T0142329

Lecture Notes in Computer Science 13157

More information about this subseries at https://link.springer.com/bookseries/7407

Yongxuan Lai · Tian Wang · Min Jiang ·
Guangquan Xu · Wei Liang ·
Aniello Castiglione (Eds.)

Algorithms and Architectures for Parallel Processing

21st International Conference, ICA3PP 2021
Virtual Event, December 3–5, 2021
Proceedings, Part III

Editors
Yongxuan Lai
Xiamen University
Xiamen, China

Min Jiang
Xiamen University
Xiamen, China

Wei Liang
Hunan University
Changsha, China

Tian Wang
Beijing Normal University
Zhuhai, China

Guangquan Xu
Tianjin University
Tianjin, China

Aniello Castiglione
University of Naples Parthenope
Naples, Italy

ISSN 0302-9743 ISSN 1611-3349 (electronic)
Lecture Notes in Computer Science
ISBN 978-3-030-95390-4 ISBN 978-3-030-95391-1 (eBook)
https://doi.org/10.1007/978-3-030-95391-1

LNCS Sublibrary: SL1 – Theoretical Computer Science and General Issues

This Springer imprint is published by the registered company Springer Nature Switzerland AG
The registered company address is: Gewerbestrasse 11, 6330 Cham, Switzerland

Preface

On behalf of the Conference Committee we welcome you to the proceedings of the 2021 International Conference on Algorithms and Architectures for Parallel Processing (ICA3PP 2021), which was held virtually during December 3–5, 2021. ICA3PP 2021 was the 21st in this series of conferences (started in 1995) that are devoted to algorithms and architectures for parallel processing. ICA3PP is now recognized as the main regular international event that covers the many dimensions of parallel algorithms and architectures, encompassing fundamental theoretical approaches, practical experimental projects, and commercial components and systems. This conference provides a forum for academics and practitioners from countries around the world to exchange ideas for improving the efficiency, performance, reliability, security, and interoperability of computing systems and applications.

A successful conference would not be possible without the high-quality contributions made by the authors. This year, ICA3PP received a total of 403 submissions from authors in 28 countries and regions. Based on rigorous peer reviews by the Program Committee members and reviewers, 145 high-quality papers were accepted to be included in the conference proceedings and submitted for EI indexing. In addition to the contributed papers, eight distinguished scholars, Yi Pan, Daqing Zhang, Yan Zhang, Shuai Ma, Weijia Jia, Keqiu Liu, Yang Yang, and Peng Cheng, were invited to give keynote lectures, providing us with the recent developments in diversified areas in algorithms and architectures for parallel processing and applications.

We would like to take this opportunity to express our sincere gratitude to the Program Committee members and 160 reviewers for their dedicated and professional service. We highly appreciate the six track chairs, Ding Wang, Songwen Pei, Zhiming Luo, Shigeng Zhang, Longbiao Chen, and Feng Wang, for their hard work in promoting this conference and organizing the reviews for the papers submitted to their tracks. We are so grateful to the publication chairs, Yang Wang, Carmen De Maio, Donglong Chen, and Yinglong Zhang, and the publication assistants for their tedious work in editing the conference proceedings. We must also say "thank you" to all the volunteers who helped us in various stages of this conference. Moreover, we are so honored to have many renowned scholars be part of this conference. Finally, we would like to thank all speakers, authors, and participants for their great contribution and support to make ICA3PP 2021 a success!

December 2021

Min Jiang
Aniello Castiglione
Guangquan Xu
Wei Liang
Jean-Luc Gaudiot
Yongxuan Lai
Tian Wang

Organization

General Co-chairs

Jean-Luc Gaudiot	University of California, Irvine, USA
Yongxuan Lai	Xiamen University, China
Tian Wang	Beijing Normal University and UIC, China

Program Co-chairs

Min Jiang	Xiamen University, China
Aniello Castiglione	University of Naples Parthenope, Italy
Guangquan Xu	Tianjin University, China
Wei Liang	Hunan University, China

Track Chairs

Ding Wang	Nankai University, China
Songwen Pei	University of Shanghai for Science and Technology, China
Zhiming Luo	Xiamen University, China
Shigeng Zhang	Central South University, China
Longbiao Chen	Xiamen University, China
Feng Wang	Wuhan University, China

Local Co-chairs

Cheng Wang	Huaqiao University, China
Liang Song	Xiamen University, China

Publication Chairs

Yan Wang	Xiamen University of Technology, China
Carmen De Maio	University of Salerno, Italy
Donglong Chen	Beijing Normal University-Hong Kong Baptist University United International College (UIC), China
Yinglong Zhang	Minnan Normal University, China

Publicity Co-chairs

Fan Lin	Xiamen University, China
Longbiao Chen	Xiamen University, China
Saiqin Long	Xiangtan University, China
Zhetao Li	Xiangtan University, China

Steering Committee

Yang Xiang	Swinburne University of Technology, Australia
Kuan-Ching Li	Providence University, Taiwan, China

Program Committee

A. M. A. Elman Bashar	Plymouth State University, USA
Jiahong Cai	Hunan University of Science and Technology, China
Yuanzheng Cai	Xiamen University, China
Jingjing Cao	Wuhan University of Technology, China
Zhihan Cao	Huaqiao University, China
Arcangelo Castiglione	University of Salerno, Italy
Lei Chai	Beihang University, China
Chao Chen	Chongqiing University, China
Donglong Chen	Beijing Normal University-Hong Kong Baptist University United International College (UIC), China
Haiming Chen	Ningbo University, China
Juan Chen	Hunan University, China
Kai Chen	Institute of Information Engineering, Chinese Academy of Sciences, China
Lifei Chen	Fujian Normal University, China
Rongmao Chen	National University of Defense Technology, China
Shuhong Chen	Guangzhou University, China
Xiaoyan Chen	Xiamen University of Technology, China
Xu Chen	Sun Yat-sen University, China
Yu Chen	Wuhan University of Technology, China
Yuanyi Chen	Zhejiang University, China
Yuxiang Chen	Huaqiao University, China
Lin Cui	Jinan University, China
Haipeng Dai	Nanjing University, China
Hong-Ning Dai	Lingnan University, China
Xia Daoxun	Guizhou Normal University, China

Dezhi Han	Shanghai Maritime University, China
Jinsong Han	Zhejiang University, China
Song Han	Zhejiang Gongshang University, China
Yulin He	Shenzhen University, China
Alan Hong	Xiamen University of Technology, China
Haokai Hong	Xiamen University, China
Zhenzhuo Hou	Peking University, China
Donghui Hu	Hefei University of Technology, China
Yupeng Hu	Hunan University, China
Yuping Hu	Guangdong University of Finance and Economics, China
Qiang-Sheng Hua	Huazhong University of Science and Technology, China
Weh Hua	University of Queensland, China
Yu Hua	Huazhong University of Science and Technology, China
Chenxi Huang	Xiamen University, China
Haiyang Huang	Huaqiao University, China
Jiawei Huang	Central South University, China
Jing Huang	Hunan University, China
Weihong Huang	Hunan University of Science and Technology, China
Xinyi Huang	Fujian Normal University, China
Zhou Jian	Nanjing University of Posts and Telecommunications, China
Fengliang Jiang	Longyan University, China
Wanchun Jiang	Central South University, China
Wenjun Jiang	Hunan University, China
He Jiezhou	Xiamen University, China
Zhengjun Jing	Jiangsu University of Technology, China
Xiaoyan Kui	Central South University, China
Xia Lei	China University of Petroleum, China
Chao Li	Shanghai Jiao Tong University, China
Dingding Li	South China Normal University, China
Fagen Li	University of Electronic Science and Technology of China, China
Fuliang Li	Northeastern University, China
Hui Li	Guizhou University, China
Jiliang Li	University of Goettingen, Germany
Lei Li	Xiamen University, China
Tao Li	Nankai University, China
Tong Li	Nankai University, China

Wei Li	Jiangxi University of Science and Technology, China
Wei Li	Nanchang University, China
Xiaoming Li	Tianjin University, China
Yang Li	East China Normal University, China
Yidong Li	Beijing Jiaotong University, China
Sheng Lian	Xiamen University, China
Junbin Liang	Guangxi University, China
Kaitai Liang	Delft University of Technology, The Netherlands
Wei Liang	Hunan University, China
Zhuofan Liao	Changsha University of Science and Technology, China
Deyu Lin	Nanchang University, China
Fan Lin	Xiamen University, China
Jingqiang Lin	University of Science and Technology of China, China
Yongguo Ling	Xiamen University, China
Guanfeng Liu	Macquarie University, Australia
Jia Liu	Nanjing University, China
Kai Liu	Chongqing University, China
Kunhong Liu	Xiamen University, China
Peng Liu	Hangzhou Dianzi University, China
Tong Liu	Shanghai University, China
Wei Liu	East China Jiaotong University, China
Ximeng Liu	Singapore Management University, Singapore
Xuan Liu	Hunan University, China
Xuxun Liu	South China University of Technology, China
Yan Liu	Huaqiao University, China
Yaqin Liu	Hunan University, China
Yong Liu	Beijing University of Chemical Technology, China
Zhaobin Liu	Dalian Maritime University, China
Zheli Liu	Nankai University, China
Jing Long	Hunan Normal University, China
Wei Lu	Renmin University of China, China
Ye Lu	Nankai University, China
Hao Luo	Beijing Normal University-Hong Kong Baptist University United International College (UIC), China
Zhiming Luo	Xiamen University, China
Chao Ma	Hong Kong Polytechnic University, China
Haoyu Ma	Xidian University, China

Chengying Mao	Jiangxi University of Finance and Economics, China
Mario Donato Marino	Leeds Beckett University, UK
Yaxin Mei	Huaqiao University, China
Hao Peng	Zhejiang Normal University, China
Hua Peng	Shaoxing University, China
Kai Peng	Huaqiao University, China
Li Peng	Hunan University of Science and Technology, China
Yao Peng	Northwest University, China
Zhaohui Peng	Shandong University, China
Aneta Poniszewska-Maranda	Lodz University of Technology, Poland
Honggang Qi	University of Chinese Academy of Sciences, China
Tie Qiu	Tianjin University, China
Dapeng Qu	Liaoning University, China
Zhihao Qu	Hohai University and Hong Kong Polytechnic University, China
Yang Quan	Huaqiao University, China
Abdul Razaque	International IT University, Kazakhstan
Chunyan Sang	Chongqing University of Posts and Telecommunications, China
Arun Kumar Sangaiah	VIT University, India
Shanchen Pang	China University of Petroleum, China
Yin Shaoyi	Paul Sabatier University, France
Hua Shen	Hubei University of Technology, China
Meng Shen	Beijing Institute of Technology, China
Huibin Shi	Nanjing University of Aeronautics and Astronautics, China
Liang Shi	East China Normal University, China
Peichang Shi	National University of Defense Technology, China
Liang Song	Xiamen University, China
Tao Song	China University of Petroleum, China
Song Han	Zhejiang Gongshang University, China
Riccardo Spolaor	University of Oxford, UK
Chunhua Su	Osaka University, Japan
Bingcai Sui	National University of Defense Technology, China
Nitin Sukhija	Slippery Rock University of Pennsylvania, USA
Bing Sun	Huaqiao University, China
Yu Sun	Guangxi University, China

Guisong Yang	University of Shanghai for Science and Technology, China
Hao Yang	Yancheng Teachers University, China
Hui Yang	National University of Defense Technology, China
Lan Yang	Quanzhou University of Information Engineering, China
Lvqing Yang	Xiamen University, China
Mujun Yin	Huaqiao University, China
Haitao Yu	Guilin University of Technology, China
Sheng Yu	Shaoguan University, China
Shuai Yu	Sun Yat-sen University, China
Zhiyong Yu	Fuzhou University, China
Liang Yuzhu	Huaqiao University, China
Tao Zan	Longyan University, China
Yingpei Zeng	Hangzhou Dianzi University, China
Bingxue Zhang	University of Shanghai for Science and Technology, China
Bo Zhang	Shanghai Normal University, China
Chongsheng Zhang	Henan University, China, China
Haibo Zhang	University of Otago, New Zealand
Hong-Bo Zhang	Huaqiao University, China
Jia Zhang	Jinan University, China
Jingwei Zhang	Guilin University of Electronic Technology, China
Jun Zhang	Dalian Maritime University, China
Mingwu Zhang	Hubei University of Technology, China
Qiang Zhang	Central South University, China
Shaobo Zhang	Hunan University of Science and Technology, China
Shengchuan Zhang	Xiamen University, China
Shiwen Zhang	Hunan University, China
Tianzhu Zhang	Nokia Bell Labs, USA
Yi Zhang	Xiamen University, China
Yilin Zhang	Huaqiao University, China
Baokang Zhao	National University of Defense Technology, China
Bowen Zhao	Singapore Management University, Singapore
Jinyuan Zhao	Changsha Normal University, China
Liang Zhao	Shenyang Aerospace University, China
Sha Zhao	Zhejiang University, China
Wan-Lei Zhao	Xiamen University, China
Yongxin Zhao	East China Normal University, China

Reviewers

Hanbin Hong
Min Wu
Siyao Chen
Kaiyue Zhang
Xiaohai Cai
Zhiwen Zhang
Tieqi Shou
Liqiang Xu
Chenhui Lu
Hang Zhu
Jiannan Gao
Hang Zhou
Haoyang An
Yang Liu
Peiwei Hu
Xiaotong Zhou
Qiang Tang
Chang Yue
Miaoqian Lin
Yudong Li
Zijing Ma
Zhankai Li
Chengyao Hua
Qingxuan Wang
Yi Wang
Mingyue Li
Songsong Zhang
Jingwei Jiang
Meijia Xu
Shuhong Hong
Anqi Yin
Shaoqiang Wu
Shuangjiao Zhai
Fei Ma
Shiya Peng
Kedong Xiu
Shengnan Zhao
Chuan Zhao
Bo Zhang
Baozhu Li
Ning Liu
Wang Yang
Xiaohui Yang
Zihao Dong
Sijie Niu
Kun Ma
Dianjie Lu
Ziqiang Yu
Lizhi Peng
Yilei Wang
Zhang Jing
Tian Jie
Jian Zhao
Hui Li
Yan Jiang
Minghao Zhao
Lei Lyu
Yanbin Han
Yanlin Wu
Lingang Huang
Mingwei Lin
Wenxiang Wang
Xinqiang Ye
Songyi Yang
Cancan Wang
Xingbao Zhang
Yafeng Sun
Li Lin
Jinxian Lei
Wentao Liu

Contents – Part III

Blockchain Systems

Data Science

Distributed and Network-Based Computing

Edge Computing and Edge Intelligence

Service Dependability and Security Algorithms

Software Systems and Efficient Algorithms

Blockchain Systems

StateSnap: A State-Aware P2P Storage Network for Blockchain NFT Content Data

Siqi Feng[1,3], Wenquan Li[1], Lanju Kong[1,2](✉), Lijin Liu[1], Fuqi Jin[1], and Xinping Min[1,2]

[1] School of Software, Shandong University, Jinan, China
klj@sdu.edu.cn
[2] Research Center Dareway Software Co., Ltd., Jinan, China
[3] Key Laboratory of Shandong Province for Software Engineering, Jinan, China

Abstract. Non-Fungible Token (NFT) has gained worldwide attention, relying on their superior data representation, as a solution for describing complex digital assets. However, the scale of NFT content data is so huge that we have to use P2P storage network to store it. Due to the lack of management capabilities of global resource allocation, retrieving content in existing P2P storage network relies heavily on distributed collaboration and cannot support low-latency off-line retrieval. Besides if an accident occurs on the blockchain or storage network that is different from the expectation, such as fork, the data state on-chain and off-chain would may be inconsistent. Therefore, we propose the concept of State for P2P storage network to represent the resource allocation of data and assist in the off-line retrieval of resources. We propose a state-aware model based on the permissioned blockchain and P2P storage network, which core is a data structure called StateSnap. The model supports the verification of the consistency of the on-chain and off-chain data and support State rollback and switching. Through experiments, we show that our model reduces the data retrieval time by about 78% compared to the traditional IPFS and performs well in terms of scalability, robustness, and State switch efficiency.

Keywords: NFT · Blockchain · P2P storage network · Retrieval · MPT

1 Introdution

Non-Fungible Token (NFT), which is a unique, non-detachable token originating from the digital world [1]. By using NFT, asset issuers can easily prove the existence and ownership of digital assets. An NFT consists of two parts: metadata and content data. Metadata, the descriptive information of the asset, is usually stored on-chain, while the content data representing the real content is stored off-chain due to its excessive content. With advantages of decentralization and non-tampering like the blockchain, P2P storage networks such as IPFS [2], Storj

Y. Lai et al. (Eds.): ICA3PP 2021, LNCS 13157, pp. 3–18, 2022.
https://doi.org/10.1007/978-3-030-95391-1_1

[3], Sia [4] are generally chosen as off-chain storage. The different colours used in Fig. 1 show that issuers of NFTs can choose different P2P storage network to store content data according to their needs. But there are two flaws when P2P storage network storing the NFT content data.

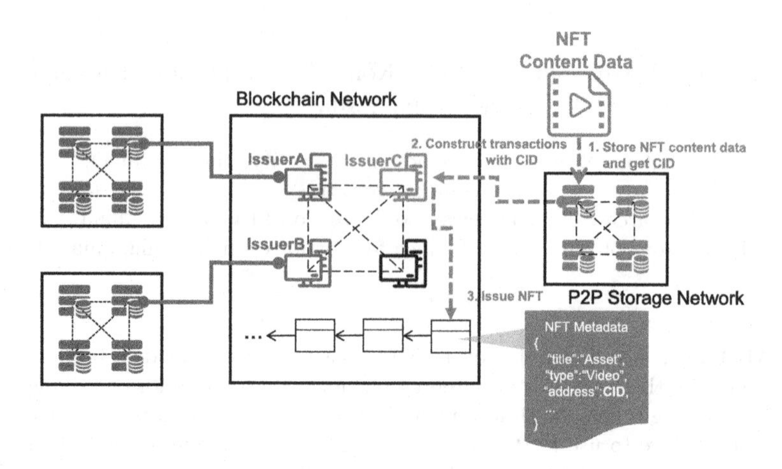

Fig. 1. Red blue and green indicate the different P2P storage networks chosen by issuers and the steps to publish NFT. (Color figure online)

On the one hand, the external quality of service is limited, which is mainly reflected in the inefficiency of information retrieval. Earlier routing algorithms, such as the Flooding algorithm, broadcast redundant information for data retrieval. Without paying extra attention to service efficiency. Later algorithms such as DHT, Coral DSHT, etc., are suited to the characteristics of distributed systems, but they require collaboration among nodes and on-line retrieval, which are also highly affected by network performance and are inefficient [5]. A better way is to allow each member to perform accurate off-line retrieval locally.

On the other hand, data in blockchain and storage network are prone to be inconsistent, because the metadata and content data of NFT are operated independently by users on different networks. Whether the inconsistency caused by the metadata on the blockchain (such as a fork occurs or a transaction fails), or the content data on the storage network (such as downtime when publishing data), the data changes cannot be detected in time, which makes the NFT security and integrity cannot be guaranteed.

Therefore, we designed a state-aware P2P storage network for NFT content data. By extracting the membership information and data information of the storage network into ***State***, efficient off-line retrieval can be achieved. Recording the digest of the State in the blockchain not only reaches a consensus on the State of the storage network but also establishes a unique association between a block and the State of storage network, which can quickly realize the verification of data consistency. To correct the data inconsistency, we allow a storage network can be switch from one State to another. In summary, our contributions are as follows.

- We propose the concept of State for the P2P storage network storing network composition State, data distribution State, and data attribute State. It highly summarizing the characteristics of the P2P storage network and the usage of data within the network, provides a way to achieve efficient off-line retrieval.
- We propose a state-aware P2P storage network model. By using a novel data structure, StateSnap, the historical State of the P2P storage network can be stored in the blockchain, enabling the storage network to switch among different States recorded in different blocks.
- A state-aware P2P storage network with StateSnap is implemented and compared with the IPFS. Experiments show that our proposed model can reduce the time of data retrieval to about 22% under general conditions and performs well in terms of scalability, robustness, and State-Switch.

2 Related Work

Blockchain blocks often store small and limited data. With the continuous expansion of blockchain applications, the storage demand for NFT content data is increasing, due to the highly storage cost, we usually store NFT metadata on blockchain to verify, and the real data in off-chain P2P storage network.

Traditional distributed storage networks generally take a network-wide broadcast approach, which brings huge communication overhead and low retrieval efficiency. Many scholars have also improved the resource search algorithms, such as reference [6–8]. However, their overall idea is to construct tree of node to reduce the resource search range, but the range is still so large that the retrieval efficiency can not significantly improved.

In addition to improvements in the search mechanism, some scholars have investigated the scheme of resource placement in expectation of providing faster services. Reference [9] proposes a DHT-based data allocation strategy achieves $O(logn)$ retrieval efficiency, but this increases the storage overhead on the chain. The IPFS network [10] uses a DHT algorithm based on the KAD protocol [11], which is capable of accomplishing file distribution and retrieval without a central node and can theoretically guarantee $O(logn)$ retrieval efficiency [12]. However, because nodes in P2P networks can leave and join the network at any time, frequent updates of DHT will reduce the retrieval efficiency or even fail to determine the storage tasks of nodes. In addition, snapshot technology is usually supported in databases to support switching [13] at different moments. Because blockchain and storage networks are difficult to keep consistent, snapshot technology is also highly needed in blockchain-P2P collaborative storage networks to help storage networks switch between different correlation State.

3 Theoretical Model

3.1 State-Unaware

Figure 2 shows that the behavior of issuing an NFT can divide into two operations: (1) The P2P storage network layer stores the content data and obtains the CID. (2) The blockchain layer publishes metadata indicating an NFT is generated. The issuer guarantees the integrity of the operation, and any mistake of above operation will break the data consistency between the two layers.

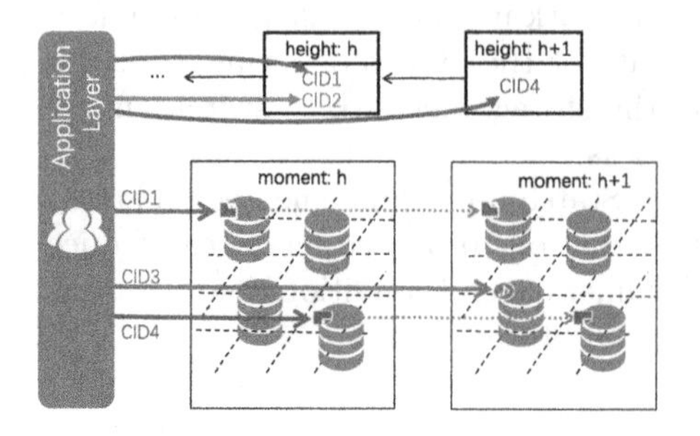

Fig. 2. Issuance of NFT

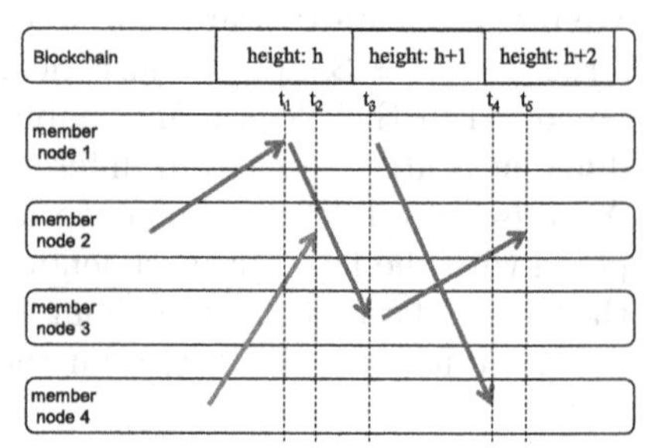

Fig. 3. Data transfer within a P2P storage network.

Figure 3 shows an example that the data transmission in assert circulation. Each member node gets data on demand just like the arrow, and the moment t at the end of the arrow is the completion of a data transmission. Then the data distribution of the whole network enters a new state. Due to the uncontrolled access to data, nodes are hard to perceive a reliable and continuously available distribution of network resources. Data retrieval have to rely on slow routing search in a distributed system, which greatly limits network efficiency.

We summarize the above characteristics of the P2P storage network as state-unaware. Only by solving them can we provide an efficient resource management platform to ensure the consistency of data between the P2P storage network layer

Table 1. Notations in the paper

Notation	Description
t, S^t	*A definite time point t, and the State of P2P Storage Network at t*
NS^t	*The network composition State at t*
CS^t	*The content data distribution State of the network at t*
CAS^t	*The content data attribute State at t*
$Data_c$,c	*An NFT content data, and its reference*
$Node_n$, n, PK_n	*A storage network member node, its reference, and its public key*
$\mathcal{C}^t$, $\mathcal{C}^t_n$	*The set of content data stored in the network, and the counterpart in n*
$\mathcal{N}^t$	*The set of all member nodes at t*
NL^t	*Operation logs of all the network member nodes before t*
nl^t_n	*A single operation record for member node n at t*
CL^t	*Operation logs of all* $Data_c$ *before t*
cl^t_n	*A single operation record for* $Data_c$ *at t*
$Info_c$, $Info_n$	*Descriptive attribute information about* $Data_c$*, and that about* $Node_n$
h, S^h	*A definite block height h, and the State of P2P Storage Network at h*

and the blockchain layer. Thus, we propose the theoretical model of state-aware to ensure it. The StateSnap structure which takes snapshots of the network is also proposed to state-aware.

3.2 State and State-Aware

Definition 1. *(State of P2P Storage Network, S^t). S^t is the extraction of network composition State, data distribution State, and the data attribute State at a certain time t.*

The notations used in the paper are listed in Table 1. At a certain time t, the S^t can be represented as a triplet consisting of the following three elements:

$$S^t = (NS^t, CS^t, CAS^t) \tag{1}$$

$$NS^t = \{(PK_n, Info_n) | n \in \mathcal{N}^t\} \tag{2}$$

$$CS^t = \{(Data_c, c \rightarrow PK_n) | n \in \mathcal{N}^t, c \in \mathcal{C}^t\} \tag{3}$$

$$CAS^t = \{(c, Info_c) | c \in \mathcal{C}^t\} \tag{4}$$

Operations on content data will cause data attribute State change. Therefore, before moment t, all logs of the storage network in terms of membership management and content data operating (such as authorization, migration, etc.), and the State transition function $StateTransition$ to the next Δt are noted as:

$$NL^t = \{nl_n^i | n \in \mathcal{N}^t, i \in [0, t]\} \tag{5}$$

$$CL^t = \{cl_n^i | c \in \mathcal{C}^t, i \in [0, t]\} \tag{6}$$

$$S^{t+\Delta t} = StateTransition(S^t, NL^{t+\Delta t} - NL^t, CL^{t+\Delta t} - CL^t) \tag{7}$$

If regard the block packing time as the standard time point, the State transition of neighboring blocks can be noted as:

$$\Delta NL^{h+1} = NL^{h+1} - NL^h \tag{8}$$

$$\Delta CL^{h+1} = CL^{h+1} - CL^h \tag{9}$$

$$S^{h+1} = StateTransition(S^h, \Delta NL^{h+1}, \Delta CL^{h+1}) \tag{10}$$

Figure 4 is a resource placement case describes how S^h is linked to the blockchain. At the time of publication, NFT content data is initially placed randomly on member node. At this moment the content data is not included in the records of State. Users can only manipulate the data through the blockchain. These operations are recorded on block h as ΔNL^h. In addition, all member nodes' joining and exiting are only managed by the blockchain with a unique identity PK_n, and the related operations recorded on block h are recorded as ΔCL^h. ΔNL^h and ΔCL^h are broadcasted along with the block generation.

Each member node of the storage network, performing as a light node, only maintains block header and does not participate in consensus. However it can

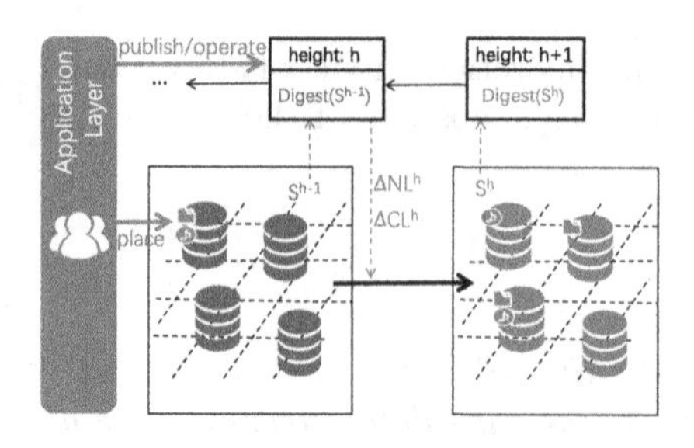

Fig. 4. S^h is anchored to the blockchain.

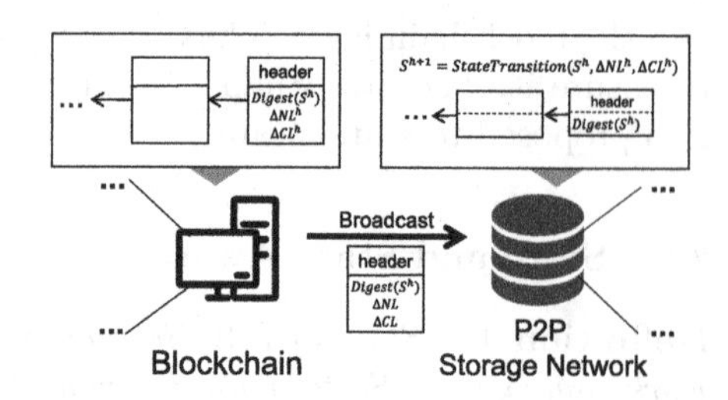

Fig. 5. Collaborations between blockchain and P2P storage network.

identify the main chain and fork and also verify the authenticity of information from the blockchain through SPV (Simple Payment Verification). The difference is that although the member node only keeps the block header, it receives the complete block, then gets the ΔNL^h and ΔCL^h in it, and performs the corresponding operation to complete the State progression, as shown in Fig. 5. For example, if the storage task of this node is recorded in ΔCL^h, then it needs to request a replica of the data from the present owner. In addition, the member node also needs to keep the corresponding S^h to complete the anchoring of the State to the block.

The above design is called state-aware. Any changes in data content, data location and data-related information are uniformly represented by State. State is only triggered by the block, and is determined at the moment of block generation. And the historical path of this State transfer is anchored to the blockchain. The storage network can use the comparison of State digests in order to quickly discover and perform a State switch when a rollback or a failed transaction occurs in the blockchain, and re-achieve data consistency with the blockchain. However, State is only kept locally for each member while the digest of State stores on blockchain due to the huge on-chain storage cost.

It is means that the state-aware network must achieve two functions: (1) The ability to organize State and ensures member nodes can quickly compare State. (2) The ability to quickly restore the digest to State with the help of historical information or information from other nodes. In addition, we define the conditions for the application of state-aware P2P storage networks:

conditions 1: The network is a fully connected network.
conditions 2: The P2P network operates in a permissioned environment, which can prevent sybil attacks by malicious nodes.
conditions 3: Assume that NFT content data is not allowed to be deleted, and can only be updated in the form of append.

3.3 MemberState and State-Switch

Since State is anchored to the blockchain, then State has three variation rules.

(1) State-Progression: Under S^h, transfer to the higher S^{h+i} of the corresponding block height.
(2) State-Rollback: Under S^h, move to S^{h-i} which corresponds to a lower block height.
(3) State-Switch: Under S^h, transfer to the corresponding S^j of the block on the other fork. A more rigorous conversion method is to rollback State to the nearest common block before fork and then progress to the target block, including State-Progression as well as State-Rollback. This synchronization method check for data correctness.

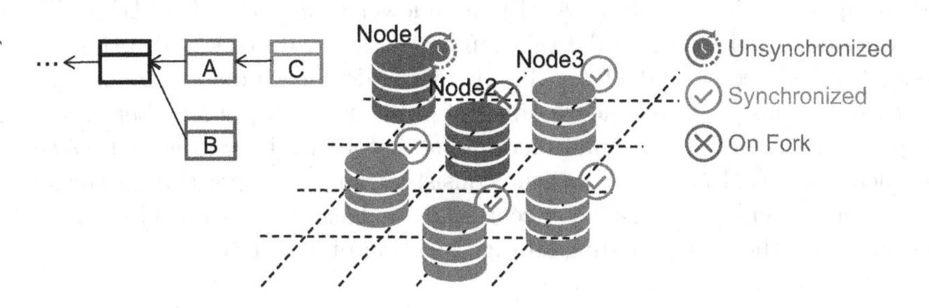

Fig. 6. Distance of member node state variation.

The above three variations correspond to A->C, C->A, B->C in Fig. 6. No matter which variation is made, it cannot be strongly consistent. This is because there will be some member nodes with untimely and incorrect State variations. Besides, because data can be migrated, as well as the withdrawal and joining of member nodes, NS^t is irreversible, so S^t is also irreversible. To solve these problems, we propose the concept of *MemberState*, it refers to the subset of S^t and a member n at some moment t. The relationship is as follows.

$$S^t = \bigcup_{n \in \mathcal{N}^t} \{MS_n^t\} \tag{11}$$

The transfer of *MemberState* implies the conversion of CS_n^t, which is an essential part of *MemberState* and contains $Data_c$. The transfer of a member State is successful only if the specified data is stored locally.

For the most complex State-Switch, assume that the target time is T. The essence of the State-Switch is that $S^t = \bigcup_{n \in \mathcal{N}^t} \{MS_n^t\}$ to $S^T = \bigcup_{n \in \mathcal{N}^T} \{MS_n^T\}$. The network State reaches final consistency when all MS_n^T are successfully restored on a member node basis.

The irreversible problem of S^t is actually the inability to restore $\mathcal{N}^t$ such that $S^T \neq \bigcup_{n \in \mathcal{N}^t} \{MS_n^T\}$. If $\mathcal{N}^T \subseteq \mathcal{N}^t$, then S^T can be completely restored, otherwise only some members can be restored. If $n \in (\mathcal{N}^t \cap \mathcal{N}^T)$, then MS_n^T can be restored; if $n \in (\mathcal{N}^t - \mathcal{N}^T)$, then MS_n^T is cleared; if $n \in (\mathcal{N}^T - \mathcal{N}^t)$, then MS_n^T cannot be restorable and needs to correct S^t by inputting ΔCL^{T-t}, ΔNL^{T-t}. Although S^t may be irreversible, P2P networks usually have data replicas, and data is generally not lost unless a major node exit occurs.

4 Implementation

4.1 Representation of StateSnap

We propose the *StateSnap*, a tree-like snapshot of the current network, as a description and digest algorithm of S, to enable the storage network to quickly switch and verify between States. Each member node maintains a detailed copy of StateSnap locally and uploads *Root* periodically to blockchain. The realization of StateSnap was facilitated by an original combination of the *Merkle Patricia Tree* [14] and the *B+ tree*. As shown in Fig. 7, its data structure can be divided into an upper *Addressing Tree* (ADT) and a lower *Consistent Hash Loop* (CHL). We re-encode the hash of all data files in a constant way to establish the unique association between c and $Data_c$. In the leaf node corresponding to c, $Info_c$ is recorded. Similarly, we use the encoding of public key PK_n of member node n as the path, and insert the information of member node in the tree as well, $Info_n$ is recorded in the leaf node. We use the consistent hashing algorithm as the siting algorithm for replica placement, accordingly, a hash loop is established in the tree to record the corresponding storage location of each data, i.e., $c \rightarrow PK_n$.

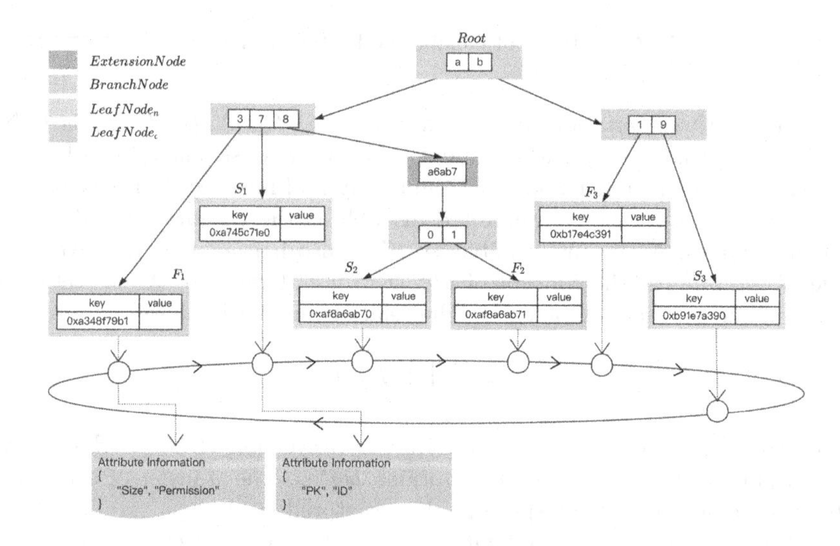

Fig. 7. The structure of StateSnap.

4.2 Data Retrieval

The resource allocation information in StateSnap is an essential assistance to rapid data retrieval, through which off-line retrieval can obviate significant response time for remote requests.

The ADT is similar to the MPT in that the indexing of key-value pairs are instructed by the path from the root to leaf nodes. The tree consists of three types of nodes: extension nodes, leaf nodes and branch nodes. The first two of them are short nodes, consisting of hash pointers to child nodes. Branch nodes are long nodes, with the first 16 items corresponding to the 16 half-byte characters that may appear in the traversal and the last item storing the value. Extension nodes and branch nodes are used to preserve the common prefix, thus reducing the height of the tree. Table 2 list the notations used in this paper with their specific meanings.

Table 2. Notations in ADT

Notation	Description
LeafNode	*The leaf nodes in StateSnap are collectively referred to and can be specifically divided into two categories*
LeafNode_c	*A category of leaf nodes representing NFT content data*
LeafNode_n	*A category of leaf nodes representing member nodes that provide storage*
ExensionNode	*Nodes in StateSnap that store public prefixes*
BranchNode	*Nodes in StateSnap that contains 16 branches*

Addressing key consists of two types: CCID and CNID. CCID is the value obtained by coding the CID of the NFT content data. The CNID has the same length as the CCID which is obtained by coding the PK_n of the member node. Key is the unique retrieval path and identifier for each LeafNode, and all NFT content data and member nodes in P2P storage network have a unique corresponding LeafNode, which can be obtained by the index down from Root following the sequence of characters in key. Therefore, depending on the content of the representation, the LeafNode is divided into LeafNode_c and LeafNode_n. When data items are added to the tree, additional encoding is performed to distinguish the types of LeafNode. We add 4 bits after the key, the least of which encodes a special token, 1 if the node is LeafNode_c and 0 if it is LeafNode_n. When the tree needs to reach consensus, deterministic cryptographic hashes for child nodes is computed from the bottom up and stored in their parents' hash pointer fields until a Root are generated. That Root will reach consensus on blockchain.

To ensure all member nodes get a uniform storage location off-line, our model uses the idea of consistent hashing and embeds it in StateSnap. CCID and CNID can be mapped evenly and randomly into the loop space. We propose the structure named CHL. We add a horizontal pointer to the next LeafNode clockwise in the value part of each LeafNode (the last node points to the first node), making a loop structure as shown in Fig. 7. And we also add a set of pointers to the local database log file for the LeafNode, which records member attributes and data attributes.

To storage a new content data, if its CCID is the same as one of a CNID, this data should be placed on the corresponding node. Otherwise, the data should be placed on the node closest to it in clockwise distance. To reduce the query burden on the data owner and improve fault tolerance, we select the nearest K (an independent fixed value) node clockwise down from the current location to store these replicas. For example, if $K = 3$ the NFT content data represented by F_1 will be stored by the corresponding member nodes of S_1, S_2, S_3.

Note that, if CHL is implemented using content-based hash pointer, then any modification of a LeafNode will result in the change of the content and hash of all leaf nodes. So the horizontal pointer in CHL should utilize the CCID or CNID of the next LeafNode. When CHL is read into memory, a physical pointer can be appended for quick direction.

Algorithm 1: Data Retrieval Algorithm

```
   input : CCID, CID, count=0
   output: Data_c
 1 while CCID is not empty do
 2     if thisNode is BranchNode then
 3         thisNode = thisNode [CCID];// Point to matching child node.
 4         CCID = CCID [1:];// Truncate the first character.
 5     else if thisNode is ExtensionNode then
 6         thisNode = thisNode.nextNode;// Point to unique child node.
 7         CCID= CCID [len(thisNode.key) : ];// Truncate identical prefix.
 8     end
 9 end
10 while count<3 do
11     while thisNode.type == 'data' do
12         thisNode = thisNode.next; // Find the next LeafNode_n
              horizontally.
13     end
14     Send (CID) to thisNode require Data_c;
15     if hash(Data_c)=CID then
16         return Data_c; // Verify CID.
17     else
18         thisNode = thisNode.next;
19     end
20     count++;
21 end
```

Algorithm 1 shows the detailed approach to data retrieval. Matching the CCID from the Root of the ADT (line 1–9), then we horizontally search for the LeafNode_n along CHL (line 11–13). Finally, remote data confirmation is conducted (line 14–20). The retrieval algorithm is a pure algorithm and never perform data modification. So the retrieval can be executed concurrently and there is no performance loss caused by thread blocking. In the subsequent experimental session, we demonstrate a significant performance improvement in the retrieval when using StateSnap compared to traditional methods.

4.3 State Progression and Switch

When it is to complete a State-Progression, in addition to the migration of content data between nodes, the StateSnap is necessarily updated. If it is a change of an existing LeafNode, instead of doing on the original LeafNode, it needs to modify a new copy, then rebuild the merkle path of ADT from the LeafNode up to Root. However, if a new content data or member node is to be added, the insertion of a new LeafNode is inevitable.

The insertion method consist of two steps. The first is insert into the ADT. Secondly, when a LeafNode of the data has been constructed, we also need to execute the insertion operation on the CHL, which is to record the previous and next neighbor LeafNode to rebuild the loop. We mark the previous neighbor LeafNode on CHL as $preLeafNode$.

Algorithm 2: Node Backtracking Algorithm

---

input : pNode = None
output: preNode
`// When inserting into the ADT, starting from the Root node, whenever a BranchNode is encountered`
1 When meet a BranchNode:
2 **if** *currentBranch is not the leftmost branch of BranchNode* **then**
3 | set tmpBranch = nearest left branch of currentBranch;
4 | set pNode = BranchNode[tmpBranch];
5 **end**
`// When the new LeafNode is created`
6 **if** *New leaf node is init* **then**
7 | preNode = right most leaf node of pNode
8 **end**

But to find the $preLeafNode$ requires traversing the whole tree to find it, which imposes an $O(n)$ time overhead. Therefore, we propose a novel **Node Backtracking Algorithm** to achieve fast finding of $preLeafNode$ in the tree, which runs with the insertion algorithm and uses a temporary node pNode to save the information during insertion and finds $preLeafNode$. Based on this information when the node is created LeafNode the space complexity is only $O(1)$.

In the algorithm, for each BranchNode passed, if the currentBranch is not the leftmost branch of the BranchNode, the pNode points to the left BranchNode of currentBranch (line 2–5). When a new LeafNode is created, the pNode always keeps the nearest common parent of the new LeafNode and its $preLeafNode$, so that we can find the $preLeafNode$ simply by finding the rightmost child of the pNode (line 6–8). Once we have solved how to find the $preLeafNode$, we can easily find the next LeafNode based on the knowledge of the link list, and how to insert a node in a directed loop will be solved.

Once the State of a storage network is anchored to blockchain, and so on fork becomes a problem. StateSnap can switch state with zero latency like MPT. This is because hash values can associate parent and child nodes, whenever any LeafNode changes (e.g., adds or updates), all changes are extracted in hash index value for the parent node. The contents of the parent node are finally changed, creating a new parent node that recurs this effect to the Root node. Eventually, a change corresponds to the creation of a new path from the changed LeafNode to the Root node, as shown in the gray path in Fig. 8. The old LeafNode is still accessible via the old path based on the old Root node. For the horizontal pointers present in the CHL under StateSnap, if hash pointers are used, the hash value of all LeafNodes will change when any change occurs in the LeafNode. Therefore, as mentioned in Sect. 4.2, our lateral pointers can be traced and restored by storing the key of the next LeafNode when they are stored.

The specific **Switch Algorithm** is as follows, firstly, we obtain the Root before the fork according to the block header $block_T$, switch the StateSnap locally to the pre-fork period, then request the blocks from the blockchain nodes during the fork period, and get all the stored transactions (line 2). Then code them according to the CID and insert them into the tree to update the StateSnap (line 4–6). Finally comparing the current recalculated Root with the consensus Root in the next block, and verify the stored transactions in $Block_{h+1}$ (line 7), which is because the stored transactions in $Block_h$ are validated in $Block_{h+1}$.

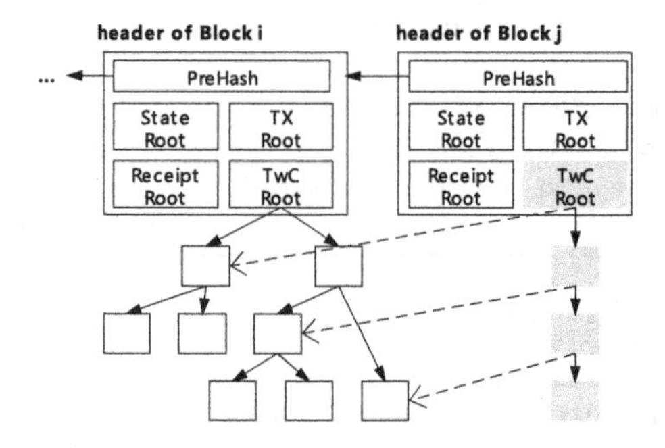

Fig. 8. When a node is added or changed, a new path is created.

Algorithm 3: Switch Algorithm

input : $block_T$

1 **for** $block_T$ *in (Rollback block, Current block)* **do**
2 Send ($block_T$) to blockchain Node and require block and Root;
3 assert block is True;
4 **for** *CID in block* **do**
5 insert Encode (CID) in Tree;
6 **end**
7 assert curRoot == next(block).Root
8 **end**

5 Experiment

We builds a private Ethereum blockchain using geth to store the metadata and implement StateSnap in a P2P storage network to provide the storage of NFT content data. We use docker to simulate a medium-scale network. Our proposed model is compared with the KAD algorithm (used by IPFS) and the search flooding algorithm in terms of retrieval efficiency, scalability, resistance to malicious members, and State-Switch efficiency.

5.1 Retrieval Efficiency

The retrieval performance is measured in time. The search time T_{search} of an NFT content data is the sum of the time T_1 for finding and the time T_2 for transferring. Without malicious nodes, we fix the number of member nodes in storage network as 100 and pre-store 15000 replicas of NFT content data. Search time of our model, KAD and search flooding is measured respectively when the number of search requests rises from 1000 to 15000. Since the T_2 time is the same in all three methods, we only plot the analysis for the T_1 time. The experimental results are shown as Fig. 9.

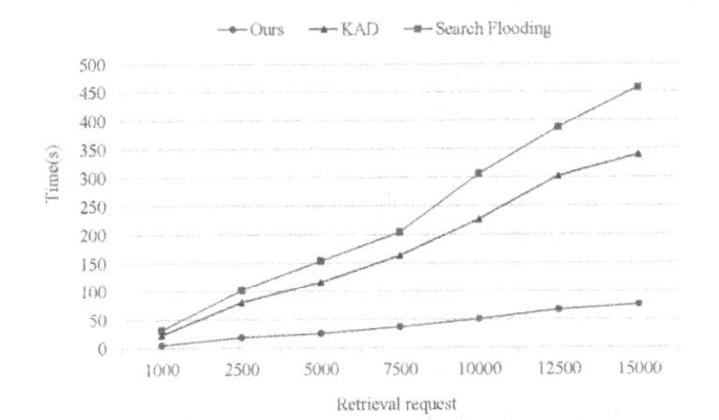

Fig. 9. Retrieval efficiency

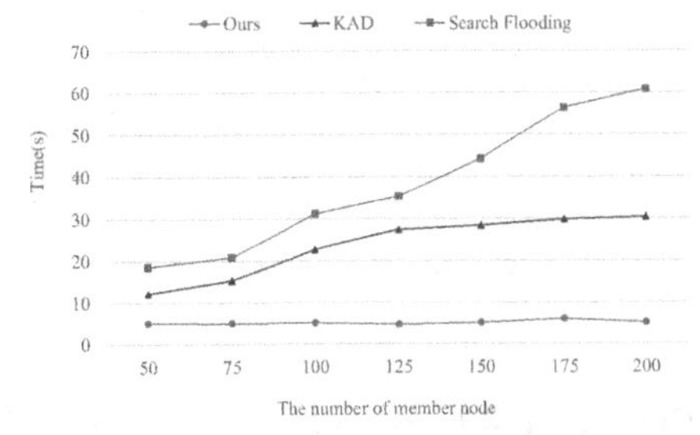

Fig. 10. System scalability

As we can see from the results, the time of three solutions increases linearly with requests increase. However, our model's retrieval time is about 22% of the KAD and 16% of search flooding. This is due to the member nodes maintaining StateSnaps can quickly and accurately find the node with $Data_c$, which greatly improves the retrieval efficiency compared to traditional member nodes that constantly request in the network.

5.2 Scalability

The time for retrieving data is also recorded to measure the scalability of network when the number of nodes changes. The experiment results are shown in Fig. 10, when the number of queries is fixed at 1000.

From the experimental results, it can be seen that with search flooding, the difficulty of searching grows as the number of nodes increases. The retrieval time of KAD grows in a logarithmic manner as the number of nodes increases, because each node in KAD needs to store the information of the 2^i nodes that are behind it. In our model, all node information is recorded in StateSnap. As the number of nodes increases, the mapped positions of LeafNode_n will become closer. Therefore, the time doesn't increase as the number of nodes increases, indicating that our model has good scalability.

5.3 Resistance to Malicious Members

Even in a permissioned environment, there is still no guarantee of complete honesty and trust of the member nodes, which greatly reduces the efficiency of services. Therefore, we compare the performance of our model, KAD and search flooding by retrieval time when the proportion of malicious nodes in the storage network varies. We randomly select some nodes as malicious nodes to keep them silent and not send messages or forge messages. The results are shown in Fig. 11.

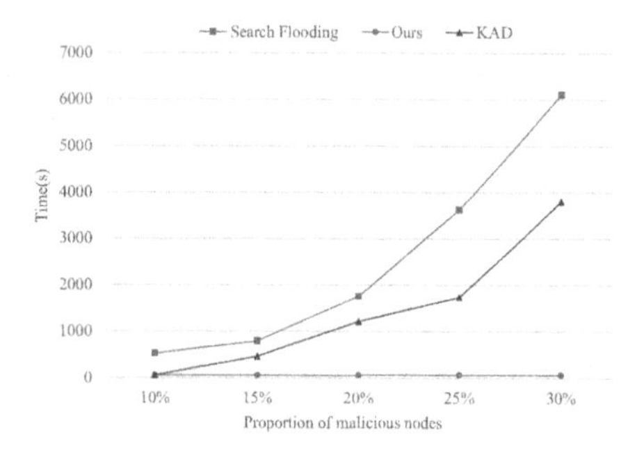

Fig. 11. Resistance to malicious members

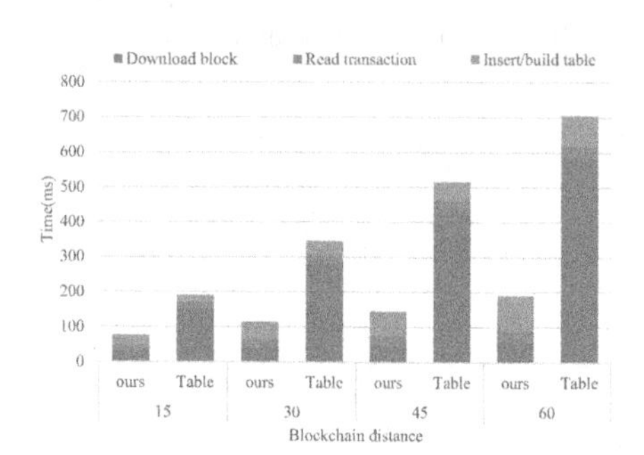

Fig. 12. State-switch efficiency

As seen in Fig. 11, the retrieval time of KAD and search flooding grows in an exponential manner as the proportion of malicious nodes increases. This is because these schemes can only keeping retrieving until the user's waiting time runs out. However, our model requests the second replica node as soon as the first replica node fails to obtain $Data_c$. When none of the K replica nodes responds, it assumes that $Data_c$ is lost and stops retrieval. Other schemes, such as the erasure code technique, can be used later to recover it [15].

5.4 State-Switch Efficiency

In order to compare the efficiency of StateSnap when forks occur, we used a tree structure and a table structure to store all the information in StateSnap separately and compared their performance. The experimental results are shown in Fig. 12. The vertical coordinate is the time required to switch to the main chain. 15, 30, 45, 60 indicates the height of the blockchain after the fork occurs. The total time is divided into three parts: the time to download blocks, the time to read transactions and the time to insert transactions.

The results show that the time of the two structures increases with the blockchain height grows, where the table has a slight advantage in the insert/build operations because the tree requires extra calculation steps in the insert operation, while the table just stores them sequentially. The time of our model to download blocks does not increase obviously with the blockchain height, while the time of tables is proportional to the blockchain height. This is because a traditional P2P network cannot adapt to a blockchain fork without a reliable trust mechanism, and it must download all blocks in order to guarantee the trustworthiness of the content. In contrast, when it is necessary to switch State, our model only needs to locate the block where the fork occurred to obtain the trusted Root and start a new computation from then on.

6 Conclusion

We propose the concept of State, which describes network composition State, data distribution State, and data attribute State, and provides a way to achieve efficient off-line retrieval locally by members. We establish a state-aware P2P storage network with StateSnap for blockchain NFT content data to help us store and quickly compare the historical State of the P2P storage network, and enable the storage network to switch between the States associated with different blocks. Experiments show that our model has better performance in terms of retrieval, robustness and State-Switch compared to IPFS and search flooding.

Acknowledgement. This work is supported by the National Natural Science Foundation, China (No.61772316); the major Science and Technology Innovation of Shandong Province (No. 2019JZZY010109, 2020CXGC010106); the Industrial Experts Program of Spring City; National Social Science Fund (No. 20BJY131); the Special Project of Science and Technology Innovation Base of Key Laboratory of Shandong Province for Software Engineering (Project ID:11480004042015).

References

1. Wang, Q., Li, R., Wang, Q., Chen, S.: Non-fungible token (NFT): overview, evaluation, opportunities and challenges. arXiv preprint arXiv:2105.07447 (2021)
2. Benet, J.: IPFS-content addressed, versioned, P2P file system. arXiv preprint arXiv:1407.3561 (2014)
3. Storj: A decentralized cloud storage network framework. Whitepaper. https://github.com/storj/whitepaper. Accessed 4 Sept 2021
4. Vorick, D., Champine, L.: Sia: Simple decentralized storage. Nebulous Inc (2014)
5. Dabek, F., Li, J., Sit, E., Robertson, J., Kaashoek, M.F., Morris, R.T.: Designing a DHT for low latency and high throughput. In: NSDI, vol. 4, pp. 85–98 (2004)

6. Schoenmakers, B.: A simple publicly verifiable secret sharing scheme and its application to electronic voting. In: Wiener, M. (ed.) CRYPTO 1999. LNCS, vol. 1666, pp. 148–164. Springer, Heidelberg (1999). https://doi.org/10.1007/3-540-48405-1_10
7. He, K., Wu, X., Zhang, Y.: Topology research of unstructured P2P network based on node of interest. Comput. Eng. Appl. 09 (2016)
8. Dahan, S., Nicod, J.M., Philippe, L.: The distributed spanning tree: a scalable interconnection topology for efficient and equitable traversal. In: CCGrid 2005, vol. 1, pp. 243–250. IEEE (2005)
9. Hassanzadeh-Nazarabadi, Y., Küpçü, A., Ozkasap, O.: Decentralized utility-and locality-aware replication for heterogeneous DHT-based P2P cloud storage systems. IEEE Trans. Parallel Distrib. Syst. **31**(5), 1183–1193 (2019)
10. Psaras, Y., Dias, D.: The interplanetary file system and the filecoin network. In: 2020 50th Annual IEEE-IFIP International Conference on Dependable Systems and Networks-Supplemental Volume (DSN-S), pp. 80–80. IEEE (2020)
11. Khudhur, N., Fujita, S.: Siva-the IPFS search engine. In: 2019 Seventh International Symposium on Computing and Networking (CANDAR), pp. 150–156. IEEE (2019)
12. Sharma, P., Jindal, R., Borah, M.D.: Blockchain technology for cloud storage: a systematic literature review. ACM Comput. Surv. (CSUR) **53**(4), 1–32 (2020)
13. Li, L., Wang, G., Wu, G., Yuan, Y., Chen, L., Lian, X.: A comparative study of consistent snapshot algorithms for main-memory database systems. IEEE Trans. Knowl. Data Eng. **33**(2), 316–330 (2019)
14. Bonneau, J.: EthIKS: using ethereum to audit a CONIKS key transparency log. In: Clark, J., Meiklejohn, S., Ryan, P.Y.A., Wallach, D., Brenner, M., Rohloff, K. (eds.) FC 2016. LNCS, vol. 9604, pp. 95–105. Springer, Heidelberg (2016). https://doi.org/10.1007/978-3-662-53357-4_7
15. Qi, X., Zhang, Z., Jin, C., Zhou, A.: BFT-store: storage partition for permissioned blockchain via erasure coding. In: 2020 IEEE 36th International Conference on Data Engineering (ICDE), pp. 1926–1929. IEEE (2020)

Towards Requester-Provider Bilateral Utility Maximization and Collision Resistance in Blockchain-Based Microgrid Energy Trading

Hailun Wang[1], Kai Zhang[1(✉)], Lifei Wei[2], and Lei Zhang[2]

[1] College of Computer Science and Technology, Shanghai University of Electric Power, Shanghai, China
kzhang@shiep.edu.cn

[2] College of Information Technology, Shanghai Ocean University, Shanghai, China

Abstract. Microgrid is a promising system for coordinating distributed energy in the future Energy Internet, where energy market is crucial for facilitating multi-directional trading. Nevertheless, the traditional solutions for energy trading usually rely on a centralized framework, which is vulnerable to high operation cost and low transparency. To address the problem, the blockchain strategy has been widely deployed in microgrid energy trading. However, existing blockchain-based systems have not formally considered: (i) collusion attacks launched by a pair of dishonest consumer and microgrid; (ii) and the demands of maximizing bilateral utility. Therefore, in this work, we propose a blockchain-based microgrid energy trading system that allows a consumer to dynamically change its service quality quotes for a microgrid. In particular, bilateral utility of requester-provider is almost maximized, with employing a Bayesian Nash equilibrium strategy to model the interaction between consumers and microgrids. Moreover, collusion attacks during the bidding process are certainly prevented from a dishonest pair of consumer and microgrid. To achieve this, we employ a request-based comparable encryption technique to compare two microgrids' revenues in a privacy-preserving way, in which each microgrid should submit an encrypted expected revenue. In addition, a general security analysis of our system is formally discussed. To illustrate the effectiveness, we take a blockchain-based framework to build the experiment platform, where the structure of nodes and transaction process are well formalized.

Keywords: Blockchain · Microgrid · Energy trading · Bilateral utility · Collision resistance

Y. Lai et al. (Eds.): ICA3PP 2021, LNCS 13157, pp. 19–34, 2022.
https://doi.org/10.1007/978-3-030-95391-1_2

1 Introduction

Microgrid is a self-sufficient energy system that includes a variety of distributed energy resources, such as solar panels, wind turbines and combined heat and power [13]. Through an energy trading system, the service provider (microgrids) can provide energy resources for service requesters (consumers) in a pay-as-you-go manner [21]. Different from classic electricity trading system where a power generator delegates a power supplier to sell its electricity to consumers, the microgrid trading system is to sell its energy directly with a consumer [14]. Hence, the balancing energy distribution and minimizing the operation costs can be achieved, since the cost-expensive intermediate trading process are not involved in the energy trading systems.

The traditional solutions for microgrid transaction managements usually rely on a centralized framework, but some problems are surfaced, such as high operation cost, low transparency, and transaction data modification [17]. To tackle the problems, the blockchain framework is introduced into the field of energy trading systems, which enables us to conduct, verify and record microgrid energy transaction [12]. Moreover, it can also provide transparent and user-friendly applications for trading process [3], since the blockchain can effectively provide trustworthy, enhanced security and the traceability of records for a decentralized network.

The blockchain-based methodology has been widely deployed in microgrid energy markets [10,18], since it can effectively reduce operation cost and realize fairness of trading transaction. In particular, the work [19] gave a blockchain-based framework for peer-to-peer electricity sales in microgrid market, and [9] introduced a distributed supply bidding algorithm to solve the global welfare optimization via a smart power contract. Nevertheless, only basic privacy and security guarantees are achieved in the solutions. Moreover, [6] developed a blockchain based approach with introducing differential privacy technique for microgrid energy auction, in which an adversary cannot infer private information of any participant. Although Wu et al. [18] proposed a user-centric P2P energy trading for residential microgrids, the proposed strategies can only maximize the profit of small-scale distributed microgrids.

Motivation. Existing blockchain-based solutions have not formally considered: (i) collusion attacks launched by a pair of dishonest consumer and microgrid [9,19]; (ii) and the demands of maximizing bilateral utility [6,18]. Therefore, enabling requester-provider bilateral utility maximization and collision resistance is highly motivated in blockchain-based microgrid energy trading systems.

1.1 Our Contributions

To address the problems, in this work, we propose a blockchain-based low-carbon microgrid energy trading system. Generally speaking, the system enables requester-provider (almost) bilateral utility maximization and collision resistance. The main contributions can be summarized as follows:

1. To realize a secure and effective decentralized energy trading for microgrid, we take a blockchain-based framework to build the trading system. After a well definition of structure nodes and transaction process, the records written on the blockchain can be not tampered by malicious outsider's attacks.
2. To maximize the bilateral utility of requester-provider, we employ a Bayesian Nash equilibrium [16] strategy to model the interaction between consumers and microgrids. In particular, a microgrid actively determines a transaction object based on a variety of judgment conditions, while a consumer is able to specify a candidate microgrid after receiving a set of expected revenues.

Table 1. Feature comparison

	Ref. [18]	Ref. [19]	Ref. [9]	Ref. [6]	Ref. [8]	Ours
P2P Trading	✓	✓	✓	✓	✓	✓
Smart Contracts	✓	✓	✓	✓	×	✓
Scoring Strategy	×	×	×	×	×	✓
Collusion-Resistance	×	×	×	×	×	✓
Utility Maximization	×	✓	✓	×	✓	✓
No SPOF	✓	✓	✓	✓	✓	✓
Anonymity	✓	✓	✓	✓	×	✓

3. To prevent collusion attacks launched by a consumer and a microgrid, we employ a request-based comparable encryption methodology to enable comparing the bidding in a privacy-preserving way.
4. To achieve quality score incentives, we build a new scoring approach to enable a microgrid to optimize its service quality. Moreover, the misbehavior of a dishonest microgrid or a consumer is forbidden.

Comparison. Table 1 gives a feature comparison between state-of-art-work and our solution. Moreover, we build a real blockchain platform with a well definition of data structure of nodes and transaction process. The results show that the system can effectively (almost) maximize the bilateral utility and enjoys high running efficiency. In particular, the forecasted electricity price is almost the same as the transaction price, while the consumed gas cost still retains low.

Organization. We review some background knowledge in Sect. 2 and describe the problem formulation in Sect. 3. Section 4 presents a blockchain-based energy trading system and Sect. 5 gives its security analysis. We implement the system and show its performance in Sect. 6 and finally conclude the work in Sect. 7.

2 Background Knowledge

2.1 Blockchain

A blockchain is a distributed ledger whose data are shared among peer-to-peer networks. It is a chain data structure that combines data blocks together in chronological order and guarantees a non-modifiable, falsifiable, decentralized sharing, which securely stores verifiable simple sequential data within the system [20]. Blockchain technology has been applied in many practical applications, such as financial services, supply chain financial management, education, and health care.

2.2 Smart Contracts

Smart contracts, a "computer agreement to enforce the terms of the contract", was first proposed as by Nick Szabo [11]. In essence, smart contracts are usually run on a shareable distributed database [7], which can be consistently executed by a network of mutually distrusting nodes, without the arbitration of a trusted authority. Due to the resistance towards tampered attacks, smart contracts have been widely employed in a number of practical applications [4]. Particularly, Ethereum is a major blockchain platform for smart contracts - Turing complete programs that are executed in a decentralized network and usually manipulate digital units of value [15].

2.3 Request-Based Comparable Encryption

Request-based comparable encryption(RBCE) [5] is an extension notion of order-preserving encryption(OPE) [2], which is a system that securely gives a comparison about two encrypted values. In RBCE, the numerical order of the encrypted numbers is preserved, while the original value is hidden. Hence, only the parties who have a comparison "token" can compare the encrypted values. In particular, the RBCE has been widely deployed in practical applications, in which the size of the generated ciphertext of an underlying confidential data is not too long.

3 Problem Formulation

In this section, we formalize the system model and system running flow, define the design goals and threat model of the system.

3.1 System Model

There are three entities included in the proposed system: authority, microgrids, and consumers, which are illustrated in Fig. 1.

- **Authority**: The authority initializes the system and distributes public parameter and secret key for microgrids. Moreover, it records and publishes the service quality score of microgrids.
- **Microgrids**: The microgrids generate energy resources and sell them to the consumers for gaining profits, where the transactions between request and feedback are conducted on the blockchain platform.
- **Consumers**: The consumers include a variety of clients that submit service requests. And a client may choose one microgrid as a target from a number of feedback that come from different microgrids.

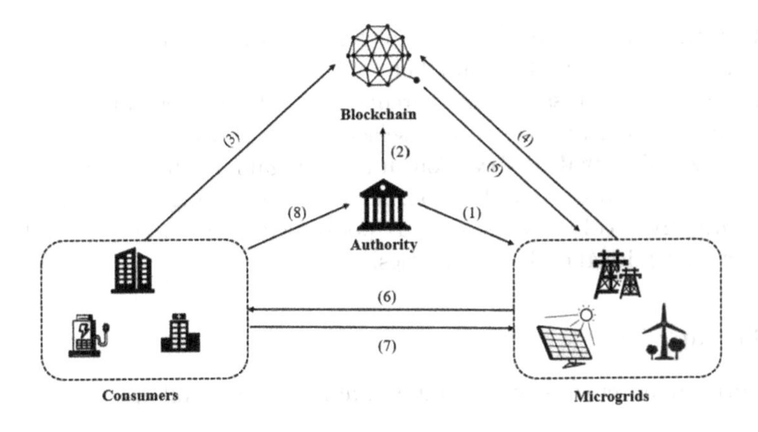

Fig. 1. Overview of the architecture of the system.

3.2 System Running Flow

The system running flow of the system is formalized as follows:

1. The authority publishes system public parameter by initializing the system, and generates secret key for each microgrid (1). It also uploads the initial service quality score of microgrids and CO_2 intensity for electricity produced of microgrids to the blockchain (2).
2. A consumer sends an electricity demand order request to the blockchain (3), which includes the time period of electricity consumption T, an electricity consumption W and an afforded the maximum electricity pricing Y. Note that the microgrid may decline the request order if Y is much lower than expected ($Y \ll \mathsf{expected}$). In addition, the consumer may also submit the margin of this transaction to the contract account of the authority.
3. Based on the request in blockchain, each microgrid does:
   (3.1) Compute its power generation cost price. Based on the electricity load, power generation cost price and electricity demand, it bids for one request with giving a request feedback (4).
   (3.2) For a same request order, each microgrid can receive other's service quality score of microgrids and CO_2 intensity for electricity produced from the blockchain (5). Later, they compute and encrypt consumer expected revenue, and send an encrypted result CIPH(ciphertext of consumer expected revenue) to the consumer (6).
4. Receiving a number of encrypted results CIPH, the consumer conducts a comparison to determine the transaction microgrid (7). For this transaction, the consumer submits the service quality score of microgrids to the authority (8). However, a consumer's margin may be deducted when it maliciously conducted a service quality score for a microgrid.

3.3 Design Goals

The design goals of a blockchain-based energy trading system are as follows:

- **Autonomy.** The transaction is completely negotiated between the consumers and the microgrids, without any help of a third party.
- **Transaction fairness.** Both the requester and provider parties participate in the pricing and realize two-way selection in a transaction.
- **Maximize bilateral utility.** Consumers can gain maximum profits, and the microgrid can choose a "right" consumer for conducting transactions.
- **Running Securely.** The system is performed safely and effectively while still retaining designed functionalities.

3.4 Threat Model

The security requirements of the system are required as follows:

- **Anonymity:** There is no established connection between the consumer electricity demands order and its consumer identity, otherwise some privacy information may be leaked.
- **Data integrity.** The data integrity should be protected during the transaction running, that is, the data can not be modified.
- **Collusion Resistance:** A malicious consumer may conclude with a malicious microgrid when it receives quoted prices from a number of microgrids. Hence, the quoted prices of honest microgrids are leaked to dishonest microgrids, which brings about severe loss for honest microgrids.
- **Security against dishonest consumers/microgrids:** A dishonest consumer may submit incorrect service quality score of microgrids to the authority, or refuse to pay for request order. Moreover, a delivered electricity from a dishonest microgrid to consumers may be less than the required in the transaction.

4 The Proposed Scheme

In this section, we present the formal description of our blockchain-based energy trading system and its running process.

4.1 Blockchain Structure

- **Consumer node:** The consumer node sends an electricity demand order $\mathsf{ED}(T, W, Y)$ to the blockchain, and obtains an encrypted consumer expected revenue CIPH from a microgrid node.
- **Microgrid node:** The microgrid node obtains ED from the blockchain, computes and sends CIPH to the consumer node.
- **Authority node:** The authority node distributes the key MK to a microgrid node, it still uploads and updates the service quality score and CO_2 intensity for electricity production of a microgrid.

4.2 Running Flow

The running process of our energy trading system is divided into the following phases: *electricity demand order initialization, expected revenue generation, transaction objects determination,* and *service quality score of microgrids generation,* which is shown in Fig. 2. Moreover, the smart contracts provide functional interactions for energy transactions and the records contract execution results.

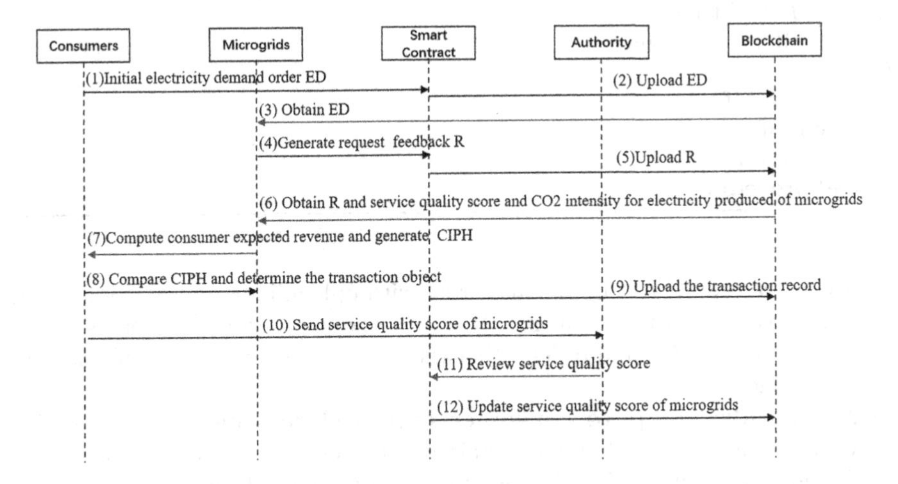

Fig. 2. Running process of blockchain-based energy trading system.

4.2.1 Electricity Demand Order Initialization A consumer sends its electricity demand order $\mathsf{ED}(T, W, Y)$ to the blockchain. If Y is much lower than expected, there will be no microgrid to accept the order. Hence, the consumer is required to re-send the electricity demand order. At the same time, the consumer submits corresponding deposit to a contract account of authority to maintain its credibility.

4.2.2 Expected Revenue Generation The microgrid's expected revenue for a consumer's request order is generated as follows:

(1) **Microgrids: participate in bidding.** The microgrid uses the random forest algorithm to estimate power generation cost price, and obtains customer electricity demand order $\mathsf{ED}(T, W, Y)$ from the blockchain. Later, the microgrid determines whether the electricity load can meet the electricity demand order during the time period T, and whether the estimated power generation cost price Y' is less than or equals to Y. Set the total electricity load during the time period T as L, and the microgrid's trading volume before the time period T is N. If we have $L \geq \sum_{i=1}^{N} W_i + W$,where $Y \geq Y'$, the microgrid may

Algorithm 1 : Bidding

Require: Electricity demand order $\mathsf{ED}(T, W, Y)$, power generation cost price Y', total electricity load L and trading volume N.
Ensure: Request feedback R
1: **procedure** BID(ED, Y', L, N)
2: $L' = \sum_{i=1}^{N} W_i + W$ //*Total electricity consumption during the time period T*
3: $R \leftarrow 0$
4: **if** $L \geq L'$ **then**
5: **if** $Y \geq Y'$ **then**
6: $R \leftarrow 1$
7: **end if**
8: **end if**
9: Return R
10: **end procedure**

decide to respond to this demand order with uploading a request feedback R to the blockchain for bidding it. If $R = 1$, the microgrid participates in the bidding for the demand order; otherwise, the microgrid does not participate in this bidding.

(2) **Microgrids: compute consumer expected revenue.** The microgrids who bid for the same ED may obtain other microgrids' service quality score G and corresponding CO_2 intensity for electricity produced C. The G has a positive impact on whether the microgrid is selected by consumers; while C has a negative impact on whether the microgrid is selected by consumers. For an electricity demand order ED, we assume that there are M microgrids participating in bidding and denote q as the performance index of each microgrid. Denote the probability of each microgrid being selected by a consumer as P, thus we have

$$q_j = \frac{G_j}{C_j}, \quad \text{and} \quad P_j = \frac{q_j}{\sum_{k=1}^{M} q_k}.$$

Here, the probability distribution of a selected microgrid is certainly obtained. Suppose the quotation for microgrids is X, the expected revenue for consumers is E, where $E_j = P_j \times (-X_j), j = 1, \cdots, M$. Moreover, the consumers tend to choose a microgrid with high consumer revenue expectation for transaction. During the process of prices determination, the probability distribution of the microgrid that can be selected is known, and the utility of each consumer and microgrid has been maximized. As a result, no participant is willing to change their strategy, thus the price game can reach Bayesian Nash Equilibrium (BNE).

(3) **Microgrids: encrypt the consumer expected revenue.** If a consumer directly obtains the consumer revenue expectation E, the real quoted price of the microgrid is certainly obtained. The dishonest consumers may collude with a malicious microgrid and thus leak the quoted pricing information

Algorithm 2 : Expectation

Require: The microgrid service quality score$G$, the CO_2 intensity for electricity produced C, the number of microgrids that participate in bid M
Ensure: Consumer expected revenue E_j
1: **procedure** EXPECTATION(G, C, M)
2: $j \leftarrow 1$
3: **while** $j \neq M + 1$ **do**
4: $q_j = G_j / C_j$
5: $P_j = \frac{q_j}{\sum_{k=1}^{M} q_k}$
6: $E_j = P_j \times (-X_j)$
7: $j = j + 1$
8: **end while**
9: Return E_j
10: **end procedure**

to the microgrid. To prevent the requester-provider collusion attacks, we employ an request-based comparable encryption strategy to first encrypt E. Hence, we give a privacy-preserving consumer expected revenue generation scheme, where the microgrid needs to send an encrypted version CIPH to a consumer for the following comparison. The process is shown in Algorithm 3.

4.2.3 Transaction Objects Determination Receiving a number of CIPHs from microgrids, the consumer compares them (actually compares the expected revenue) and determines the transaction object. After a comparison, the consumer publishes the token and the encrypted version of the highest consumer revenue expectation. Moreover, the microgrid who wins the bidding should submit its quoted price and the value I^* as a proof. Finally, the consumer checks whether the microgrid is the winner as $H(I^*) = H(I)$. If the transaction is concluded, the consumer can announce the bid-winning microgrid's quoted price, and each microgrid is able to verify the bid-winning result. Note that the shown interactions are conducted via transactions between consumers and microgrids.

4.2.4 Service Quality Score of Microgrids Generation For each transaction, the consumer sends a service quality score g of microgrids to the authority. Then, the authority reviews the authenticity of the service quality score of microgrids. If it is not a malicious/incorrect one, the authority runs the smart contract to update the service quality score of microgrids on the blockchain $G = \mathsf{AVERAGE}(G', g)$, where G' is the service quality score of microgrids before this transaction. Otherwise, the consumer's deposit will be deducted.

5 Security Analysis

In this section, we analyze the following security guarantees of our system based on the addressed threat models as follows.

Algorithm 3 : Encrypted expected revenue comparison

Require: The authority generates and distributes MK to microgrids
Ensure: $\{-1, 1\}$
1: **procedure** TOKEN GENERATION(E)
2: $E' = |E|, E' = \sum_{i=1}^{n-1} 2^i e_i = (e_{n-1}, \cdots, e_1, e_0)$
3: **while** $i \leq n$ **do**
4: $d_n = Hash(MK, (0,0,0)), d_i = Hash(MK, (1, d_{i+1}, e_i))$
5: $i++$
6: **end while**
7: Return $token = (d_0, d_1, \cdots, d_n)$
8: **end procedure**
9: **procedure** ENCRYPTION($I, token$)
10: **while** $j \leq n-1$ **do**
11: $b_i = Hash(d_i, (2, I, 0))$
12: $a_i = Hash(MK, (4, d_{i+1}, 0)) + e_i \; mod3$
13: $f_i = Hash(d_{i+1}, (5, I, 0)) + a_i \; mod3$
14: $i++$
15: **end while**
16: Return $\mathsf{CIPH} = (I, (b_0, \cdots, b_{n-1}), (f_0, \cdots, f_{n-1}))$
17: **end procedure**
18: **procedure** COMPARISON($\mathsf{CIPH}, \mathsf{CIPH}', token$)
19: **while** $j \leq n-1$ **do**
20: **while** $k \leq n$ **do**
21: **if** $b'_k = Hash(d_k, (2, I', 0)))$ **then**
22: **if** $(c'_j \neq Hash(d_j, (2, I', 0)))$ **then**
23: $k++$
24: **end if**
25: **end if**
26: **end while**
27: $b'_k = Hash(d_k, (2, I', 0))$
28: $j++$
29: $a_j = f_j - Hash(d_{j+1}, (5, I, 0)) \; mod \; 3$
30: $a'_j = f'_j - Hash(d_{j+1}, (5, I', 0)) \; mod \; 3$
31: **end while**
32: **if** $a_j - a'_j = 1 \; mod \; 3$ **then**
33: Return 1
34: **end if**
35: **if** $a_j - a'_j = 2 \; mod \; 3$ **then**
36: Return -1
37: **end if**
38: **end procedure**

- **Anonymity:** Based on inherent entity anonymity in blockchain, the contract addresses and personal identity information of consumers involved in the transactions can be certainly hidden from the public.

- **Data integrity:** Since blockchain has provided a security guarantee that the data records cannot be modified. That is, if a data addition or transaction has been completed, it cannot be edited or deleted again.
- **Collusion Resistance:** Based on request-based comparable encryption, the consumer can only receive an encrypted version CIPH of consumer revenue expectation rather a plaintext. They can still compute-then-compare the real expected revenue. Hence, the requester-provider collusion attacks can be prevented in the literature.
- **Security against dishonest consumers:** Based on our introduced smart contract, the dishonest consumers who update incorrect service quality score of microgrids to the authority will be deducted the deposit. Note that the payment function is implemented by the underlying smart contract, thus the dishonest consumer can be prevented.
- **Security against dishonest microgrids:** Similarly, the illegal behaviors of dishonest microgrids are supervised by designed smart contract. Note that the smart contract is run by the authority.

Hence, our proposed blockchain-based energy trading system is functionality-maintaining and secure under the introduced security model.

6 Performance Analysis

In this section, we give a careful comparison with related work, and also implement our system and conduct a feature comparison with related work.

6.1 Theoretical Analysis

Note that the works [6,8,9,18,19] have not formally considered the collusion attacks between dishonest consumers and microgrids; and the work of [8] was not blockchain-based solution that is vulnerable to the problem of a single point of failure. Moreover, the work [18] only considers for maximizing user benefits while not supporting bilateral utility maximization.

Our blockchain-based energy trading system has several enjoyable features than related work. All the user nodes in the blockchain network conduct the transactions via addresses (thereby hiding the user's personal identity information). Moreover, the Fee is paid through smart contracts. Even if some nodes fail, the system can still run normally that effectively avoids the problems of a single point of failure (SPOF). At the same time, the pricing method employs the Bayesian Nash equilibrium, with (nearly) maximizing the utility of both requester and providers. A request-based comparable encryption is used to effectively prevent dishonest consumers from colluding with the microgrid. To realize an effective scoring incentive mechanism, the misbehavior of dishonest microgrid can be certainly prevented. Furthermore, The consumers submit margin to the authority for addressing the threats from dishonest consumers.

6.2 Performance Analysis

Our experiments use the dataset *Electricity Prices* in Kaggle [1], which contains 38,015 electricity price data. We analyze the performance of the proposed model by building an Ethereum private chain environment, and deploy a test network. Concretely, the experimental environment is as follows: Intel(R) Core(TM) i5-8500 CPU @ 3.00GHz 3.00GHz and 16GB of RAM, Intel(R) Core(TM) i5-10210U CPU @ 1.60GHz 2.11GHz and 12GB of RAM.

1) Cost of Smart Contract. We test the consumption of gas during the execution of the function in the smart contract, where the price of gas is about 20Gwei. Table 2 shows the cost compare of smart contracts, where we can find that the cost of enforcing smart contracts is not high.

Table 2. Cost of running smart contracts

	Gas used	Gas cost (Ether)
Deploy Electricity contract	1059672	0.02119
InitalDemand	116522	0.00233
GridsBid	82820	0.00166
WriteCarbon	45947	0.0092
QueryPrice	25690	0.00051
Confirm Transaction	112643	0.00225
TransferElectrity	46453	0.00093
Submit service quality score	46960	0.00062
Upload service quality score	44488	0.00089
Pay	31380	0.00063

2) The impact of service quality score on transaction volume. Set the initial service quality score of microgrids as 5 and the initial number of transactions as 0. We observe that the service quality score of microgrids can be updated after each transaction. For this issue, we simulated 10-day transactions in the Ethereum private chain environment. Concretely, Fig. 3 shows the transaction volume of 3 microgrids in 10 cycles. And Fig. 4 shows the service quality score of 3 microgrids changes. It can be seen from Fig. 3 and Fig. 4 that our system has restored the market rules of survival of the fittest in real trading scenarios.

It can be seen from Fig. 3 and Fig. 4 that the transaction behavior of microgrid m_1 has been stable. In general, the transaction volume and service quality score of microgrids are rising, indicating that microgrid m_1 has a good service. Microgrid m_2's transaction behavior has a certain degree of inertia. For a period of time, the transaction volume and the service quality score of microgrids continue to decline, indicating that microgrid m_2 has malicious transactions. In the next few cycles, due to good transaction behaviors, the trust of consumers is

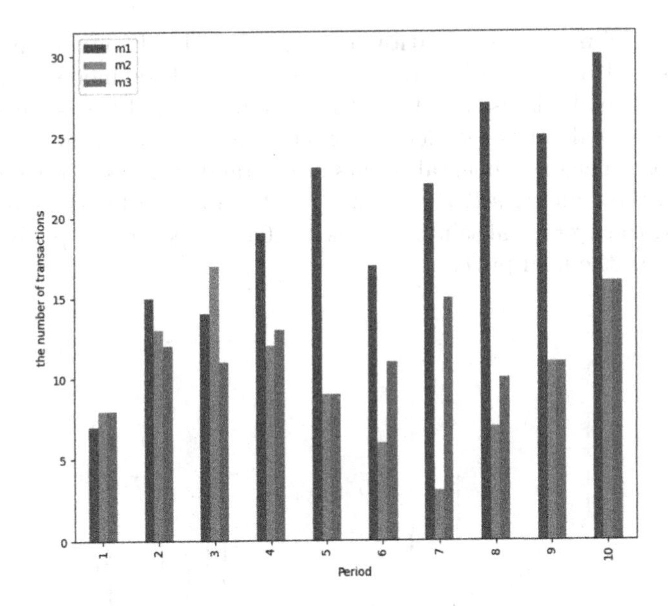

Fig. 3. Transaction volume comparison.

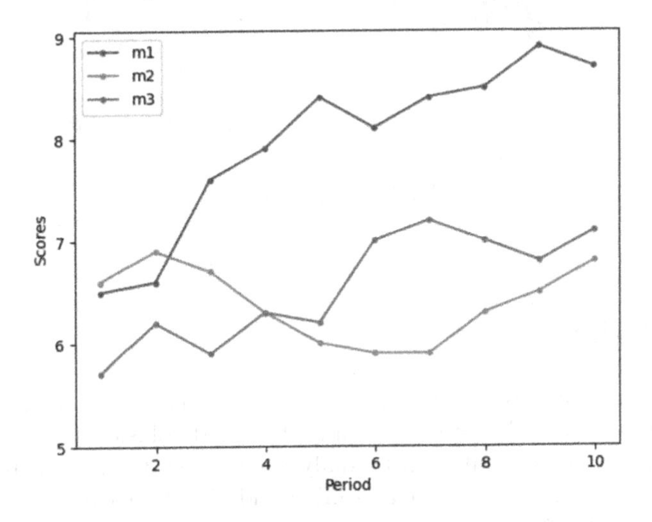

Fig. 4. Service quality score of microgrids changes.

regained, the service quality score of microgrids has also picked up to a certain extent. The transaction behavior of microgrid m_3 has always been in a negative state, the transaction volume has not changed much, and consumers' ratings are relatively stable. This shows that our system has restored the market rules of survival of the fittest in real trading scenarios.

3) Forecast of power generation cost price. The dataset is divided into training set and test set according to the form of 7:3 for model training, the random forest model is used to train the training set and leave-one-out cross-validation. We randomly select 300 samples to compare the actual price with the predicted cost. The experimental results of the random forest model are shown in Fig. 5. In the training set, the accuracy of the random forest model reaches 81.31%, and the average absolute error is 12.105. Moreover, the predicted value is very close to the final price.

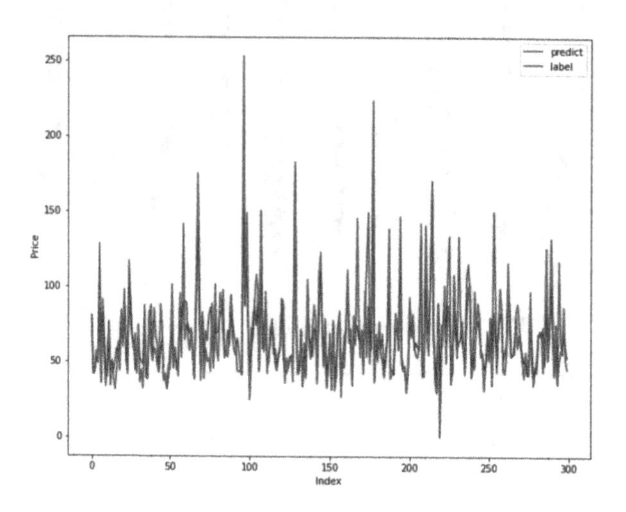

Fig. 5. Price forecast.

7 Conclusion

In this paper, we have proposed a blockchain-based energy trading system, which enables requester-provider utility maximization and collision resistance. Moreover, we have given a careful security analysis for the system, particularly under the case of a dishonest consumer and microgrid. A number of smart contracts are introduced to enhance the security, fairness, two-way price bargaining, and efficiency of the system. After an implementation of the energy trading system, the experiment results show the practical utility of our system.

Acknowledgements. This work was supported by the National Natural Science Foundation of China (61802248, 61972241, U1936213), the "Chenguang Program" supported by Shanghai Municipal Education Commission (No. 18CG62) and the Natural Science Foundation of Shanghai (18ZR1417300).

References

1. https://www.kaggle.com/salilchoubey/electrity-prices
2. Agrawal, R., Kiernan, J., Srikant, R., Xu, Y.: Order preserving encryption for numeric data. In: Proceedings of the 2004 ACM SIGMOD International Conference on Management of Data, pp. 563–574 (2004)
3. Alam, M.T., Li, H., Patidar, A.: Bitcoin for smart trading in smart grid. In: The 21st IEEE International Workshop on Local and Metropolitan Area Networks (2015)
4. Bartoletti, M., Pompianu, L.: An empirical analysis of smart contracts: platforms, applications, and design patterns. In: Brenner, M., et al. (eds.) FC 2017. LNCS, vol. 10323, pp. 494–509. Springer, Cham (2017). https://doi.org/10.1007/978-3-319-70278-0_31
5. Furukawa, J.: Request-based comparable encryption. In: Crampton, J., Jajodia, S., Mayes, K. (eds.) ESORICS 2013. LNCS, vol. 8134, pp. 129–146. Springer, Heidelberg (2013). https://doi.org/10.1007/978-3-642-40203-6_8
6. Hassan, M.U., Rehmani, M.H., Chen, J.: Deal: differentially private auction for blockchain-based microgrids energy trading. IEEE Trans. Serv. Comput. **13**(2), 263–275 (2019)
7. Hu, D., Li, Y., Pan, L., Li, M., Zheng, S.: A blockchain-based trading system for big data. Comput. Netw. **191**, 107994 (2021)
8. Kim, H., Lee, J., Bahrami, S., Wong, V.W.: Direct energy trading of microgrids in distribution energy market. IEEE Trans. Power Syst. **35**(1), 639–651 (2019)
9. Li, H.A., Nair, N.K.C.: Blockchain-based microgrid market and trading mechanism. In: 2018 Australasian Universities Power Engineering Conference (AUPEC), pp. 1–5. IEEE (2018)
10. Mengelkamp, E., Gärttner, J., Rock, K., Kessler, S., Orsini, L., Weinhardt, C.: Designing microgrid energy markets: a case study: the brooklyn microgrid. Appl. Energy **210**, 870–880 (2018)
11. Miller, M.S., Stiegler, M.: The Digital Path: Smart Contracts and the Third World. Routledge (2003)
12. Nakamoto, S.: Bitcoin: a peer-to-peer electronic cash system. Decentralized Bus. Rev. 21260 (2008)
13. Olivares, D.E., et al.: Trends in microgrid control. IEEE Trans. Smart Grid **5**(4), 1905–1919 (2014)
14. Soto, E.A., Bosman, L.B., Wollega, E., Leon-Salas, W.D.: Peer-to-peer energy trading: a review of the literature. Appl. Energy **283**, 116268 (2021)
15. Tikhomirov, S., Voskresenskaya, E., Ivanitskiy, I., Takhaviev, R., Marchenko, E., Alexandrov, Y.: Smartcheck: static analysis of ethereum smart contracts. In: Proceedings of the 1st International Workshop on Emerging Trends in Software Engineering for Blockchain, pp. 9–16 (2018)
16. Ui, T.: Bayesian nash equilibrium and variational inequalities. J. Math. Econ. **63**, 139–146 (2016)
17. Wang, J., Wang, Q., Zhou, N., Chi, Y.: A novel electricity transaction mode of microgrids based on blockchain and continuous double auction. Energies **10**(12), 1971 (2017)
18. Wu, S., Zhang, F., Li, D.: User-centric peer-to-peer energy trading mechanisms for residential microgrids. In: 2018 2nd IEEE Conference on Energy Internet and Energy System Integration (EI2), pp. 1–6. IEEE (2018)

19. Xue, L., Teng, Y., Zhang, Z., Li, J., Wang, K., Huang, Q.: Blockchain technology for electricity market in microgrid. In: 2017 2nd International Conference on Power and Renewable Energy (ICPRE), pp. 704–708. IEEE (2017)
20. Yong, Y., Feiyue, W.: Development status and prospect of blockchain technology. Acta Automatica Sinica **42**(4), 481–494 (2016)
21. Zhang, C., Wu, J., Zhou, Y., Cheng, M., Long, C.: Peer-to-peer energy trading in a microgrid. Appl. Energy **220**, 1–12 (2018)

Evaluating the Parallel Execution Schemes of Smart Contract Transactions in Different Blockchains: An Empirical Study

Jianfeng Shi[1,2], Chengzhi Li[2], Heng Wu[2], Heran Gao[1,2], Songchang Jin[4,5], Tao Huang[2], and Wenbo Zhang[2,3](✉)

[1] University of Chinese Academy of Sciences, Beijing, China
[2] Institute of Software, Chinese Academy of Sciences, Beijing, China
zhangwenbo@otcaix.iscas.ac.cn
[3] State Key Laboratory of Computer Sciences, Institute of Software, Chinese Academy of Sciences, Beijing, China
[4] Defense Innovation Institute, Beijing, China
[5] Tianjin Artificial Intelligence Innovation Center, Tianjin, China

Abstract. In order to increase throughput, more and more blockchains begin to provide the ability to execute smart contract transactions in parallel. However, there is currently no research work on evaluating parallel execution schemes in different blockchains, which makes it difficult for developers to find a blockchain technology suitable for their application scenarios. In this paper, we firstly summarize existing parallel execution schemes of smart contract transactions. Then, based on their characteristics, we propose a comprehensive evaluation framework for parallel execution schemes of smart contract transactions and implement a benchmark tool, which can compare different parallel execution schemes and discover their limitations. Finally, we utilize the evaluation framework and the tool to evaluate several excellent blockchains in China and obtain some useful findings.

Keywords: Blockchain · Smart contract · Parallel execution · Benchmark · Empirical study

1 Introduction

Blockchains, such as Ethereum [2], introduce smart contracts to further expand the application scope of blockchains beyond cryptocurrencies [1]. Therefore, blockchains are expected to have a huge impact on many industries in the future [3].

However, some blockchains, such as Ethereum, serially execute a batch of smart contract transactions within a block, leading to extremely low throughput about 15–20 transactions per second (TPS) [4–6]. Compared with VISA's 1700 TPS [7] and Alipay's 256K TPS [8], the performance of blockchains in handling transactions is obviously not enough to meet actual needs. Therefore, the serial execution of smart contract transactions has become a serious bottleneck and hinders the widespread adoption of blockchains.

Y. Lai et al. (Eds.): ICA3PP 2021, LNCS 13157, pp. 35–51, 2022.
https://doi.org/10.1007/978-3-030-95391-1_3

At present, more and more blockchains begin to provide the ability to execute smart contract transactions in parallel [22–26], however their parallel execution schemes are different. Even though we can know the best throughput from their documents, it is difficult to figure out their advantages and disadvantages, requirements and applicable scenarios. What's more, the existing blockchain testing tools only focus on some basic metrics, such as throughput and latency, which are difficult to apply to testing parallel execution schemes of blockchains. For these reasons, smart contract developers cannot easily evaluate various parallel execution schemes to select the most suitable blockchain technology for their own scenarios.

In this paper, we propose an evaluation framework and implement a benchmark tool to help researchers and developers to analyze the strengths and limitations of parallel execution schemes, so as to obtain a comprehensive overview of blockchains. The contributions in our paper are as follows.

- We classify and summarize the existing parallel execution schemes of smart contract transactions, and elaborate on their representatives.
- We propose an evaluation framework for parallel execution schemes of smart contract transactions. This evaluation framework is conducive to our in-depth understanding of parallel execution schemes of smart contract transactions in blockchains.
- We design and implement a benchmark tool named PEbench, which can test the scalability, the sensitivity to conflict rates and the correctness of parallel execution schemes in blockchains.
- We conduct a comprehensive evaluation on parallel execution schemes of smart contract transactions of FISCO BCOS [9], XuperChain [18] and ChainMaker [10]. The results of our experiments can be used to guide the further research on blockchain technologies.

The rest of this paper is structured as follows. Section 2 summarizes the existing parallel execution schemes for smart contract transactions. Section 3 describes our proposed evaluation framework for parallel execution schemes. Section 4 introduces the design and implementation of PEbench. Section 5 comprehensively evaluates the parallel execution schemes of FISCO BCOS, XuperChain and ChainMaker. Section 6 concludes this paper.

2 Parallel Execution Schemes of Smart Contract Transactions

The transaction processing flow of blockchains usually includes three steps: "execute transactions", "order transactions" and "validate transactions". According to the order of "execute transactions" and "order transactions", we can classify parallel execution schemes into two categories: parallel execution schemes based on the execute-order-validate architecture and parallel execution schemes based on the order-execute-validate architecture.

2.1 Parallel Execution Schemes Based on the Order-Execute-Validate Architecture

In the blockchains that adopt the order-execute-validate architecture, a blockchain node usually selects and orders a batch of transactions from the transaction pool, then uses a deterministic or random scheme to execute them in parallel, and finally generates a complete block.

Deterministic Parallel Execution Schemes. Whether it is a miner node or a validator node, before executing smart contract transactions, as shown in Fig. 1, they will use the source code or other information to analyze the dependencies between smart contract transactions in advance, and then generate a transaction dependency graph so that different threads can execute smart contract transactions in parallel without locks. Obviously, the transaction dependency graph generated by these schemes for the same transaction are deterministic, so there is no need to include the transaction dependency graph in the block. This paper will take FISCO BCOS as an example to elaborate.

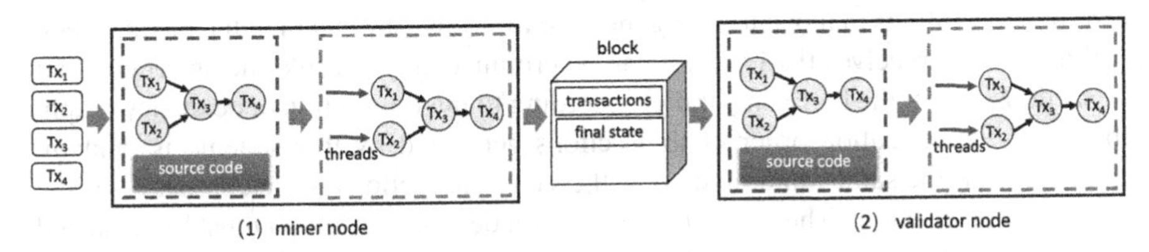

Fig. 1. Deterministic parallel execution scheme.

FISCO BCOS is an enterprise-level blockchain technology in China, with more than 1,000 companies and institutions participating in its community. Developers need to define mutually exclusive parameters when writing smart contracts, and the values of these mutually exclusive parameters will be treated as shared state objects which cannot be accessed concurrently. As shown in Table 1, *transfer()* is registered as a function that can be executed in parallel, and the first two parameters (*from* and *to*) of the function are defined as mutually exclusive parameters.

Table 1. Register functions that can be executed in parallel

```
function enableParallel() public
{
  registerParallelFunction("transfer(string, string, uint256)", 2);
}
```

Before executing a batch of smart contract transactions, the miner node constructs a transaction dependency graph (DAG) based on the values of mutually exclusive parameters, where a vertex represents a smart contract transaction and a directed edge represents the dependency relationship between two transactions.

Based on the transaction dependency graph, the miner node can execute smart contract transactions without conflict in parallel. (1) The miner node initializes a thread pool according to the number of the machine's CPU cores. (2) In the DAG, all transactions with in-degree of 0 can be executed in parallel. After executing a transaction, the in-degree of subsequent vertices of the transaction will be reduced by 1. (3) Repeat step 2 until all smart contract transactions have been executed [9].

The miner node does not store the transaction dependency graph in the new block, that is, when a validator node receives the new block, it needs to recalculate to obtain the transaction dependency graph.

Random Parallel Execution Schemes. The miner node uses concurrency control technologies in the database field to tentatively execute all smart contract transactions in parallel, and dynamically record the dependencies between smart contract transactions according to conflicts that occur during the execution, and save them in the block. When a validator node receives the block, it can deterministically execute the smart contract transactions within the block in parallel according to the transaction dependency graph. Obviously, the execution order of transactions generated by this scheme is random, because even if the same miner node runs the same transactions twice continuously, the results may be different. Therefore, the transaction dependency graph must be packaged into the block. This paper will take ChainMaker as an example to elaborate.

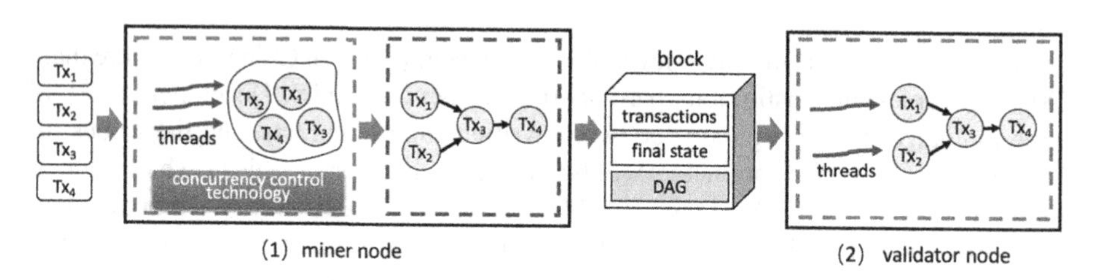

Fig. 2. Random parallel execution scheme.

ChainMaker is a new generation of open source blockchain technology. The miner node divides the execution of each smart contract transaction into two steps: execute and commit. All transactions can be executed directly in parallel at the same time, but the write operations of each transaction will be cached in memory, which does not really take effect. Only when the execution of each transaction is completed, the system will check whether the read set of the transaction has been modified by other transactions. If the read set has not been modified, let the write set of the transaction take effect. Otherwise, the transaction will be rolled back and re-executed [10].

As shown in Fig. 2, when the execution of all transactions within the block are completed, the miner node will analyze the read-write sets of all transactions to construct a DAG. If there is a read-write, write-read, or write-write conflict between any two transactions, an edge will be constructed between the vertices representing them in the DAG according to the execution order.

2.2 Parallel Execution Schemes Based on the Execute-Order-Validate Architecture

In the blockchains that adopt the execute-order-validate architecture, on the one hand, "Execute transactions" is placed before "Order transactions". On the other hand, "Execute transactions" and "Order transactions" can be handled by different nodes respectively. That is to say, multiple nodes can be allowed to execute transactions from different clients in parallel at the same time. Both Hyperledger Fabric [16] and XuperChain adopt this architecture, and this paper will take XuperChain as an example to elaborate.

XuperChain is an open source blockchain technology with a highly flexible architecture and excellent performance. As shown in Fig. 3, A client triggers the simulation execution of the smart contract by sending a transaction proposal to any node, and a read set and a write set will be recorded during the simulation execution. As the name implies, the simulation execution of smart contract based on the current state will not change the current node's blockchain state in any way. The read set consists of tuples (variable name, data version), and the write set consists of tuples (variable name, data value). Finally, the read-write set are returned to the client.

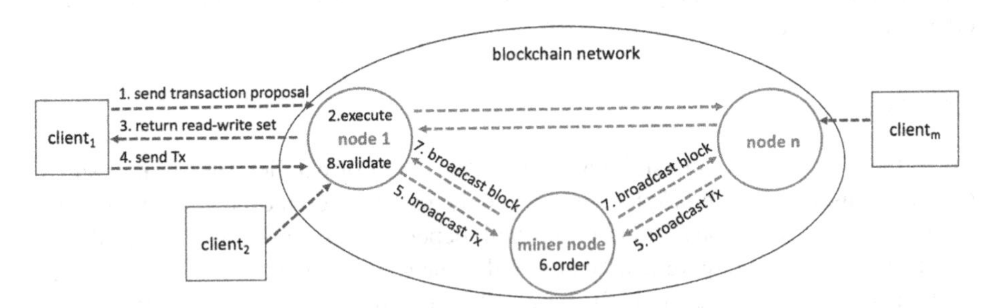

Fig. 3. Parallel execution scheme of XuperChain.

The client uses the read-write set to assemble a transaction, attach its signature, and submit it to any node. And the transaction will be broadcast on the XuperChain's network.

The miner node will collect and order a batch of transactions to generate a new block, and broadcast it to all other nodes for validation.

It can be found that the simulation execution is independent of each other and does not affect each other, so the simulation executions of smart contract transactions can be performed in parallel within a node. At the same time, different nodes can process more simulation execution requests from different clients in parallel, which is an amplification of parallel capabilities.

3 An Evaluation Framework for Parallel Execution Schemes

Based on the characteristics of the parallel execution schemes summarized in Sect. 2, we propose an evaluation framework that includes the following dimensions to help developers have a deeper understanding of the parallel execution schemes of smart contract transactions in blockchains.

3.1 Scalability

Scalability refers to the ability to increase the throughput of the blockchain system by changing the number of CPU cores of each blockchain node or increasing the number of blockchain nodes. This capability will enable blockchain system operators to dynamically adjust the hardware resources of the blockchain system according to their business development, so as to reduce the operation cost.

3.2 Sensitivity to Conflict Rates

Sensitivity to conflict rates refers to the change in throughput when the blockchain system processes transactions with different conflict rates. Transactions generated by different business scenarios have different conflict rates. At the same time, different parallel execution schemes have different sensitivity to conflict rates. Some parallel execution schemes are highly sensitive to changes in transaction conflict rates, while others are less affected.

3.3 Correctness

Correctness refers to the consistency between the result of parallel execution of smart contract transactions by blockchain nodes and the result of serial execution of the same smart contract transactions. Transactions with dependencies may cause data conflicts, so it is the most basic requirement to ensure that the result of parallel execution is equal to the result of serial execution. If the result of parallel execution of smart contract transactions may be wrong, it is meaningless to use this parallel execution scheme to improve the throughput no matter how high the throughput is.

3.4 Altruism

Altruism refers to whether the inter-transaction dependencies or read-write sets found by the miner node after the parallel execution of smart contract transactions will be passed to validator nodes, so that the validator nodes do not need to repeat some work when validating the new block, which improves the efficiency of the blockchain system to verify new blocks.

3.5 Extra Cost

For some blockchains, the parallel execution schemes of smart contract transactions require extra time cost or extra storage cost. Some blockchains even need to roll back and re-execute transactions that are terminated due to conflicts.

Extra time cost means that before executing smart contracts, blockchain nodes need to spend extra time to do some preparatory work, so that they can execute smart contract transactions in parallel without conflict.

Extra storage cost means that a block not only contains smart contract transactions themselves, but also needs to store the dependencies between these transactions.

Rollback and retry refers to the need to roll back and re-execute transactions that are terminated due to conflicts during execution.

3.6 Requirements for Smart Contracts

Requirements for smart contracts mean that smart contracts need to meet some restrictions before they can be executed in parallel correctly. For example, some parallel execution schemes of smart contracts require that the functions of smart contracts that need to be executed in parallel do not call other functions of the same smart contract or the functions of other smart contracts.

3.7 Requirements for Programmers

Requirements for programmers mean that programmers need to provide some information to the blockchain system when writing smart contract code, such as which functions in smart contracts can be executed in parallel and which parameter values need to be mutually exclusive. Therefore, the blockchain system can statically analyze the dependencies between transactions before executing them.

4 PEbench

4.1 Motivation

Faced with so many parallel execution schemes of smart contracts, developers not only want to know the best throughput of these blockchains, but also want to know the strengths and limitations of each parallel execution scheme and their applicable scenarios. However, as shown in Table 2, the existing testing tools for blockchains usually only pay attention to obtaining the best throughput of the whole blockchain system. Therefore, in order to help developers understand each blockchain in depth, we design and implement a benchmarking tool named PEbench, focusing on obtaining the scalability, the sensitivity for conflict rates and the correctness of parallel execution schemes.

Table 2. Comparison of testing tools for blockchains

Testing tool	Purpose	Performance indicators	Supported blockchains
Hyperledger Caliper [14]	Overall performance	Success rate, transaction throughput, transaction latency, resource consumption	Hyperledger Besu, Fabric, Ethereum, FISCO BCOS
Xbench [17]	Overall performance	Same as Hyperledger Caliper	XuperChain
FISCO BCOS java-sdk-demo [19]	Stress testing for serial execution and parallel execution	Throughput, delay distribution and correctness	FISCO BCOS
Truffle [20]	Functional correctness	Accuracy, time consuming and passing rate	Ethereum
BLOCKBENCH [21]	Use a set of micro and macro workloads to compare blockchains	Throughput, latency, scalability, fault tolerance and security	Ethereum, Parity, Hyperledger Fabric
DAGBENCH [30]	evaluate the performance of DAG implementations	Throughput, latency, scalability, success indicator, resource consumption, transaction data size and transaction fee	IOTA, Nano and Byteball
PEbench (our work)	Specific to parallel execution schemes of smart contracts	Scalability, sensitivity for conflict rates, correctness	FISCO BCOS, XuperChain, ChainMaker

4.2 Architecture

The software architecture of PEbench is shown in Fig. 4. The smart contract library contains all smart contracts for testing. The parameters that PEbench allows to set include the blockchain type, total amount of transactions, the conflict rate, the sending rate and the number of CPU cores. The transaction generator will generate a batch of transactions based on the values of these parameters in the configuration. The transaction transmitter is responsible for sending the transactions to the tested blockchain system at the specified rate through the adapter of the blockchain.

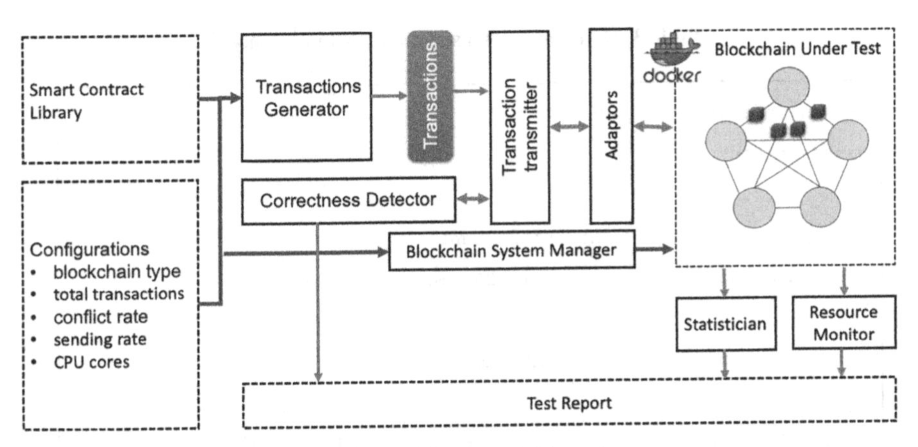

Fig. 4. Architecture of PEbench.

We run each blockchain node in the form of Docker container, so it is easy for the blockchain system operator to control the maximum number of CPU cores that each blockchain node can use. During the execution of transactions, the resource monitor will record the computer resource usage of each blockchain node. The correctness detector will verify the correctness of the parallel execution of smart contract transactions. The statistician will finally calculate the throughput based on the log of a blockchain node, which is more accurate than the client's.

4.3 Generating the Conflicting Transaction Set

We will use the smart contract shown in Table 3 to illustrate how to construct a transaction set with a specific conflict rate. The mapping *account* that contains the data about the amount of money owned by users is a global variable (aka replicated state) of the smart contract.

For the function *add()*, the following two transactions conflict:

Tx_1: add('Tom', 100)

Tx_2: add('Tom', 200)

Therefore, Tx_1 and Tx_2 cannot be executed at the same time in parallel. Only after the execution of Tx_1 is completed, can the execution of Tx_2 be start.

For the function *transfer*(), the following two transactions conflict:

Tx_3: transfer('Tom', 'Jeff', 100)

Tx_4: transfer('Tom', 'Lily', 100)

Therefore, Tx_3 and Tx_4 cannot be executed at the same time in parallel. Only after the execution of Tx_3 is completed, can the execution of Tx_4 be start.

We divide the total transaction set to be sent into a conflicting transaction set and a non-conflicting transaction set. All transactions in the conflicting transaction set are

Table 3. A fragment of a smart contract in solidity

```
pragma solidity ^0.4.25;
contract Bank
{
  mapping (string => uint256) account;
  function add(string name, uint256 money) public
  {
       account[name] = account[name] + money;
  }
  function transfer(string from, string to, uint256 money) public
  {
       account[from] = account[from] - money;
       account[to] = account[to] + money;
  }
}
```

in conflict with each other, that is, blockchain nodes can only execute them serially, and the parallel execution will cause conflicts. All transactions in the non-conflicting transaction set are conflict-free, that is, blockchain nodes can execute them in parallel without conflict.

4.4 Controlling the Number of CPU Cores

Each blockchain node will run in a Docker container. At the beginning of each test, PEbench will restart these Docker containers and set the number of CPU cores allowed for each Docker container through the parameter *cpuset-cpus*, so as to control the maximum number of CPU cores that can be used by each blockchain node when executing smart contract transactions in parallel. It is worth mentioning that we use *sshpass* to help PEbench control different machines.

4.5 Checking the Correctness of Parallel Execution

Ensuring the correctness of execution results is the minimum requirement for parallel execution schemes of smart contracts. After each round of testing, PEbench will compare the result of local serial execution with the result of a blockchain node's parallel execution to ensure that there are no problems such as dirty reads or lost updates in the process of parallel execution of smart contract transactions.

5 Evaluation and Discussion

We select three excellent open source blockchain technologies in China for evaluation and discussion, namely FISCO BCOS(v2.7.2), ChainMaker(v1.2.0) and XuperChain(master branch on 2021/05/27) introduced in Sect. 2, which use different types of parallel execution schemes. Each node of each blockchain runs on a virtual machine equipped with 8 CPU cores, 16 GB memory and 100GB disk, and PEbench runs separately on another virtual machine with the same configuration. All blockchains adopt *solidity* smart contracts and *EVM* virtual machines.

5.1 General Findings

Finding 1. The more CPU cores of each node, the higher throughput the blockchain can obtain. However, when the number of CPU cores increases to a certain number, it is no longer the main factor to improve performance.

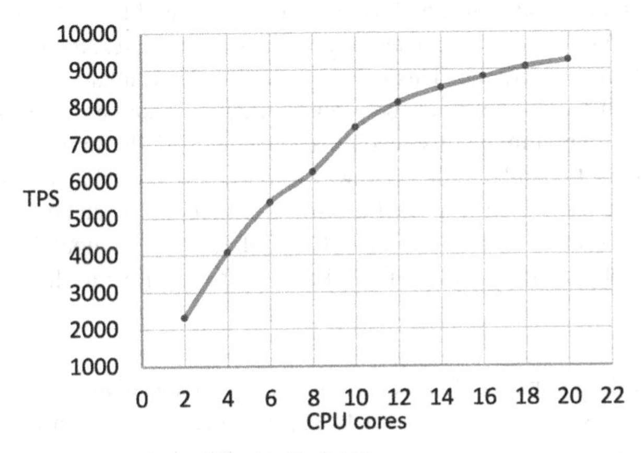

Fig. 5. Scalability test.

By changing the number of CPU cores allowed for each Docker container running a blockchain node, the throughput of different parallel execution schemes is measured. The test result of FISCO BCOS is shown in Fig. 5. The function called by all transactions is *transfer()*. A total of 40000 transactions are generated in each round of testing, and the conflict rate is 0. These transactions are sent to the blockchain node at the rate of 15000 transactions per second. We find that when the number of CPU cores is less than 10, the throughput increases almost linearly with the increase of CPU cores. However, when it exceeds 10, the throughput increases slowly. This is because the execution time accounts for a smaller and smaller proportion of the total time (ordering, execution, consensus, validation, storage, etc.), that is, the execution of smart contract transactions is no longer the major factor limiting the blockchain's throughput.

Finding 2. Random parallel execution schemes are more sensitive to the conflict rate of transactions than deterministic parallel execution schemes.

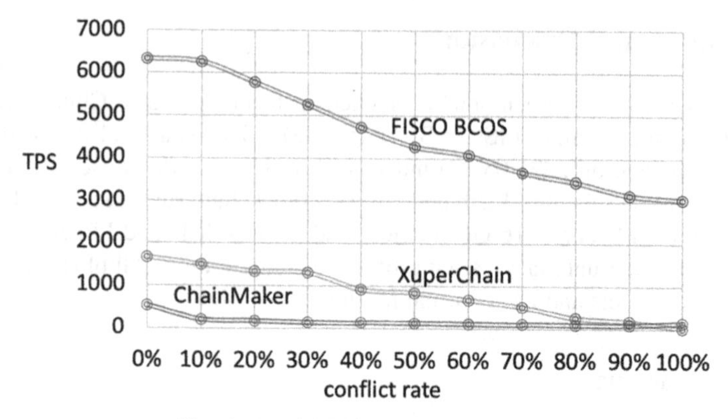

Fig. 6. Sensitivity test for conflict rate.

FISCO BCOS is selected as the representative of the deterministic parallel execution schemes, and ChainMaker is selected as the representative of the random parallel execution schemes. As shown in Fig. 6, as the conflict rate increases, the throughput of all blockchain systems is declining. This is because in the deterministic parallel execution schemes, although the dependencies between transactions can be obtained before execution, more and more transactions need to be executed serially. In the random parallel execution schemes, the high conflict rate will lead to more terminated transactions that need to be rolled back, which reduces the system throughput. It can be found that when the conflict rate increases from 0% to 10%, ChainMaker's throughput drops faster than FISCO BCOS, which shows that ChainMaker is more sensitive to the conflict rate of transactions.

Finding 3. The throughput of a blockchain system is very sensitive to the number of read and write operations in their smart contracts, so it is necessary to optimize the functional logic of smart contracts to reduce the read and write operations on the blockchain ledger (Fig. 7).

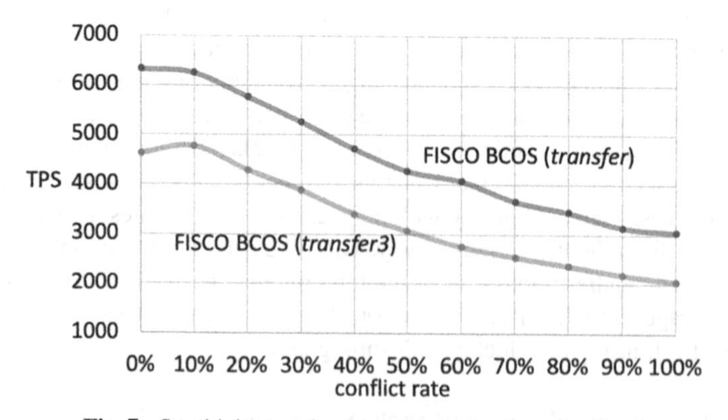

Fig. 7. Sensitivity test for the number of write operations.

As shown in Table 3, function *transfer()* has 2 read operations and 2 write operations, while function *transfer3()*, which transfers money from one user to the other three users, has 4 read operations and 4 write operations. The more write operations to the blockchain ledger in the smart contract, the lower the throughput of the blockchain system.

Finding 4. For a blockchain that contains a transaction dependency graph or a read-write set in each of its blocks, the maximum transaction capacity of the block is usually set lower in the configuration file of the blockchain system.

If the block size is too large, it will increase the communication overhead of the blockchain, which is not conducive to the overall performance. In FISCO BCOS that does not need to store the DAG of transactions, the default maximum transaction capacity of a block is 1000, while in ChainMaker that needs to store the DAG of transactions, the default maximum transaction capacity of a block is 100.

5.2 Evaluation Result

We summarize the evaluation result of the three blockchains as follows (Table 4).

Table 4. Comparison of parallel execution schemes of smart contract transactions

	FISCO BCOS	ChainMaker	XuperChain
Scalability	Middle	Middle	High
Sensitivity to conflict rates	Low	High	Middle
Correctness	Without error	Without error	Without error
Altruism	No	Yes, other nodes execute transactions in parallel according to the DAG in the new block	Yes, other nodes perform parallel verification according to the read-write sets of transactions
Extra time cost	Yes, generate the DAG	No	No
Extra storage cost	No	Yes, the DAG	Yes, the read-write set
Rollback and retry	No	Yes, rollback conflicting transactions and retry	Yes, discard conflicting transactions directly
Requirements for smart contracts	Yes, require no calls to other functions	No	No
Requirements for programmers	Yes, declare parallelizable functions and their mutually exclusive parameters	No	No

5.3 Determinism vs. Randomness

The advantage of deterministic parallel execution schemes is that different nodes executing the same batch of smart contract transactions can produce the same schedule. Therefore, it is not necessary to store a DAG in each block, thereby saving the communication and storage overhead of the block. In addition, this solution does not require locks, and is more tolerant of changes in the conflict rate because the dependencies between transactions have been obtained through static analysis before execution.

In random parallel execution schemes, the DAG generated by executing the same batch of transactions multiple times may be different. Therefore, in order to ensure the consistency of the order in which the validator nodes and the miner node execute this batch of transactions, the miner node needs to store the DAG generated by itself in the new block. Although this scheme increases the communication and storage burden, it speeds up the validation speed of the validator nodes. In addition, this solution may require using locks and is very sensitive to changes in the conflict rate.

5.4 Order-Execute-Validate vs. Execute-Order-Validate

In a blockchain with the "order-execute-validate" architecture, "order transactions" and "execute transactions" are usually completed on one node. Obviously, under this architecture, the throughput of a blockchain system is limited by the processing capacity of the miner node, and the performance of the blockchain system will not be improved as the number of nodes increases.

In a blockchain that adopts the "execute-order-validate" architecture, "order transactions" and "execute transactions" are separated, and they can be on different nodes. In addition, the "execute transactions" is placed before the "order transactions". Obviously, the processing capacity of the blockchain system can be improved by adding blockchain nodes and allowing different blockchain nodes to execute transactions from different clients in parallel. The disadvantage of this scheme is that it is not suitable for scenarios with high conflict rates [28, 29], because the execution of each transaction is independent, and it is only checked for conflicts with other transactions when submitting for storage. If the data version referenced by a transaction has expired, the execution of the transaction will fail directly.

5.5 Sharding

Sharding [12, 13] is another parallel execution scheme that is not described in this paper. Sharding divides the whole blockchain network into several subnetworks, and each subnetwork named shard has its own state and ledger. All new transactions will be allocated to different shards, and nodes in different shards can independently execute and validate its own smart contract transactions in parallel.

Obviously, each shard is actually a smaller blockchain, so sharding is an inter-chain parallel scheme, and the schemes described in this paper are the intra-chain parallel schemes, which really make full use of the parallel processing capabilities of the machines' multi-core CPU.

5.6 Are Our Examples Enough

Aside from the parallel execution schemes of smart contract transactions mentioned above, there are mainly sidechain, multi-chain and DAG implementations [34]. Sidechain and multi-chain are the same as sharding. Their principle is divide-and-conquer, that is, all smart contract transactions are divided into different small groups of nodes according to a certain rule to achieve parallelism. As described in Sect. 5.5, they are all inter-chain parallel execution schemes, while our paper discusses intra-chain parallel execution schemes. DAG implementation is a new form of distributed ledger technology. It uses DAG data structure instead of block-based data structure that has problems such as poor performance and poor scalability [30]. According to its definition, DAG implementations do not belong to the blockchains discussed in our paper.

In addition, we did not evaluate Ethereum because it executes smart contract transactions serially, and did not consider Hyperleger fabric because it uses a similar parallel execution scheme with XuperChain.

This paper categorizes parallel execution schemes of smart contracts in Sect. 2, where FISCO BCOS is the representative of deterministic parallel execution Schemes, ChainMaker is the representative of random parallel execution schemes, and XuperChain is the representative of parallel execution schemes based on the execute-order-validate architecture. Therefore, we choose to evaluate FISCO BCOS, ChainMaker and XuperChain is reasonable.

6 Related Work

Kuzlu [27] analyzed the performance of Hyperledger fabric and Wang [15] evaluated Ethereum, Hyperledger fabric, Sawtooth and FISCO BCOS in terms of throughput, latency and scalability. Dabbagh [33] proposed a comparison framework containing 10 criteria, including blockchain type, consensus algorithm and performance metrics etc. In addition, Alsunaidi [31] and Bamakan [32] evaluated the consensus algorithms in the blockchain in terms of mining efficiency, incentive, performance and security etc. As far as we know, there is currently no work to evaluate blockchains from the perspective of parallel execution of smart contract transactions.

7 Conclusion

Parallel execution of smart contract transactions is an important way to increase the throughput of the blockchain system. In order to evaluate emerging parallel execution schemes for smart contract transactions, we first summarize the existing parallel execution schemes, and then propose a comprehensive evaluation framework and implement a benchmark tool named PEbench. With the help of PEbench, we conduct an in-depth evaluation of three outstanding open source blockchain technologies in terms of parallel execution schemes, and obtain some useful findings.

Acknowledgements. Thanks to the developers of XuperChain, FISCO BCOS and ChainMaker for their selfless guidance and help, as well as their open source communities. This work is supported by National Key Research and Development Program of China (Grant No. 2018YFB1402803) and National Natural Science Foundation of China (Grant No. 61872344, 61972386).

References

1. Nakamoto, S.: Bitcoin: A peer-to-peer electronic cash system. Decentral. Bus. Rev. 21260 (2008)
2. Ethereum. https://www.ethereum.org/zh/. Accessed 15 Sept 2021
3. Yao, Q., Zhang, DW.: Survey on identity management in blockchain. J. Softw. (2021)
4. Pîrlea, G., Kumar, A., Sergey, I.: Practical smart contract sharding with ownership and commutativity analysis. In: Proceedings of the 42nd ACM SIGPLAN International Conference on Programming Language Design and Implementation, pp. 1327–1341 (2021)
5. Meneghetti, A., Parise, T., Sala, M., Taufer, D.: A survey on efficient parallelization of blockchain-based smart contracts. Ann. Emerg. Technol. Comput. (AETiC) (2019). Print ISSN 2516-0281
6. Shihab, Shahriar, Hazari, Qusay, H., Mahmoud.: A parallel proof of work to improve transaction speed and scalability in blockchain systems. In: 2019 IEEE 9th Annual Computing and Communication Workshop and Conference (CCWC), IEEE (2019)
7. Shevkar, R.: Performance-based analysis of blockchain scalability metric. Tehnički glasnik **15**(1), 133–142 (2021)
8. Qin, C., Guo, B., Shen, Y., Li, T., Zhang, Y., Zhang, Z.: A secure and effective construction scheme for blockchain networks. Secur. Commun. Networks (2020)
9. FISCO BCOS. https://fisco-bcos-documentation.readthedocs.io. Accessed 15 Sept 2021/
10. ChainMaker. https://docs.chainmaker.org.cn/index.html. Accessed 15 Sept 2021
11. Lian, Y., Tsai, W. T., Li, G., Yao, Y., Deng, E.: Smart-contract execution with concurrent block building. In: 2017 IEEE Symposium on Service-Oriented System Engineering (SOSE), IEEE, pp. 160–167 (2017)
12. Wang, G., Shi, Z. J., Nixon, M., Han, S.: Sok: Sharding on blockchain. In: Proceedings of the 1st ACM Conference on Advances in Financial Technologies, pp. 41–6 (2019)
13. Yu, G., Wang, X., Yu, K., Ni, W., Liu, R.P.: Survey: sharding in blockchains. In: IEEE Access, vol. 1–1 pp. 99, (2020)
14. Hyperledger Caliper. https://github.com/hyperledger/caliper. Accessed 15 Sept 2021
15. Wang, R., Ye, K., Meng, T., Xu, C.Z.: Performance evaluation on blockchain systems: a case study on Ethereum, Fabric, Sawtooth and Fisco-Bcos. In: International Conference on Services Computing, pp. 120–134, Springer, Cham. https://doi.org/10.1007/978-3-030-59592-0_8
16. Hyperledger Fabric. https://github.com/hyperledger/fabric. Accessed 15 Sept 2021
17. Xbench. https://github.com/xuperchain/xbench. Accessed 15 Sept 2021
18. XuperChain. https://xuper.baidu.com/n/xuperdoc/index.htm. Accessed 9 Sept 2021
19. FISCO BCOS's stress test program. https://github.com/FISCO-BCOS/java-sdk-demo. Accessed 15 Sept 2021
20. Truffle. https://www.trufflesuite.com/docs/truffle/testing/testing-your-contracts. Accessed 15 Sept 2021
21. Dinh, T., Wang, J., Chen, G., Liu, R., Ooi, B. C., Tan, K. L.: Blockbench: a framework for analyzing private blockchains. In: Proceedings of the 2017 ACM International Conference on Management of Data, pp. 1085–1100 (2017)

22. Anjana, P.S., A Tt Iya, H., Kumari, S., Peri, S., Somani, A.: Efficient concurrent execution of smart contracts in blockchains using object-based transactional memory. In: International Conference on Networked Systems, pp. 77–93. Springer, Cham (2020). https://doi.org/10.1007/978-3-030-67087-0_6
23. Anjana, P.S., Kumari, S., Peri, S., Rathor, S., Somani, A.: Optsmart: a space efficient optimistic concurrent execution of smart contracts. arXiv preprint arXiv:2102.04875 (2021)
24. Amiri, M.J., Agrawal, D., Abbadi, A.E.: ParBlockchain: Leveraging transaction parallelism in permissioned blockchain systems. In: 2019 IEEE 39th International Conference on Distributed Computing Systems (ICDCS), IEEE, pp. 1337–1347 (2019)
25. Yu, W., Luo, K., Ding, Y., You, G., Hu, K.: A parallel smart contract model. In: Proceedings of the 2018 International Conference on Machine Learning and Machine Intelligence, pp. 72–77 (2018)
26. Dickerson, T., Gazzillo, P., Herlihy, M., Koskinen, E.: Adding concurrency to smart contracts. Distrib. Comput. **33**(3–4), 209–225 (2019). https://doi.org/10.1007/s00446-019-00357-z
27. Kuzlu, M., Pipattanasomporn, M., Gurses, L., Rahman, S.: Performance analysis of a hyperledger fabric blockchain framework: throughput, latency and scalability. In: 2019 IEEE International Conference on Blockchain (Blockchain), pp. 536–540, IEEE (2019)
28. Ruan, P., Loghin, D., Ta, Q.T., Zhang, M., Chen, G., Ooi, B.C.: A transactional perspective on execute-order-validate blockchains. In: Proceedings of the 2020 ACM SIGMOD International Conference on Management of Data, pp. 543–557 (2020)
29. Sharma, A., Schuhknecht, F.M., Agrawal, D., Dittrich, J.: Blurring the lines between blockchains and database systems: the case of hyperledger fabric. In: Proceedings of the 2019 International Conference on Management of Data, pp. 105–122 (2019)
30. Dong, Z., Zheng, E., Choon, Y., Zomaya, A.Y.: Dagbench: A performance evaluation framework for DAG distributed ledgers. In: 2019 IEEE 12th International Conference on Cloud Computing (CLOUD), IEEE, pp. 264–271 (2019)
31. Alsunaidi, S.J., Alhaidari, F.A.: A survey of consensus algorithms for blockchain technology. In: 2019 International Conference on Computer and Information Sciences (ICCIS), IEEE, pp. 1–6 (2019)
32. Bamakan, S.M.H., Motavali, A., Bondarti, A.B.: A survey of blockchain consensus algorithms performance evaluation criteria. Exp. Syst. Appl. **154**, 113385
33. Dabbagh, M., Choo, K.K.R., Beheshti, A., Tahir, M., Safa, N.S.: A survey of empirical performance evaluation of permissioned blockchain platforms: Challenges and opportunities. Comput. Security **100**, 102078
34. Zhou, Q., Huang, H., Zheng, Z., Bian, J.: Solutions to scalability of blockchain: a survey. IEEE Access **8**, 16440–16455 (2020)

Misbehavior Detection in VANET Based on Federated Learning and Blockchain

Pin Lv[1,3], Linyan Xie[1], Jia Xu[1,3](✉), and Taoshen Li[1,2]

[1] School of Computer Electronics and Information, Guangxi University, Nanning 530004, China
{lvpin,xujia}@gxu.edu.cn

[2] China-ASEAN International Join Laboratory of Integrated Transport, Nanning University, Nanning 541699, China

[3] Guangxi Key Laboratory of Multimedia Communications and Network Technology, Nanning 530004, China

Abstract. As an irreversible trend, connected vehicles become increasingly more popular. They depend on the generation and sharing of data between vehicles to improve safety and efficiency of the transportation system. However, due to the open nature of the vehicle network, dishonest and misbehaving vehicles may exist in the vehicular network. Misbehavior detection has been studied using machine learning in recent years. Existing misbehavior detection approaches require network equipment with powerful computing capabilities to constantly train and update sophisticated network models, which reduces the efficiency of the misbehavior detection system due to limited resources and untimely model updates. In this paper, we propose a new federated learning scheme based on blockchain, which can reduce resource utilization while ensuring data security and privacy. Further, we also design a blockchain-based reward mechanism for participants by automatically executing smart contracts. Common data falsification attacks are studied in this paper, and the experimental results show that our proposed scheme is feasible and effective.

Keywords: VANET · Federated learning · Blockchain · Smart contract · Misbehavior detection

1 Introduction

Vehicular ad-hoc network (VANET) is a promising solution to reduce traffic accidents, improve traffic efficiency, and promote infotainment via vehicle-to-vehicle (V2V) and vehicle-to-infrastructure (V2I) communications. With the increase

Supported by National Natural Science Foundation of China (NSFC) under Grant Nos. 62062008 and 62062006; Special Funds for Guangxi BaGui Scholars; Guangxi Natural Science Foundation under Grant No. 2019JJA170045; Guangxi Science and Technology Plan Project of China under Grant No. AD20297125.

Y. Lai et al. (Eds.): ICA3PP 2021, LNCS 13157, pp. 52–64, 2022.
https://doi.org/10.1007/978-3-030-95391-1_4

of connected and automated vehicles, VANET is an indispensable part of the future transportation systems, which is safer, smarter and more efficient.

However, VANET is vulnerable to a range of external or internal attacks which can pose a danger to drivers and passengers. External attacks can be defended against using cryptographic schemes [10]. An effective way to detect internal attacks is to use misbehavior detection systems (MDSs). Data-centric misbehavior detection solutions have been widely suggested due to suitability for critical applications, privacy protection and distributed environments [3]. Data falsification attack is one of the highly concerned internal attacks. Attackers can disrupt VANET and spoof other vehicles by broadcasting false position information [12].

In recent years, a number of machine learning-based methods have been proposed to improve the accuracy of attack detection. Traditional machine learning algorithms include K-nearest neighbor (KNN), random forests, decision trees, etc. Deep neural networks such as convolutional neural networks (CNN) and multi-layer perceptrons (MLP) are also widely used in this area. Nevertheless, the above solutions require centralized training which adds computing and storage burden to the central server. These solutions are also prone to suffer from data privacy leakage issues [9].

To address the above issues, this paper proposes a scheme combining federated learning and blockchain to accurately detect data falsification attacks. The convergence and complementarity of the two technologies can effectively address the need for data sharing in this field under the protection of privacy. The main contributions of this research work are as follows:

- An efficient and reliable collaboration mechanism based on federated learning is proposed for participants who lack trust when sharing data for detection.
- A data falsification detection model based on blockchain and federated learning is proposed. This model is compared with three other deep learning models (multilayer perceptrons, convolutional neural networks, recurrent neural networks).
- A smart contract is designed to record contributions of participants and reward participants based on the quality of model training. The effectiveness of the proposed method is on real dataset.

The remainder of this paper is organized as follows. In Sect. 2, related work is summarized. The system model is described in Sect. 3. In Sect. 4, the effectiveness of the detection framework is verified by experiments. Finally, conclusions are drawn in Sect. 5.

2 Related Work

Due to various security threats that have emerged in recent years in VANET, many research efforts have been proposed to detect and mitigate these threats.

In [15], the authors propose a scheme which uses plausibility checks as a feature vector for machine learning model to detect position falsification attacks.

A feature vector with six dimensions is generated, and experiments are conducted using CNN and supported vector machine (SVM).

Grover *et al.* propose a method for detecting position falsification attacks based on machine learning [4]. They create a framework for binary and multi-class classification that distinguishes between various legitimate and improper behavior.

Rawat *et al.* present a solution for detecting data falsification attack based on hash chains by regulating the size of contention window in order to transmit accurate information to the neighbor vehicles promptly [11].

Gyawali *et al.* suggest a machine learning-based misbehavior detection system to detect false alert attack and position falsification attack. However, this work mainly focuses on false alert attack [5].

Singh *et al.* propose a machine learning-based model to detect false position information transmitted by the misbehaving vehicles [14]. Two classifiers, including logistic regression and support vector machine, are used to build the model.

Sharma and Liu propose a data-centric misbehavior detection model using supervised machine learning [13]. This work combines plausibility checks with machine learning techniques as well, and instantiates the model with six algorithms to verify their comparative validity.

As it can be noticed, most existing methods employ machine learning to detect position falsification attacks, requiring a large number of data sets to generate accurate models. However, considering communication overheads, privacy concerns and single points of failure, it is not feasible to collect a large number of data sets and store them in a centralized server.

3 System Model

Our model is designed to detect misbehaving vehicles in VANET by training with local data generated by each vehicle. Parameters in the locally trained model are uploaded. A model aggregator collects and aggregates the model parameters, and uploads the updated parameters again. Each vehicle updates the local model and starts the next iteration. The above process is repeated until the whole training process converges.

3.1 Entities in the Model

The misbehavior detection system in vehicular networks is generally composed of the following four types of nodes:

- *Participants*: Participants are vehicles with high mobility, and they move constantly in the road network. Each vehicle is represented as a node in the vehicular networks. Vehicles can communicate with other vehicles by sending messages. Due to its dynamic character, each moving vehicle has different misbehavior detection data.

- *Model aggregators*: Roadside units (RSUs) are typically regarded as model aggregators. RSUs normally have a large amount of processing, storage, and communication capacities, and they are stationary. They update the new aggregation model by using machine learning trained models submitted by vehicles in the region.
- *Blockchain*: Blockchain provides a secure and reliable environment for participants who lack mutual trust. The request and initialization model for shared data generates a transaction in the blockchain, which is visible to all nodes. Participants who satisfy the request upload their locally trained model to the blockchain, so that all model aggregators can retrieve the models. The model is constantly updated until the required task is completed.
- *Smart contract*: In the process of misbehavior detection, smart contracts record the contributions of participants involved in model updates and reward them according to the quality of model training.

3.2 Approach

The data generated by shared vehicles for training models is extremely accessible to attackers. Attackers can also use corresponding techniques to analyze these data to obtain private information about vehicle owners. Cyber attacks such as Sybil attacks and denial of service attacks are well-researched topics.

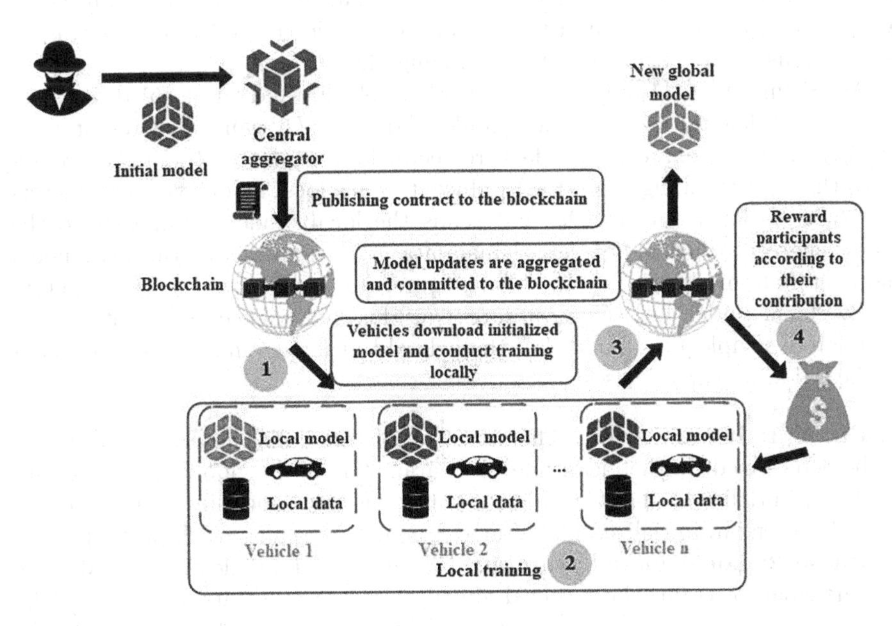

Fig. 1. An overview of proposed framework.

The proposed framework is illustrated in Fig. 1. Federated learning and blockchain are used in the framework, and the vehicles do not need to share data to the central authority. The relevant data is stored on the blockchain, and participants are later rewarded based on their contributions using smart contract. The higher the quality of the parameters uploaded by the participants, the better the effect on model aggregation. In general, the process is described as follows:

1. When a vehicle travels into the area of a RSU, it can request to share local data and update the initialization model. These operations are packaged into a transaction that is visible to all participants.
2. The vehicle obtains the pre-trained misbehavior detection model, and uses local data to train further on the basis of the existing model.
3. The vehicle uploads the trained model to the blockchain. The model aggregator performs a federated averaging on the participant-trained model, and uploads the aggregated model to the blockchain. Each participant can update the model to the local and continue training locally, which is an iterative process.
4. After completing the model aggregation task, the participants can be rewarded for their contributions.

3.3 Federated Averaging

After training the detection model using local data from the vehicle, all model weights need to be aggregated to generate a new global model. This averaging is accomplished by using federated averaging algorithm [8].

We define $N = \{1,...,n\}$ as the set of n vehicles and the local dataset on vehicle k is denoted as D_k. The sample of data in D_k can be represented by (x_j, y_j), where x_j expresses the feature vector for sample $j \in D_k$, and y_j represents the corresponding label. Each vehicle trains a local model with parameters based on its local dataset D_k, and sends the local model parameters to the model aggregator instead of the original data [7]. After receiving the local model parameters from all vehicles, the model aggregator aggregates them to generate a new global model. Let $f(x_j, y_j; \omega)$, or $f_j(\omega)$ for convenience, represent the loss function of sample j. In particular, the federal learning training process consists of the following three phases:

- **Phase 1: Initialization.** The structure of the global model first needs to be determined. In the meantime, the parameters of the global model are randomly initialized or pre-trained on a public dataset, depending on the training task. Then, an initialized global model ω_0 will be sent to each participant.
- **Phase 2: Local model training.** In the t-th communication round, each participant k trains the received global model based on its local dataset by letting $\omega_t^k \leftarrow \omega_t$, and gets the local model parameters ω_t^k. In other words, the goal of participant k is to minimize the empirical loss $F(\omega_t^k)$ by training the local model, i.e.,

$$\omega_t^k = \arg\min_{\omega_t^k} F(\omega_t^k) \tag{1}$$

$$F(\omega_t^k) = \frac{1}{|D_k|} \sum_{j \in D_k} f_j(\omega_t^k) \tag{2}$$

where $|D_k|$ represents the number of samples in dataset D_k, and $f_j(\omega_t^k)$ denotes a local loss function.
In the updating process, at each iteration, each client maintains a local model by local stochastic gradient descent (SGD) [17] as

$$\omega_{t+1}^k \leftarrow \omega_t^k - \eta \nabla F(\omega_t^k) \tag{3}$$

where $\nabla F(\omega_t^k)$ expresses the gradient of loss function, η denotes the learning rate.
- **Phase 3: Global aggregation.** The model aggregator aggregates all participants' local updated parameters to generate the global model ω_{t+1} using typical aggregation algorithms. Specifically, the objective is to minimize the global loss function, which can be represented as follows:

$$F(\omega_t) = \frac{1}{|D|} \sum_{k=1}^{N} |D_k| F(\omega_t^k), k \in 1, 2, ..., N \tag{4}$$

The above processes are repeated until the required accuracy is achieved.

3.4 Contribution Smart Contract

For models shared by different participants, different contribution values are generated, taking into account differences in dataset size or hardware. The smart contract is deployed in the blockchain, and is invoked via its address. The smart contract rewards participants accordingly based on their contribution values. When a contract reaches a trigger condition, the smart contract can automatically execute the contract without the intervention of a third party, providing greater reliability and higher efficiency.

4 Implementation and Evaluation

4.1 VeReMi Dataset

We experiment with the VeReMi dataset [6], a publicly available labeled dataset for analyzing misbehavior detection mechanisms in VANET.

The VeReMi dataset provides a wide range of traffic behaviors and attacker implementations. The simulations are conducted in the LuST scenario [1], which is designed to supply a comprehensive scenario for evaluation in the VEINS simulator [16]. The dataset is generated by simulating it under different attacker densities and traffic densities, which contains five different types of attackers. Detailed descriptions of the attacker types are shown in Table 1.

Table 1. VeReMi attack types

Attack ID	Type of attack	Description	Parameters
1	Constant	Attacker transmits a fixed location	$x = 5560$, $y = 5820$
2	Constant Offset	Attacker transmits a fixed offset added to the real position	$\Delta x = 250$, $\Delta y = -150$
4	Random	Attacker transmits a random position inside the simulation area	uniformly random in playground
8	Random Offset	Attacker transmits a random position in a rectangle around the vehicle	Δx, Δy are uniformly random from [−300, 300]
16	Eventual Stop	Attacker vehicle behaves normal for some time and then transmits a constant current position	Stop probability increases by 0.025 each position update

4.2 Evaluation Metrics

The confusion matrix is commonly used in misbehavior detection as metrics to assess classification performance, and it is a visual display tool to evaluate the strengths, weaknesses and differences of classification models. As shown in Table 2, based on the elements of the confusion matrix, the following metrics are used to evaluate the detection: accuracy, precision, recall and F1-score. These performance metrics are widely used in the machine learning and classification [2].

Table 2. Confusion matrix

		Predicted class	
		Negative	Positive
Actual class	Negative	True Negative (TN)	False Positive (FP)
	Positive	False Negative (FN)	True Positive (TP)

- Accuracy is the ratio of correctly predicted detection to the total detection.

$$Accuracy = \frac{TP + TN}{TP + FP + TN + TF} \tag{5}$$

- Precision represents the ratio of correctly predicted attackers to the total predicted attackers.

$$Precision = \frac{TP}{TP + FP} \tag{6}$$

- Recall shows the ratio of correctly predicted attackers to the total actual attackers.

$$Recall = \frac{TP}{TP + FN} \tag{7}$$

- F1-score is the weighted average of precision and recall.

$$F1 - score = 2 \times \frac{Precision \times Recall}{Precision + Recall} \tag{8}$$

The formulas of these metrics are shown above, with TP representing attacker vehicle are detected as attacker, TN indicating legitimate vehicle are detected as legitimate, FP meaning legitimate vehicle are detected as attacker, and FN denoting attacker vehicle are not detected as attacker.

4.3 Performance Analysis

The third-party libraries Pytorch for deep learning and Syft for federated learning are used to implement the proposed model. The system model is simulated on a Mac OS Catalina workstation with the Quad-Core Intel Core i5, 64-bit, 2 GHz CPU based on Python programs.

In federated learning, we compare the performance of multi-layer perceptron (MLP) with two other well-known deep learning models, namely, recurrent neural network (RNN) and convolutional neural network (CNN).

All the models are all multi-class classifiers, and the available dataset is split into two parts. One part has 80% of the data, and it is used as the training set. The other 20% of the data is for testing. The Adam optimizer is used with a learning rate of 0.01, and CrossEntropyLoss is chosen as the loss function in the training models.

Table 3. Performance comparison among MLP, RNN and CNN

Model	Accuracy	Precision	Recall	F1-score
MLP	**98.24%**	**98.24%**	**98.27%**	**98.25%**
RNN	97.27%	97.26%	97.23%	97.24%
CNN	97.75%	97.76%	97.75%	97.75%

It can be seen from Table 3, the three deep learning models perform similarly on the four metrics. In contrast, the MLP significantly outperforms the other two deep learning models in terms of training time costs. This is because the structure of the connections between MLP nodes is simpler compared to

RNN. Connections between the nodes in RNN form a cycle, while connections between the nodes in MLP do not form a directed connectivity. In addition, RNN converges more slowly. As for CNN, the use of convolutional functions typically increases an additional time overhead than the threshold functions typically used by MLP for hidden layers. Therefore, we consider that MLP is a deep learning model more suitable for federated learning as it has a higher accuracy rate and lower training time cost.

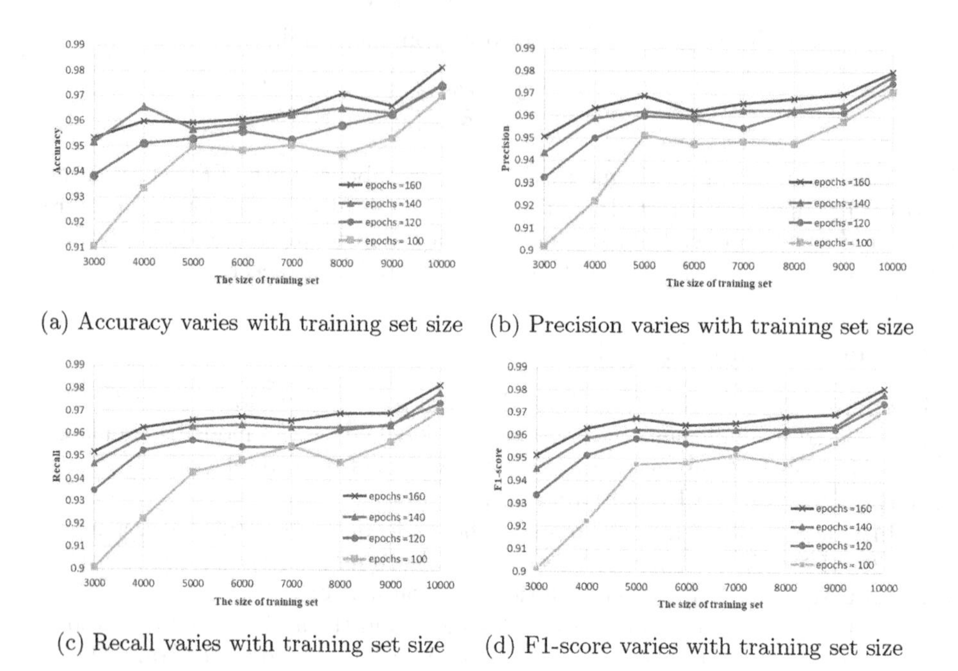

(a) Accuracy varies with training set size (b) Precision varies with training set size

(c) Recall varies with training set size (d) F1-score varies with training set size

Fig. 2. Comparison of accuracy, precision, recall and f1-score under different size training sets and different epochs.

Then, we test the accuracy under different training dataset sizes and different epochs. An epoch is the process of training all the training dataset once. Figure 2(a) shows the accuracy increases when training dataset size increases. The accuracy also increases with epoch for the same size of training dataset. The increase is not significant when the epoch is in the range of 140–160, indicating that the model basically reaches convergence when the epoch is close to 140.

Figure 2(b), Fig. 2(c) and Fig. 2(d) show that under the same epoch, precision, recall and F1-score increase with the increment of the training dataset size. The experimental results indicated that with sufficient training dataset and epochs conditions, the precision, recall and F1-score of the algorithm could achieve a high level.

Figure 3 presents that variations in the accuracy of the aggregation model when different numbers of vehicles collaborate during the federated learning training process. As it can be seen from Fig. 3 that the changing trend of the model accuracy is consistent under different numbers of vehicles. This is because, with federated learning, each collaborative vehicle has the same status as a contributor. After uploading the model parameters to the model aggregator, the model aggregator uses average arithmetic to aggregate the model parameters uploaded by several uploaders. Thus, the trend of the aggregation model accuracy is the same when the number of vehicles involved in the training varies.

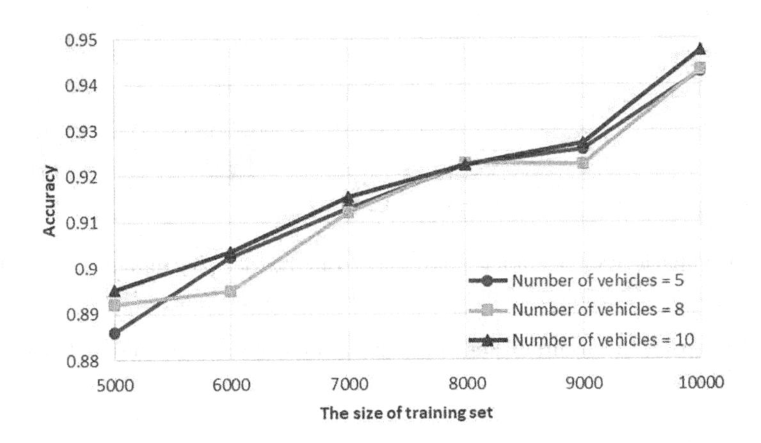

Fig. 3. Accuracy in various number of vehicles.

In addition, we compare both federated training and centralized training methods, and analyze the precision, recall and F1-score value. It can be discovered in Table 4 that, the precision and F1-score values of centralized training are generally better than those of federated training. For attacks of Type 2, the recall value for centralized training is slightly lower than for federated training.

Table 4. Attack type detection results

Attack		Type 1	Type 2	Type 4	Type 8	Type 16
Precision	Federated training	0.9819	0.9490	1.0000	0.9974	0.9862
	Centralized training	0.9918	0.9841	1.0000	0.9975	0.9866
Recall	Federated training	1.0000	0.9920	0.9976	0.9750	0.9619
	Centralized training	1.0000	0.9893	0.9976	0.9850	0.9888
F1-score	Federated training	0.9945	0.9700	0.9988	0.9861	0.9739
	Centralized training	0.9959	0.9867	0.9988	0.9912	0.9877

Figure 4 illustrates the accuracy and communication overhead of federated versus centralized training. As depicted in Fig. 4(a), the accuracy of centralized training is slightly higher than that of federated training, with 98.24% accuracy for federated training and 99.08% accuracy for centralized training. However, in terms of communication overhead, federated training is much higher than centralized training. We compare the communication overhead mainly by the amount of data exchanged between the vehicle and the server. In centralized training, all training data need to be uploaded. On the contrary, federated training only needs to upload the locally trained parameters.

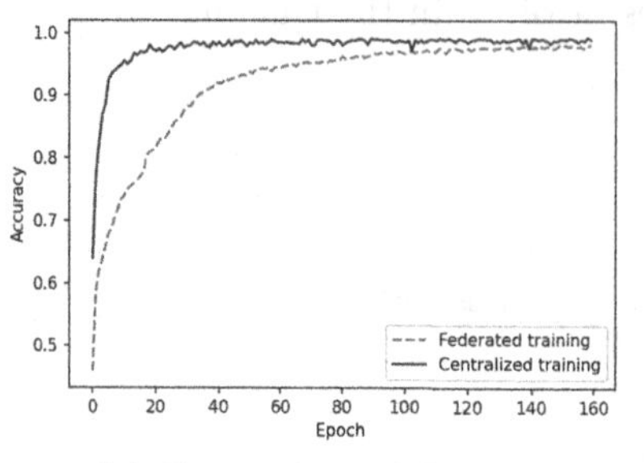

(a) Comparison of Accuracy

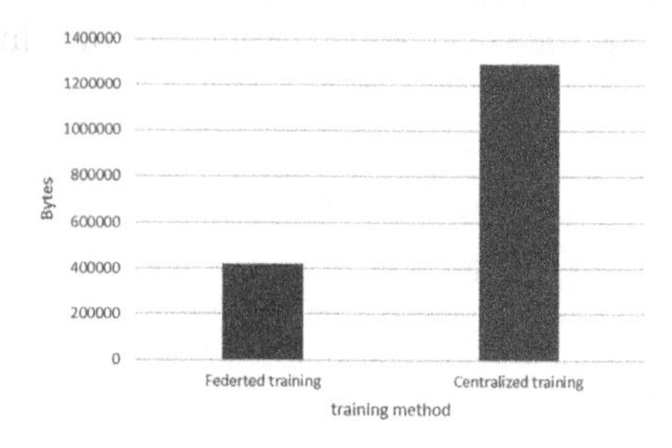

(b) Comparison of Communication Overhead

Fig. 4. Comparing the accuracy and communication overhead of federated versus centralized training.

Assume that T and D represent the amount of data to be uploaded and downloaded and the size of the data set, respectively. The sizes of the federated (S_F) and centralized (S_C) models can be represented as follows:

$$S_F = \sum_{1}^{E} \sum_{k=1}^{N} (T_k) \tag{9}$$

$$S_C = \sum_{k=1}^{N} (D_k) \tag{10}$$

where N denotes the number of vehicles and E is the epoch. We experimentally compare the communication overhead of the two training methods. For centralized training, its communication overhead is the size of all the VeReMi datasets we use for training. In reality, the data for each node is usually collected by the central server, and the sum of these data is the size of the data for centralized training. In our work, we record the amount of data communicated between the server and the node in 160 epochs. It can be found in Fig. 4(b) that the communication overhead for federated training is 422.4 KB, while the communication overhead for centralized training is 1.3 MB. With the increment of traffic flow, the communication overhead of centralized training continues to increase. The accuracy of federated training is only 0.84% lower than centralized training, but the communication overhead is much less than it, which indicates that our solution is feasible.

5 Conclusion

In VANET, malicious vehicles disrupt the VANET environment through data falsification attack. We proposed a collaborative misbehavior detection based on blockchain and federated learning. In this paper, the publicly available dataset VeReMi was used to train a misbehavior detection model. The proposed scheme eventually achieved the higher accuracy of model aggregation by multi-party aggregation training. Numerical results proved the effectiveness of the method in terms of efficiency and accuracy. At the same time, participants were rewarded using blockchain-based smart contracts to incentivize them to deliver higher quality models. Considering the timeliness of vehicle data and the limitation of the participants' local equipment, the model achieved low overhead and high efficiency, which can be better applied to misbehavior detection.

In the future, we will add more attack types to extend the detection coverage of misbehavior and use encryption schemes such as differential privacy and homomorphic encryption to make the training process more secure.

References

1. Codecá, L., Frank, R., Faye, S., Engel, T.: Luxembourg SUMO traffic (LuST) scenario: traffic demand evaluation. IEEE Intell. Transp. Syst. Mag. **9**(2), 52–63 (2017)
2. Davis, J., Goadrich, M.: The relationship between precision-recall and ROC curves. In: The 23rd International Conference on Machine Learning, pp. 233–240 (2006)
3. Dietzel, S., Petit, J., Heijenk, G., Kargl, F.: Graph-based metrics for insider attack detection in vanet multihop data dissemination protocols. IEEE Trans. Veh. Technol. **62**(4), 1505–1518 (2012)
4. Grover, J., Prajapati, N.K., Laxmi, V., Gaur, M.S.: Machine learning approach for multiple misbehavior detection in VANET. In: Abraham, A., Mauri, J.L., Buford, J.F., Suzuki, J., Thampi, S.M. (eds.) ACC 2011. CCIS, vol. 192, pp. 644–653. Springer, Heidelberg (2011). https://doi.org/10.1007/978-3-642-22720-2_68
5. Gyawali, S., Qian, Y.: Misbehavior detection using machine learning in vehicular communication networks. In: IEEE International Conference on Communications (ICC), pp. 1–6. IEEE (2019)
6. Kamel, J., Wolf, M., van der Hei, R.W., Kaiser, A., Urien, P., Kargl, F.: VeReMi extension: a dataset for comparable evaluation of misbehavior detection in VANETs. In: IEEE International Conference on Communications (ICC), pp. 1–6 (2020)
7. Kang, J., Xiong, Z., Niyato, D., Zou, Y., Zhang, Y., Guizani, M.: Reliable federated learning for mobile networks. IEEE Wirel. Commun. **27**(2), 72–80 (2020)
8. McMahan, B., Moore, E., Ramage, D., Hampson, S., Arcas, B.A.: Communication-efficient learning of deep networks from decentralized data. In: Artificial Intelligence and Statistics, pp. 1273–1282. PMLR (2017)
9. Parno, B., Perrig, A.: Challenges in securing vehicular networks. In: Workshop on Hot Topics in Networks (HotNets-IV), Maryland, USA, pp. 1–6 (2005)
10. Qu, F., Wu, Z., Wang, F.Y., Cho, W.: A security and privacy review of VANETs. IEEE Trans. Intell. Transp. Syst. **16**(6), 2985–2996 (2015)

11. Rawat, D.B., Bista, B.B., Yan, G.: Securing vehicular ad-hoc networks from data falsification attacks. In: IEEE Region 10 Conference (TENCON), pp. 99–102. IEEE (2016)
12. Ruj, S., Cavenaghi, M.A., Huang, Z., Nayak, A., Stojmenovic, I.: On data-centric misbehavior detection in VANETs. In: IEEE Vehicular Technology Conference (VTC Fall), pp. 1–5. IEEE (2011)
13. Sharma, P., Liu, H.: A machine-learning-based data-centric misbehavior detection model for internet of vehicles. IEEE Internet Things J. **8**(6), 4991–4999 (2020)
14. Singh, P.K., Gupta, S., Vashistha, R., Nandi, S.K., Nandi, S.: Machine learning based approach to detect position falsification attack in VANETs. In: Nandi, S., Jinwala, D., Singh, V., Laxmi, V., Gaur, M.S., Faruki, P. (eds.) ISEA-ISAP 2019. CCIS, vol. 939, pp. 166–178. Springer, Singapore (2019). https://doi.org/10.1007/978-981-13-7561-3_13
15. So, S., Sharma, P., Petit, J.: Integrating plausibility checks and machine learning for misbehavior detection in VANET. In: The 17th IEEE International Conference on Machine Learning and Applications (ICMLA), pp. 564–571. IEEE (2018)
16. Sommer, C., German, R., Dressler, F.: Bidirectionally coupled network and road traffic simulation for improved IVC analysis. IEEE Trans. Mob. Comput. **10**(1), 3–15 (2010)
17. Zinkevich, M., Weimer, M., Smola, A.J., Li, L.: Parallelized stochastic gradient descent. In: NIPS, vol. 4, p. 4. Citeseer (2010)

Data Science

ABE-AC4DDS: An Access Control Scheme Based on Attribute-Based Encryption for Data Distribution Service

Peng Gao[1,2] and Zhuowei Shen[1,2(✉)]

[1] School of Cyber Science and Engineering, Southeast University, Nanjing, China
zwshen@seu.edu.cn
[2] Key Laboratory of Computer Networks and Information Integration (Southeast University), Ministry of Education, Nanjing, China

Abstract. In response to the security threats faced by distributed real-time applications based on DDS, a fine-grained data access control scheme is proposed, which is based on attribute-based encryption theory and suitable for topic-based publish/subscribe communication model. The scheme takes the topic as the unit of data access control and integrates the access control process with the DDS communication process, In the discovery phase of DDS, the digital signature is used to verify the publication permission for a topic, and in the publish/subscribe phase of DDS, the CP-ABE is used to verify the subscription permission for a topic. The scheme ensures not only the privacy of users but also the confidentiality and authenticity of data. Theoretical analysis shows that this scheme can resist security threats such as unauthorized publication and unauthorized subscription. Moreover, the performance test of the prototype system shows that it matches the loose coupling and one to many characteristics of the publish/subscribe communication model and has good scalability in multi-subscriber scenarios while adjusting key parameters.

Keywords: Access control · Data distribution service · Attribute-based encryption · Publish/Subscribe

1 Introduction

Data Distribution Service (DDS) adopts a data-centric publish-subscribe model to realize the real-time transmission of massive data with features such as loose coupling, high-performance reliability and good scalability [1]. It provides a decentralized, QoS-guaranteed middleware architecture in a distributed heterogeneous environment [2], widely used in critical industrial fields in recent years [3]. However, the DDS specification [4] has certain information security risks [5, 6]. The loose coupling characteristics in time, space and data transmission make data transmission between entities in a flexible manner, which is effortless to introduce security risks and seriously threatens the security of data communication. For example, publishers may illegally publish topics

Y. Lai et al. (Eds.): ICA3PP 2021, LNCS 13157, pp. 67–80, 2022.
https://doi.org/10.1007/978-3-030-95391-1_5

and transmit confusing data; subscribers may illegally subscribe to topics and intercept private data. Therefore, it is necessary to control data access in DDS systems.

Attribute-Based Encryption (ABE) [7] is a public-key encryption system, which is a data encryption method that supports access control. The public key and private key of ABE are not one-to-one, and one public key can correspond to multiple private keys. No matter how many users the data is shared with, it only needs to be encrypted once. The encryptor does not need to know who decrypts the data, and the decryptor only needs to meet the corresponding conditions to decrypt. In this way, data access control is realized while encrypting. It has a significant advantage, that is, it is consistent with the loose coupling characteristics of DDS in the scenario of one publisher and numerous subscribers. With Ciphertext-Policy Attribute-Based Encryption (CP-ABE, a kind of ABE) [8], the private key is associated with the user attribute set, and the ciphertext is related to an access control structure. Ciphertext can be decrypted if and only if the attributes in the user's attribute set meet the access control structure.

Inspired by the above motivations, this paper proposes An Access Control Scheme based on Attribute-Based Encryption for Data Distribution Service, which is abbreviated as ABE-AC4DDS. The rest of the paper is organized as follows. Section 2 presents the related works. Section 3 models the system of ABE-AC4DDS. Section 4 describes the primary access control process of ABE-AC4DDS. Analysis and performance evaluation of ABE-AC4DDS is given in Sect. 5 and Sect. 6 separately. Finally, Sect. 7 concludes this paper.

2 Related Works

In response to the security threats faced by data publishing and subscribing in DDS, previous works have studied access control on DDS. The main defects of them are as follows.

(1) Most schemes, such as [9–11], used encryption technology to implement access control on the subscription side. A general approach is that published data is encrypted using a shared key. However, there is no proxy or central server in a pure distributed publish/subscribe middleware system. It is easy for unauthorized publishers to publish data, making it more difficult to control access of the publishers and more likely to pose a security threat.
(2) Some researchers, such as [12] and [13], focused on the identity authentication mechanism of DDS data access control and controlled data access from the user's perspective. In these coarse-grained access control schemes, each pair of publisher and subscriber need to exchange identity information in advance and negotiate encryption secrets. Once matched, the subscriber could subscribe to all the data published by the publisher, no matter which topic it belongs to, which not only easily caused information leakage between the publisher and subscriber, but also lacked access control capabilities for different topic data resources;
(3) In 2018, the OMG officially announced DDS Security specification [14] that defined the architecture of the security model and service plug-in interface compatible with

DDS implementation. [15] used attribute-based access control (ABAC) to implement DDS data access control based on permission files Access control. [16] proposed a certificate-based dynamic access control method for DDS participants, which provides flexible and configurable access control. However, some problems still exist, such as complicated access control authority configuration and cumbersome information exchange for security negotiation.

3 System Model of ABE-AC4DDS

To solve the problems of previous studies, a data access control model is abstracted from DDS publish/subscribe systems. The model does not need to specify the identities of the legal publishers and subscribers of a topic. On the contrary, it is to clarify the attributes that the legal publishers and subscribers of the topic need to have, to achieve access control to the topic data and to prevent unauthorized publication and unauthorized subscription.

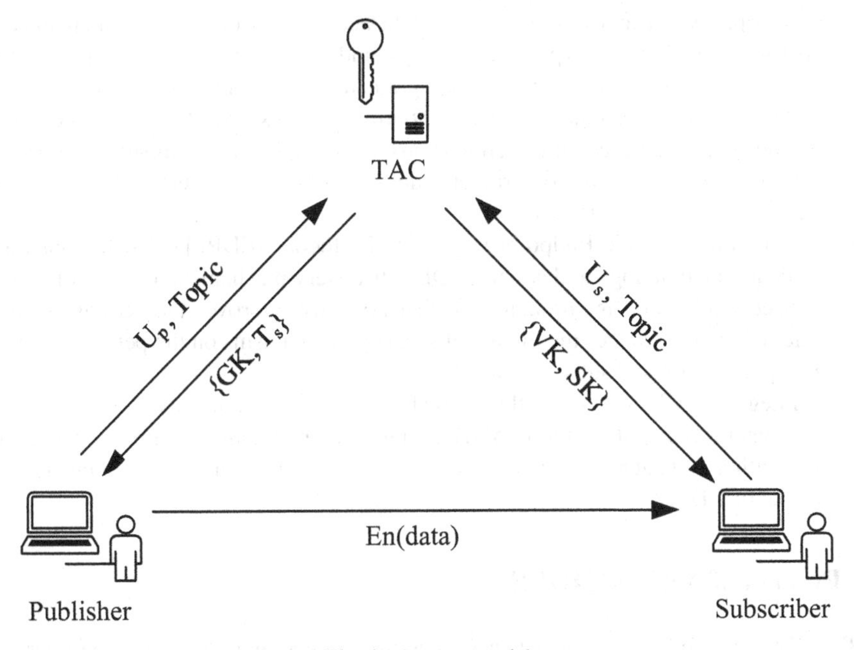

Fig. 1. System model

ABE-AC4DDS model is shown in Fig. 1. It is constructed of three entities: Trusted Authorization Center (TAC), Publisher, and Subscriber.

(1) TAC
The management organization of access control authority in the data distribution domain is mainly responsible for analyzing the data authority of a topic in the

system, generating system public key PK, master private key MSK, the publication access control structure T_p, the subscription access control structure T_s, the topic signature verification asymmetric key pair $\{GK, VK\}$, and the permission key pair related to user topic access control($\{GK, T_s\}$ or $\{VK, SK\}$). Then the permission key pair is distributed to the related users.

(2) Publisher
The publisher of the topic applies to TAC for the publication permission key pair $\{GK, T_s\}$, through its own attribute set U_p and publication topic, and uses the publication permission key pair to perform authority signature and data encryption.

(3) Subscriber
The subscriber of the topic applies for the subscription permission key pair $\{VK, SK\}$ from TAC through its own attribute set U_s and subscription topic, and uses the subscription permission key pair for signature verification and data decryption.

Compared with the traditional access control on DDS, ABE-AC4DDS has the following advantages.

(1) It's a topic-based and attribute-based data access control model, which makes full use of attribute encryption technology and the multi-dimensionality of user attributes to achieve one-to-many, fine-grained access control in DDS. It makes up for the coarse granularity of access control in the existing DDS access control schemes (based on user, single-dimension), the complex and cumbersome configuration of the access control structure (defined by the publishers and subscribers one by one), and other issues.
(2) By combining Simple Endpoint Discovery Protocol (SEDP) [17] with signature verification technology and adopting DDS discovery mechanism, the risk of unauthorized publication is eliminated, which not only improves the security of the system, but also reduces the impact of security mechanisms on the performance of DDS publication and subscription.
(3) By integrating CP-ABE with DDS publish/subscribe communication process, it not only ensures the confidentiality of data transmission, but also eliminates the threat of unauthorized subscription, and realizes loose coupling and one-to-many access control of DDS.

4 Process of ABE-AC4DDS

ABE-AC4DDS divides the topic data access control process into two stages: permission distribution and permission control. Permission distribution which TAC executes is the basis of permission control, and both publisher and subscriber implement permission control.

4.1 Permission Distribution

The data access control permission of a topic is divided into two parts: publication permission and subscription permission. By parsing the topic permission file set by the administrator user of the system, TAC determines the publication access control policy and subscription access control policy of a topic, and distributes the corresponding permission key pairs to the publishers or subscribers according to their attribute set. The specific process is shown in Fig. 2.

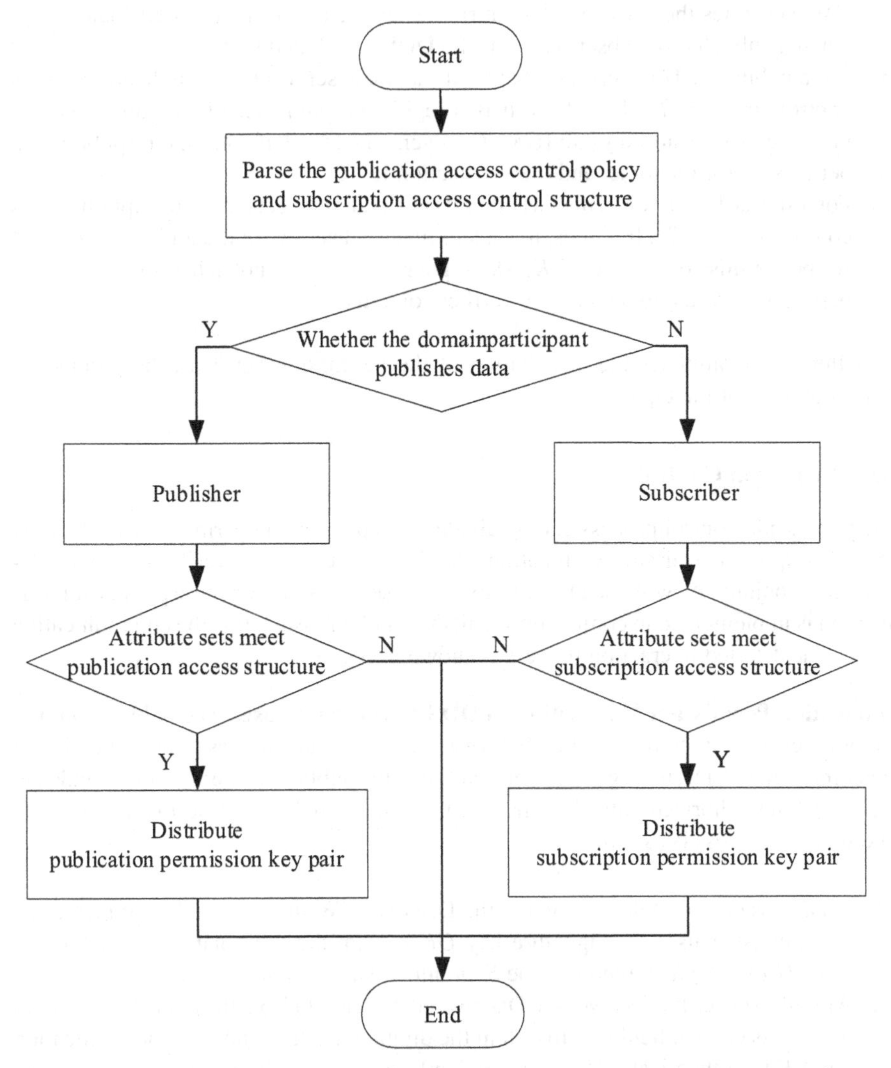

Fig. 2. Permission analysis and key distribution process

The primary process includes:

(1) TAC initializes the system and generates $\{PK, MSK\}$. PK is public in the entire system, and MSK is kept secretly by TAC;

(2) TAC determines the publication access control policy and subscription access control policy of a topic and generates the publication access control structure T_p and the subscription access control structure T_s of the topic;
(3) TAC generates an asymmetric key pair $\{GK, VK\}$ for signature verification for the topic;
(4) TAC generates the corresponding attribute key SK according to its attribute set U, when a publisher or subscriber joins the DDS distributed system;
(5) For a publisher, TAC verifies whether its attribute set meets the publication access control structure T_p. If it does, it is a legitimate publisher of the topic, and the publisher permission key pair $\{GK, T_s\}$ is generated for it; if not, it has no publication permission for the topic, and the verification fails.
(6) For a subscriber, TAC verifies whether its attribute set meets the subscription access control structure T_s. If it does, it is a legitimate subscriber of the topic, and the subscriber permission key pair $\{VK, SK\}$ is generated for it; if not, it has no subscription permission for the topic, and the verification fails.

After TAC completes the distribution of the permission key pair, the permission control process of the topic begins.

4.2 Permission Control

The permission control process can be divided into publication permission verification and subscription permission verification. Publication permission verification is implemented in conjunction with the DDS discovery phase, while subscription permission verification is implemented in conjunction with the DDS publish/subscribe communication phase. The detailed interaction process is shown in Fig. 3.

Publication Permission Verification in DDS Discovery Phase. The publication permission verification combines the DDS discovery mechanism. Based on the endpoint matching in the SEDP, the signature verification of the publication permission is added to prevent the unauthorized publishers from establishing a publish-subscribe relationship. The specific process is as follows:

(1) An access control field is added to the DiscoveredWriterData (in SEDP message). The publisher uses the signature key GK to sign the topic name *Topic*: $Sign = En_{GK}(Hash(Topic))$, then fills the *Sign* into the access control field;
(2) After the subscriber receives the DiscoveredWriterData from the publisher, it parses the access control field in it to obtain the digital signature and uses the verification key VK to authenticate: $H_1 = De_{VK}(Sign),\ H_2 = Hash(Topic);$;
(3) Compare whether the hash value H_1 and H_2 are equal. If the two are equal, it indicates that the verification is successful, and the subscriber successfully verifies the publication permission of the remote publisher in SEDP and continues to establish a publish-subscribe relationship between them; otherwise, it means that the remote publisher have no publication permissions, and the establishment of the publish-subscribe relationship fails in the SEDP.

After the DDS discovery phase, only legitimate publishers can establish a publish-subscribe relationship with the corresponding subscribers, thus entering the DDS publish/subscribe communication phase.

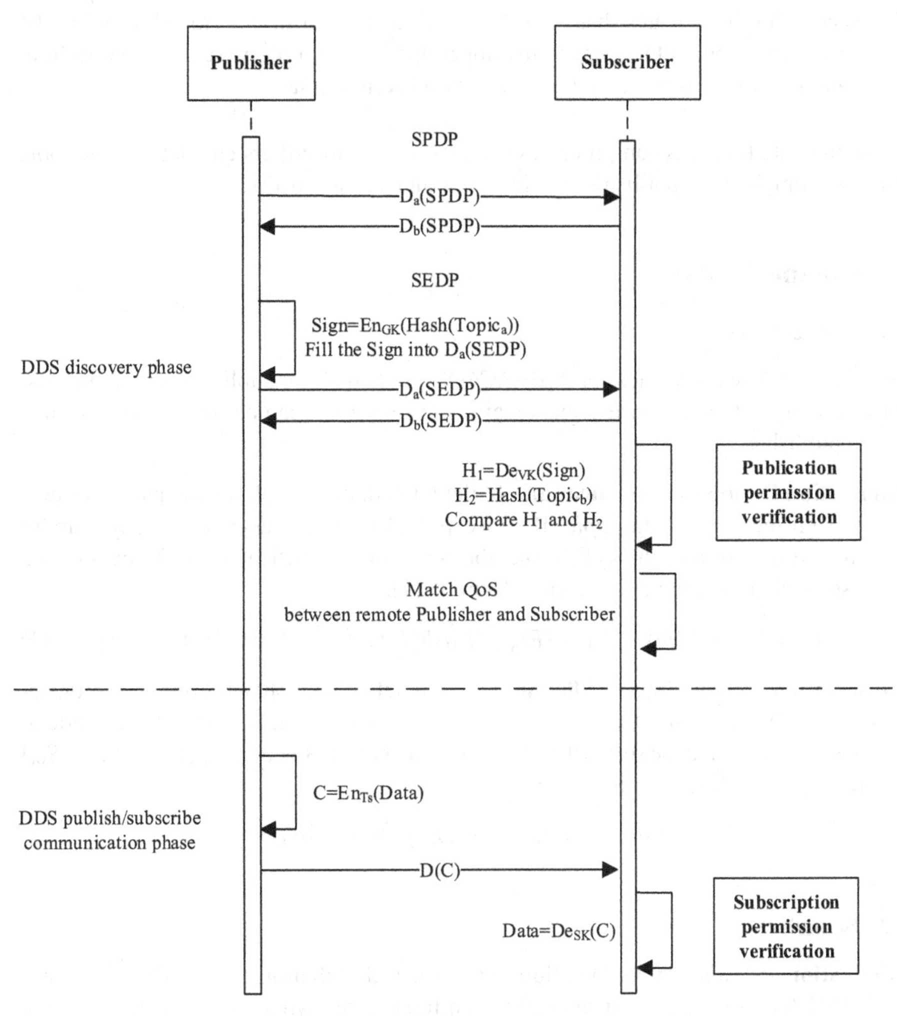

Fig. 3. Permission control process

Subscription Permission Verification in DDS Publish/Subscribe Communication Phase. Subscription permission verification combines the DDS publish/subscribe communication process. The publisher encrypts the published topic data with CP-ABE, and legitimate subscribers use the attribute key *SK* distributed by TAC to decrypt the ciphertext, thereby preventing unauthorized subscription. The specific process is as follows:

(1) The legitimate publisher uses T_s received during the permission distribution stage to perform CP-ABE encryption on the topic data: $C = En_{Ts}(Data)$, publish the ciphertext C to the Domain;
(2) When subscribing to the topic, the legal subscriber decrypts the received topic ciphertext data C with its attribute key SK: $Data = De_{SK}(C)$. If the decryption is successful, it indicates that the subscriber has the permission to subscribe to the topic, and submits the *Data* to the upper application; otherwise, it has no right to subscribe to the topic, and the topic data subscription fails.

In the data transmission process, only legitimate subscribers can decrypt the topic data and complete the publish/subscribe communication process.

5 Scheme Analysis

5.1 Correctness

Definition 1: The correctness of ABE-AC4DDS: a legitimate publisher can be successfully verified by the subscriber, and a legitimate subscriber can be successfully verified by the publisher.

Publication Permission Verification. In the DDS discovery phase, the publisher uses GK to sign the topic name *topic*. Because $\{GK, VK\}$ is the asymmetric key pair for signature verification of the same topic, the subscriber's verification of the publication permission about the topic is shown in formula (1).

$$H_1 = De_{VK}(Sign) = De_{VK}(En_{GK}(Hash(Topic))) = Hash(Topic) = H_2 \quad (1)$$

Subscription Permission Verification. In the DDS publish/subscribe communication phase, the publisher uses T_s to perform CP-ABE encryption into ciphertext: $C = En_{Ts}(Data)$, then the subscriber uses SK to decrypt it. Since $U_s = \{S_{s1}, S_{s2} \ldots, S_{sn}\}$ satisfies T_s, according to Formula (2).

$$Data = De_{SK}(C) = De_{SK}(En_{Ts}(Data)) \quad (2)$$

5.2 Security

Publication Permission Verification. In the key distribution phase, TAC generates $\{GK, VK\}$ for signature verification for each topic. Only when the attribute set of the publisher meets the publication access control structure T_p, the publisher can obtain GK. And only when the attribute set of the subscriber meets the subscription access control structure T_s, the subscriber can obtain the verification key VK. In the DDS discovery stage, due to the unforgeability of the signature verification technology [18], an illegal publisher cannot forge a message *Sign* that can be verified with VK, so that the subscriber and the illegal publisher cannot establish a publish-subscribe relationship. Hence the topic-based publication permission control is implemented. The publication permission control of the scheme satisfies Unforgeability against Chosen Message Attacks (UF-CMA).

Subscription Permission Verification. The subscription permission verification of the scheme uses an interactive game between the challenger and the attacker of the adversary model to define its security model. TAC is the challenger, and the unauthorized subscriber is the adversary.

(1) Init stage: the adversary chooses an access structure A^* to publicly announce the challenge;
(2) Setup stage: The challenger randomly selects $\alpha, \beta \in Z_p$, and executes the system CP-ABE Setup algorithm to generate a public key PK and send it to the adversary;
(3) Query stage: The adversary repeatedly asks the challenger about the user decryption private key related to the attribute set P^*, and the user attribute set P^* does not satisfy the access structure A^*. After receiving the inquiry, the challenger randomly selects $r_j \in Z_p$ for each attribute j in the user attribute set A_u, and the challenger generates the private key SK and sends it to the adversary;
(4) Challenge stage: The adversary selects two plaintext data M_1 and M_2 of equal length and sends them to the challenger. The challenger generates a random value $\mu \in \{0, 1\}$ with equal probability fairly, and encrypts the plaintext M_μ by accessing the structure A^* to generate a ciphertext CT*, and return CT* to the adversary;
(5) Guess stage: After the adversary obtains CT*, give a guess μ' about μ.

The probability of the adversary winning in this game is defined as:

$$Adv = \mid Pr\left[\mu' = \mu\right] - \frac{1}{2} \mid \tag{3}$$

Theorem 1. Assuming that the DBDH Assumption [19] of CP-ABE encryption technology is established, if there is no probability polynomial time, the adversary can break the subscription permission verification in our scheme with a non-negligible advantage, then the subscription permission verification satisfies the Indistinguishability under Chosen-Plaintext Attack (IND-CPA) [20].

Proof. The challenger randomly selects the generator g and the bilinear group (e, p, G_0, G_1). If the adversary can break the subscription permission verification of the scheme, which is equivalent to the existence of an algorithm A can solve the DBDH with the advantage of ε in polynomial time. That is, input $\left(g, g^a, g^b, g^c, T\right)$, algorithm A can judge whether T is equal to $e(g, g)^{abc}$ in polynomial time.

(1) If $\mu = 1$, the adversary succeeds, it means $T = e(g, g)^{abc}$. According to the DBDH Assumption, the adversary obtains the effective ciphertext and its advantage

$$\Pr\left[\mu' = \mu|\mu = 1\right] \leq \varepsilon + \frac{1}{2} \tag{4}$$

(2) If $\mu = 0$, it means that the adversary cannot obtain any information about μ. Because of random ciphertext, the probability of guessing success and failure is both $1/2$.

$$Pr\left[\mu' \neq \mu | \mu = 0\right] = Pr\left[\mu' = \mu | \mu = 0\right] = \frac{1}{2} \quad (5)$$

According to Formula (3), (4), (5), the advantage of the adversary's successful attack subscription permission verification is

$$\begin{aligned} Adv &= \left| Pr[\mu' = \mu] - \frac{1}{2} \right| \\ &= \left| Pr[\mu = 0] \cdot Pr[\mu' = \mu | \mu = 0] + Pr[\mu = 1] \cdot Pr[\mu' = \mu | \mu = 1] - \frac{1}{2} \right| \\ &= \left| \frac{1}{2} \cdot Pr[\mu' = \mu | \mu = 0] + \frac{1}{2} \cdot Pr[\mu' = \mu | \mu = 1] - \frac{1}{2} \right| \leq \frac{1}{2} \cdot \frac{1}{2} + \frac{1}{2}\left(\frac{1}{2} + \varepsilon\right) - \frac{1}{2} \leq \frac{\varepsilon}{2} \end{aligned}$$

In summary, because it is assumed that the adversary's advantage in solving the DBDH problem is not greater than the negligible value ε, then the adversary's advantage in winning is not greater than $\varepsilon/2$ is also negligible. Therefore, the subscription permission verification meets IND-CPA security.

6 Experiment and Evaluation

Our scheme adds an access control mechanism to the DDS specification, which has a certain impact on the real-time performance of DDS communication. This paper quantitatively tests the time characteristics of ABE-AC4DDS and determines the key factors that affect the performance index, so as to reduce the impact of access control on real-time in engineering practice application scenarios.

Based on the uDDS software [21] and the open-source attribute-based encryption library OpenABE [22], the prototype system in this paper is implemented. The system is deployed on a cloud platform, and the hosts are connected by a gigabit virtual switch. The detailed configuration of each host is shown in Table 1.

Table 1. Host configuration table

Resource type	Configuration
CPU	4*Intel® Core™2 DuoT7700@ 2.40 GHz
Memory	DDR4 8 GB
Disk	40 G
Operating system	64-bit CentOS Linux 7

Compared with the original DDS, the prototype system adds the publication permission verification in the DDS discovery phase and the subscription permission verification in the DDS publish/subscribe communication phase, which is a key factor that affects the real-time performance of DDS. Therefore, by testing the discovery delay and

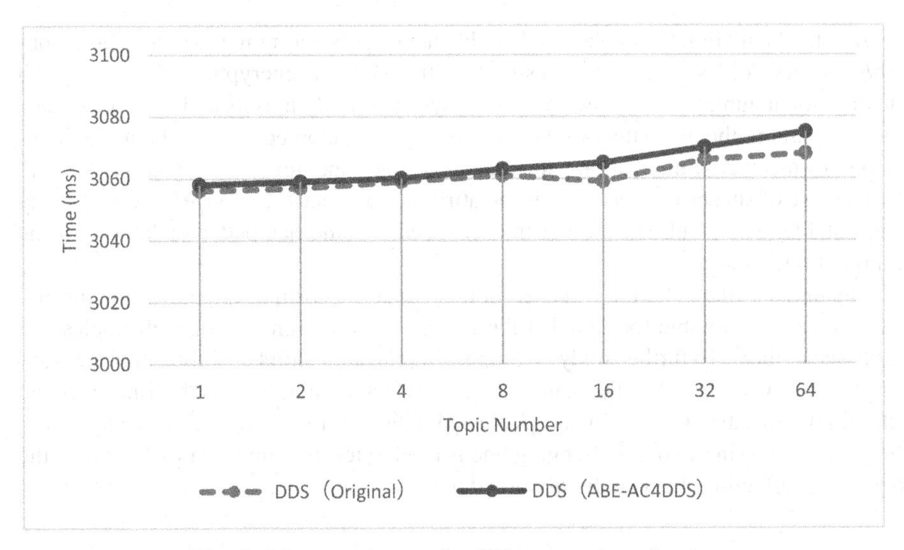

Fig. 4. Discovery latency for different number of topics

the publish/subscribe delay, the time performance index of the scheme in the practical application can be obtained.

As shown in Fig. 4, in the discovery phase, the subscriber in ABE-AC4DDS needs to verify the publication permission of the publisher. Compared with the DDS without access control, the delay increases to a certain extent. As the number of topics published in the system increases, both delays have increased slightly, but the maximum difference is 0.2% (7 ms).

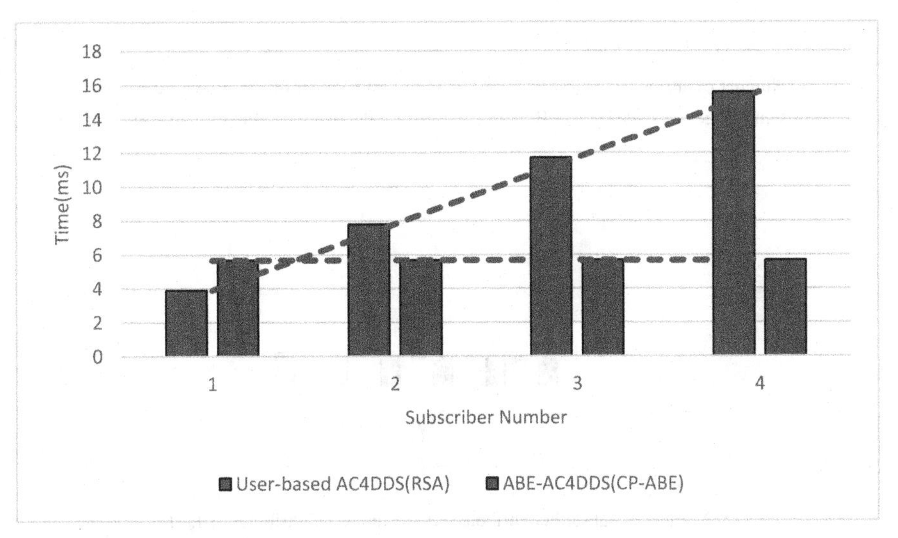

Fig. 5. Publish/subscribe latency of the same topic with different numbers of subscribers under different encryption systems (publishing 100 bytes)

As shown in Fig. 5, in the DDS publish/subscribe communication phase, publish/subscribe delays have increased significantly due to the encryption. Compared with the traditional public-private key encryption system based on users and one-to-one data communication, the attribute-based encryption system adopted in our scheme can realize one-to-many communication with encryption once and multi-party decryption. With the increase of subscribers, as long as the attribute set meet T_s, publish/subscribe delay keeps stable. The result shows that the proposed scheme has better scalability in the multi-subscriber scenario.

In summary, the solution proposed in this paper has a small impact on the discovery delay and is suitable for data distribution application scenarios of multi-topics and multi-subscribers. To further study the impact of relevant factors on publication and subscription delay in ABE-AC4DDS, including ABE factors (number of subscriber attribute set, number of attributes in T_s) and published data factors (publish message payload), the method of fixing two and changing one is used to test the time cost to determine the key factors affecting the data distribution delay.

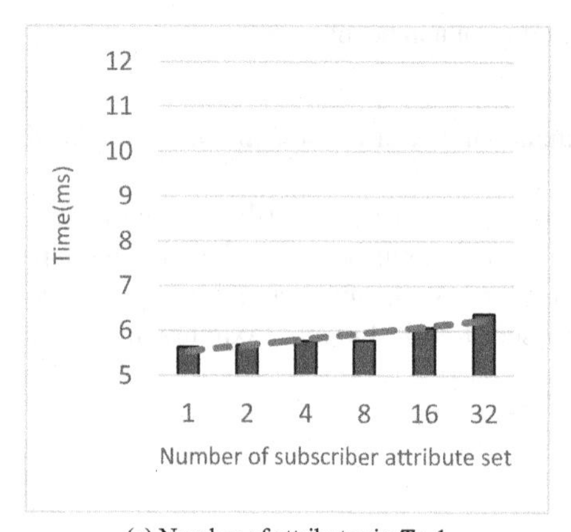

(a) Number of attributes in T_s: 1
Publish message payload size: 100B

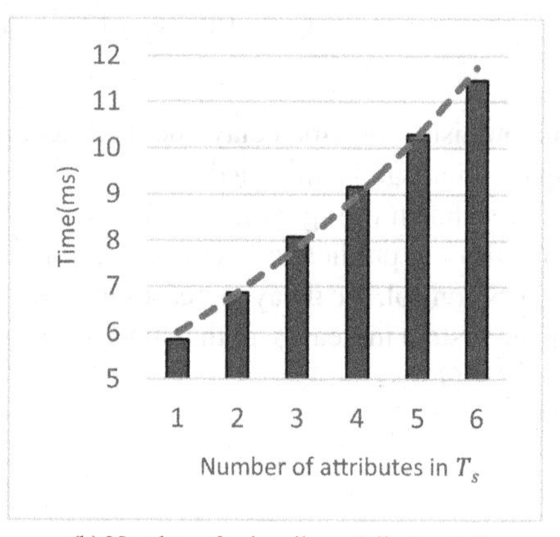

(b) Number of subscriber attribute set: 8
Publish message payload size: 100B

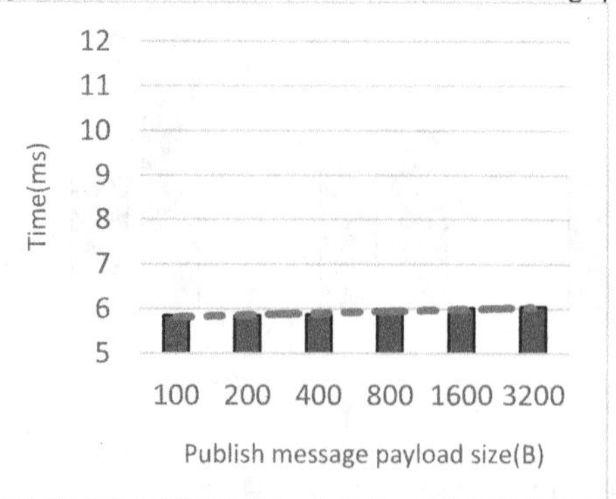

(c) Number of subscriber attribute set: 8, Number of attributes in T_s: 1

Fig. 6. The influence of different factors on publish/subscribe latency in ABE-AC4DDS

It can be seen from Fig. 6 that the attribute set of subscribers and publish message payload increased by 32 times, the delay increased by 1.13 times and 1.03 times, while the number of the subscription access control structure increased by 6 times, and the delay increased by 1.96 times. It indicates that it can effectively reduce the data publication and subscription time delay in the DDS publish/subscribe communication process in ABE-AC4DDS by adjusting the complexity of the topic subscription access control structure.

7 Conclusion

Aiming at the security threats in the DDS distributed system such as unauthorized publication/subscription, and the lack of fine-grained access control capabilities in DDS, we propose An Access Control Scheme based on Attribute-Based Encryption for Data Distribution Service (ABE-AC4DDS). The scheme makes full use of the DDS communication process, combined with attribute-based encryption and signature verification technology to realize one-to-many dynamic publish and subscribe authority control, and ensures the authenticity and confidentiality of the publish/subscribe process on the basis of certain security and privacy protection. We implemented a prototype system for the proposed scheme and conducted security analysis and tests for performance evaluation. The scheme is not suitable for real-time data distribution scenarios due to the relatively high computational complexity of using local online users to independently perform attribute-based encryption or decryption. In the future, we will propose an optimization and improvement scheme for the efficiency of ABE by using partial offline method of encryption and decryption without affecting its security, and further expand the practical application scenarios of ABE-AC4DDS.

References

1. Zou, G., Liu, Y.F.: Automatic discovery technology of real-time data. Comput. Technol. Dev. **27**(1), 25–29 (2017)
2. David, L., Vasconcelos, R., Alves, L., André, R., Endler, M.: A DDS-based middleware for scalable tracking, communication and collaboration of mobile nodes. J. Internet Serv. Appl. **4**, 16 (2013). https://doi.org/10.1186/1869-0238-4-16
3. Leigh, B., Duwe, R.: Designing autonomous vehicles for a future of unknowns. ATZelectron. Worldwide **16**(3), 44–47 (2021)
4. Object Management Group. Data Distribution Service (DDS) [EB/OL]. https://www.omg.org/spec/DDS
5. He, Z.Y., Liang, Y.: Study on the DDS network information security technology. Appl. Mech. Mater. **738–739**, 1213–1216 (2015). https://doi.org/10.4028/www.scientific.net/AMM.738-739.1213
6. White, T., Johnstone, M.N., Peacock, M.: An investigation into some security issues in the DDS messaging protocol. In: 15th Australian Information Security Management Conference, vol. 132, pp. 132–139. Edith Cowan University, Perth, Western Australia (2017). https://doi.org/10.4225/75/5a84fcff95b52
7. Sahai, A., Waters, B.: Fuzzy Identity-Based Encryption. In: Cramer, R., (eds.) Advances in Cryptology – Eurocrypt 2005. Eurocrypt 2005. Lecture Notes in Computer Science, vol. 3494, pp. 457–473. Springer, Berlin, Heidelberg (2005)https://doi.org/10.1007/11426639_27

8. Bethencourt, J., Sahai, A., Waters, B.: Ciphertext-policy attribute-based encryption. In: IEEE Symposium on Security & Privacy, vol. 321, pp. 321–334. IEEE Computer Society, Washington, USA (2007). https://doi.org/10.1109/SP.2007.11
9. Han, J.: Message encryption methods for dds security performance improvement. J. Korea Inst. Inf. Commun. Eng. **22**(11), 1554–1561 (2018). https://doi.org/10.6109/JKIICE.2018.22.11.1554
10. Shen, Z.W., Gao, P., Xu, X.Y.: Design of DDS secure communication middleware based on security negotiation. Netinfo Secur. **21**(6), 19–25 (2021). https://doi.org/10.3969/j.issn.1671-1122.2021.06.003
11. Tariq, M.A., Koldehofe, B., Rothermel, K.: Securing broker-less publish/subscribe systems using identity-based encryption. IEEE Trans. Parallel Distrib. Syst. **25**(2), 518–528 (2014). https://doi.org/10.1109/TPDS.2013.256
12. Li, M.J., Ye, H., Wang, L., et al.: Design of authentication protocol for high-security data distribution service. Aeronaut. Comput. Tech. **45**(1), 103–107 (2015)
13. Zhen, C., Di, H.T., Guo, Q.L.: Research on identity authenticationmethod for data distribution service. Electron Technol. **44**(6), 44–48 (2015). https://doi.org/10.3969/j.issn.1000-0755.2015.06.013
14. Object Management Group. DDS Security [EB/OL]. https://www.omg.org/spec/DDS-SECURITY/1.0
15. Kim, H., Kim, D.-K., Alaerjan, A.: ABAC-based security model for DDS. IEEE Trans. Depend. Secure Comput. **1**, 1 (2021). https://doi.org/10.1109/TDSC.2021.3085475
16. Zhen, C., DI, H.T., Guo, Q.L., et al.: Research on access control method of data distribution service. Inform. Commun. **2019**(5), 96–98 (2019)
17. Object Management Group. The Real-time Publish-subscribe Protocol DDS Interoperability Wire Protocol [EB/OL]. https://www.omg.org/spec/DDSI-RTPS
18. Bellare, M., Namprempre, C., Neven, G.: Security proofs for identity-based identification and signature schemes. J. Cryptol. **22**, 1–61 (2009). https://doi.org/10.1007/s00145-008-9028-8
19. Emura, K., Miyaji, A., Nomura, A., Omote, K., Soshi, M.: A ciphertext-policy attribute-based encryption scheme with constant ciphertext length. In: Bao, F., Li, H., Wang, G. (eds.) Information Security Practice and Experience. ISPEC 2009. Lecture Notes in Computer Science, vol. 5451, pp. 13–23. Springer, Berlin, Heidelberg (2009)https://doi.org/10.1007/978-3-642-00843-6_2
20. Goldwasser, S., Micali, S.: Probabilistic encryption. J. Comput. Syst. Sci. **28**(2), 270–299 (1984). https://doi.org/10.1016/0022-0000(84)90070-9
21. uDDS Homepage. https://udds.cn/. Accessed 21 July 2021
22. OpenABE Homepage. https://github.com/zeutro/openabe. Accessed 21 July 2021

An Interactive Visual System for Data Analytics of Social Media

Yi Zhang[1], Zhaohui Li[2(✉)], and Dewei Xi[1]

[1] College of Intelligence and Computing, Tianjin University, Tianjin, China
yizhang@tju.edu.cn
[2] School of New Media and Communication, Tianjin University, Tianjin, China
zhaohuil@tju.edu.cn

Abstract. With the development of the Internet, Big Data analysis of social media has become a hot research topic in recent years. However, a challenging problem in social media analysis is how to intelligently detect event information from massive media data and how to help users quickly understand event content. Therefore, we propose a method to extract important time periods of events from media data to analyze the media event information, and develop an interactive visual system based on the "5W" principle. The system provides an interactive analysis platform for exploring media Big Data (e.g., Twitter data). The system uses a dynamic topic model to extract topics, a Naive Bayesian classifier to distinguish emotions, and explores events based on the evolution of time, topics, and emotions. In terms of visualization, the system allows users to add some annotations based on segmentation of complex information and highlight important events by adding different story cards to the timeline so that users can get a quick overview of events. Users can also interactively modify the story cards during the exploration process. Finally, several case studies show that our system can effectively reduce the time required to understand social media data and allow users to quickly explore the full picture of an event through an interactive approach.

Keywords: Social media analysis · Big Data analysis · Important time period extraction · Interactive visualization · Social media event

1 Introduction

How to extract and analyze information from complex Big Data is a hot topic among researchers, especially in the field of social media data analytics.

Information in social media has long changed from text-based to view-based forms. Views display information through combinations and interactions of colors and shapes that appeal to the visual organs as much as possible and make the information more accessible to people.

The significance of our research is to propose a method to automatically extract important time periods from the large amount of media text data, which

Y. Lai et al. (Eds.): ICA3PP 2021, LNCS 13157, pp. 81–100, 2022.
https://doi.org/10.1007/978-3-030-95391-1_6

helps users to better understand and localize events. We also design an interactive visual system to support the analysis and exploration of social media Big Data through some visualization techniques. The system can display the content and causes of events in social media data, so that users without background knowledge in Big Data analysis can gain a general understanding of events and information in a short time.

Here are three innovations in our study.

- We propose a new method to extract the important time periods in social media data using topic words and emotions. Dynamic Topic Model (DTM) is used to process temporal data and analyze changes in topic words [2]. Naive Bayesian classifier is adapted to categorize emotions on different topics (Details in Sect. 5).
- We propose a visualization method to analyze the possible reasons for the turn of events in social media Big Data. By analyzing the change of topic words, emotions, and text volume in important time periods, we automatically show the user some possible reasons for the change of events in story cards (Details in Sect. 6.2).
- We designed an interactive visual system based on the "5W" principle: "When, Where, What, Who, and Why". The system can automatically generate lots of story cards that help users quickly understand the important information of the event in a short social text. Moreover, users can interactively explore the details and modify the story cards (Details in Sect. 6).

The rest of this paper is organized as follows. Section 2 introduces related work, Sect. 3 discusses the tasks of Big Data analysis in social media, Sect. 4 shows the system pipeline, Sect. 5 describes the method for extracting important time periods, Sect. 6 describes the details of the visual system, in Sect. 7 we invite volunteers to evaluate and do some case studies, and Sect. 8 is the limitation and conclusion.

2 Related Work

2.1 Researches on Socialized Short Text

Social media analysis is an important field of data analysis, and socialized short text is a special category of text data.

Socialized short texts appear on social platforms such as Twitter, Weibo, Facebook, etc. They are different from traditional long texts such as news and blogs. Liu Dexi et al. [7] did a survey on automatic summarization of socialized short texts. In the paper, they pointed out that socialized short texts have typical features, such as each short text is short but can contain many texts at the same time; there are irregular words and even misspellings; there are complex "social relations" and various topics; also, socialized short texts are emotional and their temporal features are more distinct.

More and more scholars have been studying this field. D-map [4], E-map [5] and R-map [3] analyze Weibo data from different perspectives. D-map focuses on forwarding time and user influence, and shows how a message spreads across different social groups. E-map and R-map both use a map-like visual metaphor. They use things like towns, bridges, rivers, and islands to visualize the forwarding process of socialized short texts, which is impressive.

2.2 Researches on Emotion Analysis

Emotion is important not only in the analysis of socialized short texts, but also in the important time period extraction.

Barbosa et al. [1] attempted to explore the emotions in Twitter texts. They divided the emotion detection process of Twitter into two steps by first dividing messages into primary and secondary levels, and then determining whether the text's emotion at the primary level is positive or negative. They argue that the primary information can represent the overall emotion, which is a great inspiration for this study. Marcus et al. [12] also proposed a new method to appropriately normalize the overall emotion to account for the different recognition rates of classifiers for positive and negative emotions. Recently, the model proposed by Devlin et al. [6] BERT showed better results by modeling bidirectional contexts and invalidating autoencoding-based pretraining.

2.3 Researches on Temporal Data Visualization

For temporal texts, the most common method is to design a timeline. Li J et al. [11] designed a kind of semantic-spatial-temporal cube that can easily explore data with temporal, spatial, and textual components; Eccles. R et al. [9] designed a 3D timeline combined with geolocation information; Nguyen. P et al. [13] proposed a timeline named SchemaLine that allows interactive construction of temporal patterns and seamless integration with visual data exploration and annotation; Guo, L et al. [10] show their work on event sequence data, which clusters and visualizes event sequences into threads based on tensor analysis.

3 Tasks of Social Media Data Analytics

In investigative news [8], we found that an event can be analyzed according to the "4Ws" principle (Who, When, Where, and What). Moreover, C. Tong et al. [15] used the "WHW" (Who, How and Why) scheme in their study. And at the same time, the "how" can also be divided into "where", "when" and "what". So in our study, we can actually design a system according to the "5Ws" principle (Who, When, Where, What and Why).

To overcome the above challenges, we propose the following social media data analysis tasks in conjunction with the "5Ws" principles.

- Task 1: Filter out the important time periods and determine the start and end time of the event. If users are not satisfied with the results of the analysis, they should be able to view those results, review the details, and make changes if necessary. (When)
- Task 2: Determine the location and key people. The location information shows the spatial distribution of the users who posted the message, while the key people are important for analyzing the reasons for the turn of events. (Where and who)
- Task 3: Users should be able to get an overview of the whole event based on the story, quickly understand the main information of the event, and know what the hot topics and emotional changes are. (What)
- Task 4: Analyze the reasons for the turn of events. In complex views, root cause analysis is often done manually by the user based on a complex view. However, when designing social media data analytics, this information should be automatically analyzed by the machine and presented directly to the user. The user is primarily a listener and secondarily a discoverer. Therefore, the story presented to the user should be as complete as possible. (Why)

4 Study Pipeline

The whole process of our study can be divided into five parts, which are shown in Fig. 1. They are (a) Data Collection, (b) Data Preprocessing, (c) Important Time Periods Extraction, (d) Data Analysis, and (e) Interactive Visual System.

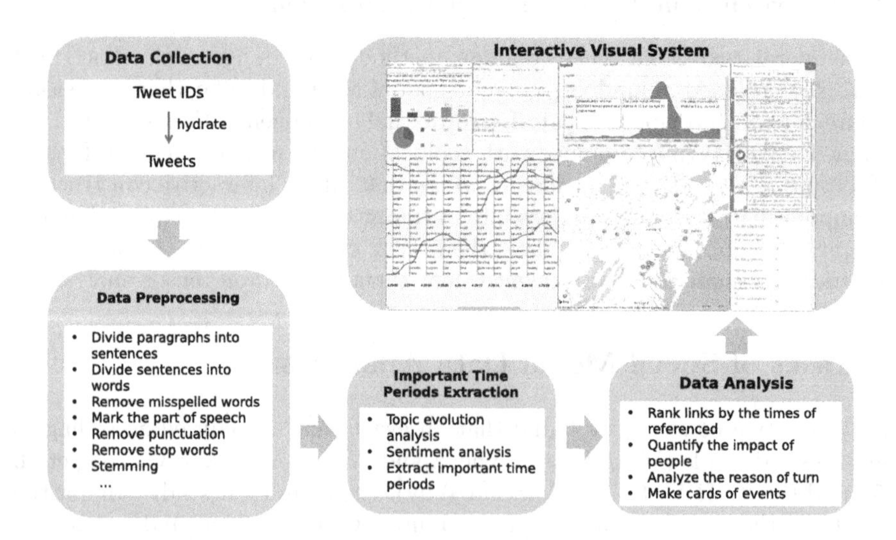

Fig. 1. The whole process

Firstly, in Data Collection part, we use Twitter data as study material. The dataset ID comes from an open source website, while the full text of tweets comes from the official Twitter API.

In the Data Preprocessing part, more rigorous preprocessing is required to match the characteristics of socialized short texts. This includes removing interference, splitting paragraphs and sentences, removing punctuation, removing stop words, removing misspellings, and extracting word stems.

In the Important Time Periods Extraction part and Data Analysis part, we come up with a method to filter out the importance of time periods based on topic evolution and emotion transformation, since key turning points are most likely to occur in these time periods. The method extracts topics by DTM and identifies the emotions of each socialized short text through a Naive Bayesian classifier. If both emotions and topic words change significantly within a time period, it could be an important time period. At the same time, we quantify the influence of people, mark the influential people, and summarize the results of the analysis.

Finally, we design a visual system that provides multiple views and interactions that help users analyze and explore information about events in social media.

5 Extract Important Time Periods

An event is chronological, so extracting important time periods is a key step in organizing the event. In this section, we describe the method for extracting the important time periods by following the evolution of topic words and emotions.

5.1 Track the Evolution of Topics

A topic is a collection of discourse words. In a temporal text, topics change as events evolve, so tracking changes in discourse words is important for understanding the overall development of events. After preprocessing the text, we use DTM [2] to extract topics and obtain a trained topic model. This comprises several topics, each of which contains a list of topic words that can be used to extract important time periods based on their level of confusion.

The follows are the details of analyzing one of the topics.

- Step 1: Divide the original time period. These original time periods will be merged to form new time periods. In our study, we divide the whole event time by every two hours.
- Step 2: Analyze the trend of the popularity of one topic word. As events continue to evolve, the popularity of each topic word has also been changing. The higher the ranking of the word, the more it is mentioned during that time period. For each topic in each time period, the top 20 topic words are taken and the rank sequence of each topic word is calculated. The topic words are denoted as w_1, w_2, $w_3, \ldots, w_{20}$ and the ranks of each topic word are denoted as $r_{w_k}^{t_1}$, $r_{w_k}^{t_2}$, $r_{w_k}^{t_3}$ ..., in which t_1, t_2, t_3 represent the different original time periods. Formula 1 calculates the difference in ranking between the ith and the $i+1$th original time period of the kth topic word.

$$D_{w_k}^{t_{i,j+1}} = r_{w_k}^{t_{i+1}} - r_{w_k}^{t_i} \tag{1}$$

For the kth topic word, $D_{w_k}^{t_{i,j+1}} < 0$ means it is less popular in the $i+1$th period, instead $D_{w_k}^{t_{i,j+1}} > 0$ means it is more popular.

- Step 3: Find the monotone intervals from the original time periods. Starting from the first period, if $D_{w_k}^{t_{i+1,i+2}} * D_{w_k}^{t_{i,i+1}} \geq 0$, which means they are in the same monotone interval, these two periods will be merged into one monotone interval. And this monotone interval will continue to be compared with $D_{w_k}^{t_{i+2,i+3}}$. Until the end, we can get some monotone intervals in this way.
- Step 4: Remove fluctuations. There may be some fluctuations in the monotone intervals, which can be ignored in the analysis. The trend of fluctuations is slightly different from the overall trend of the whole time period and the fluctuations. So we use Formula 2 to measure the difference in the rank of each topic word.

$$ED_{w_k}^{m} = \max_{1 \leq i \leq L_c} r_{w_k}^{t_i} - \min_{1 \leq i \leq L_c} r_{w_k}^{t_i} \tag{2}$$

In Formula 2, L_c represents the number of original time periods in the mth monotone interval, $\max_{1 \leq i \leq L_c} r_{w_k}^{t_i}$ is the highest rank of the kth topic word in the mth monotone interval, while $\min_{1 \leq i \leq L_c} r_{w_k}^{t_i}$ is the lowest rank of the kth topic word in the mth monotone interval, $ED_{w_k}^{m}$ is the difference between them. After calculating the values of $ED_{w_k}^{m}$ by Formula 2, some monotone intervals can be considered as a fluctuation and be merged with the adjacent monotone intervals to avoid the effect of fluctuaions by setting a threshold. For example, for the pth monotone interval, we set a threshold that if $ED_{w_k}^{p+1}$ is not greater than 2 and these two monotone intervals have opposite trends, the $p+1$th monotone interval can be considered as fluctuation and merged with the pth monotone interval, otherwise the $p+1$th monotone interval is not fluctuation and continue to look at the $p+1$th monotone interval and the $p+2$th monotone interval, until all monotone intervals are considered.
- Step 5: Calculate the rate of the change in the topic word. After the previous step, the number of new monotone intervals can be denoted as n. At this time, some adjacent monotone intervals may have the same or opposite trend and they can continue to be merged. So we use the rate of change of the kth topic word in the mth period to determine to combine them, which is calculated by Formula 3.

$$S_{w_k}^{n} = \left| \frac{r_{w_k}^{t_{last}} - r_{w_k}^{t_{first}}}{\Delta t} \right| \tag{3}$$

Here Δt represents the distance between the first original time period and the last original time period in the nth monotone interval, $r_{w_k}^{t_{\text{last}}}$ and $r_{w_k}^{t_{\text{first}}}$ represent this topic word's ranks in the first and last original time period in the nth monotone interval. If the rate of change in the former period is not lower than the rate of change in the latter period, we can continue to merge these two monotone intervals.

Algorithm 1. find the common monotone interval

Input: all monotone interval of every topic words: $\text{set}_{w_k}^{y}$
where $k = 1, 2, 3..., v$ and $y = 1, 2, 3...$
Output: the common monotone interval of all topic words: u_a
where $a = 1, 2, 3...$

1: $a = 1$
2: **while** $\cap_{k=1}^{v} \text{set}_{w_k}^{1} \neq \emptyset$ **do**
3: $\quad u_a = \cap_{k=1}^{v} \text{set}_{w_k}^{1}$
4: $\quad$ **for** k from 1 to v **do**
5: $\quad\quad \text{set}_{w_k}^{all} = \text{set}_{w_k}^{all} \cap u_i$
6: $\quad a = a + 1$

- Step 6: Find the common monotone interval for all topic words. The common monotone interval is the time period in which the trend of all topic words is monotonous. The yth monotone interval of the kth topic word is denoted as $\text{set}_{w_k}^{y}$, and the whole monotone interval of the kth topic word are denoted as $\text{set}_{w_k}^{all}$. Then the intersection of the monotone intervals of all topic words can be taken by Formula 4.

$$[u_1, u_2, ..., u_a] = \cap_{k=1}^{v} \text{set}_{w_k}^{all} \tag{4}$$

  u_a means the ath common monotone interval of all topic words. v means the number of topic words in each original time period, where k is from 1 to v. And we can get the result by Algorithm 1. The loop will be end until each $\text{set}_{w_k}^{all}$ is an empty set, then we can get lots of common monotone intervals, they are $u_1, u_2, \ldots, u_a$.
  For example, assuming there are six original time periods at the beginning, they can be denoted as $t_1 \sim t_6$. There are three topic words: A, B and C, the monotone intervals of word A are $t_1 \sim t_4$ and $t_4 \sim t_6$, the monotone intervals of word B is $t_1 \sim t_6$, the monotone intervals of word C are $t_1 \sim t_5$ and $t_5 \sim t_6$. We can get $u_1 = (t_1 \sim t_4) \cap (t_1 \sim t_6) \cap (t_1 \sim t_5) = (t_1 \sim t_4)$. Then the left intervals of A is $(t_4 \sim t_6)$, because $(t_1 \sim t_4) - (t_1 \sim t_4) = \emptyset$. The left intervals of B is $(t_1 \sim t_6) - (t_1 \sim t_4) = t_4 \sim t_6$. The left intervals of C are $(t_4 \sim t_5)$ and $(t_5 \sim t_6)$, because $(t_1 \sim t_5) - (t_1 \sim t_4) = (t_4 \sim t_5)$.
  Next, $u_2 = (t_4 \sim t_6) \cap (t_4 \sim t_6) \cap (t_4 \sim t_5) = (t_4 \sim t_5)$. And the left intervals of A is $(t_4 \sim t_6) - (t_4 \sim t_5) = (t_5 \sim t_6)$. The left intervals of B is $(t_4 \sim t_6) - (t_4 \sim t_5) = (t_5 \sim t_6)$. The left of intervals of C is $(t_5 \sim t_6)$, because $(t_4 \sim t_5) - (t_4 \sim t_5) = \emptyset$.
  Finally, $u_3 = (t_5 \sim t_6) \cap (t_5 \sim t_6) \cap (t_5 \sim t_6) = (t_5 \sim t_6)$ and the result of $(t_5 \sim t_6) - (t_5 \sim t_6)$ is an empty set. So there are three intersections, $(t_1 \sim t_4)$, $(t_4 \sim t_5)$ and $(t_5 \sim t_6)$.
- Step 7: Select the important time periods. In order to select the important time periods, We firstly calculate the chaos degree of each common monotone interval by Formula 5. For each common monotone interval in our work, there are 20 topic words, and the degree of chaos here is the sum of the range of

variation of all topic words in the common monotone interval.

$$\sum_{\mathrm{k}=1}^{v} \left| ED_{w_k}^{\mathrm{u}_a} \right| \tag{5}$$

And then, all the common monotone intervals were ranked and the top 10 common monotone intervals can be selected as the degree of chaos of the common monotone intervals. These are identified as important time periods from the perspective of topic evolution.

5.2 Track the Transition of Emotions

Beside topics, emotions are also important factors in analyzing socialized short text. We use Naive Bayesian classifier method to distinguish the emotion. The classifier will give each text an emotion tag according to the calculated probability.

For the division of emotions and the labeling of text data, due to the special characteristic of socialized short texts which often have many irony and neutral phrases, so we finally divide the emotions of the information into two types, positive and negative, which can show the best results.

For emotions, important time periods are extracted in the following steps.

- Step 1: Divide the original time period. We also divide the whole event by every two hours.
- Step 2: Calculate the range of variation for each emotion. By Naive Bayesian classifier, the proportion of each emotion can be marked as $p_{e_b}^{t_j}$, where e_b means the bth emotion and t_j denotes the jth original time period, then the range of variation of the proportion of each emotion in the adjacent original time period can be calculated according to Formula 6.

$$SD_{e_b}^{t_{j,j+1}} = p_{e_b}^{t_{j+1}} - p_{e_b}^{t_j} \tag{6}$$

- Step 3: Determine the state of each emotion. The status here is whether the emotion is in the leading position or not. Suppose there are x emotions and the status can be calculated by Formula 7.

$$\mathrm{sign}_{e_b}^{t_j} = p_{e_b}^{t_j} - \frac{1}{x} \tag{7}$$

For the bth emotion, $\mathrm{sign}_{e_b}^{t_j} > 0$ indicates that this emotion stands in the leading position at the jth original time period, and the emotion is considered as a large change when the following two conditions are satisfied.

- $SD_{e_b}^{t_{j,j+1}} \geq TH$, here TH is a manually set threshold, in this paper we use 15%;
- $\mathrm{sign}_{e_b}^{t_j}$ and $\mathrm{sign}_{e_b}^{t_{j+1}}$ have different signs.

While both two conditions are satisfied, the state of this emotion is thought changed, then the jth and $j+1$th original time periods can be merged as a union.

When both conditions are satisfied, the state of this emotion is considered to have changed, and then the jth and $j+1$th original time periods can be merged into one union.

- Step 4: Calculate the range of variation and select the important time periods.

$$\sum_{b=1}^{x} \left| SD_{e_b}^{t_{jj+1}} \right| \tag{8}$$

We calculate the range of variation for all unions by using Formula 8. After ranking all the unions, we can select the top 10 unions by the range of union changes. These are important time periods determined from the perspective of emotion transition.

5.3 Important Time Periods

With the above steps, we can get 10 important time periods determined from the perspective of topic evolution and 10 important time periods determined from the perspective of emotion transformation. Then, we combined them into the final important time periods for topic evolution and emotion transformation. These final important time periods most likely contain important information about the whole event.

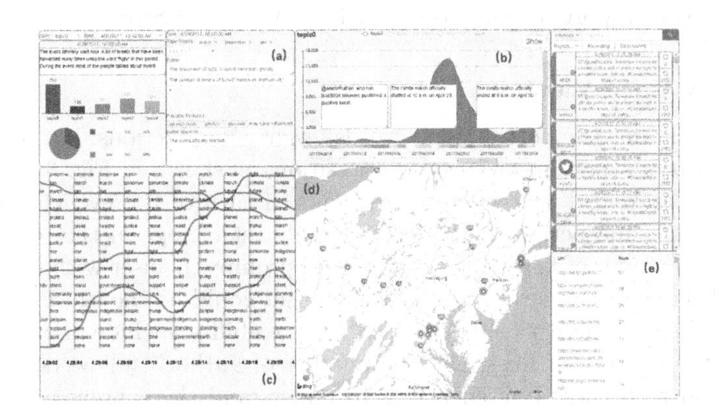

Fig. 2. The interactive visual system

6 Design of Interactive Visual System

The visual system has five views, as shown in Fig. 2. The Story Card View(a) shows topic words, event introductions, key people and the causes of their transition. Emotion Transition View(b) allows the user to quickly locate important time periods and provides the function to interactively add annotations.

Topic Evolution View(c) shows the evolution of topic words. Spatial Distribution Map(d) shows the location of the tweets and provides the function to filter the information interactively. Details View(e) shows the details of tweets. All these views work together to perform the above functions.

We summarize some necessary functions of the visual system based on the "5W" principle.

- Function 1: The system should display important time periods in the view so that users can clearly identify them and then explore them. (Support by Emotion Transition View)
- Function 2: The system should display multiple story information. The story information should include not only the current hottest topic words and key people, but also the causes of event changes, and the story cards should be generated automatically. (Support by Emotion Transition View and Story Card View)
- Function 3: The system should be able to generate story cards. After reading these story cards, the user should be able to explore the event. (Support by Details View, Spatial Distribution Map and Topic Evolution View)
- Function 4: The system should be able to display the changes in popularity, emotion and text volume of each topic. (Support by Emotion Transition View and Topic Evolution View)
- Function 5: The system should be able to refine the event. After exploring an event, if the users are not satisfied with the event generated by the system, they can be able to modify the story card and add annotations to improve the event. (Support by Emotion Transition View and Story Card View)

6.1 Story Card View

This view contains the hottest topic words, event introductions, key people and causes of topics evolution or emotion transition, as shown in Fig. 3. The Story Card View can detect whether a reason for the turn of the event occurred in the important time period. If there is a significant change in the topic words, emotions or number of tweets, the system will prompt the user with some preset tags. And the view can also filter out tweets that could be the cause by ranking the possibility of those tweets. Metrics for possibility include the number of retweets, the number of likes, and the number of author's followers. This way, the user can read these filtered tweets to understand the content and cause of the event instead of reading them one by one from the huge text data.

There are three main parts in this view. The first part is a drop-down list, located in the upper left corner, where the user can select different topics and time periods. Once the user changes the drop-down list, all the views associated with it will be changed together.

The second part is a structure view, below the drop-down list, which includes a histogram, a pie chart, and a text box. The histogram shows the number of tweets on each topic and the pie chart shows the percentage of each emotion. The text box above displays the story generated by the system. The story contains several preset sentences. For example:

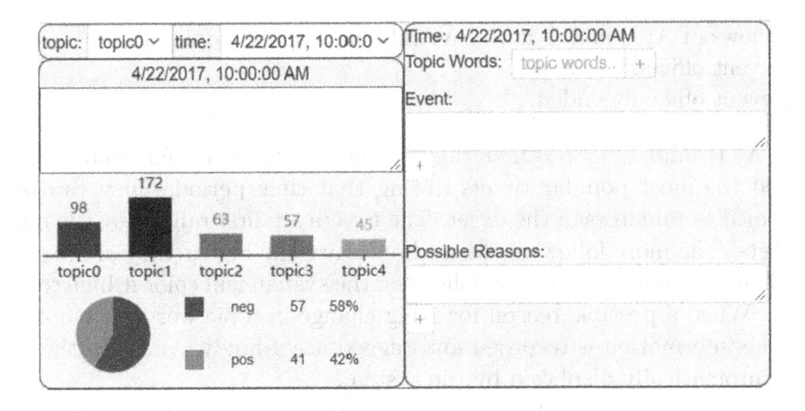

Fig. 3. Story card view

- "The event officially started now."
- "A lot of tweets that have been forwarded many times used the word'(some word)' in this period."
- "During the event, most of the people talked about (some topic)."
- "The most three influential people in this period are (@somebody, @somebody, @somebody)."
- "(some emotion) in (some topic) has a significant (increase or decrease)."
- ...

These preset sentences can make the generated stories more natural and optimize the user's reading experience.

The third part is a card, which is located on the right half. This card shows the introduction of the event, possible causes, and the three hottest topic words of the period. Both the introduction of the event and the possible causes consist of several preset tags that are automatically generated by the system.

We have set three types of preset tags for event introduction when analyzing tweets:

- The proportion of (some emotion) in (some topic) has a significant increase/decrease.
- The discussion of (some topic word) in (some topic) increases/decreases greatly.
- The number of tweets of (some topic) increases/decreases dramatically.

The parenthesis parts here can be replaced by a specific topic, emotions or topic words. The system first finds the combined important time periods generated by the method in Sect. 5, and then analyzes whether a significant change has occurred in the topic words, emotions, or number of tweets. If so, the view will select the matching tags to display after replacing the parenthesis part, which can notify the user that there might be a reason for the turn of events.

We have also preset three tags for possible reasons:

- The views of A, B and C may have influenced public opinions.
- The event officially started.
- The event officially ended.

The A, B, and C in the first tag are the top three people who posted or retweeted the most popular tweets during that time period. They can be the authors or disseminators of the tweet. The tweets are first ranked by the number of retweets. The more followers the authors have, the higher they are ranked. If one of them has more than 10,000 followers, the system will color it blue to notify the user. When a possible reason for a big change in topic words or emotions is found, the information is recorded and tagged accordingly. And also these tags will be automatically displayed by the system.

If there are other possible reasons, users can add them as tags and modify the tags as they like.

6.2 Emotion Transition View

The Emotion Transition View consists of four parts, as shown in Fig. 4, the user can see the trend of event changes as a whole. (a) is a flow view, the height represents the set of tweets at each time point, which shares the same x-axis with (b); (b) is an emotion square view; (c) is a series of annotations that provide a basic overview of the event. In addition to these three parts, there is a hidden list of important time periods. It is located below the emotion square view and can be accessed by clicking the "show" button in the upper right corner of this view. The list of important time periods is shown in Fig. 5.

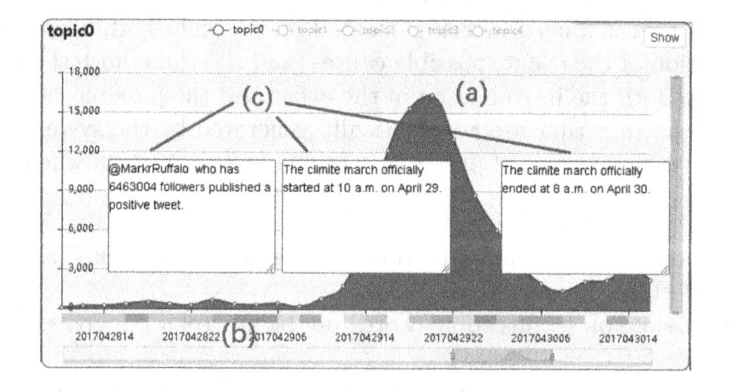

Fig. 4. Emotion transition view

In the Emotion Transition View, for important time periods, the annotation pops up when the mouse is moved over the corresponding square, while for unimportant time periods, the annotation pops up only when the square is clicked. After system analysis, most of the annotations are blank, but for the start and end time periods of the event, the following words are automatically added to their annotations, where the words in brackets will be replaced by the appropriate content:

- The event officially started (or ended) at (some time) on (some day).

Apply	Add	Delete	Ascending	Descending
2017042014	2017042016		2017042020	2017042022
2017042206	2017042208		2017042614	2017042616
2017042710	2017042712		2017042718	2017042720
2017042808	2017042810		2017042910	2017042912
2017042916	2017042918		2017042922	2017043000

Fig. 5. The list of important time periods

The start and end time periods are determined automatically based on the change in the amount of text. We first locate the time period with the largest text volume, and in most cases the time period with the largest text volume lies between the start and end time periods. The difference between the text volume of two adjacent time periods is calculated as the vertical difference, and the horizontal difference between two adjacent time periods is two hours, and the horizontal difference can be set to 1. Then from the time period with the largest text volume, looking forward and backward, we can find a time period with the largest slope difference, which is the start and end time period.

The colors of the emotion squares mean different emotions. As described in Sect. 5.2, we have divided the emotions of the text into positive and negative. So in this view, the green color represents positive emotions and the red color represents negative emotions. The higher the percentage of emotions, the closer the color of the corresponding square is to green, otherwise the color of the corresponding square is closer to red. In this way, the user can easily track the change of emotions.

6.3 Topic Evolution View

The Topic Evolution View is consistent with Function 4. We have drawn inspiration from the work of Jie Li et al. [11]. As shown in Fig. 2(c).

In this view, each vertical gray line represents an original time period, and there is a time marker under each line. We take the first 20 topic words and display them in order from top to bottom. Different original time periods for the same topic word are connected by a red line. If there is a time period without that topic word, then the topic for that time period is connected with the word "none" at the bottom. Users can interactively look up the change in rank of each topic word.

For the important time periods, the range of change is greater than for the unimportant time periods. So in the view, the important time periods look more confusing because there will be more intersections between the lines.

6.4 Details View

The Details View is shown in Fig. 6. It consists of two parts. The left part shows the details of each tweet in the time period. On the left side of each tweet is a sidebar whose color represents different emotions. Users can filter tweets by keywords or nicknames, and can also sort tweets by number of retweets, likes, or followers the user has. If a user has more than 10,000 followers, we classify them as an influencer and put a "v" icon in the corner of their avatar. The right part shows all the links in the tweets and the number of mentions per link.

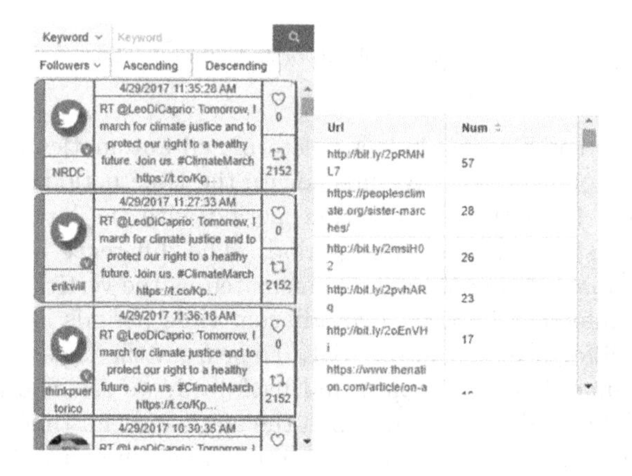

Fig. 6. Details view

6.5 Spatial Distribution Map

We used Microsoft's Bing Map as the main component of the Spatial Distribution Map. The overview of the Spatial Distribution Map is shown in Fig. 7.

All tweets with geographic information are marked with blue dots on the map. Users can quickly get an overview of the spatial distribution of each tweet. It also shows where the events happen and where people are particularly active. Clicking on each point can display its detailed information. These details include the avatar and nickname of the person who sent the tweet, as well as the content of the tweet. In our system, this diagram presents information for one time period at a time. Users can select time periods to change the graph and analyze the spatial evolution.

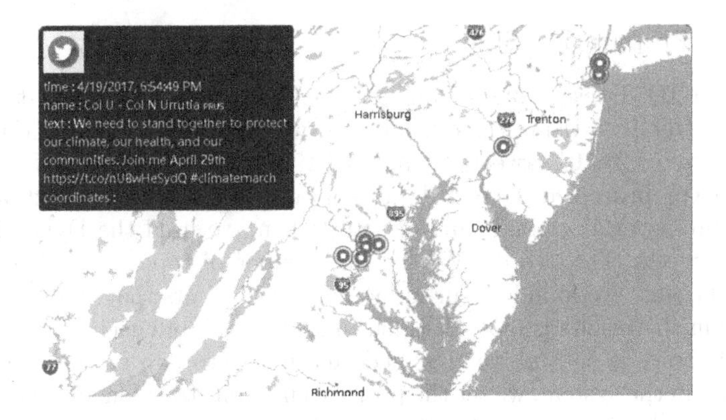

Fig. 7. Spatial distribution map

7 Case Study and Evaluation

To test the effectiveness and feasibility of the previous method and system, we analyzed a real Twitter case and invited some volunteers to evaluate the system.

The dataset used for the evaluation is a Twitter dataset about the climate march. The march took place on April 29, 2017, when Trump was in power for 100 days. The main purpose of the march was to express people's dissatisfaction with current environmental policies and to call on the government to care about and protect the environment.

First, we obtained the record of events and pre-processed the data. Then we filtered the important time periods according to the method in Sect. 5. Next we analyzed the time, topic words, causes, key people, etc. Finally, we used the system to display the results.

For the visualization analysis, we selected 16 volunteers. Half of these volunteers had relevant knowledge of visualization, and the other half did not.

None of the 16 volunteers had ever known anything about this competitive event. Before the evaluation, we first introduced them to the functions of each view in the system and then gave them four basic tasks. These basic tasks were not difficult and were designed to get the volunteers familiar with the system as quickly as possible. After completing these basic tasks, they read the automatically generated story in the Story Card View to get a quick overview of the event and then further explore the information in the story. These tasks are listed below:

1. Find out which are the important time periods that are automatically analyzed by the system.
2. Find out the start and end time periods of the event.
3. Find out when the heat of the event peaked.
4. Analyze where most of the participants gathered during this event.

Following are the details of our case study.

7.1 Data Preparation Stage

The Twitter ID dataset comes from an open source website [14]. It contains 681,668 Twitter IDs. Using the official API, we finally obtained a dataset containing a total of 284,012 tweets from April 19 to May 3, 2017.

First, we preprocessed the text data to obtain a set of cleaned words and split the data into several original time periods. Then, we used the Dynamic Topic Model to process these words. We set the number of topics to 5, from topic0 to topic4, because we found in our numerous experiments that the semantics of each topic in the results is most distinct and different from each other when the number of topics is 5. And we classified emotions into two types: positive and negative. So in the system, we set the positive emotions to green and the negative emotions to red. Finally, we identify a number of important time periods and combine the given sentences into coherent stories.

7.2 Get Automatically Generated Stories

First, the volunteers completed the four basic tasks in a similar way after reading them.

For Task 1, volunteers clicked the "Show" button in the Emotion Transition View and found the important time periods divided by the system automatically.

For Task 2, they moved the mouse over the Emotion Square View and looked at its annotations. As shown in the Fig. 4, they found that the annotations for most of the important time periods were blank, except for April 29, which contained the annotation "The climite march officially started at 10:00 a.m. on April 29". And the annotation for April 30 was "The climite march officially ended at 8:00 a.m. on April 30". These indicate that the start time of the event analyzed by the system was at 10:00 a.m. on April 29 and the end time was at 8:00 a.m. on April 30. (At the beginning, the annotation is "The event officially started at ...", and it was manually modified by the previous user by the "climite march".)

For Task 3, they found that the flux graph peaked at 8:00 p.m. on April 29, so they concluded that the heat of this event was greatest during that time period.

For Task 4, using the Spatial Distribution Map, they found that most of the participants in the event were on the east and west coasts of the United States, with more participants congregating in Washington as time went on.

After completing these four tasks, the volunteers had a good understanding of how to use the system. Next, they read the stories on the Story Card View. The story card for topic0 on April 29, 2017 at 10:00 a.m. is shown in Fig. 8.

In the Story Card View, the system automatically generated a story and some tags. The story on the left consists of three sentences: (1) This event officially started now. (2) Many tweets that were retweeted multiple times used the word "fight" in this period. (3) During the event, most people were talking about topic0.

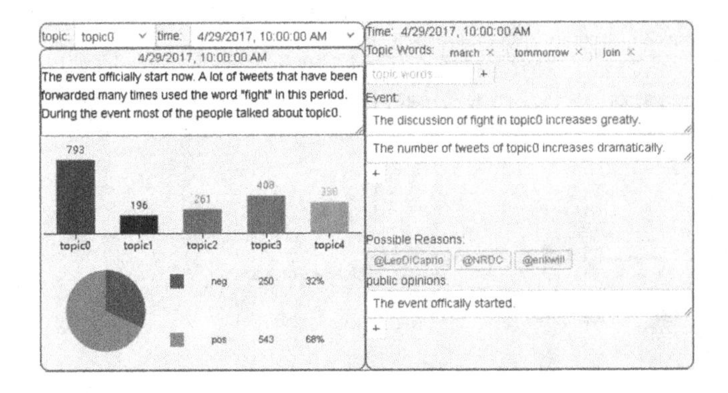

Fig. 8. A story card on April 29, 10 a.m.

In the right half of the card, the three most popular keywords are "march", "tomorrow" and "join". Among them, "march" could mean parade or third month. However, because the first letter is not capitalized, some volunteers guessed that it meant parade. In the "Event" column, the system automatically generated two tags. One was that the number of discussions about the topic word "fight" in topic0 increased sharply during this time period, which verified the previous guess that march is a meaning of parade. The other tag is that the number of tweets about topic0 increased greatly during this time period. The possible reasons for these two situations are shown below. One is that the actions of three users @LeoDiCaprio, @NRDC and @erikwill might have influenced public opinion during this period, and the other is that the system assumes that this is when the event officially began.

Next, volunteers simply read the story cards for all important time periods and then form a basic story in their minds. This story includes: when the event began and ended; when the emotion of the event changed significantly; when the number of texts changed significantly; when the discussion about a topic word changed significantly; and who may have influenced the event.

7.3 Analysis of the Celebrity Effect

One volunteer noted that the system considers influential people to be one of the main causes of change in events. So she wanted to find some examples to see if there was a "celebrity effect" to support this.

By switching time periods and topics, volunteers found that the story card in topic2 automatically generated a sentence at 4 a.m. on April 29: "A lot of tweets that have been retweeted many times used the word 'trump' in this period". As can be seen in the Fig. 9. The system pointed out two key people here: @SierraClub and @350. It then examined them in detail using the system's filtering and sorting features.

The volunteer found that a user named @SierraClub retweeted a lot of tweets about the march at 7 p.m. on April 27. And since @SierraClub has 365,404

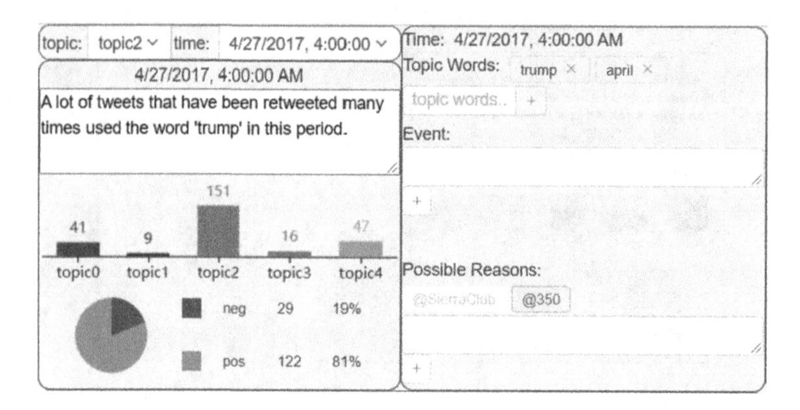

Fig. 9. A story card on April 29, 4 a.m.

followers (the user can look up the number of followers by hovering over the username), the number of retweets of that tweet increased rapidly, as shown in Fig. 10. And a user named @350, who had over 380,000 followers, said they were taking more than 30 buses to Washington, DC. This proves that 10 a.m. on April 29 can indeed be marked as the official start time of the event, which is the same result in the system.

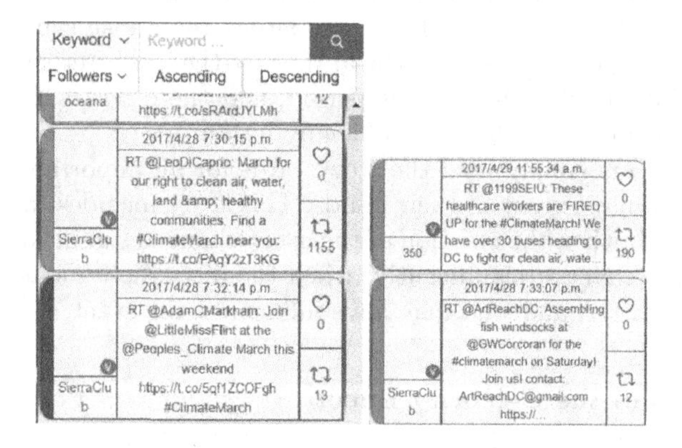

Fig. 10. The details for the case

Through a similar exploration, this volunteer has found examples that prove there is indeed a "celebrity effect". Therefore, she thoughts that our method to analyzing the causes is appropriate.

7.4 Effectiveness Evaluation

Finally, a questionnaire was designed to ask the volunteers whether the system was effective and feasible. 75.0% (12/16) of them thought that the system automatically generated stories could help them understand the events quickly; 18.75% (3/16) thought that the system was not smart enough and there were some omissions in analyzing the possible causes of changes in events; 6.25% (1/16) thought the system's interactions needed to be improved and that some buttons were too small to be easily clicked.

8 Limitation and Conclusion

According to the case study and the problems we encountered during development, we consider that the system has the following limitations and problems that need improvement.

(1) The impact of other events on the target event is not considered. In real life, there are many events that occur simultaneously, and there may be many relationships between different events. And this system only analyzes the internal evolution of one event without considering the external factors of other events. We believe that it can be solved in the future work.
(2) It might be better to use a more advanced method to analyze the emotions of the text. Our work mainly focuses on the extraction of important time periods and the design of the system for social media data analytics, so we only use DTM to extract topics and Naive Bayesian classifier to distinguish emotions. We believe that using more advanced algorithms or models to analyze the topics and emotions can lead to more effective results, and that is the next step we will try.
(3) Not consider the multi-view data. In the real world, social media data is often generated from different sources or observed from multiple viewpoints. We believe that applying our method and design to the multi-view social media data may reveal more interesting information.

In summary, we propose a new method to extract important time periods in socialized short texts through a combination of topic word analysis and emotion analysis. We propose a visual design with an interactive story card that allows users to quickly understand the what and why of events. And we design an interactive visual system based on the "5W" principle, which can work well in data analytics of social media. The system can quantify the importance of time periods, selects the most important ones to explore in detail, and analyzes the evolution of topics and emotional transitions. The results of the analysis are presented in the form of story cards, which users can not only read, but also modify according to their own opinions. And we invited some volunteers to evaluate the system, and most of them agreed that the interaction is easy and that the story card can help save time in understanding an event. So our system can help well in understanding, analyzing and exploring the information in social media data.

References

1. Barbosa, L., Feng, J.: Robust sentiment detection on twitter from biased and noisy data. In: Coling 2010: Posters, pp. 36–44 (2010)
2. Blei, D.M., Lafferty, J.D.: Dynamic topic models. In: Machine Learning, Proceedings of the Twenty-Third International Conference (ICML 2006), Pittsburgh, Pennsylvania, USA, 25–29 June 2006 (2006)
3. Chen, S., Li, S., Chen, S., Yuan, X.: R-map: a map metaphor for visualizing information reposting process in social media. IEEE Trans. Visual Comput. Graph. **26**(1), 1204–1214 (2020)
4. Chen, S., Chen, S., Wang, Z., Liang, J., Wu, Y.: D-map: visual analysis of egocentric information diffusion patterns in social media. In: 2016 IEEE Conference on Visual Analytics Science and Technology (VAST) (2016)
5. Chen, S., Chen, S., Yuan, X., Lin, L., Zhang, X.L.: E-map: a visual analytics approach for exploring significant event evolutions in social media. In: IEEE Vast (2017)
6. Devlin, J., Chang, M.W., Lee, K., Toutanova, K.: Bert: pre-training of deep bidirectional transformers for language understanding. arXiv preprint arXiv:1810.04805 (2018)
7. Dexi, L., Changxuan, W.: Survey on automatic summarization of socialized short text. J. Chin. Comput. Syst. **034**(012), 2764–2771 (2013)
8. Dou, W., Wang, X., Skau, D., Ribarsky, W., Zhou, M.X.: Leadline: interactive visual analysis of text data through event identification and exploration. In: 2012 IEEE Conference on Visual Analytics Science and Technology (VAST) (2012)
9. Eccles, R., Kapler, T., Harper, R., Wright, W.: Stories in geotime. Inf. Vis. **7**(1), 3–17 (2008)
10. Guo, S., Xu, K., Zhao, R., Gotz, D., Zha, H., Cao, N.: Eventthread: visual summarization and stage analysis of event sequence data. IEEE Trans. Vis. Comput. Graph. **24**(1), 56–65 (2018)
11. Li, J., Chen, S., Chen, W., Andrienko, G., Andrienko, N.: Semantics-space-time cube. A conceptual framework for systematic analysis of texts in space and time. IEEE Trans. Vis. Comput. Graph. **26**(4), 1789–1806 (2018)
12. Marcus, A., Bernstein, M.S., Badar, O., Karger, D.R., Madden, S., Miller, R.C.: Twitinfo: aggregating and visualizing microblogs for event exploration. In: Proceedings of the SIGCHI Conference on Human Factors in Computing Systems, pp. 227–236 (2011)
13. Nguyen, P.H., Xu, K., Walker, R., Wong, B.L.W.: Schemaline: timeline visualization for sensemaking. In: 2014 18th International Conference on Information Visualisation (2014)
14. Ruest, N.: climatemarch tweets April 19–May 3, 2017 (2017). https://doi.org/10.5683/SP/KZZVZW
15. Tong, C., et al.: Storytelling and visualization: an extended survey. Information **9**(3), 65 (2018)

Auto-Recon: An Automated Network Reconnaissance System Based on Knowledge Graph

Xiang Zhang[1,2], Qingli Guo[1(✉)], Yuan Liu[1], and Xiaorui Gong[1,2]

[1] Institute of Information Engineering, Chinese Academy of Sciences, Beijing, China
{zhangxiang,guoqingli,liuyuan,gongxiaorui}@iie.ac.cn

[2] School of Cyber Security, University of Chinese Academy of Sciences, Beijing, China

Abstract. Effective network security management is usually based on comprehensive, accurate and real-time control of enterprise network. With the rapid growth of enterprise network information, their exposure on the Internet is also expanding rapidly, which raises many security issues. It is hard for the existing approaches of network security management to cover dynamic and hidden assets. Moreover, black-box-based network reconnaissance is highly dependent on expert experience and time-consuming, which cannot meet the needs of enterprises to perform testing periodically. Therefore, target-oriented automated reconnaissance of network information becomes an urgent problem to be solved.

We proposed and constructed a knowledge graph based network reconnaissance model, NRG, from a method-level perspective which describes the relationship between different network information and the way to reconnaissance them. Based on NRG, we have designed and implemented Auto-Recon, an automated network reconnaissance system using the distributed architecture. The purpose of Auto-Recon is to automatically find exposed surfaces of targets on the Internet using the primary domain as initial information. The system reduces strong dependence on network reconnaissance knowledge and experience. We conducted an experiment and the result shows that Auto-Recon has better performance in terms of efficiency, effectiveness and automation than existing tools.

Keywords: Penetration test · Reconnaissance · Knowledge graph · Automated

1 Introduction

The rapid development of the Internet has brought great changes to human life. Thanks to the Internet, enterprises are able to operate at a high speed and continuously increase their invest in the construction of network system. However, the growth in complexity and scale of network leads to the expansion of attack surfaces, and causes serious threat to the security of enterprises [5], such as computer virus, data breaches, phishing, and so on.

Y. Lai et al. (Eds.): ICA3PP 2021, LNCS 13157, pp. 101–115, 2022.
https://doi.org/10.1007/978-3-030-95391-1_7

Penetration test is a popular way for enterprises to assess their defensive ability against these threat. It simulates an authorized cyberattack to evaluate the security of the enterprise [13,25,30]. Different from a passive situation, penetration test allows defenders to identify their weaknesses and strengths in a timely and comprehensive manner.

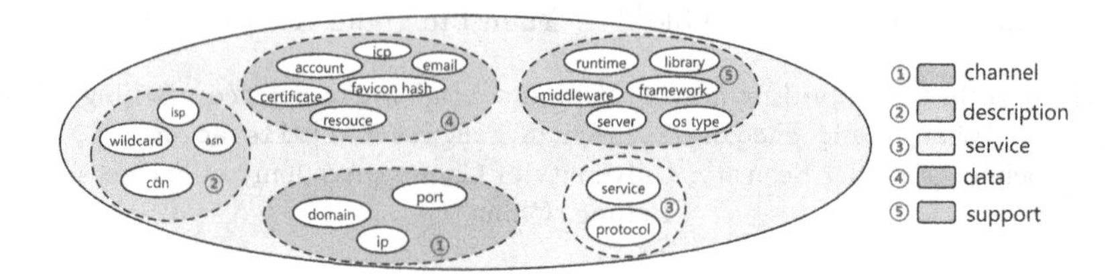

Fig. 1. Network information and their classification.

In penetration test, network information is the most concerned part of penetration testers. As shown in Fig. 1, network information usually includes communication channels, related attributes and descriptions, open services, runtime support, and data. Usually, from the very beginning of penetration test, penetration testers only know the primary domain of target. Therefore, reconnaissance of network information against the target is extremely critical in penetration testing [9]. The quality of the gathered information directly affect the understanding of the target.

Existing works related to network reconnaissance are usually conducted manually. The selection of tools and the process of data are mainly dependent on expert experience. Besides, the tools are usually customized for specific tasks, so the data formats are varied. Due to the large amount and wide variety of network information, manual reconnaissance usually takes considerable time, but produces low-quality results. To remedy this situation, some reconnaissance automation tools, such as reNgine [22] and ARL [2], are developed. However, these automation tools just combine single tools and serialize them without considering the details. Therefore, the efficiency and effectiveness can not be guaranteed.

To improve automation in network reconnaissance, this paper applies knowledge graph to information gathering. The data model of knowledge graph defines the information to be collected, the operations to be executed and the relationship between them. The data model split the subtasks into more fine-grained units, and add necessary constraints to operations. Based on the knowledge graph, we implement an automated reconnaissance system: Auto-Recon, which effectively improves the efficiency and effectiveness of network reconnaissance.

Our contributions are as follows:

1. This paper proposes an automated reconnaissance system, Auto-Recon, based on knowledge graph.

2. The data model of the knowledge graph is defined from the method perspective, and is embedded with necessary constraints to the execution of methods. Methods ena the parallization in methods improves the efficiency. The constraints guarantee the effectiveness of the results.
3. Auto-Recon significantly reduce the dependence on expert experiences. Experiments shows that, compared with ARL and rEngine, Auto-Recon achieves significant increase in efficiency and effectiveness.

The rest of the paper is organized as follows. Section 2 summarizes related works. Section 3 describes the creation of the knowledge graph. Section 4 introduces the implementation of Auto-Recon. Next, The performance of Auto-Recon is evaluated in Sect. 5. Finally, Sect. 6 concludes this paper.

2 Background and Related Work

Network reconnaissance is a fundamental and core part of penetration test. Effective network reconnaissance can improve the chances of discovering attack surfaces. Therefore, it has been a hot research topic in both the academy and industry.

2.1 Network Reconnaissance in Academy

Network reconnaissance can be divided into direct and indirect reconnaissance according to whether there is direct contact with target.

Direct reconnaissance requires connection to the target. For example, subdomain enumeration, port scanning, application layer protocol identification, web scanning, etc.

Subdomain enumeration is a technique that use a high-frequency subdomain dictionary to make enumeration attempts on DNS server [21]. Because of the large number of DNS queries that need to be sent, subdomain enumeration has high requirements for DNS query speed. However, DNS will refuse to respond to sources that exceed the request rate threshold [19].

Port scanning technique is trying to determine which ports are open for a network device that has an IP address [20,28]. There are 65536 standard ports on a computer which can be categorized into three large ranges: (i) well-known ports (0-1023), (ii) registered ports (1024-49151) and (iii) dynamic and/or private ports (49152-65535) while only a small fraction of these ports are commonly used [17]. There are many methods of port scanning have been proposed, such as stealth scanner, socks port probe, bounce scanner, tcp scanner, udp scanner, etc. [4]. However they usually focus only on the rate of detection and not on the strategy of detection.

Application layer protocol identification help penetration testers identify the type of protocol running on the port which allows for precise detection and attack. Traditional application layer protocol identification use stable ports and can be easily identified by the port number registered in Internet Assigned Numbers Authority (IANA) [6,14]. However, this simple method fails when the protocol uses dynamic ports [8,23]. In order to improve the effectiveness, researchers proposed a protocol identification method based on load differences which is done by finding the fields that differ from other protocols [7,24,29].

Web is an information system where documents and other web resources are identified by Uniform Resource Locators (URLs). The resources of the Web are transferred via the HTTP or HTTPS. Web scanning techniques are usually web server identification, web application identification, web directory scanning, etc.

Indirect reconnaissance does not require a direct connection to the target. The target data is obtained by making queries to third-party platforms such as traditional search engines, cyberspace mapping platforms, certificate transparency, etc.

Traditional search engines like Google, Bing and Baidu provide website crawling and inclusion functions, so we can retrieve the same type of websites with specified characteristics such as keyword, urls, etc. Google Hacking Database is a database of Google hacking search query commands which can help reconnaissance with Google search engine [31].

The development of fast network probing technology makes it possible to scan the entire ipv4 address space in a short period of time. Therefore, cyberspace mapping is developing rapidly. In 2009, Matherly created the Shodan search engine, focusing on the search of information about all networked devices and their component types [12]. In 2015, Durumeric combined Zmap with Google cloud platform and developed Censys system [3]. Compared to traditional search engines, cyberspace mapping is more concerned with ip, port, protocol and other elements in cyberspace.

2.2 Network Reconnaissance in Industry

Based on the research in academy, various tools have been developed to assist network reconnaissance in penetration testing.

- Subdomain Enumeration: sublist3r, subfinder, oneforall are the most used tools in enumerating subdomain [18,26,27].
- Port Scanning: Nmap is a tool can not only perform network discovery but also OS identification [16]. Zmap uses the idea of stateless scanning to greatly improve the speed of scanning [32].
- Web Scanning: Dirsearch is a web path scanner trying to find sensitive url endpoints and backup files. HMAP is a tool that recognizing what web server are running [10].
- CyberSpace Mapping Platform: such as Shodan, Censys, Fofa, Zoomeye, etc. These platform acquires assets through periodic scanning, identificating and passive traffic detection [1,3,11,12].
- Other Third-party Platform: such as crt.sh, wayback machine, etc. crt.sh is a database that aggregates the major logs of transparency logs. The Wayback Machine is a digital archive of the World Wide Web. They provide the penetration testers with a way to search related information.

Most tools implement only a few of reconnaissance methods, and penetration testers usually need to manually combine various tools to complete the network reconnaissance, extracting useful information from the output of one tool and integrate that information in another tool, which is not only time-consuming but also requires lot of professional knowledge. In order to simplify the network reconnaissance, few tools stitch these tools together in relation to their input and output, thus enabling simple automation, such as reNgine (star 3.2k from github) [22] and ARL (star 1.8k from github) [2].

In penetration testing, network reconnaissance should be performed through the cooperation of various sub-tasks including subdomain enumeration, port discovery, protocol identification, web scanning. The automation of network reconnaissance will not only increse speed and effectiveness but also save significant time and energy. However, neither the researchers in academy nor developers in industry have proposed an effective automatic strategy. On one hand, in academy, existing researches only focus on a specific sub-task. For example, [4] only concerns port scanning, and [7] only concerns protocol identification. On the other, in industry, although many tools for network reconnaissance have been proposed, they are a simple combination of existing tools that perform a specific sub-task. Few developers take the efficiency and coverage into consideration. Therefore, there is still a long way to go in network reconnaissance automation.

3 Network Reconnaissance Graph Model

This section explains the inspiration of the graph model for network reconnaissance in penetration test, and introduces the model structure and the role it plays in network reconnaissance by analyzing several scenarios.

3.1 Inspiration

Learning from the long-term experience of network reconnaissance in penetration test, it can be summarized that good network reconnaissance tools should meet the following requirements.

- **High efficiency.** It is usually required that penetration test should be completed within limited time. The faster the information is gathered, the more the time left for vulnerability detection, and the more likely a hidden attack surface will be discovered.
- **High Effectiveness.** It is inevitable that network reconnaissance will encounter invalid results. Invalid results are not helpful, and introduce unnecessary operations which will reduce the efficiency. Therefore, tools should guarantee the effectiveness of results.
- **Low dependence on humans.** Network reconnaissance tools need to free humans from complex and tedious operations. That means, the tools should reduce the interaction with humans and the dependence on expert experience.

Knowledge graph use a graph structure to store interlinked descriptions of entities such as objects, events, situations or abstract concepts. It provides structured information that can be interpreted by computers and has become the base for information systems that require structured knowledge. The graph structure can model the relationship between information and operations in network reconnaissance. Therefore, in this paper, knowledge graph is leveraged to identify useful information and develop effective reconnaissance strategy.

3.2 Our Approaches

Subtask Decomposition. Existing works study network reconnaissance from the perspective of subtask. As shown in Fig. 2, a subtask represents a class of information to be gathered such as Sub-domains gathering, IP gathering, Port gathering, etc. To achieve parallization and improve the efficiency, in this paper, subtasks are decomposed into multiple methods. Considering network reconnaissance from the perspective of methods, we can easily identify the dependencies between methods and extract methods that can be executed in parallel. As shown in Table 1, $method_1$, $method_2$ and $method_3$ can be executed in parallel. $method_n$ depends on the result of $method_{n-1}$. Besides, for some direct reconnaissance methods, we can distribute the method execution in order to bypass network rate limits and security strategy. Take port scanning as example, the port to be scanned can be randomly split and execute by different executors, in order to reduce the probability of being blocked.

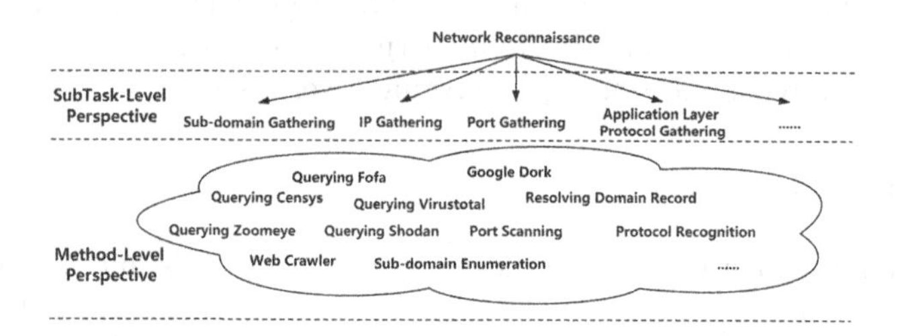

Fig. 2. Three perspective of network reconnaissance.

Method Constraint. In network reconnaissance, invalid results will result in unnecessary work and decrease the efficiency. Invalid results are caused by two factors. On one hand, before methods are executed, the required conditions are not met. Take Wildcard Domain Record and Content Delivery Network (CDN) as example. Wildcard Domain Record is a record in a DNS zone that will match requests for non-existent domain names which means all the requests of sub-domain enumeration will return true. If the wildcard status is not checked before subdomain enumeration, all the enumeration are invalid (see Fig. 3(a)). CDN is a technique that accelerating the Internet browsing. If the CDN status is not checked before resolving it will lead to invalid ip record (see Fig. 3(b)).

Table 1. Example of methods and relative subtasks.

	Method-level perspective	Subtask-level perspective
1	Get sub-domains by sub-domain enumerating	Sub-domains gathering
2	Get sub-domains by querying censys	Sub-domains gathering
3	Get sub-domains by querying fofa	Sub-domains gathering
......	......	......
n-2	Get ip by resolving domain	IP gathering
n-1	Get available ports by port scanning	Port gathering
n	Get application layer protocol by protocol recognition	Application layer protocol gathering

On the other, the results of methods are not validated. For example, information gathered by indirect reconnaissance is usually not real-time (such as querying shodan, zoomeye, etc.) and not filtered, thus producing invalid results.

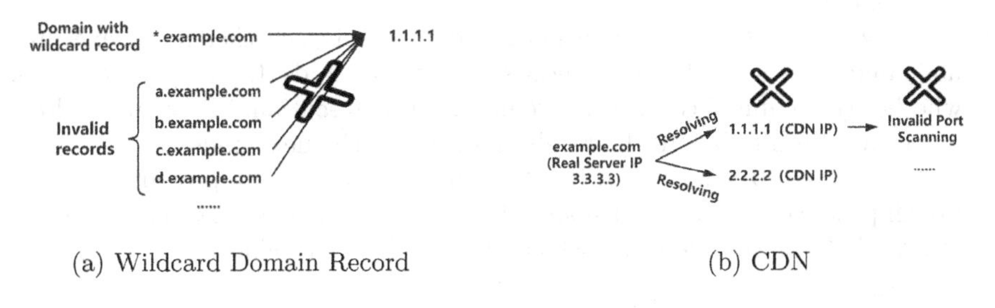

(a) Wildcard Domain Record (b) CDN

Fig. 3. Invalid results lead to pointless reconnaissance.

To improve the effectiveness of network reconnaissance, checkers and filters are designed. The checker checks whether the conditions of a method are satisfied. If the conditions are satisfied, the method will be executed, otherwise it will be abandoned. The filter helps validate the results and filters out the invalid ones. Only the results of a specific method processed by the filter can be input to other methods. For a method, if certain conditions are required before its execution, a checker will be added before it; if the results are not real-time, a filter will be added after it. Considering all the possible scenarios, four execution patterns are summarised as shown in Fig. 4.

Data Model. Combining knowledge graph with network reconnaissance practice, we construct a graph-structured data model to guide network reconnaissance. From method-level perspective, we first define the information to be gathered, the operations to be taken, and the corresponding attributes. Then, we embed the execution patterns into the graph. The ontology of the knowledge graph is shown in Fig. 5. The key components are listed as follows.

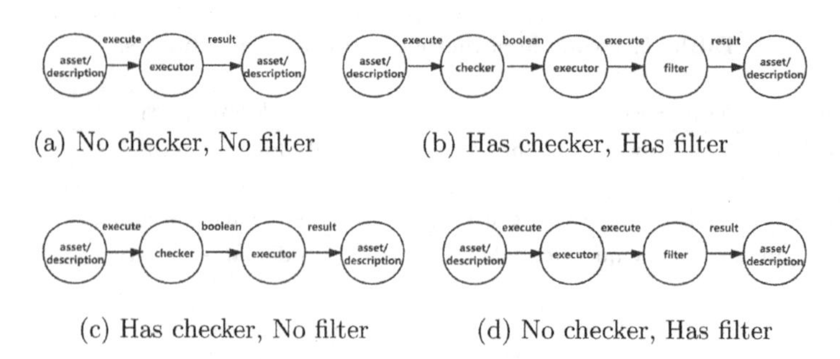

(a) No checker, No filter (b) Has checker, Has filter

(c) Has checker, No filter (d) No checker, Has filter

Fig. 4. Four patterns of reconnaissance methods.

1. $A = \{A_0, A_1, ..., A_{n-1}\}$, Information Nodes, including asset node and description node. Asset node represents those elements that help locating in cyberspace, such as ipv4 address, ipv6 address, primary domain, sub-domain, port, etc. Description node represents the attributes or status of the asset, such as Internet Service Provider (ISP), IP Location, Wildcard Status, Certificate, etc.
2. $O = \{O_0, O_1, ..., O_{n-1}\}$, Operation Nodes, including checker node, executor node and filter node. Executor nodes execute methods. Checker nodes check whether the required conditions are met for the execution. Filter nodes filters out invalid results before the results are sent to the next node..
3. $E = \{E_{ij} | i, j = 0, 1, ..., n-1\}$, Relationship collection, E_{ij} represents the relationship between $node_i$ and $node_j$. Relationship includes *Execute, Boolean, Result, RelyOn* and *Describe*, which are described in Table 2

Table 2. Relationship and example.

Relationship	Introduction	Example
Execute	E_{ij} represents $node_i$ can execute $node_j$	Primary domain can execute the operation of getting domain wildcard status
Result	E_{ij} represents that $node_j$ can be gathered by executing $node_j$	Domain Wildcard status can be gathered by checking domain wildcard status
Boolean	E_{ij} represents a branch. If $node_i$ results true, then execute $node_j$. Otherwise give up executing	If checking cdn status return true, then resolving ipv4. Otherwise, give up resolving
RelyOn	E_{ij} represents $node_i$ rely on $node_j$	Subdomain relyon primary domain
Describe	E_{ij} represents $node_i$ can describe $node_j$	Primary domain cdn status can describe primary domain

The ontology is a generalized data model which only model generalized types of information, but does not contains specific information in the domain of network reconnaissance. Using the ontology as a framework, we then add in real

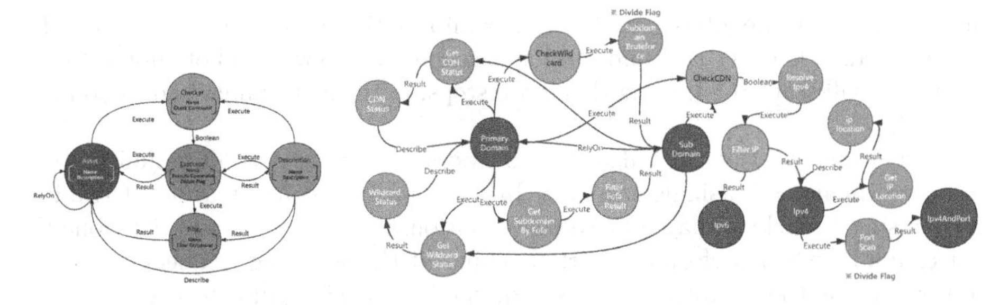

Fig. 5. Ontology of the model

Fig. 6. Partial instance of the model

data of individual assets, descriptions, operations, and relationships to create a knowledge graph. A portion of the knowledge graph is shown in Fig. 6. Whenever a reconnaissance task comes, we query the knowledge graph and the knowledge graph will return multiple execution paths. A path is a group of operations that are arranged in order, with checkers and filters placed in the appropriate positions.

4 Target-Oriented Automated Reconnaissance System

This section introduces Auto-Recon, a target-oriented automated reconnaissance system which is implemented under the guidance of the knowledge graph. Auto-Recon accepts the target and the knowledge graph as input, and output the information gathered for the target.

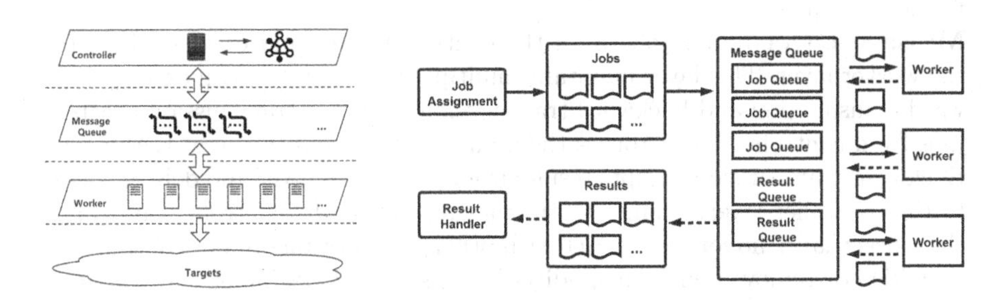

Fig. 7. Three layer model of distributed reconnaissance system.

Fig. 8. Workflow of the message layer.

Auto-Recon uses a graph database–Neo4j [15], to store the knowledge graph. Whenever a new target is input, query statements will be generated and organized according to the four patterns shown in Fig. 4. Then the query results converted into jobs. A job is a piece of Json structured information that represents the method which contains check operation, execute operation and filter operation. When the job is completed, new information related to the target will be gathered. Whenever new information is gathered, a new round of information gathering will started following the above steps, until the terminate condition is reached.

The efficiency of parallel algorithm introduced in Sect. 3 is often limited by the performance of a single machine. In addition, the concentration of traffic on one single machine may lead to the violation of the target's security policy. For example, when performing port scanning, if the scanning frequency to the server exceeds the established threshold, the server will refuse to connect. To speed up reconnaissance and avoid network concentration, Auto-Recon adopts a distributed architecture in which the parallel tasks are executed in a group of machines. For example, the scanning of different ports can be executed by several workers.

The architecture of the system is shown in Fig. 7. Auto-Recon use a three-layer architecture: the controller layer, the message layer, and the worker layer. The controller layer is the core of the system, which is responsible for job generation, registration, segmentation and distribution. The message layer is responsible for job routing. The worker layer is the actual executor of the task, fetching job message from the message queue and parsing it, calling external modules to execute it. With the benefits of distribution, we can avoid the network rate limitations and traffic concentration problems caused by single-unit deployments (Fig. 8).

1. **Controller Layer** receives information and queries the knowledge graph for executable methods based on the information, and then converts the query results into jobs. The jobs are registered into a database. If the job can be divided, it will be divided into several jobs, otherwise it will be directly sent to message queue.
2. **Message Layer** routes jobs from the controller layer and delivers them to the worker layer. This layer maintains multiple job queues. It can execute forward transmission and backward transmission. Forward transmission accepts jobs from controller layer, routes them and deliver them to the worker layer. Backward transmission accepts results from the worker layer and deliver them to the controller layer. Thanks to the message layer, the controller layer and the worker layer no longer restrict each other, and working nodes can be easily added or removed without modifying the source code. Thus, the system's scalability is improved.
3. **Worker layer** fetches jobs from the message queue and executes them. Jobs are executed by external program. The results of jobs will be stored in a database and returned to the controller layer via the message layer.

5 Experimental Results and Analysis

This section first presents the experimental setup, and then evaluates the effectiveness and efficiency of Auto-Recon (Figs. 9 and 10).

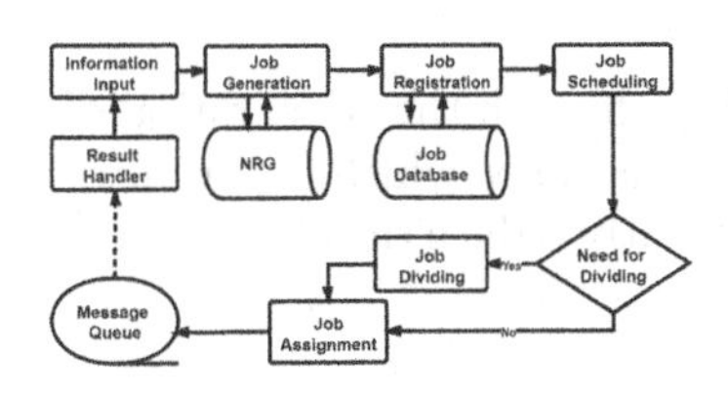

Fig. 9. Workflow of the controller layer.

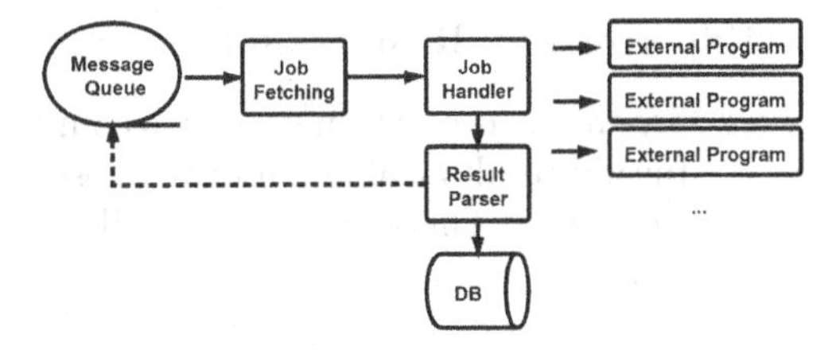

Fig. 10. Workflow of the worker layer.

5.1 Experimental Setup

To validate the advantage of Auto-Recon, we compare it with reNgine [22] and ARL [2] from the aspects of effectiveness and efficiency. reNgine and ARL are both automated reconnaissance framework for web applications with less manual involvement. Getting the most stars, reNgine and ARL are the most two popular automated reconnaissance tools in the github. All the experiments are conducted on servers with the same configuration. Each server is configured with 2 CPU, 2048 MB Memory, 80 GB NVMe. Since Auto-Recon uses a distributed architecture, we need 3 worker servers for Auto-Recon. reNgine and ARL use standalone deployment.

Since the tasks that reNgine and ARL can perform are different, for fair comparison, Auto-Recon, reNgine, and ARL are planned to perform the same sub-task: sub-domain gathering and port scanning. For sub-domain gathering sub-task, they gather sub-domains for 40 primary domains divided into 4 reference groups, with 10 randomly chosen from Aleax top 300. For port scanning sub-task, they scan all the ip address collected from one of the 4 group of primary domains. The three tools are compared based on the effectiveness and efficiency.

We first introduce several key concepts.

- **Effective sub-domains.** Effective sub-domains refer to the sub-domains that has the DNS record such as A, AAAA, TXT, etc.
- **Ineffective sub-domains.** Ineffective sub-domains refer to the sub-domains that do not exist, including the sub-domains that has no DNS record and the sub-domains gathered because of the domain has wildcard dns record.
- **Effective ports.** Effective ports refer to the ports that actually open.
- **Ineffective ports.** Ineffective ports refer to the ports are not open but are misrecognized as open.
- **Effective percentage.** Effective percentage refers to the percentage of effective sub-domains or ports in total sub-domains.
- **Effectiveness.**The effectiveness of one tool is mainly measured by effective percentage. The higher the effective percentage, the more effective the tool.
- **Efficiency.** The efficiency is measured by the time cost. The lower the time cost, the higher the efficiency

5.2 Experimental Results and Analysis

Network reconnaissance automation is accomplished through the cooperation of methods. Invalid results will be input to subsequent methods, which in turn produces invalid results. This cyclic process will significantly reduce the effectiveness and efficiency.

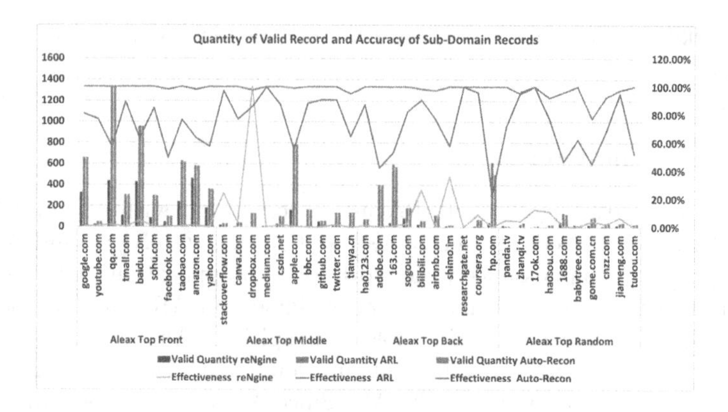

Fig. 11. Effectiveness of reconnaissance tools in sub-domain gathering.

Sub-domain Gathering. Figure 11 presents the number and percentage of effective sub-domains for Auto-Recon, reNgine, and ARL. For reNgine, both the effective number and effective percentage are extremely low. The reason for this is twofold. On one hand, reNgine performs sub-domain enumeration without checking if the domain has a wildcard record. On the other, reNgine did not validate the results, which often contains large amount of non-existent sub-domains. Therefore, reNgine's performance on effectiveness can not meet the requirements in Sect. 3. The effective number of ARL and Auto-Recon are similar. Because they use the same reconnaissance methods of sub-domain gathering. However, all the effective percentages of ARL are obviously lower than that of Auto-Recon. The average difference is as high as 24.67%. This is because ARL does not check the records when using indirect reconnaissance methods, but Auto-Recon does. Auto-Recon's effectiveness is the highest for the same reconnaissance mission and reconnaissance method.

The time cost of the three tools are presented in Fig. 12. reNgine costs the most time, while Auto-Recon cost the least time. On average, the time cost of Auto-Recon is 75.04% less than the second place ARL. Therefore, so the efficiency of Auto-Recon 3 times higher than it. This is because reNgine and ARL simply reuse existing tools without any constraint, resulting in inefficient reconnaissance. For example, reNgine use a combination of amass-passive, assetfinder, sublist3r, subfinder and oneforall. The same method exists in the implementation of these tools. Compared with reNgine and ARL, Auto-Recon analyzes reconnaissance from a method-level perspective and does not have the problem of duplicate method execution. In addition, Auto-Recon add constrains to the methods which effective guarantee of the validity of the results. When performing sub-domain gathering, we found that if the request rate exceeds a certain threshold, the DNS server will limit the query from the same source, which will result in a portion of the sub-domains not being gathered. Auto-Recon uses a distributed deployment to bypass network rate limitation and the shortcomings of traffic concentration. Therefore, Auto-Recon achieves the highest effectiveness.

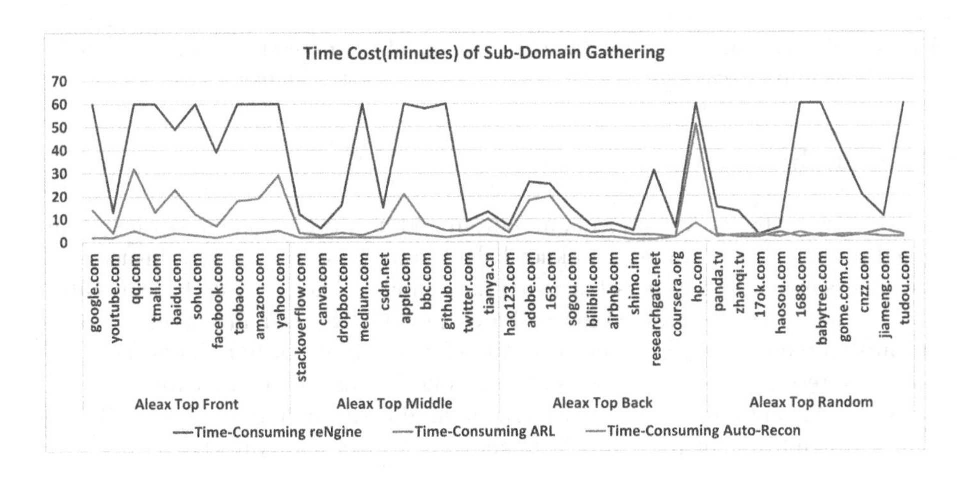

Fig. 12. Time cost of reconnaissance tools in sub-domain gathering.

Port Scanning. Figure 13 presents the number and percentage of effective ports gathered by Auto-Recon, reNgine, and ARL. For reNgine, both the effective number and effective percentage are extremely low. This is because reNgine gathers many invalid sub-domains, which leads to many invalid ips. Therefore, reNgine's performance on effectiveness can not meet the requirements in Sect. 3. Most of the effective percentages of ARL are obviously lower than that of Auto-Recon. This is because ARL does not check the records, but Auto-Recon does. The time cost of the three tools are shown in Fig. 12. reNgine costs the most time, while Auto-Recon cost the least time. On average, the time cost of Auto-Recon is 33.33% less than the second place ARL. The significant increase in effectiveness and effectiveness of Auto-Recon is mainly due to the distributed architecture and method constraint (Fig. 14).

In terms of scalability, if the reconnaissance method needs to be adjusted, reNgine and ARL need to modify the main program. Auto-Recon has a modular design, which only needs to adjust the knowledge graph and load the corresponding modules to achieve functional expansion. Therefore, Auto-Recon is better than existing automated reconnaissance tools.

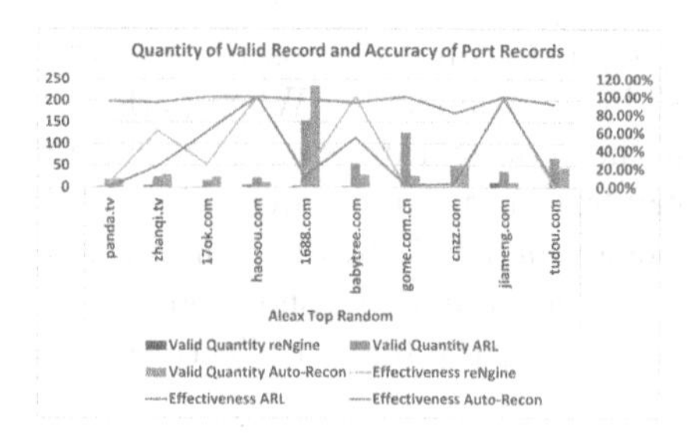

Fig. 13. Effectiveness of reconnaissance tools in port scanning.

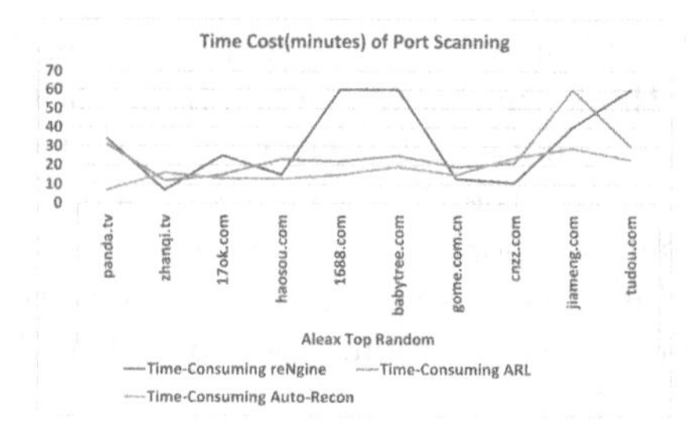

Fig. 14. Time cost of reconnaissance tools in port scanning.

6 Conclusion

This paper proposes Auto-Recon, an automated network reconnaissance system based on knowledge graph. The knowledge graph encodes the data in network reconnaissance into a graph-structured data model. The data model structures data from a more fine-grained method-level perspective, and add necessary constraints to the execution of methods. Auto-Recon significantly reduces the complexity of reconnaissance activities by reducing the dependence on expert experience. Experimental results shows that, compared with existing tools, Auto-Recon achieves significant increase in effectiveness, efficiency and Auto-Recon.

Acknowledgement. We thank anonymous reviewers for their invaluable comments and suggestions. This research was supported in part by Key Laboratory of Network Assessment Technology (Chinese Academy of Science) and Beijing Key Laboratory of Network Security and Protection Technology.

References

1. TeamfromKnownsec: Zoomeyesearchengine. https://www.zoomeye.org
2. ARL: ARL. https://github.com/TophantTechnology/ARL
3. Arnaert, M., Bertrand, Y., Boudaoud, K.: Modeling vulnerable internet of things on SHODAN and CENSYS: an ontology for cyber security. In: Proceedings of the Tenth International Conference on Emerging Security Information, Systems and Technologies (SECUREWARE 2016), pp. 299–302 (2016)
4. Bhuyan, M.H., Bhattacharyya, D.K., Kalita, J.K.: Surveying port scans and their detection methodologies. Comput. J. **54**(10), 1565–1581 (2011)
5. CNCERT: 2020 CNCERT cybersecurity analysis (2021). https://www.cert.org.cn/publish/main/upload/File/2020CNCERTCybersecurityAnalysis.pdf
6. IANA: Service name and transport protocol port number registry. https://www.iana.org/assignments/service-names-port-numbers/service-names-port-numbers.xhtml
7. Kang, H.-J., Kim, M.-S., Hong, J.W.-K.: A method on multimedia service traffic monitoring and analysis. In: Brunner, M., Keller, A. (eds.) DSOM 2003. LNCS, vol. 2867, pp. 93–105. Springer, Heidelberg (2003). https://doi.org/10.1007/978-3-540-39671-0_9

8. Kim, M.S., Won, Y.J., Hong, J.W.K.: Application-level traffic monitoring and an analysis on IP networks. ETRI J. **27**(1), 22–42 (2005)
9. Kowta, A.S.L., Bhowmick, K., Kaur, J.R., Jeyanthi, N.: Analysis and overview of information gathering & tools for pentesting. In: 2021 International Conference on Computer Communication and Informatics (ICCCI), pp. 1–13. IEEE (2021)
10. Lee, D., Rowe, J., Ko, C., Levitt, K.: Detecting and defending against web-server fingerprinting. In: Proceedings of the 18th Annual Computer Security Applications Conference, pp. 321–330. IEEE (2002)
11. Li, R., Shen, M., Yu, H., Li, C., Duan, P., Zhu, L.: A survey on cyberspace search engines. In: Lu, W., et al. (eds.) CNCERT 2020. CCIS, vol. 1299, pp. 206–214. Springer, Singapore (2020). https://doi.org/10.1007/978-981-33-4922-3_15
12. Matherly, J.: Complete guide to shodan. Shodan LLC (2016-02-25), vol. 1 (2015)
13. Mirjalili, M., Nowroozi, A., Alidoosti, M.: A survey on web penetration test. Int. J. Adv. Comput. Sci. **3** (2014)
14. Moore, A.W., Papagiannaki, K.: Toward the accurate identification of network applications. In: Dovrolis, C. (ed.) PAM 2005. LNCS, vol. 3431, pp. 41–54. Springer, Heidelberg (2005). https://doi.org/10.1007/978-3-540-31966-5_4
15. Neo4j: Neo4j. https://neo4j.com/download/
16. Nmap: Nmap. https://nmap.org/
17. nmap: Port selection data and strategies. https://nmap.org/book/performance-port-selection.html
18. OneForAll: Oneforall. https://github.com/shmilylty/OneForAll
19. Rate-limiting queries: Rate-limiting queries. https://developers.google.com/speed/public-dns/docs/security#rate_limit
20. Radhakrishnan, S.V.: A framework for system fingerprinting. Ph.D. thesis, Georgia Institute of Technology (2013)
21. Ramadhan, R.A., Aresta, R.M., Hariyadi, D.: Sudomy: information gathering tools for subdomain enumeration and analysis. IOP Conf. Ser. Mater. Sci. Eng. **771**, 012019 (2020). https://doi.org/10.1088/1757-899x/771/1/012019
22. reNgine: rengine. https://github.com/yogeshojha/rengine
23. Roughan, M., Sen, S., Spatscheck, O., Duffield, N.: Class-of-service mapping for QOS: a statistical signature-based approach to IP traffic classification. In: Proceedings of the 4th ACM SIGCOMM Conference on Internet Measurement, pp. 135–148 (2004)
24. Sen, S., Spatscheck, O., Wang, D.: Accurate, scalable in-network identification of P2P traffic using application signatures. In: Proceedings of the 13th International Conference on World Wide Web, pp. 512–521 (2004)
25. Shivayogimath, C.N.: An overview of network penetration testing. Int. J. Res. Eng. Technol. **3**(07), 5 (2014)
26. subfinder: subfinder. https://github.com/projectdiscovery/subfinder
27. Sublist3r: Sublist3r. https://github.com/aboul3la/Sublist3r
28. Trowbridge, C.: An overview of remote operating system fingerprinting. SANS InfoSec Reading Room-Penetration Testing (2003)
29. Van Der Merwe, J., Caceres, R., Chu, Y.h., Sreenan, C.: Mmdump: a tool for monitoring internet multimedia traffic. ACM SIGCOMM Comput. Commun. Rev. **30**(5), 48–59 (2000)
30. Yeo, J.: Using penetration testing to enhance your company's security. Comput. Fraud Secur. **2013**(4), 17–20 (2013)
31. Yerrapragada, K.P.: Google hacking !! (2007). Accessed 20 Jan 2010
32. zmap: zmap. https://github.com/zmap/zmap

INGCF: An Improved Recommendation Algorithm Based on NGCF

Weifeng Sun(✉), Kangkang Chang, Lijun Zhang, and Kelong Meng

School of Software, Dalian University of Technology, Dalian, China
wfsun@dlut.edu.cn,
{ckk19970825,ljzhang,2357806126}@mail.dlut.edu.cn

Abstract. Strengthening the representation and learning of user vector and item vector is the key of recommendation system. Neural Graph Collaborative Filtering (NGCF) has the problem of insufficient feature extraction of user vector and item vector. In addition, it uses the linear combination of the final embedding vectors of user and item to calculate the inner product, which is difficult to accurately obtain the user-item prediction score. In this article, we propose a graph neural network collaborative filtering algorithm based on NGCF, named INGCF. This new algorithm uses a 4-layer IndRNN layer and a feature extraction layer constructed by a self-attention mechanism to enhance the feature extraction capabilities of NGCF. And INGCF optimizes the user-item prediction score method to capture the complex structure of user-item interaction data to improve the accuracy of score calculations. We compare several indicators of INGCF, NGCF, PinSage and GCN by using same data sets and perform ablation experiments. The experimental results show that the INGCF algorithm has a better recommendation effect.

Keywords: Graph neural network · IndRNN · Collaborative filtering

1 Introduction

Traditional deep learning models such as GAN [1] have made great progress in Euclidean spatial data such as language, image, video, etc. However, there are great limitations in processing non-Euclidean spatial data such as social networks and information networks. Graph neural networks can handle some non-Euclidean data well. Graph neural networks have been widely used in many fields, especially in recommendation systems. Most of the information essentially has a graph structure, such as social relationships, knowledge graphs, bipartite graphs composed of user product interactions, etc. And because the graph neural network can learn and represent the connection between graph nodes and graph data well, some studies focus on the application of graph neural networks to recommendation systems. For example, Sankar et al. [10] propose the Grafrank method, which uses the neighbor aggregator of a specific model and the cross-model attention mechanism to learn multi-faceted user representation.

Y. Lai et al. (Eds.): ICA3PP 2021, LNCS 13157, pp. 116–129, 2022.
https://doi.org/10.1007/978-3-030-95391-1_8

You et al. [17] propose a graph comparison learning framework GraphCL to learn the unsupervised representation of graph data, and enhance the graph data.

The NGCF proposed by Wang et al. [13] carries out information dissemination at the graph structure and the affine of the central node. It uses different levels of representation to obtain the embedding of the final node and takes advantage of the residual network. NGCF has done a good job in processing graph data and performing collaborative filtering recommendations, and it has many advantages. For example, the message-passing architecture that builds GNN strengthens high-order user-item information and helps to embed high-order connectivity relationships of users and items. However, it relies on the overall graph structure and has low generalization ability. In addition, it can not fully extract the vector features of users and items. It is difficult to accurately calculate the user-item correlation score by using the inner product of the linear combination of the final user embedding vector and item embedding vector. In order to solve the above problems, we propose an improved algorithm INGCF based on NGCF. INGCF uses a 4-layer independently recurrent neural networks (IndRNN) [6] to enhance feature extraction after the concatenate operation of the user vector and item vector. IndRNN alleviates the problem of gradient explosion and gradient disappearance within and between layers of the model network to a certain extent, allows the network to learn long-term dependencies, and improves the robustness of the model. The self-attention mechanism [11] can further extract the explicit features of the data and improve the generalization of the model. To calculate user-item score more accurately, we use multi-layer perceptrons to solve the difficulty of matrix decomposition of complex user-item interaction data in the low-dimensional latent space.

The following parts of this article are organized as follows. The second section introduces the related research work of recommendation algorithm based on graph neural network in recent years. The third section describes the ideas of NGCF and some of our evaluations. In the fourth section, we describe the structure of INGCF in detail, and explain the parts and reasons we retained and improved NGCF. The fifth section introduces the experimental environment, data set, baseline, experimental data and analysis of experimental results. Finally, we summarize this article and propose future work.

2 Related Work

Recommendation algorithms based on graph neural networks can be divided into two categories, general recommendation and serialized recommendation. General recommendation includes a recommendation algorithm based on walking algorithm, such as PinSage [16]. PinSage combines random walk strategy and graph convolution to learn node embedding. It uses a random walk-based method to sample a fixed number of neighbors, so that it can be extended to a network-scale recommendation task of millions of users and items. However, the neighbor's sampling strategy may affect the performance of the model. Some general recommendation algorithms are based on representative algorithms, such as STAR-GCN [18], LightGCN [2] and NGCF. STAR-GCN stacks multiple GCN blocks, introduces a reconstruction mechanism to connect adjacent blocks, and solves the problem of data label leakage by masking some edges. LightGCN learns user and item embeddings by linear propagation on the user-item interaction graph, and uses the weighted sum of embeddings learned in all layers as the final embedding. LightGCN simplifies the design of GCN [5], makes the model structure simple and reduces the difficulty of training.

Serialized recommendation recommends the next item of interest to the user by capturing the sequence pattern in the item sequence and the user's behavior. Most works only focus on the timing preferences in the sequence. They first convert the serialized data into a sequence diagram, and then use GNN to capture the knowledge of serialization, such as SR-GNN [15], FGNN [9] and MA-GNN [8]. SR-GNN uses graph data to model sessions, uses graph neural networks to capture complex transitions between items, and uses attention networks to express each session as a combination of global preferences and current interests. FGNN uses GAT to capture the transfer of items in the sequence diagram, and assigns different weights to different neighbors. The weights depend on the adjacency matrix and the connected edges and nodes at the same time. It replaces the complex gate mechanism with weighting and updating the central node. MA-GNN integrates static, dynamic short-term and dynamic long-term user interests and item co-occurrence models for serialized recommendation. It uses the gate mechanism to balance short-term and long-term interests.

In recent years, there have been more researches on the comparative learning and pre-training of graph neural networks. Wang et al. [12] propose a new GCN-based SSL algorithm. It uses data similarity and graph structure to enrich the monitoring signal, which helps to improve the subsequent classification results. They designed a semi-supervised comparison loss model to generate improved node representations by maximizing the consistency between different views of the same type of data. Lu et al. [7] propose a self-supervised pre-training strategy L2P-GNN for GNNs, which alleviates the difference between pre-training and fine-tuning. In order to encode local and global information into the prior, L2P-GNN designs a dual adaptive mechanism at the two levels of nodes and graphs. There are also some researches on attention in graph neural networks. Wang et al. [14] propose KGIN on this issue and design a new information aggregation model to recursively aggregate long connection relationship sequences by treating each user's intention as an attention combination of a knowledge graph relationship. Cheng et al. [4] propose a relational temporal attentive graph neural networks RetaGNN. In order to have the ability of induction and transfer, the author trains relational attention GNN through the user-item pairs on the local subgraph, and proposes a sequential self-attention mechanism to encode the long-term and short-term time patterns that the user prefers.

3 Questions Raised

NGCF doesn't strengthen the ability of model feature extraction after embedding Propagation Layers, which will make the training effect lower than the optimal.

It uses a simple interactive function of calculating the inner product of e_u^* and e_i^* as the method of calculating user-item prediction score. The NGCF team don't try more complicated methods and don't deny that other methods may have better results.

NGCF is composed of three parts: the embedding layer, the embedding propagation layer and the prediction layer. In the embedding propagation layer, it uses the message-passing architecture of GNN to capture collaborative filtering signals along the graph structure while optimizing user and item embedding. The message-passing architecture can be divided into message construction and message aggregation.

Message Construction: For the connected user u and item i, define the message from i to u and the neighbor of u to u as Eq. (1):

$$\begin{cases} m_{u\leftarrow i}^{(l)} = p_{ui}(W_1^{(l)} e_i^{(l-1)} + W_2^{(l)}(e_i^{(l-1)} \odot e_u^{l-1})), \\ m_{u\leftarrow u}^{(l)} = W_1^{(l)} e_u^{(l-1)} \end{cases} \tag{1}$$

where $m_{u\leftarrow i}$ represents the information to be spread from i to u. $p_{ui} = \frac{1}{\sqrt{|N_u||N_i|}}$ is the attention factor, N_u and N_i represent the first-hop neighbors of user u and item i. $W_i^{(l)}$, $W_2^{(l)}$, $\in R^{d_1 \times d_{l-1}}$ are the trainable transformation matrixs, d_l is the transformation size. $e_i^{(l-1)}$ is the item representation generated from the previous message passing step, and is used to store messages from its (l-1)-hop neighbors. It helps the user u's representation at the i-th level. With the same principle, the item i can be represented in the l-th layer.

Message Aggregation: Aggregate messages propagated from user $u's$ neighbors to user u's representation. The aggregate function can be expressed as Eq. (2):

$$e_u^{(l)} = LeakyReLU(m_{u\leftarrow u}^l + \sum_{i\in N_u} m_{u\leftarrow i}^{(l)}), \tag{2}$$

where e_u^1 represents the representation of user u obtained after the first embedding propagation layer. $m_{u\leftarrow u} = W_1 e_u$ is the self-connection of u, which is used to retain the information of the original characteristics. The embedded propagation layer can explicitly use first-order connectivity information to associate users with items representations.

The prediction layer aggregates the refined embeddings from different propagation layers and outputs the prediction scores of user-items. After spreading with the L layer, multiple representations of the user vector e_u are obtained. NGCF connects them to form the user's final embedding. It performs the same operation on the item vector, and obtain the final user and item embedding: $e_u^* = e_u^0||...||e_u^L, e_i^* = e_i^0||...||e_i^L$, where $||$ is a series operation. In doing so, not only can the initial embedding be enriched by embedding the propagation layer, but also the propagation range can be controlled by adjusting L. However, after this layer, NGCF does not do further feature extraction for user vector and item vector, which is not conducive to better training effect.

NGCF processes e_u^* and e_i^* to calculate the correlation score of user-item by calculating the inner product of two vectors. We do not support this part and think it needs improvement. In fact, simply the inner product of the linear combination of the product of the latent features is not enough to capture the complex structure of the user item interaction data in this algorithm.

4 INGCF Algorithm

INGCF's improvements to NGCF include the following two points. First, after the concatenate operation, a feature extraction layer is designed to further extract the user-item features. Second, a multi-layer neural network is used to process the final user and item embedding vector to calculate the user item prediction score between user and item.

The network structure of INGCF is shown in Fig. 1. It is composed of five parts: Embedding Layer, Embedding Propagation Layers, Feature Processing Layers, Feature Extraction Enhancement Layers and Score Calculation Layers. The improvements of INGCF compared with NGCF will be described below.

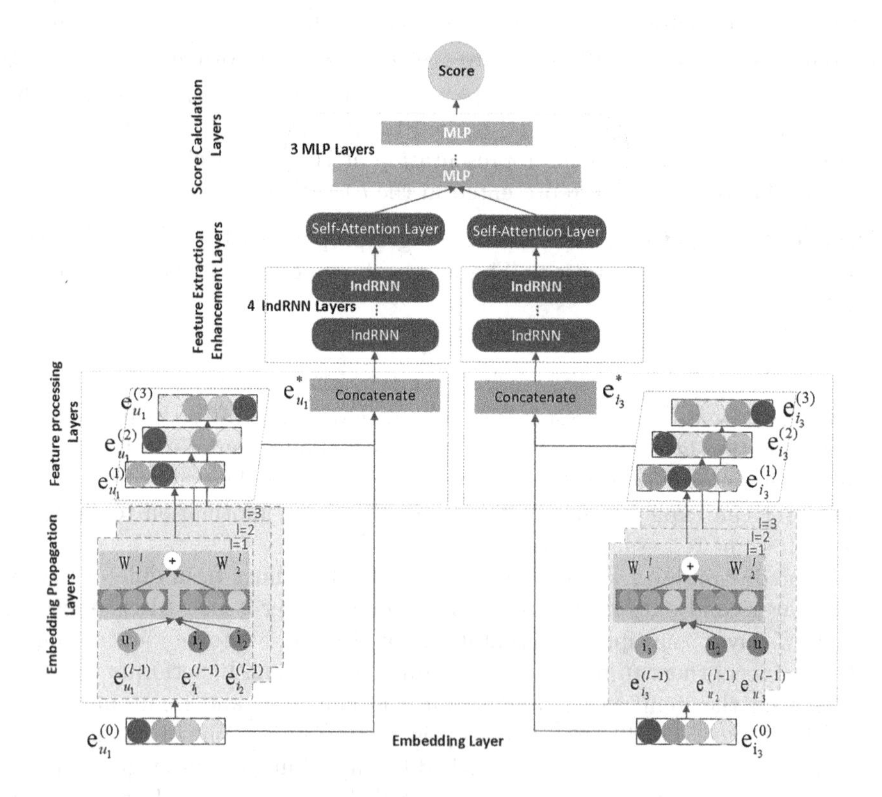

Fig. 1. INGCF network structure diagram

4.1 Feature Extraction Enhancement Layer

In order to enhance the model's ability to extract features, we add a 4-layer IndRNN layer and a self-attention mechanism layer after the Concatenate operation to form a feature extraction enhancement layer. After the Concatenate operation mentioned above, the input enters IndRNN. The status update of IndRNN can be described as Eq. (3):

$$h_t = \text{ReLU}\left(W q_t + u \odot h_{t-1} + b\right), \tag{3}$$

where h_t and q_t are the hidden state and input vector of the time step t, W is the input weight, and u is the recursive weight. $\odot$ is the product of Hadamard, and b is the offset value. ReLU is selected as the activation function. Each layer of neurons in IndRNN is independent, and the connections between neurons are realized by superimposing more layers. For the n-th layer, the hidden state G can be obtained by Eq. (4):

$$h_{n,t} = \text{ReLU}\left(W_n q_t + u_n h_{n,t-1} + b_n\right), \tag{4}$$

where $h_{n,t}$ is the hidden state of the n-th layer, W_n and u_n are the input weights and recursive weights of the n-th layer, and b_n is the offset value of the n-th layer. Each neuron only receives information from the input and its own hidden state in the previous step. Because the neurons in each layer of IndRNN are independent of each other, it can retain long-term memory, which greatly enhances the ability to extract features. IndRNN can also effectively solve the problem of gradient disappearance and gradient explosion by adjusting the time-based gradient backpropagation, thereby supporting the learning of long-term dependencies. This is also the reason why the general RNN builds 2 to 4 layers, and the training effect becomes worse if the number of layers is too much, and the high-level effect of IndRNN is still very good. In addition, IndRNN uses non-saturated functions such as ReLU as the activation function, which makes the network robust. We use 4-layer IndRNN, so that the extraction effect is better and the overall complexity of the model is not too large, so as to avoid excessive training burden. The neuron structure of IndRNN is shown in Fig. 2. BN means Batch Normalization. ReLU means ReLU activation function. Recurrent means that the layer is cyclic.

The output results from IndRNN enter the self-attention mechanism layer. The self-attention mechanism can reduce the dependence on external information, and is good at capturing the internal correlation of data or features and the linear characteristics of data. The expression of self-attention mechanism is as Eq. (5):

$$\text{Attention}\left(Q, K, V\right) = \text{softmax}\left(\frac{QK^T}{\sqrt{d_k}}\right)V, \tag{5}$$

where Q stands for Query vector, K stands for Key, and V stands for Value. d_k represents the dimensionality of each head of the self-attention mechanism, and $\sqrt{d_k}$ is the scale factor, which is used to avoid excessive inner product values

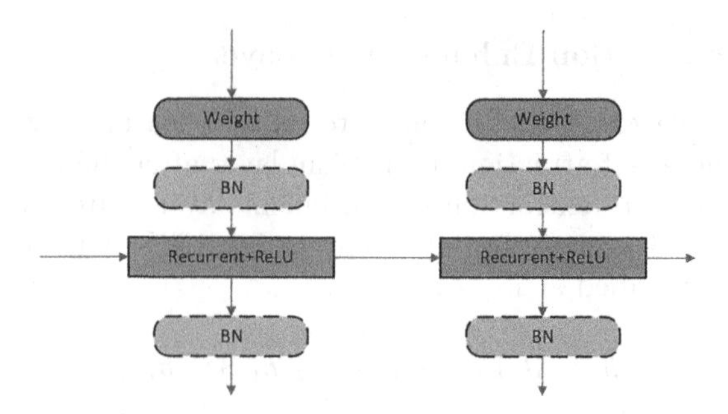

Fig. 2. Basic IndRNN structure diagram

and help the training to converge more smoothly. The algorithm uses the same feature extraction method for vector e_i^*.

The basic calculations of the self-attention mechanism are all matrix calculations, which can solve the problem of sequence length dependence, obtain explicit characteristics of data, and perform calculations in parallel. It has a wide range of applications in the field of image vision and recommendation systems. In this paper, the self-attention mechanism is used as the last step of feature extraction to strengthen the characterization ability of feature vectors. The output vectors $e_u^{'}$and $e_i^{'}$ have fully extracted features.

4.2 Optimize User-Item Prediction Score Method

NGCF uses the final embedded vectors of users and items to calculate the inner product as the prediction score of user items, but our experiments show that the prediction calculation using neural network is more effective. He et al. [3] proved this by examples. The following is an example.

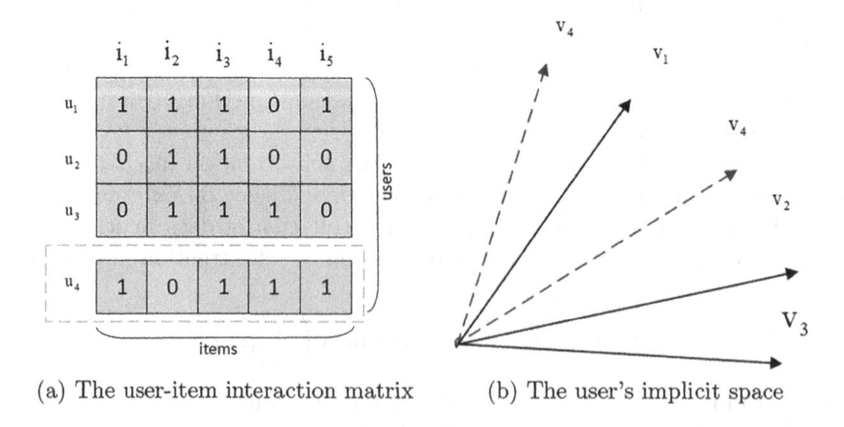

(a) The user-item interaction matrix (b) The user's implicit space

Fig. 3. An example

Figure 3(a) is the user-item interaction matrix, and Fig. 3(b) is the user's implicit space. Generally speaking, matrix decomposition distributes users and items into the same implicit space, so the similarity between two users can be calculated by the inner product of the two. Without considering the loss, the Jaccard coefficient is used as the real user similarity. The Jaccard coefficient of user u and user j is defined as Eq. (6):

$$s_{ij} = \frac{|\boldsymbol{R}_i| \cap |\boldsymbol{R}_j|}{|\boldsymbol{R}_i| \cup |\boldsymbol{R}_j|}, \tag{6}$$

where R_u represents the set of items interacting with user u.

According to the first three rows in Fig. 3(a), you can calculate the magnitude relationship between the Jaccard similarity coefficients: $S_{23}(0.66) > S_{12}(0.5) > S_{12}(0.4)$. We use V_i to represent the vector of u_i in the latent space, then the geometric relationship of V_1, V_2 and V_3 in the latent space can be drawn as shown in Fig. 3(b). Then considering a new user G whose input is given as the dashed line in Fig. 3(a), we can calculate the result:$S_{41}(0.6) > S_{43}(0.4) > S_{42}(0.2)$. The result shows that u_4 is the most similar to u_1, followed by u_3, and finally u_2. A matrix factorization model puts L representing M close to N, and the two possible positions are shown by dotted lines in Fig. 3(b). It can be seen from Fig. 3(b) that in both cases, V_4 is closer to V_2 than V_3, which results in a larger ranking loss. This example shows the limitations of matrix factorization due to the use of simple fixed inner products to estimate complex user-item interactions in the low-dimensional latent space.

Our improvement plan is to use the vector inner product to calculate the prediction score and replace it with a 3-layer MLP network as the interaction function between user and item, and use a large number of potential factors to solve the above problems. User embedding and item embedding are fed back to the MLP network, and the user's score for the item is predicted through the latent vector. MLP is chosen here because it is flexible and can better capture the data characteristics of the interaction between user and item, and the setting of 3 layers makes it possible to learn the hidden vectors of user and item well without making the model excessively complicated.

The user-item relevance score model of our algorithm can be expressed as Eq. (7):

$$\hat{y}_{ui} = f\left(P^T e_u^{\prime}, Q^T e_i^{\prime} \mid P, Q, \Theta_f\right), \tag{7}$$

where P^T and Q^T respectively represent the potential factor matrix of users and items, $e_u^{\prime}$ and $e_i^{\prime}$ respectively represent the above-mentioned user and item vectors processed by the 4-layer IndRNN and self-attention mechanism layer, and Θ is the interaction model parameter. The f function is defined as Eq. (8):

$$f\left(P^T e_u^{\prime}, Q^T e_i^{\prime}\right) = \Phi_{\text{out}}\left(\Phi_3\left(\Phi_2\left(\Phi_1\left(P^T v_u^U, Q^T, v_i^I\right)\right)\right)\right), \tag{8}$$

where $\Phi_o ut$ and Φ_3, Φ_2, and Φ_1 represent the mapping function of the output layer and the 3rd, 2nd and 1st MLP layer respectively.

4.3 Model Optimization

After experimental research, we feel that the objective loss function of NGCF is very reasonable for the overall model. It optimizes the BPR loss with a moderate amount of calculation. We also use this method, and the expression is as in Eq. (9):

$$\text{Loss} = \sum_{(u,i,j)\in O} -\ln\sigma\left(\hat{y}_{ui} - \hat{y}_{uj}\right) + \lambda\|\Theta\|_2^2, \tag{9}$$

where $O = (u,i,j)|(u,i) \in \mathrm{R}^+$ represents the paired training data, R^+ represents the observed interaction, and R^- represents the unobserved interaction. σ is the Sigmoid function, and $\Theta = \left\{E, \left\{W_1^{(l)}, W_2^{(i)}\right\}_{l=1}^{(l)}\right\}$ represents all trainable model parameters.λ is a coefficient, which controls the intensity of L2 adjustment to prevent overfitting. We use Adam to optimize the prediction model and update the model parameters.

5 Evaluation

5.1 Experimental Settings

Experimental Environment. We use Tensorflow1.4 framework, Python3.6 and GPU RTX2070. When using these three data sets for experiments, we randomly select 80% of each user's historical interaction data to form the training set, and use the remaining 20% as the test set. We randomly select 10% of the historical user interaction data in the training set as the validation set. The embedding size of the model is 64, the batch size is set to 1024, and the learning rate is set to 0.001. Experiments show that the parameter update is relatively stable.

Datasets. We use three public data sets: Gowalla, Yelp2018 and Amazon-Book in our experiments. Datasets information is shown in Table 1.

Gowalla: This data set is a record of 16,79245 social friend relationships and visited poi among 329,839 users on gowalla from September 2011 to December 2011, as well as the type classification of the places visited.

Yelp2018: This data set is taken from the 2018 Yelp Challenge. It contains business name, business address, business ID and other information. Businesses such as restaurants and bars are considered projects.

Amazon-Book: A dataset of Amazon best-selling books. It contains the top 50 bestsellers every hour. It contains information such as time, book ranking, book name, user reviews and so on.

Table 1. Datasets information

Dataset	Users	Items	Interactions	Density
Gowalla	29858	40981	1027370	0.00084
Yelp2018	31831	40841	1666869	0.00128
Amazon-Book	52643	91599	2984108	0.00062

Baselines. The baselines selected in this article are NGCF, PinSage, and GCN. Our algorithm has improved NGCF, and it is necessary to compare with NGCF. The comparison result can reflect our improvement effect. PinSage and GCN are both excellent graph neural network algorithms.

NGCF: The algorithm utilizes the structure of the user item graph by propagating embedding, which enables the expressive modeling of high-order connections in the user item graph to effectively inject collaboration signals into the embedding process in an explicit manner.

PinSage: A data-efficient graph convolution network algorithm that combines efficient random walk and graph convolution to generate node embeddings containing graph structure and node feature information.

GCN: A scalable semi-supervised learning method for graph structure data. It motivates the choice of convolution structure through the local first-order approximation of spectral graph convolution, linearly scales the number of graph edges, and learns the hidden layer representation that encodes the local graph structure and node features. GCN is the basis of STAR-GCN, NGCF, etc.

Evaluation Index. The evaluation indicators selected in our experiment are Recall and NDCG. Among them, Recall is the proportion of the positive classes in the sample that are correctly predicted. NDCG (Normalized Discounted Cumulative Gain) is usually used to measure and evaluate recommendation and ranking related algorithms. These two indicators can well evaluate the performance of our algorithm.

5.2 Result of NDCG and Recall

We experiment and calculate the data of Recall@20 and NDCG@20 for four algorithms in three data sets. The data of Recall@20 obtained in the experiment is shown in Table 2, and the data of NDCG@20 is shown in Fig. 4.

We run each of the four algorithms for 20 rounds by using these three data sets, and calculate the average Recall@20. In the Gowalla data set, INGCF is 13.34% higher than NGCF's Recall@20. By using the Yelp2018 data set, INGCF is 12.32% higher than NGCF's Recall@20. In the Amazon-Book data set, INGCF is 11.6% higher than NGCF's Recall@20. The result of our analysis

Table 2. Comparison of Recall@20 by using the three data sets.

	GCN	PinSage	NGCF	INGCF
Gowalla	0.1394	0.1380	0.1547	**0.1754**
Yelp2018	0.0359	0.0372	0.0438	**0.0492**
Amazon-Book	0.0285	0.0283	0.0344	**0.0384**

is that INGCF uses multi-layer neural network to process end-user item embedding vector, calculate user item prediction score, and fully capture the complex structure of user item interaction data.

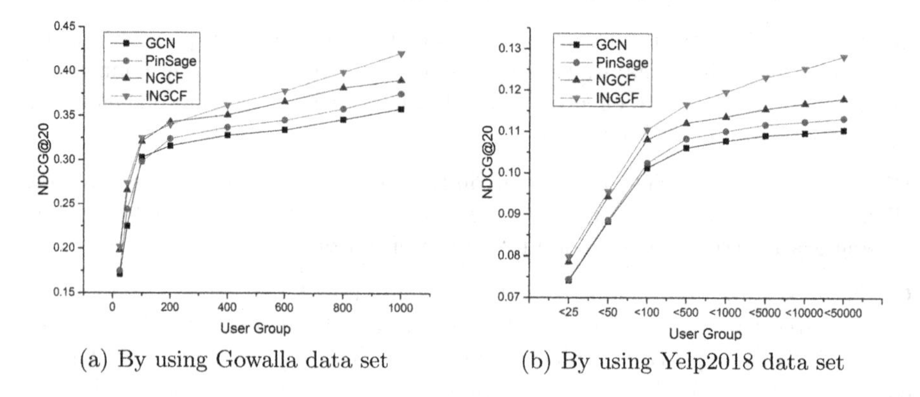

(a) By using Gowalla data set

(b) By using Yelp2018 data set

Fig. 4. Comparison of NDCG@20 by using two data sets

Figure 4(a) and Fig. 4(b) are the average NDCG@20 comparison of the four algorithms in the two data sets. User Group represents the number of interactions of each user. We select User Group less than 25,50,100,200,200,400,600,800,1000 for statistics, and calculate the corresponding NDCG@20.

It can be seen from Fig. 4(a) and Fig. 4(b) that by using the Gowalla and Yelp2018 data sets, when the number of user interactions is less than 100, the four algorithms NDCG@20 have little difference. In the Gowalla data set, INGCF is 12.3% higher than PinSage's NDCG@20 when User Group<1000, while INGCF is 7.9% higher than NGCF. In the Yelp2018 data set, INGCF is 13.3% higher than PinSage's NDCG@20 when User Group<50000, while INGCF is 8.6% higher than NGCF. It can be seen that from <100 and after User Group, the gap in NDCG@20 indicators among some algorithms has gradually become apparent. Among them, INGCF and NGCF are relatively close, and they have opened a gap with PinSage and GCN. We believe that the message passing structure of INGCF and NGCF effectively injects the high-order connection modeling of the user-item diagram into the embedding process, so that the user-item information obtained is more accurate. When using the Gowalla data set, the User Group changes from <200 to <1000, and the gap between INGCF and NGCF gradually increases. When using the Yelp2018 data set, the gap between INGCF and NGCF is that the User Group gradually increases from <100 to <50000. We believe that the feature extraction enhancement layer makes INGCF stronger in feature extraction. When dealing with user interaction data with low sparseness, it can still handle the data well.

5.3 Ablation Experiment

This section mainly discusses the influence of the feature extraction enhancement layer proposed by INGCF and the optimized user item prediction scoring method on the model. The experimental results are shown in Table 3. **FU** means the full version of INGCF,**DI** means IndRNN is not used, **DS** means the self-attention mechanism is not used, and **DM** means the user-item prediction score method is not optimized.

Table 3. Results of ablation experiments by using three data sets.

		DI	DS	DM	FU
Gowalla	Recall@20	0.1654	0.1592	0.1786	**0.1754**
	NDCG@20	0.234	0.2385	0.2386	**0.2496**
Yelp2018	Recall@20	0.0451	0.0471	0.0473	**0.0492**
	NDCG@20	0.0967	0.104	0.1008	**0.1035**
Amazon-Book	Recall@20	0.0358	0.0364	0.0362	**0.0384**
	NDCG@20	0.0681	0.0699	0.0705	**0.0739**

Through the results of the overall ablation experiment, it can be seen that the feature extraction enhancement layer proposed by INGCF and the optimized user-item prediction score method have a significant impact on the improvement of the overall algorithm. The following is the analysis part.

DI: Instead of using a 4-layer IndRNN, it uses a self-attention mechanism layer to extract features and then calculates the relevant score using an optimized user item prediction score method. As Table 3 shown, Recall@20 dropped by 5.7% and NDCG@20 dropped by 6.3% in the Gowalla dataset. By using the Yelp2018 data set, Recall@20 dropped by 8.3%, and NDCG@20 dropped by 6.6%. By using the Amazon-Book data set, Recall@20 dropped by 6.77%, and NDCG@20 dropped by 7.8%. It can be seen that under the Yelp2018 data set, the indicators of the 4-layer IndRNN have the most comprehensive decline, and Yelp2018 has the highest data density in the three data sets. This also proves that the 4-layer IndRNN is good at processing low-sparse data and obtaining the characteristics of high-density data.

DS: It does not use a self-attention mechanism. After the 4-layer IndRNN, we choose to use the optimized user project prediction score method to calculate the relevant score. It can be seen from Table 3 that after the self-attention mechanism is not adopted, Recall@20 has an average drop of 5.3% in the three data sets. NDCG@20 has an average drop of 5.4% in the three data sets. This proves that the self-attention mechanism can effectively improve the discriminative ability of the model and capture the obvious characteristics of the data.

DM: It does not use the optimized user-item prediction score method, but still uses the vector inner product to calculate the relevant score. It can be obtained from Table 3 that Recall@20 has an average drop of 4.8% in the three data sets. NDCG@20 has an average drop of 3.2% in the three data sets. This shows that the user-item prediction score method composed of three-layer MLP can better capture the interactive data characteristics between the user and the item, and can more accurately calculate the relevant score, which is very helpful to improve the model training results.

6 Conclusions and Future Work

This paper proposes an enhanced feature extraction graph neural network collaborative filtering algorithm based on NGCF, INGCF. This algorithm retains the advantages of NGCF's strong ability to capture high-level connection features of graph data. INGCF has strengthened NGCF's feature extraction capabilities for user vectors and item vectors. It also solves the problem that the evaluation function is difficult to capture the complex structure of user project interaction data. INGCF adds 4 layers of IndRNN layer and 1 layer of self-attention mechanism layer after the Concatenate operation of the user-item vector of each level of NGCF to enhance the feature extraction ability. The combination of IndRNN and the self-attention mechanism also makes the model more robust and generalized to a certain extent. We use 3-layer MLP to better capture the complex structure of user-item interaction data in the algorithm to calculate the user-item prediction score. We compare the NDCG@20 and Recall@20 indicators of the four algorithms of INGCF, NGCF, PinSage and GCN by using the Gowalla, Yelp2018 and Amazon-Book datasets. We also conduct ablation experiments and achieve good results.

In future work, we will use more data sets to conduct experiments on INGCF to evaluate its performance in more detail. We also plan to test the cold start effect of INGCF, and if it is poor, we will try to further improve the algorithm.

Acknowledgments. This work is supported by National Key R&D Program of China (2018YFB1700100), CERNET Innovation Project (NGII20190801) and the Fundamental Research Funds for the Central Universities under Grants (DUT21LAB115).

References

1. Goodfellow, I., et al.: Generative adversarial nets. In: Advances in Neural Information Processing Systems 27 (2014)

2. He, X., Deng, K., Wang, X., Li, Y., Zhang, Y., Wang, M.: LightGCN: simplifying and powering graph convolution network for recommendation. In: Proceedings of the 43rd International ACM SIGIR Conference on Research and Development in Information Retrieval, pp. 639–648 (2020)
3. He, X., Liao, L., Zhang, H., Nie, L., Hu, X., Chua, T.S.: Neural collaborative filtering. In: Proceedings of the 26th International Conference on World Wide Web, pp. 173–182 (2017)
4. Hsu, C., Li, C.T.: RetaGNN: relational temporal attentive graph neural networks for holistic sequential recommendation. In: Proceedings of the Web Conference 2021, pp. 2968–2979 (2021)
5. Kipf, T.N., Welling, M.: Semi-supervised classification with graph convolutional networks. arXiv preprint arXiv:1609.02907 (2016)
6. Li, S., Li, W., Cook, C., Zhu, C., Gao, Y.: Independently recurrent neural network (IndRNN): building a longer and deeper RNN. In: Proceedings of the IEEE Conference on Computer Vision and Pattern Recognition, pp. 5457–5466 (2018)
7. Lu, Y., Jiang, X., Fang, Y., Shi, C.: Learning to pre-train graph neural networks. In: Proceedings of the AAAI Conference on Artificial Intelligence, vol. 35, pp. 4276–4284 (2021)
8. Ma, C., Ma, L., Zhang, Y., Sun, J., Liu, X., Coates, M.: Memory augmented graph neural networks for sequential recommendation. In: Proceedings of the AAAI Conference on Artificial Intelligence, vol. 34, pp. 5045–5052 (2020)
9. Qiu, R., Li, J., Huang, Z., Yin, H.: Rethinking the item order in session-based recommendation with graph neural networks. In: Proceedings of the 28th ACM International Conference on Information and Knowledge Management, pp. 579–588 (2019)
10. Sankar, A., Liu, Y., Yu, J., Shah, N.: Graph neural networks for friend ranking in large-scale social platforms. In: Proceedings of the Web Conference 2021, pp. 2535–2546 (2021)
11. Vaswani, A., et al.: Attention is all you need. In: Advances in Neural Information Processing Systems, pp. 5998–6008 (2017)
12. Wan, S., Pan, S., Yang, J., Gong, C.: Contrastive and generative graph convolutional networks for graph-based semi-supervised learning. arXiv preprint arXiv:2009.07111 (2020)
13. Wang, X., He, X., Wang, M., Feng, F., Chua, T.S.: Neural graph collaborative filtering. In: Proceedings of the 42nd International ACM SIGIR Conference on Research and Development in Information Retrieval, pp. 165–174 (2019)
14. Wang, X., et al.: Learning intents behind interactions with knowledge graph for recommendation. In: Proceedings of the Web Conference 2021, pp. 878–887 (2021)
15. Wu, S., Tang, Y., Zhu, Y., Wang, L., Xie, X., Tan, T.: Session-based recommendation with graph neural networks. In: Proceedings of the AAAI Conference on Artificial Intelligence, vol. 33, pp. 346–353 (2019)
16. Ying, R., He, R., Chen, K., Eksombatchai, P., Hamilton, W.L., Leskovec, J.: Graph convolutional neural networks for web-scale recommender systems. In: Proceedings of the 24th ACM SIGKDD International Conference on Knowledge Discovery and Data Mining, pp. 974–983 (2018)
17. You, Y., Chen, T., Sui, Y., Chen, T., Wang, Z., Shen, Y.: Graph contrastive learning with augmentations. Adv. Neural. Inf. Process. Syst. **33**, 5812–5823 (2020)
18. Zhang, J., Shi, X., Zhao, S., King, I.: Star-GCN: stacked and reconstructed graph convolutional networks for recommender systems. arXiv preprint arXiv:1905.13129 (2019)

Distributed and Network-Based Computing

AutoFlow: Hotspot-Aware, Dynamic Load Balancing for Distributed Stream Processing

Pengqi Lu[1,2], Yue Yue[1,2], Liang Yuan[1](✉), and Yunquan Zhang[1]

[1] State Key Laboratory of Computer Architecture, Institute of Computing Technology, Chinese Academy of Sciences, Beijing, China
yuanliang@ict.ac.cn

[2] University of Chinese Academy of Sciences, Beijing, China

Abstract. Stream applications are widely deployed on the cloud. While modern distributed streaming systems like Flink and Spark Streaming can schedule and execute them efficiently, streaming dataflows are often dynamically changing, which may cause computation imbalance and back-pressure.

We introduce AutoFlow, an automatic, hotspot-aware dynamic load balance system for streaming dataflows. It incorporates a centralized scheduler that monitors the load balance in the entire dataflow dynamically and implements state migrations correspondingly. The scheduler achieves these two tasks using a simple asynchronous distributed control message mechanism and a hotspot-diminishing algorithm. The timing mechanism supports implicit barriers and a highly efficient state-migration without global barriers or pauses to operators. It also supports a time-window based load-balance measurement and feeds them to the hotspot-diminishing algorithm without user interference. We implemented AutoFlow on top of Ray, an actor-based distributed execution framework. Our evaluation based on various streaming benchmark datasets shows that AutoFlow achieves good load-balance and incurs a low latency overhead in a highly data-skew workload.

Keywords: Stream processing · Big data · Data skewness · Control message · Load balance

1 Introduction

Streaming dataflows are widely used ranging from AI applications to website unique visitors (UV) counting nowadays. These streaming jobs are generally formed as directed acyclic graphs (DAGs) by modern streaming systems to be deployed on clusters, each node in the graph represents a streaming operator defined by users. Streaming framework then wraps them in each task [7] or actor [18], and initializes network channels, in-memory key-value store, and other components, and schedules them to physical nodes.

Y. Lai et al. (Eds.): ICA3PP 2021, LNCS 13157, pp. 133–151, 2022.
https://doi.org/10.1007/978-3-030-95391-1_9

Streaming systems are designed to achieve both low latency and high throughput, while latency and throughput can be affected in different ways. An unexpectedly high input source rate that exceeds the processing capacity of a physical node in the graph will cause latency spikes [15]. Some physical nodes in the graph may become bottlenecks when there are inappropriate settings of parallel instances to operators [13], which will cause backpressure. Different fault-tolerance mechanisms may affect the system's throughput and latency in different ways [20]. The data-skewness caused by highly uneven access on a small portion of the state will lead to a hotspot issue in the streaming job, which often cannot be solved well at runtime.

In this paper, we focus on the data-skewness issue in distributed streaming dataflows. There are two general techniques to handle hotspot issues statically. One is pre-aggregation (Fig. 1), in which the map operators apply the same reduce logic as the downstream reduce operators on their portion of data before they send results to downstream. The pre-aggregation technique reduces pressure to access the state of reduce operator when there are hotspot issues, while it has obvious drawbacks. Firstly, pre-aggregation requires grouping a small batch of data, which will cause latency overhead. In some real-time applications, this overhead may not be acceptable. Secondly, pre-aggregation is only applicable to reduce operations (Count, Sum, Min, Max), and is not fit for other operations like Join and messageTimeWindow. The other technique is rehashing (Fig. 1), which adds an extra layer of reduce operators to scatter out the potential hotspots and merge results in the second layer. The rehashing method has similar drawbacks to pre-aggregation because it aggregates data twice.

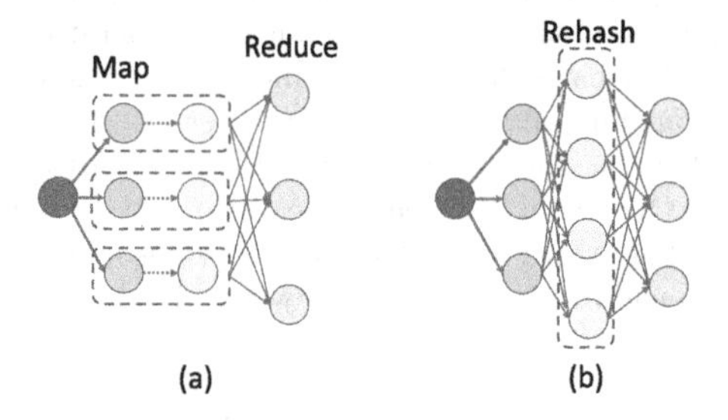

Fig. 1. (a) Pre-aggregation: insert a local operator that reduces the output of map operator. (b) Rehashing: add an extra layer of reduce operators to scatter out the potential hotspots

Pre-aggregation and rehashing are static methods that mitigate hotspots issues, which means the physical graph built at compile time cannot be changed at run time. While streaming dataflows are long-running jobs in general, and the access pattern of data to stateful operators is often unpredictable, the static method is inherently not the best choice to solve the problem. Flink [7] supports rescaling the streaming jobs, which we can change the parallelism of the operator instances and redistribute the state of stateful operators. However, this requires manual operations such as discovering when and where the hotspot issues raise, and how many instances to rescale. Furthermore, Spark [22] and Flink [7] have to halt the whole job graph until the rescaling job finishes, which is the overhead users have to tolerate. SnailTrail [13] adopts Critical Path Analysis to successfully detect data-skewness issues and other computation imbalance problem, while this method has to be integrated with a streaming system like Flink or Spark. Chi [17] uses distributed control messages to coordinate dataflow reconfiguration, and builds a programmable control plane on top of a streaming system. Although this is an innovation that firstly integrates control plane and data plane in a system, when we focus on a specific issue like data-skewness, it's a bit heavy because it requires explicit barrier on parts of the dataflow graphs when doing reconfiguration. Megaphone [14] takes control stream and data stream as inputs in each worker simultaneously, and let the control inputs to decide which downstream operator to route. When the control input modifies the routing table of map operators, it will start the state-migration between stateful operators. Megaphone supports fluid state migration unlike Flink or Chi, but its operators require buffering when timestamps of data messages are in advance of that of the control stream.

We propose a novel framework AutoFlow to handle the data-skewness problem efficiently. AutoFlow adopts a similar but lighter control message mechanism. The timing mechanism supports implicit barriers and a highly efficient state-migration without global barriers or pauses to operators. It also supports a time-window based load-balance measurement and feeds them to the hotspot-diminishing algorithm without user interference. These functions are integrated in a centralized scheduler that monitors the load balance in the entire dataflow dynamically and implements state migrations correspondingly.

We implemented AutoFlow on top of Ray, an actor-based distributed execution framework. Our evaluation based on various streaming benchmark datasets shows that AutoFlow achieves good load-balance and incurs a low latency overhead in a highly data-skew workload. Our contributions are the following:

- A lightweight centralized timing mechanism that generates total ordered control messages and induces implicit barriers.
- An efficient state-migration method that only buffers minimal data messages.
- An effective metric-collection scheme and hotspot-diminishing algorithm to reduce the load-imbalance.
- A prototype system AutoFlow integrating the above methods.

The remainder of this paper is organized as follows. Section 2 provides the related work. The AutoFlow is described in Sect. 3. We present the performance results in Sect. 4. Section 5 concludes the paper.

2 Related Work

Modern big data systems can be divided into two categories, one of them is MapReduce [10], Hadoop [21], and Spark [22] that treat the input data as batches of records, while the other are Flink [7], Structured Streaming [5], Google Dataflow [4] and others [3,16,19] that treat the data as infinite data streams. A key difference between streaming and batching is that the streaming systems support message-time operation and watermarks. Most of the streaming systems focus on how to parallelize the operators to physical nodes in the clusters and schedule the tasks efficiently. This work focuses on hotspot issues, which are often less concerned or well handled in existing streaming systems.

Modern streaming systems are more and more deployed on clusters, their goals of design first look at the ease of use, performance, scalability, and fault tolerance. Therefore most of their dataflow architectures are often static, that is, they can act in streaming mode when processing data but act discretely when reconfiguring the dataflow graphs. Currently, streaming systems like Flink can only redistribute state by halting the whole dataflow graphs, which is a discrete way that incurs latency overhead and opposite to its streaming nature.

Furthermore, existing methods often considers a wide range of areas such as rescaling, checkpointing, and other user-defined functions. Therefore, current profiling tools in streaming systems like Flink [7] requires human efforts. Users may be required to monitor the screen and adjust it when they detect a data-skewness issue, thus an inevitable latency comes between the detection and repair.

Asynchronous Barrier Snapshot [6] is another technique that has been successfully applied to Flink [7] as a checkpointing mechanism. Its core concept has also been used by Chi [17] as distributed control messages to coordinate the reconfiguration stage in stream processing systems. Distributed control messages act like an asynchronous barrier that coordinate the reconfiguration process, there may just be a few of the operators are halting at the same time and other operators work as usual like Chi [17]. In the case of checkpointing and rescaling, if without the control messages, we will have to halt all the physical nodes in the dataflow and resume until the reconfiguration job finishes. The distributed control message mechanism reduces the overhead by making the control operations asynchronous.

There are two types of work on building more reacting, more efficient streaming systems. One is building a monitoring system or algorithms that can detect issues or failures in the system faster and more accurately [9,12,13,15,19]. The other is finding a more efficient way of doing specific things like checkpointing, rescaling [6,8,11,23].

Reconfiguration. Flink [7] currently only supports rescaling when the whole dataflow graph is stopped. Megaphone [14] employ a similar fluid- migration scheme but it lacks a controller and needs to block more the data message, which also incurs overhead. SEEP [8] integrated a checkpointing mechanism that can act asynchronously with the rescaling operation to reduce the overhead. Dhalion [11] provided a control policy mechanism that allows users to define their own policy and self-tunes a streaming job with their needs. Chi [17] integrated control messages with data messages on a programmable control plane in distributed stream processing. However, our AutoFlow differs from the above methods in: 1.

we focus on tackling the data-skewness issue through distributed control message mechanism, we use control message only for state-migration but the other control operations. 2. Our AutoFlow only requires buffering the migrated part of the data messages when doing the migration, other methods either require halting the whole dataflow graph or blocking channels of some operators.

Monitoring. DS2 [15] provided an external controller that supports both automatic scaling and monitoring, SnailTrail [13] adopted Critical Path Analysis (CPA) to analyze the bottleneck of the streaming dataflows. AutoFlow embedded a lightweight scheduler as an operator to continuously analyze hotspots in the dataflows and sends control messages to perform state-migration between stragglers and non-stragglers. Thus, our model is an integrated approach to tackling a specific problem.

3 AutoFlow Design

3.1 Motivation

For the data skewness issue in stream processing, it's desirable to design a lightweight adjusting scheme that only requires buffering the related data. The challenges are how to detect and react to data-skewness issues efficiently, and how to migrate state between straggler and non-straggler without a large overhead. The key idea is to combine data-skewness detecting and reacting as components embedded in the systems without user interference. For detecting the data-skewness issues, we adopt a scheduler to continuously collect metrics from stateful operators. For reacting to the data-skewness issues, we let the scheduler to send distributed control messages to start a state-migration between operators. These two tasks adopt a control message mechanism in our AutoFlow to determine and adjust the state migration behaviors between the stragglers and non-stragglers efficiently.

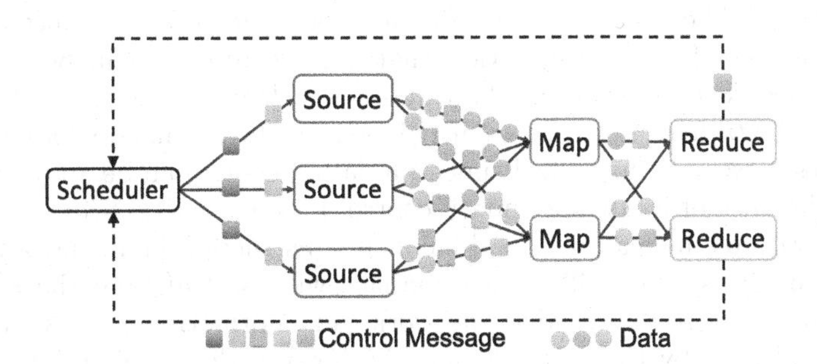

Fig. 2. AutoFlow model. The scheduler is embedded as an operator in the dataflow graph, and continuously broadcasts control messages with increasing timestamps to source operators. The source operators flush data messages until their timestamps exceed the control message, and then propagate the control message indicated a data boundary.

3.2 Overview

AutoFlow's dataflow model is similar to most of the streaming systems like Flink [7], Spark Streaming [23] and Google Dataflow [4]. It consists of two categories of operators. The first one is the stateless operator such as map, flatmap, and filter, which does not hold any local mutable state. The second is the stateful operator such as join, reduce, time-window. The stateful operator is one of the core concepts in streaming systems because many functionalities like checkpointing, state management are built for it.

The AutoFlow model is depicted in Fig. 2. It abstracts the dataflow model as a DAG graph. The major part is a centralized lightweight scheduler that is embedded in the dataflow graph as a new operator. The scheduler is connected to all source operators and all stateful operators. The scheduler carries out two tasks: dynamically monitor the load balance in the entire dataflow and implement state migrations correspondingly. To achieve these goals, the scheduler incorporates a simple timing mechanism and a hotspot diminishing algorithm.

3.3 Timing Mechanism

The scheduler performs as a timer. It periodically generates control messages that are sent to all downstream operators, in particular, all source operators. Each message is attached with a logical timestamp. In Fig. 2, the squares are control messages and the same color indicates the same timestamp. After receiving a control message, the source operator inserts it in its dataflow. In Fig. 2, the circles are data messages and control messages indicate boundaries among the data messages in a connect channel. The control messages spread over the entire dataflow graph in a similar way, i.e. after receiving control messages with the same logical timestamp from all of its input channels, one operator broadcast this control message to all of its output channels. Finally, reducers attach load-imbalance information to control messages and return them back to the scheduler.

We uniformly treat the control message and data messages. Table 1 lists the message fields. The *type* field indicates a data, control, or slot message. The slot type is used for state-migration that will be discussed in Sect. 3.4. The *time* attached to the control or data message is logical timestamp. For easily discussion, the *time* can be continuously increasing even numbers for continuous control messages, and the logical time of a data message can be an odd number between the ones of its two boundary control messages. The *migration* field is a boolean variable indicating a load-balancing instruction in a control message. If it is true, the system will transit the slot with *slot_id* from the *sender* to the *receiver*. The *metric* field is the workload information that is written by reducers and read by the scheduler for the hotspot diminishing. The *payload* field in a data or slot message is the real data that will be processed.

The function of these messages is three-fold. First, previous schemes often utilize timing schemes with partially ordered timestamps. On the contrary, control messages are launched by a centralized scheduler and totally ordered in AutoFlow. This feature serves as a base of the following functions.

Table 1. Message fields

Field	Exist in	Description
type	D, C, S	Data (D), Control (C) or Slot (S) message
time	C,(D)	The logical timestamp
migration	C	Trigger state-migration or not
sender	C	The sender name of the state
receiver	C	The receiver name of the state
slot_id	C, S	State slots to send
metric	C	Information for hotspot decision
payload	D, S	Data that will be processed

Second, control messages indicate implicit barriers. Control messages come from all upstream operators that have an equal timestamp identify an implicit control barrier. Specifically, when a downstream operator reducer r_0 receives a control message from one of its input channels that has the content ($time = 200$, $migration = True$, $sender = r\_0$, $receiver = r\_1$, $slots = \{2, 3\}$). The control message implies that: (1) there will be no data messages with timestamps $time < 200$ coming from this channel, and r_0 can safely migrate the state of the 2nd and 3rd slots to r_1 after it receives the same control messages from all its input channel. (2) the data messages with timestamps $time > 200$ will be routed according to the updated routing table in the map operators, the r_0 doesn't have to worry if it will receive coming data messages that are routed to the migrated slot it has sent. These two implications are critical to designing an efficient state-migration scheme. The detailed description is presented in Sect. 3.4.

Third, a control message associated with a timestamp indicates a boundary of data records in a period of time. The source operator simply injects the control message in the data message flow and send it to all downstream operators. It is equal to group the data messages between two continuous control messages to an atomic data set that is processed identically. Since a control barrier message will be eventually broadcasted to all stateful operators, this property provides an implication for measuring the load-imbalance. Specifically, each stateful operator keeps collecting statistics data such as processing counts of each key slot. When a stateful operator receives all the empty control messages that carry the same timestamp from upstream, it will attach statistics results to the control message and send it back to the scheduler for further analysis. The scheduler keeps collecting metrics from stateful operators, analyzes the collected profiling data, and employs an algorithm to optimize the load-balancing dynamically. It attaches a state-migration instruction to the next control message when detecting a hotspot issue. The detailed description is presented in Sect. 3.5.

Another significant difference between AutoFlow and other systems is that we actually do not necessitate a timestamp associated with the data message. The reason is that AutoFlow only models a DAG graph without loops and assumes a FIFO communication channel in this work. Then the orders between data and control messages are obvious. However, we can extend the timing scheme and support complex scenarios like dataflow graphs with loops. Specifically, we can utilize a simple logical timestamp scheme that is coordinated between the control messages and data messages. When a source operator receives an empty control message from the scheduler, it will continuously send data messages that have smaller timestamps than that of the control message to downstream operators. When the data left in the source have timestamps that are larger than that of the control message, the source operator will propagate the control message to downstream operators and fetch another control message that has a larger timestamp. Other operators can be augmented in a similar way. We leave this extension as future work.

3.4 State Migration

We now present an efficient state-migration implementation with the assistance of the timing mechanism. When an operator receives a control message with a state-migration instruction from one of its input channels, it can expect the same control messages from all other channels. With the implicit barrier induced by the timing mechanism, it just needs to buffer subsequent data messages which are routed to the migrated slot indicated by the state-migration instruction and from the same channel. Other data messages can be immediately processed as usual. Moreover, we can process multiple state-migration control messages at the same time. AutoFlow does not need to stop the data input channel or require global barriers. The implementation details for stateless and stateful operators are discussed in the following two parts, respectively.

Stateless Operator. Although stateless operators like map, filter do not hold any state of the dataflow, it's their job to decide which downstream stateful operator to send when they processed a data record. There are often many parallel instances of operators in a streaming job for horizontal scalability, so it's a common way that each map operator has the same immutable hashing function that acts as a static routing table shared across a cluster of workers. We split the distribution of state into many slots like the virtual nodes in consistency-hashing and each slot has a unique id. The slot is the atomic unit in our state-migration process. We only migrate slots between stateful operators in the state-migration.

Algorithm 1 shows how to process control messages in map operators. If the *migration* field is false, we just record the message *msg* in the *num* object that counts the number of received messages with the same *time*. When the map operator receives all the same control messages from its upstream channels (Line 12), it will broadcast it to all its downstream operators (Line 19). Thus, if there's no migration message happens, the map operator does nothing different from when there's no control message mechanism.

Algorithm 1 registers the migrated slots (Line 6) and initiates a buffer (Line 7) when the *migration* is true. It then buffers records (Line 23) from channels that have already received the control message if the corresponding slot will be migrated (Line 22). Finally, when the map operator receives control messages from all its input channels, it means an implicit barrier, and all data messages in the buffer should be routed with respect to the updated routing table. So it is safe to change the routed table, delete the migrated information, and flush all records in the message buffer (Line 14–16). In summary, we only have to buffer data messages that are routed to the migrated slot temporarily.

Algorithm 1. Message Processing in Stateless Operator

```
1: function PROCESS(msg)
2:     if msg.type == "Control" then
3:         if num.get(msg.time) == 0 then
4:             num.init(msg.time, 1)
5:             if msg.migration == True then
6:                 migrated_slot.init(msg.slot_id)
7:                 buffer.init(msg.slot_id)
8:             end if
9:         else
10:            num.add(msg.time, 1)
11:        end if
12:        if num.get(msg.time) == num_input then
13:            if msg.migration == True then
14:                routing_table.update(msg)
15:                migrated_slot.delete(msg.slot_id)
16:                buffer.flush(msg.slot_id)
17:            end if
18:            num.delete(msg.time)
19:            broadcast_control_msg(msg)
20:        end if
21:    else if msg.type == "Data" then
22:        if migrated_slot.check(msg) then
23:            buffer.append(msg.payload)
24:        else
25:            process_data(msg.payload)
26:        end if
27:    end if
28: end function
```

We demonstrate an example shown in Fig. 3 where the map operator receives state-migration messages interleaved with other data messages from three input channels. The slot named r_x_y means the y-th slot of reducer r_x. The bottom of the figure is the hashing results of all data messages and the routing table tells the slots owned by which reducer instance. In Fig. 3 (i), when the map operator receives a square control message with timestamp 8 from one uppermost channel, it parses the message and gets the name of the migrated slot r_0_3, which will be migrated from reducer r_0 to reducer r_1. Then the map operator writes down

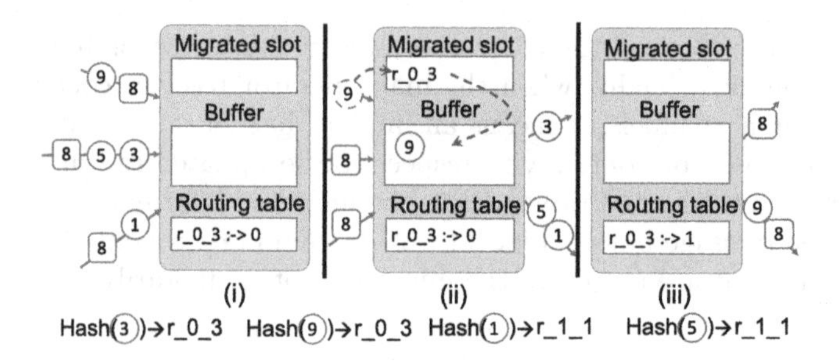

Fig. 3. Modifying the router table and buffering data messages in map operators. (i) Map operator receives a control message from the uppermost channel (ii) mapper buffers the data messages following the control message (iii) mapper finishes processing control messages and broadcasts them to downstream operators.

the slot id on its migrated slot list to indicate which data messages need to be buffered. The map operator buffers the circle data message from the uppermost channel in Fig. 3 (ii) according to the migration instruction. The data message hashed to the slot r_0_3 in the middle input channel is sent to reducer r_0 as usual since the map operator has not received the control. Figure 3 (iii) illustrates that after the map operator receives the control messages with timestamp 8 from all its input channels. it changes the routed table, broadcasts the control messages, and flushes all records in the message buffer to reducer r_1.

Stateful Operator. Many core concepts in streaming systems like state-management, flow-control, and checkpointing are highly related to stateful operators that hold locally mutable state in a key-value store. Unlike stateless operator where we barely encounter a hotspot issue because we can easily adapt schemes like random shuffle, round-robin and rebalance between operators to ensure load-balance, stateful operators often face the data-skewness problem that requires state-migration.

Cooperated with the stateless operator, specifically Algorithm 1, we know that (1) if a stateful operator has received a migrated control message, it will not receive the following message that routed to the slot. Our mechanism ensures the upstream stateless operators will route the message to the new operator. (2) the stateful operator will trigger a state-migration process after it receives all the control messages. With these two features, we present a similar algorithm for stateful operators (Algorithm 2) and describe it with an example in Fig. 4.

In Fig. 4 (i) the left reducer firstly receives a control message 6 from the uppermost channel and record it in the num (Line 4). Although the data message 4 from the middle channel is routed to the migrated slot is r_0_1, the reducer has not received the control message from it yet, so it will process the message as usual and change the state in r_0_1. In the meantime, the right reducer receives the same control message and initializes buffer and migrated-slot list (Line 7–8), and launches asynchronous receiving of migrated slots (Line 9). When there

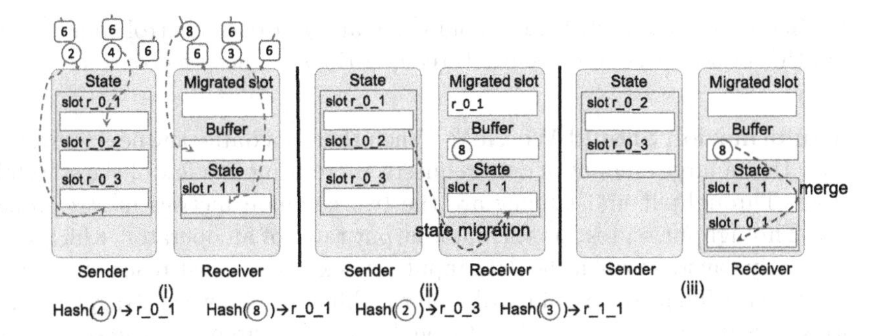

Fig. 4. Migrating states between reduce operators. (i) Reduce operators receive data messages interleaved with control messages (ii) Reduce operator(right) adds slot"r_0_1" in the migrated slot and initializes the message buffer when processing the control message. In the mean time reduce operator (left) triggers a state-migration to the right. (iii) After reduce operator (right) receives state from the left, it then merge the messages in the message buffer.

comes a data message 8 right after the control message that routed to slot r_0_1, the reducer just buffers it (Line 28). In Fig. 4 (ii) when the left reducer receives all control messages, it will start a migration and send slot data r_0_1 to the receiving reducer (Line 18). In Fig. 4 (iii) when the right reducer receives the slot, it will merge the data messages in the buffer to the slot (Line 23–25). It's worth mentioning that the migration process is asynchronous which means that the reducers are processing other data messages concurrently.

3.5 Hotspot-Diminishing

External controllers in existing big data systems are commonly used for task scheduling, resource management, and monitoring the runtime status of the cluster. It's a centralized way that can coordinates the tasks in a cluster directly. In AutoFlow we use scheduler as an operator in a more fine-grained way to tackle the data-skewness issue. Specifically, in the scheduler operator, we integrate monitoring of the runtime status in stateful operators with a feedback control component that reacts to the dataflow efficiently.

There are lots of ways to monitor a streaming system such as collecting the throughput and latency, monitoring the queue size of the input channels and probing the data input or output rates of operators. Some of them either do not reflect the situation accurately [13] or are not efficient due to the natures of distributed systems such as network latency. As the control messages act like watermarks, we let the stateful operators send metrics such as the processing counts of each slot, and the total processing count to the scheduler operator when they reach an implicit barrier triggered by the empty control messages.

Common monitoring systems collect metrics such as memory consumption in each task, throughput and latency per record in each operator, size of the network buffers and so on. These collected metrics are often shown in the dashboard for users. When we build an integrated monitoring and feedback control component tackling the data-skewness, there are three questions we need to consider:

What Information Should We Get? When there's a data-skewness issue, we will know that a large amount of data requests are routing to a few of the stateful operators. Throughput and latency are the two common metrics in streaming systems. Throughput represents the data output rates of an operator, which will reach a nearly constant when the data input rate exceeds or just reached its processing capacity. That is, we can hardly tell the difference between stragglers and non-stragglers just according to their throughputs. The latency of the stragglers will grow high beyond the non-stragglers. But these metrics are not enough for a feedback algorithm to make decisions of state-migration, in particular, determine

Algorithm 2. Message Processing in Stateful Operator

```
 1: function PROCESS(msg)
 2:     if msg.type == "Control" then
 3:         if num.get(msg.time) == 0 then
 4:             num.init(msg.time, 1)
 5:             if msg.migration == True then
 6:                 if msg.receiver == my_id then
 7:                     migrated_slot.init(msg.slot_id)
 8:                     buffer.init(msg.slot_id)
 9:                     async_recv_migrated_slot(msg)
10:                 end if
11:             end if
12:         else
13:             num.add(msg.time, 1)
14:         end if
15:         if num.get(msg.time) == num_input then
16:             if msg.migration == True then
17:                 if msg.sender == my_id then
18:                     async_send_migrated_slot(msg)
19:                 end if
20:             end if
21:             broadcast_control_msg(msg)
22:         end if
23:     else if msg.type == "Migration" then
24:         migrated_slot.deleteslot(msg.slot_id)
25:         buffer.merge(msg.slot_id)
26:     else if msg.type == "Data" then
27:         if migrated_slot.check(msg) then
28:             buffer.append(msg.payload)
29:         else
30:             process_data(msg.payload)
31:         end if
32:     end if
33: end function
```

the migrated slots. In AutoFlow, we collect the number of processing records of each slot in a period of time from every stateful operator as the input of our hotspot decision algorithm.

How Do We Get? A stateful operator sends control messages with metrics when it has received the same control messages from all channels. In the scheduler operator, we start a thread that keeps listening from stateful operators asynchronously. When the scheduler receives metrics messages from all stateful operators, it indicates a boundary of time. It then feeds those metrics to our hotspot-diminishing algorithm.

As discussed above, the scheduler operator receives the processing count of each slot in every interval determined by two continuous control messages. While the dataflows are often changing dynamically, the current metrics data it receives might not always predict the situation in the next second correctly. For example, in a flash-sale activity online there might just be a latency spike in a task and vanished in a few seconds, while in other case the stragglers might last for a long time. Setting a better window-length parameter can help to predict the situation more accurately. For a wider range of application, we adopt a time-window based feedback algorithm in AutoFlow, where the window length is an adjustable parameter.

Algorithm 3. Hotspot Diminishing

```
Input: slot : {reducer_id, {slot_id, slot_count}}
Output: migrated_slot : {sender, receiver, {slot_id}}
 1: function PROCESS_WINDOW(slot)
 2:     total_count = slot.sum_by_reducer_id()
 3:     sender = total_count.max_id()
 4:     sender_count = total_count.getcount(sender)
 5:     receiver = total_count.min_id()
 6:     receiver_count = total_count.getcount(receiver)
 7:     diff = sender_count − receiver_count
 8:     if diff/sender_count >= factor then
 9:         migrated_slot.init(sender, receiver)
10:         gap = diff/2
11:         while gap > 0 && slot.hasslot(sender) do
12:             slot_id = slot.getmaxslot(sender)
13:             slot_count = slot.getcount(sender, slot_id)
14:             if gap >= slot_count then
15:                 migrated_slot.add(slot_id)
16:                 gap = gap − slot_count
17:             end if
18:             slot.pop(sender, slot_id)
19:         end while
20:         return migrated_slot
21:     end if
22:     return NULL
23: end function
```

How to React? For static workloads, it is often satisfied to shuffle the state between all workers with a global barrier to reach a perfect load-balance. However, this scheme will incur a large overhead for long-running and dynamically changing dataflows. In AutoFlow, we tend to act dynamically and continuously that we do not need to migrate a large amount of state.

Our hotspot-diminishing algorithm is presented in Algorithm 3. It only adjusts the workloads between the operators with the maximum and minimum metric, in particular, the total number of processing records in a time-window (Line 2). Specifically, when the gap between the max-operator (Line 3–4) and the min-operator (Line 5–6) exceeds a predefined value (Line 8), we initiate a *migrated_slot* and utilize the first-fit heuristic (Line 11–19) of the knapsack problem to pack slots whose total size approximately equals to half of the gap. Finally, the scheduler operator will serialize the migration meta-info in the control message and propagate it in the dataflow. Although we illustrate one algorithm implemented in AutoFlow, there are also many other load-balance schemes that can be adopted in our framework. Introducing a user-defined function as a load-balance API is one of its future work.

3.6 Implementation

AutoFlow is built on top of Ray [18]—an actor-based distributed execution framework. We implemented the core logic of operators and transform the logical dataflow graph to a physical graph through Ray's Python bindings, in which each operator is wrapped in an actor. Each actor is connecting with each other through grpc [1] in Ray, when an operator sends a message to its downstream channel, a *remote* function will be called through the backend of Ray. Each operator in the logic graph of dataflow will be deployed in a cluster combined with a Python process. Currently, we implemented the (de)serialization of data messages and the state of stateful operators in Python's standard library. Moving them to a more efficient programming language's backend and adopting a high-performance key-value store is in our future work. In AutoFlow we bypassed the scheduling process of Ray's backend by turning on the *direct_call* mode in the *remote* function's API. In *direct_call* mode, Each *remote* function call is a Remote Procedure Call (RPC) to another process. At the start of the deployment, Ray will connect each actor through gRPC [1].

4 Evaluation

We evaluated the AutoFlow model on the generated Nexmark benchmark dataset. To measure the benefits of AutoFlow's dynamic load-balance scheme, our evaluation falls into four categories. First, we ran a stateful query on various workloads with fixed skewness percentages to show the AutoFlow can achieve load-balance through our control message mechanism (Sect. 4.2). Second, we tested our model on workloads with a suddenly changed extent of skewness to show the reaction speed of our algorithm (Sect. 4.3). Third, To an unpredictable, rapidly changing workload, our algorithm can handle it to what extent? (Sect. 4.4).

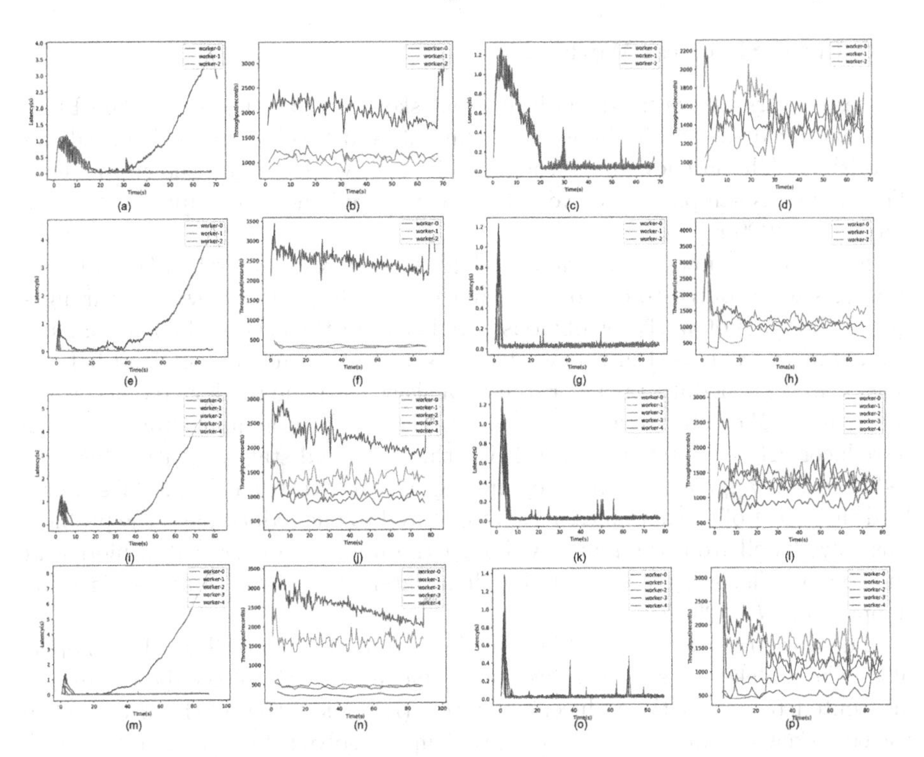

Fig. 5. (a)(b) 50% skewness in 3 reducers without dynamic load-balance. (c)(d) 50% skewness in 3 reducers with dynamic load-balance. (e)(f) 80% skewness in 3 reducers without dynamic load-balance. (g)(h) 80% skewness in 3 reducers with dynamic load-balance. (i)(j) 60% skewness in 5 reducers without dynamic load-balance. (k)(l) 60% skewness in 5 reducers with dynamic load-balance. (m)(n) 80% skewness in 5 reducers without dynamic load-balance. (o)(p) 80% skewness in 5 reducers with dynamic load-balance.

4.1 Experimental Setup

We run all our experiments on up to three machines on the Aliyun ECS cluster, each is an m5.2xlarge instance with 8 vCPUs and 32 GB of RAM, running on Ubuntu 18.04. The schema in Nexmark benchmark is a model that consists of three entities: *Person, Auction, Bid*, which simulates a public auction. We generated the dataset according to the opened source code from [2]. There many standard queries in Nexmark benchmark, such as a simple map (**Q1**), a simple filter (**Q2**), an incremental join (**Q3**). We chose to test a source-map-reduce model on our workloads as the following stateful query:

```
stream = env.source(record)
           .map(record => bids)
           .keyby(bids.id)
           .reduce(state[bids.id]
           += bids.price)
```

And each operator has the same parallel instances in our settings.

4.2 Fixed Skewness Workload

We tested various workloads with different skewness percentages on AutoFlow, and compared the results between turning off and turning on the load-balance algorithms. The latency and throughput in each workload are shown in Fig. 5. The latency is sampled once every 100 records, and the throughput is recorded once every 1000 records.

Figure 5 (a) shows an increasingly high latency overhead when 50% of data messages were routed to one of the reducers, and Fig. 5 (b) shows the throughput in the hotspot worker is obviously higher than the others, the load-balance algorithm was not used in this workload. When we turned on the load-balance scheme in the scheduler operator, the data-skewness issue is alleviated according to Fig. 5 (d). We detected a slight latency spike and a throughput dropping down of reducer r_1 (t = 30 s in Fig. 5(c)(d)), this indicated state-migration between reducers has happened. Figure 5 (e) and Fig. 5 (f) illustrate a high skewness workload that most of the data messages were routed to one reducer, and the problem was solved well from the start. Although the migration overhead is hidden at the start of Fig. 5 (g), we detected a slight latency overhead of state-migration happened at $t = 25$ (Fig. 5(h)).

The workloads depicted in Fig. 5 (i)–(o) is running on 5 workers of each operator. The x%-skewness in those workloads means that x% of the data messages are routed to two of the reducers. Figure 5 (i) depicts the overhead caused in the 60% skewness workload, and the problem was solved through migration with some slight latency overhead ($t = 50$ in Fig. 5(k)). The rest of the figures also show the alleviation of data skewness in AutoFlow.

In the above workloads, we illustrate that the AutoFlow model reaches better a load-balance. While in real-world applications the dataflows are often dynamically changing, we need to generalize our algorithm to fit in the unpredictable environment.

4.3 Spiking Changing Workloads

In real-world scenarios, we often encounter spiking messages in which a large number of requests are flushed to servers for just a few seconds. These also cause hotspot issues most of the time. Solving them requires a faster reactive method to detect the issues from the beginning.

The feedback control algorithm proposed in Sect. 3.5 has a *window-length* parameter exposed to users that can generalize to these spiking messages. The *window-length* parameter indicates the duration the scheduler operator collects metrics for. When the dataflow is stable or slowly changing, we can collect the metrics from a broad range of time by setting the *window-length* parameter bigger, and the scheduler operator can make decisions better according to those metrics. However, when the dataflow has spiking messages, the effective metrics are concentrated on a short range of time. In this case, it's wise to set the *window-length* smaller to better suited to the fast-changing workloads.

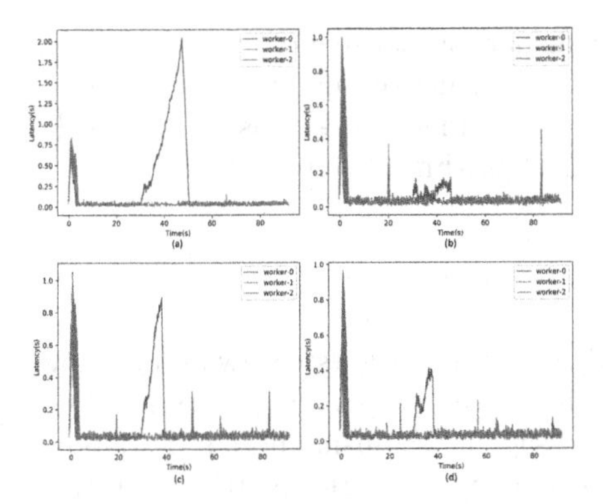

Fig. 6. (a) A spiking message occurred and lasted for 20 s. (b) The AutoFlow reduced the latency overhead caused by the spiking message with a small *window-length* parameter. (c) A spiking message occurred and lasted for 10 s occurred. (d) The AutoFlow relaxed the issue with a smaller *window-length* parameter than that of (b).

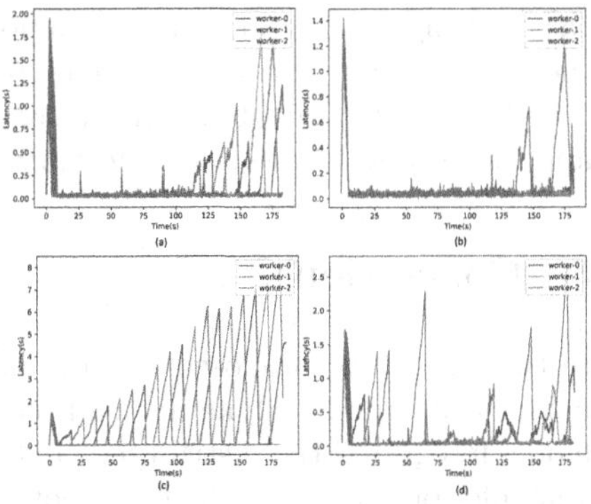

Fig. 7. (a) 50% skewness dynamically changing workload. (b) The AutoFlow reduced the overhead to some extent. (c) 80%, round-robin skewness workload. Each reducer took turns being a hotspot. (d) The AutoFlow fixed some hotspots but some others still existed.

As depicted in Fig. 6, we showed workloads that have spiking messages lasted for 20 s (Fig. 6(a)) and 10 s (Fig. 6(c)). In Fig. 6 (b), we set a smaller *window-length* and got better results compared to Fig. 6 (a). The latency overhead is greatly reduced by our feedback control algorithm. However, to the workload with shorter reaction time in Fig. 6 (c), our algorithm can only relax it to some extent. There are some reasons for that: (1) our algorithm is not efficient enough for this workload, (2) there's a delay from sending and receiving metrics, and a delay from sending control messages and performing state-migration. Building a more reactive scheme is in our future work.

4.4 Dynamically Changing Workloads

Streaming dataflows are long-running jobs and they often change dynamically. To see if our algorithm can fit into the dynamically changing environment, we generated the workloads that have hotspots in every moment and the location of the hotspot changed dynamically. At any time, there was a reducer that becomes the hotspot in the dataflow depicted in Fig. 7 (a) and Fig. 7 (c), It's worth mentioning that the situation in Fig. 7 (c) was far worse than that in Fig. 7 (a), because the time when the latency overhead rises depends on both the skewness percentage and the source input rate. Figure 7 (b) shows fixed the hotspot issues but left for one, this may due to the efficiency of our algorithm. In Fig. 7 (d), AutoFlow detected some of the hotspots and reduced the overhead greatly compared to Fig. 7 (c), but some other hotspots still existed. Imagine a case, a straggler migrates state to a non-straggler for load-balance, after a while, the non-straggler becomes the hotspot, and the situation may become worse.

Although we might not meet the above extremely workloads in real-world applications, we shows the generality of our algorithm to some extent.

5 Conclusion

In this paper, we proposed and evaluated AutoFlow, a hotspot-aware dataflow model that supports dynamic load-balance in distributed stream processing. Our model integrated the distributed control message mechanism and a self-adapted scheduler operator to detect hotspot issues quickly and perform state-migration between operators efficiently. The experimental result shows that our model achieved a better load balance on various data-skewness workloads, and the hotspot operator is diminished according to its latency and throughput. Our model also got better results under spiking changing situation due to the elastic time-window based algorithm.

In future work, we seek to adapt our AutoFlow model to more scenarios like dynamic rescaling and automatic parallelization of dataflows.

Acknowlegement. The authors would like to thank all the reviewers for their valuable comments. This work is supported by National Key R&D Program of China under Grant No. 2016YFB0200803; the National Natural Science Foundation of China under Grant No. 61972376, No. 62072431, No. 62032023; the Science Foundation of Beijing No. L182053.

References

1. gRPC. https://grpc.io/
2. NEXMark benchmark. http://datalab.cs.pdx.edu/niagara/NEXMark/
3. Akidau, T., et al.: MillWheel: fault-tolerant stream processing at internet scale. Proc. VLDB Endow. **6**(11), 1033–1044 (2013)
4. Akidau, T., et al.: The dataflow model: a practical approach to balancing correctness, latency, and cost in massive-scale, unbounded, out-of-order data processing (2015)
5. Armbrust, M., et al.: Structured streaming: a declarative API for real-time applications in apache spark. In: International Conference on Management of Data, pp. 601–613 (2018)
6. Carbone, P., Fóra, G., Ewen, S., Haridi, S., Tzoumas, K.: Lightweight asynchronous snapshots for distributed dataflows. arXiv preprint arXiv:1506.08603 (2015)
7. Carbone, P., Katsifodimos, A., Ewen, S., Markl, V., Haridi, S., Tzoumas, K.: Apache Flink: stream and batch processing in a single engine. Bull. IEEE Comput. Soc. Tech. Committee Data Eng. **36**(4) (2015)
8. Castro Fernandez, R., Migliavacca, M., Kalyvianaki, E., Pietzuch, P.: Integrating scale out and fault tolerance in stream processing using operator state management. In: ACM SIGMOD International Conference on Management of Data, pp. 725–736 (2013)

9. Dai, J., Huang, J., Huang, S., Huang, B., Liu, Y.: HiTune: dataflow-based performance analysis for big data cloud. In: ATC, pp. 87–100 (2011)
10. Dean, J., Ghemawat, S.: MapReduce: simplified data processing on large clusters. Commun. ACM **51**(1), 107–113 (2008)
11. Floratou, A., Agrawal, A., Graham, B., Rao, S., Ramasamy, K.: Dhalion: self-regulating stream processing in heron. Proc. VLDB Endow. **10**(12), 1825–1836 (2017)
12. Garduno, E., Kavulya, S.P., Tan, J., Gandhi, R., Narasimhan, P.: Theia: visual signatures for problem diagnosis in large Hadoop clusters. In: LISA'12, pp. 33–42 (2012)
13. Hoffmann, M., et al.: SnailTrail: generalizing critical paths for online analysis of distributed dataflows. In: NSDI'18, pp. 95–110 (2018)
14. Hoffmann, M., Lattuada, A., McSherry, F.: Megaphone: latency-conscious state migration for distributed streaming dataflows. Proc. VLDB Endow. **12**(9), 1002–1015 (2019)
15. Kalavri, V., Liagouris, J., Hoffmann, M., Dimitrova, D., Forshaw, M., Roscoe, T.: Three steps is all you need: fast, accurate, automatic scaling decisions for distributed streaming dataflows. In: OSDI'18, pp. 783–798 (2018)
16. Kulkarni, S., et al.: Twitter Heron: stream processing at scale. In: the 2015 ACM SIGMOD International Conference on Management of Data, pp. 239–250 (2015)
17. Mai, L., et al.: Chi: a scalable and programmable control plane for distributed stream processing systems. Proc. VLDB Endow. **11**(10), 1303–1316 (2018)
18. Moritz, P., et al.: Ray: a distributed framework for emerging {AI} applications. In: OSDI'18, pp. 561–577 (2018)
19. Toshniwal, A., et al.: Storm@ twitter. In: ACM SIGMOD International Conference on Management of Data, pp. 147–156 (2014)
20. Wang, S., et al.: Lineage stash: fault tolerance off the critical path. In: SOSP'19, pp. 338–352 (2019)
21. White, T.: Hadoop: The Definitive Guide. O'Reilly Media, Inc., Newton (2012)
22. Zaharia, M., et al.: Resilient distributed datasets: a fault-tolerant abstraction for in-memory cluster computing. In: NSDI'12, pp. 15–28 (2012)
23. Zaharia, M., Das, T., Li, H., Hunter, T., Shenker, S., Stoica, I.: Discretized streams: fault-tolerant streaming computation at scale. In: SOSP'13, pp. 423–438 (2013)

BMTP: Combining Backward Matching with Tree-Based Pruning for Large-Scale Content-Based Pub/Sub Systems

Zhengyu Liao[1], Shiyou Qian[2(✉)], Zhonglong Zheng[1(✉)], Jian Cao[2], Guangtao Xue[2], and Minglu Li[1]

[1] Zhejiang Normal University, Zhejiang, China
{zhengyu,zhonglong,mlli}@zjnu.edu.cn
[2] Shanghai Jiao Tong University, Shanghai, China
{qshiyou,cao-jian,gt_xue}@sjtu.edu.cn

Abstract. Content-based publish/subscribe systems are widely deployed in many fields to realize selective data distribution. Event matching is a potential performance bottleneck in large-scale systems that maintain tens of millions of subscriptions, such as stock data updates. To deal with the large-scale subscription problem, in this paper, we propose a new matching algorithm called BMTP, which achieves high and robust matching performance. BMTP combines a Backward Matching method with a Tree-based Pruning method. The idea behind BMTP is that after filtering out most unmatching subscriptions, the efficiency of the backward matching method can be greatly improved. To evaluate the performance of BMTP, we conducted extensive experiments based on a real-world stock market dataset. The experiment results indicate that the matching performance of BMTP is improved by up to 96.6% than its counterparts.

Keywords: Event matching · Large-scale · Performance · Robustness

1 Introduction

The publish/subscribe (pub/sub) system is widely deployed in many fields, such as system monitoring and management [13,22], real-time stock update [2,3,6], online gaming [1,17], online advertising [15,16], business process execution [11,14] and social media messaging [4,5]. It facilitates the decoupling of the communication parties in time, space and synchronization [8]. In particular, the content-based pub/sub systems [10,23] can achieve fine-grained data distribution by allowing subscribers to define Boolean expressions to express their interest in events.

For large-scale content-based pub/sub systems, event matching is a key component. Researchers have proposed many matching algorithms to achieve efficient content-based data distribution. Although each matching algorithm has its unique design, we can roughly divide existing algorithms into three categories.

Y. Lai et al. (Eds.): ICA3PP 2021, LNCS 13157, pp. 152–166, 2022.
https://doi.org/10.1007/978-3-030-95391-1_10

First, some algorithms use counting-based methods to match, such as TAMA [25] and OpIndex [24]. Second, to efficiently filter unmatching subscriptions, several algorithms adopt tree-based pruning methods, such as H-Tree [20], PS-Tree [12] and MO-Tree [7]. Third, there are also some algorithms which use backward matching methods which target unmatching subscriptions to indirectly obtain the matching ones, such as REIN [19], Ada-REIN [21] and GEM [9].

Generally, each matching algorithm has its advantages and disadvantages. There are many parameters that affect the performance of the matching algorithm [18]. Through a detailed analysis of existing work, we divide the parameters that affect the matching algorithm into two categories: explicit parameters and implicit parameters. Explicit parameters directly affect the total workload, such as the number of subscriptions and the number of predicates contained in subscriptions. Almost all algorithms are affected by explicit parameters. On the other hand, implicit parameters do not change the total workload, but indirectly affect the efficiency or robustness of the matching algorithm, such as the distribution of subscriptions and the matching probability of subscriptions. The design of different matching algorithms makes them behave differently in the face of implicit parameters.

As the amount of data continues to increase, there are at least two main reasons why it is challenging to avoid the impact of explicit and implicit parameters on the matching algorithm to ensure the QoS of content-based pub/sub systems. First, the growth in the number of subscriptions or predicates means more workload needs to be processed. Therefore, it is difficult to avoid the additional overhead caused by explicit parameters. Second, during the matching process, changes in implicit parameters may cause the performance of the matching algorithm to fluctuate. Therefore, in addition to high performance, it also has robustness requirements for the design of matching algorithms.

To reduce the matching time under large-scale subscriptions and maintain robustness under different subscription matching probabilities, in this paper, we propose a new matching algorithm called BMTP. The basic idea of BMTP is to filter out most unmatching subscriptions through the tree-based pruning method, and then use the backward matching method to determine the matching subscriptions in the set of candidate subscriptions. The combination of the two methods can fully employ their respective advantages and achieve better performance than the single method in large-scale pub/sub systems.

We conducted extensive experiments to evaluate the performance of BMTP on a real-world stock market dataset. We compare our algorithm with four counterparts, namely REIN [19], TAMA [25], H-Tree [20] and MO-Tree [7]. The experiment results show that BMTP outperforms its counterparts by up to 96.6% in terms of matching time. In addition, the performance of BMTP is more robust by up to 89.3% in terms of the standard deviation of the matching time.

The rest of the paper is organized as follows. The matching semantics are introduced in Sect. 2. In Sect. 3, we review the related work. In Sect. 4, we describe the design of BMTP. Section 5 reports the extensive experiment results and Sect. 6 concludes the paper.

2 Definitions of Matching Semantics

In this section, we define predicates, subscriptions, events and event matching.

Predicate. In this paper, we focus on the interval predicates defined as $P = \{att, low, high\}$, where $P.att$, $P.low$ and $P.high$ represent the attribute, low value and high value of P respectively. $P.low$ is not greater than $P.high$. Other types of predicates defined by the relational operators $\{>, \geq, =, <, \leq\}$ can be converted into interval predicates, similar to TAMA [25] and REIN [19].

Subscription. A subscription S has a unique ID, denoted by $S.ID$, which comprises one or more interval predicates in conjunctive form: $S = \{P_1 \wedge P_2 \wedge ... \wedge P_K\}$. The subscription size is defined as the number of predicates K in S. Each attribute appears at most once in the subscription.

Event. $E = \{R_1, R_2, ..., R_M\}$ is an event containing M attribute-value pairs. The number of pairs in E is defined as the event size. The attribute-value pair is expressed as $R = \langle att, val \rangle$, where $R.att$ and $R.val$ represent the attribute and value respectively. Each attribute appears once in an event.

Matching Semantic. Given an interval predicate P and an attribute-value pair R, $P \sim R$ denotes that P matches R, when $\{P.att = R.att\} \wedge \{P(R.val) = true\}$. The subscription S is a match of event E, denoted by $S \sim E$, when $\forall P \in S \rightarrow \{\exists R \in E\} \wedge \{P \sim R\}$. Given the event E, the matching algorithm retrieves all matches in the subscription set.

3 Related Work

A large body of work focuses on researching the matching algorithm of content-based pub/sub systems. In this section, we first review the existing matching algorithms, namely counting-based methods, tree-based methods and backward methods. Then, we explain the difference between BMTP and the existing algorithms.

(1) Counting-based methods usually identify matching subscriptions by counting the number of satisfied predicates of each subscription. Typical counting-based matching algorithms include TAMA [25] and OpIndex [24]. For example, TAMA establishes a multi-level index for each attribute. Each index level divides the value domain of the attribute into multiple cells. An interval predicate maps to multiple cells at different levels. Given an event value, the matching process of TAMA involves counting the satisfied predicates in the cells that cover the event value at all levels. When the value of the counter is equal to the number of predicates contained in the corresponding subscription, the subscription is a

match. Although TAMA has a higher matching efficiency when there are fewer satisfied predicates, it has two disadvantages. First, to improve matching efficiency, TAMA adopts an approximate matching mode, so there are certain false positive subscriptions in the matching results. Second, when the number of subscriptions or the matching probability of subscriptions increases, the matching performance of TAMA will be greatly reduced.

(2) The tree-based method aims to filter out as much unmatching subscriptions as possible layer by layer. Compared to the counting-based method, the tree-based method first identifies fewer candidate subscriptions and then checks them one by one. Generally, to achieve excellent filtering performance, an expensive filtering process needs to be performed and the memory consumption is also high. H-Tree [20] and MO-Tree [7] are representative tree-based methods. H-Tree selects a set of attributes to construct a multi-layer index. For each layer, the value domain of the attribute is divided into overlapping cells. The cells of different layers cascade to form an H-Tree. Essentially, H-Tree divides the subscription space into a set of subspaces to reduce the search cost. When the matching probability of subscriptions is low, H-Tree has good performance because it only checks subscriptions in the candidate subspaces by pruning most of the unmatching subscriptions. However, to obtain an excellent filtering effect, H-Tree needs to construct more index layers, which makes the number of divided subspaces increase exponentially.

(3) The backward matching method uses reverse thinking. Different from the counting-based and tree-based methods, algorithms which adopt the backward method aim to indirectly find matching subscriptions by quickly marking all unmatching subscriptions. The representative algorithms are REIN [19] and GEM [9]. REIN indexes subscriptions based on the low and high values of each predicate in the subscription. The first index level is built on attributes. For each attribute, there are two sets of buckets corresponding to the low and high value of the predicate. Each predicate is inserted into two buckets based on its low value and high value. The matching process of REIN is to mark the corresponding subscriptions based on the unsatisfied predicates in the target buckets. Due to the low cost of marking operations, REIN has efficient performance. However, REIN is not suitable for scenarios where the matching probability of subscriptions is low.

The unique design of each algorithm makes it perform well in certain situations. However, due to the use of a single method, it is difficult for most existing matching algorithms to achieve excellent performance in the face of changes in explicit and implicit parameters. Different from most existing algorithms, BMTP combines the backward matching method and the tree-based pruning method, which has two advantages. First, under large-scale subscriptions, the number of candidate subscriptions after filtering may still be large. Therefore, the backward matching method is more efficient than the one-by-one naive inspection method adopted by most existing tree-based algorithms. Second, the tree-based pruning method is used to filter out a large number of unmatched subscriptions, which mitigates the impact of explicit parameters. In addition, the candidate

Table 1. Sample of subscriptions

S_0	$A_0 \in [0.1, 0.2] \wedge A_2 \in [0.4, 0.7]$
S_1	$A_1 \in [0.1, 0.7] \wedge A_2 \in [0.1, 0.2]$
S_2	$A_0 \in [0.1, 0.4] \wedge A_1 \in [0.2, 0.8] \wedge A_2 \in [0.2, 0.6]$
S_3	$A_0 \in [0.2, 0.3]$
S_4	$A_0 \in [0.3, 0.4] \wedge A_1 \in [0.3, 0, 9]$

subscriptions have a high matching probability, thereby reducing the impact of implicit parameters on the backward matching method. These advantages result in BMTP having an excellent and robust matching performance in large-scale content-based pub/sub systems.

4 Design of BMTP

4.1 Overview

As discussed in Sect. 3, explicit or implicit parameters have a great impact on the performance of the matching algorithm. Therefore, how to design a high-performance and robust matching algorithm has always been one of the pursuits of large-scale content-based pub/sub systems. However, it is difficult to propose a robust matching algorithm with excellent performance in all scenarios. Through the analysis of existing work, we found that the number of subscriptions and subscription matching probability are the two main parameters that affect the performance of the matching algorithm.

To alleviate the matching cost caused by the increase in the number of subscriptions, one solution is to effectively filter out a large number of unmatching subscriptions through a tree structure. Most tree-based methods pursue good filtering effects, thereby greatly reducing the size of the candidate subscription set. However, in the face of large-scale subscriptions, building a higher tree to achieve better filtering effects usually increases memory consumption exponentially. This means that the single tree-based method does not adapt well to systems that maintain large-scale subscriptions.

Another important factor that affects the performance of the matching algorithm is the matching probability of the subscription. The performance of most existing matching methods varies monotonically with changes in subscription matching probability. For example, the performance of REIN [19] and GEM [9] increases as the matching probability increases, while TAMA [25], H-Tree [20], and MO-Tree [7] show a downward trend. This makes their matching performance fluctuate greatly when the matching probability changes dynamically. However, to ensure the QoS of the real-time data distribution, such as a stock data update, it is essential to achieve a high and robust matching performance in content-based pub/sub systems.

In order to solve these problems, we propose a new matching algorithm called BMTP. The basic idea of BMTP is embodied in two aspects. First of all, unlike

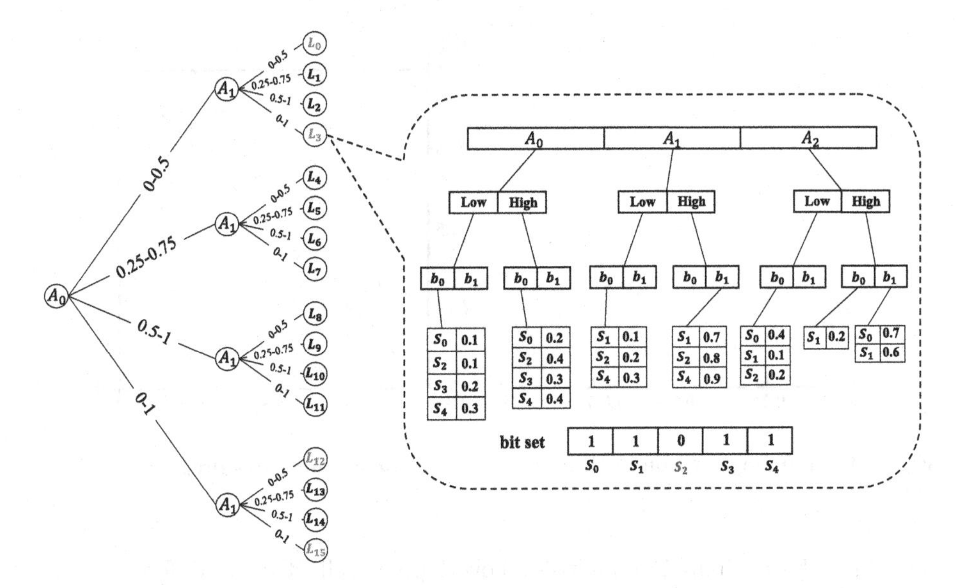

Fig. 1. The structure of BMTP for indexing subscriptions in Table 1 and the status when matching event $E = \{\langle A_0, 0.12\rangle, \langle A_1, 0.24\rangle, \langle A_2, 0.55\rangle\}$

most existing tree-based algorithms, BMTP does not pursue the best filtering effect, but filters out most unmatching subscriptions quickly and efficiently by constructing a tree with a limited height. Using the tree-based pruning method, BMTP can alleviate the impact of the number of subscriptions on the matching performance. Second, it is obvious that candidate subscriptions have a high matching probability after filtering. Therefore, BMTP adopts the backward matching method, focusing on quickly marking all unmatching subscriptions in the candidate set to indirectly determine the matches. BMTP combines the tree-based pruning method with the backward matching method to overcome the monotonic performance caused by a single method. Therefore, BMTP achieves efficient and robust event matching performance in large-scale and dynamic content-based pub/sub systems.

4.2 Structure of BMTP

BMTP is essentially a space index tree built on multiple attributes, as shown in Fig. 1. BMTP divides the total subscription space into multiple subspaces (leaf nodes). Each leaf node contains a subset of subscriptions. Considering large-scale subscriptions, a sub-matcher is constructed in each leaf node. According to the number of subscriptions in the leaf node, the sub-matcher can use the backward matching method or the naive inspection method.

The construction of the space index tree is similar to H-Tree [20] and MO-Tree [7]. We select multiple popular attributes to construct the tree, and each layer of the tree corresponds to an attribute. For each layer, we divide the value

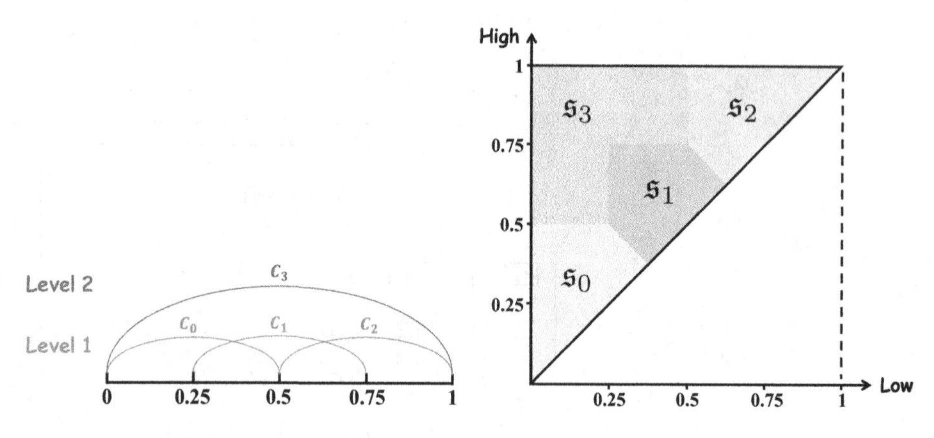

Fig. 2. Overlapping division in H-Tree [20]

Fig. 3. Subspaces division

domain of the attribute into multi-level overlapping cells. Cells at different levels have different widths. At each level, the size of the overlapping range is half the cell width. The cell width of the upper level is twice that of the lower level. For example, the cells in Fig. 2 are divided into two levels. C_0, C_1 and C_2 are the first (lowest) level cells with a width of 0.5, and C_3 is the second level cell with a width of 1. Note that for each layer of the tree, there is only one cell with a width 1 at the highest level.

Based on this partition scheme, by constructing the mapping rules from predicates to cells described in Sect. 4.3, a subscription is mapped to a cell at each layer. Therefore, the overlapping cells at each layer essentially divide the attribute space into mutually exclusive subspaces as shown in Fig. 3. For example, $\mathfrak{s}_0$ represents the subspace of C_0, and so on. By cascading the cells of each layer, a space index tree is formed. Figure 1 shows a two-layer space index tree built on A_0 and A_1. The attribute space at each layer is divided into four cells.

If the number of subscriptions in the leaf node is greater than the threshold, we build a REIN-based sub-matcher in the leaf node. REIN [19] adopts a backward matching method based on a data structure that indexes the predicates based on attributes and predicate values. REIN divides the value domain of the attribute into multiple disjoint cells (buckets) to realize mapping from predicate values to buckets. For example, as shown in Fig. 1, the value domain of A_0, A_1 and A_2 is divided into two buckets b_0 and b_1 for both low and high predicate values. So, given the interval predicate $A_1 \in [0.1, 0.7]$ in S_1 as listed in Table 1, its low value and high value are inserted to b_0 and b_1 respectively on attribute A_1. The subscription ID is inserted into the bucket together with each predicate value.

4.3 Insertion and Deletion

The subscription insertion process consists of two steps. First, we map the subscription layer by layer to a leaf node on the space index tree. Specifically, at each layer, by checking the cells from the lowest level to the highest level, the corresponding interval predicate in the subscription is mapped to the covering cell with the smallest width. The specific mapping rules are as follows:

(1) If the subscription does not contain a predicate defined on the attribute of a certain layer, the subscription is mapped to the highest-level cell. Otherwise, the mapping cell is determined by the width and center of the interval predicate.

(2) If the interval predicate width is less than half of the cell width of a certain level, according to the distance between the predicate center and the center of the cells, the predicate is mapped to the cell with the smallest distance.

(3) If the predicate width is greater than half of the cell width of a certain level but less than the cell width, the interval predicate is mapped to the cell that completely covers it.

Then, if the number of subscriptions maintained by the mapped leaf node is greater than the threshold, we insert the predicates of the subscription into the index structure of the corresponding sub-matcher, similar to REIN [19]. Specifically, for each interval predicate P contained in the subscription S, the low value and high value form two key-value pairs with $S.ID$, namely $< P.low, S.ID >$ and $< P.high, S.ID >$. Each pair is inserted into the corresponding bucket of REIN.

Table 1 lists a set of sample subscriptions. As shown in Fig. 1, for subscription S_2, we first determine the leaf node on the space index tree by A_0 and A_1. At the A_0 layer, by rule (3), S_2 first is mapped to subspace $\mathfrak{s}_0$ (corresponding to C_0). Then, also by rule (3), S_2 is mapped to subspace $\mathfrak{s}_3$ at the A_1 layer. So, S_2 is stored in leaf node L_3. After this, we insert the three predicates of S_2 into the buckets of the sub-matcher. The predicate $A_0 \in [0.3, 0.4]$ in S_4 provides an example of mapping rule (2). Although this predicate is also covered by C_1 whose domain is $[0.25, 0.75]$, the center of the predicate is closer to the center of C_0. Thus, S_4 is mapped to subspace $\mathfrak{s}_0$ of the A_0 layer. According to mapping rule (1), at the A_1 layer, S_0 is mapped to subspace $\mathfrak{s}_3$, because S_0 does not contain a predicate defined on A_1.

Deleting a subscription is similar to inserting. We first find the leaf node that the subscription maps to on the tree, and then delete each predicate in the subscription in the index structure of the sub-matcher.

4.4 Matching Algorithm

The matching process of BMTP consists of three steps, namely locating the leaf nodes that contain candidate subscriptions, marking the unmatching subscriptions in the bitset by running the sub-matchers, and finally obtaining the matching subscriptions represented by the unmarked bits.

It is easy to determine which leaf nodes contain candidate subscriptions. At each layer, cells containing the event value are candidate subspaces. For example, given an event $E = \{\langle A_0, 0.12\rangle, \langle A_1, 0.24\rangle, \langle A_2, 0.55\rangle\}$, the candidate subspaces at the A_0 layer and A_1 layer are $\mathfrak{s}_0$ and $\mathfrak{s}_3$. So, the target leaf nodes are L_0, L_3, L_{12} and L_{15} which are marked in blue, as shown in Fig. 1.

The number of subscriptions in the leaf nodes may be different. To achieve a better matching effect, we can use two matching methods to implement the sub-matcher of each leaf node. When the number of subscriptions in a leaf node is small, the one-by-one inspection is more efficient. Therefore, we set a threshold for the leaf nodes. If the number of subscriptions of a leaf node is larger than the threshold, the backward matching method is used. Otherwise, the candidate subscriptions can be checked one by one to determine the matches.

The backward matching method is like REIN [19]. Given an interval predicate P and an event value v, since $P.high$ must be greater than $P.low$, when $P.low$ is greater than v or $P.high$ is less than v, the predicate is not satisfied at v. Therefore, for each REIN-based sub-matcher, we traverse all the attributes to mark unmatching subscriptions. Specifically, on each attribute, we mark all subscriptions whose predicate low/high value is greater/smaller than v. After marking, all unmarked subscriptions are matches.

Algorithm 1 gives the pseudo code of BMTP. First, BMTP initializes a set $\mathbb{S}$ to store matching subscription and a two-dimensional array $SpaceIDs$ to record the mapping subspace IDs of each layer (lines 1–2). The size of the first dimension of $SpaceIDs$ equals the number of layers H of the space index tree. Then, at each layer, we insert the ID of the subspaces covering the event value into $SpaceIDs$ (lines 3–4). After obtaining the mapping subspaces of each layer, the target leaf nodes containing candidate subscriptions are determined. Finally, for each leaf node, we perform the sub-matcher to add the matching subscriptions to $\mathbb{S}$ (lines 5–6).

Algorithm 1. BMTP

Input: An event E
Output: The set of matching subscriptions $\mathbb{S}$
1: Initialize $\mathbb{S}$
2: Initialize $SpaceIDs$
3: **for** $i = 0$ to $H - 1$ **do**
4:     Add the ID of subspaces covering the event value at the $i - th$ layer to $SpaceIDs$[i];
5: **for** each target leaf node **do**
6:     Perform the sub-matcher and add the matching subscriptions to $\mathbb{S}$;

---

5 Experiments

In this section, we evaluate BMTP and compare it with REIN [19], TAMA [25], H-Tree [20] and MO-Tree [7] on a real-world stock dataset. The four baselines are reviewed in Sect. 3. The number of buckets in REIN is set to 500 and the

Table 2. The settings of the parameters in the experiments

Parameter	Description	Values
H	Height of BMTP	**3**, 4
η	Number of cells at each layer of BMTP	**4**, 11
ζ	Number of buckets in REIN sub-matchers	10, **50**, 100, 150, 200
θ	Threshold for different matching method	0, **5K**, 9K, 15K, 20K
N	Number of subscriptions	3M, 5M, **10M**, 15M, 20M
K	Subscription size	5, **10**, 15, 20, 25
M	Event size	10, 20, 30, **40**, 50
α	Parameter of Zipf	**0**, 0.5, 1, 1.5, 2
τ	Predicate matching probability	0.1, 0.2, 0.3, 0.4, 0.5, 0.6, 0.7

discretization level of TAMA is set to 13, as suggested in their papers. For H-Tree, the number of indexing attributes and the number of divided cells for each attribute are set to 8. The height of MO-Tree is 6 and the number of levels at one layer is set to 4. All algorithms are implemented in C++ and complied by GCC 7.5.0 with O3 optimization level on the Ubuntu 18.04.5 system. All experiments were run on a 6-core Intel 3.6 GHz physical machine with 64 GB RAM.

5.1 Setup

Table 2 lists the parameters and their settings in the experiments. The bold values represent the default settings. We first used a synthetic 5M-subscription workload to tune the parameters of BMTP. Then, we compared BMTP with the four baselines on the real-world stock dataset which was collected from the Chinese stock market on June 8, 2018. Each stock record contains 56 fields, such as stock id, transaction price, trading volume, etc. Events and subscriptions are randomly generated based on the stock records and the dimensions are set to be the same as the event size. To evaluate the impact of explicit parameters, we vary the number of subscriptions from 3M to 20M and the subscription size from 5 to 25. Implicit parameters include event size M, attribute skewness α and predicate matching probability τ. We use Zipf to simulate the skewness of attribute popularity. In each experiment, we match 1000 events and calculate their average matching time.

5.2 Results

Parameter Tuning. BMTP has four important parameters: the number of layers H in the tree, the number of cells η at each layer, the number of buckets ζ in REIN-based sub-matchers and the threshold θ for different matching methods in leaf nodes. We tune these four parameters with different settings on a synthetic 5M subscription workload.

Figure 4 shows the insertion time and deletion time of BMTP with different numbers of buckets ζ. The insertion time of the four combinations of H and η

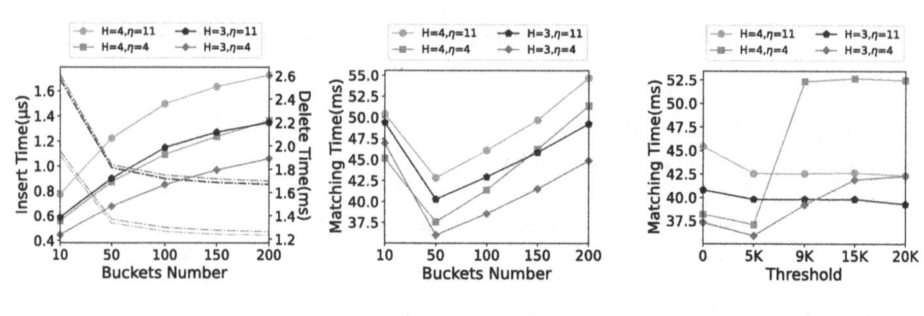

Fig. 4. Tuning ζ

Fig. 5. Tuning ζ

Fig. 6. Tuning θ

increases with an increase of ζ, while the deletion time is the opposite. Increasing H and η also increases the insertion time of BMTP. The deletion time of BMTP only decreases as the height of the index tree increases.

Figure 5 shows the matching time of BMTP with different settings of ζ. The matching time of the four combinations of H and η first decreases and then increases. This is because increasing ζ reduces the number of predicates that need to be checked in the target buckets, thus resulting in a downward trend. However, when ζ exceeds a certain value, the cost of traversing more buckets overwhelms this benefit, thereby increasing the total matching time.

Figure 6 shows the matching time of BMTP with different settings of θ. The results show that the matching time of BMTP decreases first and then increases. This is because when the threshold is low, using the backward matching method in sub-matchers with fewer subscriptions is not better than checking subscriptions one by one directly. On the contrary, when the threshold is high, the leaf nodes contain more subscriptions, which is efficient in adopting the backward matching method.

Through the analysis of the experiment results and considering the performance of insertion, deletion and matching time with different parameter settings, we set $H = 3$, $\eta = 4$, $\zeta = 20$ and $\theta = 5k$ as the default setting of BMTP.

Matching Time. Matching time is an important metric to evaluate the performance of the matching algorithm. We measure the matching time of the five algorithms with different settings of explicit and implicit parameters.

The Effect of N. Figure 7 shows the matching time with different numbers of subscriptions N. Intuitively, the matching time of all algorithms increases with more subscriptions. However, the growth rate of BMTP is slower than the four baselines. For BMTP, the matching time of $N = 20$M is 7.5 times that of $N = 3$M, which reduces the growth rate by 52.5% compared to REIN. When $N = 20$M, the matching time of BMTP is improved by 58.5%, 63.8%, 69.7% and 93.8% compared with REIN, TAMA, H-Tree and MO-Tree, respectively. This verifies that by combining the pruning strategy and the backward matching method, BMTP can alleviate the impact of N on matching performance, achieving excellent performance under large-scale subscriptions.

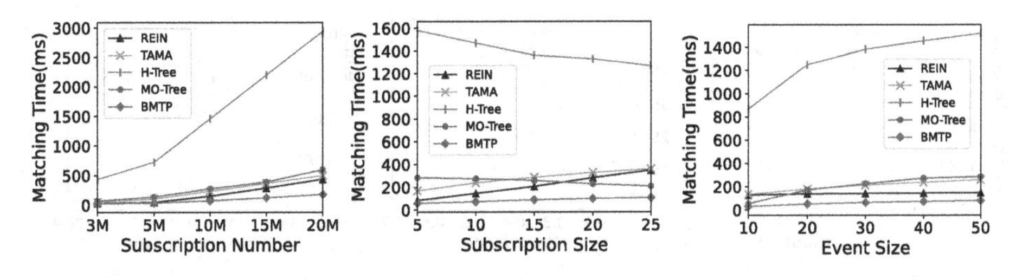

Fig. 7. Different N

Fig. 8. Different K

Fig. 9. Different M

The Effect of K. Subscription size K is also an explicit parameter. Figure 8 shows the matching time of the five algorithms with different settings of K. Compared with REIN, TAMA, H-Tree and MO-Tree, BMTP reduces the matching time by up to 68.8%, 69.9%, 72.5% and 96.3%, respectively. For H-Tree and MO-Tree that store a subscription as a whole in the leaf node, increasing K decreases the matching probability of subscriptions, which means less candidate subscriptions need to be checked. This tree-based pruning strategy mitigates the impact of K to some degree. Therefore, the matching time of H-Tree and MO-Tree decreases with the increase of K. On the other hand, BMTP, REIN and TAMA show an increasing trend when K increases. The reason is that increasing K causes the three algorithms to evaluate more predicates, which costs more time. However, BMTP has the slowest growth rate in matching time because BMTP uses the tree-based pruning strategy to filter a large number of unmatching subscriptions.

The Effect of M. Figure 9 shows the matching time with different settings of event size M. The matching time of all the tested algorithms is on the rise. For REIN and TAMA, given N and K, increasing M means traversing more attributes, which increases the total matching time. For H-Tree, MO-Tree and BMTP, when M is large and the attributes are uniformly distributed, the filtering effect of several indexing attributes decreases, which results in a large number of candidate subscriptions. However, BMTP uses the backward matching method in the sub-matchers, which alleviates the impact of M to a certain degree. The matching time of BMTP is reduced by up to 76.3%, 77.3%, 73.2% and 96.5%, compared with REIN, TAMA, H-Tree and MO-Tree, respectively.

The Effect of α. Figure 10 shows the matching time with a different distribution of attributes. Suppose that attributes follow the Zipf distribution with parameter α. The larger the α, the more serious is the skewness of the data. Skewness results in some popular attributes. The distribution of attributes has different effects on the five algorithms. First, REIN and TAMA are not sensitive to the distribution of attributes. Given N and K, the skewness of attributes has little effect on the matching probability of predicates. Therefore, their matching time remains unchanged. Second, the matching time of BMTP, H-Tree and MO-Tree exhibits a downward trend. This is because when there are some popular attributes, it is beneficial for these algorithms to choose the indexing attributes, which improves

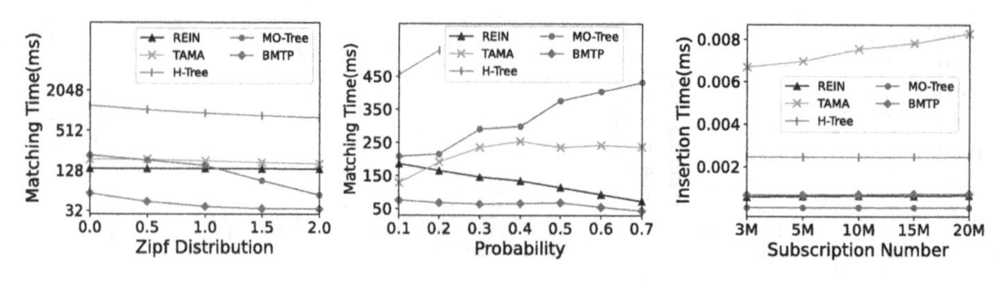

Fig. 10. Different α **Fig. 11.** Different τ **Fig. 12.** Different N

the filtering effect. BMTP outperforms REIN, TAMA, H-Tree and MO-Tree by up to 74.9%, 82.4%, 76.5% and 96.6%, respectively.

The Effect of τ. Figure 11 shows the matching time with different settings of predicate matching probability τ. The matching performance under different matching probabilities reflects the robustness of the matching algorithm in a dynamic environment. H-Tree is not suitable for situations with high matching probability, so we truncated the case where $\tau > 0.2$. The performance of the algorithms using a single matching method changes monotonically. On the contrary, BMTP overcomes this problem by combining the backward matching method and the tree-based pruning strategy, showing good robustness. The standard deviation of the matching time of BMTP from $\tau = 0.1$ to $\tau = 0.7$ is reduced by up to 76.2%, 78.4% and 89.3%, compared with REIN, TAMA and MO-Tree, respectively.

Insertion Time Figure 12 shows the average time of inserting a subscription with different settings of N. First, MO-Tree is faster than H-Tree because the height of MO-tree is lower than that of H-Tree. Second, the insertion time of BMTP is similar to REIN but larger than MO-Tree. This is because BMTP and REIN need to insert each predicate of the subscription, but MO-Tree just inserts the subscription as a whole into the leaf node. Third, TAMA has the longest insertion time because it inserts each predicate into multiple cells.

6 Conclusion

By reviewing the existing work, we found that the efficiency and robustness of the matching algorithms adopting a single matching method can be affected by explicit and implicit parameters, such as the number of subscriptions and the matching probability of subscriptions. In this paper, we propose a new matching algorithm called BMTP which combines the backward matching method and the tree-based pruning method. When facing a large number of subscriptions in a dynamic environment, BMTP can alleviate the impacts of explicit and implicit parameters, overcoming the weakness of single matching methods. We conducted comprehensive experiments to evaluate the performance of our proposed algorithm. The experiment results verify that BMTP achieves a high and robust matching performance in large-scale content-based pub/sub systems.

Acknowledgments. This work was supported by the National Key Research and Development Program of China (2019YFB1704400), the National Natural Science Foundation of China (61772334, 61702151), and the Special Fund for Scientific Instruments of the National Natural Science Foundation of China (61827810).

References

1. Cañas, C., Zhang, K., Kemme, B., Kienzle, J., Jacobsen, H.A.: Publish/subscribe network designs for multiplayer games. In: Proceedings of the 15th International Middleware Conference, pp. 241–252. ACM (2014). https://doi.org/10.1145/2663165.2663337
2. Chandramouli, B., Yang, J.: End-to-end support for joins in large-scale publish/subscribe systems. Proc. VLDB Endow. **1**(1), 434–450 (2008). https://doi.org/10.14778/1453856.1453905
3. Chen, C., Vitenberg, R., Jacobsen, H.A.: Building fault-tolerant overlays with low node degrees for topic-based publish/subscribe. IEEE Trans. Dependable Secure Comput. (Early Access), 1–17 (2021). https://doi.org/10.1109/TDSC.2021.3080281
4. Chen, L., et al.: Top-k term publish/subscribe for geo-textual data streams. VLDB J. **29**(5), 1101–1128 (2020). https://doi.org/10.1007/s00778-020-00607-8
5. Demers, A., Gehrke, J., Hong, M., Riedewald, M., White, W.: Towards expressive publish/subscribe systems. In: International Conference on Extending Database Technology, pp. 627–644. Springer (2006) https://doi.org/10.1007/11687238_38
6. Ding, T., Qian, S., Cao, J., Xue, G., Li, M.: Scsl: optimizing matching algorithms to improve real-time for content-based pub/sub systems. In: IEEE International Parallel and Distributed Processing Symposium (IPDPS), pp. 148–157 (2020)
7. Ding, T., et al.: Mo-tree: an efficient forwarding engine for spatiotemporal-aware pub/sub systems. IEEE Trans. Parallel Distrib. Syst. **32**(4), 855–866 (2021). https://doi.org/10.1109/TPDS.2020.3036014
8. Eugster, P.T., Felber, P.A., Guerraoui, R., Kermarrec, A.M.: The many faces of publish/subscribe. ACM Comput. Surv. **35**(2), 114–131 (2003). https://doi.org/10.1145/857076.857078
9. Fan, W., Liu, Y., Tang, B.: Gem: an analytic geometrical approach to fast event matching for multi-dimensional content-based publish/subscribe services. In: IEEE International Conference on Computer Communications (INFOCOM), pp. 1–9 (2016). https://doi.org/10.1109/INFOCOM.2016.7524338
10. Gupta, A., Sahin, O.D., Agrawal, D., El Abbadi, A.: Meghdoot: content-based publish/subscribe over p2p networks. In: ACM/IFIP/USENIX International Conference on Distributed Systems Platforms and Open Distributed Processing, pp. 254–273. Springer, Berlin Heidelberg (2004) https://doi.org/10.1007/978-3-540-30229-2_14
11. Jergler, M., Zhang, K., Jacobsen, H.A.: Multi-client transactions in distributed publish/subscribe systems. In: 2018 IEEE 38th International Conference on Distributed Computing Systems (ICDCS), pp. 120–131 (2018). https://doi.org/10.1109/ICDCS.2018.00022
12. Ji, S., Jacobsen, H.A.: Ps-tree-based efficient boolean expression matching for high-dimensional and dense workloads. Proc. VLDB Endow. **12**(3), 251–264 (2018). https://doi.org/10.14778/3291264.3291270

13. Ji, S., Jacobsen, H.A.: A-tree: a dynamic data structure for efficiently indexing arbitrary boolean expressions. In: Proceedings of the 2021 International Conference on Management of Data. SIGMOD/PODS '21, pp. 817–829 (2021). https://doi.org/10.1145/3448016.3457266
14. Li, G., Muthusamy, V., Jacobsen, H.A.: A distributed service-oriented architecture for business process execution. ACM Trans. Web **4**(1), 1–33 (2010). https://doi.org/10.1145/1658373.1658375
15. Machanavajjhala, A., Vee, E., Garofalakis, M., Shanmugasundaram, J.: Scalable ranked publish/subscribe. Proc. VLDB Endow. **1**(1), 451–462 (2008). https://doi.org/10.14778/1453856.1453906
16. Mahmood, A.R., et al.: Adaptive processing of spatial-keyword data over a distributed streaming cluster. In: Proceedings of the 26th ACM SIGSPATIAL International Conference on Advances in Geographic Information Systems, pp. 219–228 (2018). https://doi.org/10.1145/3274895.3274932
17. Moll, P., Isak, S., Hellwagner, H., Burke, J.: A quadtree-based synchronization protocol for inter-server game state synchronization. Comput. Netw. **185**, 107723 (2021). https://doi.org/10.1016/j.comnet.2020.107723
18. Qian, S., et al.: A fast and anti-matchability matching algorithm for content-based publish/subscribe systems. Comput. Netw. **149**, 213–225 (2019). https://doi.org/10.1016/j.comnet.2018.12.001
19. Qian, S., Cao, J., Zhu, Y., Li, M.: Rein: a fast event matching approach for content-based publish/subscribe systems. In: IEEE Conference on Computer Communications (INFOCOM), pp. 2058–2066 (2014). https://doi.org/10.1109/INFOCOM.2014.6848147
20. Qian, S., Cao, J., Zhu, Y., Li, M., Wang, J.: H-tree: an efficient index structurefor event matching in content-basedpublish/subscribe systems. IEEE Trans. Parallel Distrib. Syst. **26**(6), 1622–1632 (2015). https://doi.org/10.1109/TPDS.2014.2323262
21. Qian, S., Mao, W., Cao, J., Mouël, F.L., Li, M.: Adjusting matching algorithm to adapt to workload fluctuations in content-based publish/subscribe systems. In: IEEE Conference on Computer Communications (INFOCOM), pp. 1936–1944 (2019). https://doi.org/10.1109/INFOCOM.2019.8737647
22. Sadoghi, M., Jergler, M., Jacobsen, H.A., Hull, R., Vaculín, R.: Safe distribution and parallel execution of data-centric workflows over the publish/subscribe abstraction. IEEE Trans. Knowl. Data Eng. **27**(10), 2824–2838 (2015). https://doi.org/10.1109/TKDE.2015.2421331
23. Shen, H.: Content-based publish/subscribe systems. In: Handbook of Peer-to-peer Networking, pp. 1333–1366. Springer (2010) https://doi.org/10.1007/978-3-540-24629-9_11
24. Zhang, D., Chan, C.Y., Tan, K.L.: An efficient publish/subscribe index for e-commerce databases. Proc. VLDB Endow. **7**(8), 613–624 (2014). https://doi.org/10.14778/2732296.2732298
25. Zhao, Y., Wu, J.: Towards approximate event processing in a large-scale content-based network. In: 2011 31st International Conference on Distributed Computing Systems, pp. 790–799 (2011). https://doi.org/10.1109/ICDCS.2011.67

AF-TCP: Traffic Congestion Prediction at Arbitrary Road Segment and Flexible Future Time

Xuefeng Xie[1], Jie Zhao[2], Chao Chen[2(✉)], and Lin Wang[3]

[1] State Key Laboratory of Mechanical Transmission, Chongqing University, Chongqing, China
[2] College of Computer Science, Chongqing University, Chongqing, China
cschaochen@cqu.edu.cn
[3] School of Management Science and Real Estate, Chongqing University, Chongqing, China

Abstract. Traffic congestion prediction is a fundamental yet challenging problem in Intelligent Transportation Systems (ITS). Due to the large scale of the road network, the high nonlinearity and complexity of traffic data, few methods are well-suited for citywide traffic congestion prediction. In this paper, we propose a novel deep model named AF-TCP for Traffic Congestion Prediction at Arbitrary road segment and Flexible future time. For the model to achieve the prediction at flexible future time, we construct all time representations in a unified vector space and further improve the model's perception ability to different horizons. On the other hand, to realize the congestion prediction of arbitrary road segment within the city, we utilize road attributes and local neighbor structure to build the road segment representation, and design a deep model to fuse it with the corresponding historical traffic data. Extensive experiments on the real-world dataset demonstrate that our model exhibits stable performance at different prediction horizons and outperforms the baselines.

Keywords: Traffic congestion prediction · Time representation · Road segment representation · Intelligent transportation systems

1 Introduction

Traffic congestion is a primary issue for urban transportation systems due to its significant impact on the human/goods mobility in cities. The prediction of traffic congestion is essential to the studies of urban mobility efficiency. For example, the accurate prediction of forthcoming congestion can greatly facilitate congestion propagation analysis [1], route planning [2,3], and travel time estimation [4]. However, traffic congestion is complicatedly affected by the mobility of road users and the topology of the road network. On the one hand, the mobility of road users presents strong spatiotemporal evolving nature [5]. On the other hand, traffic flows at neighboring roads are highly interrelated yet heterogeneous according to the roads' attributes. Hence, the basic challenge of traffic congestion prediction lies in capturing the spatiotemporal dependencies of evolving human mobility in road networks.

In general, the temporal dependencies of human mobility are twofold, namely the periodic and recent evolving patterns, while the spatial dependence mainly refers to

Y. Lai et al. (Eds.): ICA3PP 2021, LNCS 13157, pp. 167–181, 2022.
https://doi.org/10.1007/978-3-030-95391-1_11

the correlations among neighboring roads or areas. To overcome the mentioned challenge, traditional approaches for traffic prediction usually conduct time series analysis and employ different machine learning techniques such as Autoregressive Integrated Moving Average (ARIMA) [6], Support Vector Machine (SVM) [7] and K-Nearest Neighbors (KNN) [8]. However, these methods fail to capture the underlying spatial and temporal dependencies completely and simultaneously. In recent years, deep learning techniques have attracted extensive attention from researchers due to their superior performance in modelling the high nonlinearity and complexity of traffic data. For instance, Recurrent Neural Networks (RNN) have been widely adopted to model the temporal dependence from the sequential traffic data [6,9,10]. Convolutional Neural Networks (CNN) have been used to model the spatial dependency after converting the urban space into a grid image [11,12]. However, by using these techniques, it is still difficult to achieve fine-grained predictions and make use of variations of roads within each grid. The most recent studies show that graph-based techniques such as graph convolutional networks (GCN) and graph attention networks (GAT) have achieved state-of-the-art performance in some traffic prediction applications [13–16]. It is because such techniques are capable of modeling data in the non-Euclidean space (e.g., road networks). Although existing studies have showed promising results, there are still two significant issues that need to be further explored.

The dilemma of prediction horizon: short-term or long-term prediction? Existing studies often adopt a horizon in their short/long-term prediction tasks, but there is no consensus on the boundary between the short-term and long-term. Moreover, the horizon in these studies has to be fixed in the model setting, in this case, one can only predict the future traffic by a time window with fixed size. However, in real-world scenarios, the needs of traffic predictions are often *flexible* and *on-demand*. For example, predictions with a short-term horizon enable drivers to avoid congested roads immediately, while predictions with a long-term horizon are informative for travelers to make efficient route planning. Hence, how to choose the horizon becomes a common controversial issue. To break the bottleneck, in this paper, we aim to provide an inclusive approach with *flexible* horizon to satisfy both the short-term and long-term prediction requirements.

The inefficiency of fine-grained traffic prediction on the city scale. Generally, it is difficult to make traffic predictions for the entire city. On one hand, the scale and complexity of the road network in a city set up a great obstacle for many methods. For example, existing graph-based methods are not able to deal with the entire road network effectively. On the other hand, very few data sources can grantee the adequate spatial and temporal coverage of observations on all road segments in the city [17]. Unfortunately, that is the foundation for most data-driven methods to model the underlying spatiotemporal correlations. Hence, we aim to establish a unified model to enable congestion prediction for all road segments in a *one-for-all* fashion.

To sum up, we propose a novel deep model named AF-TCP for the traffic congestion prediction at arbitrary road segments and flexible future time. Specifically, we design a time aware module that maps the temporal information of traffic data into a unified vector space. In the space, the relative distance between two vectors indicates their time difference. Besides, we construct the road segment representation using road segment attributes and the local neighbor structure, then encode it with the corresponding traffic data to support the prediction at arbitrary road segment. The main contributions of this paper are summarized as follows:

- We invent a novel deep model for the traffic congestion prediction at a flexible time in the future. The model can automatically capture multi-resolution time dependencies of traffic observations with the flexible prediction horizon.
- We construct a discrete representation of the road network using the road network's inherent properties, namely road attributes and road structure. Then we combine this fine-grained spatial information with historical traffic observations to achieve the congestion prediction at arbitrary road segment in the city.
- We evaluate our model on a real-world traffic dataset collected in Xi'an, China. Results demonstrate that our model is effective and robust to the prediction of traffic congestion at different horizons, and outperforms all benchmarks in terms of accuracy and F1-score.

The rest of this paper is organized as follows. In Sect. 2, some definitions and a formal statement of the study are provided. The proposed model is demonstrated in Sect. 3. Experiments and results are presented in Sect. 4. In Sect. 5, we briefly introduce the related studies on traffic prediction. Finally, in Sect. 6, we give conclusions and directions for future work.

2 Preliminary

2.1 Basic Concepts

Definition 1 (Road segment). *A road can be divided into several road segments through intersections or split points. Each road segment* $r_i (i = 1, 2, \dots)$ *is usually associated with some attributes (e.g., length, direction, speed limit, the number of lanes, etc.).*

Definition 2 (Time Slice). *Time slices are the result of discretizing the time of a day. For example, given a 2-minute interval, one day can be divided into 720 time slices (i.e.,* $t_0 \sim t_{719}$*).*

Definition 3 (Traffic Observations). *For each segment, the traffic observations are derived from the raw sensor data, including traffic flow, traffic speed, occupancy, travel time, etc. These observations are multivariate time series in essence.*

Definition 4 (Congestion Level). *Congestion level* $c_{i,t} \in \{1, 2, 3\}$ *is an integer related to the road segment* r_i *in time slice* t*, and a higher value indicates more congested traffic.*

2.2 Problem Statement

For a specific road segment r_i, given the current time slice t_{cur} and the predicted time slice $t_{pred} = t_{cur} + \Delta t$ of the day, where the time interval Δt is a flexible prediction horizon. We select the historical traffic data $P^k(k = 4, 3, 2, 1)$ and each P^k represents the traffic observations of period $[t_{pred}, t_{pred+4}]$ before k weeks. Traffic observations for each time slice here contain four attributes: *road speed*, *eta speed* (i.e., estimation of speed), *congestion level* and *number of vehicles*. Similarly, we can obtain the traffic observations X for recent period $[t_{cur-4}, t_{cur}]$. Then, the road segment attributes S_i is also joined, to predict anticipated traffic congestion level at time slice t_{pred}. At last, the problem can be formalized as:

$$c_{i,t+\Delta t} = f\left(P^4, P^3, P^2, P^1; X; S_i\right), \tag{1}$$

where the function f corresponds to our designed predictor based on deep model.

3 Methodology

Here, we elaborate on the framework of our model for the traffic congestion prediction. Figure 1 gives an overview of the model, consisting of three components, i.e., the embedding layer, the encoder-decoder and the time aware module.

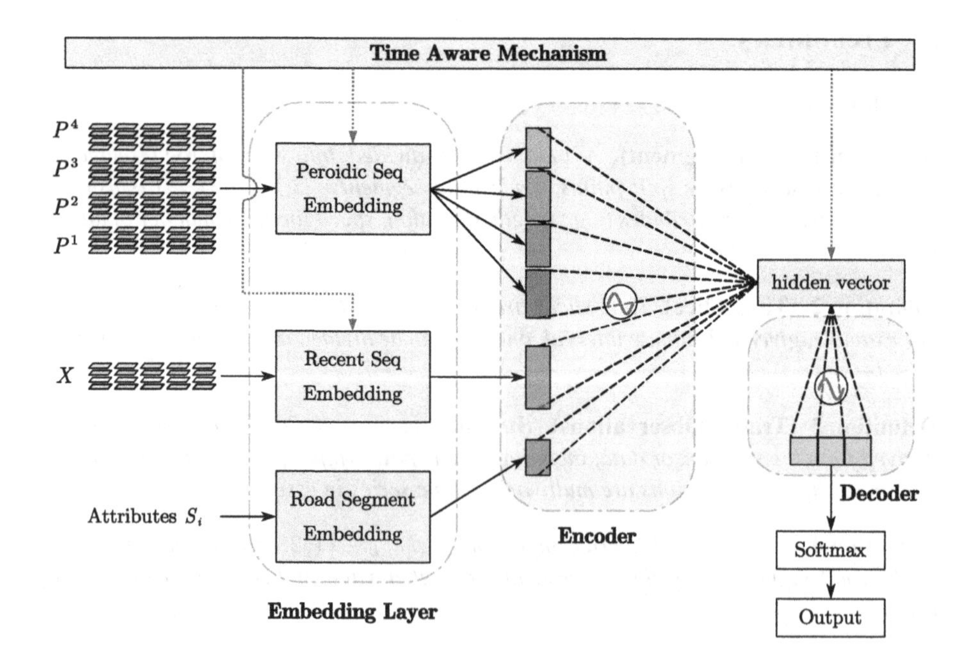

Fig. 1. The framework of AF-TCP.

3.1 Overview

The embedding layer is employed to extract initial features for the followed encoder-decoder, including periodic and recent features of road traffic, and segment features. Its inputs include periodic observation sequences P^k, recent observation sequence X, and road segment attributes S_i.

For all the obtained features, a simple and straightforward handling is to concatenate them together. However, such manner usually cannot guarantee the effectiveness of feature fusion. To circumvent this, we design an encoder to perform the deep fusion of all available features. Specifically, the encoder maps all features to a low-dimensional vector space through non-linear transformations and outputs a hidden vector. Then we construct a decoder to extract information related to traffic congestion from the hidden vector. Finally, the output of the decoder is further fed into a *Softmax* function to get the final prediction result (i.e., a probability distribution).

In addition, we also propose a time-ware module as an auxiliary service that provides temporal context to the other two components, as shown in the blue dotted line in Fig. 1.

3.2 Embedding Layer

Sequence Embedding for Modelling Temporal Dependencies. General speaking, traffic congestion always displays local coherence and periodicity. We embed the periodic and recent sequence to capture these time dependencies. $P^k(k = 4 \sim 1)$ are four periodic sequences mentioned before to be handle here. In order to learn the similar patterns of traffic observations between weeks, we set the four periodic sequences to share the same embedding network. Next, we will demonstrate the details of embedding for any P^k, regardless of the superscripts.

As shown in Fig. 2, the input $P^k \in \mathbb{R}^{5\times4}$ is a length-5 sequence with 4 channels. We first feed P^k into a compression block consisting of a 1-dimensional convolution (Conv1d), a batch normalization (BN) and a nonlinear activation function (ReLU). This block can shorten the length of the input sequence by 1. At last, we compress the features within five time slices into a vector by stacking four compression blocks as Eq. 2:

$$P^k_{(l+1)} = ReLU(BN(W_l \star P^k_{(l)} + b_l)), l = 0, \ldots, 3, \tag{2}$$

where $\star$ refers to the convolution operation, and W_l and b_l are trainable parameters in l-th layer. $P^k_{(l)}$ is equal to P^k when $l = 0$.

Then, we obtain the final embedding through a fully-connected network (FCN) with one hidden-layer:

$$\mathbf{Z}_k = ReLU(W_{f2} \cdot ReLU(W_{f1} \cdot P^k_{(4)} + b_{f1}) + b_{f2}), \tag{3}$$

where $W_{f\cdot}$ and $b_{f\cdot}$ are trainable parameters of the fully-connected network. $\mathbf{Z}_k \in \mathbb{R}^{C_k}$ indicates the embedding dimension of the periodic sequence P^k is C_k.

Note that we establish a new embedding network to map the recent sequence X into a vectorized representation with the same structure as the periodic sequence embedding. The output is denoted as $\mathbf{Z}_x \in \mathbb{R}^{C_x}$.

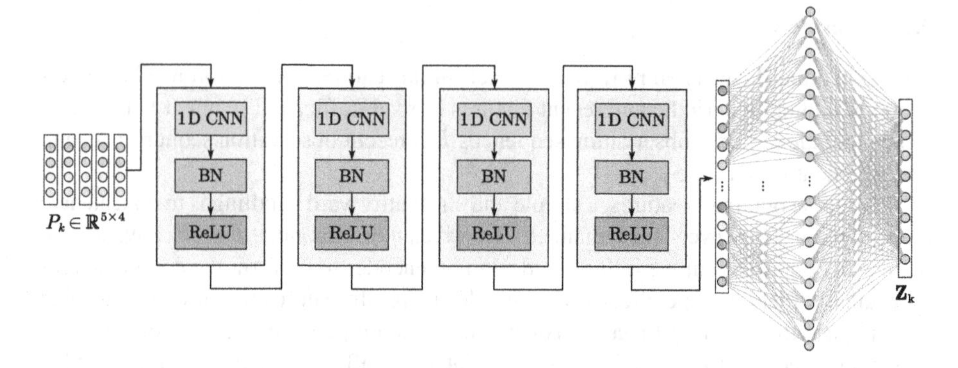

Fig. 2. The architecture of periodic sequence embedding.

Road Segment Embedding for Extracting Fine-Grained Spatial Feature. It is well known that traffic flows have inherent relationship with road networks. Specifically, the structure and attributes of road networks have potential impact on the change of traffic conditions [18]. Hence, we introduce how to utilize road networks' properties for the traffic congestion prediction in this part.

For a road segment r_i, we extract its nine attributes S_i (as listed in Table 1) to construct a representative vector. Then, we directly employ the vector to represent the road segment. In terms of the network structure, it is time-consuming and unnecessary to capture the global topology of the entire road network due to the spatial locality of traffic congestion. Thus, we focus on modelling the local neighborhood structure, that is, to extract the neighborhood number of each segment and append it to the representative vector. Following the common data preprocessing pipline, we normalize the contiguous data in the representation vector such as the length and width, and handcraft the discrete data such as the direction, path class, number of neighbors into one-hot expression. After the value transformation, each segment is represented by a 30-dim vector called segment embedding $\mathbf{Z}_r$. The detailed meaning of each dimension in the embedding vector can be founded in Table 1.

Table 1. The meaning of $\mathbf{Z}_r$ in each dimension.

Dim	Meaning	Dim	Meaning
1	Length	21	Speed_limit
2–4	Direction	22–26	Level
5–9	Path_class	27	width
10–17	Speed_class	28–30	neighbor
18–20	Lane_num		

3.3 Time Aware Module

Traffic conditions are time-dependent, and the time aware module is proposed to precisely assist the deep model in perceiving different temporal information. To that end, the key idea of this module is to obtain embeddings of time slices from a uniform vector space then integrate them into the model. Traditionally, there are two time encoding strategies: relative position encoding and absolute position encoding. The former strategy will assign a unique vector to each time slice, so that the generated embedding could be high-dimensional and sparse. We thus adopt the second strategy, namely the position encoding proposed in Transformer [19]. The detailed formula can be seen in Eq. 4:

$$\begin{aligned} PE(pos)_{2i} &= sin(pos/10000^{2i/d}) \\ PE(pos)_{2i+1} &= cos(pos/10000^{2i/d}) \end{aligned} \tag{4}$$

where "pos" represents the relative position of a time slice on a timeline, "d" represents the pre-defined dimension of the embedding, "$2i$" and "$2i + 1$" represent even and odd indexes in that embedding respectively. Note that in such a representation, $PE(pos+k)$ is a linear transformation of $PE(pos)$ given any fixed offset k [19], which can offer clues to capture temporal dependence in traffic.

In order to incorporate temporal information into the sequence embedding in advance, we concatenate the position embedding of each time slice with the corresponding traffic observations:

$$\begin{aligned} P^k_{t,:} &= \text{concat}(P^k_{t,:}, PE(t)), t \in [t_{pred}, t_{pred+4}] \\ X_{t,:} &= \text{concat}(X_{t,:}, PE(t)), t \in [t_{cur-4}, t_{cur}] \end{aligned} \tag{5}$$

The updated P^k and X are the standard inputs for the periodic/recent sequence embedding, respectively.

To sum up, the benefits of this time aware module are twofold: 1) It provides the model with the temporal context in the traffic data as auxiliary information throughout the process. 2) it provides a different perspective for the model to understand the prediction horizon, that is, the distance between t_{cur} and t_{pred} in the vector space. Following, we will describe how to utilize temporal information in the encoder-decoder component.

3.4 Encoder and Decoder

Based on the features automatically extracted by the embedding layer, we further follow an end-to-end design to conduct a deep fusion of these features in the latent space. Specifically, our model adopts the decoder-encoder architecture that is widely used in sequence-to-sequence models and Generative Adversarial Networks (GANs).

The encoder-decoder component is described as follows:

$$\begin{aligned} \mathbf{Z}'_x &= \frac{1}{\mathbf{dist}(PE(t_{pred}), PE(t_{cur}))}\mathbf{Z}_x \\ \mathbf{h} &= \mathbf{encoder}(\mathbf{Z}_4 \oplus \mathbf{Z}_3 \oplus \mathbf{Z}_2 \oplus \mathbf{Z}_1 \oplus \mathbf{Z}'_x \oplus \mathbf{Z}_r\}) \\ \mathbf{y} &= \mathbf{decoder}(h \oplus PE(t_{pred})) \end{aligned} \tag{6}$$

where **encoder** and **decoder** consist of several stacks of linear and non-linear layers, respectively. To be more specific, the encoder takes the feature set $\{\mathbf{Z}_k, \mathbf{Z}_x, \mathbf{Z}_r\}$ and map them into a hidden vector $\mathbf{h}$. The hidden vector can be viewed as a fusion of temporal traffic patterns and spatial segment representation. Then the decoder takes $\mathbf{h}$ together with the temporal embedding $PE(t_{pred})$ as its input by concatenating these two vectors and outputs a probability distribution $\mathbf{y}$. In particular, the importance of the recent feature $\mathbf{Z}_x$ is uncertain when the prediction horizon is not fixed. More precisely, the dependence between recent traffic condition and future traffic condition would decrease with the increasing horizon. To model such decay nature, we employ the reciprocal of the Euclidean distance between t_{pred} and t_{cur} in the vector space as the discount factor of $\mathbf{Z}_x$.

4 Evaluation

4.1 Dataset

We conduct experiments on a real-world dataset collected from Xi'an, China, which was released by Didi Chuxing[1]. This dataset contains real-time and historical traffic condition data on the Didi platform from July 1, 2019 to July 30, 2019, as well as road attributes and topology of the road network. The traffic condition for a segment at a given time slice is a 4-tuple: ⟨*road speed, eta speed, congestion level, number of vehicles*⟩. The dataset includes all road segments' traffic conditions for the past four weeks and recent periods, current and predicted time slices (i.e., t_{cur}, t_{pred}), as well as the congestion level at t_{pred}. Each segment contains 8 attributes, including length, direction, function level, speed limit level, number of lanes, speed limit, level, and width. The topological structure of the whole road network is expressed by list of upstream and downstream relationship. The length of the time slice is 2 min, which is the minimum time unit in the dataset.

4.2 Experimental Settings

Evaluation Protocols. We split the dataset into training set, validation set and test set according to a proportion of 0.7, 0.1 and 0.2. Our model is implemented in Pytorch on a Linux server (GPU: GeForce RTX 2080 Ti). We adopt Cross-Entropy as the loss function and use Adam optimizer to minimize it. The learning rate is set to 0.0002 and the batchsize is 256. We apply early stopping to the validation set to prevent the overfitting problem.

We set the prediction horizon as a scalable value ($\Delta t \in \{1, 2, \ldots 30\}$) to test the robustness. It means that the model will randomly receive prediction tasks with different horizons in training. We set the dimension of the sequence embedding (i.e., C_k, C_x) equal to 50. The embedding dimension d of all time slices is set to 20. More detailed settings of the hyperparameters are given in Table 2.

[1] https://gaia.didichuxing.com.

Table 2. Hyperparamter setting.

Component	Setting
Compression block × 4	Conv1d (25, 75) + BN + ReLU Conv1d (75, 100) + BN + ReLU Conv1d (100, 75) + BN + ReLU Conv1d (75, 50) + BN + ReLU.
FCN	Linear (50, 125, 50).
Encoder	Linear (280, 300) + BN + ReLU + Linear(300, 100) + BN + ReLU.
Decoder	Linear(120, 75) + BN + ReLU + Linear(75, 3)

Baselines and Metrics

- **HA** (Historic Average): HA is a traditional statistical algorithm that uses the average state of previous time slices to estimate the state of future time slice.
- **SVM** (Support Vector Machine) [20]: SVM is a traditional classification algorithm that finds a decision plane maximizing the margin between classes. Consider its time complexity, we use linear kernel function in this comparison.
- **LightGBM** (Light Gradient Boosting Machine) [21]: LightGBM is an effective implementation of Gradient Boosting Decision Tree (GBDT). It uses Gradient-based One-Side Sampling (GOSS) and Exclusive Feature Bundling (EFB) to boost the training process while reserve a relative same accuracy with traditional GBDT algorithms.
- **ANN** (Artificial Neural Network): MultiLayer Perceptron is a traditional Artificial Neural Network that uses multiple non-linear layers to form a decision plane.
- **LSTM** (Long-Short Term Memory): LSTM is a component of Deep Neural Networks (DNNs) that learns information over extented time interval. Here, we use the LSTM network proposed in [22].

We employ two metrics: Accuracy and F1-score to evaluate the prediction results of different methods.

4.3 Experiment Results

Prediction Performance. Figure 3 reports the prediction performance of our model at different horizons (i.e., $t_{pred} - t_{cur}$). We can first observe that the accuracy decreases when the interval starts to increase. Generally, the traffic conditions of the few time slices close to t_{pred} are more important for congestion prediction, so the accuracy would decrease when this nearest information is not available within the horizon. Besides, as the prediction interval increases further, we can find the model gradually achieve stable performance. It indicates that our model is robust to the dynamic changes of the horizon (i.e., flexible predicted time) when performing the prediction.

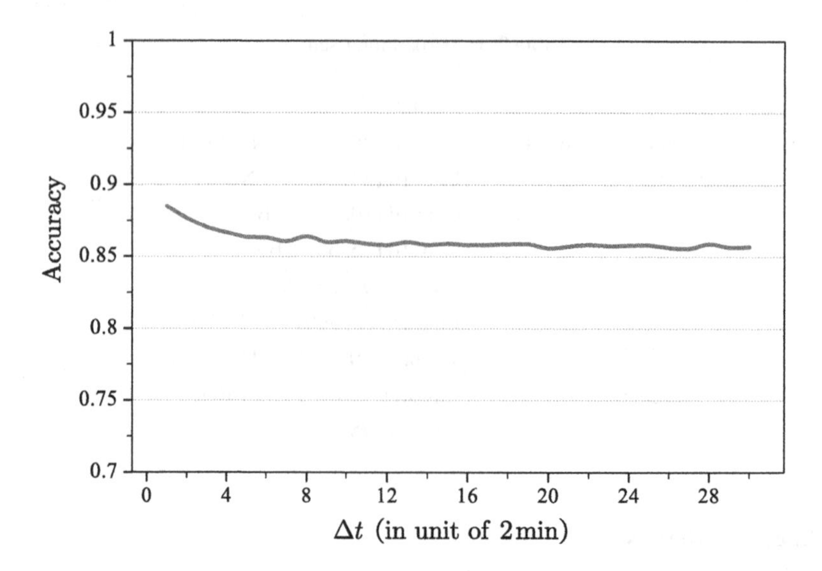

Fig. 3. The accuracy of AF-TCP for different prediction horizons.

For the sake of more detailed analysis, we also calculate the accuracy and macro-F1 score on different t_{pred} of a day. The results are shown in Fig. 4. We can observe that the prediction accuracy is close to 100% from 0 am to 6 am (i.e., the first 180 time slice), but the corresponding Macro-F1 is unstable. We infer that the model can accurately predict that most road segments' congestion level is low due to the lighter traffic in the early morning. However, there may be a few congested road segments that are not identified, resulting in fluctuations in F1 scores. One the other hand, after 6 am, the accuracy begins to decline, and F1-score starts to rise and remain stable. This could be that as the traffic volume on the road network gradually increases with the arrival of morning peak, the model begins to identify different road congestion levels (i.e., the rise and stability of F1-score). Meanwhile, the accuracy of the model tends to decrease due to the complexity of the traffic situation.

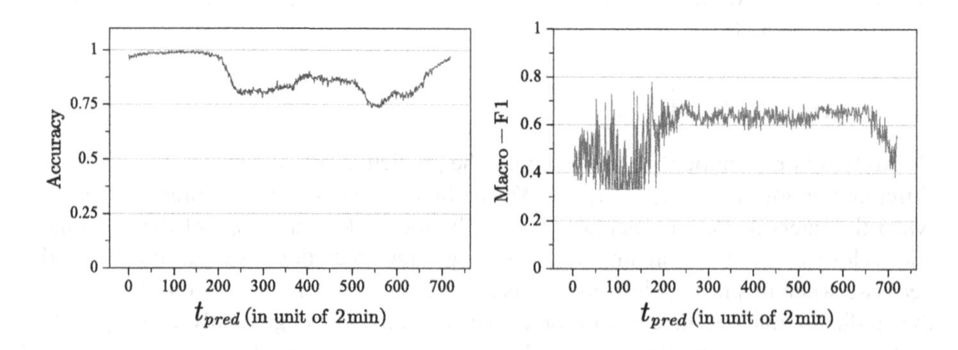

Fig. 4. The performance at different predicted time slices.

In addition, we can observe there are two obvious troughs corresponding to the morning and evening peaks on the accuracy curve in Fig. 4. One possible reason is that the road traffic becomes complicated due to the urban commuting, thus it is difficult for the model to maintain high accuracy predictions for all road segments.

To demonstrate the effectiveness of our prediction model for arbitrary segment, we first randomly choose four road segments and select their historical traffic data and road attributes. Then we train four independent ANN models for four road segments respectively. Note that some complex models that require large-scale data are not applicable due to the data sparseness of the single road segment. Thus, we only choose a simple ANN method for comparison. At last, we compare their performance with AF-TCP which designed in an one-for-all fashion. The results are shown in Fig. 5. Some interesting observations can be noted:

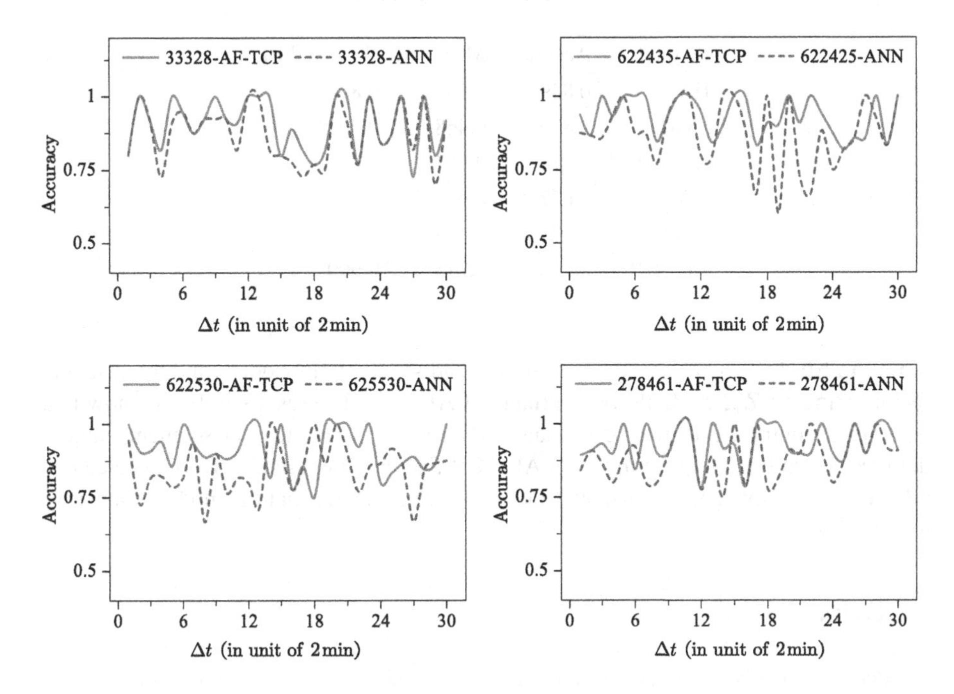

Fig. 5. Performance comparison of AF-TCP and ANN on segment 33328, 622435, 622530, and 278461.

1. In the subfigure in the upper left of Fig. 5, AF-TCP can achieve an accuracy comparable to ANN, and has better performance in the other three subfigures.
2. In each subfigure, the performance of AF-TCP over multiple segments is more stable than that of the segment-based method.

We also offer some insights gained from the above observations: 1) our model can better accomplish the prediction of multiple road segments simultaneously. 2) our model can learn more general pattern of citywide road congestion with the one-for-all fashion which would improve the model generalization.

Comparison with State-of-the-Art Methods. In order to accommodate the capabilities of different baselines, we reduced the size of the original data set. Concretely, we randomly select 10,000 records per day from nearly one month of data. The following performance comparisons are based on this downsampling data.

We compare the proposed model with baselines, and the results are shown in Table 3. Our model achieves the best performance in three metrics: Accuracy, Macro-F1 score, and Weighted-F1 score. The average improvement in these three metrics is around 5%, 24%, and 6%, respectively. Specifically, AF-TCP shows the largest improvement in the Macro-F1 score, indicating that our model is significantly better at identifying different congestion levels than other methods.

Table 3. Comparison of different baselines.

Methods	Accuracy	Macro F1	Weighted F1
HA	0.849	0.587	0.839
SVM	0.820	0.468	0.801
LightGBM	0.878	0.664	0.868
ANN	0.872	0.646	0.864
LSTM	0.871	0.639	0.861
AF-TCP	**0.906**	**0.745**	**0.900**

Ablation Study. To analyze the validity of road segment embedding and the discount factor acting on $\mathbf{Z}_x$, we build two variants of AF-TCP. The results in Fig. 6 show that both parts contribute to the performance improvement. First, road segment embedding has a relatively larger impact on AF-TCP. Second, the model performance can be affected by simply setting a discount factor, which indicates that the model is sensitive to the horizon.

5 Related Work

For decades, plenty of studies have explored the problem of traffic prediction [23,24]. Most early attempts are parametric methods based on the empirical data and theoretical assumptions, such as Kalman Filter [25] and ARIMA [26]. Although these methods can complete the traffic prediction with low model complexity, there are still some limitations for congestion prediction. For instance, ARIMA mainly targets univariable time series, while the congestion of the road segment is usually associated with multiple traffic-related variables (e.g., speed, volume).

Recent advances in deep learning have motivated its application in traffic prediction [27]. [28] employ the long short term memory (LSTM) network to capture nonlinear traffic dynamics for road speed prediction. [11] transform the city into a grid image, and use the CNN to capture spatial correlations. This type of method can only provide a coarse granularity prediction, but not suitable for congestion prediction at the road

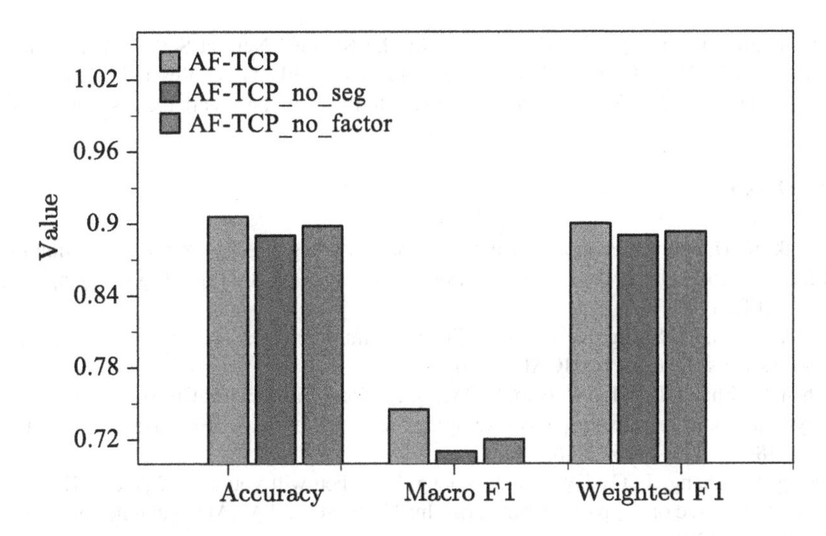

Fig. 6. The effects of road segment embedding and discount factor

segment-level. On the other hand, since many traffic data are graph-structured, it is natural to explore the graph neural networks in traffic domain to improve the prediction accuracy [29]. For example, [30] model the traffic flow dynamics as a diffusion process on a directed graph, and propose a diffusion convolutional recurrent neural network that captures the spatiotemporal dependencies. Instead of applying regular convolutional and recurrent units, [13] combine graph convolutional layers with convolutional sequence learning layers to model spatial and temporal dependencies. To model the dynamic spatiotemporal correlation of traffic data, [14] extend graph convolution by a spatiotemporal attention mechanism. However, the large number of road segments in the road network imposes a tremendous burden to existing graph-based approaches, and most of which can only handle a portion of the road network. In this paper, we propose a deep model named AF-TCP focusing on congestion prediction at arbitrary road segments and flexible future time.

6 Conclusions and Future Work

In this paper, based on the historical traffic data and road networks, we propose a novel deep model named AF-TCP for road traffic congestion prediction at arbitrary road segments and flexible future time. we combine the road attributes and road structure, with historical traffic observations to achieve the congestion prediction at arbitrary road segment. We also design a time aware module for learning different temporal contexts and improving the robustness of the model to different horizons. Extensive experiments on a real-world dataset demonstrate the effectiveness and superiority of AF-TCP. In the future, we will consider the impact of more complex road network structures on traffic, and extend our proposed model for more spatiotemporal prediction tasks.

Acknowledgments. The work was supported by the National Natural Science Foundation of China (No. 61872050 and No. 62172066), and sponsored by DiDi GAIA Research Collaboration Plan. Xuefeng Xie and Jie Zhao contributed equally to this work and share the first authorship.

References

1. Basak, S., Dubey, A., Bruno, L.: Analyzing the cascading effect of traffic congestion using LSTM networks. In: 2019 IEEE International Conference on Big Data (Big Data), pp. 2144–2153. IEEE (2019)
2. Li, K., Chen, L., Shang, S.: Towards alleviating traffic congestion: optimal route planning for massive-scale trips. In: IJCAI (2020)
3. Chen, C., Zhang, D., Ma, X., Guo, B., Wang, L., Wang, Y., Sha, E.: Crowddeliver: planning city-wide package delivery paths leveraging the crowd of taxis. IEEE Trans. Intell. Trans. Syst. **18**(6), 1478–1496 (2016)
4. Wang, D., Zhang, J., Cao, W., Li, J., Zheng, Y.: When will you arrive? AAAI, Estimating travel time based on deep neural networks. In: Thirty-Second AAAI Conference on Artificial Intelligence (2018)
5. Chen, C., Ding, Y., Xie, X., Zhang, S., Wang, Z., Feng, L.: Trajcompressor: an online map-matching-based trajectory compression framework leveraging vehicle heading direction and change. IEEE Trans. Intell. Transp. Syst. **21**(5), 2012–2028 (2019)
6. Alghamdi, T., Elgazzar, K., Bayoumi, M., Sharaf, T., Shah, S.: Forecasting traffic congestion using ARIMA modeling. In: 2019 15th International Wireless Communications and Mobile Computing Conference (IWCMC), pp. 1227–1232 (2019)
7. Nguyen, H.N., Krishnakumari, P., Vu, H.L., van Lint, H.: Traffic congestion pattern classification using multi-class svm. In: 2016 IEEE 19th International Conference on Intelligent Transportation Systems (ITSC), pp. 1059–1064 (2016)
8. Harrou, F., Zeroual, A., Sun, Y.: Traffic congestion monitoring using an improved KNN strategy. Measurement **156**, 107534 (2020)
9. Zhao, Z., Chen, W., Wu, X., Chen, P.C.Y., Liu, J.: LSTM network: a deep learning approach for short-term traffic forecast. Iet Intell. Trans. Syst. **11**, 68–75 (2017)
10. Liao, C., Chen, C., Xiang, C., Huang, H., Xie, H., Guo, S.: Taxi-passenger's destination prediction via gps embedding and attention-based bilstm model. IEEE Trans. Intell. Transp. Syst. pp. 1–14 (2021)
11. Zhang, J., Zheng, Y., Qi, D.: Deep spatio-temporal residual networks for citywide crowd flows prediction. In: AAAI (2017)
12. Chen, M., Yu, X., Liu, Y.: PCNN: Deep convolutional networks for short-term traffic congestion prediction. IEEE Trans. Intell. Transp. Syst. **19**, 3550–3559 (2018)
13. Yu, T., Yin, H., Zhu, Z.: Spatio-temporal graph convolutional networks: a deep learning framework for traffic forecasting. In: IJCAI (2018)
14. Guo, S., Lin, Y., Feng, N., Song, C., Wan, H.: Attention based spatial-temporal graph convolutional networks for traffic flow forecasting. In: AAAI (2019)
15. Fang, X., Huang, J., Wang, F., Zeng, L., Liang, H., Wang, H.: ConSTGAT: contextual spatial-temporal graph attention network for travel time estimation at baidu maps. In: Proceedings of the 26th ACM SIGKDD International Conference on Knowledge Discovery and Data Mining (2020)
16. Zhao, L., Song, Y., Zhang, C., Liu, Y., Wang, P., Lin, T., Deng, M., Li, H.: T-GCN: a temporal graph convolutional network for traffic prediction. IEEE Trans. Intell. Transp. Syst. **21**, 3848–3858 (2020)

17. Hu, J., Guo, C., Yang, B., Jensen, C.S.: Stochastic weight completion for road networks using graph convolutional networks. In: 2019 IEEE 35th International Conference on Data Engineering (ICDE), pp. 1274–1285 (2019)
18. Zhang, D., Yin, J., Zhu, X., Zhang, C.: Network representation learning: a survey. IEEE Trans. Big Data **6**, 3–28 (2020)
19. Vaswani, A., et al.: Attention is all you need. Adv. Neural Inf. Process. Syst. **30**, 5998–6008 (2017)
20. Vapnik, V.: The nature of statistical learning theory. Springer science and business media (2013)
21. Ke, G., et al.: Lightgbm: a highly efficient gradient boosting decision tree. In: Advances in Neural Information Processing Systems, pp. 3146–3154 (2017)
22. Sutskever, I., Vinyals, O., Le, Q.V.: Sequence to sequence learning with neural networks. Adv. Neural Inf. Process. Syst. **27**, 3104–3112 (2014)
23. Chen, C., Zhang, D., Castro, P.S., Li, N., Sun, L., Li, S., Wang, Z.: iboat: isolation-based online anomalous trajectory detection. IEEE Trans. Intell. Transp. Syst. **14**(2), 806–818 (2013)
24. Chen, C., Liu, Q., Wang, X., Liao, C., Zhang, D.: semi-traj2graph: identifying fine-grained driving style with gps trajectory data via multi-task learning. IEEE Trans. Big Data (2021)
25. Guo, J., Huang, W., Williams, B.M.: Adaptive Kalman filter approach for stochastic short-term traffic flow rate prediction and uncertainty quantification. Transp. Res. Part C-emerging Technol. **43**, 50–64 (2014)
26. Moreira-Matias, L., Gama, J., Ferreira, M., Mendes-Moreira, J., Damas, L.: Predicting taxi-passenger demand using streaming data. IEEE Trans. Intell. Transp. Syst. **14**, 1393–1402 (2013)
27. Chen, C., Zhang, D., Wang, Y., Huang, H.: Enabling Smart Urban Services with GPS Trajectory Data. Springer (2021) https://doi.org/10.1007/978-981-16-0178-1
28. Ma, X., Tao, Z., Wang, Y., Yu, H.: Long short-term memory neural network for traffic speed prediction using remote microwave sensor data. Transp. Res. Part C-Emerging Technol. **54**, 187–197 (2015)
29. Ye, J., Zhao, J., Ye, K., Xu, C.: How to build a graph-based deep learning architecture in traffic domain: a survey. ArXiv abs/2005.11691 (2020)
30. Li, Y., Yu, R., Shahabi, C., Liu, Y.: Diffusion convolutional recurrent neural network: data-driven traffic forecasting. In: ICLR (2018)

Collaborative QoS Prediction via Context-Aware Factorization Machine

Wenyu Tang[1], Mingdong Tang[1,2(✉)], and Wei Liang[3]

[1] School of Information Science and Technology, Guangdong University of Foreign Studies, Guangzhou, China
mdtang@gdufs.edu.cn
[2] Guangdong Key Laboratory of Big Data Analysis and Processing, Guangzhou, China
[3] School of Computer Science and Engineering, Hunan University of Science and Technology, Xiangtan, China

Abstract. With the prevalence of Web/Cloud/IoT services on the Internet, to select service with high quality is of paramount importance for building reliable distributed applications. However, the accurate values of the quality of services (QoS) are usually uneasy to obtain for they are typically personalized and highly depend on the contexts of users and services such as locations, bandwidths and other network conditions. Therefore, personalized and context-aware QoS prediction methods are desirable. By exploiting the QoS records generated by a set of users on a set of services, this paper proposes a collaborative QoS prediction method based on Context-Aware Factorization Machines named CAFM, which integrates the context information of services and users with classic factorization machines. Comprehensive experiments conducted on a real-world dataset show that the proposed method significantly outperforms existing methods in prediction accuracy.

Keywords: Collaborative filtering · QoS prediction · Context-aware · Factorization machines · Service selection

1 Introduction

The past decade has witnessed a tremendous increase in the prevalence of Web/Cloud/IoT (Internet of Things) services, which are widely used as building blocks for developing various applications in common computing environments or paradigms [1, 2]. These services are basically software components that encapsulate a set of data or functionalities and possess a set of APIs (Application Programming Interfaces), through which the services can be assessed remotely. Nowadays, it has become very popular to reuse those services for developing novel or value-added applications. For example, Mashup is a popular technology to combine different services to create new services or applications, varying from Web, Cloud, Mobile to IoT environments [3, 4].

To build reliable service-based applications, it is critical to select high quality services. For Web /Cloud/IoT services, the attributes of the quality of service (QoS) typically

Y. Lai et al. (Eds.): ICA3PP 2021, LNCS 13157, pp. 182–195, 2022.
https://doi.org/10.1007/978-3-030-95391-1_12

include response time (i.e., latency), throughput, reliability, availability, etc. However, accurate QoS values of those services are usually uneasy to obtain, for both services and users are geographically distributed and their communications depend on the underlying computer networks. That is, the QoS values of a service depend highly on the environments where and when users invoke the service. Considering the vast number of services, it is also unrealistic for users to obtain QoS data and find the best service by invoking all services, which is quite time and resource consuming [5, 6].

To obtain accurate QoS data, a number of QoS prediction methods have been proposed, among which collaborative filtering (CF) is the most popular technique. CF-based QoS prediction methods which exploit the historical QoS records of services generated by invocations from different users, can produce personalized QoS values for service users. Generally, collaborative QoS prediction methods can be classified into memory-based and model-based. Memory-based CF methods focus on computing the neighborhoods of services or users, and using the QoS records of similar neighbors to make QoS predictions. Model-based CF methods such as Matrix Factorization (MF) and Factorization Machines (FM), focus on modeling each user and service with a latent feature vector, and trying to find a function that accurately maps the interactions of a service and a user to a set of QoS values [7].

Although previous studies have made much progress in collaborative QoS prediction, they are insufficient in taking into consideration the contexts of services and users, which are important for recommender systems. In QoS predictions, the contexts of services and users such as locations and network bandwidth are undoubtedly influential to the QoS values. Therefore, in addition to the historical QoS data, the contexts of services and users should also be incorporated into QoS predictions. Based on this consideration, this paper proposes a QoS prediction method based on Context-Aware Factorization Machines named CAFM. The contributions of this paper are listed as follows:

1. We take into consideration both locations and bandwidth for accurate QoS predictions. Although locations of services or users have been used by some existing works, the bandwidth factor is rarely considered.
2. We propose to integrate the contexts of services and users with the classical factorization machine model, and thus developing a context-aware QoS prediction method.
3. We conduct a series of experiments on a real dataset to evaluate the proposed QoS prediction method, and the experimental results show that the proposed method significantly outperforms the other baselines.

The remainder of this paper is organized as follows. Section 2 describes the QoS prediction problem via an example. Section 3 introduces the proposed QoS prediction method CAFM. Section 4 presents the experimental results. Section 5 surveys related work on QoS prediction. Finally, the whole paper is concluded in Sect. 6.

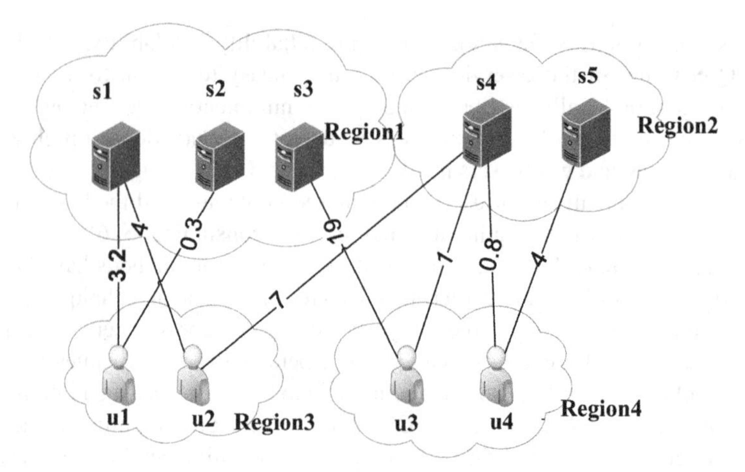

Fig. 1. A motivating example.

Table 1. QoS matrix with missing values

	s1	s2	s3	s4	s5
u1	3.2	0.3	–	–	–
u2	4	–	–	7	–
u3	–	–	19	1	–
u4	–	–	–	0.8	4

Table 2. QoS matrix after missing value predictions

	s1	s2	s3	s4	s5
u1	3.2	0.3	2	5	8
u2	4	0.4	1	7	7
u3	2.6	5.2	19	1	5
u4	3	4	19	0.8	4

2 The QoS Prediction Problem

Figure 1 describes a QoS prediction scenario. As shown in the figure, suppose there are four users and five services distributed in four different regions, namely Region1, Region 2, Region 3 and Region 4. The connections between the users and the services represent the invocation relationships, and the values on the edges represent the response time. Table 1 presents the QoS value matrix generated from the service invocation records. Let U = {u1, u2, u3, u4} be the set of users, and S = {s1, s2, s3, s4, s5} be the set of services. Each entry ru,i represents the latest invocation record between user u and service i. If user u has no invocation relationship with service i, ru,i is missing. QoS prediction aims to use the known QoS records shown in Table 1 to predict the missing QoS values. Table 2 illustrates the possible predicted QoS matrix, where the missing QoS values in the original QoS matrix is replaced by the predicted QoS values.

The challenges raised by the above QoS prediction problem rely on at least two aspects. First, the real QoS matrix is extremely sparse, for a user typically uses only a few services. This fact indicates that traditional CF methods may fail to construct similar neighborhoods of users or services. Consequently, a desirable QoS prediction method should overcome the data sparsity problem. Second, the QoS value of a service is subject to the environments where the user and the service are located. Thus, the contexts of users and services should be taken into account when making QoS predictions.

3 The Proposed Method

In this section, we first briefly introduce the model of classical factorization machines. Then, we describe the proposed QoS prediction method CAFM in detail, which revamps the classic factorization machine by incorporating the contexts of services and users such as locations and bandwidth.

3.1 Classical Factorization Machine

Factorization Machines is a new model class that combines the advantages of Support Vector Machines (SVM) with factorization models. Like SVMs, FMs are a general predictor working with any real valued feature vector, but it is able to estimate reliable parameters under very high sparsity [8]. More specifically, the factorization machine models all nested variable interactions, but uses a factorized parametrization instead of a dense parametrization like in SVMs. Another major advantage of FMs is that its model equation can be computed in linear time and it depends only on a linear number of parameters. A second-order expression of the classical factorization machine is as follows:

$$\hat{y}(\mathbf{x}) := w_0 + \sum_{i=1}^{n} w_i x_i + \sum_{i=1}^{n} \sum_{j=i+1}^{n} \langle \mathbf{v}_i, \mathbf{v}_j \rangle x_i x_j \quad (1)$$

where x is the feature vector, n is the number of variables in $\mathbf{x}$, w_0 is the global bias, w_i models the strength of the i-th variable to the target, $\mathbf{v}_i$ is the hidden vector of the i-th variable, $\mathbf{v}_j$ is the hidden vector of the j-th variable, and $\langle \mathbf{v_i}, \mathbf{v}_j \rangle$ models the interactions between the i-th variable and the j-th variable in $\mathbf{x}$. Instead of using an own model parameter $w_{i,j}$ for each interaction, the FM models the interaction by factorizing it. This is critical to allow high quality parameter estimates of higher-order interactions under sparsity.

The factorization machine employs Cholesky decomposition to convert the coefficients of the second-order term into the product of hidden vectors in each dimensional space. If a second-order polynomial regression model is implemented in QoS prediction, confronting amount of feature combinations that never interact, there is not enough data to estimate the corresponding parameters. The features of factorization machines are pairwise related, so the hidden vectors obtained by training can solve the problem of data sparsity. The gradient update formula of the factorization machine is as follows:

$$\frac{\partial}{\partial\theta}\hat{y}(\mathbf{x}) = \begin{cases} 1 & \text{if } \theta \text{ is } w_0 \\ x_i & \text{if } \theta \text{ is } w_i \\ x_i\sum_{j=1}^{n} v_{j,f}x_j - v_{i,f}{x_i}^2 & \text{if } \theta \text{ is } v_{i,f} \end{cases} \tag{2}$$

For different tasks, different loss functions can be chosen for factorization machines to train the parameters.

3.2 Context-Aware Factorization Machine

The factorization machine is a general predictor, which can easily combine a variety of features and contextual information [9]. From this aspect, this paper incorporates the following five features into the factorization machine model, namely user, service, user location, service location and user bandwidth, as shown in Fig. 2. The context-aware factorization machine can be defined as follows:

$$\begin{aligned} &\hat{y}(x(u, i, R_u, R_i, B)) = w_0 + w_u + w_i + \sum\nolimits_{m\in R_U} (w_m + \langle v_u, v_m\rangle + \langle v_i, v_m\rangle) \\ &+\sum\nolimits_{n\in R_I} (w_n + \langle v_u, v_n\rangle + \langle v_i, v_n\rangle) + \sum\nolimits_{m\in R_U} \sum\nolimits_{n\in R_I} \langle v_m, v_n\rangle \\ &+\sum\nolimits_{t\in T} (w_t x_t + \langle v_u, v_t\rangle x_t + \langle v_i, v_t\rangle x_t + \sum\nolimits_{m\in R_U} \langle v_m, v_t\rangle x_t + \sum\nolimits_{n\in R_I} \langle v_n, v_t\rangle x_t) \end{aligned} \tag{3}$$

where u represents the user ID, i represents the service ID, R_u represents the region where user u is located, R_i represents the region where service i is located, B represents the bandwidth of u, and x_i represent the i-th variable of feature x. v_u represents the latent vector of a user, v_m represents the latent vector of a user region, v_n represents the latent vector of a service region and v_t represents the latent vector of bandwidth.

Comparing with the classic factorization machine model described in Formula 1, which factorizes only the pairwise interactions between users and items, Formula 2 factorizes all pairwise interactions between the five variables, including all context variables. Please note that, since most items in the feature vector $\mathbf{x}$ are 0, as indicated by the example in Fig. 2, Formula 2 can be further simplified in computation.

The model parameters can be learned by optimizing the objective function. Due to the large number of model parameters, a regularization term is added to the loss function to prevent over-fitting:

$$OPT(S) = argmin \sum_{(x,y)\in s} L(\hat{y}(x) - y) + Reg(w_o, w_i, v_{i,f}) \tag{4}$$

$$Reg(w_0, w_i, v_{i,f}) = \lambda_0 w_0 + \lambda_1 w_i + \lambda_2 v_{i,f} \tag{5}$$

where $\hat{y}(x)$ represents the predicted QoS value, y represents the actual QoS value, L is the loss function, and Reg represents the regular term. In addition, w_0 implies the global bias term, w_i implies the first-order term coefficient of factorization machine, and $v_{i,f}$ is the hidden vector of each variables.

	Feature vector x					Target y	
	User	Service	User Region	Service Region	Bandwidth		
x_1	1000...	1000...	1000...	1000...	150	1.1	y_1
x_2	1000...	0100...	1000...	0100...	20	8.8	y_2
x_3	0100...	1000...	0100...	1000...	80	3.2	y_3
x_4	0010...	1000...	0010...	1000...	165	0.7	y_4
x_5	0010...	0100...	0010...	0100...	15	9.6	y_5
x_6	0010...	0001...	0010...	0001...	90	5	y_6
x_7	0001...	1000...	0001...	1000...	20	7.6	y_7
x_8	0001...	0100...	0001...	0100...	150	1.3	y_8

Fig. 2. Example of context-aware factorization machine.

3.3 Complexity Analysis

Here we analyze the time complexity of the proposed QoS prediction method. Smaller complexity indicates that the user can get a faster response from the QoS recommendation system. The complexity of straight forward computation of Eq. (3) is in $O(kn^2)$ because all pairwise interactions in the second-order term have to be computed, where k is the dimensionality of the factorization, and n is the size of feature vector $\mathbf{x}$. But it can be reduced to $O(kn)$ via reformulating, as demonstrated by the work [8]. The reformulating process is as follows:

$$\begin{aligned}
&\sum_{i=1}^{n}\sum_{j=i+1}^{n}\langle v_i, v_j\rangle x_i x_j \\
&= \frac{1}{2}\sum_{i=1}^{n}\sum_{j=1}^{n}\langle v_i, v_j\rangle x_i x_j - \frac{1}{2}\sum_{i=1}^{n}\langle x_i, x_i\rangle \\
&= \frac{1}{2}\left(\sum_{i=1}^{n}\sum_{j=1}^{n}\sum_{f=1}^{k} v_{i,f} v_{j,f} x_i x_j - \sum_{i=1}^{n}\sum_{f=1}^{k} v_{i,f} v_{i,f} x_i x_i\right) \\
&= \frac{1}{2}\sum_{f=1}^{k}\left(\left(\sum_{i=1}^{n} v_{i,f} x_i\right)\left(\sum_{j=1}^{n} v_{i,f} x_j\right) - \sum_{i=1}^{n} v_{i,f}^2 x_i^2\right) \\
&= \frac{1}{2}\sum_{f=1}^{k}\left(\left(\sum_{i=1}^{n} v_{i,f} x_i\right)^2 - \sum_{i=1}^{n} v_{i,f}^2 x_i^2\right)
\end{aligned} \tag{6}$$

Moreover, since most of the elements in $\mathbf{x}$ are 0 under sparsity, the sums have only to be computed over the non-zero elements. Thus, in sparse applications, the computation of the factorization machine can be further reduced to an even smaller complexity.

The above analysis indicates that our proposed method has significant scalability for large-scale datasets.

4 Experiments

4.1 Experiment Settings

The experiments are conducted on a real-world Web service QoS dataset generated from actual service invocations [10]. It contains 339 users, 5,825 services, and a 339 × 5,825 matrix consisting of QoS values that are generated by each user invoking each service. Every QoS value consist of two items: response time and throughput. The dataset also contains the location information of each user or service, such as IP address, longitude and latitude, from which we can derive the Autonomous System (AS) where each user or service locates on the Internet, and use it to divide all users and services into different regions, i.e., an AS corresponds to a region. The user bandwidth is missing in the dataset, but so some extent it can be mimicked by the throughput values in the dataset. In the experiments, we employ response time to evaluate the performance of our proposed method CAFM, and compared it with other state-of-the-art QoS prediction methods.

In real scenarios, the QoS data are likely to be very sparse since a user usually invokes only a few services. To simulate real QoS prediction tasks, we randomly remove most of the QoS values from the dataset, and decrease the data density to the range between 1% and 20%. After the data sparsification, we divide the dataset into a training set and a test set. The former is used to train the models of the proposed method and baseline methods, and the latter is used to evaluate the methods. For every experiment, we repeat it for 10 times and use the average results to do evaluation.

4.2 Evaluation Metrics

To evaluate the performance of the proposed method, we adopt two commonly used metrics, i.e., MAE (Mean Absolute Error) and Root Mean Squared Error (RMSE). The MAE is calculated with:

$$MAE = \frac{\sum_{u \in T} |y_{u,i} - \hat{y}_{u,i}|}{|T|} \tag{7}$$

And the RMSE is calculated with:

$$RMSE = \sqrt{\frac{\sum_{u \in T} |y_{u,i} - \hat{y}_{u,i}|}{|T|}} \tag{8}$$

where $y_{u,i}$ is the actual response time when the user invokes the service, $\hat{y}_{u,i}$ is the response time predicted, T is the test set, and $|T|$ represents the size of the test set. A smaller MAE or RMSE values indicates a better performance of predictions.

4.3 Performance Comparison

To validate the proposed QoS prediction method CAFM, we use the following baselines:

1) UPCC (User-based CF with PCC) [11]: UPCC is a memory-based CF method that relies only on user neighborhoods where users are similar to each other; it employs QoS values of the similar neighbors of the active user to make predictions.
2) IPCC (Item-based CF with PCC) [12]: IPCC is an item-based CF method that relies only on item neighborhoods where users are similar to each other; it employs QoS values of the similar neighbors of the target service to make predictions.
3) WSRec [13]: This is a CF-based QoS prediction method that combines the results of UPCC and IPCC with a weighted summation operator to predict QoS values.
4) PMF (Probabilistic Matrix Factorization) [14]: This method introduces probability into traditional matrix factorization methods and uses only the user-item matrix for recommendations.
5) Classical FM (Factorization Machine) [8]. This is the classical factorization machine model, which exploits only the interactions between users and items.
6) LBFM (Location-based Factorization Machine) [15]. This method revamps the classical factorization machine model by introducing the location information to predict QoS values.

Table 3. Performance comparisons of response time prediction with different data density. The results of our method CAFM are marked in bold font.

Method	Density of QoS matrix (d)									
	$d = 1\%$		$d = 2\%$		$d = 5\%$		$d = 10\%$		$d = 20\%$	
	MAE	RMSE	MAE	RMSE	MAE	RMSE	MAE	RMSE	MAE	RMSE
UPCC	0.882	1.392	0.815	1.349	0.793	1.323	0.619	1.312	0.535	1.236
IPCC	0.871	1.515	0.861	1.481	0.757	1.390	0.721	1.35	0.541	1.227
WSRec	0.810	1.332	0.775	1.303	0.675	1.273	0.605	1.276	0.532	1.239
PMF	0.679	1.497	0.663	1.469	0.496	1.235	0.431	1.207	0.395	1.018
FM	0.539	1.154	0.492	1.087	0.430	0.991	0.388	0.930	0.335	0.854
LBFM	0.534	1.143	0.482	1.061	0.417	0.955	0.372	0.891	0.325	0.836
CAFM	**0.512**	**1.090**	**0.461**	**1.005**	**0.359**	**0.869**	**0.337**	**0.825**	**0.292**	**0.771**

The performance comparisons of all baselines and CAFM are shown in Table 3. The density of the QoS matrix d is set to 1%, 2%, 5%, 10% and 20% in this experiment, so that we can compare the baseline methods and CAFM in different data sparsity. As we can see, the proposed method CAFM performs significantly better than the other comparative methods in predicting the response time in all cases of data sparsity. The model-based CF methods such as PMF, FM, LBFM and CAFM generally perform better than the memory-based CF methods such as UPCC, IPCC and WSRec. We can also observe that all the FM models (including classical FM, LBFM and CAFM) perform better than the other methods, which indicates that FM is actually a promising technology suitable for the QoS prediction issue with sparse data, and adding more context information to the FM model can further improve its performance in predictions.

4.4 Impact of Dimensionality

The dimensionality refers to the number of factors used to factorize the interactions between users, services and the other context variables. To evaluate its impact on response time predictions, we set the matrix density to 5% or 10%, and the regularization parameters ($\lambda 0$, $\lambda 1$, $\lambda 2$) to (70, 0.45, 0.002).

Figure 3 shows the MAE and RMSE values of response time predictions under the impact of dimensionality. In the beginning, when the dimensionality increases from 2 to 6, both the MAE and the RMSE values decrease rapidly, indicating that the prediction performance is being improved. However, when the dimensionality continuously increases, the MAE values increase and the RMSE values vary in a small range without a significant tendency. This observation indicates that high dimensionality is not necessary for accurate QoS predictions because it probably cause the overfitting problem. Actually, a larger dimensionality value may unnecessarily cause longer computation time, for the time complexity of the proposed method is O(kn), where k is the dimensionality.

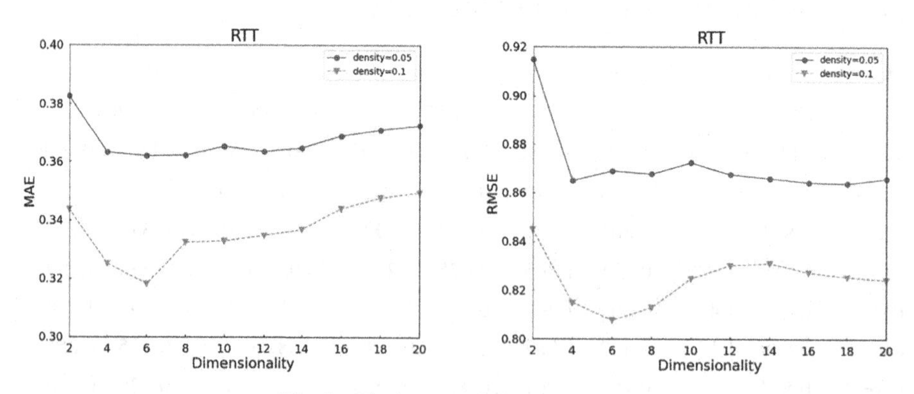

Fig. 3. The impact of dimensionality.

4.5 Impact of Regularization Parameters

Since the large number of parameters of the factorization machine, regularization parameters need to be introduced to prevent overfitting. λ0 controls the regularization weight of the global bias term, λ1 controls the regularization weight of a single feature item, and λ2 controls the regularization weight of the implicit vector used to calculate the coefficient of the feature cross term. In the experiment, we select datasets with density of 5% and 10% for the experiment, and the results illustrate that the regularization parameters will affect the prediction accuracy to a certain extent. As shown in Fig. 4–6, MAE and RMSE initially drop to a lowest point with the increase of parameters and then slowly rise. Our experiment achieves the best results when the regularization parameters are 70, 0.45, and 0.002, respectively.

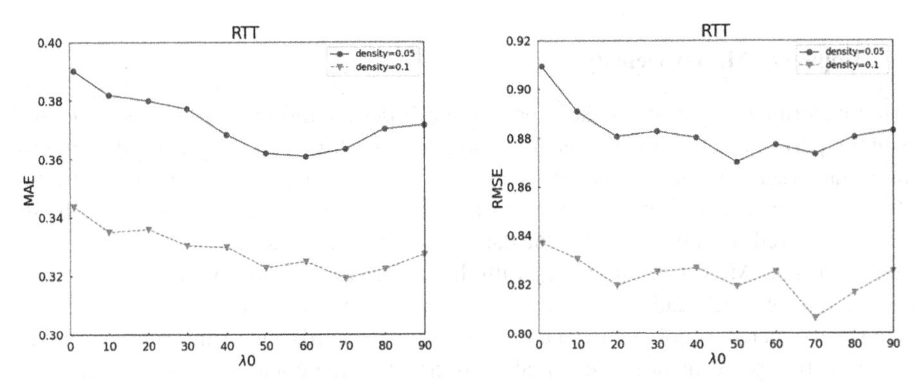

Fig. 4. The impact of parameter λ_0.

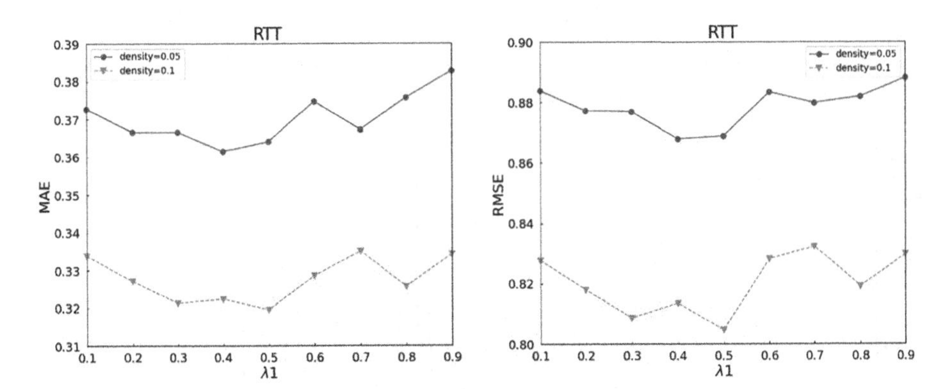

Fig. 5. The impact of parameter λ_1.

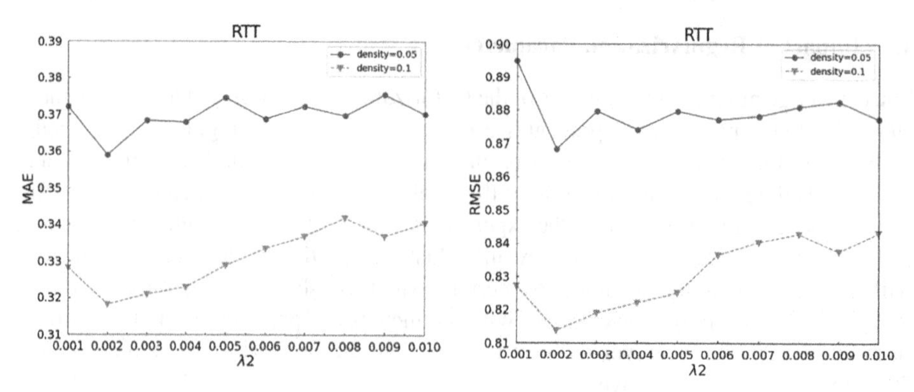

Fig. 6. The impact of parameter λ_2.

4.6 Impact of Matrix Density

In this experiment, we evaluate the impact of matrix density on QoS predictions and compare the performance of the classical factorization machine and the proposed context-aware factorization machine model. The density of the training data varies from 2% to 20%, while the regularization parameters are set to 70, 0.45 and 0.002 respectively. It can be observed from Fig. 7 that as the data density increases, MAE and RMSE decline rapidly at first. After a certain point, with the further increase of the data density, the decline rate of MAE and RMSE decreases. This observation indicates that when the data is very sparse, supplying even a little more QoS records can significantly improve the prediction performance. We can also observe that the context-aware factorization machine model is significantly better than the classical factorization machine model in the predictions.

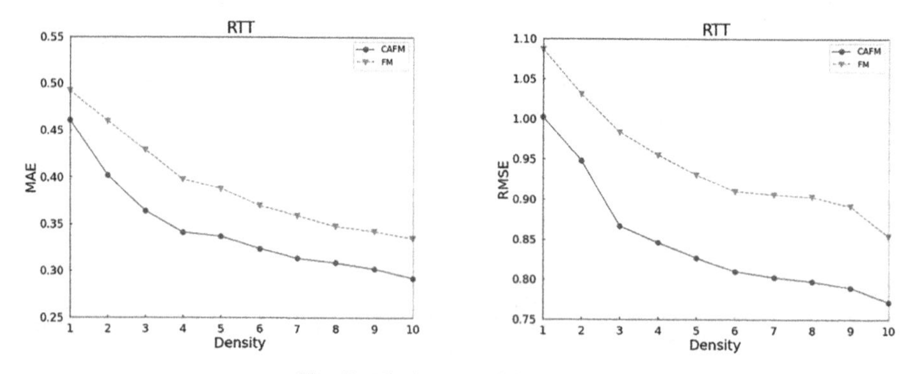

Fig. 7. The impact of data density

5 Related Work

A number of researches have been done on collaborative QoS predictions. From the perspective of technology, existing related work can roughly be divided into two categories: memory-based CF and model-based CF.

A memory-based CF method is usually carried out in three steps on QoS prediction tasks. First, it calculates the similarity between users and services based on historical QoS records. Then, it identifies the similar neighbors for the active user or target service. Finally, it aggregates the QoS information from neighbors to make QoS predictions for the active user or target service. Zheng et al. [13] proposed a memory-based CF method for QoS prediction which combines user-based prediction and service-based QoS prediction results. Tang et al. [16] took into consideration the locations of users and services to improve the similarity measurement for both users and services. In another work [17], they firstly proposed a data smoothing strategy to alleviated the data sparsity problem in the QoS prediction task, and employed a location-aware CF method to further improve QoS predictions. Ma et al. [18] refined the similarity measurement of both users and services by revealing some important characteristics such as: the magnitude of the similarity determines its stability. And based on these characteristics and memory-based CF, they proposed an accurate QoS prediction method.

Model-based CF methods can be used to address the data sparsity issue suffered by the memory-based CF methods, and are thus widely believed to be more effective for predictions under sparsity. Matrix factorization and factorization machine are popular model-based CF methods. Zhang et al. [19] used a non-negative matrix factorization to learn the hidden features of users and predict QoS values based on the user similarity. Zheng et al. [20] proposed a Neighborhood Integrated Matrix Factorization (NIMF) method, which takes into account the global and local information to perform personalized QoS prediction. Chen et al. [21] found that the prediction accuracy is greatly affected by the range of the QoS value and proposed that the application of a bias term can alleviate the problem of excessive data variance. Yang et al. [15] were the first to introduce the factorization machine model to service QoS prediction. They took into consideration the location information of users and services, and revamped the classical factorization machine model by incorporating the influence of the neighboring users and services, which are computed based on locations. However, computing user neighborhoods and service neighborhoods may be time-consuming and thus reduce the prediction efficiency. Different from [15], this paper also considered the influence of bandwidth on QoS prediction, and fused both location information and bandwidth information into the factorization machine model. Thus, our method can achieve better performance.

6 Conclusion

This paper studied the QoS prediction problem for high-quality service selection/recommendation in distributed environments. We proposed a collaborative QoS prediction method based on context-aware factorization machine, which integrates the location and bandwidth information of users and services with the classic factorization machine. Comprehensive experiments have shown that our method outperformed other prediction methods significantly. Besides, our method has linear computational complexity, and it is more suitable for applications with sparse data. In future work, we will focus on making full use of contextual information and utilizing neural networks to capture more complex interaction relationships to improve the QoS prediction accuracy.

Acknowledgement. This work is supported by National Natural Science Foundation of China under grant no. 61976061 and the Opening Project of Guangdong Key Laboratory of Big Data Analysis and Processing (202003).

References

1. Chen, X., Liang, W., Xu, J., Wang, C., Li, K. C., Qiu, M.: An efficient service recommendation algorithm for cyber-physical-social systems. In: IEEE Transactions on Network Science and Engineering (2021). https://doi.org/10.1109/TNSE.2021.3092204
2. Liang, W., Ning, Z., Xie, S., Hu, Y., Lu, S., Zhang, D.: Secure fusion approach for the internet of things in smart autonomous multi-robot systems. Inform. Sci. 1–20 (2021). https://doi.org/10.1016/j.ins.2021.08.035
3. Tang, M., Xia, Y., Tang, B., Zhou, Y., Cao, B., Hu, R.: Mining collaboration patterns between APIs for mashup creation in web of things. In: IEEE Access, vol. **7**, pp. 14206–14215 (2019)
4. Gao, H., Qin, X., Barroso, R.J.D., Hussain, W., Xu, Y., Yin, Y.: Collaborative learning-based industrial IoT API recommendation for software-defined devices: the implicit knowledge discovery perspective. In: IEEE Transactions on Emerging Topics in Computational Intelligence, pp. 1–11 (2020)
5. Tang, M., Zheng, Z., Kang, G., Liu, J., Yang, Y., Zhang, T.: Collaborative web service quality prediction via exploiting matrix factorization and network map. IEEE Trans. Netw. Serv. Manage. **13**(1), 126–137 (2016)
6. Liu, J., Tang, M., Zheng, M., Liu, X., Lyu, S.: Location-aware and personalized collaborative filtering for web service recommendation. IEEE Trans. Serv. Comput. **9**(5), 686–699 (2016)
7. Zheng, Z., Li, X., Tang, M., Xie, F., Lyu, M.R.: Web service QoS prediction via collaborative filtering: a survey. IEEE Trans. Serv. Comput. (2020)
8. Rendle, S.: Factorization machines. In: 2010 IEEE International Conference on Data Mining, IEEE. pp. 995–1000 (2010)
9. Rendle, S., Gantner, Z., Freudenthaler, C., Schmidt-Thieme, L.: Fast context-aware recommendations with factorization machines, In: Proc. 34th Int. ACM SIGIR Conf. Res. Develop. Inf. Retr., Beijing, China, Jul. 2011, pp. 635–644 (2011)
10. Zheng, Z., Zhang, Y., Lyu, M.R.: Investigating QoS of real-world web services. IEEE Trans. Serv. Comput. **7**(1), 32–39 (2014)
11. Shao, L., Zhang, J., Wei, Y., Zhao, J., Xie, B., Mei, M.: Personalized QoS prediction for web services via collaborative filtering. In: The 14th IEEE International Conference on Web Services (ICWS 2007), pp.439–446 (2007)
12. Sarwar, B.M., Karypis, G., Konstan, J.A., Riedl, J.: Item-based collaborative filtering recommendation algorithms. In: The 10th International Conference on World Wide Web (WWW 2001), pp.285–295 (2001)
13. Zheng, Z., Ma, H., Lyu, M.R., King., I.: WSRec: A collaborative filtering based web service recommender system. In: IEEE International Conference on Web Services, pp. 437–444 (2009)
14. Salakhutdinov, R., Mnih, A.: Probabilistic matrix factorization. In: The Twenty-First Annual Conference on Neural Information Processing Systems (NIPS 2007), pp.1257–1264 (2007)

15. Yang, Y., Zheng, Z., Niu, X., Tang, M., Lu, Y., Liao, X.: A location-based factorization machine model for web service QoS prediction. IEEE Trans. Serv. Comput. **14**(5), 1264–1277 (2021)
16. Tang, M., Jiang, Y., Liu, J., Liu, M.: Location-aware collaborative filtering for QoS-based service recommendation. In: 2012 IEEE 19th International Conference on Web Services (ICWS 2012), pp. 202–209 (2012)
17. Tang, M., Zhang, T., Liu, J., Chen, J.: Cloud service QoS prediction via exploiting collaborative filtering and location-based data smoothing. Concurr. Comput. Pract. Exp. **27**(18), 5826–5839 (2015)
18. Ma, Y., Wang, S., Hung, P.C.K., Hsu, C., Sun, Q., Yang, F.: A highly accurate prediction algorithm for unknown web service QoS values. IEEE Trans. Serv. Comput. **9**(4), 511–523 (2016)
19. Zhang, Y., Zheng, Z., Lyu, M.R.: Exploring latent features for memory-based QoS prediction in cloud computing. In: 2009 IEEE International Conference on Web Services, pp. 437–444 (2009)
20. Zheng, Z., Ma, H., Lyu, M.R., King, I.: Collaborative web service QoS prediction via neighborhood integrated matrix factorization. IEEE Trans. Serv. Comput. **6**(3), 289–299 (2013)
21. Chen, Z., Sun, Y., You, D., Li, F., Shen, L.: An accurate and efficient web service QoS prediction model with wide-range awareness. Future Gener. Comput. Syst. **109**, 275–292 (2020)

TDCT: Target-Driven Concolic Testing Using Extended Units by Calculating Function Relevance

Meng Fan[1,2], Wenzhi Wang[1,2](✉), Aimin Yu[1], and Dan Meng[1]

[1] Institute of Information Engineering, Chinese Academy of Sciences, Beijing, China
{fanmeng,wangwenzhi,yuaimin,mengdan}@iie.ac.cn
[2] School of Cyber Security, University of Chinese Academy of Sciences, Beijing, China

Abstract. Concolic unit testing is able to perform comprehensive analysis with a small function of the program. However, due to the following disadvantages, it cannot be widely and effectively applied to test the whole program. One is that it includes many false positives for lacking context-dependent information. The other is that it is difficult to automatically generate the whole program's inputs by unit inputs. The researchers have proposed different ways to solve the above problems, but it also causes inaccuracy or performance problems in some extent. In this paper, we present a method called Target-Driven Concolic Testing (TDCT) to meet the challenges, which combines concolic unit testing and concolic testing. TDCT is a fine-grained method based on the interprocedural control flow graph (ICFG) to construct extended unit, which could obtain a comprehensive and accurate context of target function as far as possible. We present a custom target-driven search strategy in concolic execution to automatically generate the whole program's inputs by unit inputs. It not only reduces the system performance overhead by discarding the search of irrelevant paths, but also further validates the authenticity of the potential bugs. We implement a prototype system of TDCT and apply it to 4 real-world C programs. The experiment shows that TDCT could find 83.87% of the target bugs and it possesses high precision with a true to false alarm ration is 1:5.2. It indicates that TDCT is able to effectively and accurately detect bugs and automatically generate the whole program's inputs by unit inputs.

Keywords: Unit testing · Concolic testing · Automated test generation · Search strategy

1 Introduction

Concolic unit testing [1–4] is a software testing method by which each unit of the program can be tested to determine whether they are suitable for using. It is a necessary step to ensure the product quality, with which developers can uncover and correct the defects in their programs. For a program needs to do the concolic

Y. Lai et al. (Eds.): ICA3PP 2021, LNCS 13157, pp. 196–213, 2022.
https://doi.org/10.1007/978-3-030-95391-1_13

unit testing, firstly, it requires test engineer to establish some kind of artificial environment (such as the test driver, etc.) for the tested units. Secondly, it does concolic testing in units to find potential bugs and generate unit inputs. Finally, it checks the returned values, either against known values or against the results from previous execution of the same test (regression testing).

The advantage of concolic unit testing is that it is able to perform a thorough analysis of small functions in the program. However, due to the lack of complete function context dependence in unit and it uses approximate symbolic inputs and return values to represent the context of the unit, the result of concolic unit testing include many false positives. Due to the path reachability between the path in unit and its whole program, it is difficulty to automatically generate the whole program's inputs by unit inputs. These defects make concolic unit testing difficult to be widely used to the whole program testing.

At present, many researchers propose different ways to solve the above two problems in concolic unit testing, which are divided into the following two types.

In order to reduce false positive in concolic unit testing, some state-of-the-art research exploit the filtering method [5,6], which not only brings huge labor overhead, but also is limited in accuracy. Some research uses the method of dynamically building extended unit to overcome the lack of context in the unit. This kind of method requires the program to run a period of time and dynamically collects the call relationship between functions to determine the scope of the unit. E.g. CONBRIO uses fuzzing to dynamically run program to collect the call relationships between functions [1]. However, there are some shortcomings in this kind of method. If the paths through the target function cannot be completely cover during the period of dynamically running, the collected call relationship between functions is incomplete, so that the context-dependent information in the subsequently built extended unit is inaccurate and could not effectively reduce false positive in concolic unit testing. As we all known, the path coverage through the target function in fuzzing is largely determined by the input of the program. Concolic execution also makes it difficult to fully cover all paths through the target function in a fixed time. For this reason, concolic unit testing is difficult to be widely used in practical engineering for program testing.

In order to automatically generate the whole program's inputs by unit inputs, some researchers use the method of reverse derivation. E.g. FOCAL use function summary and slice [7]. However, this kind method may exist false negative of the automatically generated inputs because the program does not run in real environment. Some researchers present the method of dynamically backward sensitive data tracing [8], which causes huge system performance overhead.

To make the concolic unit testing practical and efficient in engineering, we present a method called TDCT (Target-Driven Concolic Testing) to solve the above challenges, which combines concolic unit testing and concolic testing. At first, TDCT obtains the ICFG of the program by static analysis and uses it to construct the extended unit of the target function, which could obtain a comprehensive call relationship of target function. Then, TDCT does concolic execution in extended unit and obtains unit inputs. Finally, TDCT uses custom target-driven search strategy in concolic execution to automatically generate the whole program's inputs by unit inputs, which also could further verify whether the potential bugs really exist in the whole program. We have implement a prototype system of TDCT and apply it to 4 real-world C programs in FOCAL benchmarks [9,10]. The result of experiment shows that TDCT could find 83.87% of the target bugs and it possess high precision with a true to false alarm ration is 1:5.2. The contributions of this paper are as following:

- Present a fine-grained method based on the ICFG to construct extended unit, which could obtain a comprehensive and accurate context of target function as far as possible.
- Present a custom target-driven search strategy in concolic execution to automatically generate the whole program's inputs by unit inputs, which not only reduces the system performance overhead by discarding the search of uncorrelated paths, but also further validates the authenticity of the potential bugs.
- We implement a prototype system of TDCT and apply it to 4 real-world C programs. The result of experiment shows that TDCT could find 83.87% of the target bugs and it possess high precision with a true to false alarm ration is 1:5.2. It indicates that TDCT is able to effectively and accurately detect target bugs and automatically generate the whole program's inputs by unit inputs.

In particular, Sect. 2 discusses related work, Sect. 3 describes the detail of TDCT, Sect. 4 details the implementation, Sect. 5 discusses the experimental verifications and Sect. 6 concludes the paper with future work.

2 Related Work

TDCT is a professional detection tool based on the knowledge from multiple domains. In this section, we list some relevance knowledge of these domains. The second and third paragraphs describe concolic testing and concolic unit testing. The fourth and fifth paragraphs describe some work related to reduce false positive for concolic unit testing. The rest of this section is about related work to verify global path accessibility and automatically generate global inputs.

Concolic testing [11–13] is an automatic test generation techniques based on symbolic execution [14–16] that can systematically explore paths through a whole program. The advantage of concolic testing is automatically traversal different paths based on goal-oriented and generate test input, without relying on test cases to directly trigger vulnerabilities. However, due to the influence of the path explosion, it is unrealistic to only use concolic execution to analyze large programs, which will generate false negative [17,18].

Concolic unit testing is able to perform comprehensive analysis within small functions of the program [19]. Some researches exploit filtering method to contribute in improving accuracy [1,7] of concolic unit testing. Concolic unit testing is usually used for regression testing of patches [20–22]. In recent years, there has been a lot of research on bug search in programs based on patches [23–26], but the technical methods used are also different.

Under-constrained symbolic execution [19] is a simple and powerful combine symbolic execution with unit testing. It can obtain any function and run it without initializing any data structure or modeling the environment. UC-KLEE [27] is a typical representative of under-constrained symbolic execution in recent years. It improves scalability by directly checking a single function instead of the entire program, effectively skipping costly path prefixes, so it can be used without a main function analyze the target code, and also analyze the kernel code that is difficult to build a running environment. But preliminary experiments have shown that UC-KLEE as a tool for finding general bugs in a single version of a function results in a much higher rate of false positives.

CONBRIO [1] constructs an extended unit of a target function that consists of f and closely relevant functions to f. The relevance of a function g to f is measured by the degree of dependency of f on g. Then, CONBRIO performs concolic execution of an extended unit of f. The TDCT in this article is similar to CONBRIO to a certain extent. Both determine the range of the extended unit through the correlation between functions. However, CONBRIO obtains only the dependence of the functions through fuzzing for a period of time, which cannot accurately reflect the relationship between functions. CONBRIO implies a hypothesis, that is, there is at least one path through the target function after a period of fuzzing test. The initial inputs are generally provided by the developer, as far as possible to include the input that can hit the target function in the initial input set. But this assumption is not always true in the actual test. Therefore, the TDCT in this paper calculates the relevance between functions by using ICFG to improve the usability and accuracy of the test.

SPAIN [23] is a patch analysis framework to automatically learn the security patch patterns and vulnerability patterns, and identify them from the program binary executables. SPAIN can be useful in vulnerability and patch understanding, similar bug hunting, binary code auditing, and, eventually, the program security enhancement. But SPAIN focus on patches in which only one function is modified for one patch, but do not support patches where multiple functions are changed for one patch.

Several heuristic-based approaches have been proposed to guide an execution toward a specific branch. Symbolic Java Path Finder (Symbolic JPF) [26] describe an approach to testing complex safety critical software that combines unit-level symbolic execution and system-level concrete execution for generating test cases that satisfy user-specified testing criteria. KATCH [24] is technique for patch testing that combines symbolic execution with several novel heuristics based on program analysis which allow it to quickly reach the code of the patch. It uses existing test cases from the program's regression suite as a starting point for synthesizing new inputs. For each test case input, KATCH computes an estimated distance to the patch and then selects the closest input as the starting point for symbolic exploration. Compared with manual testing, KATCH was able to increase patch coverage more than doubled the manual. Despite the increase in coverage and the bugs found, KATCH was still unable to cover most of the targets.

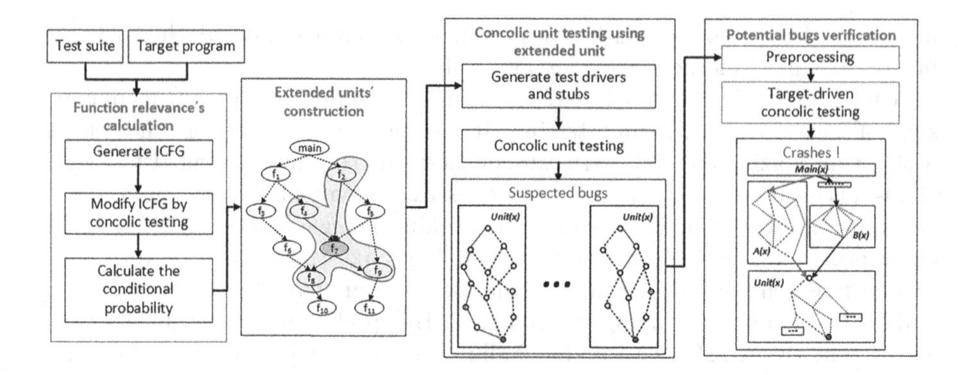

Fig. 1. The overview of TDCT.

Shadow symbolic execution [25] is a novel technique for generating inputs that trigger the new behaviors introduced by software patches. The key idea behind shadow symbolic execution is to run both versions in the same symbolic execution instance and systematically test any encountered code-level divergences. The technique unifies the two program versions via change annotations, maximizes sharing between the symbolic stores of the two versions, and focuses exactly on those paths that trigger divergences. However, Shadow is not fully automatic, while many of the annotations could be automatically added, manual assistance might still be needed.

3 Design

3.1 Overview

Figure 1 shows the overall process of TDCT. Generally, the inputs to TDCT are: 1) a program with a target function which can be a function contains bugs or a newly patched function, etc. 2) a test suite of the whole program. The aim of TDCT is to find out if there is really an error in the target function. TDCT could be divided into four steps:

Step 1: Calculating Function relevance. TDCT generates the ICFG for the whole program by static analysis. TDCT infers the relevance between functions by calculating the conditional probability based on the function calls with all the independent paths.

Step 2: Constructing extended units. With a given relevance threshold τ, we consider f has a high relevance on g if $p(g|f) \geq \tau$. Based on the calculated relevance of f on other function, TDCT constructs an extend unit of f that contains f, f's successor functions on which f has high dependency, and a calling chain of f each of which has high relevance on f.

Step 3: Concolic Unit testing using extended unit. TDCT generates test drivers and symbolic stubs for extended units constructed in step 2. Then applies concolic unit testing to an extended unit of f to explore potential bugs and provide reports of these bugs. The extended unit in this step is the expansion of the target function using function relevance in step 1, which can effectively reduce the false positive rate of concolic unit testing.

Step 4: Verification. Although step 3 has significantly reduced the false positive of concolic unit testing, the false positive still exist due to the lack of global context. TDCT takes the potential bugs reported in step 3 as the target for verification, and implements target-driven concolic testing to further confirm the actual bugs. It can reduce a lot of manual work that the test engineer needs to complete.

The following subsections will detail how each step is performed in TDCT.

3.2 Function Relevance's Calculation

In order to obtain a comprehensive and accurate context of target function as far as possible, we write an IDA script to generate the interprocedural control flow graph (ICFG) for the whole program by static analysis. Compared to CFG, ICFG increases call information between functions, which helps us collect path information for programs to calculate function relevance between.

TDCT infers the relevance between functions by calculating the conditional probability based on the ICFG. Generally, if two functions call each other frequently, they are considered to have a higher degree of interdependence. If TDCT frequently collects that f_1 calls f_2 or f_2 calls f_1 here, f_1 and f_2 will be labeled as highly relevant by TDCT. In other words, the relevance between functions indicates the degree of mutual dependence between functions.

From the ICFG, we can compute the independent paths that have a calling relationship with the target function. For example, supposing the goal is to complete the test on f in the program P. As shown in Fig. 2, the following paths that invokes f:

- $P_1 : main \rightarrow f_1 \rightarrow f_3 \rightarrow f \rightarrow f_6$
- $P_2 : main \rightarrow f_1 \rightarrow f_3 \rightarrow f \rightarrow f_5$
- $P_3 : main \rightarrow f_2 \rightarrow f \rightarrow f_6$
- $P_4 : main \rightarrow f_2 \rightarrow f \rightarrow f_5$
- $P_5 : main \rightarrow f_2 \rightarrow f_4 \rightarrow f \rightarrow f_6$
- $P_6 : main \rightarrow f_2 \rightarrow f_4 \rightarrow f \rightarrow f_5$

Suppose that a program has a target function f and other function g that invokes f. Based on function call from ICFG, we compute dependency of f on g as $p(g|f)$. Given a static call graph and we compute $p(g|f)$ as follows:

- For e which is a predecessor of f, $p(e|f)$ is calculated as N_e/N_f, N_e is a number where g calls f directly or transitively and N_f is a number of the paths that invokes f.

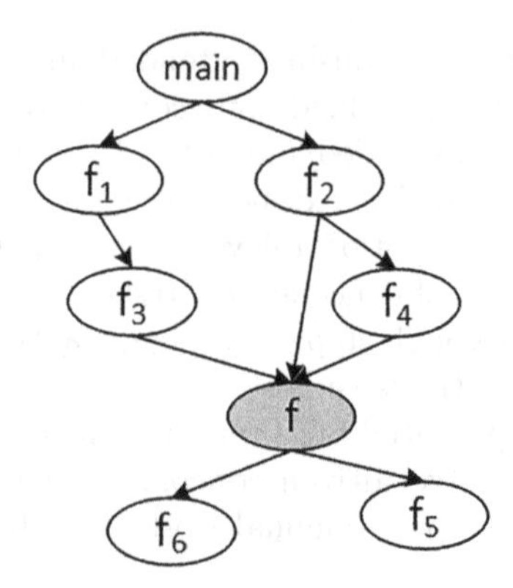

Fig. 2. Calculate the relevance of a target function f on other functions.

- For g which is a successor of f, $p(g|f)$ is calculated as N_g/N_f, N_g is a number where f calls g directly or transitively and N_f is a number of the paths that invokes f.
- For h which is a successor and predecessor of f (i.e., there exists a recursive call cycle between f and h), $p(h|f)$ is calculated as N_h/N_f, N_h is a number where f calls h or h calls f directly or transitively and N_f is a number of the paths that invokes f.

For example, in Fig. 2, we need to calculate the total number of f appear in all paths that invokes f, which is $N_f = 6$. Then we calculate the relevance of f on other functions as follows:

- $p(main|f) = 6/6 = 1.00$
- $p(f_1|f) = 2/6 = 0.33 = p(f_3|f) = p(f_4|f)$
- $p(f_2|f) = 4/6 = 0.66$
- $p(f_5|f) = 3/6 = 0.50 = p(f_6|f)$

3.3 Extended Units' Construction

Due to the lack of context, concolic unit testing usually brings a lot of false positives. In order to filter out some false alarms, TDCT expands the test object from a single function to a set of test units obtained by calculating function relevance. After having the relevance between functions, TDCT sets a threshold of the relevance according to the actual situation and constructs the extended unit related to the tested target based on this value. But thresholds will be studied in our other articles, which is not the focus of this article. For each target function f, TDCT constructs a set of extended units, including f itself, the caller e that meets the relevance threshold, and the successor g that meets the relevance threshold.

For each target function f, TDCT constructs an extended unit that contains f and f's successor functions g such that f has high dependency on all function

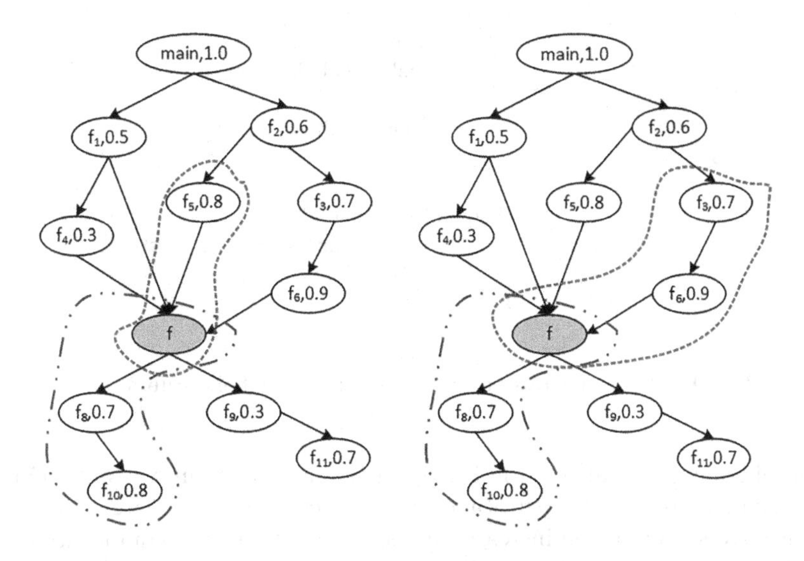

Fig. 3. Extended units of function f.

nodes in a call path from f to g (i.e. for all nodes n_i between f and g, $p(n_i|f) \geq \tau$). In other words, the relevance between all functions that directly call each other on the path from f to g exceeds the threshold. For example, assuming that the target function is f and the threshold of the function relevance is $\tau = 0.7$ in Fig. 3, this is a static call graph whose nodes are labeled with relevance of a target function f. The function relevance between f and f_5 is $p(f_5|f) = 0.8$, the relevance f and f_8 is $p(f_8|f) = 0.7$, the relevance between f_{10} and f is $p(f_10|f) = 0.8$, and the relevance between f_9 and f is $p(f_9|f) = 0.3$. Therefore, in addition to the f function itself, the functions f_8 and f_{10} will be added to the extended unit of f.

For the predecessor e, TDCT constructs an extended unit that contains f and f's predecessor functions e by traversing a static call graph from f in a reverse direction until it reaches a node labelled with low dependency of f. TDCT starts from the function f itself and searches for the upward caller function until it encounters the function e. The function relevance between the function e and the caller is lower than the preset relevance threshold. The number of function e will determine the number of extended units of the unit testing in the next stage. For example, Fig. 3 shows two callers f_5 and f_3 of the target f, so there are 2 sets of call paths for extended units will be generated $\{f_5, f\}$ and $\{f_3, f_6, f\}$. Combined with the successor, the example will generate two sets of extended units $\{f_5, f, f_8, f_10\}$ and $\{f_3, f_6, f, f_8, f_10\}$.

It is worth noting that the threshold of the relevance score is closely related to the range of the extended unit. When the threshold is higher, the extended unit will be smaller until it becomes a single target function in the extreme case. When the threshold is lower, more functions will be added to the extended unit, until the limit becomes the entire program. On one hand, the threshold

```
1   int driver_e () {
2       make_symbolic (str, 1, " str ");
3
4       char result = e (str);
5   }
6
7   char stub_g (char x) {
8       make_symbolic (res, 1, " str ");
9       return res;
10  }
```

Fig. 4. TDCT generates a test driver and a symbolic stub for f.

of the relevance score affects the false positive rate of the unit testing. On the other hand, it affects the performance and even leads to a large number of false negatives (due to the increase number of functions, the complexity of the extended unit increases, which makes it difficult to all paths are fully covered in the unit testing). We will discuss this effect in the experiment (see Sect. 5.2 for details).

3.4 Concolic Unit Testing Using Extended Unit

Concolic testing is a program analysis technique that can systematically explore paths through a program. TDCT uses concolic testing to explore the extended unit more comprehensively. For each set of extended unit related to the target function in the tested program P, TDCT automatically generates a symbolic unit test driver and some symbolic stubs in this step.

After constructing the executed units, TDCT generates a symbolic unit test *driver_e* for the entry function e to start the unit testing, and generates a symbolic stub *stub_g* for the successor function g. The test driver assigns the symbol value to the input variable of e according to the input type of the calling function e, and then calls the precursor e. As the function g is executed, it is replaced with the function *stub_g* to make the return value symbolic. For the sake of simplicity, the symbolic stub only returns the symbolic value without updating global variables and output. Test drivers and symbolic stubs usually make the actual environment of function f over approximation, which leads to infeasible execution unit paths and causes false positives. But compared to only test the target function f alone, the false positive of TDCT have been was significantly reduced.

Figure 4 shows that TDCT automatically generates a symbolic test driver *driver_e* and a symbolic stub *stub_g*. The function *driver_e* calls the function e with the symbolic variable *str* generated in the function *make_symbolic*. The function *stub_g* uses the symbolic variable *res* generated in the function *make_symbolic* to return the symbolic value instead of the concrete value.

All calls to the function g need to be replaced with $stub\_g$ to make its return value symbolic.

Because unit testing lacks context about variables, it usually bring a lot of false positive. In order to filter out most false alarms, TDCT calculates the relevance between functions by running a program over a period of time and extends the test object to an extended unit composed of multiple functions. TDCT conducts a unit testing on the extended unit to explore crash bugs related to target function f and generates a set of local inputs for each execution path of the test driver $driver\_e$. If it encounters a crash bug, TDCT will record the execution path and path constraints.

On the one hand, compared with the unit testing of a single target function, the false positive rate of TDCT is lower due to the addition of functions that are strongly depend on the target function. That is to say, the precursor constraint conditions of relevant functions are added to the extended unit. On the other hand, compared with the concolic testing with whole program that is a special unit testing with the whole program as the unit, TDCT ensures accuracy while reducing the search space and concentrating on the test for the code near the target function.

3.5 Verification

In the previous step, TDCT obtains the information about the local crash bug of the target function, such as the local input where the bug occurred and the path constraint. TDCT uses crash-related inputs and path constraints to verify whether the potential crash is real in the entire program. In this step, TDCT puts the local entry function of the potential bug as the target and marks a possible execution path in the ICFG, which can run from the function main to the above-mentioned target function entry.

In order to find the global input that can trigger the local bug as soon as possible, we design the target-driven search strategy to quickly verify the authenticity of the potential bug. See Algorithm 1 for a specific description.

When the execution of the initial state is finished, TDCT uses a path selection algorithm to select an appropriate path from the candidate pool to continue system concolic testing (line 5–7). Each state in the candidate pool is mapped to an address which is also the next execution address of the branch state. TDCT determines whether the state newly generated hits the target function by the address of the next instruction, in other words, whether the address is within the range of the target function (line 17–22). If it hits, TDCT selects the path closest to the target to continue running. This algorithm calculates the distance between the mapped address of each state in the candidate pool and the potential target address. Note that the potential target address here is the local entry address of the potential bug (that is, the entry address of function e). TDCT selects the path with the smallest D to continue execution. The calculation method of distance is: $D_i = |state_i\_addr - target\_addr|$ (line 21–27). If there are no states hit the target function, TDCT looks for the caller of the target function and confirm whether there is a state hits the caller (line 29–31). Than TDCT selects

Algorithm 1. Target-driven search strategy

Input: P-Program; T-Execution tree; Φ-Path condition; T-Execution tree; W-list of unexecuted states;
Output: Covered branches and generated input vectors

```
    //Main
 1: T ←<>;
 2: V ← initial concrete input vector;
 3: S ← initial symbolic input vector;
 4: while Termination condition are not met do
 5:    Φ_i ← selectNextState(W); //Φ = Φ_1 ∧ ... ∧ Φ_n;
 6:    Φ_i ← ConcolicExecution(P, V, S);
 7:    T ← T • Φ_i;
 8:    while State is on fork do
 9:       W ← newStateAdd(State(ℓ), Φ);
10:       addressMapping(State(ℓ));
11:    end while
12:    if ℓ_intut satisfies Φ_i then
13:       V ← V • {v};
14:    end if
15: end while
16: return pathtrace(S_i);
    //SelectNextState
17: Φ_i ← W.begin();
18: if W.size() > 1 then
19:    for p = 0, p < maxdeep do
20:       target = target_p;
21:       for state(ℓ) ∈ W do
22:          if isHitTarget(state(ℓ)) then
23:             state(ℓ).D ← |state_ℓ_addr − target_addr|;
24:          end if
25:          if state(ℓ).D is a minimum then
26:             return nextState(ℓ);
27:          end if
28:       end for
29:       target = callerSite(target_p);
30:       p++;
31:    end for
32: end if
```

the closest one among states which hit the caller to continue execution (line 19–31).

By analogy, until it finds a state which can reach the entry of a potential bug. When the execution path runs to the potential target address, that is, when it runs to the unit test entry e, TDCT will use the solver to determine whether the local input of the potential bug satisfies the path preconditions (line 12–16). If it can be satisfied, it continues to run this path until it triggers a global crash and outputs global test cases. If the local input of potential bug cannot satisfy all the path preconditions that reach the target function, TDCT discard this potential bug and think it is a false positive.

4 Implementation

We developed TDCT under the S2E [28] framework. It is mainly composed of four custom modules:

- *CalFuncRelevance*: This module generates interprocedural control flow graph (ICFG) for the whole program and infers the relevance between functions by calculating the coRnditional probability based on the function calls with all the independent paths.
- *ConsExtUnit*: Based on the calculated relevance of f on other function, TDCT constructs an extend unit of f that contains f, f's successor functions on which f has high dependency, and a calling chain of f each of which has high relevance on f.
- *ConcUnitTest*: This module automatically builds extended units according to the relevance threshold defined by user and generates test drivers and stubs for them to run concolic unit testing.
- *Verification*: Based on the bugs output by unit testing, this module completes verification with concolic testing to determine whether the local bugs exist at the whole program.

TDCT passes the function dependencies collected by *CalFuncRelevance* module to the *ConsExtUnit* module, which construct extended units by this information. The *ConcUnitTest* module performs unit testing with the extended unit and outputs potential bugs, which are used by the *Verification* module to test whether they are real bugs in the whole program.

5 Evaluation

We evaluated our approach experimentally on real-world application binaries, answering the following research questions, on a computer running Ubuntu 16.04 64-bit, equipped with a 3.4 GHz Intel Core i7-6700 CPU and 24 GB of RAM.

- Bug detecting ability: Does TDCT generate effective search heuristics and how faster TDCT detect target bugs than the current concolic testing techniques?
- Bug detecting precision: How large is the false positive ratio of TDCT with extended unit?
- Influence of the extended unit on bug detecting ability and precision: Compared with a unit with only one target function and a unit extended by static call graph, to what extent will the TDCT based on the function relevance affect the number of bugs detected and a false alarm ratio?

Table 1. Target programs and bugs.

Targets	Lines	Func.	Func cov.	Target bugs
Grep-2.0	5956	132	88.03%	6
Make-3.75	28715	555	87.28%	10
Sed-1.17	4085	73	81.73%	3
Vim-5.0	66209	1749	86.10%	12
Sum	104965	2509	#	31
Average	26241	627	85.79%	12.75

We collected 31 real-world bugs of the four FOCAL [9,10] C programs (shown in Table 1) which were continually maintained by the original developers from Dec 1996 to July 2018 and widely used in the software testing research.

We used TDCT and the following methods to conduct a comparative test: 1) Use S2E for concolic testing within a certain period of time; and 2) use static call-graph distance techniques to construct an extended unit that includes all successor functions of target f within a certain distance. The calling context of f also contains predecessor functions within a certain distance. In the second method, the distance was set as 0, 3 and 6, respectively. Specifically, when a 0 distance was selected, we had a symbolic unit test driver that contains the symbolic arguments and symbolic global variables of the target function f without any constraints on the symbol values.

Since CONBRIO [1] and FOCAL [7] are closed-source projects, we cannot directly compare these methods. But in order to fully compare the various methods, we try our best to restore their technical solutions, which called TDCT*. TDCT has been partially modified to form TDCT*, which calculates function relevance by collecting the call relationship between functions with a dynamic way.

5.1 Bug Detection Ability

In order to verify the effectiveness of bugs detecting of TDCT, as well as the other two methods were utilized to detect a total number of 31 bugs in the 4 FOCAL programs. The static call graph spent the same time as TDCT. To compare several methods as reasonable as possible, we ran all tools for 24 h. TDCT* took 2 h to dynamically collect the relationship between functions and the rest of the steps ran for 22 h. In this experiment, 0.7 was chosen as the relevance threshold. Thresholds will be studied in our other articles, which is not the focus of this article. Table 2 shows a comparison among the results of 24 h of concolic execution, static call-graph distance techniques with distance bounds 0, 3, and 6, TDCT and TDCT*.

TDCT detected 26 over 31 bugs, showing an error detection rate of 83.87%. For the method of TDCT*, the error detection rate is 77.74%. Whereas, only 11 bugs were detected in the concolic testing (S2E), with an error detection rate of 35.48%. For the static call graph distance method, the number of detected bugs was related to the selection of distance. The number of detected bugs

Table 2. Numbers of the target bugs detected.

Tar.	Tar. bugs	S2E	Static call graph distance			TDCT	TDCT*
			0	3	6		
Grep	6	2	5	3	3	5	5
Make	10	3	7	6	5	8	6
Sed	3	2	3	3	2	3	3
Vim	12	4	9	8	7	10	10
Sum	31	11	24	20	17	26	24
Rate	#	35.48%	77.74%	64.52%	54.84%	83.87%	77.74%

closest to that in TDCT when the distance was 0, but lessened to 20 and 17 respectively once the distance enlarged to 3 or 6, resulting in the error detection rate decreasing from 77.74% to 54.84%. Increasing the distance leads to increase in search range which could cause path explosions, making it is difficult to search all paths in a fixed time, consequently causes a decrease in detecting rates.

While using TDCT, 1 over 6 bugs in Grep was missed because a unit execution covered the corresponding statement but did not induce a crash. The solution is to manually add an assert function to find the error. Two target bugs in Make were omitted because the concolic execution failed to cover corresponding statements within the timeout. The strategy of using extended unit also resulted that 2 bugs in Vim were overlooked.

From this set of experiments, we can see that the number of bugs detected will decrease with the increase of the unit range. This is due to the increase in the complexity of the unit testing as functions of the unit increases. In a certain period of time, it is difficult to cover all program paths which makes false negative gradually increase. We can see that in extreme cases (when running S2E for 24 h, that is, the unit can be considered as the entire tested program), the bug discovery rate is the lowest. When the distance is 0 (that is, the unit only contains the target function f), it is easier to cover all paths in the tested unit as much as possible and more likely to find bugs in the target function.

From the above experiments, we can see that using ICFG to collect the function relevance and using the target-driven search strategy can help to increase the number of bug detection in a fixed time, that is, to achieve 83.87% of bug detection rate.

5.2 Bug Detection Precision and Effect of the Extended Unit

Table 3 shows a comparison among the false alarms number and false alarms ratio (the number of false alarms per true alarm) for static call graph distance method, TDCT and TDCT*. The reason why S2E is not added to this table is that it is a concolic testing which only has false negative and no false positive. Obviously, TDCT shows the highest bug detection precision by reporting the least number of false alarms. Interestingly, though a smaller distance could ensure the bug

Table 3. Numbers of false alarms and ratios of false alarms per true alarm.

Targets	Static call graph distance						TDCT		TDCT*	
	0		3		6					
	fal.	F/T	fal.	F/T	fal.	F/T	fal.	F/T	fal.	F/T
Grep	142	28.4	44	14.7	32	10.7	21	4.2	27	5.4
Make	504	72	126	21	79	15.8	36	4.5	53	8.8
Sed	51	17	39	13	26	13	18	6	28	9.3
Vim	657	73	356	44.5	171	24.4	59	5.9	85	8.5
Ave.	339	47.6	141	23.3	77	16	34	5.2	48.25	8.04

detecting rate of the static call graph distance method, it caused larger number of false alarms. For example, the average false alarms are 77 when the distance was 6 but raised to 339 if the distance decreased to 0. TDCT* can obtain the same bug detection capability as when the static call distance is 0, but its false positives have been significantly reduced to 48.25.

Although the static call graph method with a distance of 0 and TDCT* have the same bug detection capability as TDCT, the false alarm rate of TDCT is significantly smaller than others. Taking the testing of Grep as example, the false alarms per true alarm of TDCT is 4.2 (21/5), whereas that of the static call graph method with a 0 distance is 28.4 (142/5) and the TDCT* is 5.4 (27/5). By average, the false alarms of TDCT was even small than a half of that of the static call graph method (10% (34/339) for distance 0, 28% (34 /141) for distance 3 and 44% (34/77) for distance 6.

From this set of experiments, we can see that the number of potential bugs found in the tested unit decreases with the increase of the unit range. This is due to the increase of functions in the unit, which makes the variables in the tested unit have contextual relationships and the information is more complete, thereby reducing false positives in unit testing. We can see that in the extreme case, the number of potential bugs is highest (that is, the highest false positives) when the distance is 0 (that is, the unit only contains the target function f), which brings more workload for the confirmation of potential bugs.

Combining experiment Sect. 5.1 and experiment Sect. 5.2, it is not difficult to find that the range of the unit has a great impact on the false positive and performance. On the one hand, when the unit range is small, such as the distance is 0 (the unit only contains the objective function f in the extreme case), the unit testing can cover more paths in the unit. However, due to the lack of context, it brings a higher false positive rate, which makes it difficult to identify potential bugs. On the other hand, when the unit range is larger (for example, the distance is 6), because the context information of variables in the unit is increased, the false positive is reduced, which facilitates the confirmation of potential bugs. However, due to the increase in the scope of the unit, it is difficult for unit testing to cover all paths in a limited time, which causes false positive.

TDCT is looking for a relative balance between false positive and performance. The variable that can control the unit range in TDCT is the relevancy threshold. When the threshold is set higher, fewer functions are added to the unit. At this time, the false negative is relatively high and the performance pressure is pushed to the verification stage. When the threshold is set lower, more functions are added to the unit. At this time, the false positive is relatively low, but it takes more time to spend in the unit test stage. Therefore, the test engineer can select the appropriate relevancy threshold according to the actual situation and the previous understanding of the tested program.

From the above experiments, we can see that using ICFG to collect the function relevance can help to reduce false positives in a certain period of time, that is, TDCT has high precision with a true to false alarm ratio is 1:5.2.

6 Conclusion

To make the concolic unit testing practical and efficient in practical engineering, we present a method called TDCT (Target-Driven Concolic Testing) to solve the above challenges, which combines concolic unit testing and concolic testing. At first, TDCT obtains the ICFG of the program by static analysis and use it to construct the extended unit of the target function, which could obtain comprehensive call relationships of target function. Then, TDCT does concolic execution in extended unit and obtains unit inputs. Finally, TDCT uses custom target-driven search strategy in concolic execution to automatically generate the whole program's inputs by unit inputs, which also could further verify whether the potential bugs really exist in the whole program. The result of experiment shows that TDCT is able to effectively and accurately detect target bugs and automatically generate the whole program's inputs by unit inputs. TDCT constructs units by using ICFG to calculate the function relevance to achieve a trade-off in reducing false positive and improving performance. TDCT takes the advantages of concolic unit testing and concolic testing to improve coverage and reduce false positive.

In future work, we will further improve TDCT to make the process of obtaining global inputs easier. We will also discuss in the future whether we can better determine the relevancy threshold through certain features of the program and improve the accuracy TDCT by adding dynamic code features to machine learning.

Acknowledgment. This work is supported by the strategic Priority Research Program of Chinese Academy of Sciences, Grant No. XDC02010400.

References

1. Kim, Y., Choi, Y., Kim, M.: Precise concolic unit testing of c programs using extended units and symbolic alarm filtering. In: Proceedings of the 40th International Conference on Software Engineering (ICSE 2018). ACM, New York, NY, USA, pp. 315–326 (2018)

2. Sen, K., Marinov, D., Agha, G.: CUTE: a concolic unit testing engine for C. In: Proceedings of the 10th European Software Engineering Conference Held Jointly with 13th ACM SIGSOFT International Symposium on Foundations of Software Engineering (ESEC/FSE 2005), pp. 263–272 (2005)
3. Sen, K., Agha, G.: CUTE and jCUTE: concolic unit testing and explicit path model-checking tools. In: Ball, T., Jones, R.B. (eds.) CAV 2006. LNCS, vol. 4144, pp. 419–423. Springer, Heidelberg (2006). https://doi.org/10.1007/11817963_38
4. Ahmadi, R., Jahed, K., Dingel, J.: mCUTE: a model-level concolic unit testing engine for UML state machines. In: 2019 34th IEEE/ACM International Conference on Automated Software Engineering (ASE), pp. 1182–1185 (2019)
5. Chakrabarti, A., Godefroid, P.: Software partitioning for effective automated unit testing. In: Proceedings of the 6th International Conference on Embedded Software (EMSOFT 2006), New York, NY, USA, pp. 262–271. ACM (2006)
6. Banabic, R., Candea, G., Guerraoui, R.: Finding Trojan message vulnerabilities in distributed systems. In: Proceedings of the 19th International Conference on Architectural Support for Programming Languages and Operating Systems (ASPLOS 2014), New York, NY, USA, pp. 113–126. ACM (2014)
7. Kim, Y., Hong, S., Kim, M.: Target-driven compositional concolic testing with function summary refinement for effective bug detection, pp. 16–26 (2019)
8. Li, H., Kwon, H., Kwon, J., Lee, H.: A scalable approach for vulnerability discovery based on security patches. In: Batten, L., Li, G., Niu, W., Warren, M. (eds.) ATIS 2014. CCIS, vol. 490, pp. 109–122. Springer, Heidelberg (2014). https://doi.org/10.1007/978-3-662-45670-5_11
9. Do, H., Elbaum, S., Rothermel, G.: Supporting controlled experimentation with testing techniques: an infrastructure and its potential impact. Empirical Softw. Eng. **10**(4), 405–435 (2005)
10. FOCAL real-world crash bug benchmark. https://sites.google.com/view/focal-fse19
11. Kim, M., Kim, Y., Choi, Y.: Concolic testing of the multisector read operation for flash storage platform software. Formal Aspects Comput. **24**(3), 355–374 (2012)
12. Godefroid, P., Klarlund, N., Sen, K.: DART: directed automated random testing. In: Proceedings of the 2005 ACM SIGPLAN Conference on Programming Language Design and Implementation (PLDI 2005), pp. 213–223 (2005)
13. Kim, S.Y., et al.: CAB-fuzz: practical concolic testing techniques for COTS operating systems. In: 2017 USENIX Annual Technical Conference (USENIX ATC 2017), pp. 689–701 (2017)
14. Christakis, M., Müller, P., Wüstholz, V.: Guiding dynamic symbolic execution toward unverified program executions. In: Proceedings of the 38th International Conference on Software Engineering (ICSE 2016), pp. 144–155 (2016)
15. Stephens, N., et al.: Driller: augmenting fuzzing through selective symbolic execution. In: Proceedings of the Symposium on Network and Distributed System Security (NDSS 2016), pp. 1–16 (2016)
16. Zhang, Y., Clien, Z., Wang, J., Dong, W., Liu, Z.: Regular property guided dynamic symbolic execution. In: Proceedings of the 37th International Conference on Software Engineering (ICSE 2015), vol. 1, pp. 643–653 (2015)
17. Cadar, C., Sen, K.: Symbolic execution for software testing: three decades later. Commun. ACM **56**(2), 82–90 (2013)
18. Baldoni, R., Coppa, E., D'Elia, D.C., Demetrescu, C., Finocchi, I.: A survey of symbolic execution techniques. ACM Comput. Surv. **51**(3), 1–39 (2018). Article No. 50

19. Engler, D.R., Dunbar, D.: Under-constrained execution: making automatic code destruction easy and scalable. In: Proceedings of the of 2007 International Symposium on Software Testing and Analysis (ISSTA 2007), pp. 1–4 (2007)
20. Trabish, D., Mattavelli, A., Rinetzky, N., Cadar, C.: Chopped symbolic execution. In: 2018 IEEE/ACM 40th International Conference on Software Engineering (ICSE), Gothenburg, pp. 350–360 (2018)
21. Seo, H., Kim, S.: How we get there: a context-guided search strategy in concolic testing. In: Proceedings of the 22nd ACM SIGSOFT International Symposium on Foundations of Software Engineering (FSE 2014), New York, NY, USA, pp. 413–424. Association for Computing Machinery (2014)
22. Pham, V.-T., Ng, W.B., Rubinov, K., Roychoudhury, A.: Hercules: reproducing crashes in real-world application binaries. In: Proceedings of the 37th International Conference on Software Engineering (ICSE 2015), vol. 1, pp. 891–901. IEEE Press (2015)
23. Xu, Z., Chen, B., Chandramohan, M., Liu, Y., Song, F.: SPAIN: security patch analysis for binaries towards understanding the pain and pills. In: Proceedings of the 39th International Conference on Software Engineering, pp. 462–472. IEEE Press (2017)
24. Marinescu, P.D., Cadar, C.: KATCH: high-coverage testing of software patches. In: Proceedings of the 2013 9th Joint Meeting on Foundations of Software Engineering, pp. 235–245. ACM (2013)
25. Kuchta, T., Palikareva, H., Cadar, C.: Shadow symbolic execution for testing software patches. ACM Trans. Softw. Eng. Methodol. (TOSEM) **27**(3), 10 (2018)
26. Ramos, D.A., Engler, D.R.: Under-constrained symbolic execution: correctness checking for real code. In: Proceedings of the 24th USENIX Conference on Security Symposium (SEC 2015), pp. 49–64. USENIX Association (2015)
27. Păsăreanu, C.S., et al.: Combining unit-level symbolic execution and system-level concrete execution for testing NASA software. In: Proceedings of the 2008 International Symposium on Software Testing and Analysis (ISSTA 2008), New York, NY, USA, pp. 15–26. Association for Computing Machinery (2008)
28. Chipounov, V., Kuznetsov, V., Candea, G.: S2E: a platform for in-vivo multi-path analysis of software systems. ACM SIGARCH Comput. Archit. News **39**, 265–278 (2011)

Multi-task Allocation Based on Edge Interaction Assistance in Mobile Crowdsensing

Wenjuan Li[1], Guangsheng Feng[1(✉)], Yun Huang[2], and Yuzheng Liu[1]

[1] College of Computer Science and Technology, Harbin Engineering University, Harbin 150001, China
fengguangsheng@hrbeu.edu.cn

[2] The Second Research Institute, China Aerospace Science and Industry Corporation, Beijing, China

Abstract. This paper studies the problem of multi-task allocation in hotspots. Due to the frequent task requests in hotspots, the traditional mobile crowdsensing (MCS) cannot meet its requirements, so we adopt a distributed MCS architecture to complete this process. This paper proposes an MCS multi-task allocation framework based on edge interaction assistance, which uses edge nodes to distribute the computing load of MCS platform. In the case of short time and heavy tasks, what this paper needs to do is to minimize delay and cost while ensuring data quality, so as to express the multi-task assignment problem as a multi-objective optimization problem. Through the analysis of the process, we propose a two-phase allocation scheme (TPAS) to determine task allocation. The first phase is the task selection based on cosine similarity. The number of tasks can be reduced by considering the similarity of tasks. The second phase is task allocation based on the improved NSGA-II algorithm. The simulation results show that under the premise of satisfying the task data quality, TPAS can obtain lower task cost and delay, and significantly improve the task completion rate.

Keywords: Mobile crowdsensing · Multi-task allocation · Similarity of tasks · NSGA-II algorithm

1 Introduction

In recent years, due to the rapid growth in the number of mobile devices and wearables used, mobile crowdsensing (MCS) has become increasingly popular, becoming an attractive paradigm for sensing and collecting data [1]. Because of human intelligence, ubiquity and mobility, using crowds for data collection can lead to higher coverage and better environmental awareness. Similarly, vehicles can also be used for data collection [13]. The traditional MCS service is a cloud-based centralized architecture [2,3], in which the cloud server is responsible for task allocation. When there are a small number of tasks, traditional MCS can be done well. However, in hotspots, due to the large scale of user groups, complex information

Y. Lai et al. (Eds.): ICA3PP 2021, LNCS 13157, pp. 214–230, 2022.
https://doi.org/10.1007/978-3-030-95391-1_14

and frequent task requests from users, there may be a large number of parallel tasks at the same time, which will bring great delay and network congestion.

In order to overcome the inherent limitations of traditional MCS systems, some literatures have proposed a new MCS system architecture combining edge computing [5–8], in which the task allocation is moved from the cloud layer down to the edge layer. Because mobile edge computing (MEC) [4] has the advantages of low latency, large bandwidth, high performance and high security. Adding it to MCS can reduce the computing load of cloud computing center, relieve the pressure of network bandwidth, improve the efficiency of data processing, and reduce the delay and privacy threats. In addition, for multitasking, all tasks share the user resource pool, so there is a competitive relationship between tasks. As the number of tasks increases, resource competition between multiple tasks will become more intense, so task allocation strategies need to be used to optimize the problem. However, in order to improve the completion efficiency of tasks, it is not enough to only rely on the decentralized processing and allocation strategy of tasks, but also need to reduce the number of tasks to be performed.

Although the introduction of MEC can bring great convenience and advantages to MCS, it also brings some problems. For example, due to the mobility of users, during the task allocation process, edge nodes cannot grasp the historical information of all users in their controlled area, which causes great inconvenience to the selection and evaluation of users. Most of the literature [7] put user evaluation on the cloud, but putting all user data on the cloud will not only cause the threat of privacy leakage, but also bring some delay.

We study the problem of multi-task allocation in hotspots. Due to the large scale of user groups and the frequent real-time requests from users, it is a challenge to efficiently complete all tasks in the case of short time and heavy tasks. In addition, due to the limited capabilities of edge nodes, while reducing the computing load of the cloud server, it will also bring a certain delay and cost. Therefore, what this paper needs to do is to select fewer users to complete all tasks within the specified time, so as to reduce the required cost and time delay. Specifically, the main contributions of this paper are summarized as follows:

- For the problem of multi-task allocation in hotspots, we propose an MCS task allocation framework based on edge interaction assistance. This framework adds edge layer on the basis of the traditional MCS framework, transfers task allocation and user evaluation from the original cloud layer to the edge layer, and solves the lack of localized logs at the edge nodes by strengthening the information interaction between the edge layer.
- We express the multi-task allocation problem as a multi-objective optimization problem. Through the analysis of the task allocation process, we propose a two-phase allocation scheme (TPAS) to determine task allocation. The first phase is the task selection based on cosine similarity. The number of tasks can be reduced by considering the similarity of tasks. The second phase is task allocation based on the improved NSGA-II algorithm.
- Extensive experiments verify that the performance of TPAS is significantly better than the baseline algorithm. Using this algorithm can achieve lower task cost and time delay, and the task completion rate is significantly improved, while meeting the task data quality requirements.

2 System Model and Problem Formulation

This paper builds an MCS task assignment framework based on edge interaction assistance. As shown in Fig. 1, the specific system architecture consists of the following parts.

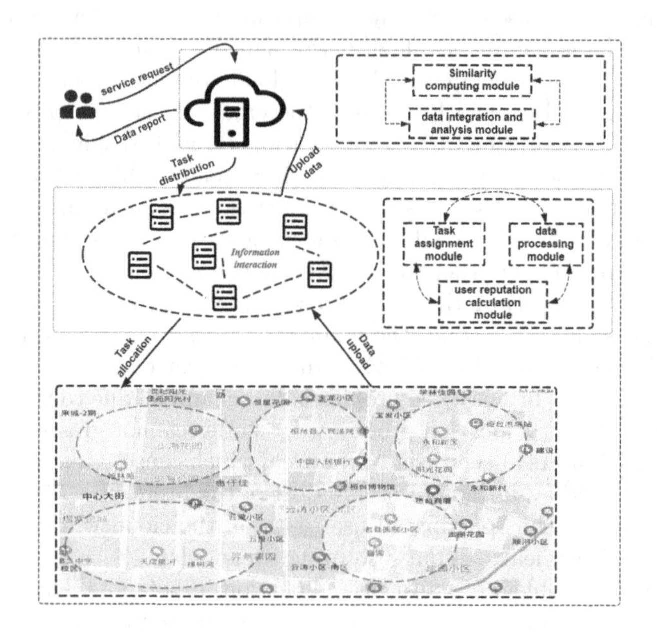

Fig. 1. MCS task allocation framework based on edge interaction assistance

2.1 Task Mode

2.1.1 Task Model Introduction

The requester sends a request task to the cloud platform according to his own needs. We consider MCS scenarios in hotspots, where perception tasks have different time and location requirements. These tasks are represented by $V = \{v_i, i = 1, 2, \cdots, N\}$. Each task is associated with a six-tuple (L_i, C_i, $[T_i^s, T_i^e]$, ϕ_i, δ_i, $type_i$), where L_i is the location of the task and C_i is the budget of the task , $[T_i^s, T_i^e]$ is the time window of the task, ϕ_i is the data size required by the task, δ_i is the data quality required by the task, and $type_i$ represents the task type.

We divide the hotspot area into M, denoted as $R = \{r_j, j = 1, 2, \cdots, M\}$. Each area selects or deploys an edge node nearby to manage all tasks in the area, and the edge node is represented by $EN = \{en_j, j = 1, 2, \cdots, M\}$.

2.1.2 Task Similarity Calculation

If the user completes it within the specified time and budget, the task will be included in the completion task pile $V^{com} = \{v_z^{com}, z = 1, 2, \cdots, N-1\}$. Next, the cloud platform judges the similarity of the two tasks by calculating the cosine similarity between the task to be performed and the task that has been completed.

First, it is necessary to determine whether each task's location, task type, and task required time are the same or very similar. If they are similar, continue to the next step. Then, calculate the cosine similarity of $\boldsymbol{v}_i$ and $\boldsymbol{v}_z^{com}$, the calculation formula is:

$$\cos\left(\overrightarrow{v_i}, \overrightarrow{v_z^{com}}\right) = \frac{\sum_1^b v_i^h \cdot v_z^{h*}}{\sqrt{\sum_1^b \left(v_i^h\right)^2}\sqrt{\sum_1^b \left(v_z^{h*}\right)^2}} \tag{1}$$

Among them, the range is [0,1]. The closer the value is to 1, the closer the directions of the two vectors are; the closer to 0, the independent of the two vectors. We set sim^* as the threshold of cosine similarity. If $\cos\left(\boldsymbol{v}_i, \overrightarrow{v_z^{com}}\right) \geq sim^*$, it means that the task v_i is similar to v_z^{com}, and the execution task decision $\chi_i = 0$ is set. If $\chi_i = 1$, it means that the task needs to be performed.

2.2 User Model

2.2.1 User Model Introduction

We think that many users can move around in this scene, represented by $U = \{u_k, k = 1, 2, \cdots, K\}$. The user also contains a lot of information, which can be represented by a six-tuple $(w_k, R_k, l_k^i, a_k^i, m_k, [t_k^s, t_k^e])$, where w_k represents the user's work ability, R_k indicates the user's reputation, l_k^i is the location of the user, a_k^i is the distance of the user from the task location, m_k is the user's moving speed, and $[t_k^s, t_k^e]$ is the user's working time period. If the edge node publishes a task, the idle workers submit their information to the edge node that manages the area according to their own interests, and the edge node selects appropriate users from the user set to perform the task. For the published task v_i, user u_k is selected to complete the task in area r_j, and the task allocation decision $p_{j,k}^i$ is set to 1.

2.2.2 User Mobility Prediction

The transfer of the user's position is mainly the movement between regions. Regarding the region as the state, the user's movement can be regarded as the transition between the states. Since the stay time in each state is generally distributed, not all time is the update point of the process, but only the state transition time is the update point, so this paper mainly uses discrete time homogeneous semi-Markov [9,10] to represent the transition probability of the user's location. The related semi-Markov kernel Z(•) is defined by the following formula

$$\begin{aligned} Z_{j,k}^i(i,j,t^*) &= P\left(L_{j,k}^{n+1} = j, t_{j,k}^{n+1} - t_{j,k}^n \leq t^* | L_{j,k}^0 \cdots L_{j,k}^n, t_{j,k}^0 \cdots t_{j,k}^n\right) \\ &= P\left(L_{j,k}^{n+1} = j, t_{j,k}^{n+1} - t_{j,k}^n \leq t^* | L_{j,k}^n = i\right) \end{aligned} \tag{2}$$

Among them, $Z_{j,k}^i(i,j,t^*)$ represents the probability that the user u_k moves from the current area i to the next area j at or before time t^*, $L_{j,k}^n$ represents the area where the user is located, and $t_{j,k}^n$ is the corresponding arrival time. It is not difficult to find that in the above equation, $L_{j,k}^{n+1}$ only depends on $L_{j,k}^{n}$, but has nothing to do with $L_{j,k}^{n-1}$. Then we get another kernel Q(•), expressed by the following equation:

$$Q_{j,k}^{i}(i,j,t^{*}) = \begin{cases} \sum_{r=1}^{M}\sum_{t=1}^{t^{*}}\left(Z_{j,k}^{i}(i,r,t) - Z_{j,k}^{i}(i,r,t-1)\right)\cdot Q_{j,k}^{i}(r,j,t^{*}-t), i \neq j \\ 1 - \sum_{j=1,j\neq i}^{M} Z_{j,k}^{i}(i,j,t^{*}) + \\ \sum_{r=1,r\neq i}^{M}\sum_{t=1}^{t^{*}}\left(Z_{j,k}^{i}(i,r,t) - Z_{j,k}^{i}(i,r,t-1)\right)\cdot Q_{j,k}^{i}(r,j,t^{*}-t), i = j \end{cases} \quad (3)$$

Among them, $Q_{j,k}^{i}(i,j,t^{*})$ represents the probability that the user stays in the area in the time slot t^{*}. When $t^{*}=0$, mobile users cannot move from one area to another, so that we can get $Q_{j,k}^{i}(i,i,0)=1$ and $Q_{j,k}^{i}(i,j,0)=0\,(i\neq j)$.

2.2.3 User Evaluation Mechanism

The quality of the collected data is affected by the user's reputation, work ability and mobility. So the user's data quality is expressed as $h_{j,k}^{i} = R_k \cdot w_k \cdot Q_{j,k}^{i}(i,j,t^{*})$, where R_k represents the user's current reputation, w_k represents the user's work ability, and $Q_{j,k}^{i}$ represents the user's mobility.

User reputation is a long-term accumulated metric, used to evaluate participants' reputation and predict their future behavior. In addition to the user's historical records and current information, what affects reputation also includes user feedback. User feedback is calculated by comparing the completion of the user with the completion of all users in the task, and the completion of the user is expressed by the ratio of data quality and user cost [12].

$$r_k^{i} = \log\left(\frac{h_{j,k}^{i}}{c_k^{user}}+1\right) - \log\left(\frac{\sum_k h_{j,k}^{i}}{\sum_k c_k^{user}}+1\right) \quad (4)$$

For the current information of users, we consider two aspects: time reliability and data reliability. Time reliability refers to the ratio of the time to complete the task and the time required by the task. The formula is defined as

$$\omega_{j,k}^{i} = \frac{e_{j,k}^{i} - s_{j,k}^{i}}{T_i^{e} - T_i^{s}} \quad (5)$$

Among them, $e_{j,k}^{i}$ represents the time when the user completes the task, and $s_{j,k}^{i}$ represents the time when the user starts the task. The larger the value, it indicates that the user spends a lot of time to complete the task. When $\omega_{j,k}^{i} > 1$, it means that the user did not complete the task on time within the specified time, so set $\omega_{j,k}^{i} = 0$. Data reliability is expressed as follows

$$\theta_{j,k}^{i} = 1 - \frac{\left|d_{j,k}^{i} - \phi_i\right|}{\phi_i} \quad (6)$$

When $\frac{|d_{j,k}^i - \phi_i|}{\phi_i} > 1$, it means that the data collected by this user is too redundant or too much useless data, so set it to $\theta_{j,k}^i = 0$.

In addition, we express the user's history as the number of real tasks n_k^{true} and fake tasks n_k^{fake} completed by the user, and we use them to describe user reliability. First, calculate the number of false and honest behaviors performed by workers in the data collection process, and then use the probability density function $Beta(\alpha, \beta)$ to evaluate the trust value.

$$Beta(\alpha,\beta) = \frac{\Gamma(\alpha+\beta)}{\Gamma(\alpha)\Gamma(\beta)}\theta^{\alpha-1}(1-\theta)^{\beta-1} = \frac{(\alpha+\beta-1)!}{(\alpha-1)!(\beta-1)!}\theta^{\alpha-1}(1-\theta)^{\beta-1} \tag{7}$$

Knowing that the expected function of $Beta(\alpha, \beta)$ is $E[Beta(\alpha, \beta)] = \frac{\alpha}{\alpha+\beta}$ [11], then user reliability is expressed as follows

$$\sigma_{j,k}^i = Beta\left(n_k^{true}+1, n_k^{fake}+1\right) = \frac{1+n_k^{true}}{2+n_k^{fake}+n_k^{true}} \tag{8}$$

Therefore, the specific definition form of the user's current reputation value is as follows

$$R_k^i = \alpha \cdot \omega_{j,k}^i + \beta \cdot \theta_{j,k}^i + (1-\alpha-\beta) \cdot \sigma_{j,k}^i \tag{9}$$

In order to prevent users from continuously submitting real data in a short period of time to improve reputation, and uploading false data in subsequent tasks, when users submit false data, the reputation should be significantly decreased exponentially. In this paper, the ratio of the number of false tasks to the number of tasks completed by users is used as the attenuation coefficient. The attenuation coefficient is defined as $\zeta_{j,k}^i = \frac{n_k^{fake}+1}{n_k^{true}+1}$, and the penalty function formula is $P_{j,k}^i = e^{-\zeta_{j,k}^i}$.

The next step is to set the update rules. When the user submits real data, the user's reputation value should increase, and when the user submits false data, the reputation should be significantly reduced. Therefore, we use arctan(•) [12] to represent the update rule.

$$R_k^{new} = \frac{1}{\pi}\arctan\left(P_{j,k}^i \cdot (R_k^i + r_k^i)\right) + \frac{1}{2} \tag{10}$$

If the user participates for the first time, set the initial value of the reputation value to 0. If $R_k \geq \eta$, it means that the reputation of the user meets the requirements, where η is the reputation threshold.

2.3 Model Goal

2.3.1 Cost of Completing the Task

The payment fee is divided into two parts, one part is the fee paid to the edge server to process data and evaluate the reputation value of the participants, and the other part is the fee paid to the user. In order to ensure the enthusiasm of users to participate, the cloud platform also needs to pay incentive costs to users participating in the task. The specific formula is as follows:

$$c_i = c_i^{ms} + \sum_k \left(c_k^{user} \cdot p_{j,k}^i\right) = (c_{ser} \sum_k (p_{j,k}^i \cdot d_{j,k}^i) + c_{eva} \cdot \gamma) \\ +(c_{per} + c_{mov} \cdot a_k^i + h_{j,k}^i \cdot c_{inc}) \tag{11}$$

Among them, c_i^{ms} is the cost paid to the edge node, and c_k^{user} is the cost paid to the user, c_{ser} represents the service price of processing unit data, c_{eva} is the fee paid to the edge server to evaluate the reputation of participants, c_{per} is the perceived cost, c_{mov} is the mobile cost, c_{inc} is the incentive cost, γ is the number of users participating in the task.

2.3.2 Time Delay and Task Completion Time

We define time delay as the time from the end of the perception phase to the completion of the data upload. Because the size of the task is negligible compared to the data, so the delay mainly includes two aspects: data upload and queuing delay (related to the processing efficiency of edge nodes). In addition, the edge node needs to perform data fusion on the acquired data before uploading the data [14].

$$t_i^{delay} = (\frac{1}{\gamma} \sum_k \frac{d_{j,k}^i}{P_{user}} \cdot p_{j,k}^i + \frac{\phi_i}{P_{en}}) + \frac{\kappa(\sum_k d_{j,k}^i + S_j^i)}{f_{en}} \tag{12}$$

Among them, P_{user} is the transmission rate when the user uploads data, and P_{en} is the transmission rate when the edge node uploads data, f_{en} represents the processing rate of the edge node, κ is the number of CPU cycles required to process one bit of data, and S_j^i is the amount of work remaining before.

The time it takes to complete each task mainly considers the time it takes to move to the task location and the time it takes to execute the task. The specific formula is as follows

$$t_i = \frac{1}{\gamma} \sum_k (e_{j,k}^i - s_{j,k}^i) \cdot p_{j,k}^i + \frac{1}{\gamma} \sum_k \frac{a_k^i}{m_k} \cdot p_{j,k}^i \tag{13}$$

2.4 Problem Formulation

Next, we focus on the multi-task allocation problem under the system. The main goal of task allocation is to minimize delay and cost while ensuring data quality. The problem is formalized as follows

$$\min_{p_{j,k}^i} \quad \sum_i c_i \cdot \chi_i \tag{14}$$

$$\min_{p_{j,k}^i} \quad \sum_i t_i^{delay} \cdot \chi_i \tag{15}$$

$$\begin{aligned} \textbf{s.t} \quad & c_i \cdot \chi_i \le C_i, i = 1, 2, \cdots, N && (16)\\ & t_i^s + \chi_i \cdot t_i \le t_i^e, i = 1, 2, \cdots, N && (17)\\ & T_i^s \le t_i \cdot \chi_i \le T_i^e, i = 1, 2, \cdots, N && (18)\\ & h_{j,k}^i \cdot p_{j,k}^i \ge \delta_i, k = 1, 2, \cdots, K && (19)\\ & \sum_k p_{j,k}^i = \gamma, i = 1, 2, \cdots, N && (20)\\ & p_{j,k}^i, \chi_i \in \{0, 1\}, i = 1, 2, \cdots, N && (21) \end{aligned}$$

Here, the first constraint ensures that the cost of completing all tasks must be within budget. The second constraint ensures that each recruited user must complete the task before the end of the participation time. The third constraint ensures that all tasks should be completed within the time window specified by the task. The fourth constraint ensures that each task performed meets the data quality requirements. The fifth constraint indicates the number of users that need to be selected in each task. The sixth constraint indicates that the task allocation decision and task execution decision are binary variables.

3 Algorithm Design

In this section, we propose a two-phase allocation scheme (TPAS) to solve this NP-hard problem. First, we first introduce the overall design of the TPAS algorithm. The TPAS algorithm is mainly divided into two parts. The first part is mainly carried out in the cloud to reduce the number of tasks. The second part is carried out at the edge layer to optimize multi-tasking.

3.1 Task Selection Based on Cosine Similarity

The similarity calculation between tasks is to calculate the cosine similarity of the information submitted by each task. Before performing the cosine similarity, in order to make the calculation more convenient, the task type needs to be coded first. Firstly, MCS tasks are divided into three categories: environmental awareness, infrastructure awareness and social awareness, which are represented by 'E', 'I' and 'S' respectively. Each sub-category in each category is coded by Gray code [6], and the coding form is shown as follows:

$$E = [E0000 \quad E0001 \quad E0011 \quad E0010 \quad E0110] \tag{22}$$

$$I = [I0111 \quad I0101 \quad I0100 \quad I1100] \tag{23}$$

$$S = [S1101 \quad S1111 \quad S1110] \tag{24}$$

After encoding, we need to perform similarity calculation and assign the result to the corresponding item in the SIM. Then we compare each item in the corresponding row of task v_i in SIM with the similarity threshold sim^* to determine whether v_i needs to be executed, and set the task execution decision χ_i.

Algorithm 1. TPAS:phase-1

Input: arriving task set V; complete task heap $V_{complete}$;
Output: the binary variable χ_i ; task queue to be executed $V_{execute}$;
1: initial $SIM = 0$ and sim^*;
2: **for** each task in V **do**
3: **for** each task in $V_{complete}$ **do**
4: **if** V and $V_{complete}$ have same task type and task location **then**
5: calculate cosine similarity;
6: add result to SIM;
7: **end if**
8: **end for**
9: **for** *each item in* SIM **do**
10: **if** *every items* $< sim^*$ **then**
11: set $\chi_i = 1$;
12: add task v_i to $V_{execute}$;
13: **else**
14: set $\chi_i = 0$;
15: skip v_i and check next task in V;
16: **end if**
17: **end for**
18: **end for**

3.2 Task Allocation Based on Improved NSGA-II Algorithm

NSGA-II retains the excellent individuals in the population through fast non-dominated sorting algorithm, introduces the elite strategy to expand the sampling space, and uses the crowding degree to maintain the diversity of the group.

(1) Initialize the main parameters of NSGA-II: The main parameters include population size, number of iterations, crossover and mutation probability, the number of objective functions M and the number of decision variables V.

(2) Initializing the population: In this paper, each chromosome represents a set of task allocation decisions, so the value of the gene must be 0 or 1. Since the number of users selected by each task is fixed, when initializing the population, we must ensure that the number of 1s in each chromosome is equal to the number of users selected by the task.

(3) Evaluation function: After initializing the population, we need to evaluate each individual with cost and delay objective functions. Note that there may be some individuals who do not satisfy the problem constraints. We call these individuals an infeasible solution. Here, we define a variable called the constraint violation value (CV), the value of CV is calculated as follows

$$\begin{aligned} CV(x) = \sum_{i=1}^{N} < c_i - C_i > + < t_i^s + t_i - t_i^e > + < T_i^s - t_i > + \\ < t_i - T_i^e > + \sum_{k=1}^{K} < \delta_i - h_{j,k}^i \cdot p_{j,k}^i > + \sum_{i=1}^{N} | \sum_{k} p_{j,k}^i - \gamma | \end{aligned} \tag{25}$$

Where $< \alpha >$ means if $\alpha \leq 0$, then $< \alpha >= 0$ otherwise $< \alpha >= |\alpha|$. Obviously, for a solution, the smaller the CV value, the better the solution. At the same time, for a feasible solution, its CV value is 0, and for an infeasible solution, its CV value is greater than 0.

(4) Fast non-dominated sorting algorithm: Assuming that the population is P, the algorithm needs to calculate two parameters n_p and S_p for each individual p in P, where n_p is the number of individuals dominating individual p, and S_p is the set of individuals dominated by individual p. First, find all individuals with $n_p = 0$ in the population and save them in the current set F_1. Then, for each individual i in F_1, the set of individuals dominated by it is S_i, traverse each individual l in S_i, execute $n_l = n_l - 1$. If $n_l = 0$, save individual l in set H. Finally, mark the individual obtained in F_1 as the individual of the first non-dominant layer, and use H as the current set, and repeat the above steps until the entire population is graded.

(5) Calculation of crowding degree: First, we need to initialize the distance of individuals in the same layer, and set $n_d = 0$. Secondly, the individuals are sorted according to the objective function. Then, we set the crowding degree of the two individuals on the boundary to be infinite, and calculate the crowding degree of the individuals.

$$n_d = n_d + \frac{f_m(i+1) - f_m(i-1)}{f_m^{max} - f_m^{min}} \tag{26}$$

(6) Improved selection operation: The selection operation in this paper uses a probability selection operator based on the logistics distribution $P = \frac{1}{1+e^{-\lambda \cdot g_c}}$ where λ is the probability selection experience parameter (this paper is set to 0.01), and g_c is the current iteration number. If the individual $X1 \prec X2$, in the early stage of the population evolution, they will give up the championship winner X1 with a higher probability and choose the inferior individual X2 instead, which can ensure the population diversity in the early stage of the algorithm. In the later stage of evolution, due to the gradual convergence of the solution set, X1 is selected with a greater probability, and the small probability of selecting X2 may also be retained at the same time.

(7) Adaptive crossover and mutation operations: We need to modify the genetic operation to adapt to the content of this paper. First, two individuals are randomly selected and a cross position is generated. The values of the two individuals' cross positions are exchanged to form a new individual and then whether the new individual is feasible is checked. If after the crossover, the number of each row is greater than or less than γ, the mutation operation is required until the requirement is met; otherwise, the mutation operation is not required.

The setting of crossover and mutation operators promotes the production of more good genes and protects them from being destroyed from the perspective of evolution. This paper divides the evolutionary stage T into three parts: the front, middle and back, which are represented as $[0, T_1]$, $(T_1, T_2]$ and $(T_2, T_3]$. This paper sets $T_1 = 0.382T$ and $T_2 = 0.618T$. The specific crossover and mutation operator formula is set as

$$P_c = \begin{cases} 0.25\frac{T_1-t}{T_1} + 0.75, & t \in [0, T_1] \\ 0.25\frac{T_2-t}{T_2-T_1} + 0.5, & t \in (T_1, T_2] \\ 0.25\frac{T-t}{T-T_2}(1-\beta) + 0.5 \cdot \beta, & t \in (T_2, T] \end{cases} \tag{27}$$

$$P_m = \begin{cases} 0.1, & t \in [0, T_1] \\ 0.1, & t \in (T_1, T_2] \\ \frac{(0.1-\frac{0.1}{V})(T-i)}{T-T_2} \cdot (1-\beta) + \frac{0.1}{V} \cdot \beta, & t \in (T_2, T] \end{cases} \tag{28}$$

(8) Generate a new population: In this step, a new population will be generated. First, the operation of steps (3) and (4) is performed on the new population. Then the whole population is classified according to their pareto rank and crowding degree, and the best chromosomes are selected to form the next generation. After that, it is judged whether the algorithm meets the termination condition. If it is satisfied, the first front end is output as the pareto optimal solution. Otherwise, the algorithm returns to step (4) and loops the process until the termination condition is met.

4 Performance Analysis

In this section, we mainly verify the performance of TPAS in task allocation. In this paper, we choose greedy algorithm (Greedy) [10] and optimal selection algorithm (OSA) as the comparison algorithm. The Greedy greedily selects a user from the set of users who meet the reputation to perform the task in each iteration until the task requirements are met. OSA selects the user with the smallest objective function from the set of users who meet the reputation to assign tasks.

4.1 Parameter Setting

This paper mainly takes task total cost, task completion rate, time delay and data quality as performance indicators, and studies the impact of task number, number of users and selected number of users on performance indicators to better reflect the performance of TPAS algorithm. In the task allocation process, the user set of the corresponding region is selected according to the location of the task. For simplicity, we only consider all tasks from 9:00 to 12:00, and each task involves at least one region. The experiment of this paper is simulated on MATLAB. The simulation parameters are shown in Table 1.

Algorithm 2. TPAS:phase-2

Input: task queue to be executed $V_{execute}$;user set U;
Output: the binary variable $p^{i}_{j,k}$; task cost,delay and user's reputation ;
1: initialize $U^{sat\_rep} = 0$ and the NSGA-II main parameters;
2: **for** each task in $V_{execute}$ **do**
3: **for** each user in U **do**
4: **if** Users who meet the reputation and data quality requirements in U **then**
5: add u_j to U^{sat_rep};
6: **else**
7: skip u_j and check next user in U;
8: **end if**
9: **end for**
10: random initial population, and calculate objective function;
11: fast non-dominated sorting and crowding distance calculation;
12: individuals are sorted according to Pareto level and crowding distance;
13: **for** i=1:gen **do**
14: produce new offspring by selection, crossover, and mutation operations;
15: calculate objective function value of offspring population;
16: form a new parent population;
17: fast non-dominated sorting and crowding distance calculation;
18: generation of the next population based on elite retention strategies;
19: **end for**
20: return Pareto optimal set and select an optimal chromosome from;
21: the value of $p^{i}_{j,k}$ is assigned according to the value in chromosome;
22: update task cost and delay, user's reputation;
23: **end for**

4.2 Performance Analysis

4.2.1 The Influence of the Number of Users

In order to see the changes in the experimental results more intuitively, we set the number of users from 500 to 1000, the number of tasks is 100, the number of selected users is 2, and the rest of the experimental parameter settings refer to Table 1. The results are shown in Fig. 2.

As can be seen from the Fig. 2 (a) to (b), with the increase of number of users, the task cost and delay increase first and then decrease. The reason is that when the number of users is small at the beginning, the number of tasks that can be completed is small, so the cost and delay are low. As the number of users increases, the number of tasks that can be completed also increases, so costs and delays gradually increase. When the number of users reaches a certain level, the more users meet the requirements, the edge nodes can more easily assign tasks to the appropriate user set, and the task cost and delay will gradually decrease.

In (c), we can observe that TPAS can meet all tasks when the number of users reaches 700. This is because when the number of users is small, many tasks fail to find users that meet their requirements, so that the task completion rate is low. As the number of users increases, the users who can meet the task requirements are also increased. When it reaches 700, all tasks can be assigned to users who meet the requirements, and the tasks can be completed on time. In addition, combined with (d), it can be found that while data quality is satisfied, the task completion rate of TPAS is higher than that of the two baseline algorithms.

Table 1. Main simulation parameter settings

Parameter	Parameter value
Number of edge nodes	5
Assign the number of tasks per task	5
Task deadline/min	5–45
The movement speed of the participants/(m $\cdot$ min^{-1})	100–150
Types of perception tasks	12
Budgets for individual tasks	15–25
Distance between participants and tasks/m	1–500
Data size required by a single task/MB	0–20
The data quality required by a single task ranges	0.5–0.7

4.2.2 The Influence of the Number of Tasks

Next, we examine the impact of the number of tasks, where we assume that the number of users is 800, the number of selected users is 2, and the number of tasks is between 100, 120, 140, 160, 180, 200. The results are shown in Fig. 3.

It can be observed from Fig. 3 (a) to (b) that as the number of tasks increases, the total task cost and delay are gradually increasing, but when the number of tasks reaches 200, there is a downward trend. This is because as the number of tasks increases, edge nodes need to recruit more users to complete the tasks, so the cost and delay will increase accordingly. However, due to the limited number of users and their limited capabilities, the number of tasks completed is also limited. As a result, as the number of tasks increases, the user's working ability is declining, and more and more tasks cannot be assigned, resulting in a continuous decrease in the completion rate, as shown in (c).

At the same time, it can be observed from (c) that the rate of decrease of the task completion rate of TPAS is lower than that of the other two algorithms, because as the number of tasks increases, the number of similar tasks is also increasing, so the number of tasks completed in TPAS is much higher than that of Greedy and OSA. In addition, it can be seen from (d) that the data quality under the three algorithms all meet the requirements of the task.

4.2.3 The Influence of the Number of Selected Users

Finally, we examine the impact of the number of selected users. This paper observes the influence of the number of selected users by setting it between 1 and 5. In addition, we assume that the number of users is 800 and the number of tasks is 100. The results are shown in Fig. 4.

It can be observed from Fig. 4 (a) to (b) that as the number of selected users increases, the total task cost and time delay under the three algorithms are in a state of rising first and then falling. At the beginning, the costs and delays increase as the number of users increases. However, as more users are selected,

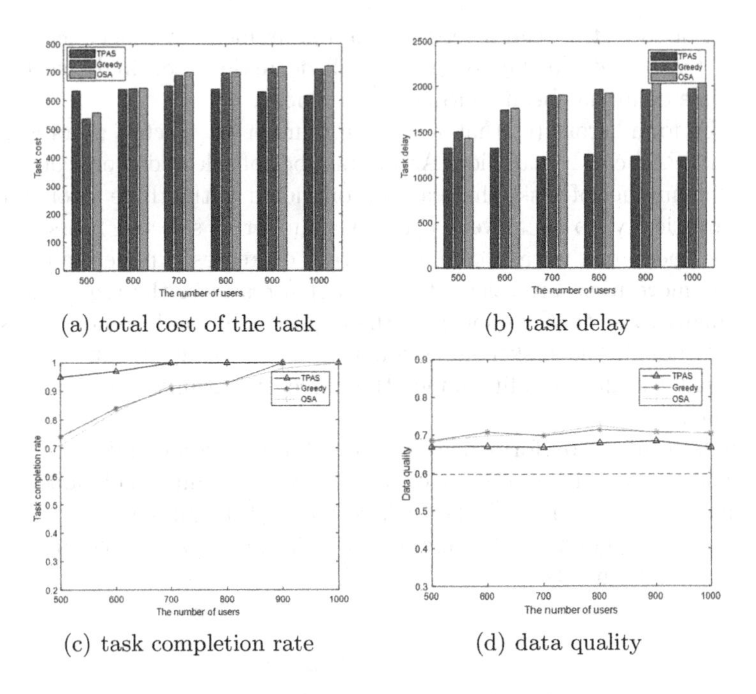

(a) total cost of the task
(b) task delay
(c) task completion rate
(d) data quality

Fig. 2. The influence of the number of users

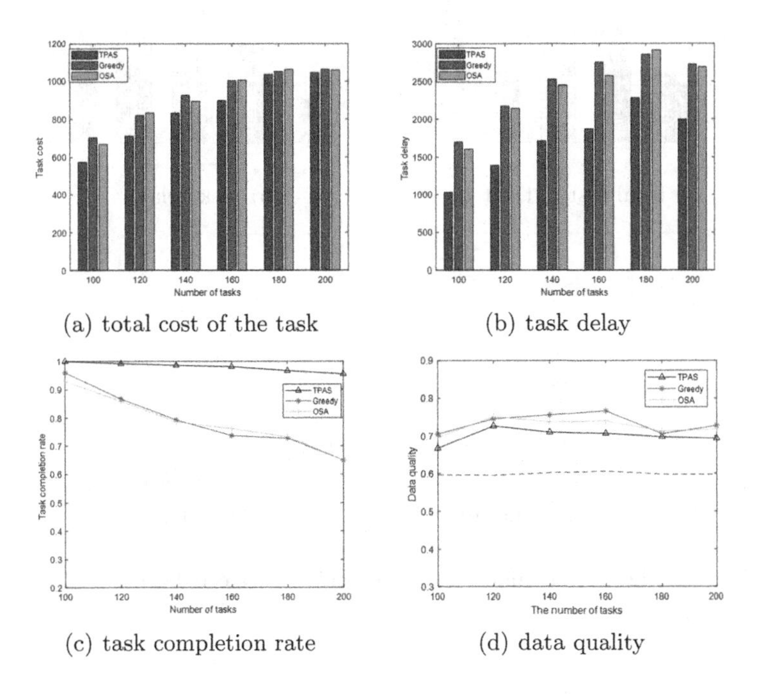

(a) total cost of the task
(b) task delay
(c) task completion rate
(d) data quality

Fig. 3. The influence of the number of tasks

the number of users that can meet the task requirements will decrease, and the number of tasks that can be completed will decrease accordingly, so that the cost and time delay has been continuously reduced.

It can be found from (c) that when the number of selected users is 1 to 2, all tasks in TPAS can be satisfied. As the number of selected users continues to increase, the number of tasks that can be completed in the three algorithms has dropped drastically, so that every time the number of selected users increases by one, the speed of the task completion rate decreases exponentially. This is because the more users are selected, the fewer users meet the requirements, so there are many tasks that cannot meet the requirements and cannot be assigned to users. Although the task completion rate is declining, it can be observed from (d) that the data quality under the three algorithms can meet the task requirements.

Through the above performance analysis of TPAS, we can find that compared with the baseline algorithm, the use of the TPAS algorithm can obviously obtain the optimal user set to reduce the task cost and delay to the greatest extent, and the task completion rate is significantly improved, while meeting the task data quality requirements.

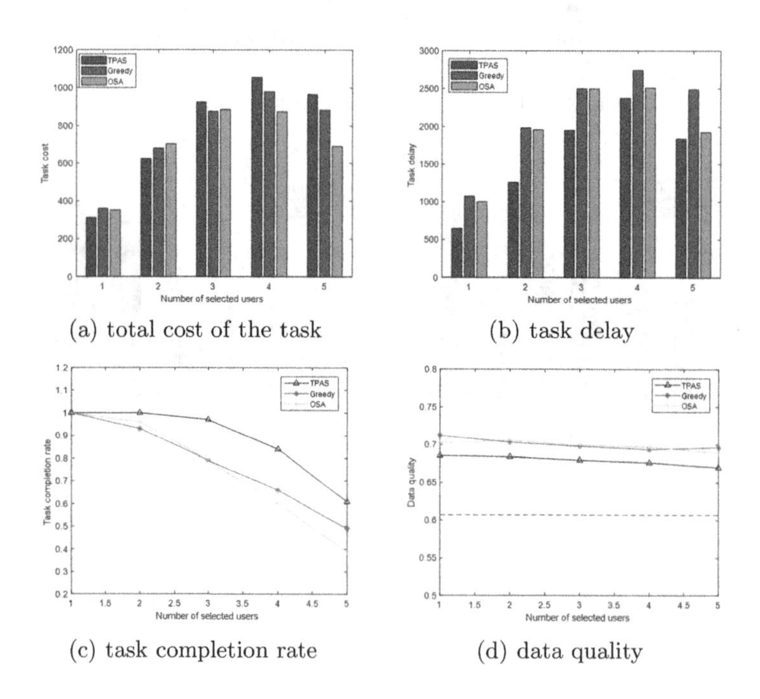

(a) total cost of the task (b) task delay

(c) task completion rate (d) data quality

Fig. 4. The influence of the number of selected users

5 Conclusion

This paper studies the problem of multi-task allocation in hotspots, and proposes an MCS task allocation framework based on edge interaction assistance. First, we introduce the system model and formalize the MCS multi-task allocation problem assisted by edge interaction. Our goal is to assign tasks to appropriate users, and to minimize task costs and delays while ensuring data quality. By analyzing the problem, we propose a TPAS algorithm to solve this problem. Finally, our simulation results show that the TPAS algorithm can obtain the optimal user set to reduce task cost and time delay, and the task completion rate is significantly improved, while meeting the task data quality requirements.

Acknowledgement. This work is supported in part by the Natural Science Foundation of China under Grant 61872104.

References

1. Capponi, A., Fiandrino, C., Kantarci, B., Foschini, L., Kliazovich, D., Bouvry, P.: A survey on mobile crowdsensing systems: challenges, solutions, and opportunities. IEEE Commun. Surv. Tutorials **21**(3), 2419–2465 (2019). https://doi.org/10.1109/COMST.2019.2914030. Thirdquarter
2. Guo, B., Yu, Z., Zhou, X., Zhang, D.: From participatory sensing to Mobile Crowd Sensing. In: 2014 IEEE International Conference on Pervasive Computing and Communication Workshops (PERCOM WORKSHOPS), Budapest, 2014, pp. 593–598 (2014). https://doi.org/10.1109/PerComW.2014.6815273
3. Zheng, Z., Wu, F., Gao, X., Zhu, H., Tang, S., Chen, G.: A budget feasible incentive mechanism for weighted coverage maximization in mobile crowdsensing. IEEE Trans. Mob. Comput. **16**(9), 2392–2407 (2017). https://doi.org/10.1109/TMC.2016.2632721
4. Abbas, N., Zhang, Y., Taherkordi, A., Skeie, T.: Mobile edge computing: a survey. IEEE Internet Things J. **5**(1), 450–465 (2018). https://doi.org/10.1109/JIOT.2017.2750180
5. Marjanović, M., Antonić, A., Žarko, I.P.: Edge computing architecture for mobile crowdsensing. IEEE Access **6**, 10662–10674 (2018). https://doi.org/10.1109/ACCESS.2018.2799707
6. Xiong, J., Chen, X., Yang, Q., Chen, L., Yao, Z.: A task-oriented user selection incentive mechanism in edge-aided mobile crowdsensing. IEEE Trans. Netw. Sci. Eng. **7**(4), 2347–2360 (2020). https://doi.org/10.1109/TNSE.2019.2940958
7. Zhou, P., Chen, W., Ji, S., Jiang, H., Yu, L., Wu, D.: Privacy-preserving online task allocation in edge-computing-enabled massive crowdsensing. IEEE Internet Things J. **6**(5), 7773–7787 (2019). https://doi.org/10.1109/JIOT.2019.2903515
8. Xia, X., Zhou, Y., Li, J., Yu, R.: Quality-aware sparse data collection in MEC-enhanced mobile crowdsensing systems. IEEE Trans. Comput. Soc. Syst. **6**(5), 1051–1062 (2019). https://doi.org/10.1109/TCSS.2019.2909265
9. Wang, E., Yang, Y., Wu, J., Liu, W., Wang, X.: An efficient prediction-based user recruitment for mobile crowdsensing. IEEE Trans. Mob. Comput. **17**(1), 16–28 (2018). https://doi.org/10.1109/TMC.2017.2702613

10. Yang, Y., Liu, W., Wang, E., Wu, J.: A prediction-based user selection framework for heterogeneous mobile crowdsensing. IEEE Trans. Mob. Comput. **18**(11), 2460–2473 (2019). https://doi.org/10.1109/TMC.2018.2879098
11. Wang, T., Wang, P., Cai, S., Ma, Y., Liu, A., Xie, M.: A unified trustworthy environment establishment based on edge computing in industrial IoT. IEEE Trans. Industr. Inf. **16**(9), 6083–6091 (2019)
12. Wang, W., Gao, H., Liu, C.H., Leung, K.K.: Credible and energy-aware participant selection with limited task budget for mobile crowd sensing. Ad Hoc Netw. **43**, 56–70 (2016). https://doi.org/10.1016/j.adhoc.2016.02.007
13. Lang, H., Liu, A., Xie, M., Wang, T.: UAVs joint vehicles as data mules for fast codes dissemination for edge networking in smart city. Peer-to-Peer Netw. Appl. **12**(6), 1550–1574 (2019)
14. Huang, M., Liu, A., Wang, T., Huang, C.: Green data gathering under delay differentiated services constraint for Internet of Things. Wirel. Commun. Mobile Comput. **2018**(3), 1–23 (2018)

A Variable-Way Address Translation Cache for the Exascale Supercomputer

Tiejun Li, Jianmin Zhang(✉), Siqing Fu, and Kefan Ma

College of Computer, National University of Defense Technology, Changsha, China
{tjli,jmzhang,fusiqingnudt,makefan14}@nudt.edu.cn

Abstract. Most of the parallel programs running at the exascale supercomputers have taken advantage of virtual address instead of physical address. Therefore the virtual and physical address translation mechanism is necessary and critical to bridge the hardware designs, driver programs and software applications. We have proposed a novel virtual and physical translation architecture. It includes the variable-way address translation Cache (VwATC), stall-hidden refresh scheme, the address validity checker and many reliability designs. The VwATC can be configured to four different tag ways and data memory capacities, and then the unused memory banks will be gated clock to reduce power dissipation. The VwATC adopts the large capacity eDRAM (embedded Dynamic Random Access Memory) to meet the high hit ratio of mapping the virtual address and physical address requirements. It also can run as a dual mode, and switch the Cache to buffer mode to reduce the accessing latency. Many experiments have been conducted on the real chip which implements the scalable address translation Cache. The results show that the VwATC has high hit ratio while running the well-known benchmarks, and save the dynamic power with different ways configuration.

Keywords: Exascale supercomputer · Address translation · Variable way · Cache

1 Introduction

From the TOP500 supercomputers list [1], each the high performance parallel computer system is composed of thousands of compute nodes. Such massive parallelism puts high pressure on the designs of the exascale supercomputers' network. It is necessary to efficiently support higher bandwidth and lower latency for data communication [2,3], because the parallel computers consist of more and more computing nodes due to the larger scale of programming software and applications. Consequently, a crucial VLSI chip called NIC (Network Interface Chip) [4] is designed and implemented to meet these requirements. In general,

This work is supported by the National Key Research and Development Program of China under grant No. 2018YFB0204301, and the National Natural Science Foundation of China under grant No. 62072464 and U19A2062.

Y. Lai et al. (Eds.): ICA3PP 2021, LNCS 13157, pp. 231–244, 2022.
https://doi.org/10.1007/978-3-030-95391-1_15

most of users are accustomed to utilize virtual address in their parallel programs, because the virtual address is consecutive and easy to use in the software. In order to fulfill RDMA (Remote Direct Memory Access) operations in virtual address mode, the NIC is necessary to support translating virtual address to physical address. However, all of the physical address data are generally stored in the main memory of the host CPU or accelerator. Since the NIC obtains physical address through the PCI-E interface to access main memory and it will introduce the large latency, the virtual and physical translation mechanism is critical for the high performance computers [5–10].

In recent years, the problem of DRAM-based Cache designs has been addressed rather frequently in the last few years, owing to its increasing importance in numerous fields. A novel DRAM set mapping policy [11] has been proposed to reduce the overall DRAM cache access latency. The experimental results have shown that the proposed policy reduces miss rate by 12.1% or hit latency by 29.3% as compared with the general set mapping policies. Another technique called an adaptive bank mapping policy [12] was also proposed to reduce inter-core cache contention in DRAM-based cache architectures. On average, the adaptive bank mapping policy increases the harmonic mean instruction-per-cycle throughput by 19.3%, compared to the state-of-the-art bank mapping policies. A new structure of small and low latency SRAM/DRAM Tag-Cache [13] can quickly determine whether an access to the large DRAM-based L3/L4 caches will be a hit or a miss. The experimental results have shown that it can improves the average harmonic mean instruction per cycle by 13.3%. However to the best of our knowledge, there is no published work in the literature devoted to integrate the DRAM-based Cache in the interconnect network chips of parallel supercomputer systems.

In this paper, we have proposed a novel virtual and physical translation architecture of a supercomputer's network. The NIC is a kernel chip in the high performance parallel computer. In the NIC chip, we have implemented the virtual and physical translation architecture which is the most important module. The key features of the virtual and physical translation mechanism include the eDRAM-based scalable address translation Cache, stall-hidden refresh scheme, checking address validity module, the dual mode Cache and many reliability designs. To reduce the power dissipation, the tag and data ways can be configured as the 1 way, 2 ways, 4 ways or full ways to work properly, and then the clocks of other ways will be gated. In other words the capacity of VwATC (Variable-way Address Translation Cache) is supported to four different sizes, including 1/2, 1/4, 1/8 and full of original size. The VwATC adopts the high capacity eDRAM even up to 39M bits, so as to improve the hit ratio and reduce the accessing latency. The non-blocking deep pipeline is implemented to improve the reading and writing memory bandwidth. The VwATC also can switch the Cache to buffer mode to shorten the readding latency. A large amount of tests on the real chip NIC have been conducted. These results show the VwATC has high hit ratio while running the well-known benchmarks for the supercomputers. The power consumption of VwATC has been evaluated during different configuration, and it is indicated that the variable-way function can save the dynamic power substantially. As compared with three typical SRAM-based Cache [14,15], the VwATC has achieved much better performance in hit ratio.

The paper is organized as follows. The next section introduces the overview of the address translation Cache architecture. Section 3 detailedly describes the design of variable-way address translation Cache. Section 4 shows and analyzes experimental results on the hit ratio and power saving of the address translation Cache, obtained by running the mostly used software on the real chip. Finally, Sect. 5 concludes the paper and outlines future research works.

2 Architecture of Variable-Way Address Translation Cache

The address translation Cache is the most important module of the virtual and physical address translation mechanism in supercomputers' interconnect network. The main function of the address translation Cache is to accumulate the address translation data, and control the reading and writing operations for the address translation items, that is the physical address. Most of users are accustomed to utilize the virtual address in their parallel programs, owing to its key advantage of consecutive address space. But in the hardware designs, the main memory is accessing by physical address. Therefore for the packets sent to the interconnect network or received from the network, it is essential to translate the virtual address in the packets to the physical address. The latency is very large to access the physical address which are stored in main memory outside of NIC chip. Therefore, in order to bring up the gap between the continual handling the packets requirements and large latency of obtaining the physical address, a high performance Cache called variable-way address translation Cache is proposed. The VwATC is to preserve those most frequently used address translation items. In general, the capacity of the Cache in chip is far more less than the main memory outside. Therefore, the VwATC cannot maintain the whole the physical address items. If the physical address items corresponding to virtual address in the packets are missed in the Cache, the VwATC will read these items from the main memory outside.

The total capacity of the VwATC on chip is 39 Mbits. It is supported to configure as 1/2, 1/4, and 1/8 of original size. Every Cache line includes 8 physical address items. Each item contains 32 bits data and 7 bits error correcting codes (ECC). Therefore, the total bit number of a Cache line is 312. The error correcting codes is adopted into tag array and data array of VwATC. This code is single error correcting and, simultaneously, double error detecting (SEC-DED). The VwATC employs 8 ways set-associative structure. Then each way is composed of 16,384 Cache lines. The eDRAM is adopted to implement the data memory banks of VwATC. Each way is an eDRAM macro with 9.5 Mbits capacity. The eDRAM is superior memory density and power over SRAM embedded in chips. Therefore, when mass storage is needed, it is much better to utilize eDRAM instead of SRAM. The replace strategy of the VwATC is least recently used (LRU) algorithm. Although the logic of LRU is more complicated, the LRU replace strategy generally has more hitting rate than other replace strategies. To avoid the unnecessary stalls of VwATC, the non-blocking deep pipeline is implemented in the VwATC. If the former accessing VwATC operation is not finished, the latter new operation will not be blocked and can enter the pipeline. This pipeline achieves strongly improvement on throughput and bandwidth of accessing the VwATC.

The architecture and interface of the VwATC is shown in the Fig. 1. The VwATC consists of 10 modules, including the input queue, the arbiter, the tag array, the data array, the miss buffer, the write back buffer, the fill buffer, the output queue, the control and state register (CSR) and memory interface.

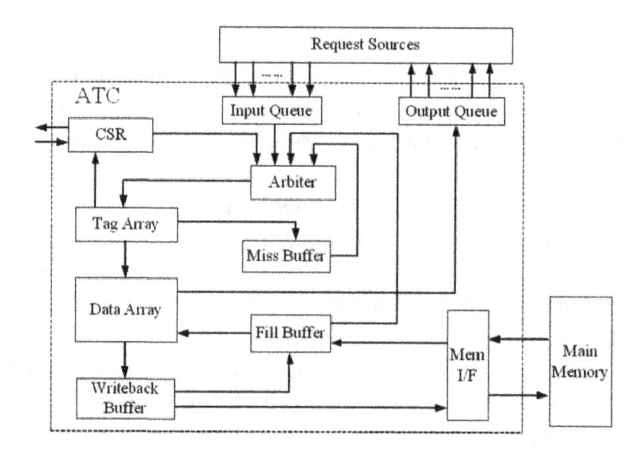

Fig. 1. Architecture and interface of the variable-way address translation Cache

3 Design of Variable-Way Address Translation Cache

First of all, we introduce how to design the variable-way Cache. To reduce the power dissipation, a variable-way address translation Cache is proposed. The VwATC supports to be configured to 1/2, 1/4, or 1/8 of total size, according to the application scenarios. While in the light workload of NIC chip, the tag memory and data memory in the VwATC are only reserved to 4 banks, 2 banks or even 1 bank, and the other banks are gated clock to save the dynamic power. In the control and state register module, the global control register contains two bits called variable way field to configure the size of the tag memory and data memory. While variable way field is two bits of 11, it is indicated that the tag and data memory are set to total size, and 10 indicate 1/2 of original size, and 01 indicate 1/4 of original size, and 00 indicate 1/8 of original size. The two bits are utilized to mask the unused ways of tag memory while the higher bits of virtual address are as compared with the tag bits of the 8 ways. Furthermore, the variable way field is also used to set the gating clock logic of the banks of the tag and data memory. For instance, if the variable way field is write to 00, the clock feeding to the tag and data memory banks number 1 to 7 are blocked to low level 0, and then the dynamic power of these 7 banks will be saved. That is to say, only the bank 0 of tag and data memory works normally. In this way, the massive memory of the VwATC can be gated clock to achieved strongly power reduction, according to the workload scenarios of the NIC chip.

Then, we detailedly describes how to implement 10 modules in VwATC. Firstly, the input queue receives the packets from several request sources outside, and encodes for each source, and then stores these packets into the FIFOs (First In First Out). The round-robin arbiter is implemented to fairly choose one from the several request FIFOs. The request packet is read from the granted FIFO, and then sent to priority arbiter module. All of the operations of the input queue works in pipeline mode. In other words it supports to accept and handle a packet from the sources at every clock cycle.

The arbiter adopts the priority arbitrating strategy. There are four request sources, including the input queue, the fill buffer, the miss buffer and the invalid engine for refresh function of the VwATC in CSR module. The priority of four sources is the invalid engine prior to the fill buffer, and the fill buffer prior to the miss buffer, the miss buffer prior to the input queue. After one source is determined, the granted request packet is transmitted to the tag array module.

The tag array is composed of three submodules, such as tag controller, tag memory bank, and replace algorithm. The tag is the higher bits of virtual address in the request packets. The relationship between the tag memory banks and the virtual address fields is illustrated in Fig. 2. The tag controller parses the virtual address, and compare the tag field with the tag reading from tag memory banks, and judges which way is hitting or all of 8 ways are missing. According to 8 ways set-associative structure, the tag memory banks are implemented by 8 SRAM memories with the depth 16,384 and width 14 bits. The 14 bits are composed of 8 bits tag, 5 bits ECC codes and 1 bit valid mark. The replace strategy adopts a typical least recently used algorithm. If all of 8 ways are used, it is necessary to evict one tag from 8 ways. Based on the LRU algorithm, the rarest used way is selected to replace by the new tag, and the physical address items in the data array are substituted simultaneously.

The data array module consists of 8 memory banks and associated control logic. Because the capacity of data array is very large, it is not suitable for implementing by the SRAM owing to its high area, and power dissipation, and more SER (Soft Error Rate) ratio. In general, the density advantage of eDRAM is approximately 2 to 3 times over SRAM, and the power advantage is approximately 2 to 10 times, and the SER advantage is approximately 250 times. Therefore, each memory bank is implemented by an eDRAM macro with depth 16,384 and width 312. The structure of 8 eDRAM banks is depicted in Fig. 2. While the tag controller determines which way is hitting, one line data in this way will be read based on the index field of virtual address. Since every line data contains 8 physical address items, one item is selected by the offset field of virtual address. Although the eDRAM has many advantages in area and power, it needs to refresh at interval. Each cell of dynamic random access memory is

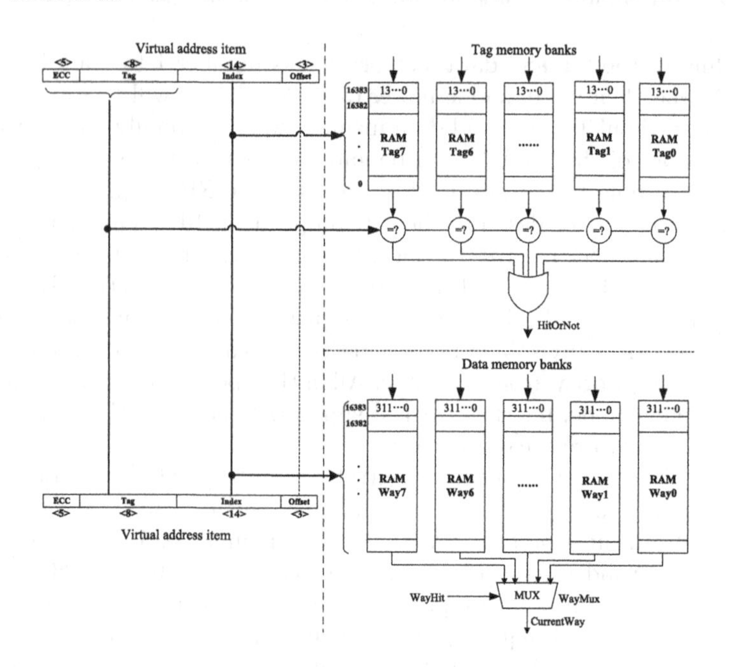

Fig. 2. Relationship between the virtual address and tag/data arrays

a transistor, which the electric charge denoted 1 will be released slowly in the process of time. Therefore, a novel eDRAM controller is proposed to hide the refresh operations into the normal read or write operations.

In general, a read, write or refresh operation for an eDRAM memory bank is completed in one cycle. Since there is only one operation occurring per cycle, a completely random address can be accessed each cycle. All other banks are available for refresh operation, making design of the eDRAM controller fairly straightforward. While refresh address and read or write address occur the collision, the refresh operation must be done firstly and a stall signal will be produced to stop the read or write operation. To avoid the stall incurring by refresh collision, an eDRAM controller is implemented. It maintains two registers, each of which points to a possible bank to be refreshed. On a given clock cycle, the controller attempts to concurrently refresh the bank indicated by its primary pointer. However, if that bank is being accessed by a read or write eDRAM operation, that is an access collision occurs, then the refresh is performed to the bank indicated by the alternate pointer. If the number of consecutive accesses to a same bank is fewer than all banks refresh cycles, that last bank will become available for at least one clock cycle, and the controller will able to perform the last refresh operation. In other words, any cycle is not need to expend for performing refreshes. Otherwise, one cycle of stall signal will be sent to the tag array to suspend a read or write operation, and then the corresponding eDRAM bank will be refresh in this cycle. Furthermore, if the user knows that he will never perform consecutive accesses to a given bank, the refresh stall will be completely hidden.

The miss buffer module collects the request packets not hitting in the tag array. After the physical address items are read from main memory and stored in the fill buffer, the corresponding request packet will be transmitted from the miss buffer to the arbiter. The fill buffer maintains the physical address items obtained from main memory through the memory interface module, and then issues a request packet with the corresponding virtual address to the arbiter. To reduce the SER ratio, the SEC-DED ECC codes are generated for the physical address items. After the missing request packet is resent to the data array, the physical address data including the ECC codes are transferred from fill buffer to the data array. The write back buffer receives the request packets missing in the tag array. Then the address is parsed from this request packet and sent to the memory interface. When the evicting operation occurs, the write back buffer also stores the physical address items from the data array selected by LRU algorithm, and then transmit them to main memory through the memory interface. The memory interface translates the internal request packets to a set of signals compatible to the interface of main memory.

The output queue receives the hitting physical address items from data array. These items include the single error correcting and double error detecting codes. Therefore the ECC codes should be verified whether there is any one bit or two bits error in the items. If there is one-bit error, the ECC checking algorithm will correct this error. If there is any two-bits error, the ECC checking algorithm only can detect this error, and mark it in the items. After ECC checking, the physical address item and ECC mark will be sent to the corresponding request source, according to the sources encoding field in the packets.

To improve the observability and controllability of the VwATC, a control and state register module is implemented. The CSR module is composed of many sets of registers, which preserve the control data set by users and error status come from other modules. The CSR has a new function. While the total number of address translation table (ATT) items used by software is smaller than 1 mega, a dual mode Cache is proposed, and the Cache can be configured as a large buffer mode. This buffer is initialized by writing all of physical address items through the CSR. It provides a channel to generate the request packets and send them to the arbiter. These request packets write the physical address items into the data array. Then all of the requests for obtaining the physical address will be hit in the tag array, and directly read the ATT items from the data array. Since removing the large latency of accessing main memory, it can achieve substantial improvement on the performance of obtaining the physical address.

4 Experimental Methodology and Results

To experimentally evaluate the efficiency of the variable-way address translation Cache, a small scale of supercomputer interconnected by the NIC chips containing the VwATC has been set up. The small parallel computer is composed of 32 compute nodes. Figure 3 gives the architecture and interconnection of 32 compute nodes. Each compute node consists of 2 multi-core processors, a NIC chip, a hard disk and 16GB main memories. A crossbar network including the router chips and optical fiber is adopted to interconnect 32 compute nodes. That is to say, every two of 32 nodes transfer data by one hop.

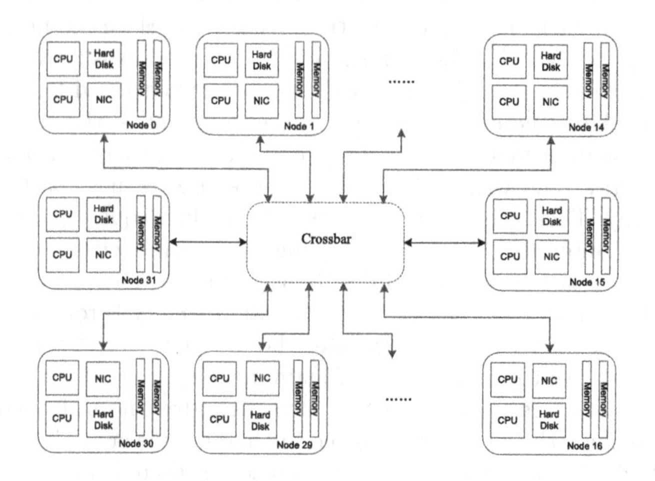

Fig. 3. Interconnection of 32 compute nodes

The NAS parallel benchmarks (NPB) [16] are well-known benchmarks for supercomputer systems. It evaluates the computation and communication performance of real user applications. That is different from the Linkpack benchmarks, which primarily evaluate the performance of floating-point computation. The NPB benchmarks include 8 classical applications, such as EP, MG, FT, IS, CG, BT, LU and SP. EP (Embarrassingly Parallel) is an application which evaluates the floating-point computing performance of mathematical function and communicates very minimally amongst the processes. MG is a 3-D multigrid solver benchmark. FT is a 3-D FFT PDE solver benchmark. IS (Integer Sort) benchmarks evaluate the performance of integer computation and collective communication. CG (conjugate gradient) benchmarks are used to evaluate the performance of unstructured set communication and point-to-point communication. BT is a block tridiagonal matrix solver, which tests the balance between communication and computation. LU (lower-upper triangular) benchmarks mainly test the fine-grained non-consecutive point-to-point communication. SP is a scalar pentadiagonal matrix solver.

The representative applications of NPB benchmarks have been compiled into four problem sizes known as the widely-used classes A, B, C, and D. The class D is the maximal size benchmarks, thus it is selected to evaluate the performance of the variable-way address translation Cache. After running the D class of NPB benchmarks on the 32 compute nodes, the hit ratio data of VwATC in each compute node are collected in Table 1. Table 1 collects the total number of hit packets, which is computed by a 64-bit counter denoted by H in the tag array of VwATC. Table 1 also shows the total number of request packets, which is provided by a 64-bit counter denoted by C in the input queue of VwATC. The last column presents the percentage of hit ratio computed through hit numbers divided by request numbers, represented by the equation $R = H/C$.

Table 1. Hit ratio of 32 compute nodes while running D class of NPB benchmarks

Nodes	Hit number	Req number	Hit ratio (%)
CN0	39483427	41226192	95.77268
CN1	40331276	42070056	95.86694
CN2	42258198	43996955	96.04801
CN3	42284018	44023433	96.04889
CN4	34947729	36685339	95.26348
CN5	34959047	36696460	95.26545
CN6	34987037	36725018	95.26758
CN7	34090802	35828671	95.14950
CN8	34869405	36607421	95.25229
CN9	35839254	37577062	95.37535
CN10	33883976	35621466	95.12235
CN11	33907842	35645747	95.12451
CN12	35799296	37536935	95.37086
CN13	35811841	37549389	95.37263
CN14	35824628	37562642	95.37303
CN15	34845864	36583313	95.25071
CN16	34898906	36637179	95.25544
CN17	35820439	37558084	95.37345
CN18	35809925	37547561	95.37217
CN19	35800341	37537660	95.37180
CN20	33912844	35650461	95.12596
CN21	33890502	35628295	95.12244
CN22	35838679	37576334	95.37567
CN23	34864811	36602500	95.25254
CN24	34085955	35823438	95.14987
CN25	34983206	36721214	95.26702
CN26	34970182	36708165	95.26540
CN27	34954375	36692557	95.26285
CN28	34989478	36727595	95.26754
CN29	34953758	36691634	95.26356
CN30	33033176	34770755	95.00276
CN31	32180045	33917795	94.87658

From Table 1, we may observe the following. Firstly, the total 8 applications of NPB benchmarks were finished in order and obtained the right results. That is to say, the virtual and physical address translation mechanism works properly and its function is correct. Secondly, for all of 32 compute nodes, the hit ratio percentage of VwATC is quite high, in most cases from 94.8% to 96.1%. The average hit ratio of VwATC in 32 compute nodes is 95.3%. For representative applications of many users' software, it is concluded that the proposed address translation Cache has high performance and is well designed.

Figure 4 shows the average hit ratio of VwATC in 32 compute nodes for each representative applications of NPB benchmarks. In Fig. 4, the hit ratio of most of the applications are more than 94%, especially 5 of them almost arrive at the 97%, including CG, MG, BT, LU, and SP. Only the hit ratio of FT benchmark is 87.9%, and less than other benchmarks. The main cause is the requests in FT of obtaining the physical address translation items from VwATC are discrete as compare with other benchmarks, and there are many missing and replacing operations occurred in the VwATC. Therefore, the hit ratio of FT application is minimum among 8 applications of the NPB benchmarks.

From Table 1 and Fig. 4, we can arrive at the hit ratio values of VwATC in 32 compute nodes are very high for NPB benchmarks. One of the main causes is the capacity of VwATC is very large. The VwATC accommodates totally 1 mega physical address translation items. The 1 mega items generally meet the requirements of most of user applications. The other cause is the VwATC implements the non-blocking deep pipeline and the stall-hidden eDRAM refresh design. It could improve the bandwidth and reduce the latency of accessing the physical address translation items.

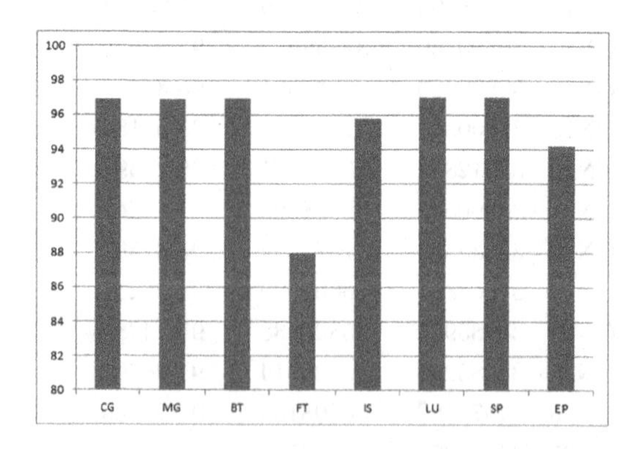

Fig. 4. Comparison of the average hit ratio of 8 NPB benchmarks

To evaluate the power reduction effect for the variable-way function, we estimate the power dissipation during different configurations of VwATC. The total power consumption of the NIC chip is 21 W. Since the VwATC is the largest module in the NIC, and occupy about 26% power of whole NIC, the power value of VwATC is approximately 5.52 W. An eDRAM macro with depth 16,384 and width 312 bits has 80 mW dynamic power and 32 mW static power. One tag SRAM memory macro with the depth 16,384 and width 14 bits has 42 mW dynamic power and 25 mW static power. The 8 ways of tag and data memory in VwATC can be configured to reserve 4 ways, 2 ways or even only 1 way. The power estimation of 4 different configurations of VwATC is depicted in Fig. 5. While the VwATC is full capacity, the dynamic power is 3.42 W, and the static power is 2.10 W. When the VwATC is set to 1/2 of original size, 4 ways of tag and data memory banks are gated clock, and then the dynamic power of these memory macros is sharply reduced, but the static power is almost same. Similarly, while the VwATC is configured to 1/4 or 1/8 of total size, 6 ways or 7 ways of eDRAM and tag SRAM memory are gated clock to save the dynamic power, and the total power values of VwATC are respectively 4.87 W and 4.74 W.

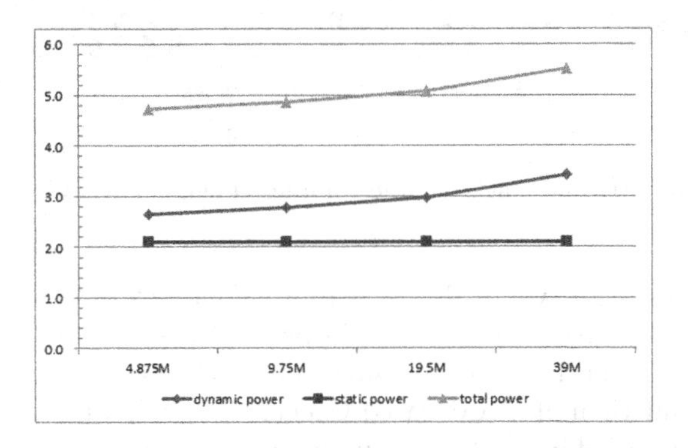

Fig. 5. Power consumption of 4 different configurations of VwATC

To test the performance of the VwATC, we have implemented three typical SRAM-based address translation Caches [14,15]. They have different capacities, respectively 4.875M bits, 9.75M bits, and 19.5M bits. Three Caches integrate by 8 SRAM macros, with the depth 2048, 4096, and 8192, and width 312 bits. They also adopt the 8 ways set-associative structure and LRU replace strategy, as same as the VwATC. Since the NICs are implemented by FPGA chips in 32 compute nodes, 4 different Caches have been implemented in NIC chip and downloaded the FPGA devices respectively. Figure 6 illustrates the hit ratio of 4 Caches in 32 nodes while running D class of NPB benchmarks. In Fig. 6, there is a column for a result obtained by each capacity of Cache, with the position along the vertical axis indicating the hit ratio in percentage, and the horizontal position indicating the 32 compute nodes. For each compute node, the first column is the hit ratio of 4.875M bits Cache, and the values are ranged from 66.32% to 68.05%, and the average ratio is 67.0%. The second column shows the hit ratio of 9.75M bits Cache, and the values are between 72.04% and 74.42%, and the average ratio is 72.9%. The third column gives the 19.5M bits Cache, and the values are from 80.35% to 81.93%, and the average ratio is 81.1%. The fourth column is the proposed eDRAM-based VwATC, and the values of hit ratio are from 94.87% to 96.04%, and the average ratio is 95.3%.

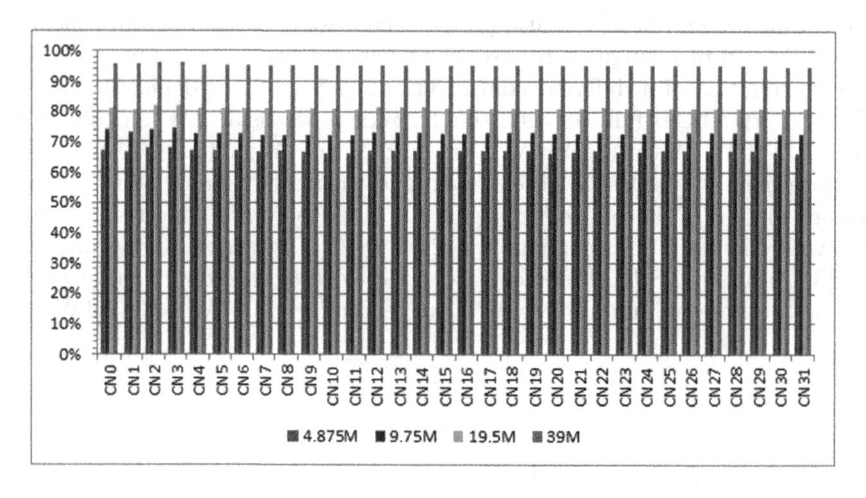

Fig. 6. Hit ratio of 4 Caches while running D class of NPB benchmarks

The density superiority of eDRAM is approximately 2 to 3 times over SRAM, thus the SRAM capacity is at most 19.5M bits in the same area. The advantage of SRAM-based Cache is prone to implement the accessing logic and achieve higher throughput than eDRAM-based Cache. However, a novel refresh scheme for eDRAM is proposed in VwATC, to improve the throughput as far as possible. If there is no consecutive access to a bank, the refresh stall will be completely hidden. Further, the VwATC can integrated the 39M bits eDRAM macro, at least twice as much as the SRAM-based Cache. Consequently, it has achieved 95.3% hit ratio and is much higher than three types of SRAM-based Cache. Furthermore, the hit ratio of 19.5M bits SRAM-based Cache is higher than 9.75M bits Cache, and the hit ratio of 9.75M bits Cache is higher than the 4.875M bits Cache. We can arrive at the conclusion that the memory capacity of a Cache is the dominant factor for hit ratio.

5 Conclusions

The virtual and physical address translation mechanism is essential to bridge the hardware designs, driver programs and software applications in the supercomputers. A novel virtual and physical address translation architecture is proposed. It includes the variable-way address translation Cache, the stall-hidden refresh scheme, the address validity checker, the dual mode Cache and many other reliability designs. The VwATC is the most important part in address translation architecture. We have conducted many tests on the real chip implemented the address translation designs, utilizing well-known benchmarks. The results show that the VwATC has high hit ratio. The power dissipation of VwATC has been estimated for different setting, and it is concluded that the variable-way function strongly reduces the dynamic power. As compared with three typical SRAM-based Cache, the VwATC has achieved much higher hit ratio. One of the future works is to exploit more aggressive techniques for shortening the latency of VwATC. The other future work is to apply the scalable address translation Cache on more real application tests.

References

1. TOP500 Supercomputer sites. http://www.top500.org/. Accessed 29 July 2021
2. Pritchard, H., Gorodetsky, I., Buntinas, D.: A uGNI-based MPICH2 nemesis network module for the cray XE. In: 18th European MPI Users' Group Conference on Recent Advances in the Message Passing Interface, pp. 110–119 (2011)
3. Xie, M., Lu, Y., Wang, K.F., et al.: TianHe-1A interconnect and message passing services. IEEE Micro **32**(2), 8–20 (2012)
4. Pang, Z., et al.: The TH express high performance interconnect networks. Front. Comp. Sci. **8**(3), 357–366 (2014). https://doi.org/10.1007/s11704-014-3500-9
5. Schoinas, I., Hill, M.D.: Address translation mechanisms in network interfaces. In: 4th International Symposium on High-Performance Computer Architecture, pp. 219–230 (1998)
6. Kostas, M.: Memory management support for multi-programmed Remote Direct Memory Access (RDMA) systems. In: 2005 IEEE International Conference on Cluster Computing, pp. 1–8 (2005)
7. Lee, M., Lee, S., Lee, J., et al.: Adopting system call based address translation into user-level communication. IEEE Comput. Archit. Lett. **5**(1), 26–29 (2006)
8. Lee, M., Lee, S., Maeng, S.: Context-aware address translation for high performance SMP cluster system. In: 2008 IEEE International Conference on Cluster Computing, pp. 292–297 (2008)
9. Mondrian, N.: Acceleration of the hardware-software interface of a communication device for parallel systems. Ph.D. thesis, University of Mannheim (2008)
10. Wang, Y., Zhang, M.: Fully memory based address translation in user-level network interface. In: IEEE 3rd International Conference on Communication Software and Networks, pp. 351–355 (2011)
11. Hameed, F., Bauer, L., Henkel, J.: Simultaneously optimizing DRAM cache hit latency and miss rate via novel set mapping policies. In: 2013 International Conference on Compilers, Architecture and Synthesis for Embedded Systems, pp. 1–10 (2013)
12. Hameed, F., Bauer, L., Henkel, J.: Reducing inter-core cache contention with an adaptive bank mapping policy in DRAM cache. In: 2013 International Conference on Hardware/Software Codesign and System Synthesis, pp. 1–8 (2013)
13. Hameed, F., Bauer, L., Henkel, J.: Reducing latency in an SRAM/DRAM cache hierarchy via a novel tag-cache architecture. In: 51st Design Automation Conference, pp. 37:1–37:6 (2014)
14. Irish, J.D., Mcbride, C.B., Ouda, I.A., et al.: Handling concurrent address translation cache misses and hits under those misses while maintaining command order. International Business Machines Corporation, United States Patent 7539840, 26 May 2009

15. Corrigan, M.J., Godtland, P., Hinojosa, J.: Selectively invalidating entries in an address translation Cache. International Business Machines Corporation, United States Patent 7822042, 26 October 2010
16. NAS parallel benchmarks. http://www.nas.nasa.gov/publications/npb.html. Accessed 20 Aug 2020

PPCTS: Performance Prediction-Based Co-located Task Scheduling in Clouds

Tianyi Yuan[1], Dongyang Ou[2], Jiwei Wang[2], Congfeng Jiang[2(✉)], Christophe Cérin[3], and Longchuan Yan[4]

[1] HDU-ITMO Joint Institute, Hangzhou Dianzi University, 310018 Hangzhou, China
tianyiyuan@hdu.edu.cn

[2] School of Computer Science and Technology, Hangzhou Dianzi University, Hangzhou 310018, China
{oudongyang,wangjiwei,cjiang}@hdu.edu.cn

[3] LIPN UMR CNRS 7030, Université Sorbonne Paris Nord, 93430 Villetaneuse, France
christophe.cerin@univ-paris13.fr

[4] Institute of Information Engineering, Chinese Academy of Sciences, Beijing 100093, China
yanlongchua@iie.ac.cn

Abstract. With increasing market competition among commercial cloud computing infrastructures, major cloud service providers are building co-located data centers to deploy online services and offline jobs in the same cluster to improve resource utilization. However, at present, related researches on the characteristics of co-location are still immature. Therefore, stable solutions for resource collaborative scheduling are still essentially absent, which restricts the further optimization of co-location technology. In this paper, we present performance prediction-based co-located task scheduling (PPCTS) model to perform fine-grained resource allocation under Quality of Service (QoS) constraints. PPCTS improves the overall cluster CPU resource utilization to 56.211%, which is competitive to the current related works. Besides, this paper fully implements a co-located system prototype. Compared with the traditional simulator-based scheduling research work, the prototype proposed in this paper is easier to be migrated to the real production environment.

Keywords: Data center · Co-located deployment · Container scheduling · Quality of Service

1 Introduction

The tasks in the data center can be divided into online services and offline jobs in terms of the type of demand [14]. Online services pursue low response delay to ensure user interaction experience, while offline jobs have stronger fault tolerance and expect higher throughput. Because the quantity of online service requests fluctuates on a daily pattern, it is difficult to determine appropriate resources

Y. Lai et al. (Eds.): ICA3PP 2021, LNCS 13157, pp. 245–257, 2022.
https://doi.org/10.1007/978-3-030-95391-1_16

allocation for these two types of tasks [5,6]. If the hardware resource is allocated based on the request peak, then there will be a huge waste of resources when the request load reaches its valley. For cloud service providers, the cost of hardware accounts for 50% to 70% of the total expense [4,8]. So that improve the efficiency of hardware resources and carrying more service requests at a lower cost is crucial for enhancing the competitiveness of data centers. In order to make full use of idle resources, co-located technology has gradually adopted by mainstream cloud service providers. By using this technology, the data centers can co-locate offline jobs and online services in the same cluster to improve resource utilization. When online services are busy, can evict offline jobs to ensure the QoS of online services.

Offline jobs and online services have different resource demand patterns. It has become a huge challenge to coordinate scheduling based on the characteristics of the two types of tasks. At present, various cloud service providers generally use container technology to achieve rapid deployment of jobs and elastic scaling of resources. Nevertheless, there is still a large optimization space for collaborative scheduling tasks. For example, the existing solutions have disadvantages such as coarse-grained scheduling, low flexibility, high resource transfer delay, high configuration cost, and monotonous metrics selection. Moreover, compared to the traditional environment, the resource imbalance in the co-located environment is more serious, which has become a bottleneck in the scheduling design. Therefore, designing an efficient resource preemption mechanism and coordinating the scheduling tasks while avoiding performance degradation has become a key issue of co-located technology.

In this paper, we present a performance prediction-based co-located task scheduling model called PPCTS to further improve resource utilization through fine-grained resource allocation under the QoS constraints. The co-located scheduling model is composed of container scheduling and offline job scheduling models. Experiment evaluation show that our model performs competitively and improves the resource efficiency and reliability of the co-located cluster.

The rest of the paper is organized as follows: In Sect. 2, we give an overview of relate previous efforts. Section 3 describes the architecture design of the PPCTS. Section 4 presents the results of our model in simulation co-located data center platform. Finally, we conclude with Sect. 5 where potential future work, as well as limitations of the model, are discussed.

2 Related Work

Since Google publicized the internal technical details of Borg in 2011, the advantage of co-located technology has prompted global cloud service vendors to deploy the co-located cloud computing infrastructure. Now Google has deployed the Borg co-located cluster for more than 10 years, and the proportion of co-located servers is reached up to 98%. Alibaba Cloud chose to integrate the original online service scheduling framework "Sigma" with the offline job scheduling framework "Fuxi" to achieve unified management [7,12]. Lu et al. [9] analyzed the Alibaba Cloud co-located log Cluster-trace-v2017 and summarized the resource imbalance as: spatial imbalance, time imbalance, resource type imbalance and requirement scale imbalance.

Chen et al. [2] proposed the Avalon co-located system, which used predictive methods to allocate the minimum number of CPU cores and cache capacity for each online service request without violating QoS, effectively decreasing resource waste. Xu et al. [13] designed a dynamic resource scheduling strategy based on game theory, incorporating QoS metrics into the resource optimization process, and achieved the balance between the fairness of scheduling and resource utilization. However, the above solutions only support the scenario of single online service deployment, which is less flexible and cannot guarantee the QoS of multiple online services. In order to design an adaptive co-located scheduling model to achieve scheduling in multiple online service co-located scenarios. Patel et al. [11] used Bayesian optimization to individually assign appropriate resources to the tasks in cluster nodes. Although this solution does not achieve the optimal effect, it simplifies the problem of high-dimensional solution space and reduces the overhead of the algorithm itself. Robert et al. [10] used the pre-test to obtain the sensitive curve of the task to measure the QoS guarantee when the task is under pressure. However, this work focuses on the performance degradation caused by the resource competition between cache and memory bandwidth, not CPU resource competition. Therefore, it has limitations in CPU-intensive workloads and co-located scenarios. Chen et al. [3] implement the scheduling of multiple resources, but each scheduling process is adjusted based on the best and worst delay performance, thus the granularity was too coarse.

In summary, the previous works mostly focus on QoS guarantee and resource distribution optimization. However, current scheduling strategies generally have poor adaptability to various types of co-located tasks, and usually use resource utilization as metrics for resource allocation, but this metrics cannot accurately reflect the true situation of resources. In response to the above issues, our solution can uniformly schedule offline jobs and online services while minimizing resource waste, and we propose new metrics to accurately perceive resource conditions: throttled rate and throttled time ratio, which improve the fine-grained resource allocation without QoS violation.

3 Task Scheduling Based on Performance Prediction

3.1 Overview

Based on the inflexibility in resource allocation of the previous work. We design a performance prediction-based co-located task scheduling (PPCTS) model. The schematic of the PPCTS is shown in Fig. 1. Which consist of container scheduling model and the cost-based multi-queue job scheduling model (CMJS). The above models will interact to achieve coordinated scheduling of different types of tasks. The container scheduling executes fine-grained resource scaling based on the results of the random forest-based container QoS violation prediction (RCVP) model. The CMJS controls the lifecycle of offline jobs from submission to completion. This model implements time-consuming prediction algorithm to estimate the cost of resources needed to complete the job.

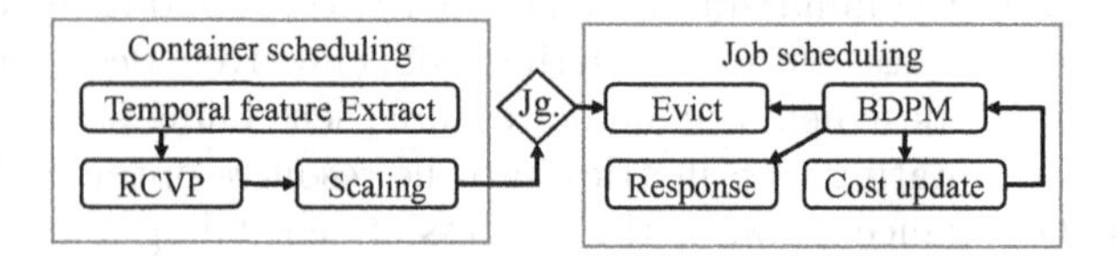

Fig. 1. Architecture of PPCTS model

3.2 Container Scheduling Based on QoS Violation Prediction

Based on the QoS violation prediction, the container scheduling model can scale container resources in advance to improve resource utilization while avoiding QoS violation. Figure 2 illustrates the workflow of the container scheduling model. In the container QoS violation prediction stage, the model predicts the QoS state of the container based on the current container state. In the container resource scaling stage, the container is scaled up or scaled down based on the prediction result. When the node resources are insufficient to allocate more resources for the container, the eviction operation of the offline job will be triggered. When the prediction result QoS is not violated and the resource utilization of the container is reached the scale-down threshold, then the container scale-down is triggered.

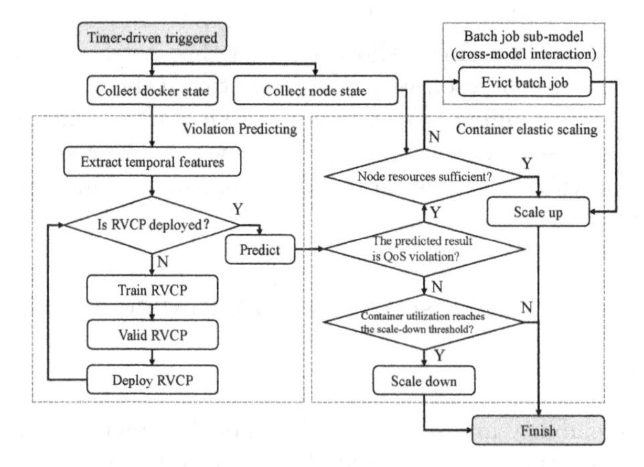

Fig. 2. The workflow of container scheduling model based on QoS violation prediction

QoS Violation Prediction. CPU utilization metrics have limitations in measuring resource starvation. In this paper, we add two metrics "throttled rate" and "throttled time ratio" as resource saturation metrics. The former is used to describe the frequency of the CPU falls into a throttled state, and the latter describes the proportion of time that the CPU falls into a throttled state. The calculation method of the two metrics is as shown in Eqs 1 and 2. Each metric prefixed with Δ represents the difference between adjacent sample values (these values are incremental values obtained from Docker low-level API response).

$$throttled\_rate = \frac{\Delta throttled\_periods}{\Delta periods} \tag{1}$$

$$throttled\_time\_ratio = \frac{\Delta throttled\_time}{\Delta timestamp} \tag{2}$$

We collect monitoring logs of online services to compare the trends of different metrics. It is illustrated in Fig. 3, the peak locations of the throttled rate and throttled time ratio is consistent with the 95th percentile response time, showing a more significant correlation than the CPU utilization. In Table 1, r refers to the Pearson correlation coefficients between each metrics and the long-tail response time of online services. Since the p-values of the three metrics are all less than 0.05, the correlation coefficients are valid. As shown in Table 1, the Pearson correlation coefficients of throttled rate and the throttled time ratio reach 0.709 and 0.781 respectively. Therefore, the above two metrics are more sensitive to the stress of container resources than CPU utilization metrics (which Pearson correlation coefficients is only 0.166).

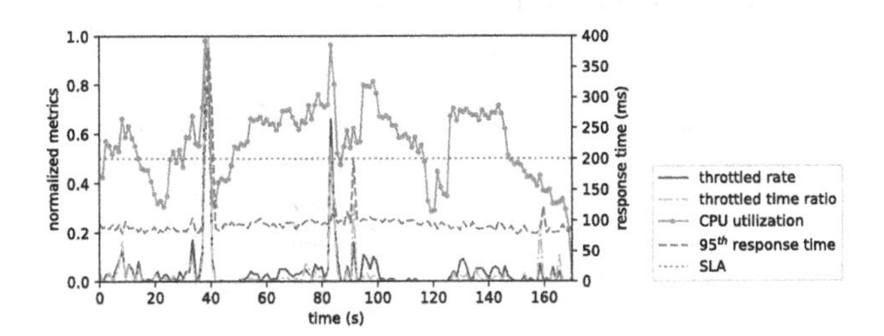

Fig. 3. The trend of CPU metrics and response time over time

Container Scheduling Model Design. In order to perform the vertical scaling of containers with the co-located cluster monitoring data, we design a random forest-based container QoS violation prediction (RCVP) model, which predicts the online service QoS violation state by using the temporal features of the historical performance data of the container. In this paper, we selected 5 statistical-based features to extract the temporal features of the CPU utilization and saturation within a given time window, which are peak_num, cidce [1], std, range and 95p. The RCVP first predicts the result of QoS violation base

Table 1. Correlation between utilization/saturation metrics and 95th percentile response time

Metrics	r	p-value
Throttled rate	0.709	0.000
Throttled time ratio	0.781	0.000
CPU utilization	0.166	0.043

on temporal features. Then the model will decide whether to perform scale-up or scale-down operation based on the QoS prediction result and the container resource utilization. This mechanism ensures the resource-starvation container can always reserve sufficient resources even when the requests fluctuate drastically, and successfully suppresses the negative effects caused by the accidental failure of the prediction model.

3.3 Cost-Based Multi-queue Job Scheduling

We design a cost-based multi-queue job scheduling (CMJS) model, which is based on the offline job execution time prediction model to perform offline job scheduling in the co-located cluster. The offline job scheduling model manages the lifecycle of offline jobs and undertakes the eviction operation triggered by the container scheduling model. Since the offline job execution pattern is relatively stable, its performance is easy to predict. By submitting 3000 offline jobs with different input data size and parallelism to the cluster nodes, the offline job execution time results illustrate in Fig. 4.

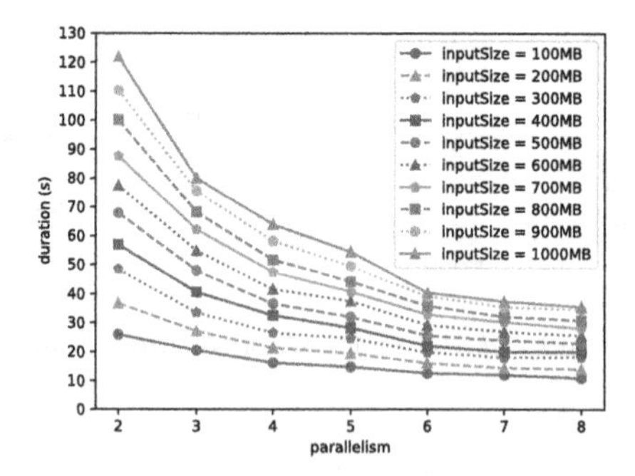

Fig. 4. Offline job execution time under different parallelism and input volume configurations

According to the curve shown in Fig. 4, the completion time of the offline job is inversely related to the degree of parallelism and linearly related to the data size. This suggests that we can model the execution time of offline jobs and calculate the cost of offline jobs with the prediction results. According to the characteristics of offline job, this paper proposes a highly interpretable offline job execution time prediction model, as shown in Eq. 3. p is the degree of execution parallelism, *inputSize* is the volume of the input file. α, β, γ, and k are the parameters of the prediction model. Each cluster node modeling based on historical execution logs to obtain its own predictive model parameters.

$$t_{batch} = \frac{\alpha + k \cdot inputSize^2}{p} + \beta \cdot inputSize + \gamma \tag{3}$$

Offline Job Scheduling Model Design. The offline job scheduling model manages the whole lifecycle of offline jobs, including submission, distribution, eviction, and result response. The management process of offline jobs is shown in Fig. 5.

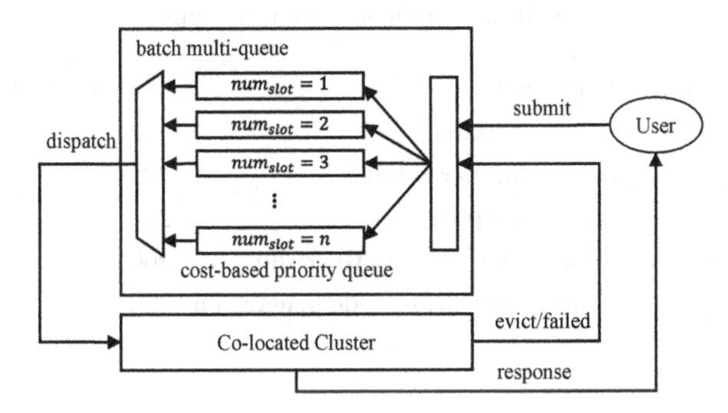

Fig. 5. The workflow of CMJS

In order to fully utilize the resources of online service clusters, offline jobs are scheduled in a multi-priority queue model. The operation of the model is mainly divided into two types: enqueue and dequeue.

Enqueue: Offline job submission and eviction will perform enqueue operations. During the enqueue operation, the offline job is first mapped to the sub-priority queue according to the parallelism specified by the user, then the offline job is inserted into a suitable position, keep the jobs in the sub-priority queue is ascending order of the total cost.

Dequeue: The distribution of offline jobs will perform dequeue operations. When there are sufficient idle resources in the cluster, the node with the most remaining resources will accept offline jobs as the target node. By comparing the heads of each sub-queue, get the target offline job with the lowest cost in the cluster. The target offline job is deployed to the target node for execution after dequeued.

The offline scheduling model will interact with the online service container scheduling model to achieve stable and reliable overall co-located scheduling operations.

4 Experiments and Performance Evaluation

To evaluate the effectiveness of the proposed PPCTS model, we build a small cluster with a master-slave structure as an experimental prototype system. The Master node is configured as AMD Ryzen5 3600, 6 cores, 3.6 GHz, hyper-threading enabled. The Slave1 node is configured as Intel Core i7-7700, 4 cores, 3.6 GHz, hyper-threading enabled. The Slave2 and Slave3 nodes are configured as Intel Core i5-8400, 6 cores, 2.8 GHz, hyper-threading disabled. The memory of all nodes is configured to 16 GB.

Based on the common online services in the existing co-located clusters, we implement 5 available online service use cases (see Table 2). We sample the load data of the online service container in Cluster-trace-v2018[1] released by Alibaba Cloud in 2018 as the request trace.

Table 2. Online service prototype

Number	Name	Implement	QoS constraints (ms)
S1	Search	Full text search	50
S2	Prime	Prime number solving	200
S3	Image	Image compression	200
S4	MD5	Information Digest Algorithm	200
S5	Verify	Image verification code generation	150

4.1 QoS Violation Prediction Evaluation

In order to evaluate the QoS violation prediction accuracy of the RCVP, we compare the proposed model with support vector machines, logistic regression, K-nearest neighbors, and CART. As shown in Table 3, CART and the proposed model successfully captured the correlation between the internal resource state of the container and the QoS violation. The two algorithms reached 0.781 and 0.803 on the *recall*. The proposed model has reached 0.849 on the F_1 score. Compared to the sub-optimal CART model, it increased by 9.267%. This suggests the RCVP further improves the prediction effect by integrating the CART model.

4.2 Container Scheduling

In order to evaluate the QoS guarantee capability and resource efficiency of the proposed container scheduling model. We compared the proposed container scheduling model with the threshold policy. The conservative and aggressive threshold policy respectively use the 70% and 90% threshold container vertical scaling strategies.

[1] https://github.com/alibaba/clusterdata.

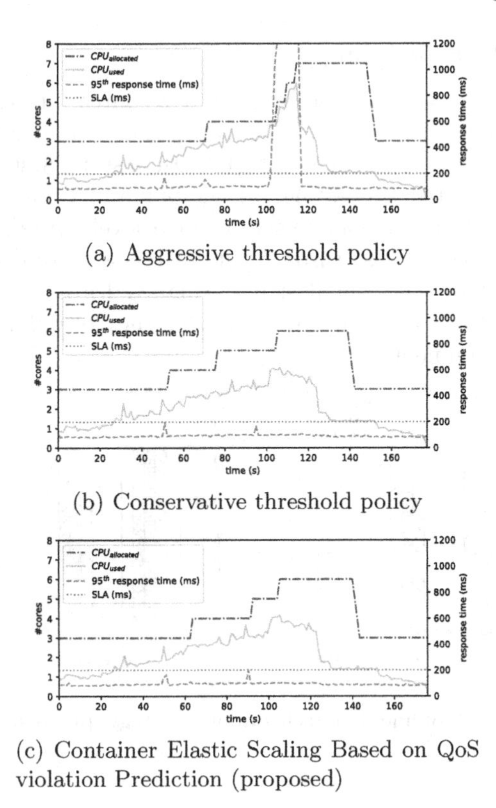

(a) Aggressive threshold policy

(b) Conservative threshold policy

(c) Container Elastic Scaling Based on QoS violation Prediction (proposed)

Fig. 6. Container CPU resource status and online service QoS status

Table 3. Comparison of QoS violation prediction of different models

Prediction model	*acc*	*Precision*	*Recall*	F_1
Support vector machines	0.607	0.667	0.192	0.298
Logistic regression	0.649	0.706	0.329	0.449
K-nearest neighbors	0.738	0.796	0.534	0.639
CART	0.806	0.774	0.781	0.777
RCVP	0.876	0.902	0.803	0.849

As shown in Fig. 6, for the aggressive threshold policy, the scale-up threshold is set too large and the scale-up operation is triggered too late, which brings about a large number of accumulated requests, hence causing serious QoS violation. The conservative threshold policy and the proposed model perform better in terms of QoS guarantee, with only sporadic violations (less than 1%), but the conservative policy caused a waste of resources due to premature scale-up operation. The proposed model performs scale-up operation according to the actual container resource status and achieves safe and cost-efficient resource allocation.

4.3 Batch Tasks Scheduling

We evaluate the execution performance and resource efficiency of the offline job scheduling model in a co-located environment. The comparison strategies are random eviction; first in first out (FIFO); last in first out (LIFO). Our experiment compares the processing throughput and resource usage of different strategies under different online service load levels. As illustrated in Fig. 7, the CMJS reaches the highest throughput. In terms of resource efficiency, the proposed model achieves the smallest performance degradation. The evaluation result proves that the model improves the efficiency of offline job execution in the co-located environment.

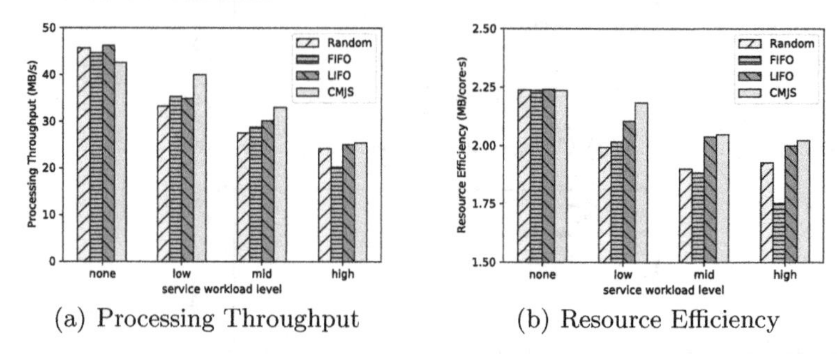

(a) Processing Throughput (b) Resource Efficiency

Fig. 7. Comparison of offline jobs scheduling strategy processing throughput and resource efficiency

4.4 CPU Utilization

To illustrate the resource utilization efficiency of the PPCTS model, we experiment with the model on the different online service loads. As shown in Fig. 8, the overall utilization of the cluster fluctuates drastically, the deployment and execution of offline jobs are the main factors that cause fluctuations. Although the overall resource utilization of the cluster is on a downward trend as the load of online services increases, resource utilization still maintains a high level. For high online service load, the overall average utilization of the cluster still reached 56.211% (compare to 39.820% of Alibaba Cloud).

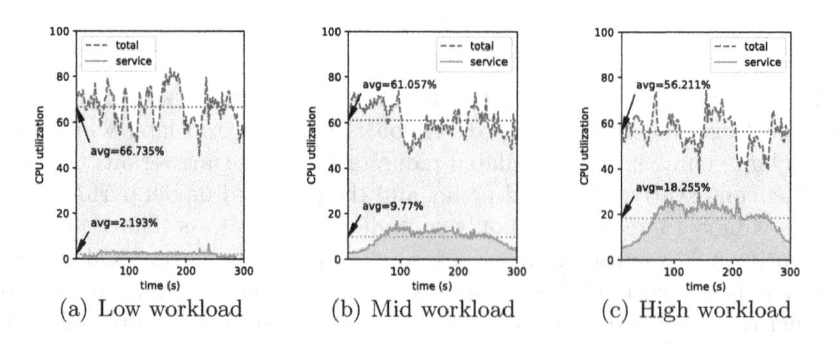

(a) Low workload (b) Mid workload (c) High workload

Fig. 8. CPU utilization of co-located clusters

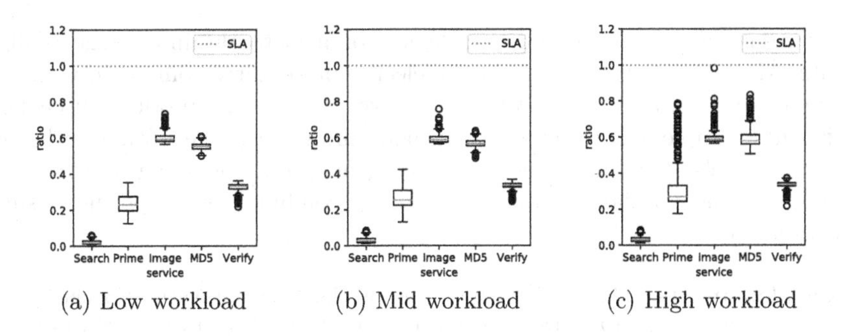

(a) Low workload (b) Mid workload (c) High workload

Fig. 9. QoS guarantee for co-located clusters

4.5 QoS Guarantee

As illustrated in Fig. 9, we count the 95th percentile response time of each online service. By using container elastic scaling strategy and the offline job eviction strategy, the performance interference suffered by the online service is always within an acceptable range, even if there are multiple response time peaks, the preset constraints in the SLA are not violated, which proves the proposed model effectively guarantees the QoS of online services. Based on the above evaluation results, the proposed PPCTS model improves the overall CPU utilization under the QoS constraints. Achieving a good balance between QoS guarantee and resource utilization in different load levels.

5 Conclusion

In this paper, we propose the PPCTS, a performance prediction-based co-located task scheduling model. Based on the proposed RCVP model, we design a container scheduling model to achieve a fine-grained adaptive allocation of container resources. The precision of the RCVP model up to 0.9 proves that the throttled rate and throttled time ratio we proposed accurately capture the degree of resource starvation. The RCVP prediction result is used as the primary indicator to guide container scale to ensure the highest priority of QoS guarantee in container scheduling. CMJS realizes the trade-off of the accuracy of model calculation complexity. Excessive model complexity will affect the performance of scheduling. Experiments show that the CMJS achieves the highest offline job throughput, which proves the effectiveness of the strategy. In the evaluation experiment of different online service load levels, the PPCTS increase the overall CPU resource utilization of the cluster to 56.211% without QoS violation. On the premise of the QoS guarantee, the utilization of resources is maximized. Compared with the 39.28% of the Alibaba Cloud co-located cluster, it shows strong competitiveness.

In this paper, we study the tasks scheduling of the co-located data center. However, many performance bottlenecks in the data center will appear only when the cluster scale is expanded to a certain scale. In the future, we will try to carry out verification in a large-scale environment to obtain more generalized results. Meanwhile, in this paper we select representative online services and offline jobs for analysis and evaluation. However, in actual production clusters, tasks will be implemented by various programming languages, frameworks, and resource models. The impact of task heterogeneity on scheduling design cannot be ignored, and special research needs to be combined with specific business characteristics.

Acknowledgments. This work is supported by the National Natural Science Foundation of China (No. 61972118), the Science and Technology Project of State Grid Corporation of China (Research and Application on Multi-Datacenters Cooperation & Intelligent Operation and Maintenance, No. 5700-202018194A-0-0-00), and the Science and Technology Project of State Grid Corporation of China (No. SGS-DXT00DKJS1900040).

References

1. Batista, G.E., Keogh, E.J., Tataw, O.M., De Souza, V.M.: CID: an efficient complexity-invariant distance for time series. Data Min. Knowl. Disc. **28**(3), 634–669 (2014)
2. Chen, Q., Wang, Z., Leng, J., Li, C., Zheng, W., Guo, M.: Avalon: towards QoS awareness and improved utilization through multi-resource management in datacenters. In: Proceedings of the ACM International Conference on Supercomputing, pp. 272–283 (2019)
3. Chen, S., Delimitrou, C., Martínez, J.F.: Parties: QoS-aware resource partitioning for multiple interactive services. In: Proceedings of the Twenty-Fourth International Conference on Architectural Support for Programming Languages and Operating Systems, pp. 107–120 (2019)
4. Jiang, C., Fan, T., Gao, H., Shi, W., Wan, J.: Energy aware edge computing: a survey. Comput. Commun. **151**, 556–580 (2020)
5. Jiang, C., Han, G., Lin, J., Jia, G., Shi, W., Wan, J.: Characteristics of co-allocated online services and batch jobs in internet data centers: a case study from Alibaba cloud. IEEE Access **7**, 22495–22508 (2019)
6. Jiang, C., et al.: Characterizing co-located workloads in Alibaba cloud datacenters. IEEE Trans. Cloud Comput. 1 (2020)
7. Liu, Q., Yu, Z.: The elasticity and plasticity in semi-containerized co-locating cloud workload: a view from Alibaba trace. In: Proceedings of the ACM Symposium on Cloud Computing, pp. 347–360 (2018)
8. Lo, D., Cheng, L., Govindaraju, R., Ranganathan, P., Kozyrakis, C.: Heracles: improving resource efficiency at scale. In: Proceedings of the 42nd Annual International Symposium on Computer Architecture, pp. 450–462 (2015)
9. Lu, C., Ye, K., Xu, G., Xu, C.Z., Bai, T.: Imbalance in the cloud: an analysis on Alibaba cluster trace. In: 2017 IEEE International Conference on Big Data (Big Data), pp. 2884–2892. IEEE (2017)
10. Mars, J., Tang, L., Hundt, R., Skadron, K., Soffa, M.L.: Bubble-up: increasing utilization in modern warehouse scale computers via sensible co-locations. In: Proceedings of the 44th annual IEEE/ACM International Symposium on Microarchitecture, pp. 248–259 (2011)

11. Patel, T., Tiwari, D.: CLITE: efficient and QoS-aware co-location of multiple latency-critical jobs for warehouse scale computers. In: 2020 IEEE International Symposium on High Performance Computer Architecture (HPCA), pp. 193–206. IEEE (2020)
12. Tian, H., Zheng, Y., Wang, W.: Characterizing and synthesizing task dependencies of data-parallel jobs in Alibaba cloud. In: Proceedings of the ACM Symposium on Cloud Computing, pp. 139–151 (2019)
13. Xu, F., Wang, S., Yang, W.: Cloud resource scheduling algorithm based on game theory. Comput. Sci. **46**(6A), 295–299 (2019)
14. Zhao, S., Xue, S., Chen, Q., Guo, M.: Characterizing and balancing the workloads of semi-containerized clouds. In: 2019 IEEE 25th International Conference on Parallel and Distributed Systems (ICPADS), pp. 145–148. IEEE Computer Society (2019)

Edge Computing and Edge Intelligence

Risk-Aware Optimization of Distribution-Based Resilient Task Assignment in Edge Computing

Hang Jin[1,2], Jiayue Liang[2], and Fang Liu[1(✉)]

[1] Hunan University, Changsha, China
fangl@hnu.edu.cn
[2] Sun Yat-sen University, Guangzhou, China
{jinh26,liangjy77}@mail2.sysu.edu.cn

Abstract. Computing resources of mobile devices are growing, and unoccupied resources can be shared to provide support for edge computing services in edge clouds. Unlike stable servers, a significant challenge is that mobile devices may exit or join an edge cloud at any time due to change of position. This dynamic nature of mobile devices may result in abortions of task execution. In this paper, a risk-aware task assignment scheme called RATA is proposed. RATA minimizes the overhead caused by potential abortions of task execution by prioritizing tasks to the edge nodes which are unlikely to exit during task execution. We first quantify the abortion risk of each task-node pair to an expected extra overhead time, and formulate a risk-aware task assignment problem that strives to minimize the average completion time of all tasks, as well as the expected extra overhead time of each task. We then design a novel task assignment scheme to solve this problem with genetic algorithm. Finally, we implement a prototype system to evaluate the performance. The experimental results show that our scheme outperforms the state-of-art task assignment schemes in terms of average completion time and deadline miss rate in most cases.

Keywords: Task assignment · Task scheduling · Edge computing · Risk-aware · Mobile device

1 Introduction

Edge computing aims to perform computation tasks by making full use of resources at the edge of the network [1,11,16]. Similar with BOINC [2], under-utilized resources of mobile devices in crowded places such as business centers, can also be fully utilized by user devices with limited resources to speed up task execution. Assigning tasks, especially latency-sensitive ones, appropriately and efficiently in edge clouds is of great importance. However, the dynamic nature of mobile devices is a great challenge, as they can dynamically exit the edge cloud, resulting in abortions of tasks assigned to these devices. Task completion can be ensured through re-assignment and re-execution, but this brings significant

Y. Lai et al. (Eds.): ICA3PP 2021, LNCS 13157, pp. 261–277, 2022.
https://doi.org/10.1007/978-3-030-95391-1_17

overhead, resulting in missing the deadline, which is a potential time limit for a device to complete an assigned task.

Most previous works on the task assignment considered bandwidth constraints and task constraints [12,22]. However, these work did not take into account the instability of edge nodes in edge clouds as well as the unstable connectivity of mobile networks, assuming that the execution process is fault-free, which is unrealistic. Femtocloud* [10] proposes a risk-controlled task allocation mechanism, which aims to minimize the risk of edge nodes leaving before tasks are completed. However, it performs worse in terms of average completion time especially when the workload is light, compared with other approaches to minimize the average completion time.

We focus on the abortion of task execution caused by unstable edge nodes, or abruptly interrupted wireless network in edge environment. The key to solve this problem is minimizing the potential overhead time caused by abortions. Main contributions of this work can be summarized as follows.

- We study the task assignment in edge cloud consisting of dynamic mobile devices. We explicitly consider the impact of abortions of task execution on task completion time. We analyze and quantify the expected extra overhead time introduced by abortions based on the probability distribution of task completion time and the remaining presence time of edge nodes.
- We propose RATA, a risk-aware task assignment scheme. With extra overhead time to be taken into account, RATA aims to minimize the average completion time as well as the extra overhead time of each task, by selecting more reliable and powerful edge nodes for each task.
- We implement a prototype system. The experiment results show that RATA performs better in terms of average completion time and deadline miss rate, compared with the state-of-the-art works.

2 Related Work

Most existing works [3,12,19] focus on task assignment problem with edge servers in edge clouds. Even though they take various factors (e.g., users' location, network capabilities, order between assigned tasks) into account, execution environments of these works are failure-free. BGTA, CoGTA and TDBU [21–23] are proposed to assign social sensing tasks to mobile edge devices. These frameworks allow edge devices to choose task with preferences by game theory so as to optimize both their payoffs and the deadline miss rate. However, these works don't consider the dynamic churn rate of mobile devices.

Deng et al. [7] present a novel offloading system to make robust offloading decisions with a trade-off fault-tolerance mechanism, which can pick a better choice between waiting for reconnection and directly restarting the service when a fault happens. Although this work considers the unstable connectivity of mobile networks, it only considers reducing the overhead brought by happened abortion failures rather than reducing the occurrence of abortion failures. Habak et al. [9] present Femtocloud, an edge system to leverage mobile devices to provide edge service with a task assignment approach which aims to maximize cluster's overall useful computations. They present an improvement on the system architecture and workload management in Femtocloud* [10]. This work presents a risk-controlled task assignment approach, which is the first to focus on reducing potential abortions. It applies a point estimate (e.g., mean of subset of historical running times) to predict task completion time. Besides, Femtocloud calculates the abortion probability to represent the abortion risk. However, the abortion probability doesn't accurately reflect the extra time consumed by abortions since duration of task execution before the abortion is different. As a result, Femtocloud actually prioritizes on reducing potential abortions more than minimizing task completion time.

Tumanov et al. [17] present a novel black-box approach called JamaisVu to predict job running time utilizing its running history. This approach is tested to perform well for predicting real-world job mixes and help to make robust scheduling decisions. Park et al. [14] conduct experiments to show that the approach derived from JamaisVu can give more accurate prediction for job running time than a point estimates. Their work develops JamaisVu and presents 3Sigma which leverages full distribution of relevant running time histories to schedule jobs by using probability density function to model the utility of jobs. With this perspective of distribution-based scheduling, we quantify the abortion risk as the expected extra overhead time based on the distributions of completion time taken by tasks and the presence time that the edge nodes exist in the edge cloud.

Checkpointing can provide distributed systems [5,6] with the ability of fault tolerance by periodically saving tasks' state for recovery. Some works use checkpointing approach to build fault-tolerance mechanisms for robust computation offloading [7,8]. Since we mainly focus on the intrinsic ability of task assignment schemes, we don't apply checkpointing into RATA and baselines for task recovery.

3 System Model and Problem Formulation

3.1 System Model

A diagram of edge cloud leveraging mobile devices is shown in Fig. 1. The model includes following components.

Edge Cloud: An edge cloud consists of a controller (a cloud server or edge server) and a certain number of edge nodes (servers or mobile devices), connected with each other via Access Points (APs) in a Local Area Network (LAN). Presence time is used to denote the duration between the moment a mobile device joins and the moment the mobile device exits. Edge nodes are modeled as $EN = \{en_1, en_2, ..., en_J\}$, where J is the total number of edge nodes in the system. Transmission rates between the controller and edge nodes are modeled as $TX = \{tx_1, tx_2, ..., tx_J\}$. Each edge node maintains a certain number of workers to execute tasks, and a waiting queue for tasks. For edge node j where $1 \leq j \leq J$, the waiting queue length of it is defined as ql_j, and the average execution time of tasks among workers is defined as T_j^{Avg}. Note that mobile devices in the figure are

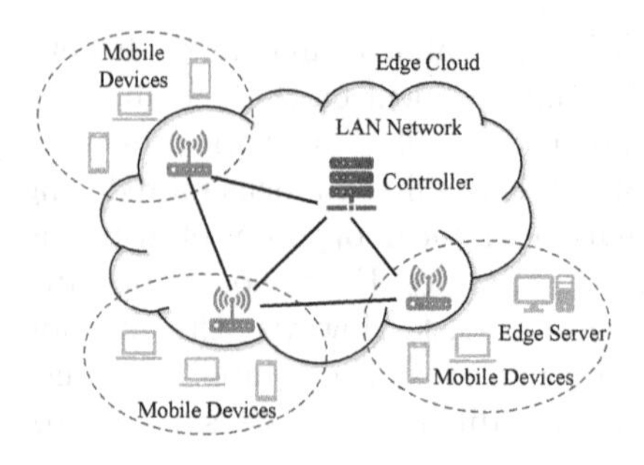

Fig. 1. An edge cloud leveraging mobile devices.

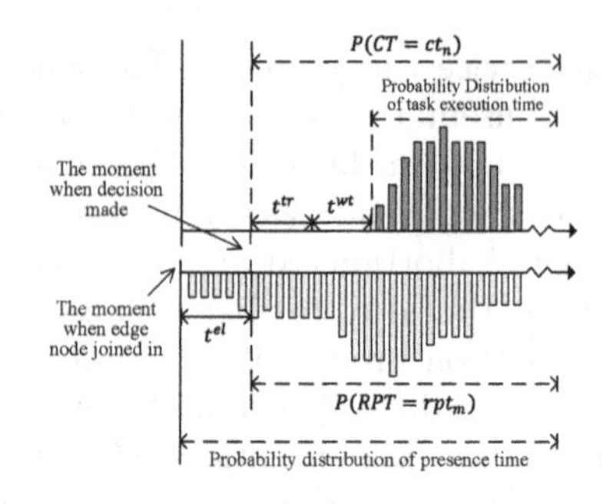

Fig. 2. Relation of related probability distributions.

the ones sharing their underutilized computation resources, rather than devices submitting tasks.

Edge Service: An edge service is invoked by a mobile user. Each edge service accepts a task with input data, and produces a result by assigning them to edge nodes for executing. All tasks in one assignment are modeled as $TK = \{tk_1, tk_2, ..., tk_I\}$, where I is the total number of tasks. For task i where $1 \le i \le I$, the size of the input data is defined as s_i. Since edge nodes are heterogeneous in terms of hardware and software, execution time of a specified task varies from node to node. We define $t^{ex}_{i,j}$ as the execution time of task i in edge node j where $1 \le i \le I, 1 \le j \le J$.

Task Assignment Strategy Profile: A task assignment strategy profile defines each task's choice of edge node, which are modeled as variable matrix $X = \{x_{1,1}, x_{1,2}, ..., x_{I,1}, x_{I,2}, ..., x_{I,j}\}$. $x_{i,j} = 1, 1 \le i \le I, 1 \le j \le J$ implies task i would be assigned to edge node j, while $x_{i,j} = 0$ means not.

3.2 Problem Formulation Without Abortion Risk

The entire process of task assignment consists of three stages. First, the task and its input data need to be transmitted from controller to the target edge node. If all the workers in the target edge node are busy, then the task will be added into the waiting queue and wait until being chosen to execute. At last, the task will be executed by one of the workers in the target edge node. Therefore, if we assume that the connection is always stable, the completion time $t^{cpt}_{i,j}$ of task i in edge node j can be calculated as

$$t^{cpt}_{i,j} = t^{tr}_{i,j} + t^{wt}_{j} + t^{ex}_{i,j} \tag{1}$$

where the transmission time $t^{tr}_{i,j}$ can be calculated by $t^{tr}_{i,j} = \dfrac{s_i}{tx_j}$. The waiting time t^{wt}_{j} can be calculated by $t^{wt}_{i,j} = ql_j \times T^{Avg}_{j}$. The execution time $t^{ex}_{i,j}$ can be predicted using the distribution of task execution time history, we will describe the approach in Sect. 4. Equation (1) ignores propagation latency as well as the transmission time of the output, because they are too small to be compared with other parts.

For Eq. (1), a frequently-used task assignment policy is assigning a task to the edge node with shortest complete time greedily. However, this policy may assign multiple tasks to a particular edge node, which results in extra waiting time among these tasks. In consideration of this issue, we formulate the task assignment problem as a 0–1 integer linear program as follows, which strives to minimize the batch completion time of all the tasks, in other words, to minimize the average completion time.

$$min \sum_{i,j} (x_{i,j} \times t_{i,j}^{cpt}) + (\sum_{i} x_{i,j} - 1) \times T_j^{Avg} \tag{2}$$

$$s.t. \sum_{j} x_{i,j} = 1 \tag{3}$$

$$x_{i,j} \in \{0, 1\}, 1 \leq i \leq I, 1 \leq j \leq J \tag{4}$$

We present a congestion-avoiding mechanism which adds up a compensation term to represent the extra waiting time cause by multiple tasks in the same edge node, in order to solve the aforementioned issue. Constraints (3) ensure that each task can be only assigned to one edge node.

3.3 Problem Formulation with Abortion Risk

To consider the dynamic nature of mobile devices and the abortion risk, we follow the perspective of distribution-based scheduling [14]. We think of the completion time of task, as well as the presence time of edge nodes using probability distribution. In fact, we focus on the remaining presence time which is equal to the difference between the presence time of an edge node and the elapsed time from the moment it joined in. When assigning a task to an edge node, we suppose the remaining presence time of the edge node as random variable RPT and its law of probability distribution is

$$p_m = P(RPT = rpt_m), m = 1, 2, ..., M \tag{5}$$

where rpt_m denotes the possible values of RPT and M denotes the total number of these values. Similarly, we suppose the completion time of the task as random variable CT and its law of probability distribution is

$$q_n = P(CT = ct_n), n = 1, 2, ..., N \tag{6}$$

where ct_n denotes the possible values of CT and N denotes the total number of these values. In fact, the two random variables are independent of each other. The relationship of related probability distribution is showed in Fig. 2, where t^{el} refers to the time elapsed from the moment edge node joined in.

Then we can calculate the abortion probability α, namely the probability that the remaining presence time is shorter than the completion time, by Eq. (7).

$$\alpha = P(RPT \leq CT) = \sum_{rpt_m \leq ct_n} (p_m \times q_n) \tag{7}$$

Further, we quantify the extra time that a potential abortion of task execution will bring up. Since the task assignment decisions at each cycle are independent with each other, in order to calculate the potential abortion overhead for one certain decision, we assume that after an abortion, the second execution of the task will be certainly completed. And we assume this task will be reassigned to an edge node which has a similar capacity with that of the origin one, so the completion time will be the same. Based on these assumptions, we suppose the total completion time under abortion risk as a random variable TCT defined by Eq. (8).

$$TCT = \begin{cases} CT & , RPT > CT \\ RPT + CT & , RPT \leq CT \end{cases} \tag{8}$$

The expectation of TCT can be calculated by:

$$\begin{aligned} E(TCT) &= \sum_{rpt_m > ct_n} (p_m \times q_n \times ct_n) + \sum_{rpt_m \leq ct_n} (p_m \times q_n \times (rpt_m + ct_n)) \\ &= \sum (p_m \times q_n \times ct_n) + \sum_{rpt_m \leq ct_n} (p_m \times q_n \times rpt_m) \\ &= E(CT) + \sum_{rpt_m \leq ct_n} (p_m \times q_n \times rpt_m) \end{aligned} \tag{9}$$

where $E(CT)$ is the expectation of CT. We define random variable ET as the extra time caused by the abortion, the expectation of ET will be

$$E(ET) = \sum_{rpt_m \leq ct_n} (p_m \times q_n \times rpt_m) \tag{10}$$

Particularly, $E(ET)$ will be zero if the abortion probability α equals to zero, when an edge node will never exit from the system or lose connection to the controller. We can treat the expected extra time as the representative of the abortion risk. The shorter the expected extra time is, the smaller the abortion risk is.

Based on the derivation above, we update the Eq. (1) to Eq. (11). We define $t_{i,j}^{et}$ as the predictions of the expected extra time ET. Therefore, the expected task completion time $t_{i,j}^{ect}$ under abortion risk, is the sum of task completion time and the expected extra time.

$$t_{i,j}^{ect} = t_{i,j}^{cpt} + t_{i,j}^{et} \tag{11}$$

We will describe the details to predict $t_{i,j}^{et}$ in Section 4. The risk-aware task assignment problem can be updated as follows.

$$min \sum_{i,j}(x_{i,j} \times t_{i,j}^{ect}) + (\sum_{i} x_{i,j} - 1) \times T_j^{Avg} \quad (12)$$

$$s.t. \sum_{j} x_{i,j} = 1 \quad (13)$$

$$x_{i,j} \in \{0,1\}, 1 \leq i \leq I, 1 \leq j \leq J \quad (14)$$

This updated problem aims at minimizing the sum of all the expected task completion time. In the optimization process, the solver will try to choose a stabler edge node to reduce the expected extra time for each task.

4 System Design of Task Assignment

4.1 System Architecture

The overview of system architecture is shown in Fig. 3. It consists of two main components: a controller and a set of edge nodes. The controller is responsible for receiving tasks from users, distributing tasks to suitable edge nodes and return the results. It consists of following modules.

- **Task Originator Interface:** receives service requests (tasks and their input data) from users, and returns back results of completed tasks.
- **Task Manager:** collects tasks, which are new or going to be reassigned, and submits these tasks to the Scheduler.
- **Node Manager:** maintains the connection with edge nodes and collects information of each edge node (e.g. IP address and waiting queue length).
- **Predictor:** responsible for generating time predictions as the input in equation (1) and (11). It consists of several Task Predictor Units and Node Predictor Units. Each Task Predictor Unit maintains distributions of execution time for tasks of a group of edge nodes with the same software and hardware configuration. Similarly, Each Node Predictor Unit maintains distributions of presence time for a group of edge nodes with the same network environment. The way to make predictions is discussed in the Subsect. 4.2.
- **Scheduler:** collects predictions of task-node pairs from the Predictor, and make assignment decisions at the start of each assignment cycle.
- **Task Tracker:** forwards tasks to target edge nodes, then tracks execution states, returns task results to the Task Manager and the Predictor. Specially, failed tasks will be kept in Task Manager for next assignment.

Edge nodes are responsible for executing their assigned tasks. Each of them consists of following modules.

- **Task Manager:** receives assigned tasks from the Controller, puts them into a waiting queue, and returns execution results after completed.
- **Work Thread:** when idle, execute a task in the waiting queue.
- **Heartbeat Thread:** responsible for keeping the connection with the Controller, report node status periodically.

4.2 Generating Predictions and Probability Distributions

We follow the black-box approach of 3σPredict [14] to generate distributions and predictions for tasks. This approach does assume that most tasks will be similar to some subsets of previous tasks, and similar tasks will have similar execution times. It does not require any task structure, or user-provided information about the execution time, but a set of features of each task. We choose three features, the user name of the task, the task name, and the logic task name (e.g. face recognition). For each feature-value pair, the approach tracks every completed tasks which have the same value of that feature and generate an empirical distribution which is stored as a histogram using a stream histogram algorithm [4] with a maximum of 40 bins. Each histogram will maintain four point estimates using four estimation techniques: average, median, moving average with decay 0.5 and average of 20 recent tasks. In addition, each histogram tracks the Normalized Mean Absolute Error (NMAE) of these four estimates. When predicting the execution time of a task with several features, this approach firstly collects all point estimates of these features. Then it chooses the point estimate with lowest NMAE as the prediction of execution time, and at last, the distribution which owns that point estimate will be chosen as the distribution of execution time for the task. Based on the approach described above, each Task Predictor Unit serves for a group of edge nodes with the same machine type. A machine type means a configuration of hardware and software. Edge nodes with the same machine type will have a similar computational capacity. What's more, we append one histogram to track all the completed tasks regardless of features in each Task Predictor Unit, in order to predict the average task execution time t^{Avg}.

As for the presence time of edge nodes, we use the same histogram algorithm to generate distributions in each Node Predictor Unit serving for a group of edge nodes with the same environment type. A environment type represents a concrete network environment. For example, edge nodes connecting to the same AP share the same type. We assume that each edge node with the same type will have similar characteristic of mobility.

We append two algorithms as follows to translate a histogram to a probability histogram, based on the histogram algorithms defined in [4]. A histogram is a set of B pairs (called bins) as an approximate representation of a set of real numbers, denoted by $\{(v_1, f_1), ..., (v_B, f_B)\}$. For each pair $(v_i, f_i), 1 \leq i \leq B$, v_i denote the value of a number, f_i denote the frequency of the number. We define a new concept called probability histogram as a variant of the histogram. A probability histogram is also a set of B pairs $\{(v_1, p_1), ..., (v_B, p_B)\}$, but for each pair $(v_i, p_i), 1 \leq i \leq B$, p_i denotes the probability of the number. In other word, a probability histogram is an approximate representation of a probability distribution. These two new algorithms show the details of two particular translation from histogram to probability histogram. Algorithm 1 first performs a right shift operation on a histogram with an offset, then translates the new histogram to be a probability histogram. Algorithm 2 first filters the pairs whose value is smaller than an offset from a histogram, then translates the rest of the histogram to a probability histogram.

Algorithm 1. Right-Shift and Probability Translation Procedure

Input: A histogram $\{(v_1, f_1), ..., (v_B, f_B)\}$, a offset δ
Output: A probability histogram $\{(u_1, p_1), ..., (u_B, p_B)\}$.
1: Set $S = \sum_{i=1}^{B} f_i$
2: **for** $i = 1$ to B **do**
3: $u_i = v_i + \delta$
4: $p_i = f_i \div S$
5: **end for**

Algorithm 2. Probability Translation Beginning with an Offset Procedure

Input: A histogram $\{(v_1, f_1), ..., (v_B, f_B)\}$, a offset δ
Output: A probability histogram $\{(u_1, p_1), ..., (u_C, p_C)\}$.
1: Find i such that $p_i \leq \delta < p_{i+1}$
2: **if** $p_i = \delta$ **then**
3: Set $S = \sum_{j=i}^{B} f_i$
4: **else**
5: Set $S = \sum_{j=i+1}^{B} f_i, i = i + 1$
6: **end if**
7: **for** $j = i$ to B, $k = 1$ to $B - i + 1$ **do**
8: $u_k = v_j$
9: $p_k = f_k \div S$
10: **end for**

The relation between probability distribution of task completion time and that of task execution time is showed in Fig. 2. To get the probability histogram of task completion time, we first calculate the sum of transmission time and waiting time as the offset, then we call Algorithm 1 with histogram of task execution time as the input. Similarly, the relation between probability distribution of remaining presence time and of presence time are also depicted in Fig. 2. To get the probability histogram of remaining presence time, we first calculate the elapsed time from the moment when the edge node joined as offset, then we call Algorithm 2 with histogram of presence time as the input.

At last, once we have generated the probability histogram of task completion time and that of the remaining time of edge node, we can calculate the expected extra time by Eq. (10), and the expected task completion time by Eq. (11).

In consideration of scalability, we employ three techniques to reduce the memory footprint. Firstly, the original design of 3σPredict can maintain a histogram using constant memory regardless of the number of points. Secondly, in spite of heterogeneity, we use one Task Predictor Unit to serve for a groups of edge nodes with the same machine type rather than just one edge node. Finally, we use one Node Predictor Unit to serve for a groups of edge nodes with the same environment type rather than just one edge node.

4.3 Task Assignment Algorithm

We choose the genetic algorithm to solve the problem, which is a heuristic method to search approximate solutions for optimization problems with evolutionary theory, and is often used to solve task assignment problems [7,13,20].

For a given task set $TK = \{tk_1, tk_2, ..., tk_I\}$ and edge node set $EN = \{en_1, en_2, ..., en_J\}$, we encode the task assignment strategy $X = \{x_{i,j}, 1 \leq i \leq I, 1 \leq j \leq J\}$ to a vector $S = \{s_i = en_j, 1 \leq i \leq I, 1 \leq j \leq J\}$ as a chromosome. $s_i = en_j$ means that task i is assigned to edge node j. We define the fitness function as follows.

$$Fitness(TK, EN, X) = \frac{1}{C(X)} \tag{15}$$

$$C(X) = \sum_{i,j}(x_{i,j} \times t_{i,j}^{ect}) + (\sum_{i} x_{i,j} - 1) \times T_j^{Avg} \tag{16}$$

Referring partly to the experiment of [7], we use the roulette-wheel method in the selection phase. In the crossover phase, we choose the standard one-point crossover operator and set the cross probability to 0.3. In the mutation phase, we choose the standard uniform mutation operator and set the mutation probability to 0.3. The population size is set to 20. We limit the maximum iteration number to 100, in order to ensure an negligible overhead of calculation compared with task completion time.

5 Experimental Evaluation

5.1 Experimental Setup

A prototype is implemented to evaluate RATA. Controller module runs on a desktop PC, while the edge node modules run on heterogeneous mobile devices such as laptops and Raspberry Pis. Modules are written in GoLang and launched in Linux environment, using gPRC for communication between modules. When a new edge node comes up, the edge node calls *JoinGroup* to register itself first, then starts calling *Heartbeat* to report its information periodically. When exiting, the edge node module calls *ExitGroup* to logout. The controller module calls *AssignTask* to forward tasks to edge node modules, which call *ReturnResult* to return results. To evaluate the intrinsic ability of reducing abortions, we don't apply any checkpointing techniques.

Hardware Configuration: Table 1 listed hardware characteristics of devices used for testbed experiment as edge nodes. Besides, a desktop PC equipped with an Intel(R) Core(TM) i5-8400 processor, 16GB RAM and Gigabit Network Adapter (connected to LAN via cable) is used as the Controller.

Workload: We use synthetic matrix multiplication Python tasks derived from the Google cluster trace [15], which is also used in experiments of previous works [12,14]. As tasks with short execution time are more suitable in edge computing with mobile devices, we filter out tasks whose execution time are larger than

Table 1. Hardware characteristics of edge node devices

Device	Computation capacity	Bandwidth	Connection
2 × Raspberry Pi 3B+	2.7 MFLOPs	13.2 Mbps	Wireless LAN
3 × Raspberry Pi 4B	15.1 MFLOPs	15.3 Mbps	Wireless LAN
Surface Pro 4(M3)	8.7 MFLOPs	95.9 Mbps	Wireless LAN

10 min. We cluster the remaining tasks using k-means clustering on their execution time, and draw tasks from each task class proportionally to generate the workload. We bind each task in the workload with three features: user name of the task, task name, and logic task name. Since the tasks derived from the Google cluster trace don't have task names, we replace their job name by their task name, so do the logic task name. We generate the size of input data file of each task using a uniform distribution U(0MB, 10 MB).

In order to take into account the impact of deadline, we use deadline slack [14] to generate deadlines for 50% of the tasks in the workload randomly. Deadline slack is defined as follows.

$$deadlineslack = \frac{(deadline - executiontime)}{executiontime} \times 100\% \tag{17}$$

By default, we use a uniform distribution U(250%,300%) to generate deadline slacks, which are experimented in [18].

We totally generate 1200 tasks for testbed experiment and 15000 tasks for simulation experiments. For each experiment, the workload will be divided into 6 groups, we randomly choose a group to pre-train the Predictor, and the rest groups to conduct evaluation. We use a Poisson arrival process to model the arrival of tasks.

Baseline: We compare the performance of our risk-aware task assignment scheme (RATA) with the following existing task assignment schemes.

- **Shortest Completion Time First (SCTF):** Dispatch a task to the edge node with shortest completion time greedily.
- **Minimize Batch Completion Time (MBCT):** Use genetic algorithm to solve the original task assignment problem as formulation (2)–(4).
- **Femtocloud:** Femtocloud* [10] first sorts edge nodes by churn probability and gets a temporary best node. Then it iterates on the sorted list of edge nodes, comparing each candidate node with the best one in terms of gain (relative difference of completion time) and risk (relative difference of churn probability). If the gain exceeds the risk, it switches these two edge nodes.

Metrics: For tasks without deadlines, we use average completion time and average number of executions to measure the performance of task assignment. For tasks with deadlines, we use deadline miss rate to measure the performance. These metrics are listed below.

- **Average Completion Time:** average time tasks consume to get completely executed.
- **Average Number of Executions:** average times that tasks are executed (include abortions).
- **Deadline Miss Rate:** the percentage of tasks missed their deadline.

5.2 Testbed Experiment

In testbed experiments, similarly with Femtocloud [10], once an edge node leaves, it will return after an OFF period which follows a normal distribution with mean equals to 20% of the average presence time. Therefore, each edge node's duty cycle is set to 80% on average. Some parameters are shown in Table 2.

Table 2. Parameters of testbed experiment

Parameter		Value
Duration of each assignment cycle		1 s
Worker thread at each edge node		3
Edge node queue management policy		First-Come-First-Serve
Average presence time	Raspberry Pi 3B+	25 min
	Raspberry Pi 4B	15 min
	Surface Pro 4(M3)	20 min

Figure 4 shows the performance of all these task assignment schemes on our testbed. Overall, RATA has the shortest average completion time and lowest deadline miss rate among listed schemes. Although RATA has a similar average number of executions with Femtocloud, the average completion time is shorter and deadline miss rate is lower. In simple terms, Femtocloud tends to select stabler edge nodes which may lead to longer task completion time to acquire lower abortion probability. As for MBCT, RATA outperforms it because the risk-aware mechanism reduces the abortions of task execution, which is reflected by the lower average number of executions. SCTF performs worst since it minimizes neither the batch completion time nor the abortion risk.

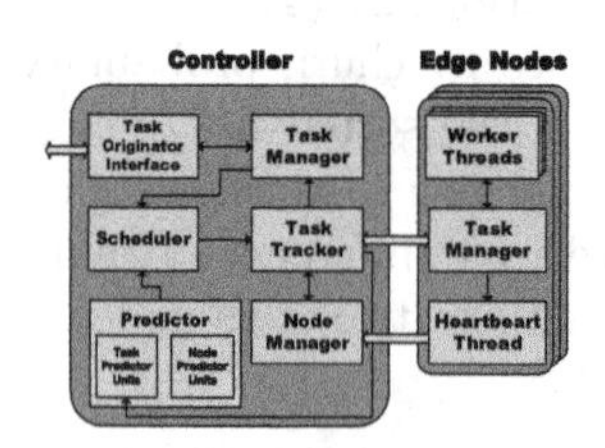

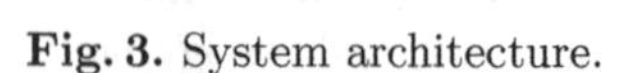
Fig. 3. System architecture.

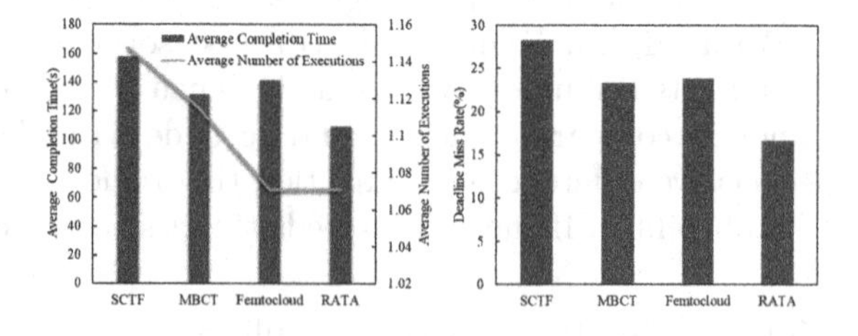

Fig. 4. Result of the testbed experiment.

5.3 Simulation Experiment

Simulation experiments are launched with larger number of simulated edge nodes. Each simulated edge node is an edge node module running on the desktop PC. We generate the bandwidth of simulated edge nodes using a uniform distribution U(8 Mbps, 80 Mbps). Transmission processes are simulated by a sleep function. A Poisson arrival process is used to model the arrival of edge nodes, and a normal distribution is used to generate presence time of each edge node. The standard deviation is set to 20% of the mean, similarly to Femtocloud. We study the impact of some parameters on task assignment performance in this experiment. Variations of these parameters can affect load status and the abortion risk. Default values of them are listed in Table 3.

Table 3. Default parameters of simulation experiments

Parameter	Value
Task arrival rate	180 tasks/min
Average presence time	20 min
Edge node arrival rate	5 nodes/min

Impact of Task Arrival Rate: Task arrival rate directly affects the waiting queue length in edge nodes. As Fig. 5 shows, all metrics increase as this rate increases. However, RATA has the shortest average completion time and lowest deadline miss rate in most cases. Compared with MBCT, RATA has a similar average completion time but a lower deadline miss rate and average number of executions. Femtocloud has a similar average number of executions but a longer average completion time compared with RATA. Since RATA can quantify the expected overhead time brought by an abortion, it will compare a potential longer overhead time with a shorter task completion time when making task assignment decisions.

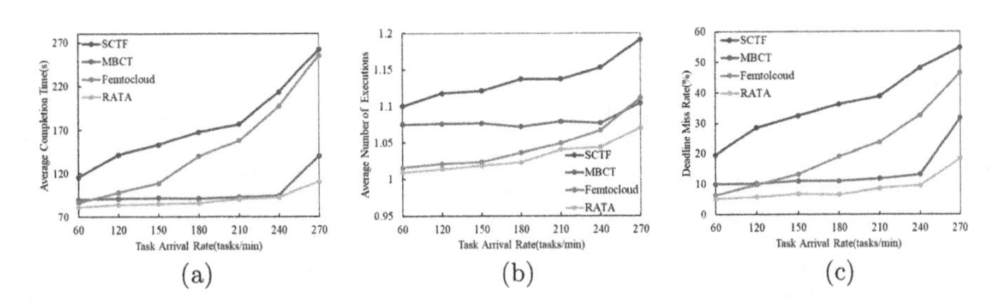

Fig. 5. Impact of task arrival rate. (a) Average completion time. (b) Average number of executions. (c) Deadline miss rate.

Impact of Edge Node Arrival Rate: Edge node arrival rate directly affects the number of edge nodes in the edge cloud. Figure 6 shows the impact of edge node arrival rate. Overall, all the metrics decrease as the arrival rate increases, and RATA outperforms other baselines in terms of three metrics in most cases. RATA has a similar performance with Femtocloud to reduce abortions of task execution in terms of average number of executions, but has a shorter average completion time and a lower deadline miss rate. Compared with MBCT, RATA has a similar performance on average completion time, but a lower average number of executions and deadline miss rate.

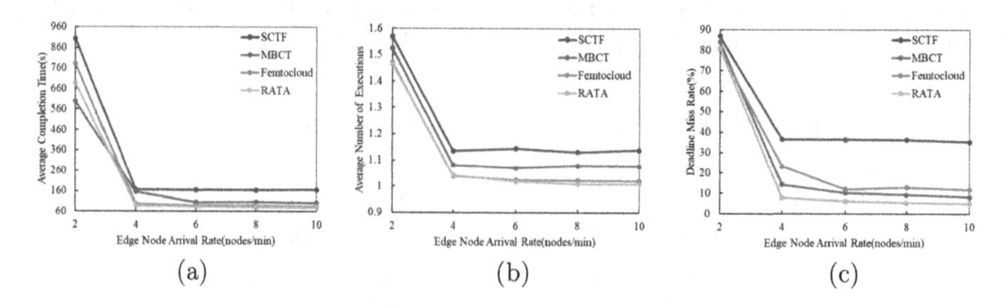

Fig. 6. Impact of edge node arrival rate. (a) Average completion time. (b) Average number of executions. (c) Deadline miss rate.

Impact of Edge Node Heterogeneity: This aims to test the ability to make full use of edge nodes with more computation capacity but shorter presence time. A new group of simulated edge nodes with this characteristic are newly introduced, modeled as a normal distribution N(5 min, 30 s). Figure 7 shows the impact of changing the arrival rate of these simulated edge nodes. Overall, all the metrics except the average number of executions decrease as the arrival rate increases. Femtocloud performs better than other schemes including RATA on keeping low average number of executions. This result shows the difference that RATA prefers to choose those edge nodes which can offer a shorter completion time for tasks despite of the higher abortion risk, while Femtocloud prefers stabler edge nodes.

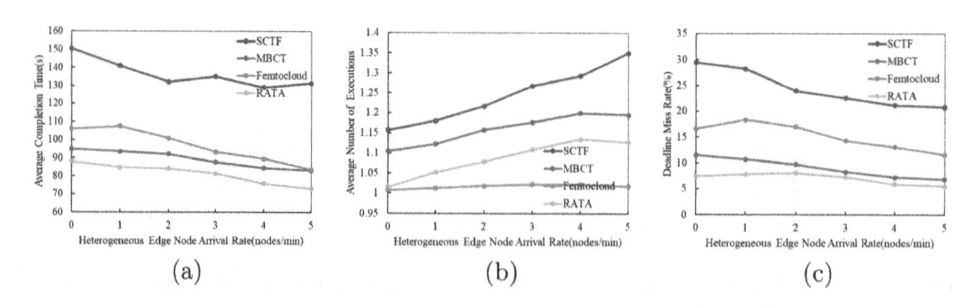

Fig. 7. Impact of edge node heterogeneity. (a) Average completion time. (b) Average number of executions. (c) Deadline miss rate.

Impact of Average Presence Time of Edge Nodes: This directly affects the abortion probability of task execution. Once an edge node leaves, it will return after an OFF period. The OFF period is modeled the same as the testbed experiment. Figure 8 shows the impact of changing the average presence time, where $+\infty$ means the edge nodes will never leave during the experiment. Overall, most metrics decrease as the presence time increases. RATA and MBCT have the shortest average completion time and lowest deadline miss rate, while RATA and Femtocloud outperform other baselines in terms of average number of executions.

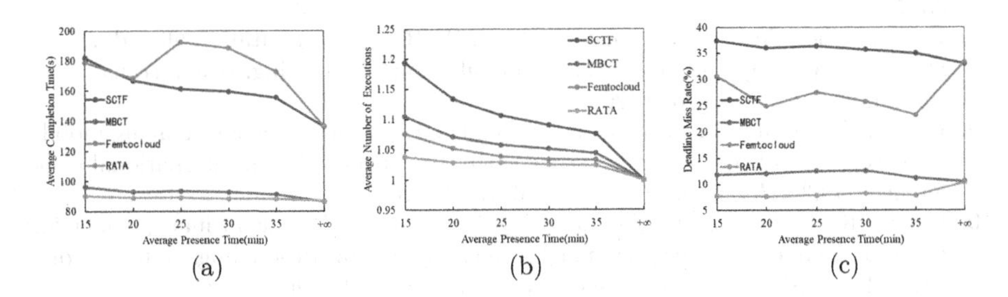

Fig. 8. Impact of average presence time of edge nodes. (a) Average completion time. (b) Average number of executions. (c) Deadline miss rate.

6 Conclusion

In this work, we study the risk-aware optimization of task assignment in edge cloud consisting of dynamic mobile devices. We formulate a risk-aware task assignment problem, which aims to reduce the average completion time as well as abortions of task execution, and give a solution to this problem. We design and implement a prototype system to evaluate the method. Experiment results show that our scheme outperforms the state-of-the-art work in terms of average completion time and deadline miss rate in most cases.

Acknowledgment. This work is supported by National Natural Science Foundation of China (62172155, 62072465).

References

1. Abbas, N., Zhang, Y., Taherkordi, A., Skeie, T.: Mobile edge computing: a survey. IEEE Internet Things J. **5**(1), 450–465 (2017)
2. Anderson, D.P.: BOINC: a system for public-resource computing and storage. In: Fifth IEEE/ACM International Workshop on Grid Computing, pp. 4–10. IEEE (2004)
3. Bahreini, T., Grosu, D.: Efficient placement of multi-component applications in edge computing systems. In: Proceedings of the Second ACM/IEEE Symposium on Edge Computing, pp. 1–11 (2017)

4. Ben-Haim, Y., Tom-Tov, E.: A streaming parallel decision tree algorithm. J. Mach. Learn. Res. **11**(Feb), 849–872 (2010)
5. Cao, G., Singhal, M.: Mutable checkpoints: a new checkpointing approach for mobile computing systems. IEEE Trans. Parallel Distrib. Syst. **12**(2), 157–172 (2001)
6. Chen, X., Lyu, M.R.: Performance and effectiveness analysis of checkpointing in mobile environments. In: 22nd International Symposium on Reliable Distributed Systems 2003, Proceedings, pp. 131–140. IEEE (2003)
7. Deng, S., Huang, L., Taheri, J., Zomaya, A.Y.: Computation offloading for service workflow in mobile cloud computing. IEEE Trans. Parallel Distrib. Syst. **26**(12), 3317–3329 (2014)
8. Guo, S., Chen, M., Liu, K., Liao, X., Xiao, B.: Robust computation offloading and resource scheduling in cloudlet-based mobile cloud computing. IEEE Trans. Mob. Comput. **20**(5), 2025–2040 (2020)
9. Habak, K., Ammar, M., Harras, K.A., Zegura, E.: Femto clouds: leveraging mobile devices to provide cloud service at the edge. In: 2015 IEEE 8th International Conference on Cloud Computing, pp. 9–16. IEEE (2015)
10. Habak, K., Zegura, E.W., Ammar, M., Harras, K.A.: Workload management for dynamic mobile device clusters in edge femtoclouds. In: Proceedings of the Second ACM/IEEE Symposium on Edge Computing, pp. 1–14 (2017)
11. Liu, F., Guo, Y., Cai, Z., Xiao, N., Zhao, Z.: Edge-enabled disaster rescue: a case study of searching for missing people. ACM Trans. Intell. Syst. Technol. (TIST) **10**(6), 1–21 (2019)
12. Meng, J., Tan, H., Xu, C., Cao, W., Liu, L., Li, B.: Dedas: online task dispatching and scheduling with bandwidth constraint in edge computing. In: IEEE INFOCOM 2019-IEEE Conference on Computer Communications, pp. 2287–2295. IEEE (2019)
13. Omara, F.A., Arafa, M.M.: Genetic algorithms for task scheduling problem. In: Abraham, A., Hassanien, AE., Siarry, P., Engelbrecht, A. (eds.) Foundations of Computational Intelligence Volume 3. Studies in Computational Intelligence, vol. 203, pp. 479–507. Springer, Heidelberg (2009). https://doi.org/10.1007/978-3-642-01085-9_16
14. Park, J.W., Tumanov, A., Jiang, A., Kozuch, M.A., Ganger, G.R.: 3sigma: distribution-based cluster scheduling for runtime uncertainty. In: Proceedings of the Thirteenth EuroSys Conference, pp. 1–17 (2018)
15. Reiss, C., Wilkes, J., Hellerstein, J.L.: Google cluster-usage traces: format+schema. Google Inc., White Paper, pp. 1–14 (2011)
16. Shi, W., Cao, J., Zhang, Q., Li, Y., Xu, L.: Edge computing: vision and challenges. IEEE Internet Things J. **3**(5), 637–646 (2016)
17. Tumanov, A., Jiang, A., Park, J.W., Kozuch, M.A., Ganger, G.R.: JamaisVu: robust scheduling with auto-estimated job runtimes. Technical report CMU-PDL-16-104. Carnegie Mellon University (2016)
18. Tumanov, A., Zhu, T., Park, J.W., Kozuch, M.A., Harchol-Balter, M., Ganger, G.R.: TetriSched: global rescheduling with adaptive plan-ahead in dynamic heterogeneous clusters. In: Proceedings of the Eleventh European Conference on Computer Systems, pp. 1–16 (2016)
19. Wu, H., et al.: Resolving multi-task competition for constrained resources in dispersed computing: a bilateral matching game. IEEE Internet Things J. **8**(23), 16972–16983 (2021)
20. Xu, Y., Li, K., Hu, J., Li, K.: A genetic algorithm for task scheduling on heterogeneous computing systems using multiple priority queues. Inf. Sci. **270**, 255–287 (2014)

21. Zhang, D., Ma, Y., Zhang, Y., Lin, S., Hu, X.S., Wang, D.: A real-time and non-cooperative task allocation framework for social sensing applications in edge computing systems. In: 2018 IEEE Real-Time and Embedded Technology and Applications Symposium (RTAS), pp. 316–326. IEEE (2018)
22. Zhang, D., Ma, Y., Zheng, C., Zhang, Y., Hu, X.S., Wang, D.: Cooperative-competitive task allocation in edge computing for delay-sensitive social sensing. In: 2018 IEEE/ACM Symposium on Edge Computing (SEC), pp. 243–259. IEEE (2018)
23. Zhang, D.Y., Wang, D.: An integrated top-down and bottom-up task allocation approach in social sensing based edge computing systems. In: IEEE INFOCOM 2019-IEEE Conference on Computer Communications, pp. 766–774. IEEE (2019)

Budget-Aware Scheduling for Hyperparameter Optimization Process in Cloud Environment

Yan Yao[1], Jiguo Yu[2,3](✉), Jian Cao[4], and Zengguang Liu[5]

[1] School of Computer Science and Technology, Qilu University of Technology (Shandong Academy of Sciences), Jinan, China
yaoyan@qlu.edu.cn

[2] Big Data Institute, Qilu University of Technology (Shandong Academy of Sciences), Jinan, China
jiguoyu@sina.com

[3] Shandong Laboratory of Computer Networks, Jinan, China

[4] Department of Computer Science and Engineering, Shanghai Jiao Tong University, Shanghai, China
cao-jian@sjtu.edu.cn

[5] College of Computer Science and Engineering, Shandong University of Science and Technology, Qingdao, China

Abstract. Hyperparameter optimization, as a necessary step for majority machine learning models, is crucial to achieving optimal model performance. Unfortunately, the process of hyperparameter optimization is usually computation-intensive and time-consuming due to the large searching space. To date, with the popularity and maturity of cloud computing, many researchers leverage public cloud services (i.e. Amazon AWS) to train machine learning models. Time and monetary cost, two contradictory targets, are what cloud machine learning users are more concerned about. In this paper, we propose *HyperWorkflow*, a workflow engine service for hyperparameter optimization execution, that coordinates between hyperparameter optimization job and cloud service instances. *HyperWorkflow* orchestrates the hyperparameter optimization process in a parallel and cost-effective manner upon heterogeneous cloud resources, and schedules hyperparameter trials using bin packing approach to make the best use of cloud resources to speed up the tuning processing under budget constraint. The evaluations show that *HyperWorkflow* can speed up hyperparameter optimization execution across a range of different budgets.

Keywords: Hyperparameter optimization · Cloud computing · Resource scheduling

Y. Lai et al. (Eds.): ICA3PP 2021, LNCS 13157, pp. 278–292, 2022.
https://doi.org/10.1007/978-3-030-95391-1_18

1 Introduction

Over the last decade, Machine Learning (ML) has become an active area for both academics and industry, who can learn insight from the collected data. Nearly every machine learning model requires hyperparameters, which are configured prior to training a model. Examples of these hyperparameters include number of clusters in k-means clustering, learning rate (in many models), batch size, and number of hidden nodes in neural networks, and many more. The task performance of trained models strongly depends on the choice of hyperparameters [13].

The process of choosing the optimal configuration of hyperparameters for a machine learning model is referred as hyperparameter optimization (HPO). A HPO job contains a large group of training trials, each with its own configuration. It get feedback from running previous trials before selecting new ones to explore until reaching a target accuracy or stopped by the user. While running hyperparameter optimization (a.k.a. hyperparameter tuning, hyperparameter search), a sizable number of hyperparameter configurations are evaluated, and one final configuration is chosen. Each configuration also referred as a *trial*. Then this final configuration is trained extensively so as to maximized model performance (i.e., accuracy).

Two common practices for the optimization of hyperparameters are manual tuning and automatic tuning. Manual methods involve tuning hyperparameters manually to obtain a good rule of thumb and default values. Automatic tuning methods (grid search, random search, Bayesian optimization [14] etc.) are adopted when the number of hyperparameters is large. In spite of the different tuning approaches, multiple configurations of hyperparameters are decided according to certain rules so that different model instances are generated which need to be tested and evaluated on the data. Typically, this process has to be repeated (trial-and-error procedure) till an optimized configuration is found. Hyperparameter optimization gets much more complex when the number of hyperparameters grow. Therefore, the process of hyperparameter optimization is usually computationally expensive and time-consuming.

To date, with the popularity and maturity of cloud computing, many researcher leverage public cloud services (i.e. Amazon AWS) to train machine learning models. To efficiently leverage distributed computation of cloud computing, each training trial is often as a single task running on a single worker, instead of planing a group of training trials as a whole [5]. However, a single training trial may not fully occupy worker, which lead to low utilization. Some recent works have proposed parallel hyperparameter optimization solution to better utilize resources [4,9].

HPO under time and monetary cost constraints is naturally a constrained optimization problem. While individual trials can execute faster given high performance or more resources, it is also critical that the time and monetary cost jointly considered. In this paper, we focus on the fundamental problem of minimizing the running time of executing a hyperparameter tuning job, subject to a budget constraint. We propose *HyperWorkflow*, and workflow engine service that coordinates between the cloud resources and hyperparameter optimization algorithms. By modeling hyperparameter optimizaiton jobs as workflows, *Hyper-Workflow* re-choreography the execution logic of the process of hyperparameter optimization algorithms.

In summary, the main contributions of this paper are as follows:

- *Execution time prediction for machine learning models under different hyper-paramenter configurations*: In order to predict the execution time of machine learning models under different settings, we propose a regression-based approach which is based on the time complexity analysis of this model.
- *Workflow-based scheduler for hyperparamater optimization*: We propose *HyperWorkfow*, an workflow-basedl scheduler for hyperparamater optimization in cloud environment to minimize execution time under a budget.
- *Generate an optimized workflow instance model for HyperWorkfow*: We transform this problem into a bin-packing problem. After an initial workflow instance model is created, it will be restructured based on deadline constraints using a Dual Variable Size Bin Packing algorithm (Dual-VSBP).

The reminder of this paper is organized as follows. Section 2 presents the related work of this paper. Section 3 formulates the scheduling problem which minimizes the execution cost with given deadline constraint. How to generate the workflow instance model for hyperparameter optimization is discussed in Sect. 4. Section 5 presents the experiment results. Finally, conclusions are drawn in Sect. 6.

2 Background and Related Work

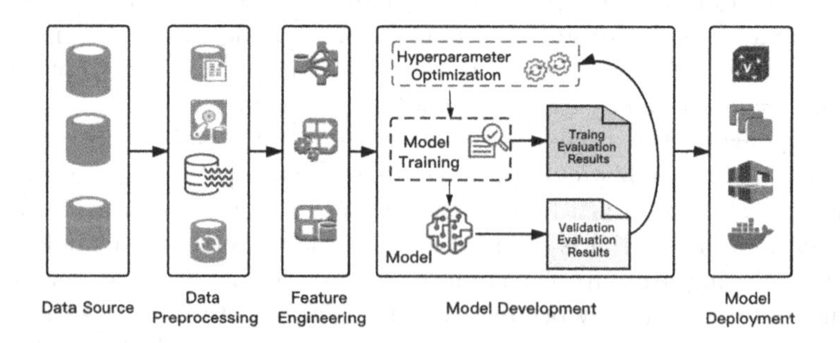

Fig. 1. Overview of the machine learning workflow

Currently, a large variety of machine learning methods exist, ranging from neural networks over kernel methods to ensemble models. A common trait of these methods is that they are parameterized by a set of hyperparameters, which must be set appropriately by the user to maximize the usefulness of the learning approach.

Figure 1 gives a high-level overview of a typical workflow for machine learning and how hyperparameter optimization algorithms work in general. First, users upload data from various sources and prepare it for the preprocessing phase. Then performing feature engineering, to create features that make machine learning algorithms work. In the first iteration, a simple and time-efficient method is selected to train the first baseline model. The evaluation process helps to find out where the baseline model stands concerning the accuracy and the business value. Deploying a model in early stages helps to monitor its performance in the production environment and to collect feedback to improve the model in the next iterations.

Historically, hyperparameter tuning has been viewed as a global optimization problem in the machine learning community. Some examples of such approaches include grid search [2], random search [3], Bayesian optimization [1,8,15] and the swarm optimization algorithm (e.g., genetic algorithm [17], particle swarm optimization algorithm [6]). Grid search is a classical hyperparameter optimization strategy in the context of machine learning. Typically, it is used to search through a manually-defined subset of hyperparameters of a learning algorithm. Random search is another approach that randomly samples parameters in a defined search space. It can also be very time-consuming when working with a large number of hyperparameters and a large number of sample points in the search space. The focus of our paper is not a hyperparameter search strategy but trying to optimize the execution process after multiple settings have been decided by a dedicated search strategy.

Li et al. [10] designed Hyperband which runs multiple instances of successive halving, with each training its configurations to different lengths before elimination. Asynchronous successive halving (ASHA) [9] adapts Hyperband to multiple parallel workers. Falkner et al. [4] propose a Bayesian version of Hyperband (BOHB) which assumes a prior on the accuracy over all hyperparameter values and chooses its recommendation based on the posterior. However, all of the above works didn't consider how to allocate resources (key factors in many fields, such as [7,16]) while training. To our knowledge, only a few works which explicitly studies resource allocation for hyperparameter tuning. HyperSched [11] is similar to ASHA; however, as the deadline approaches it allocates more resources to the promising ones using some heuristics. Rubberband [12] aims to minimize the cost of a given hyperparameter tuning job via profiling models and greedily searching for the best allocation of resources.

3 *HyperWorkflow* Scheduler

In this section, we start with the problem formulation, then we introduce the *HyperWorkflow* and the challenges that must be solved for hyperparameter optimization jobs. Finally, we present the overview of the *HyperWorkflow*.

3.1 Problem Formulation

Given a machine learning model $\mathcal{M}$ having a set of hyperparameters $\lambda = \{\lambda_1, ..., \lambda_g\}$, the hyperparameter optimization job consists of a group of training trials, each with its specific configuration. And hyperparameter optimization is an running loop of hyperpaprameter optimization job with the goal of obtaining the optimal set of hyperparameters. We named each training trial in the job as *HyperTask*, which is a g-tuple $\Lambda_i = (\lambda_1, \lambda_2, \cdots, \lambda_g)$.

The objective of *HyperWorkflow* scheduler is to provision cloud instances to process these *HyperTasks* in the hyperparameter optimization job with minimizing execution time under the budget constraint. In this paper, a cloud instance is a virtual machine (VM) hosted on cloud infrastructure, which is adopted by most cloud service providers (e.g. Google Cloud, Amazon AWS) in reality. In general, there are multiple cloud instance types comprise varying combinations of CPU, memory, storage, and networking capacity. The billing model of cloud instance is on-demand pricing model, paying for compute capacity by the hour or second (minimum of 60 s) with no long-term commitments. Pricing is per instance-hour consumed for each instance, from the time an instance is launched until it is terminated or stopped. Each partial instance-hour consumed will be billed per-second.

3.2 Hyperparameter Optimization Workflow (*HyperWorkflow*)

There are wide variety of hyperparameter optimization algorithms, each of which generates trails and evaluates performance in different ways. To generalized different hyperparameter optimization algorithms and machine learning models, we model hyperparameter optimization job as workflow, called *HyperWorkflow*.

Definition 1. ***HyperWorkflow***. *HyperWorkflow is a fully managed cloud service that coordinates the execution of HyperTasks (a group of training trials).*

We denoted the total *HyperTasks* of a hyperparameter optimization job as m, that is, $HyperTasks = \{task_1, task_2, ..., task_m\}$. We have that $\boldsymbol{m} = \prod_g x_i$, where x_i is the candidate configurations of the hyperparameter λ_i. The execution time of the tuning process refers to the time it takes to complete all the *HyperTasks*.

In *HyperWorkflow*, we can orchestrate *HyperTasks* in sequence, in parallel, or in hybrid, as illustrated in Fig. 2. The dependencies between tasks represent the logical order of execution. In other words, there is only control flow in hyperparameter optimization workflow and no data transfer or communication between tasks. In the sequential workflow scheme, all machine learning tasks are executed one by one in sequence (as shown in Fig. 2(a)). Obviously, this sequential execution method is intensive time-consuming, and the total time spent is the cumulative sum of the running time of all tasks, $Time_{\mathcal{M}}^{Seq} = \sum_{i=1}^{m} (t_{\Lambda_i})$, where t_{Λ_i} is the execution time of *HyperTasks* $task_i$.

Another scheme is parallel workflow, as shown in Fig. 2(b), all machine learning tasks are executed in parallel and independently, which can leverage the distributed characteristics of cloud computing. There is no doubt that

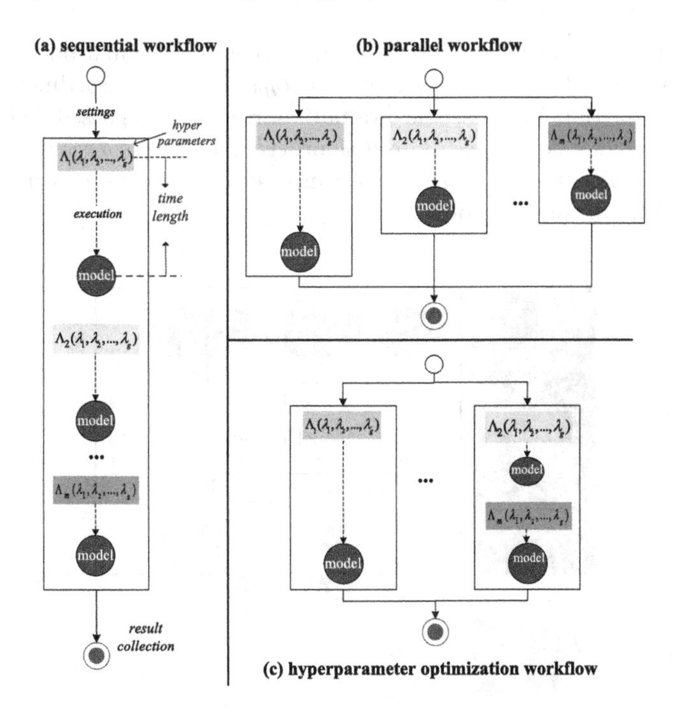

Fig. 2. Sequential, parallel and HPO workflow model

the tuning process is the fastest in this way. The total time is determined by the task with the longest running time among all machine learning tasks, $Time_{\mathcal{M}}^{Para} = \max(t_{\Lambda_i})$. However, there will be a large number of parallel branches as the search space of hyperparameter tuning of a ML model usually very large, which means a large number of cloud servers are need to be rented. In addition, different *HyperTasks* have different processing time, some run fast and some run slow. The cloud servers that running fast *HyperTasks* becomes idle, which will cause a waste of resources and a waste of money. There will be monetary and resource waste because of the idle of cloud servers that running the fast *HyperTasks.*

A good choice is to re-choreograph these *HyperTasks* (Fig. 2(c)), placing the short-running tasks into one branch, which can reduce the monetary cost and speed up the hyperparameter tuning process at the same time. How to re-choreograph these *HyperTasks* with the objective of minimizing the monetary cost as well as execution time is a combinatorial optimization problem.

3.3 *HyperWorkflow* Overview

Figure 3 illustrates the position of *HyperWorkflow* in the machine learning applications. *HyperWorkflow* coordinates between the cloud resources and HPO algorithms. The hyperparameter configureations are generated and passed on for evaluation (which can be executed parally); then the evaluation logic creates corresponding workflow model instances to *HyperWorkflow* scheduler over time. During optimization process, the evaluation results are passed back the for next trails configuration. *HyperWorkflow* provision cloud resource instances (can be virtual machine instance, docker instance, etc.) to execute workflow model instances with optimization objectives.

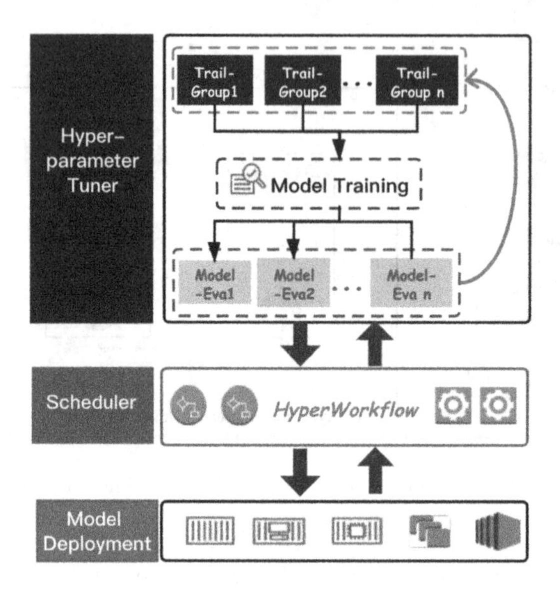

Fig. 3. Hyperparameter optimization with *Hyperworkflow* in cloud environment

4 *HyperWorkflow* Scheduling Algorithm

4.1 Upper and Lower Bounds on Execution Time of *HyperWorkflow*

Given a set of tasks, the maximum Time required to complete the task (i.e., the upper bound of the running Time, $Time^{ub}$) is all the tasks in a minimum size of the virtual machine, the calculation formula is Eq. 1, the corresponding cost (rent virtual machine cost) is Eq. 2. The minimum Time required to complete this set of tasks (i.e., the lower bound of the running Time, $Time^{lb}$) is for all the tasks to run in full parallel on the highest-spec VIRTUAL machine, as shown in 3, and the corresponding cost (the cost of renting the virtual machine) as shown in 4.

$$Time^{ub} = \sum_{i=1}^{N} time_{i,vm.small} \tag{1}$$

$$Cost^{lb} = price(vm.small) * Time^{ub} \tag{2}$$

$$Time^{lb} = max_{alltasks}(time_{i,vm.largest}) \tag{3}$$

$$Cost^{ub} = price(vm.largest) * \sum_{i=1}^{N} time_{i,vm.largest} \tag{4}$$

4.2 Scheduling Algorithm

Definition 2. ***Classical Bin Packing (BP) Problem.*** *Given an infinite supply of identical bins $B = \{b_1, b_2, ..., b_i\}$ with maximum capacity $C = 1$ and a set of n items $Item = \{item_1, item_2, ..., item_j\}$. Let a value $w_j \equiv w(item_j)$ be the size of item $item_j$ that statisfies $0 \leq w_j C$ and $1 \leq j \leq n$. The objective is to pack all the given items into the fewest bins possible such that the total item size in each bin must not exceed the bin capacity C : $\sum_{item_j \in Item_(b_i)} w_j \leq C$, $\forall b_i \in B$.*

In the classical BP problem, one can normalize $C = 1$ without loss of generality since the bin capacity is just a scale factor. Aggregating item sizes not exceeding the capacity of the corresponding bin is the only constraint. This problem is known to be an $\mathcal{NP}$-hard combinatorial optimization problem

Definition 3. ***Variable Sized Bin Packing (VSBP) Problem.*** *Given a limited collection of bin sizes such that $1 = size(b_1) \geq size(b_2) \geq ... \geq size(b_k)$, there is an infinite supply of bins for each bin type b_k. Let $L = \{b^1, b^2, ..., b^l\}$ be the list of bins needed for packing all items. Given a list of items $Item = \{item_1, item_2, ..., item_j\}$ with $size(item_j) \in (0, 1]$, the objective of the VSBP problem is to find an item-bin assignment so that the total size of the bins required $\sum_{s=1}^{l} size(b^s)$ is minimum.*

In the classical BP problem, all bins are homogeneous with a similar bin capacity. VSBP is a more general variant of the classical BP in which a limited collection of bin sizes is allowed. VSBP aims at minimizing the total size of the bins used, which is slightly different compared to the objective of the classical BP problem as discussed above.

We formulate the scheduling problem of a HyperWorkflow as a bin packing problem. Bin packing involves packing a set of items of different sizes in containers of various sizes. The size of the container shouldn't be bigger than the size of the objects. The goal is to pack as many items as possible in the least number of containers possible.

Assume there are k VM types $VM = \{vm_1, vm_2, ..., vm_k\}$, and the total *HyperTasks* of a hyperparameter optimization job as m, that is, *HyperTasks* = $\{task_1, task_2, ..., task_m\}$. Given the budget B, the optimization objective is to *minimize execution time (or makespan) with a budget constraint*, which can be reduced to a variant version of the variable sized bin packing problem (as shown in Sect. 2).

We regard the VMs as bins, the types of bins is the VM types k, and the capacity of the bins is the running time of corresponding VMs. The problem is which types of bins and at least how many are required, so that all n items can be packed into the bins, and the maximum capacity of all bins is the minimized.

There are three differences between the HyperWorkflow scheduling problem and the variable sized bin packing problem:

1. The capacity of all types bins are uncertain in this paper, while the size of bins in the problem of variable sized bin packing are known already;
2. Different with the variable sized bin packing problem, the total number of bins is not infinite, it is limited. There is a monetary cost on each bin type, and the total cost of all bins can not exceed the budget B;
3. In the HyperWorkflow scheduling problem, not only the size of bins are variable, but also the size of items. In other words, items can be placed in different types of bins and have different sizes, as a task runs at different times on different types of VMs.

Based on the above distinction from the variable size packing problem, we reduce HyperWorkflow scheduling problem as Dual Variable Size Bin Packing Algorithm, Dual-VSBP.

Algorithm 1 shows the pseudo-code of our proposed heuristic algorithm. As each task has different running times on different types of VMs, each task has multiple candidate execution times. The heuristic strategy as follows: First of all, find the task with the maximum minimum candidate running time and assign it to the type of virtual machine that makes it run the fastest. Then repeat the first step for the remaining tasks until the budget B reached. If unassigned tasks still exist while reaching budget, the unallocated tasks will be assigned to the rented VMs, with the rule of the VMs with higher performance provisioned first.

4.3 Complexity Analysis

According to the description in Algorithm 1, in order to find the task with the maximum minimum candidate running time, items need to be sorted first, the time complexity is $O(nK)$, $n$is the total number of tasks, and $k$is the types of VMs. The worst time complexity of the loop step is $O(nlogn)$. Thus, the Dual-VSB algorithm can give an approximate solution in $O(nlogn) + O(nK) = (O(nlogn + nK)) = O(nlogn)$ time.

5 Evaluation

5.1 Experimental Setup

In order to prove the effectiveness of our proposed approach for a hyperparameter search, we take three machine learning algorithms, i.e., kNN, k-means and CNN, as examples in our experiments. Without loss of generality, a server configuration with a CPU Intel core i5-6500 and memory of 16 G is used in this experiment. In real applications, the server is selected based on the total budget and deadline.

We use Greedy Algorithm (GA) as compared strategy. In this strategy, the execution time of models corresponding to the different trials of hyperparameters are first sorted in descending order. Then, the elements are taken from the sorted list, and are placed into the subset with the smallest sum of elements.

Algorithm 1 Dual-Variable Size Bin Packing Algorithm (Dual-VSBP)

Require: 1) K: VM types; 2) $task[n]$: n machine learning models; 3) B: the budget.
Ensure: $vm[\,]$: rented VMs; $mapping[\,][\,]$: the mapping between tasks and VMs.

```
1: Initialize vm[ ] ← null;
2: Initialize time[n][K];
3: Initialize Cost ← 0;
4: Initialize mapping[task][ ] ← null;
5: for i = 1 : i ≤ n; i + + do
6:     time_min[i] ← min(time[i][ ]);
7: end for
8: Sort the time_in in decease order timeSorted[ ];
9: int j ← 1, i ← 1;
10: for i = 1; i ≤ n; i + + do
11:     while Cost ≤ B do
12:         time_min[i] ← min(time[i][ ]);
13:         k ← timeSorted[i];
14:         % rent new VM.
15:         j ← j + 1;
16:         vm[j].type ← k;
17:         assign task[taskID] to the vm[j];
18:         Cost ← Cost + price[k];
19:         vm[j].time ← time[i][k];
20:     end while
21:     k' ← timeSorted[i];
22:     while k' ∈ vm[ ].type do
23:         int tmp ← 1;
24:         for j' = 1; j' ≤ size(vm[ ]); j' + + do
25:             if vm[j'].type == k' then
26:                 vm'[tmp] ← vm[j'];
27:                 tmp ← tmp + 1;
28:             end if
29:         end for
30:     end while
31:     assign task[i] to the min(vm'[ ]);
32: end for
33: return vm and mapping;
```

5.2 Execution Time Estimation of *HyperWorkflow*

To improve the initial HyperWorkflow instance model discussed in the previous section, the execution time of g tasks ($T = t_1, t_2, ..., t_g$) need to be predicted.

As a convention, there exists time complexity analysis for almost every algorithm with Big O notation. There is a linear relationship between the execution time and the expression of its hyperparameters (Eq. (5)) on a platform, which can be learned by a linear fitting process from historical execution data on the same platform.

$$T_{alg} = \alpha C + \beta \tag{5}$$

Here, we use k nearest neighbor (kNN), k-means and the CNN algorithm to verify the correctness of the time-complexity-based time prediction method.

kNN The time complexity of kNN is $O(n \cdot logn)$, where n is the size of the dataset. Then we can draw a conclusion that the execution time of kNN does not change with the value of hyperparameter k. Therefore, all parallel branches of the initial hyperparameter optimization workflow for kNN have the same execution time on the same dataset.

Table 1. Relation of different k and execution time of the kNN

k	1	3	5	7	9
Time	3848.4	3151.5	3112.2	3080.9	3104.2

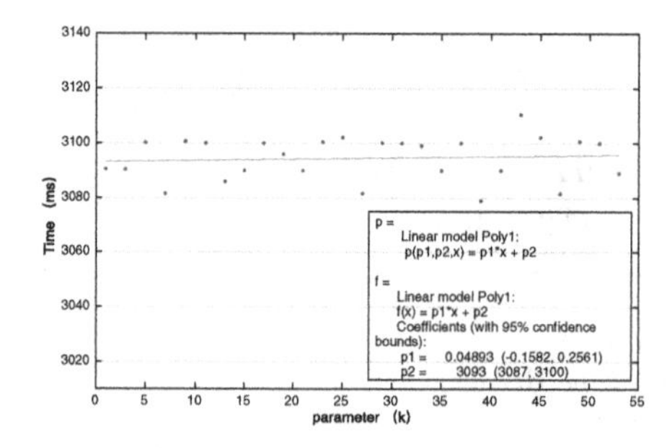

Fig. 4. Relation between *Hyperparameters Settings* and execution time (KNN)

We conducted experiments to verify this approach on the server with a CPU Intel core i5-6500 and memory of 8 G. We ran the kNN algorithm on the MINIST dataset of handwritten digits with a training set of 60,000 examples. Table 1 and Fig. 4 shows the training time spent with different k values. It can be found that the execution time of the algorithm is almost unchanged with the change of k. Therefore, the execution time of the kNN algorithm can be viewed as a constant approximately. This is consistent with the results of time estimation based on time complexity analysis.

k-means. The time complexity of the k-means algorithm is $O(n \cdot k \cdot iter)$, where n is the scale of the dataset, k is the number of clusters (centroids), and $iter$ is the number of iterations. That is, the hyperparameter tuple is $n \cdot k \cdot iter$. We can learn the linear relationship between the execution time and the hyperparameter term using several samples. First, we assign n, k, $iter$ with different values to get the data points, as listed in Table 2.

Table 2. Relation of different *Hyperparameter Settings* and execution time (second) of the k-means

n	k	iter1	Time1	iter2	Time2	iter3	Time3
1	4	300	305.32	400	396.54	500	487.88
1	5	300	348.33	400	458.76	500	574.26
1	6	300	393.90	400	524.91	500	653.62
1	7	300	463.63	400	601.94	500	752.62
1	8	300	501.55	400	666.02	500	837.43
5	4	300	1541.99	400	1992.41	500	2481.61
5	5	300	1729.34	400	2346.40	500	2906.33
5	6	300	2012.54	400	2648.23	500	3326.77
5	7	300	223.98	400	2984.32	500	3701.26
5	8	300	2478.74	400	3313.20	500	4133.45

Figure 5 shows the relation between the hyperparameter term $n \cdot k \cdot iter$ and execution time. We find that the execution time does have a linear relation with the hyperparameter term $n \cdot k \cdot iter$, which is consistent with the time complexity analysis result. After regression analysis, the linear equation is $T_{k-means} = \alpha * (n * k * iter) + \beta$, where α equals 21280 and β is 60.68.

CNN. We use a classic LeNet with MNIST as input data for the experiment. The length of the input feature map M_0 is 28, and the kernel size K is 5. Then we adjust the other hyperparameters as shown in Table 3. The settings and the execution time are both recorded in the table.

Table 3. Relation of different settings in the CNN and execution time (second)

M_0	K	$C_1.out$	$S_2.scale$	$C_3.out$	$S_4.scale$	Time
28	5	6	4	12	2	3201.22
28	5	3	4	6	2	1361.32
28	5	4	3	12	2	2308.45
28	5	4	3	6	4	1832.61
28	5	6	2	12	2	5023.75

According to the algorithm's time complexity analysis, the hyperparameter expression is $\sum_{l=1}^{D} M_l^2 \cdot K_l^2 \cdot C_{l-1} \cdot C_l$. So we plot the figure to explore the relation between the values of the hyperparameter expression and execution time. Figure 6 shows a linear relation and the fit equation is $T_{CNN} = \alpha * (\sum_{l=1}^{D} M_l^2 \cdot K_l^2 \cdot C_{l-1} \cdot C_l) + \beta$. According to the experiment, α equals 0.11427 and β is 616.8985.

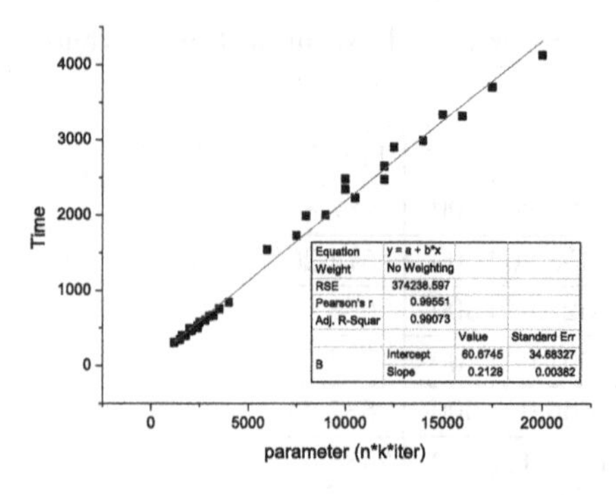

Fig. 5. Relation of *Hyperparameters Settings* and execution time (k-means)

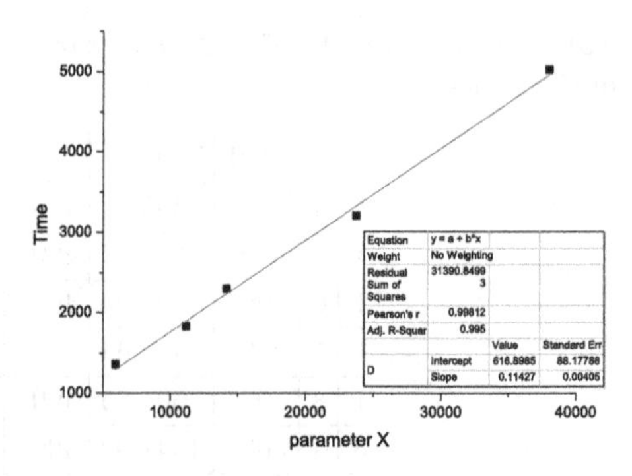

Fig. 6. Relation of *Hyperparameters Settings* and execution time (CNN)

5.3 Scheduling Algorithm Analysis

Table 4. Relative cost with different scheduling strategies

Instance Type	Core	Mem (GB)	Hourly price ($)
ecs.g5.large	2	8	0.13
ecs.g5.xlarge	4	16	0.25
ecs.g5.2xlarge	8	32	0.49
ecs.g5.3xlarge	12	48	0.73

Given a set of tasks, we can obtain upper and lower bounds based on the cost, so we set them as random values between the upper and lower bounds based on the value of the budget. In the experimental process, for part of the budget, complete parallelism cannot be realized (i.e., complete parallelism will exceed the budget and there is no effective solution). Therefore, we only compare greedy algorithm GA and Dual-VSBP algorithm proposed by us. The completion time is equal to the completion time of the slowest branch. The figure shows the average completion time of schemes with different number of tasks. It can be seen from the figure that the completion speed of scheduling schemes generated by Dual-VSBP algorithm is relatively fast (Fig. 7 and Table 4).

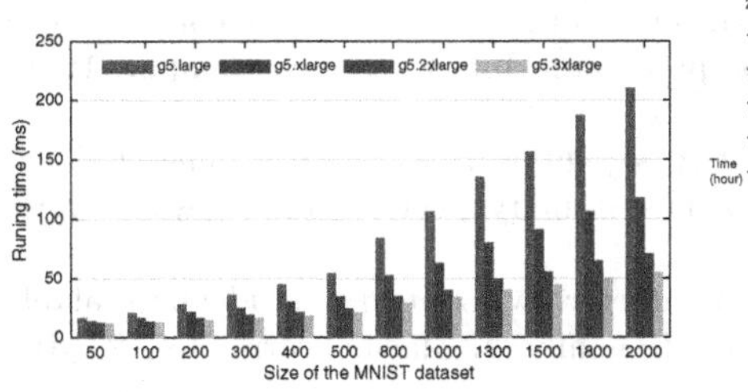

(a) number of parallel braches.

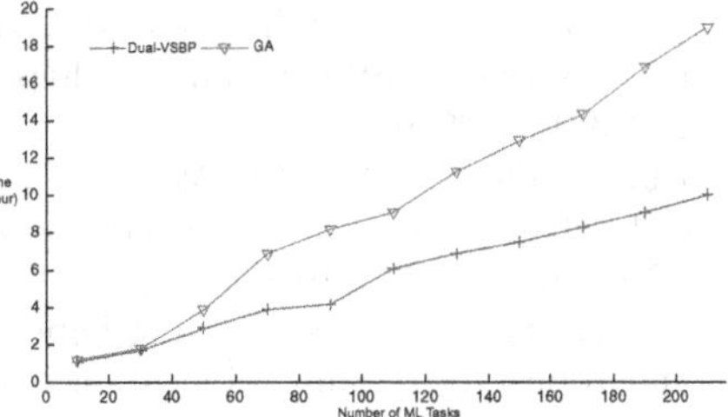

(b) percentage deviation of cost from optimal solution.

Fig. 7. Result comparision

6 Conclusions

In this paper, we described how to generate the hyperparameter optimization workflow model to ensure the cost under deadline constraints when multiple settings are specified. Specifically, we can predict the execution time of each machine learning algorithm according to its hyperparameter values based on computation complexity analysis. Then a Dual Variable Size Bin Packing algorithm is proposed to arrange the parallel branches in the workflow model.

Through the experiments, we can see that our model has stronger stability in different workflow models and reduce much more cost than GA strategy. This demonstrates that our model has the advantage of reducing time while ensuring that budget requirements can be met.

Acknowledgment. This work was supported in part by the NSF of China under Grants 61771289 and 61832012, and the Key Research and Development Program of Shandong Province under Grant 2019JZZY020124, and the Key Program of Science and Technology of Shandong under Grant No. 2020CXGC010901.

References

1. Balandat, M., et al.: BoTorch: programmable Bayesian optimization in PyTorch. arxiv e-prints, arXiv-1910 (2019)
2. Bergstra, J., Bardenet, R., Bengio, Y., Kegl, B.: Algorithms for hyper-parameter optimization, vol. 24, pp. 2546–2554 (2011)
3. Bergstra, J., Bengio, Y.: Random search for hyper-parameter optimization. J. Mach. Learn. Res. **13**(1), 281–305 (2012)
4. Falkner, S., Klein, A., Hutter, F.: BOHB: robust and efficient hyperparameter optimization at scale. Proc. Mach. Learn. Res. **80**, 1436–1445. PMLR (2018)
5. Gu, J., et al.: Tiresias: a GPU cluster manager for distributed deep learning. In: Lorch, J.R., Yu, M. (eds.) 16th USENIX Symposium on Networked Systems Design and Implementation, NSDI 2019, Boston, MA, 26–28 February 2019, pp. 485–500. USENIX Association (2019)

6. Guo, D., et al.: Image thresholding using a membrane algorithm based on enhanced particle swarm optimization with hyperparameter. Int. J. Unconv. Comput. **15**(1–2), 83–106 (2020)
7. Hu, L., Liu, A., Xie, M., Wang, T.: UAVs joint vehicles as data mules for fast codes dissemination for edge networking in smart city. Peer-to-Peer Network. Appl. **12**(6), 1550–1574 (2019)
8. Kandasamy, K., et al.: Tuning hyperparameters without grad students: scalable and robust Bayesian optimisation with dragonfly. J. Mach. Learn. Res. **21**(81), 1–27 (2020)
9. Li, L., et al.: A system for massively parallel hyperparameter tuning. In: Dhillon, I.S., Papailiopoulos, D.S., Sze, V. (eds.) Proceedings of Machine Learning and Systems 2020, MLSys 2020, Austin, TX, USA, 2–4 March 2020 (2020). mlsys.org
10. Li, L., Jamieson, K., DeSalvo, G., Rostamizadeh, A., Talwalkar, A.: Hyperband: a novel bandit-based approach to hyperparameter optimization. J. Mach. Learn. Res. **18**(1), 6765–6816 (2017)
11. Liaw, R., et al.: HyperSched: dynamic resource reallocation for model development on a deadline. In: Proceedings of the ACM Symposium on Cloud Computing, SoCC 2019, Santa Cruz, CA, USA, 20–23 November 2019, pp. 61–73. ACM (2019)
12. Misra, U., et al.: RubberBand: cloud-based hyperparameter tuning. In: Barbalace, A., Bhatotia, P., Alvisi, L., Cadar, C. (eds.) EuroSys 2021: Sixteenth European Conference on Computer Systems, Online Event, United Kingdom, 26–28 April 2021, pp. 327–342. ACM (2021)
13. Probst, P., Boulesteix, A., Bischl, B.: Tunability: importance of hyperparameters of machine learning algorithms. J. Mach. Learn. Res. **20**, 53:1–53:32 (2019)
14. Shahriari, B., Swersky, K., Wang, Z., Adams, R.P., De Freitas, N.: Taking the human out of the loop: a review of Bayesian optimization. Proc. IEEE **104**(1), 148–175 (2015)
15. Snoek, J., Larochelle, H., Adams, R.P.: Practical Bayesian optimization of machine learning algorithms. In: Neural Information Processing Systems, pp. 2951–2959 (2012)
16. Wang, T., Wang, P., Cai, S., Ma, Y., Liu, A., Xie, M.: A unified trustworthy environment establishment based on edge computing in industrial IoT. IEEE Trans. Ind. Inform. **16**(9), 6083–6091 (2020)
17. Wei, X., You, Z.: Neural network hyperparameter tuning based on improved genetic algorithm. In: ICCPR 2019: 8th International Conference on Computing and Pattern Recognition, Beijing, China, 23–25 October 2019, pp. 17–24. ACM (2019)

Workload Prediction and VM Clustering Based Server Energy Optimization in Enterprise Cloud Data Center

Longchuan Yan[1,2(✉)], Wantao Liu[1], Biyu Zhou[1], Congfeng Jiang[3], Ruixuan Li[1], and Songlin Hu[1]

[1] Institute of Information Engineering, Chinese Academy of Sciences, 100093 Beijing, China
{yanlongchua,liuwantao,zhoubiyu,liruixuan,husonglin}@iie.ac.cn
[2] State Grid Information and Telecommunication Branch, 100761 Beijing, China
[3] School of Computer Science and Technology, Hangzhou Dianzi University, Hanghzou, China
cjiang@hdu.edu.cn

Abstract. The abstract should briefly summarize the contents of the Server energy consumption of data center is an important issue of energy management. Energy optimization of server is also necessary to reduce energy consumption of data center cooling and power supply, and reduce the operation cost of whole data center. High server energy consumption is mainly caused by excessive allocation of IT resources according to the highest application workload. This paper studies the optimization algorithm of server energy consumption in enterprise cloud environment. By introducing deep learning model LSTM to predict application workload, the proposed algorithm can dynamically determine the starting up and shutting down time of virtual machines (VMs) and physical machines (PMs) according to the workload to realize the matching of application workload needs between IT resources. K-mean clustering algorithm is used to find VMs with similar starting up and shutting down time and put them on same PM group. By properly extending the running time and increasing number of VMs, the algorithm can compensate the impact of inaccurate prediction and workload fluctuation and guarantee the applications QoS. The simulation results show that the proposed method in this paper can reduce the energy consumption of servers by 45–53% with QoS guarantee when the prediction relative error is 20%, which can provide a good balance between energy saving and application QoS.

Keywords: Energy optimization · Virtual machine · Cloud computing · Resource management · Workload prediction · VM clustering

This work is supported by The Natural Key Research and Development Program of China(2017YFB1010001).

Y. Lai et al. (Eds.): ICA3PP 2021, LNCS 13157, pp. 293–312, 2022.
https://doi.org/10.1007/978-3-030-95391-1_19

1 Introduction

With the rapid development and popularization of the new generation of information technology, such as mobile applications, the Internet of Things (IoT), artificial intelligence (AI), etc., the demand for IT resources of cloud computing data center is ever-increasing. The proportion of electricity used by data centers in the United States has increased from 1.8% in 2014 to about 3% in 2019 [1,18]. For the operators of cloud computing data center, they are faced with huge challenges in resource management and energy consumption reduction [1,8]. Therefore, the issue of energy consumption is becoming increasingly prominent in modern data centers.

Currently, there are various IT devices in the cloud computing data center, including servers, network devices and security devices, among which servers account for the largest amount and cause the highest energy consumption [4,20]. The proportion of servers and energy consumption in data center is further increasing. In terms of the scale of data centers, Google, Microsoft, Alibaba and other large Internet companies only have a small number of Internet data centers but with huge scale. However, large amounts of enterprises, governments branch and public utilities around the world have a large number of small and medium-sized cloud computing data centers [15], which have much lower resource utilization and greater total energy consumption as a whole. Some energy consumption optimization methods for large data centers are difficult to apply in small data centers. Therefore, this paper focuses on the optimization of server energy consumption in the enterprise cloud data center (ECDC), specifically, the small and mid-sized ECDCs.

In cloud computing environment, many researchers use VMs migration and consolidation technology to optimize resource management and energy consumption [24]. Many researchers try to save power and energy by dynamic migrating of VMs, adjusting resource allocation and merging VMs, shutting down or keep idle servers sleeping, and etc. However, in practice, dynamic VMs migration and consolidation are faced with many challenges. Researchers model the dynamic migration and merging of VMs as knapsack problems and propose some optimization algorithms. Since knapsack problem is NP hard [6], heuristic algorithm, genetic algorithm, simulated annealing and other algorithms can be used to calculate optimal solution firstly [17]. Then the algorithm performs VMs migration and consolidation, and starts up, shuts down or hibernates physical machines (PMs) to save energy. In this scenario, global optimization may cause a large number of VMs migration. In practical deployment, it will lead to the increase of network traffic and has a negative impact on the Quality of Service (QoS) of applications [13]. Therefore, it is essentially important to decrease the times of VMs migration and reduce the impact on QoS of applications caused by the dynamic VMs migration and consolidation.

In addition, the major reason of low utilization of server resources in ECDC is that resources are allocated according to the highest peak requirement of applications, while the actual service time of applications at the peak workload is very short. In ECDC, the deployed applications mainly serve the enterprise internal business, and their workload characteristics are more regularly cyclic. This provides the opportunity to dynamically allocate IT resources according to the workload of applications, so as to match IT resources requirement of application and reduce energy consumption.

In recent years, deep learning (DL) has made remarkable progress in data prediction, natural language processing (NLP) and image classification. In this paper, DL technology is introduced to carry out workload prediction of application to assist VMs resource management of ECDC. Specifically, VMs resource management algorithm allocates and manages IT resources based on the workload of application by dynamically opening and closing VMs and PMs to realize the matching of application requirements with VMs resources. Such approaches can reduce the waste of IT resources significantly. More specifically, according to the opening and closing time of VM, this paper uses clustering algorithm to divide VMs into multiple groups, and places VMs with similar opening and closing time on the same PM groups, so that VMs and PMs can be started and shut down dynamically according to the predicted application workload. Then the dynamic migration of VMs is arranged during the low application workload time to reduce the impact on application QoS.

There are three contributions in this paper:

- A server energy management method for ECDC using DL based workload prediction is proposed.
- By grouping VMs and PMs, the migration of VMs in the peak period of business workload is avoided, and the impact of VMs migration on business application QoS is alleviated.
- Through the workload prediction and prediction error compensation, the problem of sudden workload change and inaccurate prediction is solved, and the robustness of the algorithm is improved.

The reminder of the paper are organized as follows. Section 2 and 3 introduce the relevant research progress and application characteristics of ECDC, respectively. Section 4 mainly describes the energy model and system architecture. The algorithm design is presented in Sect. 5. In Sect. 6, the experiment and simulation results are analyzed, and Sect. 7 summarizes the paper.

2 Related Work

The energy optimization technologies of server in data center can be simply grouped into two categories, i.e., the dynamic voltage frequency scaling (DVFS), and the dynamic power on/off servers (DPS). For most servers, even if CPU utilization is zero, their energy consumption is still about 60% of its peak energy consumption [9]. In this scenario, DPS can effectively reduce the energy consumption of servers significantly.

Researchers studies the energy consumption optimization of server cluster by using DPS technology. For example, Chase et al. [5] design and implement a resource management architecture of hosting center operating system MUSE.

By dynamically adjusting the number of active servers, it can automatically adapt to the workload change and improve the energy efficiency of the server cluster. Rajamani et al. [16] study the energy-saving problem of server cluster and proposed an algorithm that calculates the number of servers in the running state, shuts down some servers that can be idle to realize the energy saving based on energy consumption, QoS, system characteristics and load characteristics, as well as the historical business characteristics of the system, the processing ability of the server, the start-up delay of the server and other parameters. These early works provide a solid foundation for energy consumption optimization.

Ye et al. [22] introduce the energy consumption management of cloud computing platform from four aspects: energy consumption measurement, energy consumption modeling, energy consumption management implementation and optimization, and propose 3 energy consumption models of VMs, server consolidation and VMs online migration. The energy consumption model of VM migration assumes a linear growth relationship between energy consumption and the network overhead. In order to guarantee QoS, Xiong et al. model the server energy consumption based on CPU utilization, and propose a double threshold heuristic algorithm of energy perception to migrate the VM to reduce energy consumption [21]. The performance evaluation indicators include energy consumption, service level agreement (SLA) violation times (VM does not get enough MIPS), VM migration times. Beloglazov et al. [2] proposed the energy consumption optimization methods of computing nodes and storage nodes in the cloud video monitoring environment. Specifically, computing nodes mainly involve the DVFS, the optimization methods of VM deployment, dynamic migration, task access and scheduling, and etc. At the mean time, storage nodes mainly involve the low-power control in hardware, as well as the static data placement and dynamic data migration technology to reduce energy consumption in software.

To effectively utilize the historical information of business workload, Gaussian process regression (GPR) [3], LSTM [10], support vector regression (SVR) [19] and kernel density estimation (KDE) [11] are used to predict the demand of business applications, jobs or VMs, so as to allocate and adjust IT resources more accurately. Bui et al. propose an energy-efficient solution for orchestrating the resource in cloud computing. In nature, the proposed approach predicts the resource utilization of the upcoming period based on the GPR method [3]. Liu et al. [10]proposed a hierarchical cloud resource and energy management architecture based on deep reinforcement learning (DRL) is proposed, which includes a global layer responsible for VM allocation and a local layer responsible for local power management. In [19], a data-driven energy consumption modeling method for high-performance computers is proposed. This model is a system level energy consumption model, which includes three parts: using SVR method to predict job energy consumption, using heuristic rules to predict job completion time and idle energy consumption. Finally, three parts of information are fused by linear model to predict system level energy consumption. Mahdhi et al. [11] propose a virtual machine consolidation approach based on the estimation of requested resources and the future VM migration traffic. KDE technology is used to forecast the future resource usage of each VM. Mohiuddin et al. [12] proposed a workload aware VM consolidation method, namely, WAVMCM, and classify VM resources during VM allocation, and then select suitable PM for placement. When the threshold is reached, it migrates and merges VMs according to the workload type and demand of the VM, and turns the PM that is no longer in use into the sleep state. Zhang et al. [23] investigate the burstiness-aware server consolidation problem from the perspective of resource reservation, i.e., reserving a certain amount of extra resources on each PM to avoid live migrations, and propose a novel server consolidation algorithm, QUEUE. Nathuji et al. [14] proposed an approach to online VM power management, and make it possible to control and globally coordinate the effects of the diverse power management policies applied by these VMs to virtualized resources.

For energy optimization of data center, the VM migration and consolidation, workload analysis and prediction, and SDP are often used in various application scenarios. However, the above mentioned work do not consider the workload characteristics of real ECDC. It is necessary to analyze the access characteristics of applications in ECDC.

3 Characteristics Analysis

Compared with the Internet data center, most of the IT devices deployed in ECDC host the applications serving for the enterprise employee. The access characteristics of the enterprise applications and IT devices are obviously different from that of the Internet applications. First of all, the workload of application has obvious characteristics related with working time. Most of the applications used in enterprises have high workload in working time and low workload in nonworking time. The application workload increases rapidly, and generally the workload increases several times or dozens of times in several minutes. Some application with small business volume have low workload during business peak period. Figure 1 shows the number of online users of an application in an enterprise data center in a week, reflecting the above characteristics in detail.

Secondly, in ECDC VMs will not be created and destroyed frequently. Generally, VM will last for many years after created until the applications are no longer used. In general, the enterprise data center does not need dynamic migration of VMs in the peak period of business to avoid the impact on application QoS.

Third, the resource utilization of enterprise data center is low. The CPU utilization rate of servers is less than 5% in 70% of the time. The CPU utilization rate of some servers is less than 1% for a long time. The power difference between servers running in low power state and idle state is not significant.

The workload characteristics of applications in enterprise data centers and the running characteristics of servers and VMs inspire us:

- Through workload prediction, it is feasible to grasp the change trend of workload in advance and allocate resources in advance.

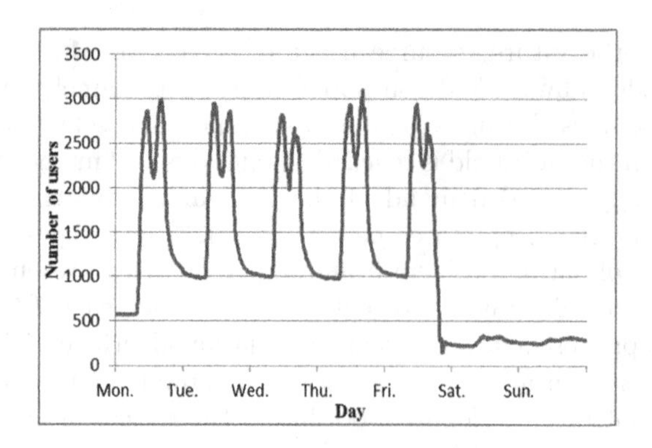

Fig. 1. Workload of application in enterprise data center (number of online users).

- The enterprise data centers need to avoid frequent VMs migration in the business peak time, so as not to affect the application QoS.
- The server utilization rate is low and the basic power is high. Shutting down the idle servers is the most effective energy-saving measure.

4 System Architecture

4.1 Energy Consumption Model

In the cloud computing environment, the server power consists of the idle PMs' and VMs' power. The server power is expressed as the sum of the idle power and the VMs power. The calculation formula is as follows:

$$P_{server} = P_{idle} + \sum_{i=1}^{n} P_{vm}^{i} \tag{1}$$

The total power of all servers is expressed as:

$$P_{total} = \sum_{j=1}^{m} s_j P_{server}^{i} \tag{2}$$

$$s_j = \begin{cases} 0, \text{if } jth \text{ server is off} \\ 1, \text{if } jth \text{ server is running} \end{cases} \tag{3}$$

Assuming that 24 h of a day is divided into k time slices, and the length of each time slice is T, the total energy consumption of all servers is expressed as:

$$E_{total} = \sum_{t=1}^{k} P_{total}^{t} \times T = T \times \sum_{t=1}^{k} \sum_{j=1}^{m} s_j^t \times (\overline{P}_{idle}^{t,j} + \sum_{i=1}^{n} \overline{P}_{vm}^{t,j,i}) \tag{4}$$

where, $\overline{P}_{idle}^{t,j}$ is the average basic power in the t_{th} time slice of the jth server, and $\overline{P}_{vm}^{t,j,i}$ is the average power in the tth time slice of the ith VM on the jth server.

In heterogeneous cloud computing environment, the basic power of servers is different. The goal of server energy consumption optimization is to allocate VMs on servers, dynamically select the running time s_j^t of servers and VMs, and minimize the total energy consumption under the condition of guaranteeing the QoS.

$$\begin{aligned} &\min_{s_j^t} E_{total} \\ &s.t.\ QoS_i \geq Q_i^T \\ &s_j^t = 0\ or\ 1 \end{aligned} \tag{5}$$

where, Q_i^T is the QoS threshold of the ith application.

The QoS of application is defined as the ratio between the time when the number of VMs provided by the algorithm is greater than the number of VMs required by users and the actual running time of the application. The formula is as follows:

$$QoS_{app} = 1 - \frac{UT_{app}}{T_{app}} \times 100\% \tag{6}$$

in which, UT_{app} is the time that the number of VMs provided by the algorithm is less than the number of VMs required by users, T_{app} is the running time of application.

4.2 Solutions

Considering the important periodicity of workload of application in enterprise data center, we use clustering algorithm and heuristic algorithm to solve the problem of energy consumption optimization. Specifically, according to the predicted application workload, the VM opening and closing time is determined. VMs with similar opening and closing time are placed on the same PM by clustering method. When the application workload drops, the related VM is closed. When all VMs on a PM are closed, the PM is closed too. When the application workload rises, the related VMs and PMs are opened.

When the utilization of some PMs are low, it is necessary to migrate the VMs on this PM to other PMs to reduce energy consumption. In order to measure the utilization efficiency of PM during operation, we define the utilization rate of PM η in operation time as follows:

$$\eta = \frac{1}{N} \sum_{i=1}^{n} \frac{T_{vm}^i}{T_{pm}} \tag{7}$$

in which, n is the number of VMs on the PM, N is the maximum number of VMs that can run on the PM, T_{vm}^i is the running time length of the i_{th} VM, and T_{pm} is the running time length of the PM. The value range of PM running time utilization rate η is $[0, 1]$. The higher η value means less energy waste and vice versa.

4.3 System Description

This paper proposes a server energy optimization management architecture based on the opening and closing time of the VMs and PMs. Energy optimizer is the core of server energy consumption optimization management. It dynamically schedules and manages PMs and VMs, and is responsible for the VM allocation and adjustment the opening and closing time of VMs and PMs, etc. Its main functional modules include resource scheduler, workload prediction, VM allocation, VM adjustment, PM and VM control, and VM recycle. The system architecture is shown in Fig. 2.

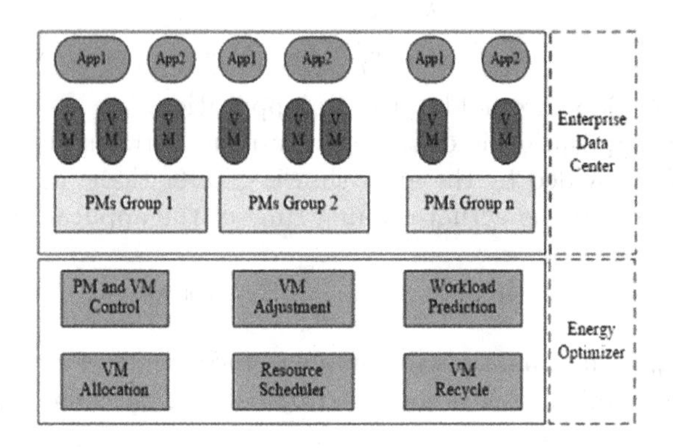

Fig. 2. PM and VM energy management framework of cloud data center.

The resource scheduler module schedules the VMs and PMs resources every certain time interval. According to the application workload, the module is responsible for the control of whole VMs management process which includes VMs running time optimization, opening and closing VMs and PMs determination, VM allocation for new application and release unused VMs, and etc. The workload prediction module predicts the 24-h workload of each application, and provides one-day workload data for the resource scheduler module. VM allocation module allocates VMs for new application and selects PMs. The VM recycle module releases the occupied PM resources of VMs of the application that is no longer running. The VM adjustment module determines the opening and closing time of VM according to the change of application workload and the change of server resource utilization. Similar VMs can be found by clustering algorithm, and VMs with large change of opening and closing time can be dynamically migrated to the new servers for the, so as to improve the server resource utilization. The PM and VM control module is responsible for the opening and closing of the PMs and VMs under the management of resource scheduler.

5 Algorithm Design and Description

5.1 Resource Scheduling Algorithm

Resource scheduling algorithm controls the whole process of PMs and VMs management and energy consumption optimization. In this paper, fixed time slot T is used for resource scheduling and management. The resource scheduling algorithm is implemented every T minutes to allocate VMs for new business applications, release the VMs of applications that are no longer running, and carry out the opening and closing scheduling of PMs and VMs.

In order to avoid the impact of VM migration on application QoS when the application is under high workload, the resource scheduling algorithm tries to minimize times of VM migration. Only at the beginning of every day (generally, at that time the application is in a low workload), the workload is predicted, and VMs migration and PMs adjustment are performed according to the workload. The pseudo code of the algorithm is shown in Algorithm 1.

Algorithm 1. Resource Scheduling

```
Require: history workload workload_h
    day=0;
 1: while day++ do
 2:    workload = Workload_Predict(workload_h);
 3:    pmtimeList, vmtimeList = VM_Adjust(workload);
       // M is number of time slots in a day
 4:    for i = 1 : M do
 5:       if isNotEmpty(newApp) then
 6:          VM_Allocate(newApp);
 7:       end if
 8:       if isNotEmpty(retirdApp) then
 9:          VM_Release(retirdApp);
10:       end if
11:       //start and close VM and PM according to run time
12:       startPM(pmtimeList)
13:       startVM(vmtimeList)
14:       stopVM(vmtimeList)
15:       stopPM(pmtimeList)
16:    end for
17: end while
18: return ;
```

The algorithm adjusts VMs only through VMs migration in the low workload time in a day to avoid the impact of VMs migration on application QoS.

5.2 Workload Prediction Algorithm

Due to the excellent performance of DL in prediction, we use LSTM to predict workload of the application. LSTM is an improvement of RNN, which solves the problem of long dependence of RNN [7]. LSTM, like other RNN, is a neural network with repeated module chain structure. LSTM network can delete or add information to cell state through a structure called gate. These three gates are called forgetting gate, input gate and output gate.

In this paper, the multi-dimensional single step method is used to predict the business workload, which is to use the workload of previous days to predict the workload of next day. The neural network we used is a stacked LSTM with three LSTM layers and one Dense layer. The input dimension of the neural network is 288, the number of neurons in each layer is 128, the window length is 5–8, the window step length is 1, the prediction length is 1, and the neural network training algorithm is Adagrad. The prediction model in this article is implemented by Python Keras. Before the prediction model training, we normalize the application workload data to [0, 1].

Workload prediction error affects the QoS of application. To compensate the decline of application QoS caused by prediction error, a QoS compensation mechanism is proposed in the VM adjustment algorithm. By increasing the predicted workload by times of prediction error, that is more VMs are allocated to the application, so as to eliminate the negative impact of prediction error on QoS. The workload ($workload_a$) used for VM allocation is calculated as follows:

$$Workload_a = Workload_p \times (1 + e_{max}) \tag{8}$$

where, $Workload_p$ is predicted workload, and e_{max} is the maximal relative error of predicted workload.

5.3 VM Adjustment Algorithm

The VM adjustment algorithm selects the appropriate PMs according to the opening and closing time of the VMs, and places the VMs with similar opening and closing time characteristics on the same PM. Firstly, K-mean clustering method is used to cluster VMs, and the same number of PM groups are set according to the number of categories. Then, the same kind of VMs are placed on the same PM group. In order to reduce VMs migration, only PMs are reselected for VMs with different VM categories after clustering. VMs placement problem is a typical knapsack problem. In this paper, greedy algorithm is used to solve the problem. The resource requirements of CPU, memory and network of VMs are sorted in descending order to form three queues, and then a VM is selected from each queue to place on the available PM in turn. If the available PMs cannot be carried, a PM is added until all VMs are placed. Finally, the opening and closing time of the PM is determined according to the opening and closing time of the VMs on the PM. The algorithm is described in Algorithm 2.

As a part of the adjustment algorithm, when the utilization rate of the PM is lower than the threshold u, the VM on the PM is tried to migrate to other PMs. If there is no VM on the PM, remove the PM from the its group. The algorithm is described shown in Algorithm 3.

5.4 VM Allocation and Release Algorithm

The VM allocation algorithm allocates VMs for the new application according to the VM resource requirements set by the administrator. The VM recovery algorithm first shuts down the VMs that is no longer running, and then deletes the VMs from the cloud platform to release the occupied PMs resources.

Algorithm 2. *VM_Adjust*

Require: VM list *vmlist*,PM list *pmList*,Workload *workload*;
Ensure: VM time list *vmtimeList* PM time list *pmtimeList*;

```
1: //workload compensation
2: workload = workload × (1 + e_max);
3: vmtimeList = Determine_RunTime(workload);
4: vmCluster = K − Mean(vmtimeList);
5: vmClass = getClassUnchangedVM(vmCluster);
6: vmClassNum = getClassNum(vmClass);
7: for i = 1 : vmClassNum do
8:    vmNum = getVMNum(vmClass(i));
9:    for j = 1:vmNum do
10:      vm = selectVM(vmClass(i));
11:      pmNum = getPMNum(pmGroup(i));
12:      pm = getPM(pmGroup(i));
13:      migrateFlag = False;
14:      for n = 1:pmNum and pm(n).Support(vm(j)) do
15:         migrateVM(vm(j), pm(n));
16:         addPM(pmGroup(i), newPM);
17:         Break;
18:      end for
19:      if migrateFlag==False then
20:         newpm = getNewPM();
21:         migrateVM(vm(j), newpm);
22:         addPM(pmGroup(i), newpm);
23:      end if
24:   end for
25: end for
26: //Merge PM if its utilization rate is low
27: VM_Merge(pmGroup, u);
28: pmtimeList = Determine_RunTime(vmtimeList);
29: return pmtimeList,vmtimeList;
```

Algorithm 3. *VM_Merge*

```
Require: PM group pmGroup,PM utilization u
Ensure: number of VM migration mTimes
 1: mTime = 0;
 2: groupNum = getGroupNUm(pmGroup);
 3: for i = 1:groupNum do
 4:    pmNum = getPMNum(pmGroup(i));
 5:    for j = 1 : pmNum do
 6:       if pmGroupi, j.u < u then
 7:          vm = getVM(pmGroup(i, j));
 8:          vmNum = getVMNum(vm);
 9:          for n = 1:vmNum do
10:             for m = 1:pmNUm and m! = j do
11:                if pmGroup(i, m).Support(vm(n)) then
12:                   migrateVM(vm(n), pmGroup(i, m));
13:                   mTime + +;
14:                   break;
15:                end if
16:             end for
17:          end for
18:       end if
19:    end for
20: end for
21: return mTimes;
```

6 Simulation Experiments and Analysis

6.1 Description and Evaluation of Simulation Experiment

Java and Python are used to implement programs for the simulation experiment of server energy optimization in ECDC. Assume that at the beginning of the experiment, there are 4 applications, 65 servers with 5 models in cloud data center. The server power parameters is shown in Table 1.

The power of VM is determined by the number of CPU cores, size of memory and disk in this paper, and its formula is as follows:

$$P_{vm} = \alpha \times Cores_{cpu} + \beta \times Size_{memory} + \gamma \times Size_{disk} \tag{9}$$

in which, α, β and γ are power coefficients of CPU, memory and disk. The number of new coming applications per day conforms to the Poisson distribution. The probability function of Poisson distribution is as follows:

$$P(X = k) = \frac{\lambda^k}{k!}e^{-\lambda}, k = 1, 2, 3, ... \tag{10}$$

where, the parameter λ is the average number of random events in unit time (or unit area).

In this paper, the scheduling interval is 5 min. The experiment is divided into 2 time periods, the first period is 7 days, which is used to test the energy consumption of servers without energy consumption optimization. The second period is 22 days, which is used to test the performance of the proposed energy consumption optimization algorithm.

Table 1. Server parameters in simulation

No.	Idle power (W)	CPU number	Memory size (GB)	Server number
1	150	72	256	8
2	180	80	256	12
3	220	96	384	10
4	130	72	256	15
5	120	72	256	20

6.2 Energy Consumption Analysis

According to the simulation results, 35 PMs are used to carry 4 applications, and the energy consumption of PMs and VMs is 267.26 kWh. When the optimization algorithm of energy consumption is turned on, the energy consumption of cloud platform system is reduced to 67.05–136.25 kWh without workload prediction error, and the maximum energy consumption of nonworking days is reduced to the original 25.08%. The energy consumption curve of the cloud data center server every day is shown in Fig. 3.

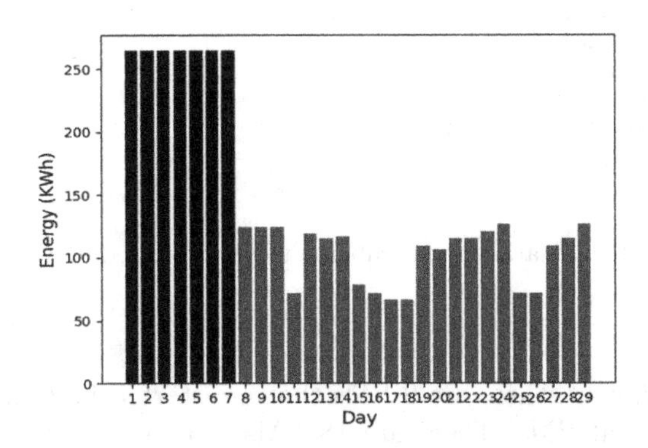

Fig. 3. Energy consumed by servers in data center.

Figure 4 and 5 show the workload of an application and the number of users supported by the VMs according to the algorithm when there is no prediction error in a workday and nonworking day, respectively. The algorithm turns on and off the VMs in time according to the workload of application to optimize the energy consumption. In nonworking hours and nonworking day, only 25% of VMs is enough to meet the needs of application. The least opening time in the VMs accounts for 4.86% of a day.

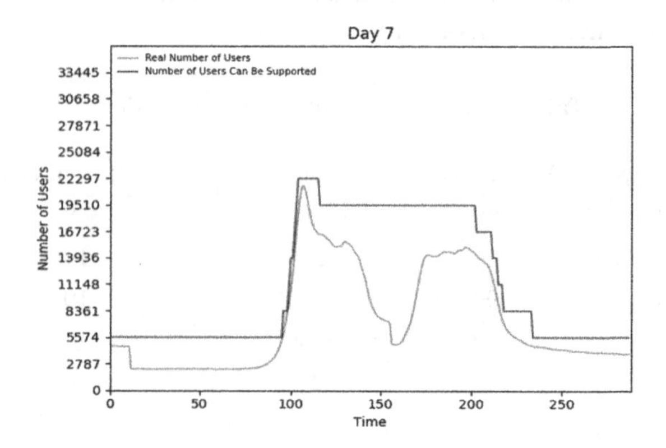

Fig. 4. Workload of application and number of users supported by VMs in a working day.

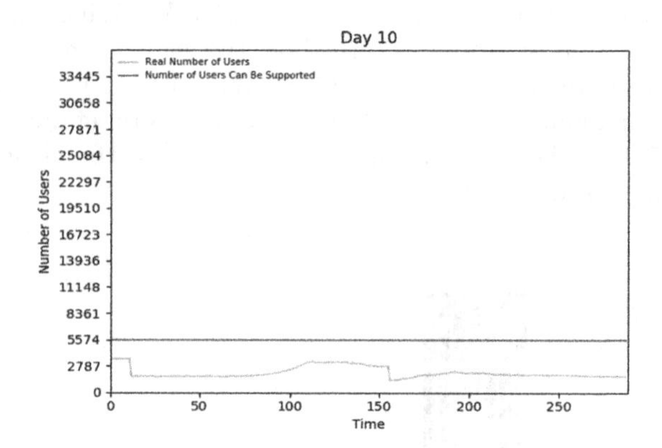

Fig. 5. Workload of application and number of users supported by VMs in a nonworking day.

In this experiment, there are 9 PMs that are started in 24 h, accounting for 25.71% of the total PMs. There are 18 PMs that are started in 8–11 hours, accounting for 51.74% of the total PMs. Other 6 PMs running time is less than 2 h, and one PM does not need to be started. The number of PMs opened on nonworking days is less than that of working days. For the running time of VMs in a working day, Only 60 VMs need to run for 24 h, and the running time of other VMs is less than 13 h.

6.3 Prediction Analysis

The mean absolute percentage error (MAPE) and maximum relative error are both used to characterize the accuracy of neural network model prediction. MAPE is defined as follows:

$$MAPE = \frac{1}{n}\sum_{i=1}^{N}\frac{|y_i - \overline{y}_i|}{y_i} \tag{11}$$

where y_i and $\overline{y}_i$ represent the real value and the predicted value respectively.

Three applications are selected for workload prediction, and the MAPE of the models is 5.14%, 2.57% and 4.52% respectively, which shows that the prediction algorithm has good accuracy. The prediction results of an application are shown in Fig. 6.

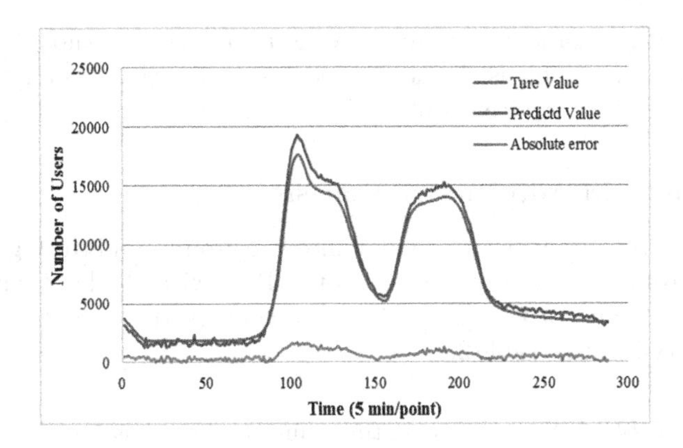

Fig. 6. Application workload prediction results.

In this paper, the predicted workload is used for VMs scheduling, so the maximum relative error is needed to calculate the number of VMs used to compensate the prediction error to guarantee the QoS. The larger the maximum relative error is, the more VMs are needed, and the higher the energy consumption is. In the experiment, the maximum relative error is between 8.63% and 19.67%.

6.4 Clustering Analysis

The K-means clustering algorithm is used to find the VMs with similar opening and closing time. The number of VMs clusters is set to 5. There are 5 clusters of VMs with different color shown in Fig. 7. The X-axis represents the opening time of VM, and the Y-axis represents the close time of VM. The point at the top left represents the VMs that are always on.

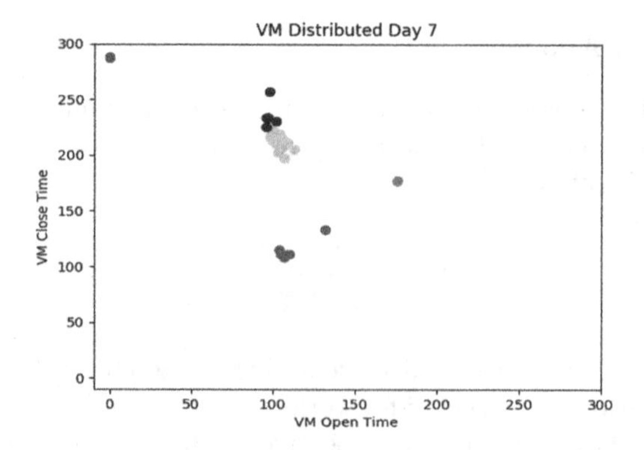

Fig. 7. VMs clusters by K-means algorithm (the coordinate interval of X and Y-axis is 5 min).

By clustering algorithm, we divide VMs into 5 groups and place them on different PM groups. This is the key to reduce the energy consumption of VMs and PMs in the cloud environment.

6.5 PM and VM Adjustment Analysis

In this experiment, the PMs are divided into five groups, and the 0 group is PMs that are always on. The second group is the PMs with the least running time, whose running time range is [103, 118]. The running time range of each group of PMs is shown in the Table 2.

Table 2. Server running time range (each vale is 5 min)

No.	Group label	Running time	Server number
0	0	[0, 288]	8
1	3	[92, 259]	2
2	4	[96, 238]	7
3	1	[8, 219]	19
4	2	[103, 118]	2

In the 22 days of the simulation experiment, VM migration occurred on the 7th, 8th, 18th, 21st and 23rd days due to the workload change. The reason is that the application workload change causes the change of the VM opening and closing time, which leads to the VM migration. Figure 8 shows number of VMs migration and PMs adding. Due to the migration of VMs, the number of PMs in the target group cannot host more VMs, so it is necessary to add PMs to the corresponding group from the PMs that have never been opened. This resulted in an increase in PMs on days 7, 8 and 23.

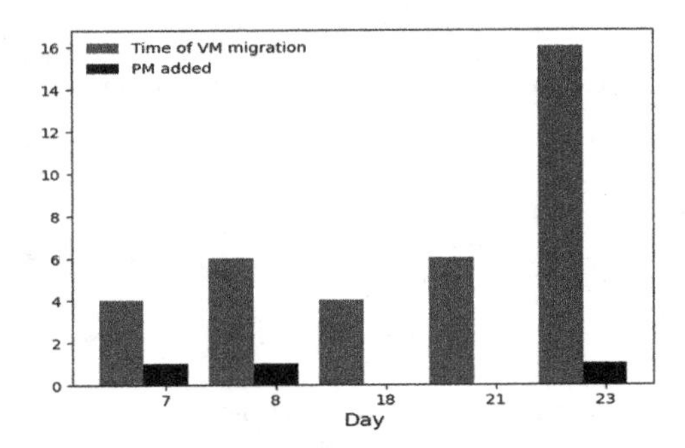

Fig. 8. Time of VM migration and number of PMs adding.

6.6 QoS Analysis and Compensation Mechanism

The impact of different workload prediction errors on application QoS has been calculated. Figure 9 shows the proportion of QoS decline of four applications when the workload has 20% prediction error.

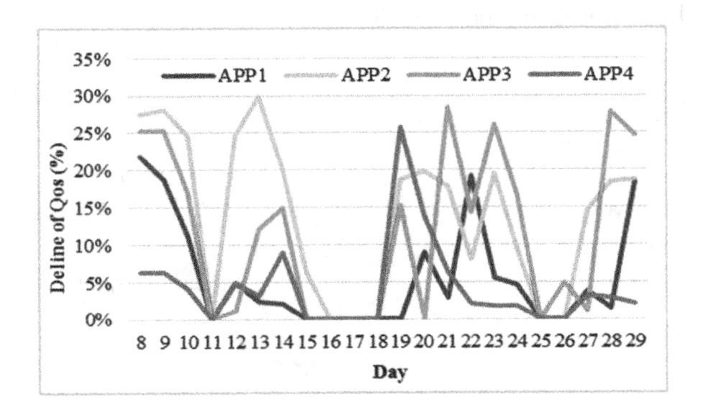

Fig. 9. QoS decline of applications with 20% prediction relative error.

From Fig. 9, it can be seen that the same prediction error has different impact on each application. There are more VM resources on nonworking days, so the prediction error has less impact on QoS. The influence of prediction error on QoS is related to the shape of workload curve (access characteristics), VM number and the time of prediction error occurrence.

Through the prediction error compensation mechanism, it can be ensure that the resources provided to the application meet the needs of QoS, but at the same time, the total energy consumption will increase slightly.

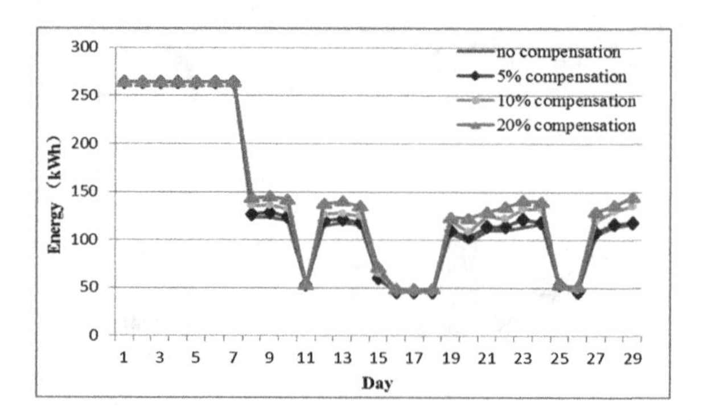

Fig. 10. Energy consumed under various prediction error compensation.

Figure 10 shows the total energy consumption without compensation and with compensation of 3 kinds of prediction error. From the figure, it can be seen that the larger the prediction error is, the more compensation energy consumption will be increased. The simulation results of energy consumption show that the algorithm can reduce 45–53% energy consumption of data center servers when the prediction relative error is 20%.

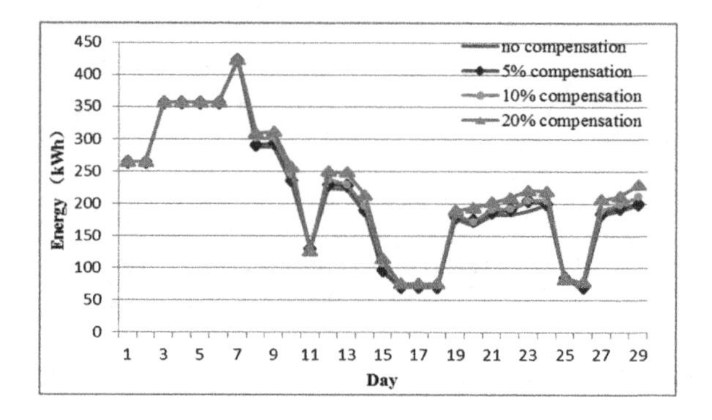

Fig. 11. Energy consumed under various prediction error compensation with new coming applicaitons.

In Fig. 11, 1 and 2 applications are added on 3rd and 7th day, respectively, which result in an energy consumption increase. After the 7 days observation period, the new applications are added to the scope of energy consumption optimization, and the total energy consumption on the 10th and 14th day decreased by about 18% and 15% respectively compared with the previous day. This shows that the proposed algorithm can effectively deal with the energy consumption optimization of new applications.

7 Conclusion

This paper analyzes the workload characteristics of applications and server usage characteristics of ECDC, and proposes an algorithm of server energy consumption optimization management based on workload prediction and VM clustering. The algorithm dynamically allocates the running time of VMs according to the application workload predicted by LSTM algorithm. For VMs placement, K-means clustering algorithm is used to place VMs with similar opening and closing time on the same PMs, so that it allows to adjust the PMs to sleep or shutdown state when the VMs does not need to run, so as to reduce the energy consumption of the PMs. The simulation results of energy consumption show that the algorithm can reduce 45–53% energy consumption of data center servers when the prediction relative error is 20%, and guarantee application QoS. By increasing running time of VMs or adding some VMs, the algorithm can achieve the balance between server energy saving and application QoS.

References

1. Energy 101: Energy Efficient Data Centers. https://www.energy.gov/eere/videos/energy-101-energy-efficient-data-centers
2. Beloglazov, A., Abawajy, J., Buyya, R.: Energy-aware resource allocation heuristics for efficient management of data centers for cloud computing. Futur. Gener. Comput. Syst. **28**(5), 755–768 (2012)
3. Bui, D.M., Yoon, Y., Huh, E.N., Jun, S., Lee, S.: Energy efficiency for cloud computing system based on predictive optimization. J. Parallel Distrib. Comput. **102**, 103–114 (2017)
4. Buyya, R., Yeo, C.S., Venugopal, S., Broberg, J., Brandic, I.: Cloud computing and emerging it platforms: vision, hype, and reality for delivering computing as the 5th utility. Futur. Gener. Comput. Syst. **25**(6), 599–616 (2009)
5. Chase, J.S., Anderson, D.C., Thakar, P.N., Vahdat, A.M., Doyle, R.P.: Managing energy and server resources in hosting centers. ACM SIGOPS Oper. Syst. Rev. **35**(5), 103–116 (2001)
6. Gary, M.R., Johnson, D.S.: Computers and Intractability: A Guide to the Theory of NP-Completeness (1979)
7. Hochreiter, S., Schmidhuber, J.: Long short-term memory. Neural Comput. **9**(8), 1735–1780 (1997)
8. Iqbal, W., Berral, J.L., Carrera, D., et al.: Adaptive sliding windows for improved estimation of data center resource utilization. Futur. Gener. Comput. Syst. **104**, 212–224 (2020)
9. Li, H., Zhu, G., Cui, C., Tang, H., Dou, Y., He, C.: Energy-efficient migration and consolidation algorithm of virtual machines in data centers for cloud computing. Computing **98**(3), 303–317 (2015). https://doi.org/10.1007/s00607-015-0467-4
10. Liu, N., et al.: A hierarchical framework of cloud resource allocation and power management using deep reinforcement learning. In: 2017 IEEE 37th International Conference on Distributed Computing Systems (ICDCS), pp. 372–382. IEEE (2017)
11. Mahdhi, T., Mezni, H.: A prediction-based VM consolidation approach in IaaS cloud data centers. J. Syst. Softw. **146**, 263–285 (2018)

12. Mohiuddin, I., Almogren, A.: Workload aware VM consolidation method in edge/cloud computing for IoT applications. J. Parallel Distrib. Comput. **123**, 204–214 (2019)
13. Najm, M., Tamarapalli, V.: VM migration for profit maximization in federated cloud data centers. In: 2020 International Conference on COMmunication Systems & NETworkS (COMSNETS), pp. 882–884. IEEE (2020)
14. Nathuji, R., Schwan, K.: VirtualPower: coordinated power management in virtualized enterprise systems. ACM SIGOPS Oper. Syst. Rev. **41**(6), 265–278 (2007)
15. Qiu, Y., Jiang, C., Wang, Y., Ou, D., Li, Y., Wan, J.: Energy aware virtual machine scheduling in data centers. Energies **12**(4), 646 (2019)
16. Rajamani, K., Lefurgy, C.: On evaluating request-distribution schemes for saving energy in server clusters. In: 2003 IEEE International Symposium on Performance Analysis of Systems and Software, ISPASS 2003, pp. 111–122. IEEE (2003)
17. Sha, J., Ebadi, A.G., Mavaluru, D., Alshehri, M., Alfarraj, O., Rajabion, L.: A method for virtual machine migration in cloud computing using a collective behavior-based metaheuristics algorithm. Concurrency Comput. Pract. Exp. **32**(2), e5441 (2020)
18. Shehabi, A., et al.: United states data center energy usage report. Technical report, Lawrence Berkeley National Lab. (LBNL), Berkeley, CA, United States (2016)
19. Sîrbu, A., Babaoglu, O.: A data-driven approach to modeling power consumption for a hybrid supercomputer. Concurrency Comput. Pract. Exp. **30**(9), e4410 (2018)
20. Varia, J.: Best practices in architecting cloud applications in the AWS cloud. In: Cloud Computing: Principles and Paradigms, vol. 18, pp. 459–490. Wiley Online Library (2011)
21. Xiong, Y., Zhang, Y., Chen, X., Wu, M.: Research of energy consumption optimization methods for cloud video surveillance system. J. Softw. **26**(03), 680–698 (2015)
22. Ye, K., Wu, C., Jiang, X., He, Q.: Power management of virtualized cloud computing platfrom. Chin. J. Comput. **35**(06), 1262–1285 (2012)
23. Zhang, S., Qian, Z., Luo, Z., Wu, J., Lu, S.: Burstiness-aware resource reservation for server consolidation in computing clouds. IEEE Trans. Parallel Distrib. Syst. **27**(4), 964–977 (2015)
24. Zhou, Q., et al.: Energy efficient algorithms based on VM consolidation for cloud computing: comparisons and evaluations. arXiv preprint arXiv:2002.04860 (2020)

Soft Actor-Critic-Based DAG Tasks Offloading in Multi-access Edge Computing with Inter-user Cooperation

Pengbo Liu(✉), Shuxin Ge, Xiaobo Zhou, Chaokun Zhang, and Keqiu Li

Tianjin Key Laboratory of Advanced Networking, College of Intelligence and Computing, Tianjin University, Tianjin 300350, China
{pengboliu,cecilge,xiaobo.zhou,zhangchaokun,keqiu}@tju.edu.cn

Abstract. Multi-access edge computing (MEC) enables mobile applications, which consists of multiple dependent subtasks, to be executed in full parallelism between edge servers and mobile devices. The idea is to offload some of the subtasks to edge servers while utilizing the directed acyclic graphs (DAGs) based subtask dependency. In some cases involving multiple users running the same application, users can cooperate to improve the application's performance (e.g., object detection accuracy in connected and autonomous vehicles) by sharing the intermediate results of the subtasks, which is referred to as cooperation gain. However, inter-user cooperation also introduces additional dependencies between the subtasks of different users, which inevitably increases the execution delay of the application with DAG tasks offloading. Therefore, how to jointly optimize execution delay and cooperation gain remains an open question. In this paper, we present an approach for DAG tasks offloading in MEC with inter-user cooperation to optimize the execution delay of the application and the cooperation gain. First, we introduce the concept of application utility, which incorporates both delay and cooperation gain. We formulate the DAG tasks offloading problem with inter-user cooperation as a Markov decision process (MDP) to maximize the total application utility. Next, we propose a quantized soft actor-critic (QSAC) algorithm to solve the formulated problem. Specifically, QSAC generates multiple potential solutions according to the order-preserving quantizing method, increasing the exploration efficiency to avoid falling into local optimal while maintaining the overall stability. Finally, simulation results validate that the proposed algorithm significantly improves the total application utility compared with the other benchmark algorithms.

Keywords: Multi-access edge computing (MEC) · Tasks offloading · Directed acyclic graphs (DAGs) · Inter-user cooperation · Markov decision process (MDP) · Soft actor-critic (SAC)

This work is support in part by National Key R&D Program of China under Grant 2019YFB2102404, and in part by the National Natural Science Foundation of China under Grant 62072330.

Y. Lai et al. (Eds.): ICA3PP 2021, LNCS 13157, pp. 313–327, 2022.
https://doi.org/10.1007/978-3-030-95391-1_20

1 Introduction

Multi-access Edge Computing (MEC) has emerged as a viable solution for relieving resource-constrained mobile devices of computation-intensive applications, such as autonomous driving and augmented reality/virtual reality (AR/VR) [1,2]. By deploying relatively resource-rich edge servers at the Base Stations (BSs) in close proximity to mobile users, users can offload their computational tasks to a nearby BS over wireless channels and receive the result sent back from the BS after task execution. In this way, MEC also alleviates the communication overhead of the backhaul network and enables the computation of delay-sensitive applications [3].

Typically, a mobile application is made of a sequence of distinct dependent subtasks, where the dependency relationships among the subtasks can be modeled by a Directed Acyclic Graph (DAG). According to the DAG, the subtasks should be executed following a prescribed order, i.e., a subtask can be started when it receives the intermediate result of its preceding subtasks. There has been a great number of DAG tasks offloading strategies which make offloading decision for each subtask by taking the execution order of the subtasks into account [4,5]. By offloading part of the subtasks to the edge servers, the mobile application is executed in full parallelism between edger servers and mobile devices to minimize the execution delay of the application [6].

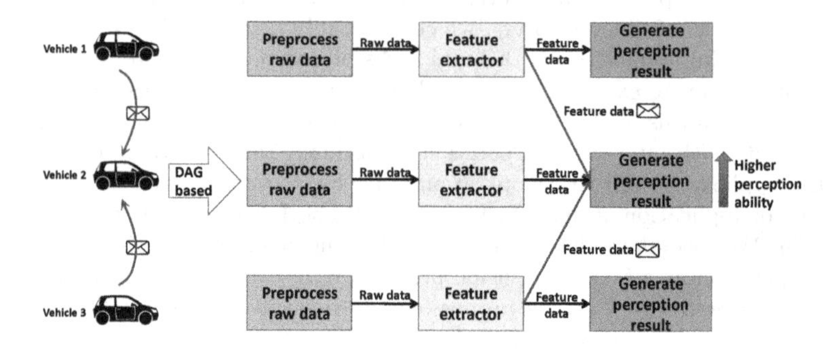

Fig. 1. An example of inter-user cooperation in connected and autonomous vehicles.

In some cases involving multiple users running the same application, users can cooperate to improve the application's performance by sharing the intermediate results of the subtasks. For example, in the cooperative perception of Connected and Autonomous Vehicles (CAVs), different vehicles share their sensing information, such as feature data extracted from Light Detection and Ranging (LiDAR) sensor input, to improve their perception ability in terms of object detection accuracy and sensing range [7,8], as shown in Fig. 1. We refer to the performance improvement brought in by inter-user cooperation as cooperation gain. Note that inter-user cooperation introduces additional dependencies between the subtasks of different users. When the users make DAG tasks offloading decision, these additional dependencies should also be taken into account [9].

While inter-user cooperation has some benefits, it may also increase the execution delay of the application due to the additional dependencies between the subtasks of different users. Specifically, given the dynamic nature of the network conditions, it is difficult to meet the stringent delay requirements of delay-sensitive applications. Notably, each user may operate independently to avoid the additional delay caused by the inter-user cooperation when the channel condition falls below a specified threshold. This drives us to make joint decisions on tasks offloading and inter-user cooperation, considering the application's execution delay, cooperation gain, available resources of the edge servers, and network conditions. However, it is pretty challenging to make joint decisions because tasks offloading and inter-user cooperation are tightly coupled, resulting in extremely high computational complexity.

To address the above issues, in this paper, we propose a soft actor-critic (SAC)-based DAG tasks offloading approach in MEC with inter-user cooperation to optimize the execution delay of the application and the cooperation gain simultaneously. The main contributions are summarized as follows.

- We introduce the concept of application utility, which incorporates both execution delay and cooperation gain. We then formulate the DAG tasks offloading problem with inter-user cooperation as a Markov decision process (MDP) to maximize the total application utility. This is the first work that jointly optimizes tasks offloading and inter-user cooperation decisions in multi-user MEC systems to the best of our knowledge.
- We propose a quantized soft actor-critic (QSAC) algorithm to solve the formulated problem. Specifically, QSAC generates multiple candidate solutions according to the order-preserving quantizing method, increasing the exploration efficiency to avoid falling into local optimal while maintaining the overall stability.
- We verify the effectiveness of the proposed QSAC algorithm with a series of simulations. The results show that the QSAC algorithm can significantly increase the application utility in dynamic network conditions, comparing with the benchmark algorithms.

The remainder of this paper is organized as follows. In Sect. 2, we present the system model and the problem formulation. The proposed QSAC algorithm is detailed in Sect. 3, following with the simulation results shown in Sect. 4. Finally, we conclude this paper in Sect. 5.

2 System Model

2.1 System Overviews

We consider a MEC system with a single BS and N users (indexed by n), as shown in Fig. 2. The BS is equipped with a k_{es}-core[1] edge server, and the CPU

[1] Each core can execute only one task at same time.

frequency of each core is f_{es}. Each user (or mobile devices) is equipped with a k_{ls}-core local server, the CPU frequency of each core is f_{ls}. Each user runs an identical application, and the users can cooperate to further improve the application's performance, e.g., detection accuracy. The applications are composed of several subtasks, and a DAG is used to model the dependencies between the subtasks.

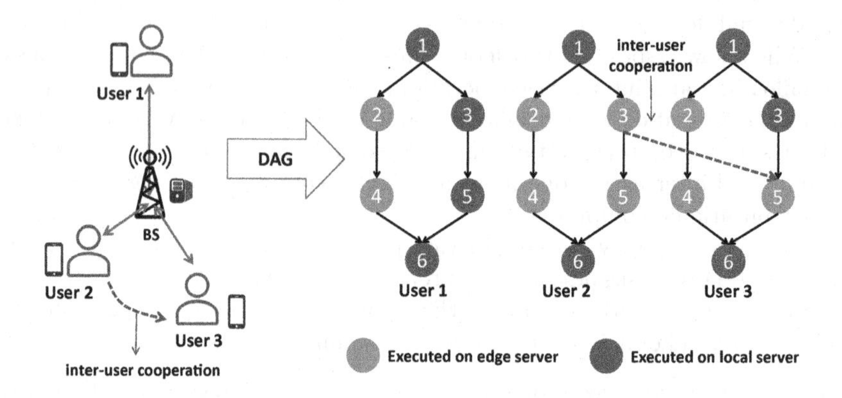

Fig. 2. An example of a multi-user MEC system with DAG tasks offloading and inter-user cooperation.

Each subtask can be executed on the user's local server or offloaded to the edge server via a wireless channel. A network operator is responsible for making offloading decision for each subtask and determining if two users should cooperate, i.e., share the intermediate outcomes of a specific subtask. Due to user mobility and network fluctuations, the operator must make decisions adaptively to reduce the application's execution delay while maximizing the cooperation gain. The total time is divided into T time slots (indexed by t) in our system. At the beginning of each time slot, the operator makes offloading and inter-user cooperation decisions. Each application has to complete all its subtasks in each time slot.

2.2 Application and DAG Model

There are M subtasks (indexed by m) in an application, and let $\mathbf{G} = (\mathcal{V}, \mathcal{E})$ denote the DAG of the applications in our system. $\mathcal{V}$ is the set of all $M * N$ subtasks, and $v_{n,m} \in \mathcal{V}$ denotes subtask m of user n, which has $C_{n,m}$ computation workload. Note that the application begins at the mobile device and the final result must be transmitted back to the device, the first and the last subtasks of each user, $v_{n,1}$ and $v_{n,M}$, are two virtual subtasks with $C_{n,1} = 0$, $C_{n,M} = 0$, $\forall n$. Let $x^t_{n,m}$ denotes the offloading decision of $v_{n,m}$ in time slot t, where $x^t_{n,m} = 0$ means $v_{n,m}$ is executed locally and $x^t_{n,m} = 1$ means $v_{n,m}$ is offloaded to the edge server in time slot t. We have $x^t_{n,1} = 0$ and $x^t_{n,M} = 0$, $\forall t, n$.

Let $\mathcal{E} = \{< v_{n',m'}, v_{n,m} > | n, n' \in N, \text{and } m, m' \in M\}$ denotes the set of dependencies among all users' subtasks, where $< v_{n',m'}, v_{n,m} >$ means that $v_{n',m'}$ must be completed before executing $v_{n,m}$. Note that in $\mathcal{E}$, there are J inter-user dependencies, which are denoted by $e^j, j = 1, 2, \cdots, J$. Let y_j^t denotes the inter-user cooperation decision in time slot t, i.e., $y_j^t = 1$ indicates two subtasks with inter-user dependency j have cooperation relationship in time slot t, and $y_j^t = 0$ otherwise. In other words, $y_j^t = 1$ means two users n and n' cooperate in time slot t, and the intermediate result of $v_{n',m'}$ will be a part of the input of $v_{n,m}$. We use $\mathcal{E}_t$ to denote the set of dependencies after the network operator makes inter-user cooperation decision in time slot t. Obviously, we have $\mathcal{E}_t \subseteq \mathcal{E}$. Let $O_{n,m}$ denotes the size of data that output from $v_{n,m}$.

2.3 Cooperation Gain

As stated above, inter-user cooperation contributes to the application's performance improvement. Depending on the practical scenarios, metrics are used to evaluate the application's performance, for example, the object identification accuracy of cooperative perception in CAVs and image/video quality in AR/VR. Therefore, it is challenging to characterize cooperation gain for all kinds of mobile applications quantitatively. Alternatively, in this paper, we define cooperation benefit as the amount of shared intermediate results between subtasks which come from different users.

Intuitively, with more intermediate results shared between users, more performance improvement of the applications can be obtained. However, the relationship between the amount of shared intermediate results and performance improvement is not always linear. With the increase of the amount of shared intermediate results, the performance improvement gradually converges to a certain level and eventually becomes negligible [10]. Thus, we use a $\log_{10}(\cdot)$ function to describe the cooperation gain, and the total cooperation gain in time slot t can be obtained as

$$G(t) = \sum_{j=1}^{J} y_j^t \log_{10} O_{n,m}, \tag{1}$$

where $v_{n,m}$ is the preceding subtask of j-th inter-user dependency.

2.4 Execution Delay of the Application

The computation delay of a subtask $v_{n,m}$ is

$$d_{n,m}^C(t) = x_{n,m}^t \frac{C_{n,m}}{f_{es}} + (1 - x_{n,m}^t) \frac{C_{n,m}}{f_{ls}}. \tag{2}$$

Let $R_n^t(a, b), a, b \in \{0,\ 1\}, a \neq b$ denotes the transmission rate between edge server and user n in time slot t, where $a = 0, b = 1$ indicate the data is transmitted from user n to the BS, and $a = 1, b = 0$ indicate the data is transmitted from the BS to user n. Thus, we have

$$R_n^t(a,b) = W \log\left[1 + \frac{(aP_{es} + bP_{ls}) \cdot h_n^t}{\sigma^2}\right], \tag{3}$$

where P_{es} and P_{ls} are the transmitting power of the BS and the user device, respectively. W, σ^2 and h_n^t are the bandwidth of the orthogonal channels, the channel noisy, and the channel gain between user n and the BS in time slot t, respectively.

For $< v_{n',m'}, v_{n,m} >\in \mathcal{E}$, the transmission delay of the intermediate result from $v_{n',m'}$ to $v_{n,m}$ can be classified into the following three cases:

- **Case 1**: If $v_{n',m'}$ and $v_{n,m}$ are executed at the same place, i.e., $n = n'$ and $x_{n,m}^t = x_{n',m'}^t$, or $n \neq n'$ and $x_{n,m}^t = x_{n',m'}^t = 1$, the transmission delay is 0.
- **Case 2**: If $v_{n',m'}$ and $v_{n,m}$ are executed at two user devices, i.e., $n \neq n'$ and $x_{n,m}^t = x_{n',m'}^t = 0$, the result of $v_{n',m'}$ is first transmitted to the BS from user n', and then is transmitted to user n from the BS.
- **Case 3**: If $v_{n',m'}$ and $v_{n,m}$ are executed at different places with one of the place is the edge server, i.e., $x_{n,m}^t \neq x_{n',m'}^t$, the result of $v_{n',m'}$ should be transmitted to the place where $v_{n,m}$ is executed.

Thus, the transmission delay between $v_{n',m'}$ and $v_{n,m}$ is

$$d_{n,m}^{n',m'}(t) = \begin{cases} 0, & \textbf{Case 1}, \\ \frac{O_{n',m'}}{R_{n'}^t(0,1)} + \frac{O_{n',m'}}{R_n^t(1,0)}, & \textbf{Case 2}, \\ \frac{O_{n',m'}}{R_{n'}^t(x_{n',m'}^t, x_{n,m}^t)}, & \textbf{Case 3}. \end{cases} \tag{4}$$

The completion time of $v_{n,m}$ in time slot t is

$$d_{n,m}(t) = d_{n,m}^R(t) + d_{n,m}^C(t), \tag{5}$$

where $d_{n,m}^R(t)$ denotes the waiting time of $v_{n,m}$, which includes the time of receiving the result of all its precedent subtasks and the time waiting for an idle CPU core. $d_{n,m}^R(t)$ can be obtained as

$$d_{n,m}^R(t) = \max\left\{\max_{<v_{n',m'},v_{n,m}>\in\mathcal{E}_t}\left\{d_{n',m'}(t) + d_{n,m}^{n',m'}(t)\right\}, d_{n,m}^W(t)\right\}, \tag{6}$$

where $d_{n,m}^W(t)$ is the earliest time that the assigned server of $v_{n,m}$ has idle CPU core to execute it. Therefore, the execution delay of the application of user n in time slot t is $d_{n,M}(t)$.

2.5 Problem Formulation

We introduce the concept of application utility to incorporate both execution delay of the application and cooperation gain, which is defined as

$$U(t) = \alpha \cdot G(t) - (1-\alpha) \cdot \sum_{n=1}^{N} d_{n,M}(t), \tag{7}$$

where $\alpha > 0$ is the weighting parameter. Our objective is to maximize the total application utility by making optimal inter-user cooperation and tasks offloading decisions, which can be formulated as the following optimization problem.

$$\begin{aligned} \max_{\mathbf{x},\mathbf{y}} \quad & \sum_{t=1}^{T} U(t) \\ s.t. \quad & x_{n,1}^t = 0, x_{n,M}^t = 0, \forall t, n, \\ & x_{n,m}^t = \{0,1\}, m \in \{2, \cdots, M-1\}, \forall t, n, \\ & y_j^t = \{0,1\}, \forall t, j. \end{aligned} \tag{8}$$

The formulated problem is an integer nonlinear optimization problem, which is NP-hard and can not be solved directly. Note that the communication channel conditions between the BS and each user are dynamically changing, and the available resources of the BS and each device will also influence the application's performance. Thus the operator has to collect a large number of environment states to make optimal decisions on tasks offloading and inter-user cooperation. Heuristic algorithms are prone to fall into local optimal, and traditional deep reinforcement learning (DRL) algorithms, such as actor-critic, are sensitive to hyperparameters and are difficult to deal with high-dimensional problems [11]. Therefore, we use a SAC-based method, QSAC, to tackle the formulated problem, which is detailed in Sect. 3.

3 Deep Reinforcement Learning Based Decision Policy

3.1 Problem Transformation

We formulate the optimization problem as a MDP $(\mathcal{S}, \mathcal{A}, \mathcal{R})$, which includes

- **Environment State:** In time slot t, the agent observes a state $\mathbf{s}_t$, $\mathbf{s}_t \in \mathcal{S}$, about environment (network condition and computation resources) to make corresponding action, the state is defined as

$$\mathbf{s_t} = (h_1^t, \cdots, h_N^t, k_{es}, k_{ls}, f_{es}, f_{ls}). \tag{9}$$

- **Decision Action:** The action space $\mathcal{A}$ consists of all the candidate tasks offloading decision and the inter-user cooperation decision. Because the first and last subtasks of each user have been set to execute locally, the action $\mathbf{a_t} \in \mathcal{A}$ in time slot t is

$$\mathbf{a}_t = (x_{1,2}^t, \cdots, x_{1,M-1}^t, \cdots, x_{N,2}^t, \cdots, x_{N,M-1}^t, y_1^t, \cdots, y_J^t) \tag{10}$$

the length of $\mathbf{a_t}$ is denoted as Z, $Z = N * (M-2) + J$.
- **Utility Reward:** After the agent takes an action $\mathbf{a}_t$ in time slot t, the environment will return an immediate reward of the application utility $r_t \in \mathcal{R}$ to the agent.

$$r_t = r(\mathbf{s}_t, \mathbf{a}_t) = U(t). \tag{11}$$

In time slot t, the agent obtains an environment state $\mathbf{s}_t$, and takes an action $\mathbf{a}_t$ involving inter-user cooperation decision and tasks offloading decision. Then, the environment will return a new state $\mathbf{s}_{t+1}$ and a reward r_t to the agent.

3.2 Soft Policy Functions

The objective of the agent is to find the optimal policy which can maximize the reward with the different environment state, an entropy term $\mathcal{H}(\pi(\mathbf{a}_t|\mathbf{s}_t)) = -\log \pi(\mathbf{a}_t|\mathbf{s}_t)$ is introduced in SAC to explore more potential solutions, where $\pi(\mathbf{a}_t|\mathbf{s}_t)$ is the corresponding policy [12]. The objective with entropy of policy $\pi(\mathbf{a}_t|\mathbf{s}_t)$ (Note that to simply the expression, we use π to represent $\pi(\mathbf{a}_t|\mathbf{s}_t)$ in the later) of SAC is

$$\psi(\pi) = \mathbb{E}_\pi \left[\sum_{t=0}^{T} \mu^t \cdot [r(\mathbf{s}_t, \mathbf{a}_t) + \kappa \mathcal{H}(\pi)] \right], \tag{12}$$

where κ and μ are the temperature parameter that controls the stochasticity of the optimal policy, and the discounting factor indicates the importance of the immediate and the future rewards, respectively.

SAC defines a soft Q-value to evaluate the state-action pair under policy π when it is given certain initial state and action, which is

$$Q^\pi(s,a) = \mathbb{E}_{\pi, a_0 = a, s_0 = s} \left[\sum_{t=0}^{T} \mu^t [r(\mathbf{s}_t, \mathbf{a}_t) + \kappa \mathcal{H}(\pi)] \right]. \tag{13}$$

SAC also defines a soft V-value to evaluate the state under policy π through repeatedly applying modified Bellman backup, which is

$$V^\pi(\mathbf{s_t}) = \frac{1}{\kappa} \mathbb{E}_{\mathbf{a_t} \sim \pi} [Q^\pi(\mathbf{s_t}, \mathbf{a_t}) + \mathcal{H}(\pi)]. \tag{14}$$

3.3 Quantized Soft Actor-Critic

The proposed QSAC algorithm is summarized in Algorithm 1. As shown in Fig. 3, in the process of generating training data, the DNN in actor-part is used to generate a relaxed action $\mathbf{a}_t'$ according to current environment $\mathbf{s}_t$ (line 4), $\mathbf{a}_t'(i) \in [0,1]$, $\forall i \in \{1, 2, \cdots, Z\}$, where $\mathbf{a}_t'(i)$ is the i-th element of $\mathbf{a}_t'$.

To further explore potential optimal action, in QSAC, we add a quantizing module into actor-part to explore $Z+2$ potential actions according to $\mathbf{a}_t'$ (line 5), which has the following 3 steps.

Step 1 : Firstly, $\mathbf{a}_t^1$ is directly generated by the output $\mathbf{a}_t'$ of the DNN in the actor-part, which is

$$\mathbf{a}_t^1(i) = \begin{cases} 1, \ \mathbf{a}_t'(i) > 0.5, \\ 0, \ \mathbf{a}_t'(i) \le 0.5. \end{cases} \tag{15}$$

Step 2 : To avoid the invalid exploration of actions, we set a special candidate action a_t^2, i.e.,

Algorithm 1. Quantized Soft Actor-Critic

1 Initialize $\mathbf{s}_0, \mathbf{a}_0$, $\mathcal{B}$, the parameters of each deep neural network (DNN).
2 **foreach** *iteration* **do**
3   **foreach** *environment step* **do**
4   Generate $\mathbf{a}'_t$ through the DNN of actor-part,
5   Explore candidate actions according to $\mathbf{a}'_t$,
6   Choose the action $\mathbf{a}_t$ from the candidate actions with maximum r_t,
7   Return state $\mathbf{s}_{t+1}$ and reward r_t.
8   Update $\mathcal{B}$ according to (17),
9   **end**
10   **foreach** *gradient step* **do**
11   Randomly sample tuple from $\mathcal{B}$,
12   Update the soft Q-value and soft V-value according to (13),(14).
13   Update parameter sets of DNNs in critic-part according to (19),(21),
14   Update parameter set of the DNN in actor-part according to (24).
15   **end**
16 **end**

$$\mathbf{a}_t^2 = \arg\max_{\mathbf{a}^*, \overline{\mathbf{a}}^*} \{r(\mathbf{s}_t, \mathbf{a}^*), r(\mathbf{s}_t, \overline{\mathbf{a}}^*)\}, \tag{16}$$

where $\mathbf{a}^* = \{1\}^Z$ and $\overline{\mathbf{a}}^* = \{0\}^Z$.

Step 3 : According to the action $\mathbf{a}_t^1$, it is easy to generate many other potential actions by randomly replacing the element $\mathbf{a}_t^1(i), \forall i \in \{1, 2, \cdots, Z\}$ with 0 or 1. However, it may lead to high uncertainty in the exploration direction. Therefore, we generate a series of potential actions from the relaxed action $\mathbf{a}'_t$, by utilizing order-preserving quantization method. Order-preserving quantization method was introduced to explore the output of the DNN in [13]. We only explore the part of tasks offloading decision for each user n, $\mathbf{a}_t^1(i), \forall i \in \{2 + (n-1) \cdot M, \cdots, n \cdot M - 1\}$ by this method at a time, while the other elements remain consistent with $\mathbf{a}_t^1$. This method maintains the overall stability and increases the exploration efficiency. Hence, we can generate N candidate action sets based on order-preserving quantization method, and each set contains $M-2$ candidate actions. We also apply the order-preserving quantization method to the part of inter-user cooperation decision of $\mathbf{a}_t^1$ to generate J candidate actions. Therefore, we finally generate Z candidate actions by step 3, denoted by $\mathbf{a}_t^{3,z}$, $z \in \{1, 2, \cdots, Z\}$.

After obtaining $Z+2$ candidate actions through the above 3 steps, we choose the action which can get the maximum reward r_t in time slot t as $\mathbf{a}_t$, which represents for the output of actor-part (line 6).

When the agent taking action $\mathbf{a}_t$, the environment will return state $\mathbf{s}_{t+1}$ and reward r_t (line 7). Then the agent update the replay buffer (line 8) by

$$\mathcal{B} = (\mathbf{s}_t, \mathbf{a}_t, r_t, \mathbf{s}_{t+1}) \cup \mathcal{B}. \tag{17}$$

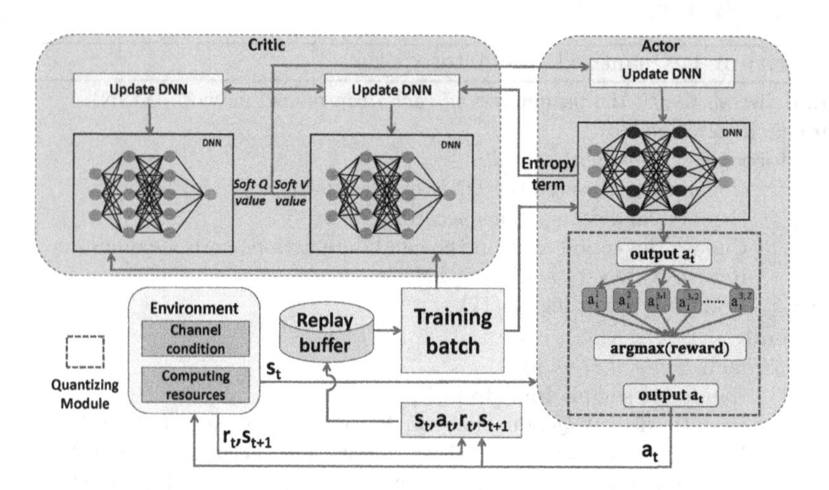

Fig. 3. The schematics of the framework of QSAC.

In the training process, QSAC will random sampling tuple in the replay buffer $\mathcal{B}$ (line 11). The critic-part first updates the soft Q-value and soft V-value (line 12). Then the critic-part updates the two DNNs in it which are used to approximate soft Q-value and soft V-value based on the trained parameter sets θ and ϑ, respectively (line 13). After that, the critic-part updates the DNN's parameter set θ through stochastic gradient descent method, this DNN is used to approximate the soft Q-value. We define the loss function of θ as

$$L(\theta) = \mathbb{E}_\pi \left[\frac{1}{2} (Q^\pi_\theta(\mathbf{s}_t, \mathbf{a}_t) - \hat{Q}^\pi_\theta(\mathbf{s}_t, \mathbf{a}_t))^2 \right], \tag{18}$$

where $\hat{Q}(\mathbf{s}_t, \mathbf{a}_t)$ is the Q target value and satisfies $\hat{Q}(\mathbf{s}_t, \mathbf{a}_t) = r_t + \mu \hat{V}_{\hat{\theta}}(\mathbf{s}_{t+1})$, the $\hat{V}_{\hat{\theta}}(\mathbf{s}_{t+1})$ is the target state value with parameter set $\hat{\theta}$. In stochastic gradient descent method, we should update the parameter set θ in the direction of gradient decent to minimize $L(\theta)$, i.e.,

$$\theta(t+1) = \theta(t) - \tau \nabla_\theta L(\theta), \tag{19}$$

where τ is the learning rate of critic part and $\nabla_\theta L(\theta)$ is the gradient of $L(\theta)$.

For the DNN which approximate the soft V-value, critic-part employs the following mean squared error to measure the performance of the parameter set ϑ, which is

$$L(\vartheta) = \mathbb{E}_{s_t} \left[\frac{1}{2} (V_\vartheta(\mathbf{s}_t) - \mathbb{E}_\pi \left[Q^\pi_\theta(\mathbf{s}_t, \mathbf{a}_t) + \kappa \mathcal{H}(\pi) \right])^2 \right]. \tag{20}$$

The corresponding update process of parameter set ϑ is

$$\vartheta(t+1) = \vartheta(t) - \tau \nabla_\vartheta L(\vartheta), \tag{21}$$

where $\nabla_\vartheta L(\vartheta)$ is the gradient of $L(\vartheta)$.

For the DNN in actor-part, the V-value and the Q-value obtained from the critic-part are used to update the policy parameter set via the stochastic gradient descent method, and the parameter set of this DNN is ϕ (line 14). The optimal policy can be learned by minimizing the Kullback-Leibler divergence expectation, which is

$$L(\phi) = \mathbb{E}_{\mathbf{s}_t}\left\{D_{KL}\left[\pi_\phi(\cdot|\mathbf{s}_t)||\frac{exp(Q_\theta(\mathbf{s}_t,\cdot))}{Z_\phi(\mathbf{s}_t)}\right]\right\}, \tag{22}$$

where $Z_\phi(\mathbf{s}_t)$ is the distribution normalization function.

We reparameterize the policy using a neural network transformation $a_t = g(\gamma_t; \mathbf{s}_t)$, where γ_t is an input Gaussian noise vector. The objective in (22) can be rewritten as

$$L(\phi) = \mathbb{E}_{s_t}\left[\kappa log\pi_\phi(g(\gamma_t;\mathbf{s}_t)|\mathbf{s}_t) - Q_\theta(\mathbf{s}_t, g(\gamma_t;\mathbf{s}_t)) + V_\vartheta(s_t)\right]. \tag{23}$$

The update process of parameter set ϕ is

$$\phi(t+1) = \phi(t) - \beta\nabla_\phi L(\phi), \tag{24}$$

where β is the learning rate of actor-part, and $\nabla_\phi L(\phi)$ is the gradient of $L(\phi)$.

4 Performance Evaluation

4.1 Simulation Setup

We emulate a MEC system consisting of one BS and 4 users that uniformly distributed around the BS (about 5–60 m away from the BS). The users run the same application with 6 subtasks. The randomly generated DAG of applications is shown in Fig. 4, where the values of the directed line indicates the data size of the intermediate results. There are 4 potential inter-user cooperation opportunities which are marked with dotted red lines. The computation workload of the 6 subtasks is randomly set to [0, 60, 80, 150, 100, 0] (M Cycles).

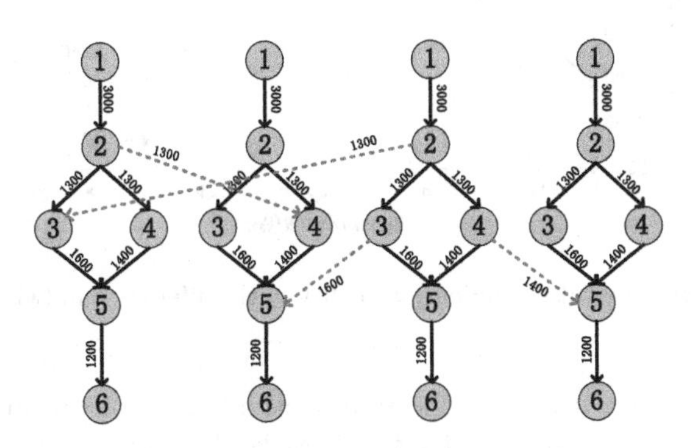

Fig. 4. The considered DAG in the simulation

In the simulations, the CPU frequency, number of cores and the transmitting power of the BS are set to $f_{es} = 2.5 * 10^9$ cycles/s, $k_{es} = 8$ and $P_{es} = 1$ W, respectively. The channel noisy is set to $\sigma^2 = 10^{-10}$. For each user device, the CPU frequency, number of cores and the transmitting power are set to $f_{ls} = 8 * 10^7$ cycles/s, $k_{ls} = 2$ and $P_{ls} = 0.1$ W [6,9]. The wireless channel gains h_n^t follow the free space path loss channel model [9]. The number of time slots is $T = 500$, and the weighting parameter α is 0.5.

We compare the performance of QSAC with the following three benchmarks approaches on DAG tasks offloading in MEC systems:

- **Never Cooperate (NC)** [6]: In this approach, each user works independently without inter-user cooperation. A greedy algorithm is used to make tasks offloading decision to minimize the delay and meet the energy constraints.
- **Always Cooperate (AC)** [9]: In this approach, it is assumed that the users always cooperate with each other. Gibbs sampling and one-climb policy are used to make tasks offloading decision to minimize the delay and energy consumption. The inter-user task dependencies are taken into account.
- **Random Cooperate (RC)**: In this approach, we first randomly make inter-user decision, i.e., we randomly remove some of the red lines in Fig. 4. Then, the tasks offloading decision is made by QSAC to maximize the total application utility.

4.2 Performance Analysis

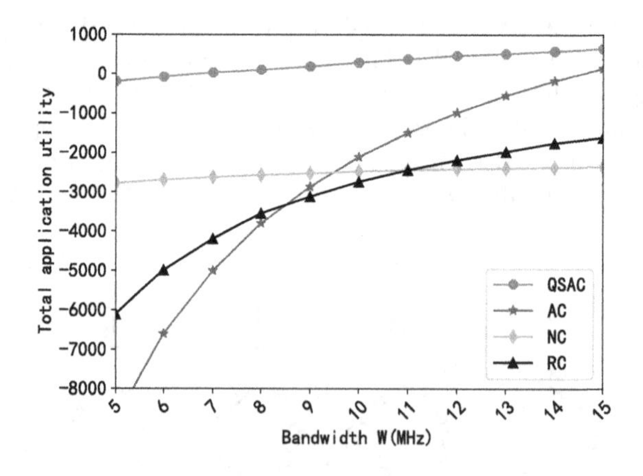

Fig. 5. The total application utility with different bandwidth.

The total application utility of the four methods is depicted in Fig. 5, where the bandwidth varies from 5 MHz to 15 MHz. It is found that AC gets the lowest total application utility when the bandwidth is lower than 8 MHz, because the additional execution delay caused by inter-user cooperation dominates the total application utility. However, when the bandwidth exceeds 10 MHz, its total application grows fast and is higher than that of NC and RC, because cooperation gain begins to dominate the total application utility. On the other hand, NC has an opposite performance as compare with AC. When the bandwidth is lower than 9 MHz, its total application utility is higher than AC and RC because there is no additional execution delay. When the bandwidth is larger than 11 MHz, its total application utility is lower than AC and RC because it can not enjoy the cooperation gain. It should be noted that the total application utility of QSAC is always the highest. This is because it jointly optimize the inter-user cooperation decision and the tasks offloading decision according to current network condition.

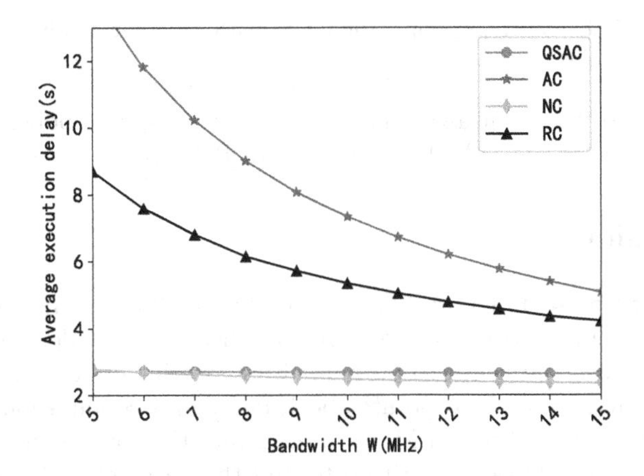

Fig. 6. The average execution delay of the application with different bandwidth.

The performance of the average execution delay of the application of the four methods are shown in Fig. 6, where the bandwidth varies from 5 MHz to 15 MHz. It can be observed that the average execution delay of AC and NC is the highest and lowest among the 4 methods, respectively. Since with RC, users randomly cooperate with others, the average execution delay is always the second highest. It is surprisingly to find that the average execution delay with QSAC is the second lowest, which is quite close to that of NC. This is because QSAC adaptively make inter-user cooperation decision according to the current network condition.

Figure 7 illustrates the average cooperation gain of the four methods, where the bandwidth varies from 5 MHz to 15 MHz. The average cooperation gain of NC is 0, while the average cooperation gain of AC is the highest. Because the RC randomly make the inter-user cooperation decision, its cooperation gain is about half of that of AC, no matter how the bandwidth changes. As mentioned above, QSAC strikes a balance between execution delay and cooperation gain.

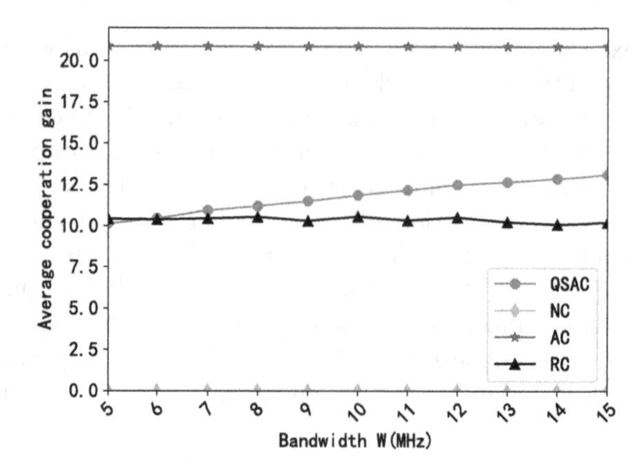

Fig. 7. The average cooperation gain with different bandwidth.

It is interesting to find that as the bandwidth increases, the cooperation gain of QSAC also increases and is higher than that of RC.

5 Conclusion

In this paper, we presented a SAC-based DAG tasks offloading approach in MEC systems with inter-user cooperation to reduce the execution delay of the application and increase the cooperation gain. First, we introduced the concept of application utility, which incorporates both delay and cooperation gain. Next, we formulated the problem as a MDP to maximize the total application utility. Then, we proposed a QSAC algorithm to find the near-optimal solution of the formulated MDP problem with fast convergence. Finally, we conducted simulations to verify the effectiveness of the proposed QSAC algorithm in terms of the total application utility, the average application execution delay, and the average cooperation gain. For future work, we will extend the current approach to the scenario with multiple heterogeneous edge servers.

References

1. Dai, B., Xu, F., Cao, Y., Xu, Y.: Hybrid sensing data fusion of cooperative perception for autonomous driving with augmented vehicular reality. IEEE Syst. J. **15**(1), 1413–1422 (2021)
2. Du, J., Yu, F.R., Lu, G., Wang, J., Jiang, J., Chu, X.: MEC-assisted immersive VR video streaming over terahertz wireless networks: a deep reinforcement learning approach. IEEE Internet Things J. **7**(10), 9517–9529 (2020)
3. Shi, W., Cao, J., Zhang, Q., Li, Y., Xu, L.: Edge computing: vision and challenges. IEEE Internet Things J. **3**(5), 637–646 (2016)
4. Song, F., Xing, H., Luo, S., Zhan, D., Dai, P., Qu, R.: A multiobjective computation offloading algorithm for mobile-edge computing. IEEE Internet Things J. **7**(9), 8780–8799 (2020)

5. Sahni, Y., Cao, J., Yang, L., Ji, Y.: Multihop offloading of multiple DAG tasks in collaborative edge computing. IEEE Internet Things J. **8**(6), 4893–4905 (2021)
6. Fu, X., Tang, B., Guo, F., Kang, L.: Priority and dependency-based DAG tasks offloading in fog/edge collaborative environment. In: 2021 IEEE 24th International Conference on Computer Supported Cooperative Work in Design (CSCWD), pp. 440–445 (2021)
7. Wang, T.-H., Manivasagam, S., Liang, M., Yang, B., Zeng, W., Urtasun, R.: V2VNet: vehicle-to-vehicle communication for joint perception and prediction. In: Vedaldi, A., Bischof, H., Brox, T., Frahm, J.-M. (eds.) ECCV 2020. LNCS, vol. 12347, pp. 605–621. Springer, Cham (2020). https://doi.org/10.1007/978-3-030-58536-5_36
8. Chen, Q., Ma, X., Tang, S., Guo, J., Yang, Q., Fu, S.: F-cooper: feature based cooperative perception for autonomous vehicle edge computing system using 3D point clouds. In: Chen, S., Onishi, R., Ananthanarayanan, G., Li, Q. (eds.) Proceedings of the 4th ACM/IEEE Symposium on Edge Computing, SEC 2019, Arlington, Virginia, USA, 7–9 November 2019, pp. 88–100. ACM (2019)
9. Yan, J., Bi, S., Zhang, Y.J., Tao, M.: Optimal task offloading and resource allocation in mobile-edge computing with inter-user task dependency. IEEE Trans. Wireless Commun. **19**(1), 235–250 (2020)
10. Chen, Q., Tang, S., Yang, Q., Fu, S.: Cooper: cooperative perception for connected autonomous vehicles based on 3D point clouds. In: 2019 IEEE 39th International Conference on Distributed Computing Systems (ICDCS), pp. 514–524 (2019)
11. Fu, F., Kang, Y., Zhang, Z., Yu, F.R., Wu, T.: Soft actor-critic DRL for live transcoding and streaming in vehicular fog-computing-enabled IoV. IEEE Internet Things J. **8**(3), 1308–1321 (2021)
12. Haarnoja, T., Zhou, A., Abbeel, P., Levine, S.: Soft actor-critic: off-policy maximum entropy deep reinforcement learning with a stochastic actor. In: Dy, J.G., Krause, A. (eds.) Proceedings of the 35th International Conference on Machine Learning, ICML 2018, Stockholmsmässan, Stockholm, Sweden, 10–15 July 2018, vol. 80. Proceedings of Machine Learning Research, pp. 1856–1865. PMLR (2018)
13. Huang, L., Bi, S., Zhang, Y.A.: Deep reinforcement learning for online computation offloading in wireless powered mobile-edge computing networks. IEEE Trans. Mob. Comput. **19**(11), 2581–2593 (2020)

Service Dependability and Security Algorithms

Sensor Data Normalization Among Heterogeneous Smartphones for Implicit Authentication

Zuodong Jin, Muyan Yao, and Dan Tao(✉)

School of Electronic and Information Engineering, Beijing Jiaotong University, Beijing 100044, China
dtao@bjtu.edu.cn

Abstract. Nowadays, smartphones have become an important part in people's life. Existing traditional explicit authentication mechanisms can hardly protect the privacy of users, which makes implicit authentication mechanisms a hot topic. Considering the non-uniform sensor data, which are collected from heterogeneous smartphones, e.g., different series or versions, a sensor data normalization solution among heterogeneous smartphones is proposed in this paper. First, the sensors used to collect data can be determined by whether they can describe users' behavior characteristics well. On this basis, we propose a sensor data normalization solution from three aspects: the proportion-based one, statistics-based one and attitude-based normalization. Particularly, we establish an adaptive complementary filter with a dynamic parameter, making it possible to adjust the proportion of accelerometer and gyroscope. Finally, based on 2,000 samples collected from 40 volunteers on three different smartphones, we perform experiments to evaluate our proposed solution. Compared with non-normalized data, the authentication efficiency through our solution can respectively increase 15% and 18% in Accuracy and AUC, and reduce FAR and FRR by 10% and 18%, respectively.

Keywords: Implicit authentication · Smartphone · Pattern unlock · Sensor data normalization · One-class support vector machine

1 Introduction

With the development of smartphones, we use them in all kinds of aspects, such as transportation, payment and so on. Since smartphones store lots of user-specific sensitive information (e.g., password, credit card), the problem of how to ensure the validity of authentication while avoiding the perceivable deterioration of user experience is gradually evolved into an important topic. As a widely used authentication method, explicit pattern password can be unlocked by anyone

This work is supported in part by the National Natural Science Foundation of China under Grant No. 61872027, Open Research Fund of the State Key Laboratory of Integrated Services Networks under Grant No. ISN21-16.

Y. Lai et al. (Eds.): ICA3PP 2021, LNCS 13157, pp. 331–345, 2022.
https://doi.org/10.1007/978-3-030-95391-1_21

who knows the password, its shortcomings are obvious [1,2]. Higher-assurance authentication through use of a second factor (e.g., a SecurID token) also falls short in the implicit authentication on smartphones, where device limitations and consumer attitude demand a more integrated, convenient, yet secure experience. So the implicit authentication mechanism has produced and developed in recent years. As a privacy security protection mechanism, implicit authentication has its own advantages in some aspects such as security, user-friendly and so on. When a user unlocks smartphones, implicit authentication mechanism needs to use built-in sensors of smartphones to collect data. Then we process the data and use them to form authentication model according to the user's characteristics. The implicit authentication mechanism compares user's model with the training phase, which requires not only the correct pattern password, but also the same characteristics to unlock. Because personal habits are difficult to imitate [3], this authentication approach has attracted more and more attention. In this way, the security of the authentication is improved.

There are different sensors (e.g., screen, pressure sensor, accelerometer and gyroscope) in a smartphone, which can collect users' unlocking data. However, the same kind of sensor in heterogeneous smartphones have differences in some aspects such as precision, range and so on. When a user unlocks by using heterogeneous smartphones, the implicit authentication results are different even though the same person using same pattern to unlock. Generally, most works collected users' behavior data on single smartphone to avoid the decrease of authentication efficiency caused by sensors among heterogeneous smartphones, without considering the normalization of the same kind of sensor on heterogeneous smartphones. To deal with the problem, this paper propose normalization solution to process the data collected from same kind of sensor on heterogeneous smartphones. Different screen sizes can mainly affect timestamp data, so the screen size and aspect ratio are used for normalization. The differences of pressure sensors affect the distribution of pressure characteristic curve, which is normalized by statistics-based solution. The differences between accelerometer and gyroscope affect the establishment of user characteristics model, so the attitude algorithm is used for normalization. Then, we introduce one-class support vector machine (OCSVM) which enables the authentication model to work only with the training data from smartphone owner. The results show that the proposed normalization solution is superior to that without normalization. And the rate of improvement on same series smartphones is higher than that on different series smartphones.

The remainder of this paper is organized as follows. Section 2 overviews the related work. In Sect. 3, we describe the framework of our normalization solution in details. Then the sensor data normalization solution is proposed and the normalized results are shown in Sect. 4. Section 5 presents the experimental evaluation results. Finally, we conclude this paper in Sect. 6.

2 Related Work

With the widespread use of smartphones, some studies concentrate implicit authentication more on smartphones. *Khan et al.* [4] attempted to challenge the device-centric approach to implicit authentication on smartphones. *Sitova et al.* [5] required volunteers to input password in vertical mode and data were collected on smartphones from sitting and walking. They almost used different methods to optimize their implicit authentication efficiency on single smartphones in some simple situation.

With the abundance of sensors, implicit authentication has been extensively studied in some more complicated situation. *Xu et al.* [6] introduced the implicit identity authentication framework based on local and network. On the basis, *Zhang et al.* [7] designed an implicit authentication mechanism to protect VR headsets. *Vhaduri et al.* [8] presented an implicit wearable device user authentication mechanism using combinations of three types of coarse-grain minute-level biometrics: behavioral (step counts), physiological (heart rate), and hybrid (calorie burn and metabolic equivalent of task).

And then, researchers have paid more attention on pattern lock based implicit authentication due to its convenience. *Shi et al.* [9] proposed a sensor fusion based implicit authentication system to enhance the protection level of the identity authentication mechanism for smartphone. *Liu et al.* [10] extracted users' behavioral characteristics and proposed a system to extract the time, pressure, size and angle keystroke feature in the process of rendering the pattern lock. Although some complicated situation are considered, they do not consider the difference on heterogeneous smartphones.

To our best knowledge, the existing research works have focused on the sensor data collected by using the same kind of smartphone in order to achieve the performance improvement of implicit authentication solution. However, the difference of sensors built-in heterogeneous smartphones hasn't been considered, which would directly affect the authentication performance. Motivated by this, we propose a sensor data normalization solution to handle with the problem of non-normalized sensor data in heterogeneous smartphones.

3 Solution Design

3.1 Framework

The framework of our work is illustrated in Fig. 1, which consisted of training phase and authentication phase. In the training phase, we use multiple smartphones for data acquisition. When users input password to unlock, their behavior data are collected through a customized collection software installed on smartphones. We normalize them and use binary classifier to form authentication model.

In the authentication phase, the user only needs to unlock the smartphone as usual to complete the authentication. Once the behavior information is matched with the trained one, the user will be regarded as legal one, who can log in any smartphone. In this process, OCSVM [11] is used to perform anomaly detection. We draw a conclusion by verifying the efficiency of identification. In the following part, we will describe the selection of sensors in three dimensions and data acquisition.

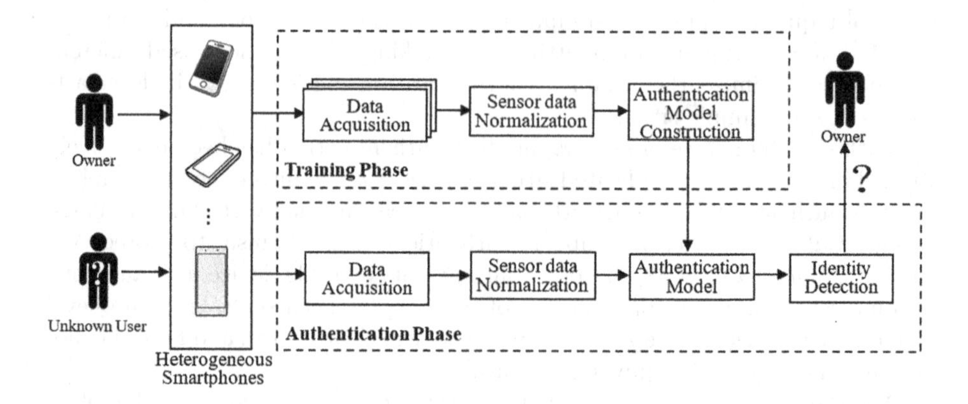

Fig. 1. The framework of our normalization solution

3.2 Sensor Selection

Since there are various sensors in smartphones, it is necessary to select some built-in sensors which involve key information of users' behavior [18]. Considering the possibility of periodic or sudden changes in users features, we process all of data and take all situations into account. In our previous work, accelerometer and gyroscope are used to obtain users' behavior data [3]. In this paper, we select four sensors (screen, pressure data, accelerometer and gyroscope) from three dimensions which can reflect users behavior well.

- For 1D features, we choose screens data since it can reflect users' time information at each pattern point. Through the time data, we can analysis the speed of user's finger on each pattern point, which represents user's speed habit.
- For 2D features, we choose pressure data. Through the pressure data, we can analyze user's finger strength on pattern path, which can reflect user's tough behavior.
- For 3D features, we choose accelerometer and gyroscope data, which can reflect users' position behavior. The position of the smartphone changes accordingly as the user enters the pattern password, and these two sensors just reflect the acceleration and angular velocity of the smartphone when the user unlocks.

After selecting sensors, we acquire sensor data, analyze and process the data by our proposed solution.

3.3 Data Acquisition

To collect users' behavior data when they draw the pattern password repeatedly, we develop an Android APP. There are existing works using IMU sensors to capture the motion characteristics of touch gestures [12,13]. Similarly, some sensors (screen, pressure sensors, gyroscopes and accelerometers) are employed to capture the motion characteristics of users during their pattern drawing process. Considering the attributes of pattern passwords, such as points, overlapping, size etc., we preset two kinds of patterns to evaluate the effectiveness of the solution, as illustrated in Fig. 2. More importantly, to show the differences of same kind of sensor in heterogeneous smartphones, three typical smartphones (HUAWEI nova7, OPPO realme x50 pro and OPPO R9s) are used to collect data, which consisted of two different series smartphones (HUAWEI and OPPO) and two same series but different versions smartphones (OPPO realme x50 pro and OPPO R9s).

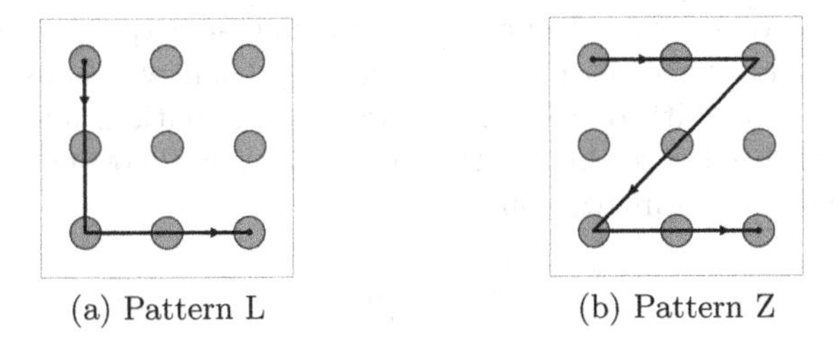

(a) Pattern L (b) Pattern Z

Fig. 2. The two pattern passwords in system (The pattern L is given in Fig. 2(a), which consists of five points and a 90-degree corner. And the pattern Z is given in Fig. 2(b), which consists of seven points and two 45-degree corners. Through analyzing the pattern L and Z, most factors (the points, overlap and length) are considered.)

4 Sensor Data Normalization

The type of components in the same model of smartphones may vary in massive production, in case of supply shortage. Thus, the specific components can respond similarly, but not exactly identical, when received the same physical signal from the outer world. This situation may cause differences in some aspects, such as precision, range, etc., in sensors' feedback. Considering the differences caused by sensors among heterogeneous smartphones can reduce the authentication efficiency, a specialized normalization solution is proposed in this section.

4.1 Proportion-Based Normalization

One of the most basic features related to smartphone screen is timestamp feature. Different screen sizes also lead to different aspect ratio. When users unlock their smartphones, the corresponding timestamp cannot be directly used to establish authentication model. To deal with this problem, we proposed a proportion-based normalization solution. Considering the physical property of smartphones are different, we choose three typical smartphones to describe the proportion-based normalization solution. Table 1 contains the screen size and ratio property of the three kinds of smartphones involved in our paper.

Table 1. Parameters of three heterogeneous smartphones

Smartphones	Screen size (inch)	Screen ratio
OPPO R9s	5.5	76:37
OPPO realme x50 pro	6.4	41:19
HUAWEI nova7	6.5	80:37

To make the data more accurate on different smartphones, we take full account of common screen size of smartphones. Then we choose 6.4 in. with a ratio of length to width of 16:9 as the standard situation, and other smartphones are normalized according to this situation. The following formulation is used to complete the normalization.

$$\frac{o_s}{n_s} \times \frac{o_r}{n_r} \times timestamp \tag{1}$$

where o_s is original size, n_s is normalization size, o_r is original length and width ratio, n_r is normalization ratio. For example, if OPPO R9s is used to acquisition data, and the timestamp of a certain contact is 200, the normalized timestamp data can be calculated as 199.59 ($= (5.5/6.4) \times (76/37)/(16/9) \times 200$). The proportion-based normalization solution is illustrated as Fig. 3.

After normalizing the screen parameters, the data from heterogeneous smartphones are simulated to have a same screen property. The influence of different screen sizes and aspect ratios on heterogeneous smartphones is thus reduced. We use the normalized timestamp data to reflect part of users' behavior and establish authentication model. However, normalizing the timestamp data is not enough if we want to get better performance.

4.2 Statistics-Based Normalization

For 2D features, pressure data can reflect users' characteristics when they input pattern passwords. The pressure sensors in heterogeneous smartphones are different, so we proposed statistics-based normalization solution to normalize the pressure data. Information of the same user drawing the same pattern on HUAWEI nova7 for six times are shown in Fig. 4.

It can be seen that the pressure sensor data distribution gauge of each user conforms to the statistical law. So, we propose a statistics-based normalization

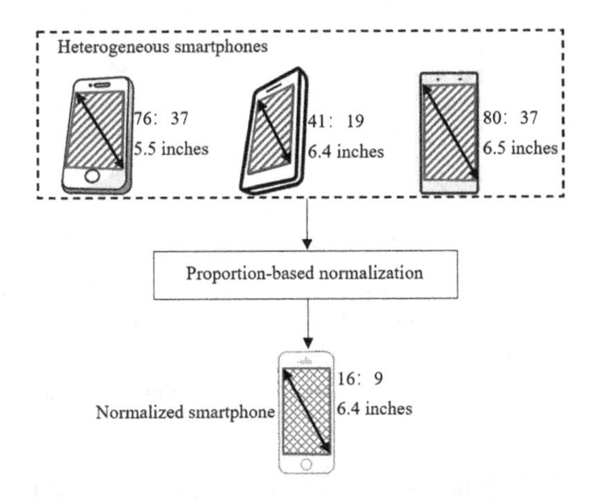

Fig. 3. The proportion-based normalization solution

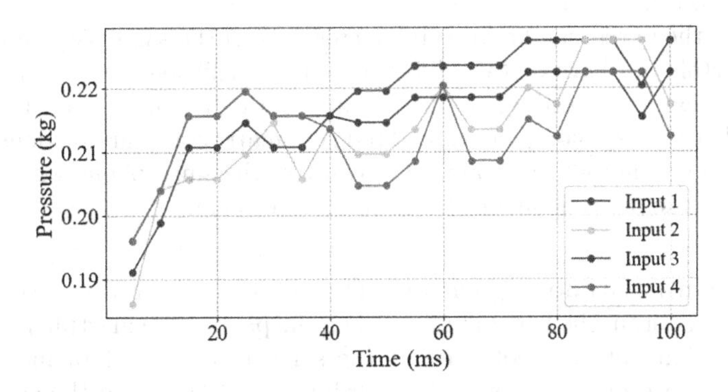

Fig. 4. Pressure data of three users on HUAWEI nova7

solution, which makes the irregular pressure sensor data distribution approximately conform to the normal distribution [14]. The pattern L unlocking pressure data of three identical users on heterogeneous smartphones are selected for analysis. The result is given in Fig. 5.

To distinguish the pressure data distribution of three kinds of smartphones easily, we normalize them to Gaussian distribution. And then, the normalized probability density function is sampled according to the number of original sample data. We add data in the sample data set as new eigenvalue.

4.3 Attitude Algorithm Normalization

For 3D features, we choose gyroscope and accelerometer data to normalize. By using these sensors, acceleration and angular velocity data are collected when users unlock their smartphones. Through attitude algorithm, the acceleration and angular velocity data are transformed into "Euler angles (pitch, roll, yaw)",

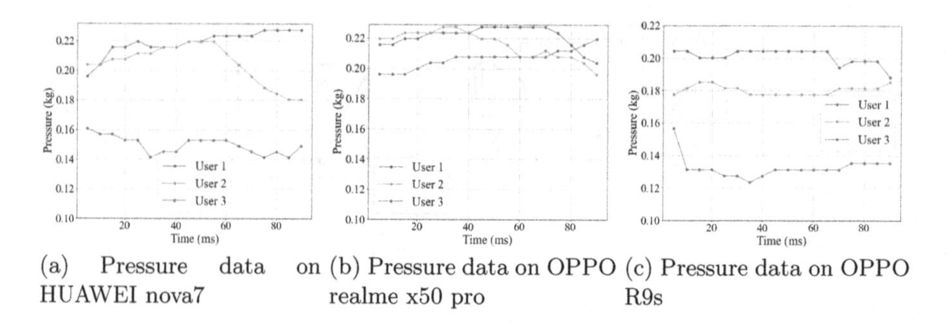

(a) Pressure data on HUAWEI nova7 (b) Pressure data on OPPO realme x50 pro (c) Pressure data on OPPO R9s

Fig. 5. Pressure data of three users on heterogeneous smartphones

which can reflect users' behavior. For observation, we present the figure of data collected from the same user with pattern L on three kinds of smartphones, which is shown in Fig. 6.

Because the gyroscope and accelerometer data of the same user on heterogeneous smartphones have certain differences, it is inefficient to directly perform the authentication according to the original data. It can be seen from Fig. 6 that the differences among the three kinds of smartphones are obvious, so it is necessary to normalize. To obtain the posture of the smartphone when the user unlocks, we take attitude algorithm to normalize them.

Acceleration Attitude Algorithm. Through the acceleration attitude algorithm, we can obtain the Euler angles. When the posture of smartphone changes, the acceleration data are collected at the same time. The data measured by accelerometer are projected onto the coordinate axis to obtain the acceleration component, and then the data are transformed into Euler angles. The corresponding relationship between accelerometer data and Euler angles is given as follows:

$$\phi = \arctan \frac{A_y}{\sqrt{A_z^2 + A_x^2}} \tag{2}$$

$$\theta = \arctan \frac{-A_x}{\sqrt{A_y^2 + A_z^2}} \tag{3}$$

$$\psi = \arctan \frac{A_z}{\sqrt{A_x^2 + A_y^2}} \tag{4}$$

where ϕ, θ and ψ represent the three "Euler angles (pitch, roll, yaw)", the A_x, A_y and A_z are the components of acceleration in the three coordinate axis. However, the Euler angles from accelerometer output have large high-frequency noise and poor reliability in short-term application. So the cooperation with gyroscope is necessary.

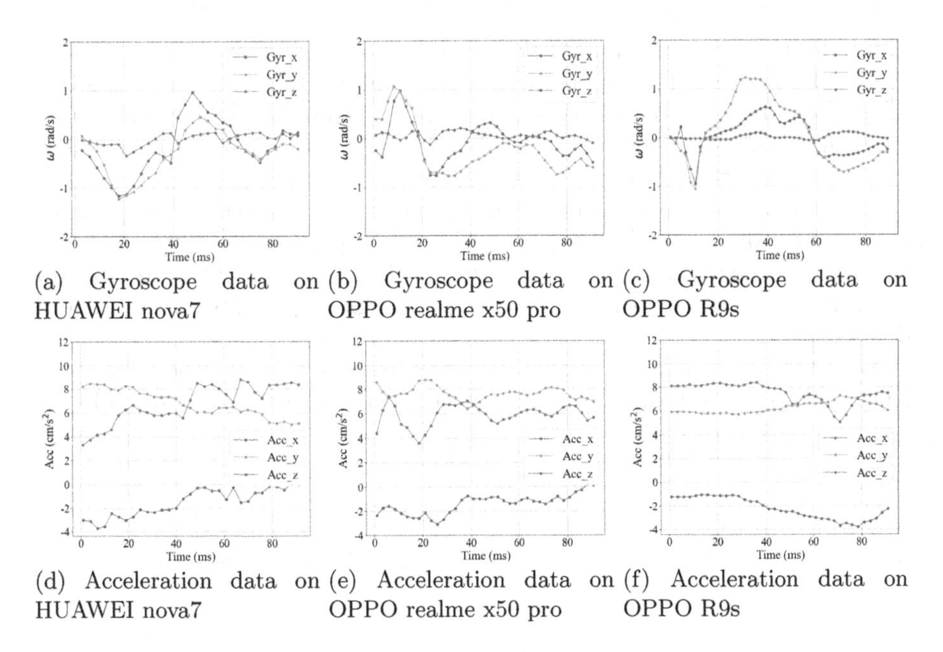

(a) Gyroscope data on HUAWEI nova7 (b) Gyroscope data on OPPO realme x50 pro (c) Gyroscope data on OPPO R9s

(d) Acceleration data on HUAWEI nova7 (e) Acceleration data on OPPO realme x50 pro (f) Acceleration data on OPPO R9s

Fig. 6. Gyroscope and acceleration data on heterogeneous smartphones

Adaptive Complementary Filtering. Accelerometer and gyroscope have their own advantages and disadvantages in attitude calculation. The attitude angle can be obtained by integrating the angular velocity data from gyroscope, but the gyroscope itself produces low-frequency attitude drift. The drift error is amplified with the process of attitude algorithm, which leads to a large attitude angle error obtained only by gyroscope integration. The attitude angle calculated by the accelerometer output value do not produce attitude drift, but the accelerometer itself has large high-frequency noise and poor reliability in short-term application. It can be clearly found that accelerometer and gyroscope can complement each other and solve their shortcomings, so we design complementary filter to normalize them. Although complementary filter can reduce the interference of two kinds of noise theoretically, the stop-band attenuation of low-pass filter is too slow to keep up with the change of smartphones. Therefore, this paper selects adaptive complementary filter and adds proportional and integration (PI) control. The core idea of adaptive complementary filter is to establish a dynamic parameter model to adjust the proportion of the two sensors. The adaptive parameter design model is defined as follows:

$$K_p = \begin{cases} K_{p0}, & 0 \leq |\omega| \leq \omega_e \\ K_{p0} + \frac{K_{p1} - K_{p0}}{\omega_{max} - \omega_e}(\omega - \omega_e), & \omega_e \leq |\omega| \leq \omega_{max} - \omega_e \\ K_{p1}, & \omega_{max} - \omega_e \leq |\omega| \leq \omega_{max} \end{cases} \tag{5}$$

where K_{p0} is the initial value of the filter scale coefficient; K_{p1} is the upper limit value of scale coefficient. ω_e indicates the cut-off angular rate of gyroscope;

ω_{max} indicates the maximum angular velocity that the gyroscope can measure. The angular velocity of gyroscope reflects the reliability of gyroscope data, so the angular velocity range of gyroscope is divided into three areas. In the area with small angular velocity, more gyroscope data are selected and the initial value of scale coefficient is adopted. In the medium speed area, if only the measured data of gyroscope is used, there will be a certain degree of error. Hence, more accelerometer data are selected. The value of proportional coefficient at this time will be linearly increased according to the proportion to better correct the deviation of gyroscope. In the high-speed region, the upper limit value is adopted [15].

After the above attitude algorithm normalization, the corresponding Euler angles of the same user unlocking heterogeneous smartphones as shown in Fig. 7.

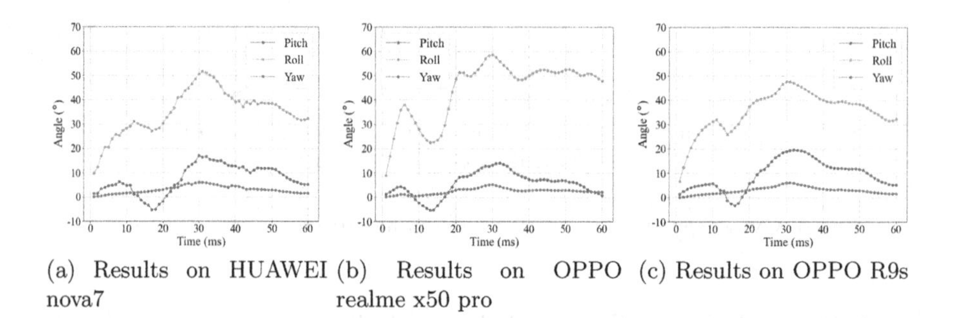

(a) Results on HUAWEI nova7 (b) Results on OPPO realme x50 pro (c) Results on OPPO R9s

Fig. 7. Normalized gyroscope and acceleration data on heterogeneous smartphones

It can be seen that the original acceleration and gyroscope data with great differences are transformed into "Euler angle (pitch, roll, yaw)" data. And the attitude algorithm results of the same user on heterogeneous smartphones are similar in images. In this way, the normalized data can reflect users' behavior characteristics.

5 Experiment and Analysis

5.1 Settings

In our experiment, three kinds of smartphones are used for data acquisition, they are HUAWEI nova7, OPPO realme x50 pro and OPPO R9s[1]. The behavior information during the unlocking process of 20 male and 20 female volunteers aged 20–50 years has been collected. Considering the difference between 90 and 45 turns and the number of inflection points can affect the results [16], the information of unlocking with pattern L and Z is collected respectively, and 50 groups of data are collected by each volunteer to form a data set.

[1] For convenience, HUAWEI nova7, OPPO realme x50 pro and OPPO R9s are simplified as H, O1 and O2 in this section, respectively.

5.2 Selection of Binary Classifier

To evaluate the performance of the binary classifiers, we compare the authentication results of three common binary classifiers. The OCSVM, Local Outlier Factor classifier and Isolation Forests classifier are selected to detect the abnormality of the pattern L and Z password unlocking data after normalization. We compare the Precision, False Acceptance Rate (FAR), False Rejection Rate (FRR) and Area Under ROC Curve (AUC), which are shown in Table 2.

Table 2. Classifier comparison

Pattern	Classifier	Precision/%	FAR/%	FRR/%	AUC
L	OCSVM	97.66	35.15	3.47	0.8070
	Isolation forest	93.42	39.81	54.33	0.5293
	Local outlier factor	92.14	62.53	51.06	0.4319
Z	OCSVM	99.13	25.79	5.20	0.8451
	Isolation Forest	98.73	33.18	13.24	0.7679
	Local outlier factor	95.89	66.04	50.59	0.4169

It can be seen from Table 2 that OCSVM is about 2% $\sim$ 5% more accurate than the other two classifiers, showing about 10% $\sim$ 30% advantages over the other two classifiers in FAR, 10% $\sim$ 40% advantages in FRR, and even 20% $\sim$ 30% in AUC. Therefore, OCSVM is chosen as the classifier to detect the anomaly of the model.

5.3 Performance Analysis

Effect on Different Series Smartphones. Firstly, we compare the normalization performance among different series smartphones (H and O1). A volunteer uses H and O1 to simulate unlocking pattern L and Z. The result is shown as Table 3.

We can see from the Table 3 that the authentication effectiveness after normalizing is improved significantly, and our normalization solution is meaningful. By comparing effect of pattern L on different series smartphones, the normalized data accuracy is improved by 27%, the FAR is reduced by 12%, the FRR is reduced by 28%, the AUC is improved by 20%, the normalization results are obvious.

In order to prove generalization of our conclusion, the pattern Z has been normalized. The results of H and O1 smartphones before and after normalization are still compared. The accuracy of normalized data is improved by 24%, the FAR is reduced by 4%, the FRR is reduced by 25%, the AUC is improved by 14%, the authentication efficiency is improved. The improvement rate of pattern L is higher than pattern Z.

Table 3. Effect on different series smartphones (H and O1)

Pattern	Data	Accuracy/%	FAR/%	FRR/%	AUC/%
L	Original data[a]	95.02	20.94	4.85	87.10
	Non-normalized data[b]	67.51	47.12	31.97	60.46
	Normalized data[c]	94.58	35.15	3.47	80.70
Z	Original data	96.70	30.52	2.29	83.60
	Non-normalized data	69.71	29.46	30.26	70.15
	Normalized data	94.12	25.79	5.20	84.51

[a]Original data is defined as the authentication data model established by using data collected by H.
[b]Non-normalized data is defined as the data model established by H and O1 without normalization.
[c]Normalized data is defined as the data model established by normalizing data collected by H and O1.

Effect on Different Versions Smartphones. Then, we compare the normalization influence on different versions smartphones (O1 and O2). A volunteer uses O1 and O2 to simulate unlocking pattern L and Z. The result is shown as Table 4.

Table 4. Effect on different versions smartphones (O1 and O2)

Pattern	Data	Accuracy/%	FAR/%	FRR/%	AUC/%
L	Original data	96.09	39.04	2.99	83.02
	Non-normalized data	81.22	53.36	16.65	64.99
	Normalized data	96.42	27.03	6.25	83.35
Z	Original data	95.71	36.25	2.58	80.58
	Non-normalized data	87.43	38.96	11.25	64.99
	Normalized data	98.31	32.95	0.11	83.52

By comparing effect of pattern L on different versions smartphones, it is found that the normalized data accuracy is improved by 15%, FAR is decreased by 26%, FRR is decreased by 10%, AUC is increased by 18%. Normalization results are obvious. As pattern Z, it can be seen that the normalized data accuracy is improved by 11%, FAR is decreased by 6%, FRR is decreased by 10%, AUC is increased by 19%, and the authentication efficiency is significantly improved. And the improvement rate of pattern L is also higher than pattern Z.

Moreover, we compare the rate of improvement between same series smartphones and different series but same version smartphones, which is listed in Fig. 8.

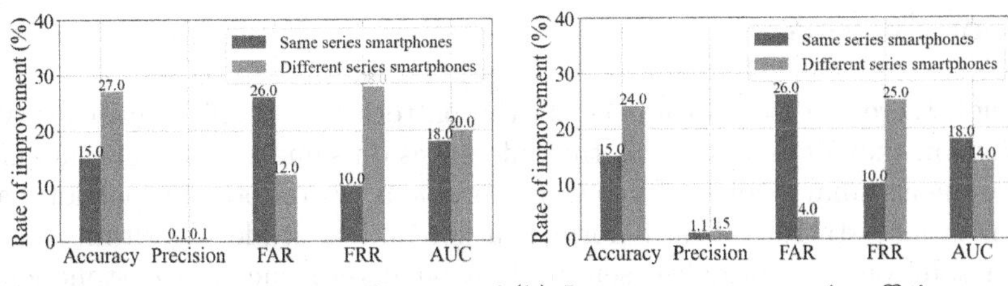

(a) Improvement rates in efficiency of pattern L

(b) Improvement rates in efficiency of pattern Z

Fig. 8. Comparison of authentication efficiency on different smartphones

After normalization, the improvement of authentication efficiency of different series smartphones (such as H and O1) is higher than that of different versions in the same series (such as O1 and O2). It can be concluded that the difference between built-in sensors of different series smartphones is greater than that of different versions in same series.

According to the improvement rate of authentication efficiency after normalization on pattern L and Z, different series smartphones (H and O1) and different versions smartphones in the same series (O1 and O2) are analyzed respectively. The results are shown in Fig. 9.

By comparison the result, improvement of authentication efficiency on pattern L after normalization is higher than that of pattern Z. The authentication rules of different pattern passwords are different, but there is a general rule: the recognition performance in straight line and inflection point is better, which means the longer the straight line, the better authentication efficiency of pattern unlock. The more inflection points exists, the better authentication efficiency of pattern unlock [17]. We compare pattern L with pattern Z in this paper, pattern L has a longer straight line distance, and it includes more information of user's behavior. So its authentication efficiency is improved more after normalization. Although the pattern Z has two inflection points, they are all 45 inflection points and occupy more points. The proportion of inflection points decreases, resulting in relatively poor information it has. Therefore, there is less improvement of authentication efficiency after normalization.

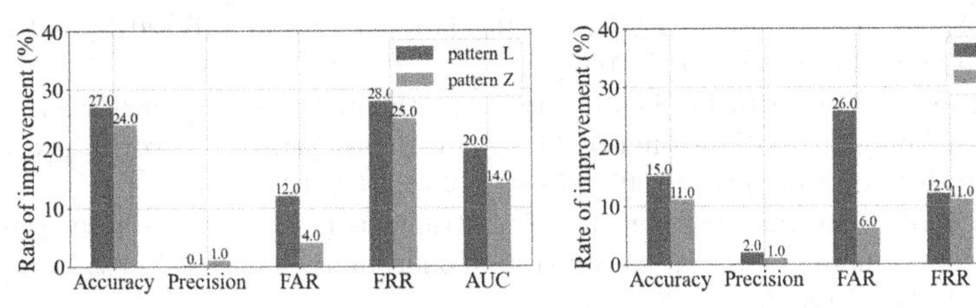

(a) Improvement rates of different series smartphones

(b) Improvement rates of different versions smartphones in same series

Fig. 9. Comparison of authentication efficiency of different patterns

6 Conclusions

Although there are many works study the pattern lock implicit authentication mechanism, they do not consider the differences on same kind of sensors among heterogeneous smartphones. To solve the problem, we propose proportion-based normalization, statistics-based normalization and attitude algorithm normalization solutions to normalize sensor data on heterogeneous smartphones i.e. HUAWEI nova7, OPPO realme x50 pro, OPPO R9s. And then, we use OCSVM to establish authentication models and detect identification. Through comparing their performance indicators, the normalization solution our proposed can improve the authentication efficiency significantly.

In our future work, we will optimize the normalization solution by adding more contextual information in our system.

References

1. Patel, V.M., et al.: Continuous user authentication on mobile devices: recent progress and remaining challenges. IEEE Signal Process. Mag. **33**(4), 49–61 (2016)
2. Ye, G., Tang, Z., et al.: A video-based attack for android pattern lock. ACM Trans. Priv. Secur. (TOPS) 2018 **21**, 1–31. https://doi.org/10.1145/3230740
3. Shi, D., Tao, D., et al.: Fine-grained and context-aware behavioral biometrics for pattern lock on smartphones. In: Proceedings of the ACM on Interactive, Mobile, Wearable and Ubiquitous Technologies (IMWUT) 2021, vol. 5, pp. 1–30 (2021). https://doi.org/10.1145/3448080
4. Khan, H., Hengartner, U.: Towards application-centric implicit authentication on smartphones. In: Proceedings of the 15th Workshop on Mobile Computing Systems and Applications 2014, Article 10, 6 p. (2014) https://doi.org/10.1145/2565585.2565590
5. Sitová, Z., Šeděnka, J., et al.: New behavioral biometric features for continuous authentication of smartphone users. IEEE Trans. Inf. Forensics Secur. **11**, 877–892 (2016). https://doi.org/10.1109/TIFS.2015.2506542
6. Xu, G.Y., et al.: Overview of implicit identity authentication mechanisms for mobile terminals. Comput. Eng. Appl. **54**(6), 19–25 (2018)
7. Zhang, Y.T., Hu, W., et al.: Continuous authentication using eye movement response of implicit visual stimuli. ACM Interact. Mob. Wearable Ubiquit. Technol. **1**, 22 (2018). https://doi.org/10.1145/3161410
8. Vhaduri, S., Poellabauer, C.: Multi-modal biometric-based implicit authentication of wearable device users. In: IEEE Trans. Inf. Forensics Secur., 10 (2018). https://doi.org/10.1109/TIFS.2019.2911170
9. Shi, D., Tao, D.: Sensor fusion based implicit authentication for smartphones. In: 14th China Conference on Internet of Things (Wireless Sensor Network), pp. 157–168 (2020). https://doi.org/10.1007/978-981-33-4214-9_12
10. Liu, C.L., et al.: Implementing multiple biometric features for a recall-based graphical keystroke dynamics authentication system on a smart phone. J. Netw. Comput. Appl. **53**, 128–139 (2015)
11. Scholkopf, B., et al.: Estimating the support of a high-dimensional distribution. Neural Comput. **13**(7), 1443–1471 (2001)

12. Ott, F., Wehbi, M., et al.: The OnHW dataset: online handwriting recognition from IMU-enhanced ballpoint pens with machine learning. In: Proceedings of the ACM on Interactive, Mobile, Wearable and Ubiquitous Technologies, vol. 4, pp. 1–20 (2020). https://doi.org/10.1145/3411842
13. Shi, Y.L., et al.: Ready, steady, touch! Sensing physical contact with a finger-mounted IMU. In: Proceedings of the ACM on Interactive Mobile Wearable and Ubiquitous Technologies, vol. 4, pp. 1–25 (2020). https://doi.org/10.1145/3397309
14. Zhang, Y.B.: Research on user behavior recognition model based on sensor data. Master dissertation, University of Electronic Science and Technology of China (UESTC), Chengdu (2020)
15. Wei, H., et al.: Implementation of improved adaptive complementary filtering MEMS-IMU attitude calculation. Electron. Meas. Technol. **43**(24), 81–86 (2020)
16. Yao, M.Y., et al.: Implicit identity authentication mechanism of smartphone pattern password based on up-sampling single classification. Comput. Sci. **46**(11), 19–24 (2020)
17. Liu, B.Y.: Smart phone identity authentication based on gesture recognition. Master dissertation, Beijing Jiaotong University, Beijing (2018)
18. Zheng, Z., et al.: A fused method of machine learning and dynamic time warping for road anomalies detection. IEEE Trans. Intell. Transp. Syst., 1–13 (2020). https://doi.org/10.1109/TITS.2020.3016288

Privacy-Preserving and Reliable Federated Learning

Yi Lu[1,2,3], Lei Zhang[1,2,3(✉)], Lulu Wang[1], and Yuanyuan Gao[1]

[1] Engineering Research Center of Software/Hardware Co-design Technology and Application, Ministry of Education, East China Normal University, Shanghai, China
leizhang@sei.ecnu.edu.cn
[2] Guangxi Key Laboratory of Cryptography and Information Security, Guilin, China
[3] Shanghai Key Laboratory of Trustworthy Computing, Shanghai, China

Abstract. In Internet of Things (IoT), it is often impossible to share datasets owned by different participants (usually IoT devices) for machine learning model training due to privacy concerns. Federated learning (FL) is a promising technique to address this challenge. However, existing FL schemes face the problem of how to avoid low-quality/malicious update. To solve this problem, we propose a privacy-preserving and reliable federated learning scheme (PPRFLS) to select reliable participants and evaluate the quality of the participants' updates. Analysis shows that the proposed scheme achieves data privacy and model reliability.

Keywords: Federated learning · Model reliability · Data privacy

1 Introduction

With the development of IoT [7,21,34], it has a large number of highly available training datasets and computing resources, which makes machine learning widely used in IoT, e.g., speech and image recognition, language translation [9,22,31]. Traditional machine learning needs to aggregate massive participants' datasets into a central server for model training, which leads to security and privacy risks, high economic cost, etc. [19,26,35]. To address the challenges, FL, as an emerging distributed machine learning paradigm, has been introduced to allow participants to collaboratively train a global model in a decentralized manner. A typical FL architecture consists of a central server (usually called aggregator) and multiple participants (who manage their respective local datasets). The participants in FL only need to send their updates trained on their local datasets to the central server instead of uploading the datasets [20].

Despite the great benefits of FL, it is facing severe challenges. The fundamental one is data privacy. In FL, since a participant's dataset is stored locally, an attacker cannot directly violate the privacy of a participant's dataset. However, if a raw update, i.e., a raw gradient in this paper, of a participant is uploaded to the central server directly, an attacker can obtain a large amount of information

Y. Lai et al. (Eds.): ICA3PP 2021, LNCS 13157, pp. 346–361, 2022.
https://doi.org/10.1007/978-3-030-95391-1_22

related to the participant's local dataset by membership inference attacks [25] and gradient analysis [2], etc. Therefore, it is important to design a privacy-preserving mechanism to protect the privacy of an update.

In addition to data privacy, another challenge is model reliability. In FL, poisoning attacks and low-quality participants may lead to low model reliability. In poisoning attacks, malicious data is injected into the training datasets of the participants by an attacker, such that the global model has a high testing error rate [14,23]. Low-quality participants are the participants with constrained computing resources, small datasets, and/or non-IID (non-independent and-identically-distributed) data. They will lead to large model convergence time, weak generalization ability and overfitting of the model [32]. We note that if privacy protection mechanism(s) is applied to FL, this will make it difficult to evaluate the model reliability [2,30]. Therefore, how to design an effective privacy-preserving model evaluation scheme is of great interest in FL.

1.1 Related Work

The virtual keyboard application developed by Google first introduced the concept of FL. Later, the concept, architecture and potential applications of FL are further discussed in [27,31,33]. However, as discussed above, to put FL into practice, it faces the challenges of data privacy and model reliability.

Multi-party security computing (MPC) and differential privacy are the main tools to solve the data privacy challenge in FL. The former is usually based on secret sharing, homomorphic encryption, etc. The existing MPC based FL schemes have high communication and/or computation overhead [6,15,28,33]. For instance, the research in [33] showed that the iteration time cost is extended by 96$\times$ if Paillier (a homomorphic encryption scheme) is applied to protect the privacy of an update when comparing with directly transferring an update to the central server. Further, since an update sent to the central server is in an encrypted form, it makes the evaluation of model reliability difficult (See the next paragraph). As to differential privacy based solutions [1,24,30], the main idea is to add noise to participant datasets which decreases model accuracy. Besides, it is also difficult to determine how much noise has to be added to a dataset or find a balance between model accuracy and data privacy.

Model reliability is threatened by data poisoning attacks and low-quality participants. To prevent poisoning attacks, a popular method is reject on negative impact (RONI) [4,11] whose idea is to discard an update if it results in a large error rate of the local model. However, in this method, the central server has to know each raw update. Thus, this method is not suitable for MPC based FL since an update is usually in an encrypted form. Although this method may be applied to differential privacy based FL, this will decrease the model accuracy since noise is added to a participant's dataset or update. The cosine similarity of participants' updates may also be used to avoid poisoning attacks [3,13,14] and low-quality participants [20,23]. However, the existing solutions [14,20] do not consider the challenge of data privacy and high computational complexity [3,14,23] due to the use of hierarchical clustering algorithm. We note that to avoid low-quality participants, a nature way is to select participants with rich local resources (e.g., rich computing resources, large datasets) [18,29,31]. However, the existing schemes cannot deal with selfish participants who aim to get more rewards by uploading wrong information.

1.2 Our Contribution

To address the data privacy and model reliability challenges in existing FL schemes, we propose a privacy-preserving and reliable federated learning scheme (PPRFLS) using a dual-server architecture and based on DK-Means and CKKS.

Our PPRFLS applies dual servers, i.e., aggregation server (AS) and platform sever (PS). In each round of an FL task, participants send their respective updates encrypted using CKKS to the AS. With the help of the PS, the AS clusters the encrypted updates from the participants into groups using homomorphism-based DK-Means with the pairwise cosine similarities as the metrics. Then, for the encrypted updates corresponding to each group, the AS aggregates them into a sub-global model based on the homomorphic addition property supported by the CKKS. The encrypted sub-global models corresponding to the groups are sent to the PS. The PS then can test the accuracy of each sub-global model using its local test dataset after the decryption of each sub-global model. The sub-global model whose accuracy is lower than a threshold is recognized as an unreliable sub-global model and is discarded. The rest reliable sub-global models are aggregated based on an accuracy-based weighted aggregation method to generate a global model corresponding to this round.

Our PPRFLS achieves the goal of data privacy because both the AS and the PS can't get the raw updates from the participants. Besides, our PPRFLS achieves the goal of model reliability because it discards the unreliable sub-global models and only aggregates the reliable sub-global models to get a reliable global model.

1.3 Organization

The rest of this paper is organized as follows. We present the background in Sect. 2. Section 3 describes our PPRFLS. The performance evaluation and the security analysis are described in Sect. 4. We conclude the paper in Sect. 5.

2 Background

In this section, we introduce our system model, threat model, design goals and homomorphic encryption. The main mathematical notations used in this paper are listed in Table 1.

Table 1. Summary of symbols and notation

Notation	Definition
c_i	The i-th participant
D_i	The dataset of c_i
$\mathcal{M}_{c_i}$	The local model of c_i
$\boldsymbol{w}$	A parameter vector
$F_{c_i}(\boldsymbol{w})$	The loss function of c_i
$\nabla F_{c_i}(\boldsymbol{w})$	The update of c_i
$\text{Cosim}(c_i, c_j)$	The cosine similarity between the updates from c_i and c_j
$\hat{M}_{in}$	An encrypted inner products matrix
$\hat{M}_{ss}$	An encrypted sum of squares matrix
m	The number of groups
$R_{g_k}^{(t)}$	The normalization factor of group g_k at round t

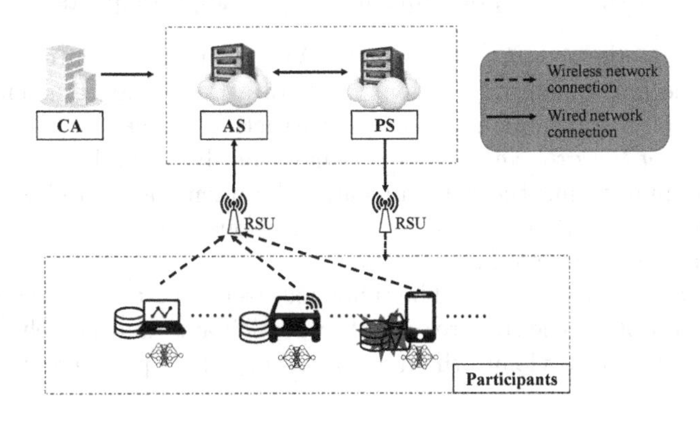

Fig. 1. System model.

2.1 System Model

Figure 1 shows our system model which consists of a certificate authority (CA), an aggregation server (AS), a platform server (PS) and participants. The main functions of each entity are as follows:

- CA: The CA is a trusted third party. It is responsible to generate the system parameters and issue certificates for the entities in the system.
- PS: It is responsible for task publishing, participant selection and global model updating. Besides, it has a local dataset that can be used as a test dataset to evaluate the global model.
- AS: It is used for global model updating. Unlike the PS, for each round, it can only generate an encrypted sub-global model (using the updates from the selected participants) which can be decrypted by the PS. Further, for

each round, it generates and sends an encrypted inner product matrix and an encrypted sum of squares matrix to the PS.
- Participants: Participants are the entities with IoT devices (such as vehicles or mobile phones). Participants will generate the updates (based on their respective datasets) and submit the updates to the AS.

We assume the PS and the AS are semi-trusted. As for the participants, we assume most of the participants are semi-trusted, while some of them are malicious.

2.2 Threat Model

An FL task may suffer threats from insiders (participants, AS, PS) and outsiders.

Threats to Privacy: The attackers aim to get the privacy information through the updates and the information uploaded from the participants.

1. *Threats from the AS and the PS:* The AS may try to violate the data privacy of the participants by various types of attacks, e.g., gradient analysis, membership inference attacks and reconstruction attacks.
2. *Threats from the external attackers:* They can obtain updates of the participants by monitoring the communication between the participants and the AS. Then, they can launch the attacks mentioned above to violate the data privacy of the participants.
3. *Threats from participants:* The malicious/semi-trusted participants may attempt to obtain the updates of other participants and launch the attacks mentioned above to violate the data privacy of other participants.

Threats to Reliability: The attackers aim to reduce the global model accuracy by sending malicious updates or low-quality updates.

1. *Threats from participants:* The malicious participants may deliberately submit updates of poisoning attacks [14] to control the training result of the global model. In addition, the low-quality participants may indeliberately send low-quality updates which may reduce the global model accuracy.
2. *Threats from the external attackers:* They may eavesdrop on the communications, tamper the updates and send malicious updates to control the training result of the global model. However, we can use an authentication mechanism to avoid this threat easily. Thus, we don't consider this threat in this paper.

2.3 Design Goals

To avoid the threats mentioned above, our PPRFLS has to satisfy the following properties:

Model Reliability: As discussed above, we just need to consider the attacks from *participants*, a scheme is required to identify the reliable updates. We can achieve the goal of getting a reliable global model by aggregating reliable updates.

Data Privacy: We want to achieve model reliability under privacy protection. The current schemes (e.g., RONI, etc.) are carried out under raw updates. In the current reliable model selection scheme, the participants need to upload local information which faces the problems that the participants are unwilling to upload local information due to privacy concerns and malicious participants uploading false information to be selected by the server. Therefore, data privacy requires the attackers can't get the raw updates from the participants directly. At the same time, the model reliability can be accurately evaluated without local information.

2.4 Homomorphic Encryption

Homomorphic encryption is a widely used tool to achieve data privacy that allows users to perform a specific form of algebraic operation on the ciphertext to get the result that is still encrypted, and the result obtained by decrypting ciphertext is the same as the result of the same operation on the plaintext. To support the operations in the PPRFLS, we choose the scheme of Cheon, Kim, Kim, and Song (hereafter referred to as the CKKS scheme) [8], a fully homomorphic encryption scheme, which can support additive homomorphism and multiplication homomorphism. The scheme consists of the following algorithms (**KeyGen**, **Encrypt**, **Decrypt**, **Add**, **Mul**), which is described as follows:

KeyGen$(\lambda) \rightarrow (sk, pk)$: generate a public key pk, a secret key sk.

Encrypt$(pk, m) \rightarrow c$: probabilistic encryption algorithm produces c, the ciphertext of message m.

Decrypt(sk, c): decryption algorithm returns message m' encrypted in c, where $m' \approx m$.

Add(c_1, c_2): for ciphertexts c_1 and c_2, the output is the encryption of plaintext addition c_{add}.

Mul(c_1, c_2): for ciphertexts c_1 and c_2, the output is the encryption of plaintext multiplication c_{mul}.

We use (**KeyGen**, **Encrypt**, **Decrypt**, **Add**, **Mul**) to define a CKKS scheme used in this paper.

3 Our Scheme

We propose the PPRFLS in this section.

3.1 High-Level Description

We notice that the direction of an update (i.e., gradient in this paper) can reflect the distribution of a dataset [10,14]. Poisoning attack [14] or a non-IID dataset [17] will lead to incorrect direction of an update, which will lead to the low accuracy of the model. Our scheme is based on this property.

We first compute the pairwise cosine similarities using participants' updates, in which cosine similarity reflects the angle between two directions of updates. We note that an update is a vector and the direction of an update is a scalar. So, cosine similarity of two directions of updates can be also computed using the corresponding updates. With the pairwise cosine similarities, we cluster the participants into groups based on the similarities, until now we achieve a secure clustering by DK-means. Then we generate the sub-global model of each group whose reliability is evaluated by using the accuracy of the sub-global model. If the reliability of a sub-global model is lower than a threshold, then the sub-global model is discarded, which implies the corresponding participants suffer from poisoning attacks and/or have non-IID datasets. Finally, we get the reliable global model generated using high reliable sub-global models.

3.2 Definitions

Assume there are n participants $(c_1, ..., c_n)$, the i-th participant c_i holds dataset $D_i = \{(x_1, y_1), (x_2, y_2), ..., (x_{s_i}, y_{s_i})\}$, where s_i is the dataset size of D_i, (x_j, y_j) is a data sample, x_j is the j-th input of a machine learning algorithm and y_j is the label value.

The global model training in FL involves multiple iteration rounds. At the beginning, the PS initializes the global model $\boldsymbol{w}$ and broadcasts to the participants. Each participant trains its local model $\mathcal{M}_{c_i} = f(\boldsymbol{w}, x)$ over training dataset D_i, where f is the model function, and $\boldsymbol{w}$ is the parameter vector. To achieve this goal, a participants c_i has to minimize the loss function defined as:

$$F_{c_i}(\boldsymbol{w}) = \frac{1}{s_i} \sum_{j=1}^{s_i} L(f(\boldsymbol{w}, x_j), y_j), \tag{1}$$

where $L(f(\boldsymbol{w}, x_j), y_j)$ is the loss value of the data sample (x_j, y_j) with $\boldsymbol{w}$. We will choose stochastic gradient descent (SGD) to solve the above optimization problem. SGD minimizes the loss function $F_{c_i}(\boldsymbol{w})$ to get the update (gradient):

$$\nabla F_{c_i}(\boldsymbol{w}) = \frac{\partial F_{c_i}(\boldsymbol{w})}{\partial \boldsymbol{w}}, \tag{2}$$

which is a vector.

For the t-th round, we assume the update generated by c_i is $\nabla F_{c_i}(\boldsymbol{w}^{(t)}) = (V_{c_i}(\boldsymbol{w}^{(t,1)}), ..., V_{c_i}(\boldsymbol{w}^{(t,\theta)}))$, where θ is the dimension of $\nabla F_{c_i}(\boldsymbol{w}^{(t)})$. Then the cosine similarity between $\nabla F_{c_i}(\boldsymbol{w}^{(t)})$ and $\nabla F_{c_j}(\boldsymbol{w}^{(t)})$ is defined as:

$$\text{Cosim}(c_i, c_j) \triangleq \frac{\langle \nabla F_{c_i}(\boldsymbol{w}^{(t)}), \nabla F_{c_j}(\boldsymbol{w}^{(t)}) \rangle}{\|\nabla F_{c_i}(\boldsymbol{w}^{(t)})\| \|\nabla F_{c_j}(\boldsymbol{w}^{(t)})\|}, \tag{3}$$

where $\langle \cdot, \cdot \rangle$ represents the inner product of two vectors, $\|\cdot\|$ represents the 2-norm of a vector.

We note that, in FL, server(s) cannot get raw updates of the participants due to privacy consideration. Therefore, raw updates have to be protected. As differential privacy based methods decrease model accuracy, here we use HE to protect update privacy. Particularly, in the PPRFLS, we use CKKS to realize update privacy. However, if CKKS is applied, it is particularly time-consuming to perform homomorphic division and homomorphic square root operations. In the PPRFLS, we present a method to compute the cosine similarity without computing these two types of operations.

3.3 The Proposal

In the PPRFLS, the PS has to generate a public-private key pair (pk, sk) according to the CKKS. When an FL task is initialized, pk has to be passed to the participants. Suppose that the global model initialized by the PS is $\boldsymbol{w}^{(0)}$ which is vector. $\boldsymbol{w}^{(0)}$ is broadcasted to the participants. For the i-th participant c_i, it initializes the local model $\boldsymbol{w}_{c_i}^{(0)} = \boldsymbol{w}^{(0)}$ based on the global model. The rest steps will describe how to achieve the rest of the steps of the PPRFLS, select reliable updates and aggregate updates under privacy protection in detail.

Step 1: Each participant that involves in the FL task has to run this step. Suppose the current round is the t-th round. For c_i, it gets update $\nabla F_{c_i}(\boldsymbol{w}^{(t)})$ by training its local model with the local dataset to minimize the loss function according to the Eqs. (1) and (2). Then, c_i sends the encrypted update $\nabla \hat{F}_{c_i}(\boldsymbol{w}^{(t)}) = (\hat{V}_{c_i}(\boldsymbol{w}^{(t,1)}), ..., \hat{V}_{c_i}(\boldsymbol{w}^{(t,\theta)}))$ to the AS , where $\hat{V}_{c_i}(\boldsymbol{w}^{(t,l)}) = \mathbf{Encrypt}(pk, V_{c_i}(\boldsymbol{w}^{(t,l)})), l \in [1, \theta]$.

Step 2: Suppose the AS receives the encrypted updates from n participants. Although the AS can calculate the pairwise encrypted cosine similarities directly using the received encrypted updates, it is costly to compute the homomorphic division and homomorphic square root operations. To avoid these costly operations, our method is to let the AS and the PS compute cosine similarities cooperatively. In this step, the AS just needs to compute the pairwise encrypted inner products and the encrypted sum of squares.

Assume the encrypted updates from c_i and c_j are $\nabla \hat{F}_{c_i}(\boldsymbol{w}^{(t)}) = (\hat{V}_{c_i}(\boldsymbol{w}^{(t,1)}), ..., \hat{V}_{c_i}(\boldsymbol{w}^{(t,\theta)}))$ and $\nabla \hat{F}_{c_j}(\boldsymbol{w}^{(t)}) = (\hat{V}_{c_j}(\boldsymbol{w}^{(t,1)}), ..., \hat{V}_{c_j}(\boldsymbol{w}^{(t,\theta)}))$. The corresponding encrypted inner product is computed by the AS as follows:

1. Compute $mul\hat{V}_{c_{i,j}}^{(t,l)} = \mathbf{Mul}(\hat{V}_{c_i}(\boldsymbol{w}^{(t,l)}), \hat{V}_{c_j}(\boldsymbol{w}^{(t,l)}))$ for $l \in [1, \theta]$.
2. Set $\hat{in}_{c_{i,j}} = 0$, compute $\hat{in}_{c_{i,j}} = \mathbf{Add}(\hat{in}_{c_{i,j}}, mul\hat{V}_{c_{i,j}}^{(t,l)})$ for $l \in [1, \theta]$.
3. Output the encrypted inner product $\hat{in}_{c_{i,j}}$.

The encrypted sum of squares for participant c_i is computed by the AS as follows:

1. Compute $mul\hat{V}_{c_i}^{(t,l)} = \mathbf{Mul}(\hat{V}_{c_i}(\boldsymbol{w}^{(t,l)}), \hat{V}_{c_i}(\boldsymbol{w}^{(t,l)}))$ for $l \in [1, \theta]$.
2. Set $\hat{ss}_{c_i} = 0$, compute $\hat{ss}_{c_i} = \mathbf{Add}(\hat{ss}_{c_i}, mul\hat{V}_{c_i}^{(t,l)})$ for $l \in [1, \theta]$.
3. Output the encrypted sum of squares $\hat{ss}_{c_i}$.

Finally, the AS sends the matrix

$$\hat{M}_{in} = \begin{bmatrix} \hat{in}_{c_{1,1}} & \dots & \hat{in}_{c_{1,n}} \\ \vdots & \vdots & \vdots \\ \hat{in}_{c_{n,1}} & \dots & \hat{in}_{c_{n,n}} \end{bmatrix}$$

and $\hat{M}_{ss} = (\hat{ss}_{c_1}, ..., \hat{ss}_{c_n})$ to the PS.

Step 3: When the PS receives $\hat{M}_{in}$ and $\hat{M}_{ss}$, for $i, j \in [1, n]$, it computes the inner product $in_{c_{i,j}} = \mathbf{Decrypt}(sk, \hat{in}_{c_{i,j}})$, the sum of squares $ss_{c_i} = \mathbf{Decrypt}(sk, \hat{ss}_{c_i})$, and the 2-norm of the updates $\sqrt{ss_{c_i}}$. Then the PS can compute the pairwise cosine similarities by using the inner products and the 2-norm of the updates according to Eq. (3), and obtains the similarity matrix

$$\mathcal{M}_s = \begin{bmatrix} Cosim(c_1, c_1) & \dots & Cosim(c_1, c_n) \\ \vdots & \ddots & \vdots \\ Cosim(c_n, c_1) & \dots & Cosim(c_n, c_n) \end{bmatrix}. \quad (4)$$

Then, the PS clusters the updates into m groups[1] using the DK-Means [16] algorithm which takes $\mathcal{M}_s$ as an input. The participants whose updates are in the same group are divided into a group. Let the groups are $\{g_1, g_2, ..., g_m\}$. The PS sends $\{g_1, g_2, ..., g_m\}$ to the AS.

Step 4: When the AS receives $\{g_1, g_2, ..., g_m\}$ from the PS, it computes the encrypted group aggregate updates corresponding to the groups in this step. Let $\nabla sum\hat{F}_{g_k}(\boldsymbol{w}^{(t)}) = 0$. For $k \in [1, m]$, $c_i \in g_k$, the AS sets the encrypted group aggregate update corresponding to g_k to be

$$\nabla sum\hat{F}_{g_k}(\boldsymbol{w}^{(t)}) = \mathbf{Add}(\nabla sum\hat{F}_{g_k}(\boldsymbol{w}^{(t)}), \nabla\hat{F}_{c_i}(\boldsymbol{w}^{(t)})).$$

Finally, the AS sends $(\nabla sum\hat{F}_{g_1}(\boldsymbol{w}^{(t)}), ..., \nabla sum\hat{F}_{g_m}(\boldsymbol{w}^{(t)}))$ to the PS.

Step 5: When the PS receives the above message from the AS, it generates a reliable global model corresponding to this round as follows:

1. For $k \in [1, m]$, compute the group aggregate update $\nabla sumF_{g_k}(\boldsymbol{w}^{(t)}) = \mathbf{Decrypt}(sk, \nabla sum\hat{F}_{g_k}(\boldsymbol{w}^{(t)}))$.
2. Because the participants belong to the same group have similar updates, for $k \in [1, m]$, the PS can use equal weights to get the sub-global update $\nabla F_{g_k}(\boldsymbol{w}^{(t)}) = \frac{1}{v_k}\nabla sumF_{g_k}(\boldsymbol{w}^{(t)})$ and hence the sub-global model

$$\boldsymbol{w}_{g_k}^{(t)} \leftarrow \boldsymbol{w}^{(t)} - \nabla F_{g_k}(\boldsymbol{w}^{(t)}) \quad (5)$$

   corresponding to g_k, where v_k is the number of participants in g_k.
3. For $k \in [1, m]$, calculate the accuracy $Acc_{g_k}^{(t)}$ of the sub-global model $\boldsymbol{w}_{g_k}^{(t)}$ using the local test dataset. If the accuracy of a sub-global model is lower than a threshold, then the sub-global model is discarded; otherwise, the sub-global model is defined to be a reliable sub-global model. Without loss of

[1] The elbow method [5] can be used to select the value m.

generality, we assume $\boldsymbol{w}_{g_1}^{(t)}, \dots, \boldsymbol{w}_{g_{m'}}^{(t)}$ are reliable sub-global models. In this paper, we define a group corresponding to a reliable sub- global model to be a reliable group, the update in a reliable group is a reliable update and the corresponding participant is a reliable participant; otherwise, the update and the participant corresponding to the unreliable group is an unreliable update and unreliable participant.

4. For the reliable sub-global models, compute the normalization factors of the sub-global updates corresponding to their sub-global models based on the accuracies of the sub-global models. The normalization factor corresponding to $\nabla F_{g_k}(\boldsymbol{w}^{(t)})$ is set to be $R_{g_k}^{(t)}$.
5. According to [12], the PS takes $\{R_{g_k}^{(t)}\}_{1 \leq k \leq m'}$ as the weights of the reliable sub-global updates and generates a reliable global model to achieve model reliability:

$$\boldsymbol{w}^{(t+1)} \leftarrow \boldsymbol{w}^{(t)} - \sum_{k=1}^{m'} R_{g_k}^{(t)} \nabla F_{g_k}(\boldsymbol{w}^{(t)}). \tag{6}$$

The above steps will be continued until the accuracy of the global model (computed using the local test dataset of the PS) is high enough or the maximum number of rounds/training time specified by the PS is reached.

We note that, in FL, it is anticipated that the participants will contribute reliable updates for global model training. On the other hand, considering FL tasks consume participants' resources (e.g., computation and communication resources), the participants will be willing to upload reliable updates, only if they can get enough payments. Therefore, for the PS, a properly designed incentive mechanism is required to improve the privacy-preserving evaluation scheme which can increase the proportion of reliable participants while achieving effective expense control. Due to page limitation, we will discuss this problem in the full version of this paper.

4 Evaluation

In this section, we evaluate the performance and security of our PPRFLS.

4.1 Simulation Setting

We chose PyTorch as the experimental platform. The simulations were performed on a PC with Inter Xeon Silver 4114 at 2.20 GHz @64G RAM. We use a two-layer neural network to train the classification model on the MNIST dataset. There are 50 participants in the FL tasks. Each reliable participant is randomly assigned a local dataset which has a uniform distribution over 10 classes as its local dataset, while each low-quality participant only receives a certain number of classes randomly as its non-IID local dataset. For the malicious participants that can lunch poisoning attacks, they are randomly assigned local datasets with 10 classes. Some labels of the local datasets are modified to mislead the global model training, e.g., the label "7" is set to be "1". We use the proportion of the labels corresponding to a local dataset that are modified to define the attack strength of a malicious participant. The proportion of the unreceived label indicates the non-IID strength of a low-quality participant, i.e., the non-IID strength is 0.2 which means the participant may don't have the data whose label is "0" and "1". Both the attack strength and the non-IID strength are called data unreliable strength in this paper. Besides, we set the proportion of reliable participants as the participant reliable strength and the proportion of unreliable participants as the participant unreliable strength. Each participant whose data size is 2000 uses a batch of 32 randomly sampled local datasets and trains the local model with 6 iterations to generate an update.

4.2 Experiment Results

For each round of an FL task, to find the relationship between the participant reliable strength and the accuracy of a reliable model, we selected different participant reliable strengths (0.4, 0.6, 0.8). In Fig. 2, we can find that the higher the participant reliable strength, the fewer rounds required for a reliable global model to achieve high accuracy. In Fig. 3, we can find that our PPRFLS can accurately distinguish reliable participants and unreliable participants. If we set the threshold to be 80%, only the group whose corresponding sub-global model with the accuracy 93.2% is a reliable group which also means the accuracy of a reliable global model is 93.2%.

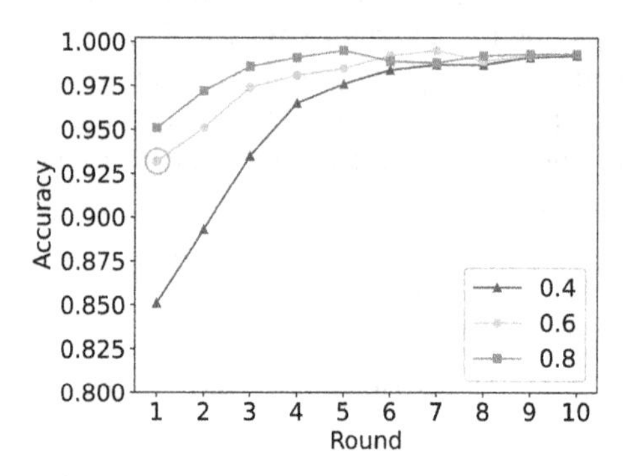

Fig. 2. The accuracy of a reliable global model with the different participant reliable strength (with data unreliable strength 0.2).

Figure 4 shows the accuracy of a global model after 10 rounds of global model training under different data unreliable strength and participant unreliable strength. Both the increase of the data unreliable strength and the participant unreliable strength reduces the accuracy of the model. Besides, we can observe that as the participant unreliable strength and data unreliable strength changes, the accuracy of the global model without unreliable sub-global models changes little. What's more, if we discard the unreliable sub-global models, the

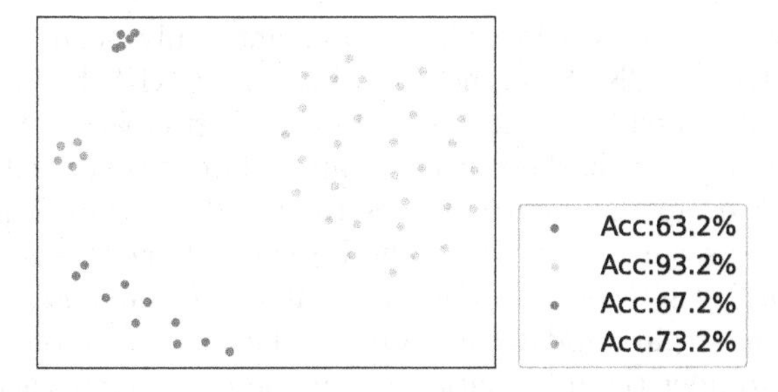

Fig. 3. The spatial distribution of the updates after clustering (with data unreliable strength 0.2, participant reliable strength 0.6).

accuracy of the global models can be improved a lot. For example, when the data unreliable strength is 0.2 and the participant unreliable strength is 0.2, the accuracy of the global model with unreliable sub-global models is 80.2% while the accuracy of the global model without unreliable sub-global models is 98.4%.

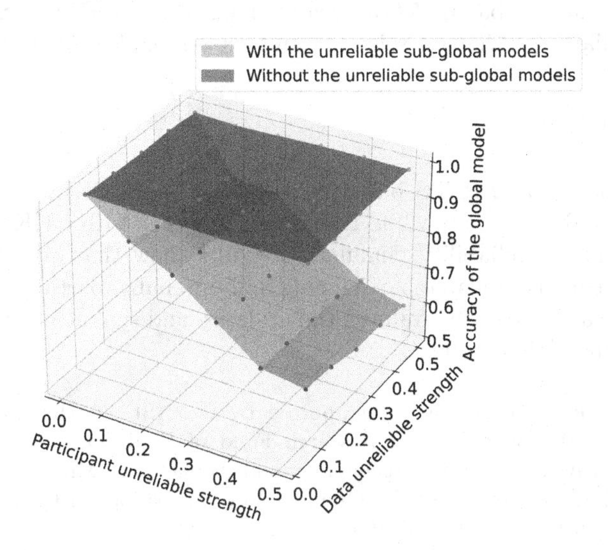

Fig. 4. The accuracy of a global model under different participant unreliable strength and different data unreliable strength.

4.3 Security Analysis

In this section, we show that our PPRFLS can achieve data privacy and model reliability defined in Sect. 2.3.

Data Privacy: In our scheme, the participants only send their respective updates encrypted by CKKS to the AS. Since the CKKS has been proved to be secure, the AS, outsider attackers and other participants cannot learn the content of an update. Thus, the privacy of an update is protected. The AS only sends encrypted group aggregate updates to the PS. The data privacy is also guaranteed by the security of CKKS. The PS can only get the group aggregate updates $\{\nabla sum\hat{F}_{g_k}(\boldsymbol{w}^{(t)})\}_{1\leq k\leq m}$ and the matrix $\hat{M}_{in}$ and $\hat{M}_{ss}$ from the AS. Since the group aggregate updates are already the aggregated ones, it is unlikely for an attacker to infer the information about the raw updates according to the definition of FL.

Model Reliability: According to the PPRFLS in Sect. 3, the AS clusters the updates of the participants into groups based on the pairwise cosine similarities and then aggregates the updates corresponding to each group into a sub-global model with the help of the PS. The PS calculates the accuracy of the sub-global models and discards the sub-global models whose accuracy is lower than a threshold. Then, the PS aggregates the reliable sub-global models using the accuracy of the sub-global models as weight. Therefore, our PPRFLS can achieve the goal of reliable participant selection and get a reliable global model.

5 Conclusion

We have proposed a PPRFLS based on DK-Means and CKKS. The CKKS is used to achieve data privacy while DK-Means together with CKKS are applied to guarantee model reliability. Simulation results show that our PPRFLS can accurately distinguish reliable participants and unreliable participants. Besides, our PPRFLS can accurately evaluate the updates, and significantly improve the reliability of the global model.

Acknowledgement. This work is supported by the NSF of China under Grants 61972159, 61572198; by the Open Research Fund of Engineering Research Center of Software/Hardware Co-design Technology and Application, Ministry of Education (East China Normal University); by Guangxi Key Laboratory of Cryptography and Information Security (No. GCIS202109).

References

1. Abadi, M., et al.: Deep learning with differential privacy. In: Proceedings of the 2016 ACM SIGSAC Conference on Computer and Communications Security, CCS'16, pp. 308–318. Association for Computing Machinery, New York (2016). https://doi.org/10.1145/2976749.2978318
2. Aono, Y., Hayashi, T., Wang, L., Moriai, S., et al.: Privacy-preserving deep learning via additively homomorphic encryption. IEEE Trans. Inf. Forensics Secur. **13**(5), 1333–1345 (2017)

3. Awan, S., Luo, B., Li, F.: CONTRA: defending against poisoning attacks in federated learning. In: Bertino, E., Shulman, H., Waidner, M. (eds.) ESORICS 2021. LNCS, vol. 12972, pp. 455–475. Springer, Cham (2021). https://doi.org/10.1007/978-3-030-88418-5_22
4. Barreno, M., Nelson, B., Joseph, A.D., Tygar, J.D.: The security of machine learning. Mach. Learn. **81**(2), 121–148 (2010). https://doi.org/10.1007/s10994-010-5188-5
5. Bholowalia, P., Kumar, A.: EBK-means: a clustering technique based on elbow method and k-means in WSN. Int. J. Comput. Appl. **105**(9), 17–24 (2014)
6. Bonawitz, K., et al.: Practical secure aggregation for privacy-preserving machine learning, pp. 1175–1191. Association for Computing Machinery, New York (2017). https://doi.org/10.1145/3133956.3133982
7. Chen, T., Zhang, L., Choo, K.K.R., Zhang, R., Meng, X.: Blockchain-based key management scheme in fog-enabled IoT systems. IEEE Internet Things J. **8**(13), 10766–10778 (2021)
8. Cheon, J.H., Kim, A., Kim, M., Song, Y.: Homomorphic encryption for arithmetic of approximate numbers. In: Takagi, T., Peyrin, T. (eds.) ASIACRYPT 2017. LNCS, vol. 10624, pp. 409–437. Springer, Cham (2017). https://doi.org/10.1007/978-3-319-70694-8_15
9. Collobert, R., Weston, J.: A unified architecture for natural language processing: deep neural networks with multitask learning. In: Proceedings of the 25th International Conference on Machine Learning, ICML'08, pp. 160–167. Association for Computing Machinery, New York (2008). https://doi.org/10.1145/1390156.1390177
10. Duan, M., et al.: FedGroup: ternary cosine similarity-based clustered federated learning framework toward high accuracy in heterogeneity data. arXiv preprint arXiv:2010.06870 (2020)
11. Fang, M., Cao, X., Jia, J., Gong, N.: Local model poisoning attacks to byzantine-robust federated learning. In: 29th USENIX Security Symposium (USENIX Security 20), pp. 1605–1622. USENIX Association (2020)
12. Freund, Y., Schapire, R.E.: A decision-theoretic generalization of on-line learning and an application to boosting. J. Comput. Syst. Sci. **55**(1), 119–139 (1997)
13. Fung, C., Yoon, C.J.M., Beschastnikh, I.: The limitations of federated learning in sybil settings. In: 23rd International Symposium on Research in Attacks, Intrusions and Defenses (RAID 2020), pp. 301–316. USENIX Association, San Sebastian (2020)
14. Fung, C., Yoon, C.J., Beschastnikh, I.: Mitigating sybils in federated learning poisoning. arXiv preprint arXiv:1808.04866 (2018)
15. Gilad-Bachrach, R., Dowlin, N., Laine, K., Lauter, K., Naehrig, M., Wernsing, J.: CryptoNets: applying neural networks to encrypted data with high throughput and accuracy. In: Balcan, M.F., Weinberger, K.Q. (eds.) Proceedings of The 33rd International Conference on Machine Learning. Proceedings of Machine Learning Research, vol. 48, pp. 201–210. PMLR, New York (2016)
16. Jothi, R., Mohanty, S.K., Ojha, A.: DK-means: a deterministic k-means clustering algorithm for gene expression analysis. Pattern Anal. Appl. **22**(2), 649–667 (2019). https://doi.org/10.1007/s10044-017-0673-0
17. Kang, J., Xiong, Z., Niyato, D., Xie, S., Zhang, J.: Incentive mechanism for reliable federated learning: a joint optimization approach to combining reputation and contract theory. IEEE Internet Things J. **6**(6), 10700–10714 (2019)
18. Kang, J., Xiong, Z., Niyato, D., Zou, Y., Zhang, Y., Guizani, M.: Reliable federated learning for mobile networks. IEEE Wirel. Commun. **27**(2), 72–80 (2020)

19. Liu, J., et al.: Secure intelligent traffic light control using fog computing. Futur. Gener. Comput. Syst. **78**, 817–824 (2018)
20. McMahan, B., Moore, E., Ramage, D., Hampson, S., Arcas, B.A.Y.: Communication-efficient learning of deep networks from decentralized data. In: Singh, A., Zhu, J. (eds.) Proceedings of the 20th International Conference on Artificial Intelligence and Statistics. Proceedings of Machine Learning Research, vol. 54, pp. 1273–1282. PMLR (2017)
21. Meng, X., Zhang, L., Kang, B.: Fast secure and anonymous key agreement against bad randomness for cloud computing. IEEE Trans. Cloud Comput. (2020). https://doi.org/10.1109/TCC.2020.3008795
22. Rehman, M.H.U., Dirir, A.M., Salah, K., Damiani, E., Svetinovic, D.: TrustFed: a framework for fair and trustworthy cross-device federated learning in IIoT. IEEE Trans. Ind. Inform. **17**(12), 8485–8494 (2021)
23. Sattler, F., Müller, K.R., Samek, W.: Clustered federated learning: model-agnostic distributed multitask optimization under privacy constraints. IEEE Trans. Neural Netw. Learn. Syst. **32**(8), 3710–3722 (2021). https://doi.org/10.1109/TNNLS.2020.3015958
24. Shayan, M., Fung, C., Yoon, C.J., Beschastnikh, I.: Biscotti: a ledger for private and secure peer-to-peer machine learning. arXiv preprint arXiv:1811.09904 (2018)
25. Shokri, R., Stronati, M., Song, C., Shmatikov, V.: Membership inference attacks against machine learning models. In: 2017 IEEE Symposium on Security and Privacy (SP), pp. 3–18 (2017). https://doi.org/10.1109/SP.2017.41
26. Song, L., Shokri, R., Mittal, P.: Privacy risks of securing machine learning models against adversarial examples. In: Proceedings of the 2019 ACM SIGSAC Conference on Computer and Communications Security, CCS'19, pp. 241–257. Association for Computing Machinery, New York (2019) . https://doi.org/10.1145/3319535.3354211
27. Tran, N.H., Bao, W., Zomaya, A., Nguyen, M.N.H., Hong, C.S.: Federated learning over wireless networks: optimization model design and analysis. In: IEEE INFOCOM 2019 - IEEE Conference on Computer Communications, pp. 1387–1395 (2019). https://doi.org/10.1109/INFOCOM.2019.8737464
28. Truex, S., et al.: A hybrid approach to privacy-preserving federated learning. In: Proceedings of the 12th ACM Workshop on Artificial Intelligence and Security, AISec'19, pp. 1–11. Association for Computing Machinery, New York (2019). https://doi.org/10.1145/3338501.3357370
29. Wang, X., Han, Y., Wang, C., Zhao, Q., Chen, X., Chen, M.: In-edge AI: intelligentizing mobile edge computing, caching and communication by federated learning. IEEE Netw. **33**(5), 156–165 (2019)
30. Wei, K., et al.: Federated learning with differential privacy: algorithms and performance analysis. IEEE Trans. Inf. Forensics Secur. **15**, 3454–3469 (2020)
31. Yao, S., et al.: Deep learning for the Internet of Things. Computer **51**(5), 32–41 (2018)
32. Yeom, S., Giacomelli, I., Fredrikson, M., Jha, S.: Privacy risk in machine learning: analyzing the connection to overfitting. In: 2018 IEEE 31st Computer Security Foundations Symposium (CSF), pp. 268–282 (2018). https://doi.org/10.1109/CSF.2018.00027
33. Zhang, C., Li, S., Xia, J., Wang, W., Yan, F., Liu, Y.: BatchCrypt: efficient homomorphic encryption for cross-silo federated learning. In: 2020 USENIX Annual Technical Conference (USENIX ATC 20), pp. 493–506. USENIX Association (2020)

34. Zhang, L.: Key management scheme for secure channel establishment in fog computing. IEEE Trans. Cloud Comput. **9**(3), 1117–1128 (2021)
35. Zhang, L., Meng, X., Choo, K.K.R., Zhang, Y., Dai, F.: Privacy-preserving cloud establishment and data dissemination scheme for vehicular cloud. IEEE Trans. Dependable Secure Comput. **17**(3), 634–647 (2020)

Security Authentication of Smart Grid Based on RFF

Yang Lei[1], Caidan Zhao[1](✉), Yilin Wang[1], Yicheng Zheng[1], and Lei Zhang[2]

[1] Department of Informatics, Xiamen University, Xiamen, China
zcd@xmu.edu.cn
[2] State Grid Fujian Electric Power Co., Ltd., Fuzhou, China

Abstract. Due to the openness of the smart grid, the traditional physical isolation technology can no longer meet the communication security requirements of the smart grid system, and the current security authentication technologies commonly used also need strong computing power, which leads to inefficiency. Based on the above problems, we propose a smart grid security authentication method based on radio frequency fingerprint (RFF). On the one hand, we propose a lightweight deep learning framework for extracting RFFs from raw received signals, which reduces the complexity of the network model and achieves high accuracy. On the other hand, we propose an improved triplet loss to perform open-set recognition. The improved triple loss could make the similar features more compact so as to effectively distinguish the unknown signals. We use radio frequency (RF) signals collected in realistic environments to conduct experiments, and the results prove the effectiveness and robustness of our method.

Keywords: Secure authentication · Deep learning · Smart grid · Radio frequency fingerprint · Signal recognition

1 Introduction

Smart grid [1] is an inevitable trend in the development of power grid technology. Based on an integrated, high-speed two-way communication network, it integrates traditional power grids with communications and computer technologies. As an important application field of the Internet of Things (IoT), the internal structure of a smart grid is equivalent to different forms of IoT application scenarios [2]. Essentially, the smart grid's success depends on the integration between communication equipment and networks, and its rapid development has promoted the establishment of smart communities and smart cities and improved people's quality of life.

For the traditional power grid, the only thing between it and users is the transmission of control information from the power grid to users with no information interaction. Therefore, physical isolation technology is often used as a security defense strategy. Physical isolation is the absolute isolation between the

Y. Lai et al. (Eds.): ICA3PP 2021, LNCS 13157, pp. 362–375, 2022.
https://doi.org/10.1007/978-3-030-95391-1_23

internal network and the public external network, both of which require protection, realizing the complete physical separation of the internal network and the external network to ensure absolute security. However, as an open system, smart grid terminals directly exposed to users may be attacked by unauthorized access equipment due to its two-way communication digital network of real-time information interaction with users, resulting in the destruction of grid integrity and private data leakage. Therefore, in the planning and construction of a smart grid supported by IoT technology, its security must be seriously considered.

In the communication process of power grid equipment, the security mechanisms of different transmission protocols and different levels of transmission data are quite different, which makes system security maintenance difficult [3]. At present, many experts and scholars have carried out research work on the access authentication of devices in the smart grid and have successfully improved and applied various authentication technologies. Literature [4] proposed a device access authentication scheme based on identity authentication and public key infrastructure (PKI) mechanism. This scheme effectively reduced the number of message exchanges with the device access authentication process, but it required an authoritative third party to provide services. Moreover, identity authentication based on the PKI mechanism also involves complex tasks such as the management, distribution, and update of certificates and keys, which directly affect the efficiency of an authentication scheme due to the limited storage and computing capabilities of power equipment. According to the error-correcting code (ECC) algorithm, Liping Zhang *et al.* [5] designed a more efficient and robust authentication protocol to achieve two-way authentication and key agreement between power equipment and substations, but the use of the ECC algorithm also had a higher demand for the computing power of power equipment. Sha. *et al.* [6] proposed a secure reading authentication framework based on smart card readers and cloud computing, requiring smart card readers' help. For complex networks of a large number of terminal types and numbers, such as smart grids, the above algorithms are difficult to implement and deploy. Literature [7] designed a lightweight message authentication scheme based on hash message authentication codes. Researchers used the Diffie-Hellman key exchange protocol as the key exchange mechanism and then proposed a communication scheme suitable for smart grids. The hash-based messages authentication code introduced by this scheme relieved the computational pressure on access devices and communication nodes to a certain extent, but it's still a one-way authentication whose security still requires improvement.

In order to solve the above-mentioned difficulties, we adopt RFF [8] technology and propose a smart grid authentication scheme based on RFF. Because there are many types of smart devices in the power grid and there are electronic component tolerances between different devices, a unique characteristic fingerprint signal can be generated as a unique identifier for each smart grid device. With the development of physical layer security authentication, on the one hand, the use of physical layer feature fingerprint security authentication reduces the calculation and communication overhead of the authentication process and does not require strong key calculation capabilities. On the other hand, the uniqueness and irreplicability of physical layer fingerprints effectively overcome the problem of key leakage of traditional upper-layer encryption algorithms. We send the RF signals generated by different power grid devices into an end-to-end deep neural network to complete the identification and authentication of smart grid access devices.

2 RFF Techniques

Each wireless device will inevitably introduce inherent differences in the manufacturing and production links, which have different effects on the electromagnetic wave signal sent. In the process of converting digital baseband signals into analog signals, the nonlinear characteristics of analog-to-digital converters, filters, and other modules will cause different distortions and distortions. This difference can be identified as the device's single feature, which is called the RFF. The RFF is only determined by the physical hardware characteristics of the device and is not affected by the information it carries, so it is unique and difficult to be copied. By analyzing the RFF of the received signals, we can accurately identify different signal transmitters that use the same frequency band, the same bandwidth, and the same modulation method, which can be employed in individual identification and identity authentication of wireless devices.

The key to the success of traditional individual identification technology based on RFF technology is to extract the unique RF signal characteristics of the device. Researchers used the transmitter's transient signal to extract the RFF feature firstly. Transient characteristics refer to the envelope characteristics of the signal during the process of the signal power changing from zero to the rated power when the working state of the wireless device changes [9]. Kennedy *et al.* [10] first proposed an RFF research method based on steady-state signals until 2008. Since then, the technology of using the steady-state signal of the transmitter to perform RFF identification has begun to emerge. After that, Hu *et al.* [11] proposed RFF extraction methods based on the constellation trajectory of RF signals, which provided new ideas for the research of RFF. Whether using the transient or steady-state characteristics of the signal. The traditional RFF identification authentication algorithm process can be summarized as follows: signal sampling, RFF extraction, training, and identification. Unfortunately, these methods need to manually calculate the signal characteristics. The parameters of the feature extraction algorithm need to be adjusted for different types of RF signals. Without prior knowledge of the signal (such as signal modulation method, SNR, etc.), these feature extraction algorithms may be difficult to work on and directly affect the performance of the classifier.

In recent years, the deep learning network has been proven to be a feature extractor, which can optimize the performance of the model through model training in order to quickly complete the automatic extraction and recognition of the features of the input data. Scholars try to introduce deep learning methods into the field of RFF recognition to solve the difficulties of feature extraction, feature selection, and transmission environmental impact in RFF recognition. In 2018, Ding L *et al.* [12] proposed an RFF identification method based on deep learning. The method selects the steady-state part of the signal and firstly uses a bispectral transformation to extract features. Finally, it uses a convolutional neural network (CNN) [13] to identify features. [14] performs framing, windowing, and short-time Fourier transform processing on the original RF signal. They put the RF signal spectrum into CNN so that CNN can better complete the extraction and identification of RFF characteristics. [15] uses long and short-term memory networks (LSTM) [16] to identify unauthorized broadcast signals and achieve a good recognition rate in the real electromagnetic environment.

3 System Design

3.1 Data Collection and Preprocessing

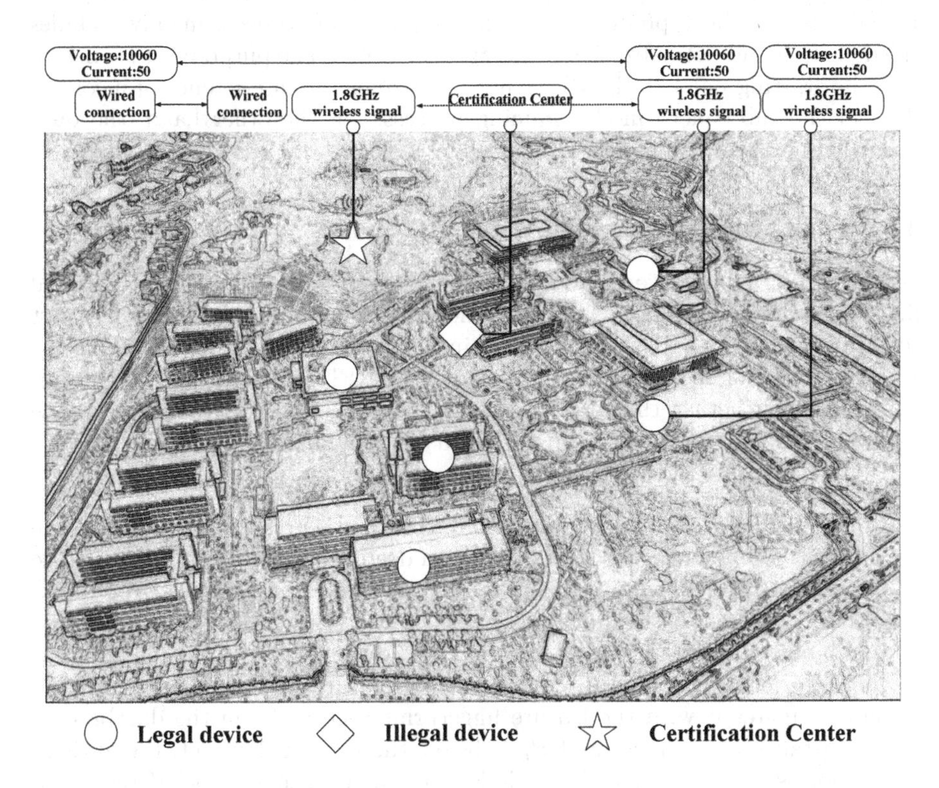

Fig. 1. Security authentication of smart grid. Our experiments are based on real-world scenarios.

In this paper, we use the universal software radio peripheral (USRP) N210 and oscilloscope to collect the RF signals generated by four smart grid devices in

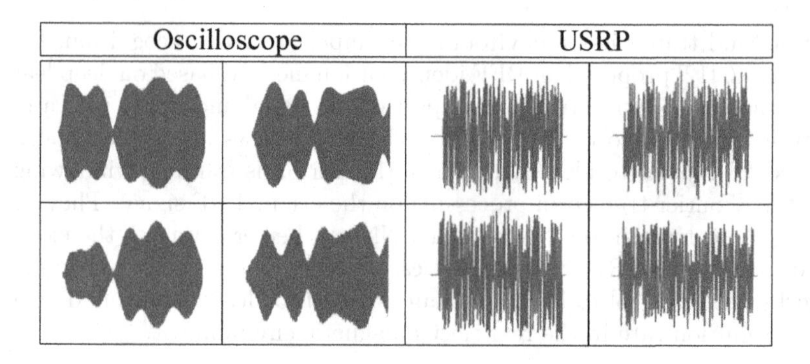

Fig. 2. Four types of RF signals after preprocessing.

realistic environments (Fig. 1). The waveforms are shown in Fig. 2. The sampling frequency and signal frequency band are 20 GHz and 1.79 GHz, respectively. All RF signals have been preprocessed. The preprocessing process mainly includes the detection and normalization of the starting point. The purpose is to synchronize the starting point of the RF signal and reduce the interference of incorrect data in the subsequent model training process. On the other hand, the data normalization also reduces the difference in the collection distance.

3.2 Security Authentication Using RF Signals

Using RFF for identity verification in smart grids is similar to that of traditional fingerprints. When the device enters the network, a certain signal is sampled first to obtain the device's RFF and record it in the database, after that, checking the RFF of all signals received in real-time with the RFF stored in the RFF library to determine whether the received signal is legal or not to avoid the device password damage to the IoT caused by hacking or identity theft. As shown in Fig. 1, the authentication process is as follows:

- Collect the RFF of all legal smart grid devices and send them to the certificate authority (CA) for registration. CA uses a deep learning model to extract the features of the RFF of the device, and then the features are encrypted and stored in the database.
- CA monitors the RFF signals sent by all access devices in real-time and uses the deep learning model to calculate their feature vectors $R_v^{'}$ in real-time, and compares it with the feature fingerprint vectors R_v in the database. If the distance between $R_v^{'}$ and R_v is below the threshold we set, the authentication is successful and the node is a legitimate access node. If there is a malicious counterfeit access node, the distance between the R_v sent by it and the $R_v^{'}$ extracted in real-time is relatively large, so the malicious nodes can be effectively distinguished.

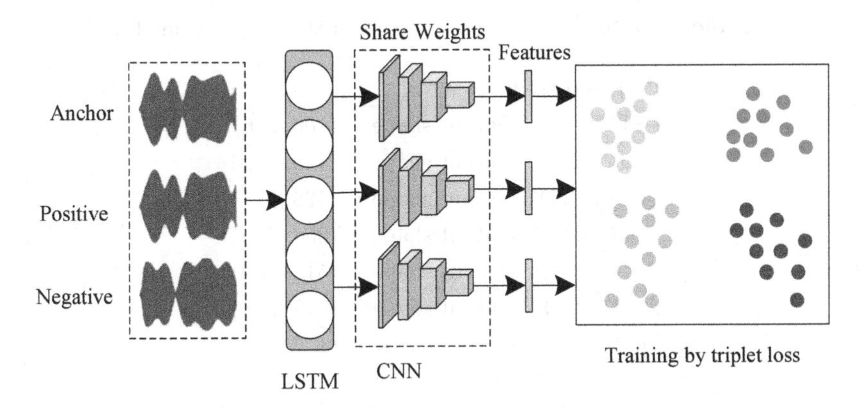

Fig. 3. Deep learning model we proposed. Each convolutional layer block also includes normalization and the rectified linear unit (ReLU) layers.

3.3 Deep Learning Model

For achieving physical fingerprint identification and authentication using wireless signals in the smart grid, many problems of traditional wireless signal identification need to be overcome. On the one hand, traditional recognition methods should establish a reliable and complete database for the data in advance. However, in the smart grid, due to the complex wireless environment and the number of nodes that keep growing, there is a problem of multi-target open recognition. It is impossible to use the closed database for identification and verification. Therefore, it is more difficult to establish a recognition database for signals. On the other hand, traditional RF signal identification methods need to select distinctive feature values as the feature fingerprint of the device according to different wireless signals [17–20], which require calculating and selecting characteristic values manually. In order to overcome the limitations of traditional clustering algorithms, in our paper, we propose a feature extraction and recognition model for wireless signals (as shown in Fig. 3 and Table 1). It is based on deep metric learning. Our model consists of LSTM and CNNs, which can adaptively capture features of RF signals. The specific structure will be described in detail below.

LSTM is a special kind of recurrent neural network, mainly to solve the problem of gradient disappearance and gradient explosion in the process of recurrent neural network (RNN) training. Due to the advantages of its unit gate design, LSTM can utilize and learn the dynamic information of process data. In a basic LSTM network architecture, given an input sequence represented by $[x_1, x_2, \ldots, x_T]$, the output sequence y_t of the LSTM can be calculated iteratively by Eqs. (1) and (2) for $t = 1, \ldots, T$:

$$h_t = \mathrm{LSTM}\left(h_{t-1}, x_t; W\right) \tag{1}$$

$$y_t = W_{hy} h_t + b_y \tag{2}$$

Table 1. Output of the various layers of the proposed model.

Proposed model	
LSTM	Output shape = (100, 576)
Conv2D	Output shape = (128, 14, 14)
MaxPool2D	Output shape = (128, 7, 7)
Res Conv2D	Output shape = (256, 7, 7)
Res Conv2D	Output shape = (512, 4, 4)
AvgPool2D	Output shape = (512, 2, 2)
Linear	Output shape = (1, 128)

Where W represents a different weight matrix, b_y represents the bias vector for the output y_t. h represents the hidden state. In the cell function LSTM(·), the hidden state is determined by the input gate i, forget gate f, output gate o. The cell state c via the following equations:

$$i_t = \sigma\left(W_{xi}x_t + W_{hi}h_{t-1} + W_{ci}c_{t-1} + b_i\right) \tag{3}$$

$$f_t = \sigma\left(W_{xf}x_t + W_{hf}h_{t-1} + W_{cf}c_{t-1} + b_f\right) \tag{4}$$

$$c_t = f_t c_{t-1} + i_t \tanh\left(W_{xc}x_t + W_{hc}h_{t-1} + b_c\right) \tag{5}$$

$$o_t = \sigma\left(W_{xo}x_t + W_{ho}h_{t-1} + W_{co}c_t + b_o\right) \tag{6}$$

$$h_t = o_t \tanh\left(c_t\right) \tag{7}$$

Where W_{ab} is the weight matrix from layers a to b; $\sigma(\bullet)$ denotes the sigmoid activation function; each b term with a subscript is the bias vector for the appropriate layer.

LSTM has complex dynamic characteristics which can capture the dependencies between sequences and is more suitable for feature extraction of one-dimensional signals. We use one LSTM layer as an encoder to extract the features of RF signals. However, though LSTM can bring some advantages, deep LSTM is difficult to train, so the features encoded by LSTM are sent to the CNNs for training. The CNN we used is ResNet [21] to solve the problem of the disappearance of gradients in the backpropagation process through the shortcut connection design. Thus the neural network can achieve continuous performance improvement when deepening the number of network layers.

Our model is trained by triplet loss [22]. Denote by (x_i^a, x_i^p, x_i^n) the three input RF signals forming the $i-th$ triplet, where x_i^a and x_i^p represent an anchor sample and a positive sample from the same individual, while x_i^n is a negative sample from a different individual. The triple loss ensures that a signal (anchor) is closer to all other signals (positive) than it is to any signal (negative) which is defined as Eq. (8).

$$L_t = \sum_{i}^{N} \left[d_i^{a,p} - d_i^{a,n} + \alpha\right]_+ \tag{8}$$

Where $[\cdot]_+$ means that when the calculated value is greater than zero, the loss is the result of the calculation, and if the calculated value is less than zero, the loss value is zero. After training, we can judge whether the unknown device is legal based on the metric difference.

The original triple loss mainly considers the relative distance between samples in different individuals. However, since this loss function does not stipulate how close the pair (x_i^a, x_i^p) should be, as a consequence, RF signals belonging to the same individual may form a large cluster with a relatively large average intra-class distance in the learned feature space. In the scene of wireless communication, because the environment is complicated and the difference between two RF signals belonging to the same individual is smaller, further consideration is needed to pull the RF signals of the same individual closer, and at the same time push the RF signals belonging to different individuals farther from each other in the learned feature space. Therefore, the improved triplet loss function is proposed, which is defined as Eq. (9):

$$L_t = \sum_i^N \left(d_i^{a,p} + [d_i^{a,p} - d_i^{a,n} + \alpha]_+\right) \tag{9}$$

We add the distance of the positive sample to the loss function so that the training result can not only distinguish the distribution of the positive and negative sample pairs in the feature space, but also ensure that the absolute distance between the positive samples is smaller and speed up the convergence of the network speed. It provides a more effective solution for the feature extraction and recognition of the wireless signal of subsequent nodes.

4 Experiments

4.1 Security Authentication

We use the above four kinds of RF signals collected by the oscilloscope to train our model. We use traditional machine learning algorithms support vector machines (SVM) [23], k-Nearest Neighbors (KNN) [24] and deep learning model LSTM, ResNet18 for comparison. Above two traditional machine learning algorithms are widely used in the field of signal recognition [25–27]. At the same time, because the wireless environment for smart grid certification is more complex, we have added varying SNRs to the signal to verify the anti-interference performance of the algorithm. Figure 4 shows the experimental results of multiple classification tasks with different models. (a) uses the signals collected by an oscilloscope, while (b) uses the signals collected by the USRP. Our experiments show that our model can achieve better performance in RFF identification tasks, and the anti-interference is stronger than SVM and KNN. What's more, compared with ResNet18 and LSTM, the advantages of the joint model of the two are more obvious.

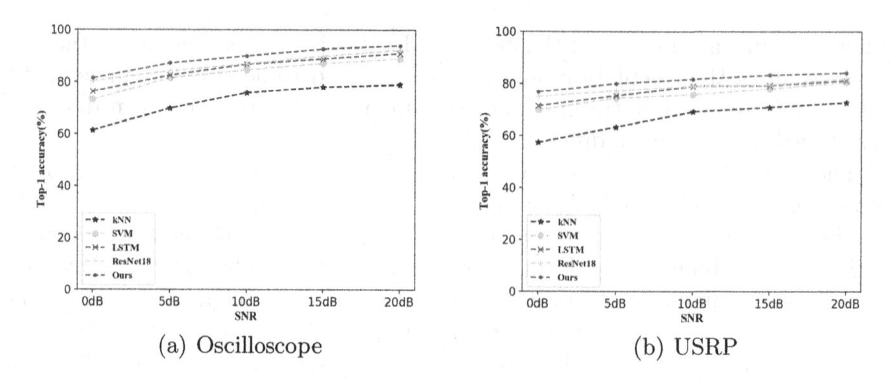

(a) Oscilloscope (b) USRP

Fig. 4. Recognition results under different SNRs.

4.2 Open-Set Recognition

In the above experiment, we assume that all signals are known, so all categories can be included in the training data to build a classifier. However, in actual smart grid security authentication tasks, since illegally connected devices are of unknown type, a classifier based on known signals may not be able to accurately detect illegal signals. Therefore, in this section, we regard any three types of RF signals as known classes and the remaining ones as an unknown class. A USRP is placed in the certification center as a receiving device. In the experiment, we use the triple loss to train the deep learning model to perform open-set recognition.

Figure 5 depicts the results of our experiment. Our method can complete the open class recognition task very well. Specifically, when SNR is equal to 0dB, the Sensitive and Specificity of the model are above 70%, and when the SNR is higher than 20 dB, our recognition rate can reach above 90%. Therefore, our method can effectively determine unknown illegal devices.

We also reproduce [28,29], which could achieve open-set identification for comparison. [28] applied minimum covariance determinant (MCD) and k-means clustering methods at the outputs of the signal classifier's convolutional layers to achieve high accuracy in classifying signals with unknown jamming signals. In [29], an open-set classification model based on a generative adversarial network (GAN) is employed to realize RF signal identification under the open-set. Based on the Table 2, it can be found that the advantages of deep metric learning are more obvious than others. What's more, our model is more suitable for triple loss because of the LSTM-CNN structure.

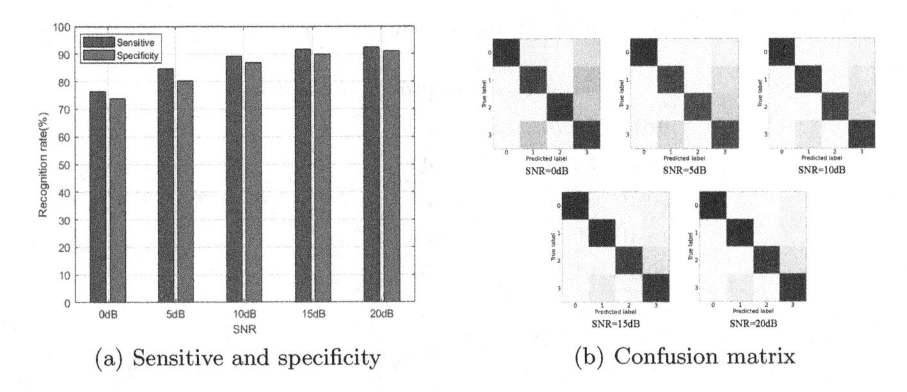

(a) Sensitive and specificity

(b) Confusion matrix

Fig. 5. Open-set recognition results.

Table 2. Experimental result (%)

SNRs	0 dB	5 dB	10 dB	15 dB	20 dB
Model [28]	65.61	69.78	73.52	75.15	77.31
Model [29]	68.43	74.81	76.96	78.29	80.04
LSTM	71.76	81.41	85.13	87.78	88.94
ResNet18	72.55	81.67	86.64	88.97	90.16
Ours	74.76	82.37	88.56	90.81	91.24

4.3 Improved Triplet Loss

In this section, we verify the performance of the improved triplet loss. Table 3 details the recognition results. It can be seen that the improved triple loss can improve the recognition performance.The lower the SNR, the more obvious the performance improvement. We believe that this is because the improved triplet loss makes the same classes closer and the interval between different classes larger, so it can reduce the interference of noise.

In order to further verify the performance, we perform an experiment for better visualizing on features. Figure 6 presents the feature distributions in Euclidean space, it can be seen that though the original triplet loss can distinguish different classes, the intra class features are still dispersed while the improved triplet loss performs much better. This kind of visualization only provides a visual explanation, and the difference between classes may be more obvious in a high-dimensional space.

Table 3. Top-1 accuracy (%)

	Triplet loss	Improved triplet loss
0 dB	74.76	77.62
5 dB	82.37	85.49
10 dB	88.56	90.04
15 dB	90.81	91.56
20 dB	91.24	92.94

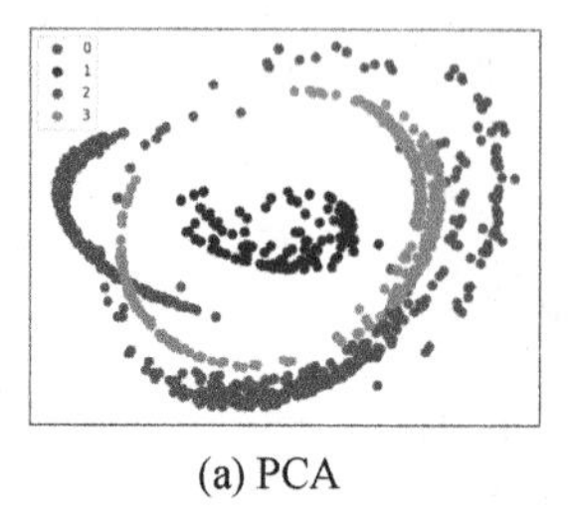

(a) PCA

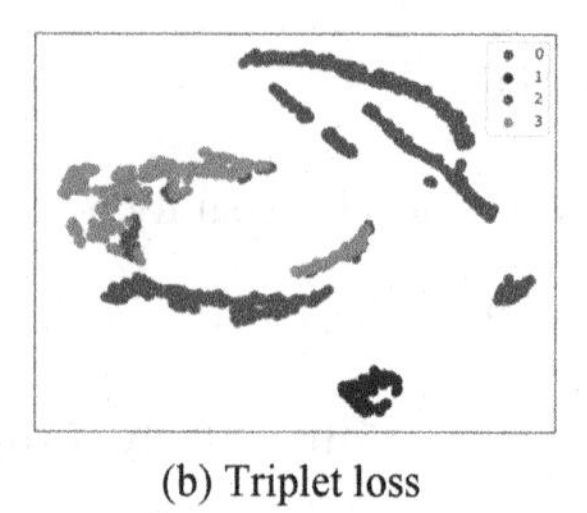

(b) Triplet loss

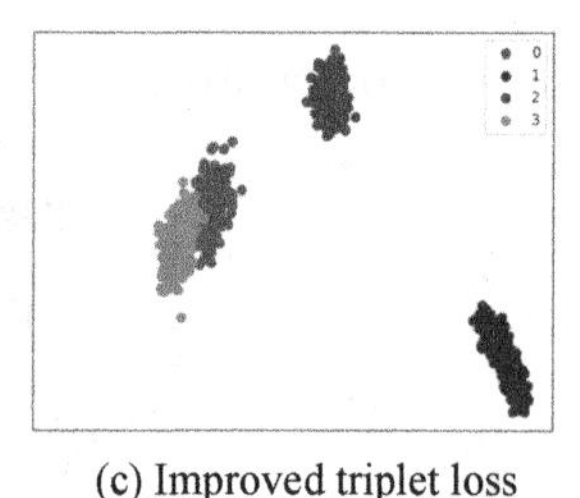

(c) Improved triplet loss

Fig. 6. The feature distributions in Euclidean space.

4.4 Real-Time Analysis

The authentication of smart grid needs to meet real-time requirements. We conducted experiments on the same server and compared the inference time of different algorithms, as shown in Table 4. Because KNN needs to traverse the dataset for each test sample and SVM needs to recalculate many times in order to find the convergent hyperplane, so their speed is much slower than that of deep learning model. We use a layer of LSTM instead of multi-layer convolution, so our model is lighter and faster than ResNet18.

Table 4. Average inference time (ms) for a single sample.

Method	Oscilloscope	USRP
SVM	31	23.6
KNN	95.2	43.9
ResNet18	6.6	5.4
Ours	**4.7**	**3.1**

5 Conclusion

Aiming at the problem of smart grid security certification, we innovatively propose a security certification method based on RFF technology, which solves the shortcoming of the high cost of encryption algorithms. Not only that, we construct a lightweight neural network composed of LSTM and ResCNN. The proposed model accelerates inference speed and performs well on RFF identification tasks. We also proposed an improved triplet loss for open-set recognition. Improved triplet loss pulls the signals of the same class closer, thus getting higher accuracy. The experiments proved the effectiveness and universality of our method. In the future, we will continue to optimize these ways to apply to different types of RF signals.

Acknowledgment. This work was supported in part by the National Natural Science Foundation of China under Grant No. 61971368, in part by the Natural Science Foundation of Fujian Province of China No. 2019J01003, in part by the National Natural Science Foundation of China under Grant No. U20A20162, in part by the National Natural Science Foundation of China (grant number 61731012).

References

1. Meng, F., Ye, W.: Summary of research on security and privacy of Smart Grid. In: 2020 International Conference on Computer Communication and Network Security (CCNS), pp. 39–42 (2020)
2. Chen, B., Chen, L., Tan, Y., Kuang, X.: Investigations on communication and management techniques for electric Internet of Things applications in Smart Grid. In: 2021 China International Conference on Electricity Distribution (CICED), pp. 515–518 (2021)
3. Yan, Y., Qian, Y., Sharif, H., Tipper, D.: A survey on Smart Grid communication infrastructures: motivations, requirements and challenges. IEEE Commun. Surv. Tutor. **15**(1), 5–20 (2013)
4. Lee, S., Bong, J., Shin, S., et al.: A security mechanism of Smart Grid AMI network through smart device mutual authentication. In: International Conference on Information Networking (ICOIN), Phuket, Thailand, pp. 592–595 (2014)
5. Zhang, L., Tang, S., Jiang, Y., et al.: Robust and efficient authentication protocol based on elliptic curve cryptography for Smart Grids. In: 2013 IEEE International Conference on Green Computing and Communications and IEEE Internet of Things and IEEE Cyber, Physical and Social Computing (GreenCom-iThings-CPSCom), Beijing, China, pp. 2089–2093 (2013)
6. Sha, K., Alatrash, N., Wang, Z.: A secure and efficient framework to read isolated Smart Grid devices. IEEE Trans. Smart Grid **8**(6), 2519–2531 (2017)
7. Fouda, M.M., Fadlullah, Z.M., Kato, N., et al.: A lightweight message authentication scheme for Smart Grid communications. Comput. Electr. Eng. **2**(4), 675–685 (2016)
8. Hall, J., Barbeau, M., Kranakis, E.: Detection of transient in radio frequency fingerprinting using signal phase. In: Proceedings of the IASTED International Conference on Wireless and Optical Communications, Banff, Canada, pp. 13–18 (2003)
9. Zhuo, F., Huang, Y., Chen, J.: Specific emitter identification based on linear polynomial fitting of the energy envelope. In: Proceedings of 2016 IEEE 6th International Conference on Electronics Information and Emergency Communication (ICEIEC), Beijing, China, pp. 278–281 (2016)

10. Kennedy, I.O., Scanlon, P., Buddhikot, M.M.: Passive steady state RF fingerprinting: a cognitive technique for scalable deployment of co-channel Femto cell underlays. In: 2008 IEEE Symposium on New Frontiers in Dynamic Spectrum Access Networks (DySPAN), Chicago, IL, United States, pp. 1–12 (2008)
11. Peng, L., Hu, A., Jiang, Y., Yan, Y., Zhu, C.: A differential constellation trace figure based device identification method for ZigBee nodes. In: 2016 8th International Conference on Wireless Communications and Signal Processing (WCSP), Yangzhou, China, pp. 1–6 (2016)
12. Ding, L., Wang, S., Wang, F., Zhang, W.: Specific emitter identification via convolutional neural networks. IEEE Commun. Lett. **22**(12), 2591–2594 (2018)
13. Xiao, L., Li, Y., Han, G., et al.: PHY-layer spoofing detection with reinforcement learning in wireless networks. IEEE Trans. Veh. Technol. **65**(12), 10037–10047 (2016)
14. Zong, L., Xu, C., Yuan, H.: A RF fingerprint recognition method based on deeply convolutional neural network. In: Proceedings of 2020 IEEE 5th Information Technology and Mechatronics Engineering Conference (ITOEC), Chongqing, China, pp. 1778–1781 (2020)
15. Ma, J., Liu, H., Peng, C., et al.: Unauthorized broadcasting identification: a deep LSTM recurrent learning approach. IEEE Trans. Instrum. Meas. **69**(9), 5981–5983 (2020)
16. Zaremba, W., Sutskever, I., Vinyals, O.: Recurrent neural network regularization. arXiv preprint arXiv:1409.2329 (2014)
17. Luo, W., et al.: A RFF access authentication technology based on K-nearest neighbor method. In: 2019 International Conference on Intelligent Computing, Automation and Systems (ICICAS), pp. 852–855 (2019)
18. Xing, Y., Hu, A., Zhang, J., Yu, J., Li, G., Wang, T.: Design of a robust radio-frequency fingerprint identification scheme for multimode LFM radar. IEEE Internet Things J. **7**(10), 10581–10593 (2020)
19. Yuan, Y., Huang, Z., Wu, H., Wang, X.: Specific emitter identification based on Hilbert-Huang transform-based time-frequency-energy distribution features. IET Commun. **8**(13), 2404–2412 (2014)
20. Nouichi, D., Abdelsalam, M., Nasir, Q., Abbas, S.: IoT devices security using RF fingerprinting. In: 2019 Advances in Science and Engineering Technology International Conferences (ASET), pp. 1–7 (2019)
21. He, K., Zhang, X., Ren, S., Sun, J.: Deep residual learning for image recognition. In: Proceedings of the IEEE Computer Society Conference on Computer Vision and Pattern Recognition (CVPR), Las Vegas, NV, United States, pp. 770–778 (2016)
22. Schroff, F., Kalenichenko, D., Philbin, J.: FaceNet: a unified embedding for face recognition and clustering. In: Proceedings of the IEEE Computer Society Conference on Computer Vision and Pattern Recognition (CVPR), Boston, MA, United States, pp. 815–823 (2015)
23. Suykens, J.A.K., Vandewalle, J.: Least squares support vector machine classifiers. Neural Process. Lett. **9**(3), 293–300 (1999). https://doi.org/10.1023/A:1018628609742
24. Keller, J.M., Gray, M.R., Givens, J.A.: A fuzzy k-nearest neighbor algorithm. IEEE Trans. Syst. Man Cybern. **SMC-15**, 580–585 (1985)
25. Reising, D., Cancelleri, J., Loveless, T.D., et al.: Radio identity verification-based IoT security using RF-DNA fingerprints and SVM. IEEE Internet Things J. **8**(10), 8356–8371 (2020)

26. Wang, X., Zhang, Y., Zhang, H., Wei, X., Wang, G.: Identification and authentication for wireless transmission security based on RF-DNA fingerprint. EURASIP J. Wirel. Commun. Netw. **2019**(1), 1–12 (2019). https://doi.org/10.1186/s13638-019-1544-8
27. Ezuma, M., Erden, F., Anjinappa, C.K., et al.: Micro-UAV detection and classification from RF fingerprints using machine learning techniques. In: IEEE Aerospace Conference Proceedings, Big Sky, MT, United States, pp. 1–13 (2019)
28. Shi, Y., Davaslioglu, K., Sagduyu, Y.E., et al.: Deep learning for RF signal classification in unknown and dynamic spectrum environments. In: 2019 IEEE International Symposium on Dynamic Spectrum Access Networks (DySPAN), Newark, NJ, United States, pp. 1–10, November 2019
29. Zhao, C., Shi, M., Cai, Z., et al.: Research on the open-categorical classification of the Internet-of-things based on generative adversarial networks. Appl. Sci. **8**(12), 2351 (2018)

SEPoW: Secure and Efficient Proof of Work Sidechains

Taotao Li[1,2(✉)], Mingsheng Wang[1,2], Zhihong Deng[1], and Dongdong Liu[1,2]

[1] State Key Laboratory of Information Security, Institute of Information Engineering, Chinese Academy of Sciences, Beijing 100093, China
{litaotao,wangmingsheng,dengzhihong,liudongdong}@iie.ac.cn
[2] School of Cyber Security, University of Chinese Academy of Sciences, Beijing 100049, China

Abstract. Since the advent of sidechains in 2014, they have been acknowledged as the key enabler of blockchain interoperability and upgradability. However, sidechains suffer from significant challenges such as centralization, inefficiency and insecurity, meaning that they are rarely used in practice. In this paper, we present SEPoW, a secure and efficient sidechains construction that is suitable for proof of work (PoW) sidechain systems. The drawbacks for the centralized exchange of cross-chain assets in the participating blockchains are overcome by our decentralized SEPoW. To reduce the size of a cross-chain proof, we introduce merged mining into our SEPoW such that the proof consists of two Merkle tree paths regardless of the size of the current blockchain. We prove that the proposed SEPoW achieves the desirable security properties that a secure sidechains construction should have. As an exemplary concrete instantiation we propose SEPoW for a PoW blockchain system consistent with Bitcoin. We evaluate the size of SEPoW proof and compare it with the state-of-the-art PoW sidechains protocols. Results demonstrate that SEPoW achieves a proof size of 416 bytes which is roughly 123×, 510× and 62000× smaller than zkRelay proof, PoW sidechains proof and BTCRelay proof, respectively.

Keywords: Sidechains · Merged mining · Decentralized construction · Succinct proof · Proof of work

1 Introduction

Since blockchain technology was introduced by Satoshi [1] in 2008, various blockchains with different characteristics and their applications have been gaining increasing attention and adoption by communities. However, blockchains suffer from several fundamental open questions [2], as organized below: (i) Interoperability: How can assets or other data be interoperated and transferred among different blockchains? (ii) Upgradability: How can a new functionality and an implementation problem, e.g., smart contract and transaction malleability [3], be supported and corrected in a deployed blockchain?

Y. Lai et al. (Eds.): ICA3PP 2021, LNCS 13157, pp. 376–396, 2022.
https://doi.org/10.1007/978-3-030-95391-1_24

Fortunately, a major approach to overcoming the above-raised questions is to use *sidechains* [4], which are a technology that can allow different blockchains to communicate with each other and react to events taking place in the other as desired. At present, sidechains have two forms. The former is parallel chains. This means that any two chains, for example, Bitcoin [1] and Ethereum [5], are treated as equals; any of them can be the sidechain of the other. The latter is parent-child chains. In this case, a sidechain can be a "child" of a parent chain; the child chain is somehow bootstrapped from the parent chain.

There are, however, concerns regarding the deficiency of security and efficiency in existing constructions for proof of work sidechains [2,4,6–13], which are explained as below:

- Some existing constructions for sidechains are highly centralized making them resistant to change, vulnerable to attacks and failures [14–16]. For example, trusted intermediaries on the involved blockchains are responsible for maintaining and managing the exchange of cross-chain assets, which will inevitably lead to a single-point-failure. Moreover, centralized constructions cast a major contradiction to the decentralized nature of blockchain.
- Cross-chain proof used for testifying the validity of cross-chain assets, consisting of a list of block headers that increases linearly with the size of the entire blockchain as well as a cryptographic proof, is fairly large, which reduces the efficiency of the constructions and limits potential scalability development.
- About security properties, many constructions for sidechains lack formal definition and rigorous proof. The security of cross-chain assets transfer can not be guaranteed. Even worse, when a sidechain is corrupted, another sidechain is unable to avoid the damage caused by that compromised sidechain.

Those challenges motivate our work.

1.1 Our Contributions

We present SEPoW, a secure and efficient construction that is suitable for proof of work sidechain systems, to complement deficiencies in the security and efficiency in existing constructions for proof of work sidechains.

In this work, we show that our SEPoW is decentralized and allows bidirectional communication between proof of work blockchains without trusted intermediaries. This means that the exchange of all cross-chain assets is managed not by a centralized party but by all honest nodes in participating blockchains.

To reduce the size of a cross-chain proof, we introduce the merged mining mechanism into our SEPoW such that a public chain of block headers is shared by the participating blockchains. Exploiting this, the proof consists of two Merkle tree paths: a transaction Merkle tree path and a merged Merkle tree path; their sizes depend on the number of transactions included in a block and the number of participating blockchains, respectively, regardless of the size of the current blockchain.

Next, we prove the security of SEPoW using a secure cross-chain proof protocol and a collision resistant hash function. Our SEPoW captures: (i) cross-chain assets can be transferred securely when the security assumptions of the participating blockchains are held, namely that an honest majority of computational power exists in the participating blockchains, and (ii) the firewall property maintained by SEPoW guarantees that any catastrophic blockchain corruption, such as a violation of the blockchain security assumptions, does not impact other blockchains.

Further, we present a concrete exemplary construction for proof of work sidechains. For conciseness our SEPoW is outlined with regard to a generic proof of work (PoW) blockchain consistent with the Bitcoin protocol [1] that underlies the Bitcoin blockchain, which is the most popular PoW blockchain so far.

We also evaluate the size of SEPoW proof and compare it with the state-of-the-art PoW sidechains protocols. Results demonstrate that SEPoW achieves a proof size of 416 bytes which is roughly 123x, 510x and 62000x smaller than zkRelay proof [12], PoW sidechains proof [7,17] and BTCRelay proof [6] for the blockchain length of 300000 and a 1 MBytes block.

1.2 Organization

The rest of this paper is organized as follows. In Sect. 2, preliminaries of SEPoW are introduced. We provide an exemplary concrete construction for proof of work sidechains in Sect. 3. The security of SEPoW and the discussion of merged mining are proved and presented in Sect. 4 and Sect. 5, respectively. In Sect. 6, we evaluate the size of SEPoW proof and compare it with the state-of-the-art work. Related works are proposed in Sect. 7. We conclude this paper in the last section.

2 Preliminaries

2.1 Cross-Chain Proof Protocol

We introduce a cross-chain proof protocol [17] that attests to the validity of cross-chain transactions into SEPoW. In this protocol the prover is a miner or full node on a chain denoted by $\mathcal{C}_1$; the verifier does not have access to $\mathcal{C}_1$, but holds a list of block headers on the chain denoted by $\mathcal{C}_2$ he/she participates in. The prover wants to convince the verifier that the predicate (i.e., the transaction tx is confirmed in $\mathcal{C}_1$) is true by sending a valid proof.

The definition of a cross-chain proof protocol is given below.

Definition 1 (Cross-chain proof protocol). *A cross-chain proof protocol for a predicate* Q *(i.e., an event occurred on blockchain) is a pair of probabilistic polynomial time (PPT) algorithms (P, V) such that:*

- ***P.*** *The algorithm takes as input a full chain $\mathcal{C}$, and outputs proof π about the predicate* Q. *We note this as $\pi \leftarrow P\ (\mathcal{C})$.*
- ***V.*** *The algorithm takes as input a proof π produced by an honest node or a malicious node, and outputs a decision $d \in \{T, F\}$. We note this as $d \leftarrow V\ (\pi)$. The predicate* Q *is true if $V(\pi) = T$ and false otherwise.*

A desired security property of a cross-chain proof protocol is shown as follows.

Definition 2 (Security). *A cross-chain proof protocol for a predicate* Q *is secure if for all environments and for all PPT adversaries* $\mathcal{A}$ *and for all rounds* $r \geq \eta k$, *if* V *receives a set of proofs* Π *at the beginning of round* r, *at least one of which has been generated by the honest prover* $\mathcal{P}$, *then the output of* V *at the end of round* r *has the following constraints:*

- *If the output of* V *is false, then the evaluation of* Q *for all honest nodes must be false at the end of round* $r - \eta k$.
- *If the output of* V *is true, then the evaluation of* Q *for all honest nodes must be true at the end of round* $r + \eta k$.

Note that the parameter η represents the rate at which new blocks are produced and k is the number of subsequent blocks. See more details in [17].

2.2 Merged Mining

To generate a succinct cross-chain proof for a predicate, we introduce a merged mining mechanism [18, 19] that allows a miner to produce multiple blocks at different chains with a single PoW solution. Exploiting this, a public chain of block headers, produced and maintained by the miners running the merged mining, is shared by the participating sidechains. Thus, the proof is only two Merkle tree paths: a transaction Merkle tree path and a merged Merkle tree path. As an exemplary concrete instantiation, in Sect. 3.1.4 we describe how a merged mining mechanism is useful for generating a succinct proof for a predicate.

Miners are allowed to perform merged mining for all involved sidechains based on their mining power. The process of the merged mining is as follows.

- Step 1 (Build merged-block-header). Miners will verify transactions for all involved c sidechains (an efficient way that verifies all transactions without monitoring all sidechains is given in Sect. 5) and collect c transaction Merkle tree root r_i, i.e., part **C** in Fig. 1a. Then, the miners calculate the root mr (part **B** in Fig. 1a) of the merged Merkle tree over the leaves encoded from the following TxRoots:
$$(r_1, r_2, ..., r_c). \tag{1}$$
- Step 2 (Search for nonce). Without merged mining, similar to Bitcoin, miners are required to search for n nonce ne that each satisfies
$$\mathrm{H}(pub, r, ne) \leq \mathrm{T}. \tag{2}$$
Where pub denotes public block parameters described in part **A** in Fig. 1a. With merged mining, miners only need to search for a nonce ne that satisfies
$$\mathrm{H}(pub, mr, ne) \leq \mathrm{T}. \tag{3}$$
Here we assume the difficulty target T is the same for all involved sidechains.

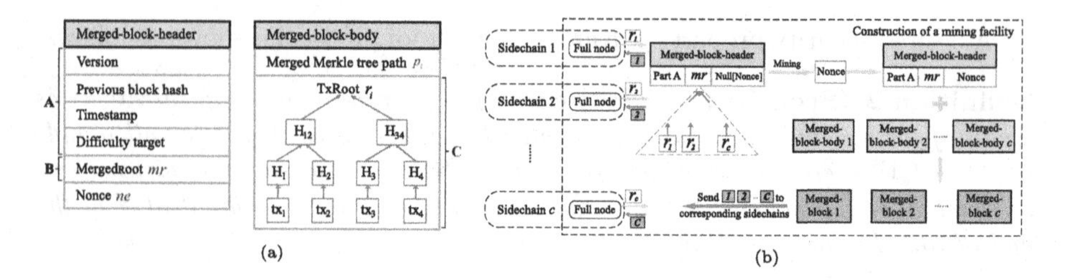

Fig. 1. The construction of a merged-block (a) and a mining facility (b).

- Step 3 (Diffuse block). Upon finding a valid ne, the miner will compose sidechain-specific merged-blocks and send them to corresponding sidechain networks, as described in Fig. 1b. It is worth pointing out that a merged-block-header is shared by these merged-blocks. As shown in Fig. 1b, for each sidechain, a merged-block includes a merged-block-header and a merged-block-body. A merged Merkle tree path p connecting r to mr is used to verify whether r is tied in the corresponding merged-block-header.
- Step 4 (Extend blockchain). In each sidechain, miners and full nodes verify the merged-block according to the specification of a block. Then the sidechain-specific merged-block will be appended in the corresponding sidechain if it is considered valid.

2.3 Security Definition of Sidechains

The first security definition of sidechains was introduced by Gaži et al. [2]. In a secure sidechains system, two fundamental properties, persistence and liveness, are necessary because a robust public transaction ledger must satisfy persistence and liveness [20]. Especially, a critical security feature that a secure sidechains system should have is the firewall property in which any catastrophic chain corruption, such as a violation of the chain security assumptions, does not impact other chains.

A formal security definition of sidechains is presented as follows.

Definition 3 (Sidechains security). *A system-of-sidechain ledgers protocol* Π *for* $\{\mathbf{L}_i\}_{i\in[c]}$ *is pegging-secure with liveness parameter* $u \in \mathbb{N}$ *with respect to:*

- *a set of assumptions* $\mathbb{A}_i$ *for ledgers* $\{\mathbf{L}_i\}_{i\in[c]}$,
- *a merge mapping* $\mathsf{merge}\,(\cdot)$,
- *validity languages* $\mathbb{V}_{\mathsf{A}}$ *for each* $\mathsf{A} \in \bigcup_{i\in[c]} \mathsf{Assets}\,(\mathbf{L}_i)$,

if for all PPT adversaries, all rounds r *and for* $\mathcal{S}_r \triangleq \{i : \mathbb{A}_i[r]\ \text{holds}\ \}$ *we have that except with negligible probability in the security parameter:*

- ***Ledger persistence:*** *For each* $i \in \mathcal{S}_r, \mathbf{L}_i$ *satisfies the persistence property.*
- ***Ledger liveness:*** *For each* $i \in \mathcal{S}_r, \mathbf{L}_i$ *satisfies the liveness property parametrized by* u.

- ***Firewall:*** *For all* $\mathsf{A} \in \bigcup_{i\in\mathcal{S}_r} \mathsf{Assets}\,(\mathbf{L}_i)$

$$\pi_{\mathsf{A}}\left(\mathrm{merge}\left(\{\mathbf{L}_i^{\cup}[r] : i \in \mathcal{S}_r\}\right)\right) \in \pi_{\mathcal{S}_r}\left(\mathbb{V}_{\mathsf{A}}\right).$$

Where $\mathbb{A}_i$ denotes the security assumption of a ledger $\mathbf{L}_i$. For instance, a majority of authority (computational power, stakeholding, or node) is never controlled by the adversary. merge$(\cdot)$ is a function that combines a set of ledger states $\mathsf{ST} = \{\mathbf{L}_1, \mathbf{L}_2, ..., \mathbf{L}_c\}$ into a single ledger state denoted by merge(ST). A concrete instantiation of merge$(\cdot)$ we will give later. For each asset denoted by A, the validity language $\mathbb{V}_{\mathsf{A}}$ can capture specific rules of behavior for A, e.g., an asset A is transferred from a chain to the other chain. Note that $\mathsf{Assets}\,(\mathbf{L})$ is the set of all assets that belong to the ledger $\mathbf{L}$. π is a ledger state projection. Specifically speaking, $\pi_{\mathsf{A}}\,(\mathbf{L})$ denotes the projection of the transactions of $\mathbf{L}$ with respect to the asset A.

2.4 An Example of an Asset A

In this part we describe an example of a fungible asset A, and present the validity language $\mathbb{V}_{\mathsf{A}}$ with respect to the asset A. This example is a modification version based on an asset example in [2].

Instantiating $\mathbb{V}_{\mathsf{A}}$. For validity language $\mathbb{V}_{\mathsf{A}}$ we consider two ledger: the main-chain ledger $\mathbf{L}_1 \triangleq \mathbf{MC}$ and the sidechain ledger $\mathbf{L}_2 \triangleq \mathbf{SC}$. For the asset A, each transaction tx has the form $\mathsf{tx} = ((\mathrm{oAddr}, \textsc{utxo}, \sigma), (\mathrm{dAddr}, a, \pi))$, here:

- oAddr is an origin address on the origin ledger $\mathbf{L}_{\mathsf{ori}}$. Note that a bitcoin address is derived from its public key.
- UTXO is an unspent transaction output. A UTXO represents the unspent amount, and is locked by the private key held by the sender.
- σ is a signature generated by the sender on the metadata ((oAddr, UTXO), (dAddr, a, π)), which is used to unlock a UTXO.
- dAddr is a destination address on the destination ledger $\mathbf{L}_{\mathsf{des}}$. We say either $\mathbf{L}_{\mathsf{ori}} = \mathbf{L}_{\mathsf{des}}$, meaning that tx is a *local transaction*, or $\mathbf{L}_{\mathsf{ori}} \neq \mathbf{L}_{\mathsf{des}}$, meaning that tx is a *cross-chain transfer transaction.*
- a is the transfer amount.
- π is the succinct proof data that validates the validity of a *cross-chain transfer transaction.* Note that π is empty iif tx is a *local transaction* or an *origin transaction.*

Instantiating Merge$(\cdot)$. A merge$(\cdot)$ function takes as input a pair of ledger states (**MC**, **SC**) outlined above, and outputs a single ledger state.

3 Implementing Sidechain Ledger

We provide SEPoW for sidechains that are based on Bitcoin. Our protocol will execute a system of sidechains with sidechain security with respect to Definition 3 under an assumption on honest hashing-power majority. Here our SEPoW adopts the form of parent-child chains.

The main challenge in SEPoW is how to ensure secure cross-chain transfers. To achieve this, we introduce a cross-chain proof protocol described above into SEPoW. Consider two sidechains $\mathcal{C}_1$ and $\mathcal{C}_2$ (the notations $\mathcal{C}_1$ and $\mathcal{C}_2$ will be used throughout the rest of this paper), as well as a predicate (e.g., a cross-chain transaction took place in the sidechain $\mathcal{C}_1$ (resp. $\mathcal{C}_2$)). A cross-chain proof protocol for the predicate means that, the prover of $\mathcal{C}_1$ (resp. $\mathcal{C}_2$) can convince the verifier of $\mathcal{C}_2$ (resp. $\mathcal{C}_1$) that the predicate is true by generating a valid proof.

It is easy to establish *valid but not succinct* cross-chain proof for any computable predicate: the prover provides the entire linearly-growing chain of block headers as proof and the verifier simply selects the longest chain. To address this problem, we also introduce a merged mining mechanism described above into SEPoW, which allows the miners of sidechains to generate multiple blocks at different sidechains with a single PoW solution. Exploiting this, a public chain of block headers is shared by the participating sidechains. In this case, our proof is composed of two Merkle tree paths and thus is succinct.

3.1 The Sidechain Construction

We now present an elaborate module design of SEPoW based on the fundamental building block described above: cross-chain proof protocol and merged mining. These interacting modules are initialization, maintenance, cross-chain transfer and generating cross-chain proof. A graphical depiction about SEPoW is shown in Fig. 2.

3.1.1 The Sidechain Initialization

The initialization of a new **SC** can be launched by any miners of **MC** deploying the configuration transaction that configures **SC** described below. This action only requires the miners to follow the rule in the configuration to support it (i.e., following merged mining). A sidechain that is successfully created will obtain a unique identifier $\mathtt{id}_{\mathbf{SC}}$.

Consider two rounds d_η, s_η on **MC**, as described in Fig. 2. d_η means that the first configuration transaction tx_0 has been included in a block on **MC**. s_η denotes the start time of the sidechain if it is successfully activated. The configuration transaction contains a set of predefined rules that describe how to activate the sidechain and determine s_η successfully. A typical example is as follows: a sidechain starts in the **MC**-round s_η that meets: (i) $\mathrm{s}_\eta - \mathrm{d}_\eta > v_1$, here v_1 denotes the number of round, which is used to determine the round s_η; (ii) at least v_2-majority mining power on **MC** is controlled by honest miners that have supported **SC**.

The process of the activation is as follows (see more details in Algorithm 1).

First, the miners of **MC** that support the sidechain mine a new block including the configuration transaction tx_0 on mainchain. If the sidechain is successfully activated, then during the round s_η the miners of **MC** that support the

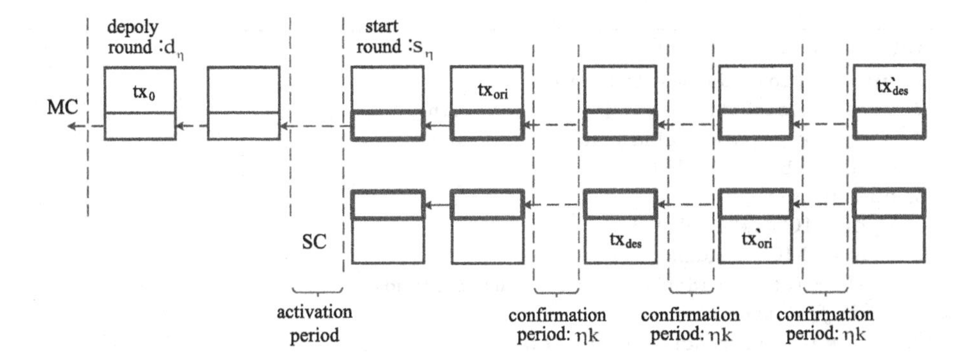

Fig. 2. An overview of SEPoW. **MC** is at the top, while **SC** is at the bottom. d_η is the round of deploying sidechain. s_η is the round of starting sidechain. Solid arrows is used to connect adjacent blocks. Dashed arrows denote some blocks are omitted. Red rectangles denote some merged-block-headers. Transactions tx of interest: tx_0. A configuration transaction; tx_{ori}. An origin transaction carrying withdraw operation from **MC** to **SC**; tx_{des}. A corresponding deposit transaction from **MC** to **SC**; tx'_{ori}. An origin transaction carrying withdraw operation from **SC** to **MC**; tx'_{des}. A corresponding deposit transaction from **SC** to **MC**.

sidechain create a genesis block GB = (merged-block-header $\triangleq$ (*pub*, *mr*, *ne*, $\texttt{id}_{\mathbf{SC}}$), merged-block-body $\triangleq$ (p, r)) for **SC**. Here there are the variables *pub*, *ne*, p, r, as described in Fig. 1a. Note that GB can be created as soon as *ne* is found. In addition, if the activation of the sidechain fails, the initialization of the sidechain is aborted.

To demonstrate that **SC** has been successfully created, the miners of **MC** that adopted **SC** broadcast a special transaction $\texttt{success\_sidechain(id}_{\mathbf{SC}}\texttt{)}$ into **MC**. Otherwise, the miners of **MC** that supported **SC** broadcast another special transaction $\texttt{failure\_sidechain(id}_{\mathbf{SC}}\texttt{)}$ into **MC**. In this case, we will deduce whether the sidechain is valid, according to the transactions of **MC** only.

3.1.2 The Sidechain Maintenance

Once **SC** is successfully created, both **MC** and **SC** will be maintained by miners and their respective full nodes.

For mainchain, its maintenance procedure is executed by **MC**-workers who are acted by the miners running merged mining and the full nodes of **MC** running the proof of work protocol, cf. Algorithm 2. For concreteness, the miners attempt to mine new blocks on top of both the longest $\mathcal{C}_{\mathbf{MC}}$ and $\mathcal{C}_{\mathbf{SC}}$ by running the merged mining mechanism described above, and then broadcast them as soon as their nonce *ne* is found. The full nodes of **MC** only maintain the longest $\mathcal{C}_{\mathbf{MC}}$.

MC-workers, on every new round after the round s_η, receive all the possible **MC**-chains $\mathcal{C}_{\text{MC-}col}$ from the other peers via Diffuse, and then check them to find the "best" chain denoted by $\bar{\mathcal{C}}$. In this case they choose the chain $\bar{\mathcal{C}}$. Adopting the chain $\bar{\mathcal{C}}$ is done for chain validity function (using Check-Chain given in Line 4 of

Algorithm 1. **SC** initialization.

```
1: upon DeploySidechain(id_SC) do
2:     state_sidechain[id_SC] ← initializing
3:     block ← pack configuration_sidechain
4:     mine block on MC
5: end upon
6: upon MC.NewRound() do
7:     s ← MC.RoundIndex()
8:     f1 ≜ (state_sidechain[id_SC]=initialized)
9:     if f1 = true then
10:        if FailureActivation then
11:            state_sidechain ← failed
12:            block ← pack failure_sidechain
13:            mine block on MC
14:        else if SuccessActivation then
15:            state_sidechain ← initialized
16:            block ← pack success_sidechain
17:            s_η ← ActivationRound()
18:            mine block on MC
19:        end if
20:    end if
21:    f2 ≜ (s = s_η)
22:    if SuccessActivation ∧ (f2 = true) then
23:        ne ← search(nonce)
24:        header ← (pub, ne, mr, id_SC)
25:        body ← (p, r)
26:        GM ← (header, body)
27:    end if
28: end upon
```

Algorithm 2. **MC** maintenance.

```
1: upon MC.NewRound do
2:     C_MC-col ← chains collected via Diffuse
3:     ▷ chain validity
4:     C_valchain ← Check-Chain(C_MC-col)
5:     ▷ chain comparison
6:     C̄ ← Check-Comparison(C_valchain)
7:     tx_s ← transactions collected via Diffuse
8:     ▷ transaction validity
9:     tx_valtx ← Check-Tx(tx_s)
10:    ▷ extend chain by invoking merged mining
11:    (B_MC, B_SC) ← Merged-PoW(tx_valtx, C̄)
12:    C̄ ← C̄ || B_MC
13:    C̃ ← receive the best chain w.r.t. SC via Diffuse
14:    C̃ ← C̃ || B_SC
15:    respective networks ← Diffuse(C̄, C̃)
16: end upon
```

Algorithm 2) and chain comparison function (using `Chain-Comparison` given in Line 6 of Algorithm 2), as well as transaction validity function (using `Check-Tx` given in Line 9 of Algorithm 2). Then, the miners try to extend $\bar{\mathcal{C}}$ by running the merged mining (using `Merged-PoW` given in Line 11 of Algorithm 2) described below.

Most importantly, the miners can simultaneously extend $\mathcal{C}_{\mathbf{MC}}$ and $\mathcal{C}_{\mathbf{SC}}$ by invoking `Merged-PoW`. In this case, a public chain of block headers is formed naturally in $\mathcal{C}_{\mathbf{MC}}$ and $\mathcal{C}_{\mathbf{SC}}$. Let us recall the work of `Merged-PoW`. First, the miners collect two valid TxRoot r from $\mathcal{C}_{\mathbf{MC}}$ and $\mathcal{C}_{\mathbf{SC}}$, and then calculate the root mr of the merged Merkle tree over the two TxRoot r. Next, the miners search for a nonce ne that satisfies $\mathrm{H}(pub, mr, ne) \leq \mathrm{T}$ (see more details in Eq. 3). Once ne is found, the miner composes specific-chain merged-blocks, consisting of a public merged-block-header and a specific-chain merged-block-body (as described in Fig. 1a), and sends them (B_{MC}, B_{SC}) to corresponding blockchain networks. Finally, workers of **MC** (resp. **SC**) verify B_{MC} (resp. B_{SC}) and append it to $\mathcal{C}_{\mathbf{MC}}$ (resp. $\mathcal{C}_{\mathbf{SC}}$) if it is considered valid. As a result, a public chain of merged-block-headers is formed naturally in $\mathcal{C}_{\mathbf{MC}}$ and $\mathcal{C}_{\mathbf{SC}}$.

Regarding the sidechain, its maintenance procedure is similar to the mainchain. Hence we only present their differences.

The main difference is that in Algorithm 2, **MC**-workers, all the possible **MC**-chains $\mathcal{C}_{\mathrm{MC}\text{-}col}$, the two new blocks B_{MC} and B_{SC}, as well as all occurrences of the best chain $\tilde{\mathcal{C}}$ with respect to **SC**, are respectively replaced by **SC**-workers, **SC**-chains $\mathcal{C}_{\mathrm{SC}\text{-}col}$, B_{SC} and B_{MC}, as well as the best chain $\mathcal{C}$ with respect to **MC**.

Algorithm 3. Transferring funds from **MC** into **SC**.

```
1: ▷ withdraw operation
2: function Withdraw(oAddr, dAddr, UTXO)
3:     σ ← Sign_sk ((oAddr, UTXO), (dAddr, a))
4:     tx_ori ← ((oAddr, UTXO, σ), (dAddr, a))
5:     mine block carrying tx_ori on MC
6: end function
7: ▷ deposit operation
8: function Deposit(oAddr, dAddr, a)
9:     ▷ tx_ori is included in the stable MC
10:    wait until block buried under k blocks
11:    π ← generate a cross-chain proof about tx_ori
12:    σ ← Sign_sk ((oAddr, UTXO), (dAddr, a, π))
13:    tx_des ← ((oAddr, UTXO, σ), (dAddr, a, π))
14:    mine block carrying tx_des on SC
15: end function
```

3.1.3 Cross-Chain Transfer

Now nodes (or clients) can move funds from **MC** to **SC** by a cross-chain transfer transaction, which consists of a transaction $\mathsf{tx}_{\mathsf{ori}}$ carrying the withdraw operation, and a transaction $\mathsf{tx_{des}}$ carrying the deposit operation. Two of them have the same fields, except for that metadata π for $\mathsf{tx_{ori}}$ is empty and metadata π for $\mathsf{tx_{des}}$ includes a cross-chain proof. The *origin transaction* $\mathsf{tx_{ori}}$ that only involves the state in **MC** is handled by **MC**-workers, while the *destination transaction* $\mathsf{tx_{des}}$ that only involves the state in **SC** is handled by **SC**-workers.

Moving funds from the mainchain **MC** into the sidechain **SC** works as follows. First, a client on **MC** diffuses $\mathsf{tx_{ori}}$ with the desired UTXO and the valid receiving address on **SC**. If $\mathsf{tx_{ori}}$ is considered valid, the corresponding block B carrying $\mathsf{tx_{ori}}$ will be generated by the miner and only be appended to **MC**; the withdraw operation will be executed.

When B has been buried under k blocks within **MC**, **MC**-workers create a cross-chain proof π about the predicate, claiming that $\mathsf{tx_{ori}}$ has been included in the stable mainchain **MC**. Here, π contains a transaction Merkle tree path for $\mathsf{tx_{ori}}$ as well as a merged Merkle tree path p. The construction of π is described below. We now suppose that π has been produced, and then received by the client that will diffuse it.

After that, the corresponding $\mathsf{tx_{des}}$ is composed by the client in **MC** and forwarded to the sidechain **SC**. If $\mathsf{tx_{des}}$ is considered valid, it contains: (i) a valid cross-chain proof π; (ii) a valid signature σ; (iii) sufficient UTXO that is not less than the transfer amount a. If included, the deposit operation will be executed, concluding the completion of transferring from **MC** to **SC**. The core of transferring from **MC** to **SC** is shown in Algorithm 3.

Withdrawing to MC. Clients can then transfer their funds from **SC** back into **MC**. They follow the reverse procedure of Algorithm 3.

3.1.4 Generating Cross-Chain Proof

In the part we present a concrete construction of a cross-chain proof π. Let us consider cross-chain transactions consisting of the origin transaction $\mathsf{tx_{ori}}$ occurring in **MC** and the destination transaction $\mathsf{tx_{des}}$ occurring in **SC**, as well as the predicate Q which claims that $\mathsf{tx_{ori}}$ has been included in block B in the stable **MC**. To maintain the transfer of cross-chain assets for the **SC** verifiers (i.e., full nodes in **SC**) that cannot evaluate Q, a cross-chain proof that deduces Q is essential.

When the block B carrying $\mathsf{tx_{ori}}$ has been buried under k blocks within **MC**, the cross-chain proof π about Q is produced by the provers on **MC**, and contains:

- **The transaction Merkle tree path.** The path for $\mathsf{tx_{ori}}$ is produced from the transaction Merkle tree located in B' body (see more details in Fig. 1a). It testifies that $\mathsf{tx_{ori}}$ is aggregated in the transaction Merkle tree root r.
- **The merged Merkle tree path** $\boldsymbol{p}$. The path is generated from the merged Merkle tree over the pair of (r_1, r_2) of the involved mainchain and sidechain. $\boldsymbol{p}$ connecting the transaction Merkle tree root r to the merged Merkle tree root mr is used to attest that r is tied in B' header.

The above two paths allow the provers to convince the verifiers that Q is true. Concrete space requirements about the proof are discussed in Sect. 6.1.

4 Proofs of Security

In this section, we first prove that SEPoW from Sect. 3 satisfies persistence and liveness, then prove that SEPoW achieves the firewall property, similar to the proof method of Gaži et al. [2].

4.1 Persistence and Liveness

Lemma 1 (Persistence and Liveness). *Consider SEPoW from Sect. 3 with respect to the assumptions $\mathbb{A}_{\mathbf{MC}}$ and $\mathbb{A}_{\mathbf{SC}}$. For all rounds r, if $\mathbb{A}_{\mathbf{MC}}[r]$ (resp. $\mathbb{A}_{\mathbf{SC}}[r]$) holds, then* **MC** *(resp.* **SC***) achieves persistence and liveness up to round r with overwhelming probability in k.*

Proof (sketch). We directly borrow previous work [20,21] to prove that both persistence and liveness hold. In [20] it was shown that the Bitcoin protocol with the honest majority of computational power provides three security properties: common prefix, chain quality and chain growth. Further, According to the work [21–23], persistence and liveness needed by a ledger can be derived from the above three properties. Therefore, SEPoW satisfies persistence and liveness, completing the proof.

4.2 Firewall Property

Lemma 2. *For all PPT adversaries $\mathcal{A}$, SEPoW from Sect. 3 using a secure cross-chain proof protocol and a collision resistant hash function achieves the firewall property with the assumptions $\mathbb{A}_{\mathbf{MC}}$ and $\mathbb{A}_{\mathbf{SC}}$ with respect to overwhelming probability in k.*

Proof (sketch). To illustrate that the firewall property holds, we employ the idea of computational reduction in our proof. The line of the proof is as follows: suppose the firewall property is broken, an insecure cross-chain proof protocol or hash function is used by our protocol; then we show the probability of using the insecure cross-chain proof protocol or hash function is negligible.
We denote by $\mathcal{A}$ an arbitrary PPT adversary attacking the firewall property, and denote by $\mathcal{Z}$ an arbitrary environment supporting the execution of $\mathcal{A}$. We will consider two PPT adversaries:

- $\mathcal{A}_1$ is an adversary attacking the cross-chain proof protocol.
- $\mathcal{A}_2$ is an adversary attacking the hash function.

Next, we start by describing the behavior of these adversaries.

The Adversary $\mathcal{A}_1$bf . First, it models the execution of $\mathcal{A}$. That is, $\mathcal{A}$ requests that a cross-chain proof about the predicate Q that an origin transaction is included in $\mathbf{MC}$ is produced (without loss of generality, in this proof we consider only the cross-chain transfers from $\mathbf{MC}$ to $\mathbf{SC}$ due to the symmetry of SEPoW), $\mathcal{A}_1$ calls its proof generation algorithm $\mathbf{P}$ (as described in Sect. 2.1) to get the corresponding proof to provide to $\mathcal{A}$.

$\mathcal{A}_1$ monitors the ledgers, $\mathbf{L_{MC}}$ and $\mathbf{L_{SC}}$, adopted by honest $\mathbf{MC}$-workers and $\mathbf{SC}$-workers, and for every round r checks the state of all honest $\mathbf{MC}$-workers and $\mathbf{SC}$-workers. To evaluate whether $\mathcal{A}$ has succeeded, the adversary inspects whether $\mathbf{L} = \text{merge}(\mathbf{L_{MC}}, \mathbf{L_{SC}}) \notin \mathbb{V}_{\mathsf{A}}$. If $\mathcal{A}_1$ can not find such a round r and entities $\mathbf{MC}$-workers, $\mathbf{SC}$-workers, it returns FAILURE.

Otherwise it exists a round r such that $\mathbf{L} = \text{merge}(\mathbf{L_{MC}}, \mathbf{L_{SC}}) \notin \mathbb{V}_{\mathsf{A}}$. Suppose that $\mathbf{L}'$ is the prefix of $\mathbf{L}$ that satisfies $\mathbf{L}' \notin \mathbb{V}_{\mathsf{A}}$ and $\mathsf{tx} = \mathbf{L}'[-1]$. If tx has $\mathsf{oAddr}(\mathsf{tx}) \notin \mathbf{MC}$ or $\mathsf{dAddr}(\mathsf{tx}) \notin \mathbf{SC}$, then $\mathcal{A}_1$ returns FAILURE. Otherwise $\mathsf{oAddr}(\mathsf{tx}) \in \mathbf{MC}$ and $\mathsf{dAddr}(\mathsf{tx}) \in \mathbf{SC}$ (and so $\mathsf{tx} \in \mathbf{L_{SC}}$). Therefore, there must exist a predicate Q that the origin transaction (tx') corresponding to tx is committed in $\mathbf{L_{MC}}$ is true.

Let Q^* be the set of $\mathbf{L_{MC}}$ including all predicates up to and containing Q. We will show that Q^* must contain a predicate attested by a forgery cross-chain proof. $\mathcal{A}_1$ inspects every predicate $\mathrm{Q}_r \in \mathrm{Q}^*$. Q_r involves a proof π_r for an origin transaction. $\mathcal{A}_1$ produces a proof π'_r for Q_r based on the view of $\mathbf{SC}$-workers, and examines whether the following *predicate violation* condition holds:

$$(\mathrm{Q}_r \in \mathsf{true}) \wedge (\neg \mathrm{V}(\pi_r) \vee (\pi_r \neq \pi'_r)). \tag{4}$$

Suppose it exists a round r^* that satisfies the condition (4) and then outputs the tuple $(\mathrm{Q}_{r^*}, \pi_{r^*})$. Otherwise $\mathcal{A}_1$ returns FAILURE.

The Adversary $\mathcal{A}_2$. Similar to $\mathcal{A}_1$, $\mathcal{A}_2$ models the execution of $\mathcal{A}$. When $\mathcal{A}$ requests that a proof of a predicate Q is created, $\mathcal{A}_2$ invokes its algorithm $\mathbf{P}$ to get the corresponding proof to provide to $\mathcal{A}$.

$\mathcal{A}_2$ monitors the ledgers, $\mathbf{L_{MC}}$ and $\mathbf{L_{SC}}$, adopted by honest $\mathbf{MC}$-workers and $\mathbf{SC}$-workers, and for every round r checks the state of all honest $\mathbf{MC}$-workers and $\mathbf{SC}$-workers. $\mathcal{A}_2$ checks whether $\mathbf{L} = \text{merge}(\mathbf{MC}, \mathbf{SC}) \notin \mathbb{V}_{\mathsf{A}}$, to evaluate whether $\mathcal{A}$ has succeeded. If the adversary can not find such a round r and entities $\mathbf{MC}$-workers, $\mathbf{SC}$-workers, it returns FAILURE. Suppose tx, tx' are as described in $\mathcal{A}_1$. If tx has $\mathsf{oAddr}(\mathsf{tx}) \notin \mathbf{MC}$ or $\mathsf{dAddr}(\mathsf{tx}) \notin \mathbf{SC}$, then $\mathcal{A}_2$ returns FAILURE. Then the predicate Q that tx' is committed in $\mathbf{L}_1$ is true, and the corresponding cross-chain proof π for tx' was created. If Q is false, then $\mathcal{A}_2$ returns FAILURE. When Q for $\mathbf{SC}$-workers is true, there exists a cross-chain proof π' that attests to tx' in the view of $\mathbf{SC}$-workers. Based on the above results, $\mathcal{A}_2$ finds a collision for hash function and returns it.

Probability Analysis. We define the following events:

- SC-FAILURE$[r]$: $\mathcal{A}$ is successful at round r, i.e., $\pi_{\mathsf{A}}(\text{merge}(\{\mathbf{L}_i[t] : i \in \mathcal{S}_t\})) \notin \pi_{\mathcal{S}_t}(\mathbb{V}_{\mathsf{A}})$.
- CCPP-BREAK: $\mathcal{A}_1$ finds such a round r^* such that the condition (4) holds.
- HASH-BREAK: $\mathcal{A}_2$ finds a collision for the hash function.

Next, we will argue that the probability Pr[SC-FAILURE[r]] is negligible for every time round r. We will demonstrate this probability in three successive claims.

The first claim shows that one of CCPP-BREAK, HASH-BREAK happens if SC-FAILURE[r] happens. As a result, according to a union bound, we have

$$\Pr[\text{SC-FAILURE}[r]] \leq \Pr[\text{CCPP-BREAK}] + \Pr[\text{HASH-BREAK}].$$

The other two claims show that Pr[CCPP-BREAK] and Pr[HASH-BREAK] are negligible.

Claim 1: SC-FAILURE[r] $\Rightarrow$ CCPP-BREAK $\vee$ HASH-BREAK.

By the Lemma 2, there exist two ledgers $\mathbf{L}_{\mathbf{MC}}$ and $\mathbf{L}_{\mathbf{SC}}$. Thus SC-FAILURE[r] is meant to

$$\text{merge}(\{\mathbf{L}_{\mathbf{MC}}[r], \mathbf{L}_{\mathbf{SC}}[r]\}) \notin \mathbb{V}_{\mathsf{A}}.$$

Without loss of generality, suppose that tx, tx' and Q are as described in $\mathcal{A}_1$. By the Lemma 2 and the Lemma 2 from [2], we deduce that tx is a *destination transaction*, and $\mathsf{oAddr}(\mathsf{tx}) \in \mathbf{MC}$ and $\mathsf{dAddr}(\mathsf{tx}) \in \mathbf{SC}$. Thus, Q is true. If $\mathcal{A}_1$ discovers such a round r^* that satisfies the condition (4), then CCPP-BREAK has happened and the claim holds. Therefore, for each predicate Q_r involving a proof π_r, the following condition holds:

$$(\mathrm{Q}_r \in \mathsf{true}) \wedge \mathrm{V}(\pi_r) \Rightarrow (\pi_r = \pi_r'). \tag{5}$$

Thus, we have a set of all predicates, each of which is true and is proved by the corresponding cross-chain proof π_r; this equation $\pi_r = \pi_r'$ holds. However, the origin transaction tx' has been committed by π_r, but not committed by π_r'. Therefore, there must exist a Merkle tree collision, meaning that a hash collision found by $\mathcal{A}_2$ occurs. As a result, HASH-BREAK happens.

Claim 2: Pr[CCPP-BREAK] is negligible.

Suppose that CCPP-BREAK happens. We can discover that the condition (4) holds. This case, however, is negligible according to the assumption that the cross-chain proof protocol in use is secure.

Claim 3: Pr[HASH-BREAK] is negligible.

It is the same as claim 2, there exists another cross-chain proof π for the origin transaction corresponding to tx is created, which contradicts with the assumption that the hash function in use is collision-resistant.

Relying on the three claims above, we conclude that for negligible probability **negl**, Pr[SC-FAILURE] $\leq$ **negl**. Therefore, $\pi_{\mathsf{A}}(\text{merge}(\{\mathbf{L}_i[r] : i \in \mathcal{S}_r\})) \in \pi_{\mathcal{S}_r}(\mathbb{V}_{\mathsf{A}})$ holds, completing the proof.

The above two Lemmas directly prove Theorem 1 presented below.

Theorem 1 (Sidechains security). *SEPoW from Sect. 3 using a secure cross-chain proof protocol and a collision resistant hash function is secure with respect to assumptions $\mathbb{A}_{\mathbf{MC}}$ and $\mathbb{A}_{\mathbf{SC}}$, and* merge *and $\mathbb{V}_{\mathsf{A}}$ defined in Sect. 2.4.*

5 Discussion of Merged Mining

Transaction Verification. In the merged mining mechanism described in Sect. 2.2, merged miners should validate all transactions in sidechains; otherwise invalid transactions may be included in some sidechain transaction Merkle Trees. To achieve this, the merged miners are required to monitoring all involved sidechains. However, this contradicts the sidechains agnosticism [7]: miners of **MC** do not need to be aware of **SC** at all. Only the entities interested in facilitating cross-chain events must be aware of both. To resolve this question, we introduce a cryptographic tool, the recursive composition of zk-SNARKs [24,25], which can create proofs that attest to the validity of other proofs. We leverage the tool to generate a succinct proof of sidechain transaction Merkle tree that attests to the correctness of all transactions at the base of the tree. By verifying the validity of the proof, the merged miners can efficiently evaluate the validity of all transactions in the tree without re-executing transaction verification and monitoring all involved sidechains.

Consider an example that a list of transactions tx_1, tx_2, tx_3, tx_4 in a sidechain transaction Merkle tree needs to be proved. First, a full node in the sidechain generates "base" proofs proving the validity of single transactions. Then two adjacent "base" proofs are merged and further generate a new "merge" proof. Finally, these "merge" proofs are recursively merged into a "merge" proof called root proof. As a result, the root proof attests to the validity of all the transactions.

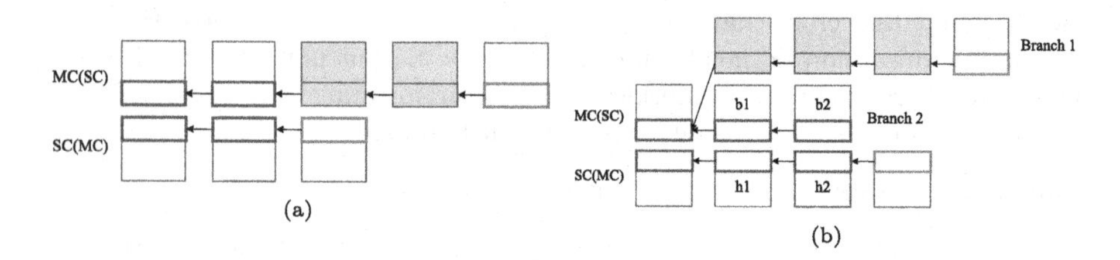

Fig. 3. Asynchronous extension of sidechains (a); Sidechain fork (b). (Color figure online)

Sidechains Extension. In SEPoW, some "regular miners" exist who do not run merged mining but only work on the chain they participate in. This will cause the computational power of each sidechain to be different. In this case, one sidechain extension will exceed the other sidechain. A typical example is shown in Fig. 3a. Where **MC** extension exceeds **SC**. A red public chain of block headers is broken due to the two gray blocks on **MC** that the "regular miners" mine. As a result, cross-chain proofs are affected.

To overcome this problem, merged miners need to rebuild a new public chain of block headers to inherit the previous public chain and backup the information testifying these gray blocks' validity into **SC**. For the former, merged miners run merged mining on the top of two chains and produce new public block headers (blue block headers in Fig. 3a). For the latter, full nodes on **MC** send gray blocks information, consisting of block headers verifying consensus and proofs attesting to the validity of transaction Merkle trees, to merged miners. Upon receiving them, merged miners check their validity by calling a hash function and the zk-SNARKs described above. Then merged miners include these hashes of valid gray block headers into subsequent merged-block-headers by merged mining (here, the hashes are seen as the leaves of a merged Merkle tree).

Fork Solution. The setting of merged mining seems to be a bit coercive in that it makes all sidechains collapse into one single blockchain. That is, the "fork" in a sidechain will affect all sidechains unstable. A typical example is shown in Fig. 3b. Where both **MC** (i.e., Branch 2) and **SC** maintain merged mining and form a red public chain of block headers. With blockchains growth, however, the length of Branch 1 exceeds Branch 2. Thus, branch 1 becomes the "main" chain according to the longest chain rule, while Branch 2 is discarded. Meanwhile, blocks b1 and b2 become "orphan" blocks. As a result, the red public chain of block headers on **MC** and **SC** is broken; cross-chain proofs are affected.

To complement this deficiency, the following efforts need to be done: 1) merged miners run merged mining on top of Branch 1 and **SC**, and create a new public chain of block headers (blue block headers in Fig. 3b); 2) the information of gray blocks on Branch 1 need to be backed up to **SC** via the way described in **Sidechains Extension**; 3) merged miners need to mark "orphan" blocks as invalid in subsequent merged-blocks. This is because adversaries may utilize the headers information of h1 and h2 blocks, including some proofs that testify to the validity of "orphan" blocks transactions, to implement the double-spend attack.

6 Performance Evaluation

In this section, we evaluate the size of SEPoW proof and compare it with the state-of-the-art work.

6.1 Size of Cross-Chain Proofs

We first evaluate the size of SEPoW proofs, consisting of a transaction Merkle tree path denoted by *path* and a merged Merkle tree path denoted by p. Similar to Bitcoin Core, we assume a 256-bit hash function is used to build the Merkle tree construction, and a block of 1 MBytes and a block header of 80 bytes are applied to SEPoW. More specifically, in 1 MB block including up to 2048 transactions (in fact, since January 1, 2020, a Bitcoin block includes 1869 transactions[1] on average), we have in bits: $|p| = |256 + 256| = 512$ bits, $|path| = |\log(2048) \times 256| = 2{,}816$ bits, and hence the size of SEPoW proofs is $|path| + |p| = 512 + 2816 = 416$ bytes.

[1] See https://blockchair.com/zh/bitcoin/charts/total-transaction-count.

6.2 Comparison with Existing Work

Without loss of generality, let us consider some work: BTCRelay [6], PoW sidechains [7,17] and zkRelay [12], which have been implemented or can be evaluated. Their cross-chain proofs sizes can be denoted as $O(n \cdot |BH| + \log_2 |BT| \cdot |H|)$ [6], $O(\log_{1/m}(2)\lambda \cdot ((\log_2(n)+1) \cdot |BH| + \log_2(n) \cdot \lceil \log_2(\log_2(n,2),2) \rceil \cdot |H|))$ [26] and $O(1/504 \cdot n \cdot |BH|)$ [12], respectively, according to their experimental results. Here, n is the length of a blockchain, $|BH|$ is the size of a block header, $|BT|$ is the number of transactions included in the block B, $|H|$ is the size of a hash, m and λ denote an attacker who controls a m fraction of honest chain's mining power succeeds with probability $2^{-\lambda}$.

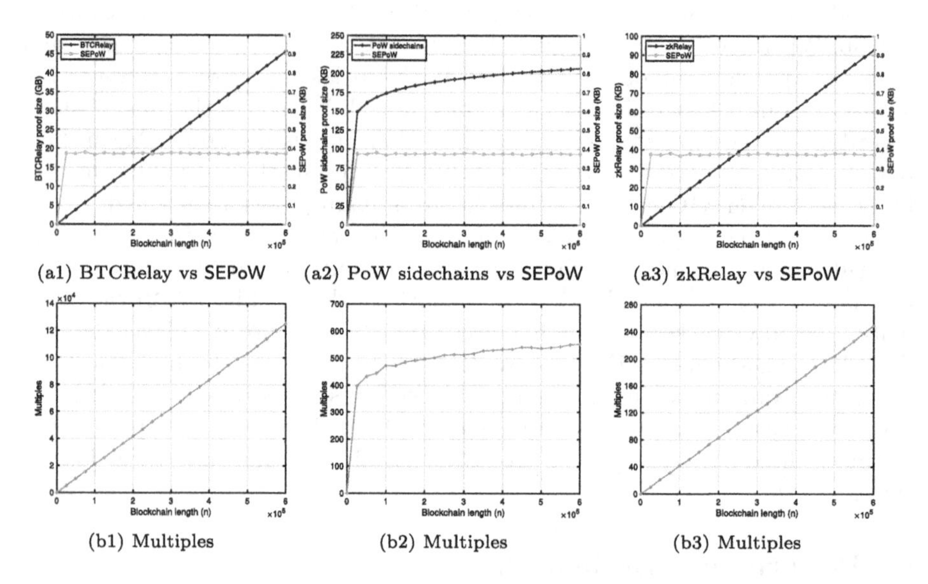

(a1) BTCRelay vs SEPoW (a2) PoW sidechains vs SEPoW (a3) zkRelay vs SEPoW

(b1) Multiples (b2) Multiples (b3) Multiples

Fig. 4. Comparison of SEPoW, BTCRelay, PoW sidechains and zkRelay at different blockchain length, $|BH| = 80$ bytes, $|BT| = 2048$, $|H| = 32$ bytes, $m = 0.5$, $\lambda = 50$.

As depicted in Fig. 4, we make comparisons of SEPoW with BTCRelay, PoW sidechains and zkRelay in terms of the size of a cross-chain proof (The source code related to this comparison experiment has been released to the Github[2]). Figure 4a(1–3) show the impact of blockchain length on a cross-chain proof. Here, BTCRelay proof and zkRelay proof, located in Fig. 4a1 and 4a3, are linear and sublinear in the length of the blockchain, respectively; PoW sidechains proof, located in Fig. 4a2, is logarithmic in the length of the blockchain. More importantly, our SEPoW proof is composed of two Merkle tree paths regardless of the blockchain length, which significantly outperforms all proofs of the above, especially for longer chains.

[2] See https://github.com/01007467319/sepow.git.

Figure 4b(1–3) show the multiple relationships between SEPoW proof and the existing work at different blockchain length. Where, the multiple that SEPoW proof is less than BTCRelay proof and zkRelay proof, indicated by Fig. 4b1 and 4b3, increases linearly; the multiple that SEPoW proof is less than PoW sidechains proof, indicated by Fig. 4b2, is exponential growth. Note that for a blockchain length (n) of 300000 and block header size ($|BH|$) of 80 bytes, SEPoW achieves a proof size of 416 bytes which is roughly 123×, 510× and 62000× smaller than zkRelay proof, PoW sidechains proof and BTCRelay proof, respectively.

7 Related Work

Recently, some researchers have focused on federation construction for sidechains. Dilley et al. [9] designed a federation construction that allows cross-chain assets transfer between disparate blockchains. In this construction cross-chain assets are managed by a trusted committee and are transferred only when the majority of the committee sign cross-chain transactions. Similarly, Back et al. [4] proposed a federation cross-chain solution, in which cross-chain assets are transferred by a trusted group of parties. However, their work has not entirely overcome the political centralization risk as the federation constructions still rely on a trusted committee to maintain and manage cross-chain assets transfer.

To prevent centralization risk, decentralized constructions for sidechains have been proposed. Kiayias et al. [7] presented the first decentralized construction for proof of work sidechains. The construction that is built based on a cryptographic primitive, Non-Interactive Proofs of Proof-of-Work (NIPoPoWs) [17], allows the passing of any information between proof of work blockchains without a trusted third party. However, its cross-chain proof has linear in the length of the blockchain and thus is fairly large. Sztorc [10] and Lerner [11] proposed Drivechains, a decentralized construction for proof of work sidechains. In Drivechains cross-chain assets moved from Bitcoin to Drivechain are authenticated by SPV proofs consisting of all Bitcoin block headers. Yet, these SPV proofs are quite large. An implemented and decentralized construction for sidechains was given in BTCRelay [6]. It supports assets transferred from Bitcoin to Ethereum but not back. To verify the validity of the transactions that took place in Bitcoin, BTCRelay requires saving the entirety of Bitcoin block headers into Ethereum; limiting any potential scalability.

Some other studies are devoted to generating succinct cross-chain proofs about the transactions that took place in a blockchain. Garoffolo et al. [8] proposed Zendoo, a decentralized construction for blockchain systems that allows communication with different sidechains without trusted intermediaries. The construction introduces zk-SNARKs [27] to produce a constant-sized proof that attests to the validity of all cross-chain transactions during a period. Westerkamp et al. [12] presented an efficient sidechains construction, which uses zkSNARK-based chain-relays to generate a succinct proof that testifies the validity of cross-chain assets. The proof size does not grow linearly with the number of block headers but is constant for any batch size. However, these work lacked a formal security definition and proof.

In some different efforts, Gaži et al. [2] presented the first formal treatment of sidechains and proof of stake sidechains construction. To attest to the validity of cross-chain assets, they introduce a trust committee chosen among sidechain block creators to generate the sidechain certificates. The first Bitcoin sidechain in production was given in RSK [13]. It supports smart contracts functionality, compatible with the Ethereum standards. Yet, RSK proofs (from Bitcoin to RBTC) have linear complexity in the length of Bitcoin blockchain and RSK lacked a formal security definition and proof.

Furthermore, different ideas regarding cross-chain transfers, such as Polkadot [28], Cosmos [29], Tendermint [29], Blockstream's Liquid [30] and Interledger [30], have been proposed. Their construction was centralized and lacked formal security definition and proof. Other related effort also included COMIT [31,32], Plasma [33,34], NOCUST [35,36], Dogethereum [37] and XCLAIM [38].

8 Conclusion

In this paper, we proposed SEPoW, which makes up for deficiencies in security and efficiency in existing PoW sidechains construction. In SEPoW the exchange of all cross-chain assets is managed by all honest nodes in participating blockchains. To generate a succinct cross-chain proof, we utilized the merged mining to produce a constant-size proof, regardless of the size of the current blockchain. The security of SEPoW for PoW sidechains is proved formally. We presented a detailed design of SEPoW and evaluated the size of SEPoW proof; SEPoW achieves a proof size of 416 bytes which significantly outperforms all proofs of the existing work, especially for longer chains.

Acknowledgment. The authors would like to thank the anonymous reviewers of ICA3PP 2021 for their insightful suggestions. This work is partially supported by the Shandong Provincial Key Research and Development Program under Grant Number 2019JZZY020127.

References

1. Nakamoto, S.: Bitcoin: a peer-to-peer electronic cash system (2008). http://bitcoin.org/bitcoin.pdf
2. Gazi, P., Kiayias, A., Zindros, D.: Proof-of-stake sidechains. In: S&P 2019, Piscataway, pp. 139–156. IEEE (2019)
3. Decker, C., Wattenhofer, R.: Bitcoin transaction malleability and MtGox. In: Kutyłowski, M., Vaidya, J. (eds.) ESORICS 2014. LNCS, vol. 8713, pp. 313–326. Springer, Cham (2014). https://doi.org/10.1007/978-3-319-11212-1_18
4. Back, A., et al.: Enabling blockchain innovations with pegged sidechains (2014). https://www.blockstream.com/sidechains.pdf
5. Buterin, V.: Ethereum: a next-generation smart contract and decentralized application platform (2014). https://www.github.com/ethereum/wiki/wiki/White-Paper

6. BTCRelay, Community: BTCRelay reference implementation (2017). https://www.github.com/ethereum/btcrelay
7. Kiayias, A., Zindros, D.: Proof-of-work sidechains. In: Bracciali, A., Clark, J., Pintore, F., Rønne, P.B., Sala, M. (eds.) FC 2019. LNCS, vol. 11599, pp. 21–34. Springer, Cham (2020). https://doi.org/10.1007/978-3-030-43725-1_3
8. Garoffolo, A., Kaidalov, D., Oliynykov, R.: Zendoo: a zk-SNARK verifiable cross-chain transfer protocol enabling decoupled and decentralized sidechains. In: ICDCS 2020, Piscataway, pp. 1257–1262. IEEE (2020)
9. Dilley, J., Poelstra, A., Wilkins, J., Piekarska, M., Gorlick, B., Friedenbach, M.: Strong federations: an interoperable blockchain solution to centralized third party risks. CoRR arXiv:1612.05491 (2016)
10. Sztorc, P.: Drivechain - the simple two way peg (2015). https://www.truthcoin.info/blog/drivechain/
11. Lerner, S.D.: Drivechains, sidechains and hybrid 2-way peg designs (2016). https://docs.rsk.co/Drivechains_Sidechains_and_Hybrid_2-way_peg_Designs_R9.pdf
12. Westerkamp, M., Eberhardt, J.: zkRelay: facilitating sidechains using zkSNARK-based chain-relays. In: Euro S&P Workshops, Piscataway, pp. 378–386. IEEE (2020)
13. Lerner, S.D.: Rootstock: smart contracts on bitcoin network (2015). https://blog.rsk.co/wp-content/uploads/2019/02/RSK_White_Paper-ORIGINAL.pdf
14. Singh, A., Click, K., Parizi, R.M., Zhang, Q., Dehghantanha, A., Choo, K.R.: Sidechain technologies in blockchain networks: an examination and state-of-the-art review. J. Netw. Comput. Appl. **149**, 102471 (2020)
15. Atzei, N., Bartoletti, M., Cimoli, T.: A survey of attacks on Ethereum Smart Contracts (SoK). In: Maffei, M., Ryan, M. (eds.) POST 2017. LNCS, vol. 10204, pp. 164–186. Springer, Heidelberg (2017). https://doi.org/10.1007/978-3-662-54455-6_8
16. Moore, T., Christin, N.: Beware the middleman: empirical analysis of bitcoin-exchange risk. In: Sadeghi, A.-R. (ed.) FC 2013. LNCS, vol. 7859, pp. 25–33. Springer, Heidelberg (2013). https://doi.org/10.1007/978-3-642-39884-1_3
17. Kiayias, A., Miller, A., Zindros, D.: Non-interactive proofs of proof-of-work. In: Bonneau, J., Heninger, N. (eds.) FC 2020. LNCS, vol. 12059, pp. 505–522. Springer, Cham (2020). https://doi.org/10.1007/978-3-030-51280-4_27
18. Judmayer, A., Zamyatin, A., Stifter, N., Voyiatzis, A.G., Weippl, E.: Merged mining: curse or cure? In: Garcia-Alfaro, J., Navarro-Arribas, G., Hartenstein, H., Herrera-Joancomartí, J. (eds.) ESORICS/DPM/CBT -2017. LNCS, vol. 10436, pp. 316–333. Springer, Cham (2017). https://doi.org/10.1007/978-3-319-67816-0_18
19. Wang, J., Wang, H.: Monoxide: scale out blockchains with asynchronous consensus zones. In: Lorch, J.R., Yu, M. (eds.) NSDI 2019, pp. 95–112. USENIX Association, Berkeley (2019)
20. Garay, J., Kiayias, A., Leonardos, N.: The bitcoin backbone protocol: analysis and applications. In: Oswald, E., Fischlin, M. (eds.) EUROCRYPT 2015. LNCS, vol. 9057, pp. 281–310. Springer, Heidelberg (2015). https://doi.org/10.1007/978-3-662-46803-6_10
21. Pass, R., Seeman, L., Shelat, A.: Analysis of the blockchain protocol in asynchronous networks. In: Coron, J.-S., Nielsen, J.B. (eds.) EUROCRYPT 2017. LNCS, vol. 10211, pp. 643–673. Springer, Cham (2017). https://doi.org/10.1007/978-3-319-56614-6_22
22. Kiayias, A., Panagiotakos, G.: Speed-security tradeoffs in blockchain protocols. IACR Cryptol. ePrint Arch. **2015**, 1019 (2015)

23. Kiayias, A., Russell, A., David, B., Oliynykov, R.: Ouroboros: a provably secure proof-of-stake blockchain protocol. In: Katz, J., Shacham, H. (eds.) CRYPTO 2017. LNCS, vol. 10401, pp. 357–388. Springer, Cham (2017). https://doi.org/10.1007/978-3-319-63688-7_12
24. Ben-Sasson, E., Chiesa, A., Tromer, E., Virza, M.: Scalable zero knowledge via cycles of elliptic curves. In: Garay, J.A., Gennaro, R. (eds.) CRYPTO 2014. LNCS, vol. 8617, pp. 276–294. Springer, Heidelberg (2014). https://doi.org/10.1007/978-3-662-44381-1_16
25. Bitansky, N., Canetti, R., Chiesa, A., Tromer, E.: Recursive composition and bootstrapping for SNARKS and proof-carrying data. In: Boneh, D., Roughgarden, T., Feigenbaum, J. (eds.) STOC 2013, pp. 111–120. ACM, New York (2013)
26. Bünz, B., Kiffer, L., Luu, L., Zamani, M.: FlyClient: super-light clients for cryptocurrencies. In: S&P 2020, Piscataway, pp. 928–946. IEEE (2020)
27. Ben-Sasson, E., Chiesa, A., Tromer, E., Virza, M.: Succinct non-interactive zero knowledge for a von Neumann architecture. In: Fu, K., Jung, J. (eds.) USENIX Security 2014, pp. 781–796. USENIX Association, Berkeley (2014)
28. Wood, G.: Polkadot: vision for a heterogeneous multi-chain framework (2016). https://www.polkadot.network
29. Buchman, E.: Tendermint: byzantine fault tolerance in the age of blockchains (2016). https://github.com/tendermint/tendermint
30. Group, T.I.P.C.: Interledger protocol v4 (2021). https://www.interledger.org
31. Hosp, J., Hoenisch, T., Kittiwongsunthorn, P.: COMIT - cryptographically-secure off-chain multi-asset instant transaction network. CoRR arXiv:1810.02174 (2018)
32. Tian, W., Pan, W., Shaobin, C., Ying, M., Anfeng, L., Mande, X.: A unified trustworthy environment establishment based on edge computing in industrial IoT. IEEE Trans. Ind. Inform. **16**(9), 6083–6091 (2020). https://doi.org/10.1109/TII.2019.2955152
33. Poon, J., Buterin, V.: Plasma: scalable autonomous smart contracts (2017). https://www.plasma.io/plasma.pdf
34. Tian, W., et al.: Propagation modeling and defending of a mobile sensor worm in wireless sensor and actuator networks. Sensors **17**(1), 139 (2017). https://doi.org/10.3390/s17010139
35. Khalil, R., Gervais, A.: NOCUST - a non-custodial 2nd-layer financial intermediary. IACR Cryptol. ePrint Arch. **2018**, 642 (2018)
36. Mingfeng, H., Anfeng, L., Tian, W., Changqin, H.: Green data gathering under delay differentiated services constraint for internet of things. Wirel. Commun. Mob. Comput. **2018**, 1–23 (2018). https://doi.org/10.1155/2018/9715428
37. Teutsch, J., Straka, M., Boneh, D.: Retrofitting a two-way peg between blockchains. CoRR arXiv:1908.03999 (2019)
38. Zamyatin, A., Harz, D., Lind, J., Panayiotou, P., Gervais, A., Knottenbelt, W.J.: XCLAIM: trustless, interoperable, cryptocurrency-backed assets. In: S&P 2019, Piscataway, pp. 193–210. IEEE (2019)

A Spectral Clustering Algorithm Based on Differential Privacy Preservation

Yuyang Cui[1], Huaming Wu[2], Yongting Zhang[1], Yonggang Gao[1], and Xiang Wu[1](✉)

[1] School of Medical Information and Engineering, Xuzhou Medical University, Xuzhou 221000, Jiangsu, China
wuxiang@xzhmu.edu.cn

[2] Center for Applied Mathematics, Tianjin University, Tianjin 300072, China

Abstract. Spectral clustering is a widely used clustering algorithm based on the advantages of simple implementation, small computational cost, and good adaptability to arbitrarily shaped data sets. However, due to the lack of data protection mechanism in spectral clustering algorithm and the fact that the processed data often contains a large amount of sensitive user information, thus an existing risk of privacy leakage. To address this potential risk, a spectral clustering algorithm based on differential privacy protection is proposed in this paper, which uses the Laplace mechanism to add noise to the input data perturbing the original data information, and then perform spectral clustering, so as to achieve the purpose of privacy protection. Experiments show that the algorithm has both stability and usability, can correctly complete the clustering task with a small loss of accuracy, and can prevent reconstruction attacks, greatly reduce the risk of sensitive information leakage, and effectively protect the model and the original data.

Keywords: Differential privacy · Spectral clustering · Privacy preservation

1 Introduction

Machine learning clustering algorithms have made disruptive breakthroughs in recent years and are widely used in computer vision, data mining and remote sensing mapping. However, most machine learning clustering algorithms are designed without considering the security and privacy issues of data and models [1]. To protect the security of private information, many industry scholars have conducted extensive research.

There are three broad directions of privacy preservation currently combined with clustering algorithms: data transformation, data anonymization, and data perturbation represented by differential privacy. Among them, Oliverira et al. [2] proposed a new spatial data transformation method RT (Rotation-based Transformation) inspired by the rotational changes of geometry in space, whose advantage lies in being able to hide the original information while maintaining the validity of the data attributes before and after the transformation. However, the computation is complex with higher dimensionality, and it is difficult to resist consistency attacks. Nayahi [3] proposed an anonymous data algorithm by distributing anonymous data in Hadoop Distributed File System (HDFS)

Y. Lai et al. (Eds.): ICA3PP 2021, LNCS 13157, pp. 397–410, 2022.
https://doi.org/10.1007/978-3-030-95391-1_25

based on the principle of clustering and resilient similarity attacks and probabilistic inference attacks, but it is difficult to resist emerging combinatorial attacks and foreground knowledge attacks. Current research on differential privacy combined with clustering algorithms focuses on the k-means algorithm. Blum et al. [4] were the first to combine differential privacy techniques with k-means clustering algorithm in 2005, proposed the DPk-means algorithm, which pioneered the research of data perturbation represented by differential privacy. Dwork [5] proposed a method of privacy budget allocation and sensitivity calculation in view of the DPk-eams algorithm's shortcomings that the large sensitivity of query function and the privacy budget allocation method are not given. FU et al. [6] improved the usability of clustering results by dynamically assigning the selection of initial centroids and iteratively updating the privacy budget during the operation of the algorithm, but they don't consider the effect of isolated points in the dataset on the clustering effect. NI et al. [7] proposed the DP-KCCM algorithm to offset the effect of differential privacy techniques by merging adjacent clusters to join of noise on the clustering results. X. W et al. [8] proposed a DP-CFMF algorithm based on differential privacy protection, which guarantees the privacy of DNA data with high recognition accuracy and high utility. W. Wu et al. [9] proposed a DP-DBS can algorithm that realizes differential privacy protection by adding laplace noise, which was experimentally shown to be able to protect privacy while being both usable. In addition, W. Wu et al. [10] integrated various privacy protection schemes and innovatively designed a data-sharing platform that can guarantee data security, practicing the previous research results.

Compared with the traditional k-means algorithm, the spectral clustering algorithm is applicable to arbitrarily shaped data sets, and does not require prior assumptions about the probability distribution of the data, it is fast in computation and simple. In recent years, it has been widely used in computer vision, data mining, image processing and natural language processing [11–14]. For example, Yang Fan et al. [15] used the spectral clustering algorithm to mine the data of chemical reagents in stock of China Institute of Petrochemical Sciences. Guo Lei et al. [16] applied the spectral clustering algorithm to cognitive diagnostic assessment (CDA), and explored the possible effects of introducing the spectral clustering algorithm on CDA in terms of attribute hierarchy, number of attributes, sample size and failure rate, and achieved good results in specific experiments.

However, the spectral clustering algorithm lacks privacy protection mechanism and there is a risk of privacy leakage. How to protect the privacy of the spectral clustering algorithm while ensuring the clustering accuracy becomes an urgent problem. While the relevant research on spectral clustering algorithms oriented to differential privacy protection are proposed. Zheng et al. [17] achieved differential privacy protection by adding laplace noise to perturb the objective function to hide the true weight values and the clustering results were more accurate compared with the spectral clustering algorithm without differential privacy protection, but the specific privacy budget allocation method was not described. The internal scale parameter and the number of clusters of the spectral clustering algorithm were optimized to further improve the accuracy of clustering [18]. The DP-CSC algorithm is proposed based on the improved spectral clustering algorithm CCL-S, and the noise is added to the compressed laplacian matrix using Wishart mechanism [19].

This paper will propose a new spectral clustering algorithm based on differential privacy to achieve privacy preservation, by adding noise conforming to the Laplace distribution to the input data to hiding the original data.

2 Theoretical Foundation

2.1 Differential Privacy

Differential privacy (DP) is a privacy-preserving model with rigorous mathematical proof, first proposed by Microsoft's DWork team [20].

Definition 1 Differential Privacy: Assume that there exists a random function A such that A in any two neighbor data sets $Q, Q^{'}$(i.e.$\left|\left|Q - Q^{'}\right|\right|_1 \leq 1$) to obtain any identical set of outputs B with probability satisfying

$$\Pr[A(Q) \in B] \leq e^{\varepsilon} \Pr\left[A\left(Q'\right) \in B\right] \tag{1}$$

Then the random function is said to A satisfies$\varepsilon - \mathit{differentialprivacy}$, abbreviated as$\varepsilon - DP$. Where the neighbor datasets are the two datasets that differ by only 1 record and $\|\|_1$ is the L_1 paradigm,$Pr[]$ is the probability of occurrence of an event, and ε is the privacy budget, the size of its value represents the degree of privacy protection, the smaller the value means the better the privacy protection.

Differential privacy is achieved mainly by adding noise to the data to perturb the data so that the output results are randomly different from the real results each time. The common mechanisms for adding noise are the Laplace mechanism for continuous data, the Exponential mechanism for discrete data, such as race, education, etc. and the Gaussian mechanism for image data. Since the data to be processed in this paper are of continuous type, the Laplace mechanism is adopted.

The Laplacian mechanism is implemented by adding random noise obeying the Laplacian distribution to the exact query result $\varepsilon - \mathit{differentialprivacy}$ protection. Let the location parameter be 0 and the scale parameter be b the Laplace distribution of $Lap(b)$, then its probability density function is

$$p(x) = \frac{1}{2b} e^{\left(-\frac{|x|}{b}\right)} \tag{2}$$

where e is the natural logarithm.

Its probability density function, as shown in Fig. 1.

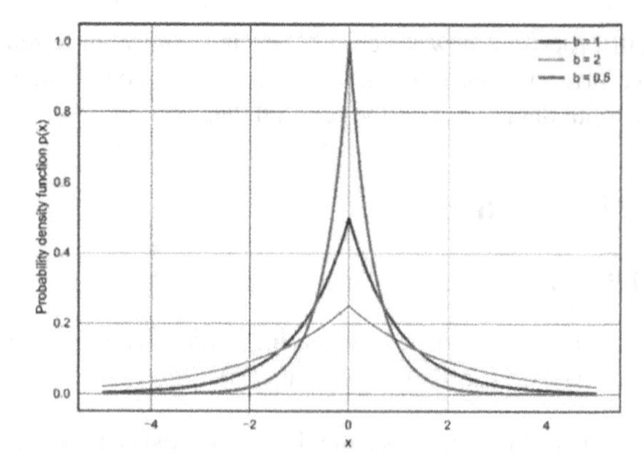

Fig. 1. Laplace probability density function

For the Laplace mechanism, there are the following definitions.

Definition 2 Laplace Mechanism: Given a data set D, there is a function $f : D- > Rd$, whose sensitivity is Δf, then the randomized algorithm $M(D) = f(D) + Y$ provides $\varepsilon - differentialprivacy$ protection, where $Y \sim Lap(\Delta f/\varepsilon)$ is the random noise, which obeys the scale parameter $\Delta f/\varepsilon$ of the Laplace distribution, and the random noise size depends mainly on the sensitivity of the function Δf[21].

Definition 3 Sensitivity Δf: With the function $f : D \rightarrow Rd$, the input is the dataset and the output is d dimensional real vector, for two neighbor datasets D, D', the sensitivities are

$$\Delta f = \max D, D' f(D) - f(D')_1 \tag{3}$$

Sensitivity measures the maximum change to the results caused by the deletion of anyone record from the dataset and is an important parameter in determining the amount of noise introduced into the data.

Spectral Clustering Algorithm

Compared with the traditional k-means algorithm and EM algorithm, the spectral clustering algorithm is applicable to data sets of arbitrary shapes that can do without prior assumptions about the probability distribution of the data, with fast computational speed and simple algorithmic ideas is simple and easy to implement.

The basic principle of the spectral clustering algorithm is that the sample data is considered as a vertex in the undirected graph, and then the similarity W_{ij} between the vertices is calculated based on the similarity function as the weight between the vertices, and finally the graph is divided according to different partitioning criteria to maximize the similarity within each subgraph and minimize the similarity between the subgraphs [22], each subgraph is equivalent to a cluster.

For a graph G to be divided, denote by A, B two subgraphs (where $A \cup B = V, A \cap B = \varnothing$, V denotes the set of all vertices in the graph G), u, v denote the points in the graph

A, B respectively, $w(u, v)$ denotes the similarity between the points u, v, and the division criteria are mainly as follows.

(1) Minimum cut [23]

$$\min\{Cut(A, B) = \sum_{u \in A, v \in B} w(u, v)\} \tag{4}$$

Where $Cut(A, B)$ is the cost function when dividing graph G into subgraphs A, B, which represents the sum of similarity between points u, v inside graphs A, B. The minimum cut set criterion divides graph G by minimizing the cost function, which proves to be effective for some image datasets in practice, but when the number of subgraphs divided exceeds 2, it is easy to cluster isolated points into a separate class. To address this situation, Shi and Malik proposed the canonical cut-set criterion and Hagen and Kahng proposed the proportional cut-set criterion, both of which solve the problem well.

(2) Normalized Cut [22]

$$\begin{gathered} \min\{\ NCut(A, B) = \tfrac{Cut(A,B)}{Vol(A,V)} + \tfrac{Cut(A,B)}{Vol(B,V)}\} \\ \text{among it}, Vol(A, V) = \sum_{i \in A, t \in V} W_{it} \end{gathered} \tag{5}$$

The minimization $NCut$ function is called the canonical cut-set criterion. This criterion measures not only the degree of similarity between samples within a class, but also the degree of dissimilarity between samples between classes.

(3) Ratio Cut [24]

$$\min\{RCut(A, B) = \frac{Cut(A, B)}{\min(|A|, |B|)}\} \tag{6}$$

where $|A|, |B|$ denote the number of vertices in subgraphs A and B, respectively. The cut of graph G according to the case when the $RCut$ function of subgraphs A, B is minimized is the proportional cut set criterion, which can minimize the similarity between subgraphs and reduce the possibility of over-cutting, but the operation speed is slow.

(4) Average Cut [25]

$$\min\{\ AvCut = \frac{Cut(A, B)}{|A|} + \frac{Cut(A, B)}{|B|}\} \tag{7}$$

It can be seen that *Avcut* uses the sum of the ratio of the cost function and the number of data points in the divided region, which can theoretically produce a more accurate division, but the same drawback is that it is easy to divide smaller subgraphs that contain only a few vertices. In addition, it is pointed out in the literature [22] that the experimental results of using Normalized cut criterion are better than Average cut criterion when dividing the same image.

(5) Minimum-Max Cut [26]

$$\min \{ MCut = \frac{Cut(A, B)}{Vol(A, A)} + \frac{Cut(A, B)}{Vol(B, B)} \} \tag{8}$$

The central idea of the minimum-maximum cut-set criterion is to minimize Cut(A, B) while maximizing $Vol(A, A)$ and $Vol(B, B)$. Minimizing this function avoids dividing a smaller subgraph containing only a few vertices, so it tends to produce balanced cut sets, but is slower to implement. *MCut* satisfies the same principle of small similarity between samples between classes and large similarity between samples within classes as *NCut*, and has similar behavior to *NCut*, but when the overlap between classes is large, *MCut* is more efficient than *NCut*.

(6) Multiway Normalized Cut [27]
The objective functions used in the above five partitioning criteria are all 2-way partitioning functions that partition the graph G into 2 subgraphs, and Meila proposes a partitioning function that can partition the graph G into k subgraphs k-ways at the same time,

$$MNCut = \frac{Cut(A_1, V-A_1)}{Vol(A_1, V)} + \frac{Cut(A_2, V-A_2)}{Vol(A_2, V)} + \cdots + \frac{Cut(A_k, V-A_k)}{Vol(A_k, V)} \tag{9}$$

The only difference between *Ncut* and *MNcut* is that the used spectral mapping is different and *MNcut* is equivalent to *Ncut* for $k = 2$. The multiplexed canonical cut-set criterion is reasonable and effective in practice, but its optimization problem is usually difficult to solve.

Although there are various classification criteria and implementations of spectral clustering algorithms, they can be summarized in the following general flow [28].

1) Calculate the similarity W_{ij} between the vertices, construct the similarity matrix W, and construct the matrix Z representing the sample set according to the different objective functions;
2) Calculate the first k eigenvectors of Z, and build the eigenvector space;
3) Clustering the eigenvectors in the eigenvector space by $K - means$ or other classical clustering algorithms.

Among them, the similarity function used to calculate the similarity between vertices W_{ij} often uses the Gaussian kernel function.

$$W_{ij} = e^{\left(-\frac{d(s_i, s_j)}{2\sigma^2}\right)} \tag{10}$$

Where s_i, s_j is the data point in the sample, and $d(s_i, s_j)$ is the distance between index data points, generally referred to as the Euclidean distance. σ is the scale parameter, the σ The value taken affects the W_{ij} the computation of the algorithm, which indirectly affects the clustering results of the algorithm. In addition, the optimal σ values are not the same, so the value in practice needs to be determined after several experiments based on the specific dataset.

The significance of using Gaussian kernel function for the similarity function is to map the data points to a high-dimensional space, add more features, highlight the differences between data points, and thus calculate the similarity between data points more accurately.

Based on the general process above, the specific steps of a standard spectral clustering algorithm [29] are given below, the input dataset X is an arbitrarily shaped dataset consisting of n points $x_1, x_2, \cdots, x_n$ consisting of an arbitrarily shaped dataset, each point can be an arbitrary object, the similarity function uses a Gaussian kernel function, the division criterion is based on *Ncut*, and the final clustering method uses $K - means$ is as follows.

Input: sample set $X = (x_1, x_2, \cdots, x_n)$, scale parameter σ , number of clusters k. Output: Cluster division result $C(c_1, c_2, \cdots, c_k)$.
Step1: Construct the similarity matrix $W_{n\times n}$. Where the element W_{ij} in $W_{n\times n}$ is the similarity between points i and j in the sample, calculated according to formula (10) . Step2: Construct the degree matrix D, The element D_{ij} in D is the sum of the i row in the $W_{n\times n}$ matrix. Step3: Based on $L = I - D^{-\frac{1}{2}}SD^{-\frac{1}{2}}$, construct the normalized Laplacian matrix L_{sym}, where I is the unit matrix. Step4: Calculate the eigenvectors f corresponding to each of the smallest k eigenvalues before L_{sym}. Step5: Normalize the eigenvectors f to finally form an $n * k_1$ -dimensional eigenmatrix F. Step6: for each row in F as a k_1-dimensional sample, a total of n samples, clustering by k-means or other clustering methods, the clustering dimension is k_2. Step7: Get the cluster division result $C(c_1, c_2, \cdots, c_k)$.

3 Analysis of Spectral Clustering Algorithm Based on Differential Privacy Preservation

3.1 Algorithm Description

This algorithm can be divided into two stages in the execution process, the first stage is to add noise conforming to the Laplace distribution to the training data to perturb the original data set; The second stage is to apply the standard spectral clustering algorithm in Table 1 to cluster the noise-added dataset. The specific algorithm is as follows.

Input: sample set $X = (x_1, x_2, \cdots, x_n)$, scale parameter σ, number of clusters k , privacy budget ε.
Output: Cluster division result $C(c_1, c_2, \cdots, c_k)$.

Add the Laplacian noise under a given privacy budget to the points in the sample set X to obtain the disturbed sample set X'.

Step1: Construct the similarity matrix $W_{n \times n}$. Where the element W_{ij} in $W_{n \times n}$. is the similarity between points i and j in the sample, calculated according to formula (10).

Step2: Construct the degree matrix D, The element D_{ij} in D is the sum of the i row in the $W_{n \times n}$ matrix.

Step3: Based on $L = I - D^{-\frac{1}{2}} S D^{-\frac{1}{2}}$, construct the normalized Laplacian matrix L_{sym}, where I is the unit matrix.

Step4: Calculate the eigenvectors f corresponding to each of the smallest k eigenvalues before L_{sym}.

Step5: Normalize the eigenvectors f to finally form an $n * k_1$-dimensional eigenmatrix F.

Step6: for each row in F as a k_1-dimensional sample, a total of n samples, clustering by k-means or other clustering methods, the clustering dimension is k_2.

Step7: Get the cluster division result $C(c_1, c_2, \cdots, c_k)$.

4 Algorithm Analysis

Scheme for Input Perturbations

Among several existing machine learning differential privacy protection schemes, the algorithm proposed in this paper uses an input perturbation scheme to protect privacy security by adding noise conforming to the Laplace distribution to the original dataset before clustering. Compared with other schemes, the input perturbation scheme has the advantages of easy implementation and less loss of clustering accuracy [30], which enables the algorithm to achieve privacy protection at source under the condition of guaranteeing clustering accuracy; from the specific implementation of this scheme in this algorithm, the disturbance of input data will lose dataset reconstruction on the one hand so that the trained model can resist model inversion attacks [31, 32] and model theft attacks [30, 33], greatly reducing the risk of model information leakage; on the other hand, the datasets contacted by the model are perturbed, hiding the true intimacy between points in the original dataset, so that this algorithm can also solve the traditional spectral clustering algorithm concerned in the literature [16], it is easy to reveal the problem of intimacy between samples.·

Sensitivity

Regarding the sensitivity, this algorithm is sensitive in the neighboring data set D, D' When any record is modified on the neighboring data set, the data sensitivity is 1 for each dimension, so the global sensitivity is n.

Privacy Budget

This algorithm adds noise that fits the Laplace distribution to each data set before the model is learned, so the total privacy budget of the algorithm ε satisfies the $\varepsilon - differentialprivacy$ the defined measure of the differential privacy model.

5 Simulation Experiments

5.1 Experimental Design

The experimental design session includes the selection of evaluation metrics for the clustering algorithm, the introduction and pre-processing of the data set, and the hardware and software environment for running the experiments.

We choose to use the Adjusted Rand Index (ARI)[34] as the evaluation index of the clustering algorithm to measure how well the algorithm clustering results match with the actual situation.

$$ARI = \frac{RI - E[RI]}{\max(RI) - E[RI]} \tag{11}$$

Where RI is the Rand Index (RI, Eq. 12), $E[RI]$ is the mathematical expectation of the Rand Index, and $max(RI)$ is the maximum value of the Rand Index. The Rand coefficient is calculated by the following formula:

$$RI = \frac{a + b}{C^2_{n\_samples}} \tag{12}$$

where a is the number of correct similar pairs, b is the number of correct dissimilar pairs, C is the combinatorial number symbol, and $n\_samples$ is the total number of data points. RI takes a range of [0, 1], and a larger value means that the clustering results match the real situation.

The Adjusted Rand Index is an improvement of the Rand Index (RI), which overcomes the shortcomings of the original Rand index for "random clustering does not guarantee that the score is close to zero", it also has a higher degree of discrimination.

(1) Experimental data set and pre-processing

The datasets used in this paper include two artificially synthetic-sized two-dimension datasets, Moon and R2, and two datasets wine and iris from the UCI Machine Learning Repository [34]. Detailed information is shown in Table 1.

Table 1. Data set

Dataset	Number of samples	Number of attributes	Number of categories
Moon	600	2	2
R2	358	2	4
Wine	178	13	3
Iris	150	4	3

In the preprocessing step, this article first normalizes all the data sets so that their attribute values all fall within the interval [0,1]. In addition, considering that this algorithm uses a Gaussian kernel function for the calculation of the similarity between samples (Eq. 4), the selection of the scale parameter ∂ affects the final clustering result. In order to eliminate this influence, the value of ∂_g in the following experiment of the differential privacy spectral clustering algorithm is the value of ∂ that makes the spectral clustering effect of the four data sets optimal. The process of determining the specific value of ∂_g is as follows: adjust the value of ∂, such as 0.1, 0.2, 0.5, 2, 6, 9, 11 and 12, perform multiple spectral clustering on the normalized data set, and cluster Calculate the ARI coefficient from the class result, and select the ∂ value corresponding to the largest coefficient as the value of ∂_g. It can be seen from Fig. 2 that for the data set moon and R2, the optimal ∂ value of the clustering effect should be selected 12. For the data set iris and wine, the optimal ∂ value of the clustering effect is maintained at about 8.

5.2 Experimental Results and Analysis

Experiment 1 compared the clustering results of spectral clustering algorithm and differential privacy spectral clustering algorithm.

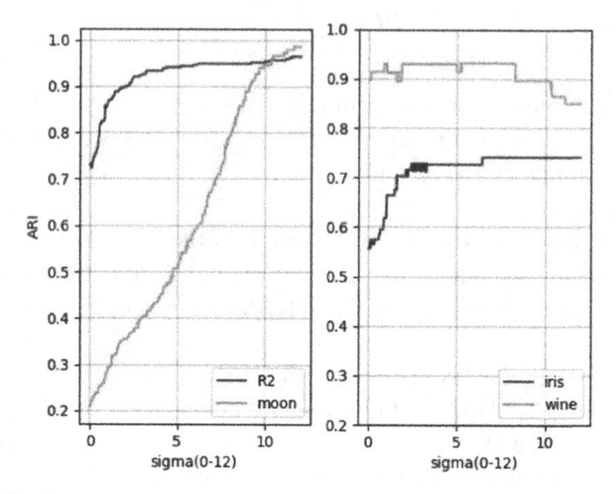

Fig. 2. Relationship between parameter ∂ and clustering effect

Figure 3 and 4 show the clustering effects on the two datasets R2 and moon, respectively. The ARI index are marked in the lower right corner of each image. Among them, Fig. 3(a) and Fig. 4(a) depict the clustering effect without the effect of differential privacy protection, while Fig. 3(b) and Fig. 4(b) correspond to the clustering effect of the differential privacy spectrum clustering. From the experimental results, the differential privacy spectral clustering algorithm proposed in this paper is able to identify the correct clustering classes with less loss of accuracy.

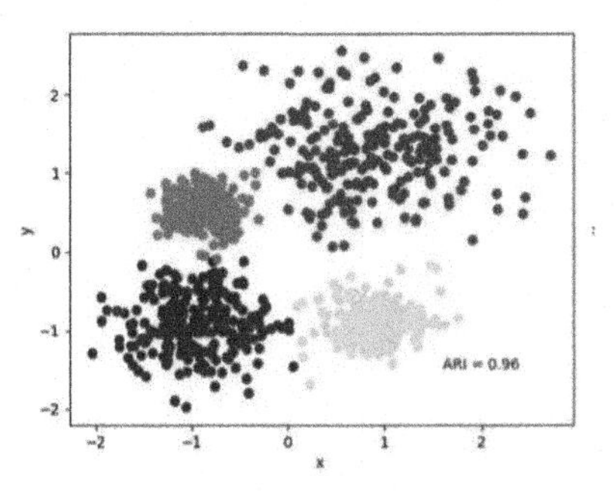

(a) Effect of spectral clustering algorithm

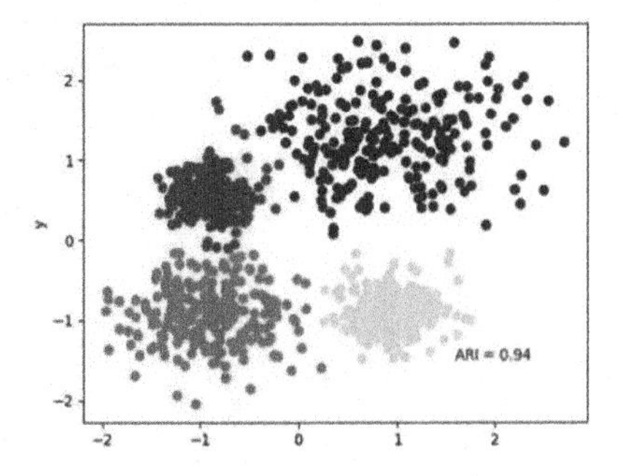

(b) Effect of differential privacy spectral clustering algorithm

Fig. 3. Clustering effect on R2 dataset

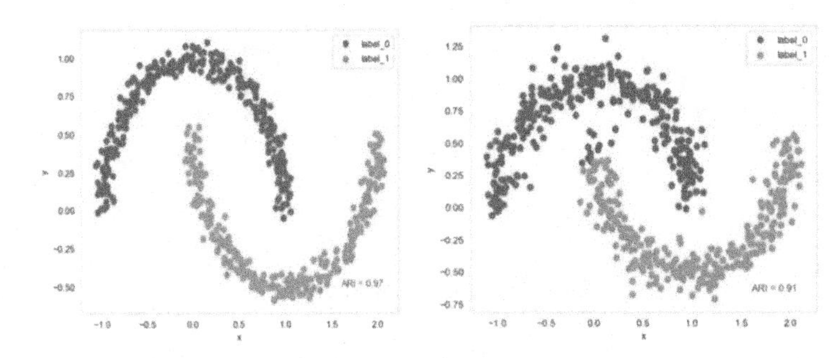

(a) Effect of spectral clustering algorithm (b) Effect of differential privacy spectral clustering algorithm

Fig. 4. Clustering effect on moon dataset

The spectral clustering algorithm and the differential privacy spectral clustering algorithm were run several times on two datasets from UCI, wine and iris, and draw a trend graph, where the x-axis is the number of runs of the algorithm was run n (each run of the algorithm was based on the original dataset), and the y-axis is the average of the n ARI index for n runs of the algorithm (for distinction, it is denoted as "Average ARI"). The aim is to observe the usability and stability of the algorithm by the trend of the average ARI with the number of runs.

As can be seen from Fig. 5 and 6, the Average ARI coefficient of the differential privacy spectral clustering algorithm is slightly lower than that of the spectral clustering algorithm, because a certain amount of perturbation is generated after adding noise to the original dataset, which causes the original dataset to lose a portion of its accuracy. From the average ARI coefficient, the scale of the reduction of the value is not large, indicating that the clustering results do not produce a large change, so the differential privacy spectrum is clustered class algorithms are still available. In addition, the average ARI index of the differential privacy spectrum clustering algorithm gradually stabilizes with the increase of the number of runs, which also reflects the stability of the algorithm.

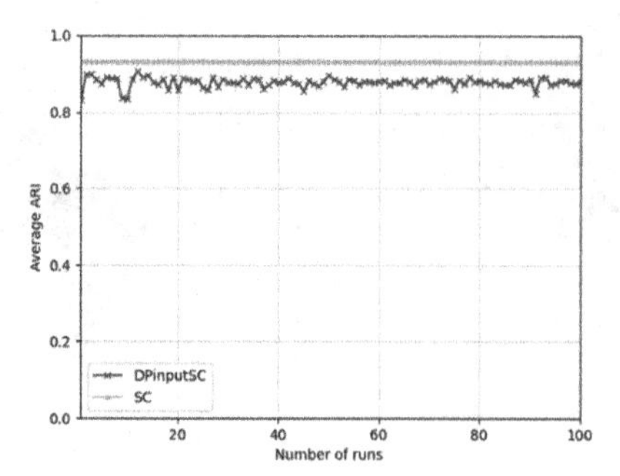

Fig. 5. Wine data clustering classes

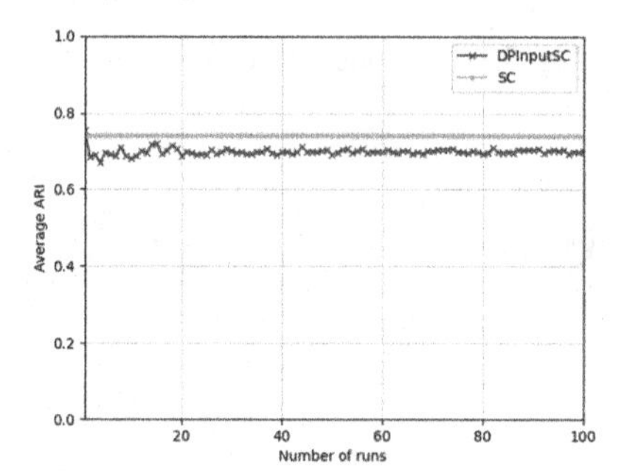

Fig. 6. Iris dataset clustering

6 Conclusion

In order to solve the problem of privacy leakage in traditional clustering algorithms, we design a spectral clustering algorithm with differential privacy mechanism. By adding noise that conforms to the laplacian distribution to the original data, to reduce the risk of sensitive information leakage and prevent reconstruction attacks against the model to achieve the purpose of protecting privacy and security. The experimental results show that the spectral clustering algorithm based on differential privacy protection proposed in this paper can not only achieve privacy protection, but also has stability and usability. Since the perturbed data can affect the clustering accuracy, the next step will be to investigate how to guarantee the differential privacy implementation while improving the accuracy of the algorithm clustering as much as possible and increasing the algorithm usability.

References

1. Xiang, W., Wang, H., Shi, M., Wang, A., Xia, K.: DNA Motif finding method without protection can leak user privacy. IEEE Access **7**, 152076–152087 (2019)
2. Achieving Privacy Preservation when Sharing Data for Clustering. Springer, Berlin Heidelberg (2004). https://doi.org/10.1007/978-3-540-30073-1_6

3. Nayahi, J., Kavitha, V.: Privacy and utility preserving data clustering for data anonymization and distribution on Hadoop. Future Gen. Comput. Syst. **74**(SEP.), 393–408 (2016)
4. Practical privacy: the SuLQ framework. In: Proceedings of the Twenty-fourth ACM SIGACT-SIGMOD-SIGART Symposium on Principles of Database Systems, June 13–15, 2005, ACM, Baltimore, Maryland, USA (2005)
5. Dwork, C.: A firm foundation for private data analysis. Commun. ACM **54**(1), 86–95 (2011)
6. Yanming, F.U., Zhenduo, L.I.: Research on k-means++ clustering algorithm based on laplace mechanism for differential privacy protection. Netinfo Sec. (2019)
7. Ni, T., Qiao, M., Chen, Z., Zhang, S., Zhong, H.: Utility-efficient differentially private K-means clustering based on cluster merging. Neurocomputing **424**, 205–214 (2021)
8. Xiang, W., Wei, Y., Mao, Y., Wang, L.: A differential privacy DNA motif finding method based on closed frequent patterns. Clust. Comput. **22**(S2), 2907–2919 (2018)
9. Wu, W.M., Huang, H.K.: Research on DP-DBScan clustering algorithm based on differential privacy preservation. Comput. Eng. Sci. **37**(4), 830–834 (2015)
10. Wu, X., Zhang, Y., Wang, A., et al.: MNSSp3: Medical big data privacy protection platform based on Internet of things. Neural Comput. Appl. **4** (2020)
11. Wang, T., Yucheng, L., Wang, J., Dai, H.-N., Zheng, X., Jia, W.: EIHDP: edge-intelligent hierarchical dynamic pricing based on cloud-edge-client collaboration for IoT systems. IEEE Trans. Comput. **70**(8), 1285–1298 (2021)
12. Youke, W., Huang, H., Ningyun, W., Yue Wang, M., Bhuiyan, Z.A., Wang, T.: An incentive-based protection and recovery strategy for secure big data in social networks. Inf. Sci. **508**, 79–91 (2020)
13. Xiang, W., Zhang, Y., Shi, M., Li, P., Li, R., Xiong, N.N.: An adaptive federated learning scheme with differential privacy reserving. Futur. Gener. Comput. Syst. **127**, 362–372 (2022). https://doi.org/10.1016/j.future.2021.09.015
14. Wang, T., Liu, Y., Zheng, X., Dai, H.-N., Jia, W., Xie, M.: Edge-based communication optimization for distributed federated learning. In: IEEE Transactions on Network Science and Engineering (2021). https://doi.org/10.1109/TNSE.2021.3083263
15. Fan, Y., Xiang, Z., Li, M.: Application of spectral clustering algorithm in chemical reagent library preparation optimization. Comput. Appl. Chem. **36**(5), 3 (2019)
16. Guo, L., Yang, J., Song, N.Q.: Application of spectral clustering algorithm in the diagnostic assessment of different attribute hierarchical structures. Psychol. Sci. **41**(3), 8 (2018)
17. Xiaoyao, Z., Dongmei, C., Yuqing, L., et al.: A spectral clustering algorithm based on differential privacy preservation. Comput. Appl. **38**(10), 5
18. Hu, B.: Research on clustering algorithm for differential privacy protection. Nanjing University of Posts and Telecommunications (2019)
19. Li, J., Wei, J., Ye, M., Liu, W., Xuexian, H.: Privacy-preserving constrained spectral clustering algorithm for large-scale data sets. IET Inf. Secur. **14**(3), 321–331 (2020). https://doi.org/10.1049/iet-ifs.2019.0255
20. Dwork, C.: Differential privacy. In: Bugliesi, M., Preneel, B., Sassone, V., Wegener, I. (eds.) Automata, Languages and Programming, pp. 1–12. Springer, Berlin, Heidelberg (2006). https://doi.org/10.1007/11787006_1
21. Liu, J.X., Meng, S.F.: A review of privacy-preserving research on machine learning. Comput. Res. Dev. **057**(002), 346–362 (2020)
22. Shi, J., Malik, J.: Normalized cuts and image segmentation. IEEE Trans. Patt. Anal. Mach. Intell. **22**(8), 888–905 (2000)
23. Wu, Z., Leahy, R.: An optimal graph theoretic approach to data clustering: theory and its application to image segmentation. IEEE Trans. Pattern Anal. Mach. Intell. **15**(11), 1101–1113 (1993)
24. Hagen, L, Kahng, A.B.: New spectral methods for ratio cut partitioning and clustering. IEEE Trans. Comput. Aided Des. Integr. Circ. Syst. **11**(9), 1074–1085 (1992)

25. Sarkar, S., Soundararajan, P.: Supervised learning of large perceptual organization: graph spectral partitioning and learning automata. IEEE Trans. Patt. Anal. Mach. Intell. **22**(5), 504–525 (2000)
26. Ding, C., He, X., Zha, H., et al.: Spectral Min-Max Cut for Graph Partitioning and Data Clustering (2001)
27. Meila, M., Xu, L.: Multiway Cuts and Spectral Clustering. U .Washingt on Tech Report (2003)
28. Cai, X.Y., Dai, G.Z., Yang, L.B.: Survey on spectral clustering algorithms. Comput. Sci. **35**(7), 14–18 (2008)
29. Bai, L., Zhao, X., Kong, Y., et al.: A review of spectral clustering algorithms. Comput. Eng. Appl. 57(14), 12
30. Zhao, Z.D., Chang, X.L., Wang, Y.X.: A review of privacy protection in machine learning. J. Inf. Secur. **4**(05), 5–17 (2019)
31. Privacy in Pharmacogenetics: An End-to-End Case Study of Personalized Warfarin Dosing. USENIX Association (2014)
32. Model inversion attacks that exploit confidence information and basic countermeasures. In: The 22nd ACM SIGSAC Conference. ACM (2015)
33. Stealing machine learning models via prediction APIs. In: 25th USENIX Security Symposium, USENIX Security 16, Austin, TX, USA, August 10–12, 2016 (2016)
34. Information Theoretic Measures for Clusterings Comparison: Variants, Properties, Normalization and Correction for Chance. JMLR.org (2010)
35. Dua, D., Graff, C.: UCI Machine Learning Repository. University of California, School of Information and Computer Science, Irvine, CA (2019)

A Compact Secret Image Sharing Scheme Based on Flexible Secret Matrix Sharing Scheme

Lingfu Wang, Jing Wang(✉), Mingwu Zhang, and Weijia Huang

School of Computer Science and Information Security,
Guilin University of Electronic Technology, Guilin, China
wjing@guet.edu.cn

Abstract. In social networks, Secret Image Sharing (SIS) provides an effective way to protect secret images. However, most existing SIS schemes only support limited access policies, which are not flexible enough for lots of scenarios. In this work, we propose an SIS scheme to solve this defect. The core contributions of our scheme can be summarized as follows. Firstly, we propose a Secret Matrix Sharing Scheme (SMSS), which is extended from the traditional Linear Secret Sharing Scheme (LSSS). Different from LSSS, SMSS shares a secret matrix instead of a single secret value. Secondly, based on our SMSS, we propose an SIS scheme named LM-SIS, which supports *monotonous access policies*. Compared with other SIS schemes, our scheme has advantages in flexibility and efficiency. Furthermore, the LM-SIS scheme is *compact*, which is reflected in its shadow size ratio is approximately $1/k$, where k denotes the root threshold of the access tree. Finally, our scheme provides lossless or approximately lossless recovery, and experimental results show that the PSNR of the recovered images is always greater than 30 dB and the SSIM usually exceeds 0.99. By sacrificing a little storage cost, our LM-SIS scheme can achieve *lossless recovery*.

Keywords: Secret image sharing · Linear secret sharing scheme · Access policy · Cauchy matrix · Lossless recovery

1 Introduction

Nowadays, with the popularization of photo devices and the development of social networks, information (particularly image information) sharing on the internet plays an important role in our daily lives. However, some images, such as sensitive personal images, and confidential military maps, need to be shared securely. Secret Image Sharing (SIS) [1] gives a secure sharing pattern of secrets among a set of participants. Take the threshold (i.e. k-out-of-n) SIS for example. A secret image is encoded into n shadows, and the secret image can be recovered iff we get k or more shadows. In some scenarios, participants may have different statuses and be assigned different privileges. Thus, Essential SIS (ESIS) [2] and Weighted SIS (WSIS) [3] are proposed with limited access policies. In an ESIS

Y. Lai et al. (Eds.): ICA3PP 2021, LNCS 13157, pp. 411–431, 2022.
https://doi.org/10.1007/978-3-030-95391-1_26

scheme, there are two kinds of shadows: essential shadows and non-essential shadows. The secret image can be recovered iff the number of shadows and the number of essential shadows are equal to or greater than the thresholds respectively. Furthermore, WSIS introduces weights for shadows to present diverse privileges of different participants. The secret image can be recovered iff the total weight of the collected shadows reaches the threshold.

However, ESIS and WSIS only implement a relatively limited access policy, which will severely restrict the flexibility of SIS in some situations. For example, in a company with many departments, each department may have subordinate relationships, the implementation of secret sharing in the company requires complex access policies. In this paper, we propose an SIS scheme that supports monotonous access policies. At the same time, considering computational efficiency and storage redundancy, the access policy should be presented in a compact form. Linear Secret Sharing Scheme (LSSS) [4] provides a smart way to share secrets through linear matrices. On the one hand, LSSS can present arbitrary monotonous access policies. On the other hand, LSSS has advantages in computational efficiency, since it can be processed by linear functions. However, LSSS can not be directly introduced into SIS, because LSSS only supports sharing a single secret value. Thus, it requires a multi-secret sharing scheme to support a flexible, efficient and secure SIS. Additionally, considering that the recovered image quality is very important in some cases. So the proposed SIS scheme should provide lossless or approximately lossless recovered images.

In this paper, we propose a Secret Matrix Sharing Scheme (SMSS) extended from LSSS. Different from LSSS, the proposed SMSS can securely share a matrix instead of a single value. The SMSS has advantages in both flexibility and efficiency. An SMSS-based SIS named LM-SIS is also proposed to achieve flexible image sharing. In the LM-SIS, an SMSS generating matrix is generated by a monotonous access policy, which is more flexible than the traditional threshold policy. Then shadows are generated by the SMSS generating matrix and the secret image matrix. Significantly, the generation process is very efficient, because it only invokes some linear transformations. During the recovery procedure, like LSSS, only the authorized participant sets can recover the secret image. Furthermore, the recovered images of LM-SIS are lossless or approximately lossless, which implies that the proposed scheme can support the secure sharing of high-quality images. In brief, the contributions can be summarized as follows:

1. We propose an SMSS, which can share a secret matrix with monotonous access policies. In detail, SMSS is an improved version of traditional LSSS.
2. We propose an SIS scheme based on the SMSS, which shares a secret image as a matrix with monotonous access policies.
3. Sufficient experimental results are given to evaluate the performance of our SIS scheme, and the results imply that our scheme is flexible, efficient and lossless.

The rest of the paper is organized as follows. Section 2 introduces the related work of the SIS scheme. Section 3 introduces the background of the proposed scheme like access policy and LSSS. The SMSS is proposed in Sect. 4. Section 5 proposes our LM-SIS scheme in detail. The experiments and analyses are shown in Sect. 6. Finally, conclusions and future work are drawn in Sect. 7.

2 Related Work

Secret Sharing. Secret sharing [5] refers to a cryptographic method that securely shares secrets, which has a wide range of applications, such as secure multiparty computation [6], SIS [7–9], and electronic voting [10]. There are many ways to realize secret sharing. In addition to the polynomial-based technique, the Chinese Remainder Theorem (CRT) [11] and LSSS can also implement secret sharing. LSSS can be regarded as a general promotion of Shamir's secret sharing scheme [5] and its formal definition was first proposed by Beimel [12]. LSSS can describe any secret sharing scheme by a LSSS matrix as long as the scheme is linear. However, Beimel did not provide a method to implement an access policy through LSSS. Thus, Lewko and Waters [4] proposed a scheme to convert any monotonous Boolean Formulas to the LSSS matrices. However, when the algorithm is applied to the Access Tree, a large LSSS matrix will be generated. To address this problem, Liu et al. [13] proposed a new algorithm, which can directly support threshold gates and obtain a smaller LSSS matrix. Nevertheless, the scalability is still ignored. Therefore, Wang et al. [14] proposed a block LSSS with strong scalability. When the access policy is updated, the LSSS matrix only needs to be partially modified.

Secret Image Sharing. SIS is an extension of secret sharing applied to images, so many methods in secret sharing can be introduced to SIS. Thien and Lin [15] embedded the secret image pixel value in the constant term of a $(k-1)$-degree polynomial over $\mathbb{Z}_{251}$ to achieve a (k, n) threshold SIS. Due to the truncation of pixel values, the recovered image will be distorted. To address this deficiency, Zhou et al. [16] regarded two adjacent pixels in the secret image as a secret value, and selected 65537 as the prime in the sharing polynomial to achieve lossless recovery. Meanwhile, the CRT-based SIS schemes have also been studied to achieve lossless recovery [1,17]. Another weakness of [15] is that part of the secret can be revealed from $k-1$ shadows, which will compromise the threshold property. To this end, Zhou et al. [18] applied the pixel encryption method. Although the shadow size is increased, the security of SIS is improved.

Extended Secret Image Sharing. All the schemes discussed above only implement a threshold access policy, in which each participant plays the same role. But in some cases, participants will be assigned different privileges based on their status. To meet this scenario, ESIS and WSIS have been proposed. Li et al. [19] firstly proposed a (t, s, k, n) ESIS, all n shadows are classified into s essential shadows and $n-s$ non-essential shadows. When recovering, k shadows included at least t essential shadows are required. However, in this scheme, the size of the essential shadows is not equal to the size of non-essential shadows, which may lead to security vulnerabilities. To solve this problem, Li et al. [20,21] proposed ESIS schemes to generate shadows with the same size and reduce the shadow size to 1/k times of the secret image respectively. However, the concatenation operation of sub-shadows will increase computational complexity. Wu et al. [22] proposed an ESIS scheme using derivative polynomials to solve this defect. To increase the flexibility of the scheme, Hu et al. [2] proposed a scalable (t, s, k, n) ESIS based on the Lagrange interpolation. Even if there is no essential shadow, k non-essential shadows can obtain the outline of the secret image. In a WSIS scheme, all shadows are assigned different weights and the secret image can be recovered iff the total weight of the collected shadows reaches the threshold. Tan et al. [23] proposed a WSIS based on the CRT which has progressive characteristics. That is, as the number of collected shadows increases, the quality of the recovered image also improves. Additionally, WSIS schemes based on other techniques have also been proposed, such as polynomial-based WSIS [3], and random grid-based WSIS [24].

However, even the access policies supported by WSIS and ESIS are relatively limited. In this paper, we propose a flexible and compact SIS scheme. Compared with the existing schemes, our scheme has the advantage of supporting monotonous access policies. Meanwhile, the quality of the recovered image is also high, so our scheme can be used for secret sharing of high-quality images.

3 Preliminaries

3.1 Access Policy

Access policy is also known as access structure, which is a core notion of secret sharing scheme. The formal definition of access policy is given as follows.

Definition 1 (Access Policy [13]**).** *Let* $\mathcal{P} = \{P_0, P_1, ..., P_{N-1}\}$ *be a set of participants.* $\forall \mathbb{A} \subset 2^{\{P_0, P_1, ..., P_{N-1}\}}$ *can be called an access policy of* P*. Furthermore, the access policy* $\mathbb{A}$ *is monotone if:*

$$\forall\, B, C \subset P : B \in \mathbb{A} \wedge B \subset C \rightarrow C \in \mathbb{A}.$$

Additionally, $\forall A \in \mathbb{A}$ is the authorized set of the access policy, otherwise, A is the unauthorized set. Furthermore, access policy can be presented in different forms, such as Access Tree, and Boolean Formula. For example, Fig. 1 shows an Access Tree, where non-leaf nodes represent threshold gates [14], and leaf nodes represent participants. Suppose the secret is denoted as $\mathbf{S}$ and the shadow set is denoted as $\bar{\mathbf{S}}$. Each $P_i \in \mathcal{P}$ gets a $\bar{\mathbf{S}}_i \in \bar{\mathbf{S}}$. The $\mathbf{S}$ can be outputted iff the recovery set $\mathbf{S}'$ is authorized. The existing SIS schemes cannot support the Access Tree, so this paper proposes an SIS scheme which can support monotonous access policies.

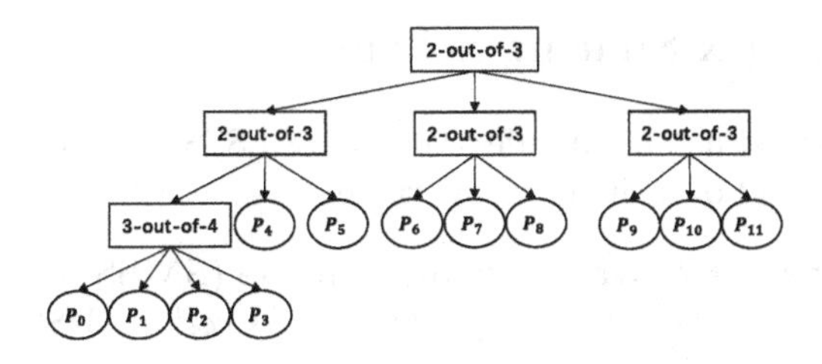

Fig. 1. Access tree

3.2 Linear Secret Sharing Schemes (LSSS)

As a commonly used form of access policy, LSSS is defined as follows:

Definition 2 (Linear Secret Sharing Schemes (LSSS) [13]). *Let Π be a secret sharing scheme over a set of participants $\mathcal{P}$, Π is linear (over $\mathbb{Z}_p$), if:*

1. *The shadows for each participant are presented as a vector over $\mathbb{Z}_p$.*
2. *There is a generating matrix $\mathbf{M}$, in which each row vector $\mathbf{M}_i$ corresponds to a participant $P_i \in \mathcal{P}$. Furthermore, there is a vector $\boldsymbol{v} = (s, r_1, ..., r_{d-1})^T$, where $s \in \mathbb{Z}_p$ is the secret, $r_1, ..., r_{d-1} \in \mathbb{Z}_p$ are random numbers and d denotes the column number of $\mathbf{M}$. Finally, the shadow of P_i is calculated as $\lambda_i = (\mathbf{M}\boldsymbol{v})_i$.*

Suppose Π is a LSSS of access policy $\mathbb{A}$, S be an authorized set and $I = \{i : P_i \in S\}$. We can find a set of constants $\{\omega_i \in \mathbb{Z}_p\}_{i \in I}$ in polynomial time such that $\sum_{i \in I} \omega_i \mathbf{M}_i = (1, 0, ..., 0)$. Furthermore, we can get $\sum_{i \in I} \omega_i \lambda_i = s$.

3.3 Image Quality Evaluation Indicator

The Peak to Signal Noise Ratio(PSNR) [25] and Structural Similarity(SSIM) [26] are two common indicators to measure the distortion of images. Let x and y be the original image and the recovered image, respectively. The first indicator PSNR is calculated as follows:

$$PSNR = 20 \cdot log_{10}(\frac{MAX_I}{\sqrt{MSE}}), \tag{1}$$

where MAX_I represents the maximum value of the image pixels, MSE is the mean square error of images x and y. The higher $PSNR$ implies y is more similar to x. The second indicator SSIM is given as the following equation:

$$SSIM(x, y) = \frac{(2\mu_x\mu_y + c_1)(2\sigma_{xy} + c_2)}{(\mu_x^2 + \mu_y^2 + c_1)(\sigma_x^2 + \sigma_y^2 + c_2)}, \tag{2}$$

where μ_x is the average of x, μ_y is the average of y, σ_x is the variance of x, σ_y is the variance of y, σ_{xy} is the covariance of x and y, c_1 and c_2 are constants, $SSIM \in [-1, 1]$. The higher $SSIM$ implies the higher similarity between x and y.

4 Secret Matrix Sharing Scheme

SMSS can be viewed as an extended version of LSSS, which is used to share a secret matrix instead of a single secret value.

Definition 3 (Secret Matrix Sharing Scheme (SMSS)). *Let Π be a secret sharing scheme over a set of participants $\mathcal{P}$, Π is called SMSS (over $\mathbb{Z}_p$), if:*

1. *The shadows for each participant are represented by a vector $S_i \in \mathbb{Z}_p^{1\times t}$.*
2. *There is a matrix $\mathbf{M}$ called the share-generating matrix with the access policy $\mathbb{T}$, and each row vector $\mathbf{M}_i$ corresponds to a participant $P_i \in \mathcal{P}$. Let matrix $\bar{\mathbf{S}} = [\mathbf{S}, \mathbf{R}]^T$, where $\mathbf{S} \in \mathbb{Z}_p^{t\times t}$ denotes the secret matrix and $\mathbf{R} \in \mathbb{Z}_p^{t\times(n-t)}$ denotes a non-zero random matrix. $\hat{\mathbf{S}} = \mathbf{M}\bar{\mathbf{S}}$ is calculated as shadow matrix, and each row vector $\hat{\mathbf{S}}_i = \mathbf{M}_i\bar{\mathbf{S}}$ denotes the shadow of P_i.*

Similar to LSSS, SMSS gets linear reconstruction property. Suppose Π is an SMSS of access policy $\mathbb{T}$, which is presented as generating matrix $\mathbf{M}$. Suppose S be a subset of $\mathcal{P}$ according with sub generating matrix $\mathbf{M}^{'}$ and shadow matrix $\mathbf{S}^{'}$, where $\mathbf{M}^{'}$ is composed of the row vector $\mathbf{M}_i, P_i \in S$ and $\mathbf{S}'$ is composed of according shadows $\hat{\mathbf{S}}_i, P_i \in S$. Iff S is an authorized set of $\mathbb{T}$, it can find matrix $\tilde{\mathbf{M}}$ in polynomial time to satisfy $\tilde{\mathbf{M}}\mathbf{M}^{'} = [\mathbf{I}_t, \mathbf{O}]$, where $\mathbf{I}_t$ denotes an identity matrix with order t and $\mathbf{O}$ denotes a zero matrix. Thus, we can get the secret matrix as $\mathbf{S}^* = \tilde{\mathbf{M}}\mathbf{S}^{'} = \mathbf{S}$. Two or more LSSS policies can be combined into one by the theorem proposed in [27]. Additionally, we introduce the theorem into SMSS to get a generating matrix with monotonous access policies.

Theorem 1. *Let $\mathbb{T}_1$ and $\mathbb{T}_2$ be the monotone access policies defined on the set of participants $\mathcal{P}_1$ and $\mathcal{P}_2$ respectively, and let $P_z \in \mathcal{P}_1$. Additionally, $\mathbb{T}_1$ and $\mathbb{T}_2$ are described by matrices $\mathbf{M}$ and $\bar{\mathbf{M}}$ respectively, where $\mathbf{M} = (m_{i,j})_{n\times k}$ and $\bar{\mathbf{M}} = (\bar{m}_{i,j})_{x\times y}$. The combined policy $\mathbb{T}_1(P_z \to \mathbb{T}_2)$ denotes the insertion of $\mathbb{T}_2$ at participant P_z in $\mathbb{T}_1$, which can be described by the following matrix:*

$$
\mathbf{M}_{\mathbb{T}_1(P_z\to\mathbb{T}_2)} = \begin{bmatrix}
m_{0,0} & \cdots & m_{0,k-1} & 0 & 0 & 0 \\
\vdots & \ddots & \vdots & 0 & 0 & 0 \\
m_{z,0}\bar{m}_{0,0} & \cdots & m_{z,k-1}\bar{m}_{0,0} & \bar{m}_{0,1} & \cdots & \bar{m}_{0,y-1} \\
\vdots & \ddots & \vdots & \vdots & \ddots & \vdots \\
m_{z,0}\bar{m}_{x-1,0} & \cdots & m_{z,k-1}\bar{m}_{x-1,0} & \bar{m}_{x-1,1} & \cdots & \bar{m}_{x-1,y-1} \\
\vdots & \ddots & \vdots & 0 & 0 & 0 \\
m_{n-1,0} & \cdots & m_{n-1,k-1} & 0 & 0 & 0
\end{bmatrix}. \quad (3)
$$

In reference [13], Liu et al. proposed a method to generate a Vandermonde matrix based LSSS matrix by Theorem 1. However, Cauchy matrix not only gets the same linear property as Vandermonde matrix, but also has advantages in efficiency. It implies that the time complexity of Cauchy matrix solving is only $O(n^2)$ while the time complexity of Vandermonde matrix is $O(n^3)$ [28]. Thus, we adopt the Cauchy matrix based SMSS to describe the access policy of SIS in this work. The definition of Cauchy matrix is given as follows.

Definition 4 (Cauchy matrix over $\mathbb{Z}_p$). *Let $X = \{x_0, ..., x_{m-1}\}$ and $Y = \{y_0, ..., y_{n-1}\}$ are two disjoint sets of elements over $\mathbb{Z}_p$, where p is a prime number. Cauchy matrix $\mathbf{C} = (c_{i,j})_{m\times n}$, where $c_{i,j} = 1/(x_i + y_j)$ and $x_i + y_j \not\equiv 0 \bmod p$ $(0 \le i \le m-1,\ 0 \le j \le n-1)$.*

In brief, a (k, n) threshold of SMSS can be described by a Cauchy matrix $\mathbf{C}_{n\times k}$. To improve computational efficiency, we provide a function *Divide*() to process the Cauchy matrix. Given a Cauchy matrix $\mathbf{C} = (c_{i,j})_{m\times n}$. The function $Divide(\mathbf{C}) = (c_{i,j})_{m\times(n-1)}$, where $c_{i,j} = (x_i+y_0)/(x_i+y_{j+1})$ $(0 \le i \le m-1,\ 0 \le j \le n-2)$.

Following [13], we propose Algorithm 1 as the SMSS generating matrix generation algorithm. Let $\mathbb{T}$ be an access policy, which is presented as an access tree. Algorithm 1 takes the access policy $\mathbb{T}$ as input, performs the depth-first traversal of $\mathbb{T}$ to get threshold sequence $TG(\mathbb{T}) = \{TG_0, ..., TG_{m-1}\}$ and generates a Cauchy matrix $\mathbf{C}_i$ $(0 \le i \le m-1)$ for each gate. Then, such Cauchy matrices are combined by Theorem 1 to describe the policy $\mathbb{T}$. Finally, the SMSS generating matrix $\mathbf{M}$ is outputted.

Algorithm 1. SMSSMatrix($\mathbb{T}$)

Input: Access policy $\mathbb{T}$
Output: The SMSS generating matrix $\mathbf{M}$
1: Perform the depth-first traversal of $\mathbb{T}$ to get $TG(\mathbb{T}) = \{TG_0, ..., TG_{m-1}\}$ and denote each TG_i as policy $\mathbb{T}_i$
2: Generate the Cauchy matrix $\mathbf{C}_i$ for each threshold-gate TG_i
3: $\mathbf{M} \leftarrow \mathbf{C}_0$
4: **for** i from 1 to $m-1$ **do**
5: $\quad \mathbf{C}_i \leftarrow Divide(\mathbf{C}_i)$
6: $\quad$ Get the row index k of TG_i
7: $\quad \mathbf{M} \leftarrow \mathbf{M}_{\mathbb{T}_{i-1}(P_k \to \mathbb{T}_i)}$
8: $\quad$ Update the row index of nodes $TG_i \in TG(\mathbb{T})$
9: **end for**
10: **return** $\mathbf{M}$

It is important that the flexibility of SMSS is stronger than the access policies of other SIS schemes. For example, Fig. 2 shows a complex access tree which can be easily described by an SMSS. However, traditional ESIS and WSIS schemes can not support such complex access policies.

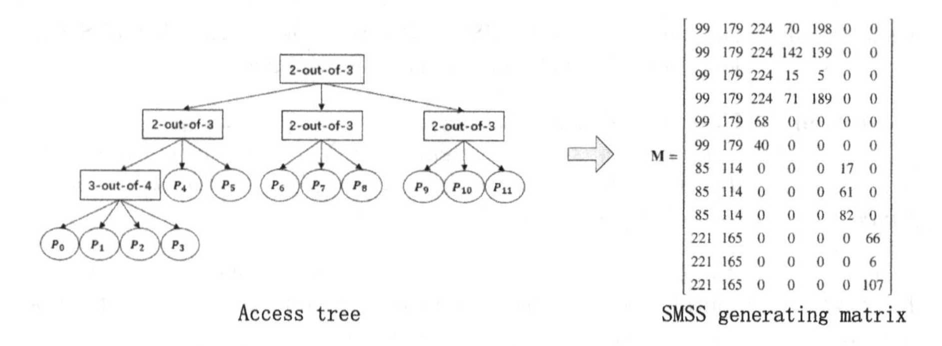

Fig. 2. Access policy to SMSS generating matrix

Furthermore, the security and correctness of SMSS are intuitive, since it is an All-or-Nothing Transformation (AONT) [29] scheme. AONT means that authorized set can get all (i.e. the original secret matrix can be reconstruct correctly), and unauthorized set can get nothing (i.e. the original secret matrix cannot be reconstruct correctly). "All" corresponds to the correctness of the scheme, and "Nothing" corresponds to the security of the scheme.

Theorem 2. *The SMSS is an AONT scheme.*

Proof. The theorem can be easily proved according to the theories of monotone span programs [27] and LSSS [13]. □

5 Linear Matrix Based Secret Image Sharing Scheme

Based on the SMSS, we propose the Linear Matrix based Secret Image Sharing (LM-SIS) scheme which supports arbitrary monotonous access policies. The LM-SIS scheme includes three procedures: secret image preprocessing, shadow generation and recovery.

5.1 Secret Image Preprocessing

Considering the property of digital images that each pixel can be viewed as an element of $GF(2^8)$, the computation of LM-SIS is performed in $\mathbb{Z}_p$, where p is a prime number close to 256 (i.e. 2^8). The LM-SIS can be divided into two cases: 251-LM-SIS and 257-LM-SIS, that is, the prime number is picked as 251 or 257. The LM-SIS secret image preprocessing procedure is shown in Fig. 3.

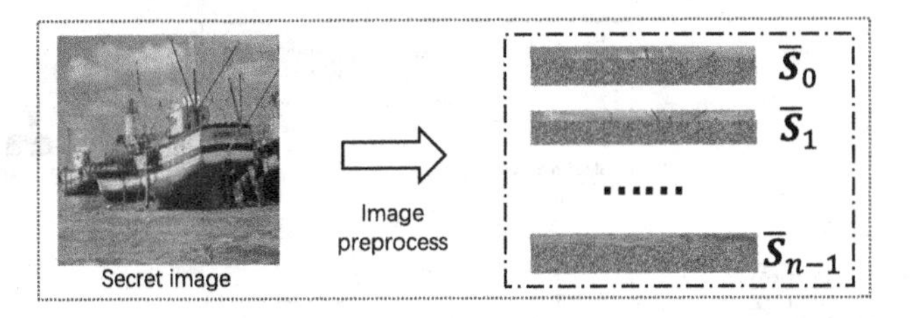

Fig. 3. Secret image preprocess

Algorithm 2. Preprocess the secret image matrix

Input: Secret image matrix $\mathbf{S}_{M\times L}$; Parameter k and T; Prime number p
Output: The set of secret matrices $\bar{\mathbf{S}} = \{\bar{\mathbf{S}}_j | \ 0 \leq j \leq n-1\}$
1: **if** $p \leq 256$ **then**
2: Truncate all pixel value larger than $p-1$ to $p-1$
3: **end if**
4: Divide each k rows of the $\mathbf{S}$ into a unit, obtained $\bar{\mathbf{S}} = \{\bar{\mathbf{S}}_j | \ 0 \leq j \leq n-1\}$
5: **for** each $\bar{\mathbf{S}}_j$ **do**
6: Generate a non-zero random matrix $\mathbf{R}^*_{L\times(T-k)}$
7: $\bar{\mathbf{S}}_j \leftarrow (\bar{\mathbf{S}}_j^T, \ \mathbf{R}^*)^T$
8: **end for**
9: **return** $\bar{\mathbf{S}}$

We propose Algorithm 2 to transform image $\mathbf{S}$ into secret matrix set $\bar{\mathbf{S}}$. Suppose the size of the secret image is $M \times L$. Algorithm 2 first maps all pixels of $\mathbf{S}$ into $\mathbb{Z}_p$, where p is a prime number. Then the matrix over $\mathbb{Z}_p$ is divided into a set of secret matrices. $\bar{\mathbf{S}} = \{\bar{\mathbf{S}}_j | \ 0 \leq j \leq n-1\}$, where the size of each $\bar{\mathbf{S}}_j \in \bar{\mathbf{S}}$ is $T \times L$, $n = \lceil M/k \rceil$, k denotes the root threshold of access policy and T denotes the column number of the SMSS generating matrix $\mathbf{M}$.

5.2 Shadow Generation

Figure 4 shows the shadow generation procedure of LM-SIS. The shadow generation requires two matrices: the SMSS generating matrix $\mathbf{M}$ and the random full rank matrix $\mathbf{R}$. We call Algorithm 1 to generate matrix $\mathbf{M}$ with access policy $\mathbb{T}$. The matrix $\mathbf{R}$ is composed of two non-zero blocks $\mathbf{R}^{'}$ and $\mathbf{R}^*$. Furthermore, $\mathbf{R}^{'}$ is a $k \times k$ matrix and $\mathbf{R}^*$ is a $(T-k) \times (T-k)$ matrix, where k denotes the root threshold of $\mathbb{T}$ and T denotes the column number of the SMSS generating matrix. For simplicity, it can take the identity matrix $\mathbf{I}_{(T-k)}$ as $\mathbf{R}^*$.

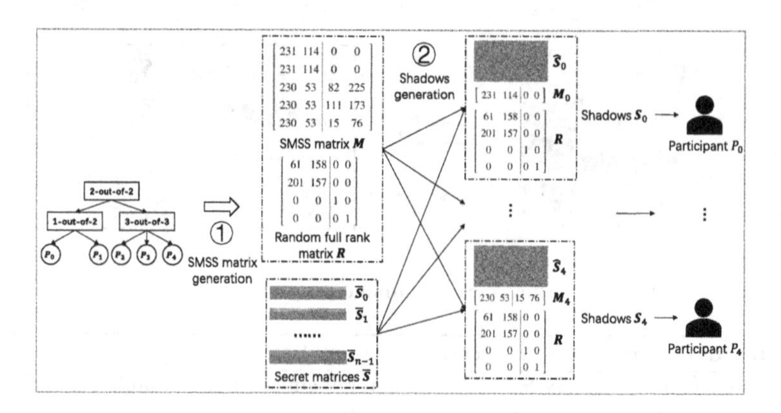

Fig. 4. Shadow generation

Algorithm 3. Generate the shadow matrices

Input: Access policy $\mathbb{T}$; Random matrix $\mathbf{R}$; The set of secret matrices $\bar{\mathbf{S}}$
Output: The set of shadow matrices $\hat{\mathbf{S}} = \{\hat{\mathbf{S}}_i | \ 0 \leq i \leq N-1\}$

1: $\mathbf{M}$ = SMSSMatrix($\mathbb{T}$) // Call Algorithm 1
2: **for** each $\mathbf{M}_i$ **do**
3: $\quad \mathbf{M}_i^{(0)} \leftarrow \mathbf{M}_i$
4: $\quad \hat{\mathbf{S}}_{i,0} \leftarrow \mathbf{M}_i^{(0)} \bar{\mathbf{S}}_0 \ mod \ p$
5: $\quad$ **for** j from 1 to $n-1$ **do**
6: $\quad\quad \mathbf{M}_i^{(j)} \leftarrow \mathbf{M}_i^{(j-1)} \mathbf{R} \ mod \ p$
7: $\quad\quad \hat{\mathbf{S}}_{i,j} \leftarrow \mathbf{M}_i^{(j)} \bar{\mathbf{S}}_j \ mod \ p$
8: $\quad$ **end for**
9: $\quad \hat{\mathbf{S}}_i \leftarrow (\hat{\mathbf{S}}_{i,0}, ..., \hat{\mathbf{S}}_{i,n-1})^T$
10: $\quad$ **if** $p > 256$ **then**
11: $\quad\quad$ Each value in $\hat{\mathbf{S}}_i$ is represented by 9 or more bits, and every 8 bits of $\hat{\mathbf{S}}_i$ is saved as a pixel of shadow images $\hat{\mathbf{S}}$
12: $\quad$ **end if**
13: **end for**
14: **return** $\hat{\mathbf{S}}$

Algorithm 3 shows the shadow generation algorithm. Each shadow matrix $\hat{\mathbf{S}}_i$ is composed of a set of shadow vectors $\hat{\mathbf{S}}_{i,j}$ $(0 \leq j \leq n-1)$. The computation of the shadow vector $\hat{\mathbf{S}}_{i,j}$ is similar to the Output Feedback (OFB) mode [30] which is usually used in block ciphers. As shown in Fig. 5, the core idea of OFB mode is getting each shadow vector $\hat{\mathbf{S}}_{i,j}$ by the vector $\mathbf{M}_i^{(j)}$, where $\mathbf{M}_i^{(j)}$ is generated by the row $\mathbf{M}_i$ of matrix $\mathbf{M}$ and the matrix $\mathbf{R}$. Finally, the matrix $\hat{\mathbf{S}}_i$, row vector $\mathbf{M}_i$ and the random matrix $\mathbf{R}$ are sent to participant P_i as the shadow.

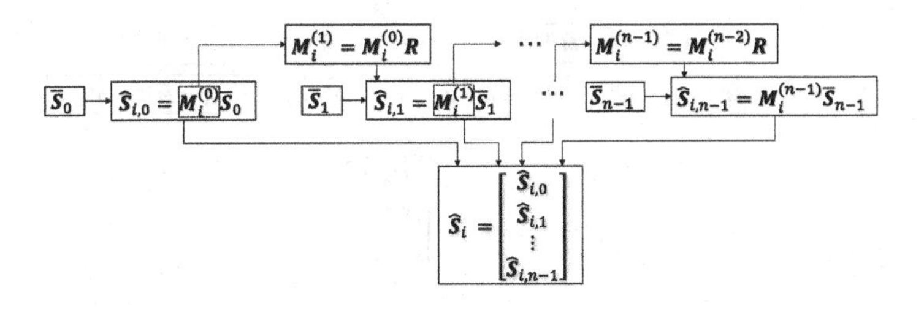

Fig. 5. Shadow generation of LM-SIS

5.3 Recovery

The recovery procedure of LM-SIS is shown in Fig. 6. Assume that $I^{'}$ is an authorized set of access policy $\mathbb{T}$ and the number of participants in the authorized set $|I^{'}| = N^{'}$. The secret image $\mathbf{S}$ can be recovered after all participants $P_i \in I^{'}$ submit their shadows. Then, it gets the intermediate matrices $\mathbf{S}_j^{'} = [\hat{\mathbf{S}}_{0,j}, ..., \hat{\mathbf{S}}_{N'-1,j}](0 \leq j \leq n-1)$, where $\hat{\mathbf{S}}_{i'} = [\hat{\mathbf{S}}_{i',0}, ..., \hat{\mathbf{S}}_{i',n-1}]$ denotes the shadow of P_i and $i' \in I'$. The recovery SMSS matrix $\mathbf{M}^{'}$ is composed of the row vector $\mathbf{M}_{i'}$, $i^{'} \in I^{'}$. Following the definition of SMSS, it can find a matrix $\tilde{\mathbf{M}}$ such that $\tilde{\mathbf{M}}\mathbf{M}^{'} = [\mathbf{I}_t, \ \mathbf{O}]$. Furthermore, we calculate the inverse matrix $\mathbf{R}^{'-1}$ of the sub-matrix $\mathbf{R}^{'}$ of the random matrix $\mathbf{R}$. Algorithm 4 takes $\mathbf{R}^{'-1}$, $\tilde{\mathbf{M}}$, $\mathbf{S}^{'}$ as input and outputs the recovered image $\mathbf{S}^*$, where $\mathbf{S}^* = [\mathbf{S}_0^*, ..., \mathbf{S}_{n-1}^*]^T$. The $\mathbf{S}^*$ is combined by the components $\mathbf{S}_j^*$ $(0 \leq j \leq n-1\})$, and the calculation of secret image recovery is shown in Fig. 7.

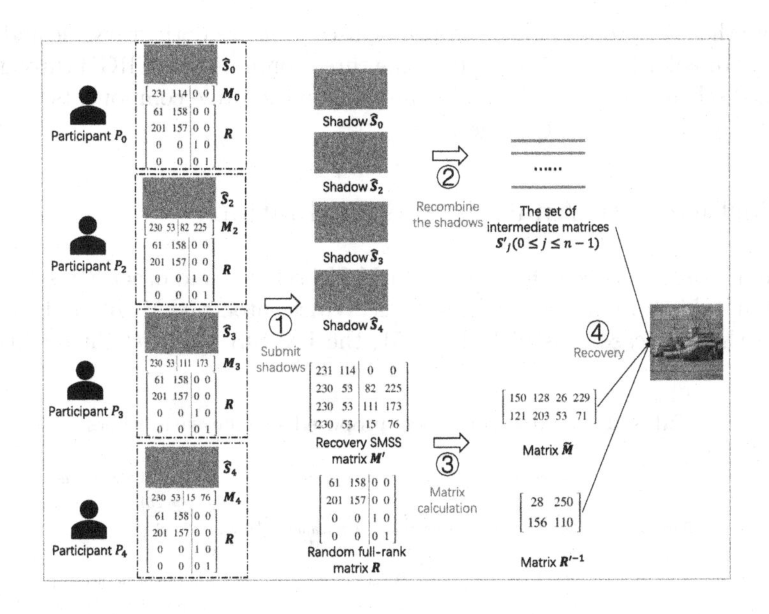

Fig. 6. Recovery

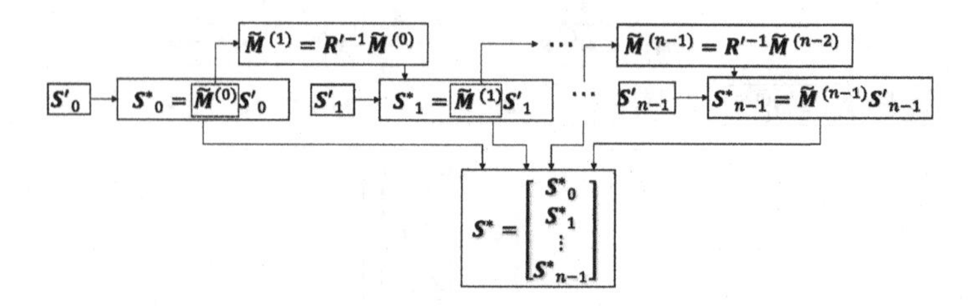

Fig. 7. Image recovery of LM-SIS

Algorithm 4. Recovery

Input: Inverse matrix $\mathbf{R}^{'-1}$; Matrix $\tilde{\mathbf{M}}$; The set of intermediate matrices $\mathbf{S}^{'} = \{\mathbf{S}_j^{'} | 0 \leq j \leq n-1\}$
Output: Recovered image matrix $\mathbf{S}^*$
1: **if** $p > 256$ **then**
2: $\quad \mathbf{S}^{'}$ is represented by a set of 8-bit pixel, and every 9 bits is saved as a pixel
3: **end if**
4: $\tilde{\mathbf{M}}^{(0)} = \tilde{\mathbf{M}}$
5: $\mathbf{S}_0^* = \tilde{\mathbf{M}}^{(0)}\mathbf{S}_0^{'} \ mod \ p$
6: **for** j from 1 to $n-1$ **do**
7: $\quad \tilde{\mathbf{M}}^{(j)} = \mathbf{R}^{'-1} \ \tilde{\mathbf{M}}^{(j-1)} \ mod \ p$
8: $\quad \mathbf{S}_j^* = \tilde{\mathbf{M}}^{(j)}\mathbf{S}_j^{'} \ mod \ p$
9: **end for**
10: $\mathbf{S}^* = (\mathbf{S}_0^*, ..., \mathbf{S}_{n-1}^*)^T$
11: **return** $\mathbf{S}^*$

Note that the proposed LM-SIS can be used to share both grayscale and color images. For color images, it can share the three components of RGB through the proposed scheme respectively, and finally merge the three components to realize the sharing of color secret images.

6 Performance Analysis and Evaluation

The comparison between the proposed LM-SIS scheme and other SIS schemes is shown in Table 1. Significantly, LM-SIS is divided into 251-LM-SIS and 257-LM-SIS. In the first case, p is picked as 251, the LM-SIS brings slight quality loss

Table 1. Comparison of the proposed scheme with others

	Yan et al.'s [9]	Wu et al.'s [22]	Tan et al.'s [23]	Our LM-SIS scheme	
				$p = 251$	$p = 257$
Method	Random grid	Derivative polynomial	Chinese remainder theorem	SMSS	
Shadow size ratio	1	$1/k$	1	$1/k$	$9/8k$
Access structure	(k, n)	(t, s, k, n)	Weighted (k, n)	Monotonous access policy	
Recovery quality	Lossy	Lossless	Lossless	Approximate lossless	Lossless

of recovered images while it gets the more compact shadow size. In the other case (i.e. $p = 257$), each value of the LM-SIS matrix should be performed as a 9-bit unit instead of an 8-bit unit. It implies that the LM-SIS scheme over $\mathbb{Z}_{257}$ can keep the secret image quality but also brings some storage redundancy. Compared with other schemes, the advantage of our scheme is the flexibility of the access policy. As shown in Table 1, lots of methods are proposed to construct SIS schemes, such as random grid [9], derivative polynomial [22] and Chinese remainder theorem [23]. However, these schemes only support limited access policies. For example, Yan et al. [9] implemented a traditional (k, n)-SIS, Wu et al. [22] implemented an ESIS and Tan et al. [23] proposed a WSIS. Our LM-SIS is more flexible than others, because SMSS provides a monotonous access policy. Meanwhile, LM-SIS also has advantages in storage costs and recovered quality. On the one hand, it gets smaller shadow size than most SIS schemes. On the other hand, it can provide lossless or approximate lossless recovered images.

We first take the images in Fig. 8 as examples to evaluate the performance of LM-SIS. The simulation runs on the PYCHARM[1] platform, and the computer used in the experiments is Intel(R) Core(TM) i5-9300H CPU 2.40 GHz, RAM 8.00 GB. As shown in Fig. 8, there are two grayscale images and two color images with size 512×512.

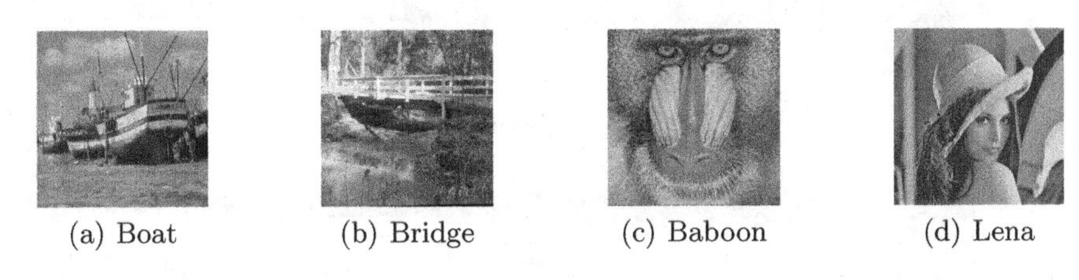

(a) Boat (b) Bridge (c) Baboon (d) Lena

Fig. 8. The secret images

Figure 9 shows the shadows of different images generated by LM-SIS. Although the secret images are different, the shadows they generate are meaningless. It implies that our scheme reduces the probability of information leakage when the shadow is obtained by attackers. Meanwhile, it can also be seen from Fig. 9, the shadow size of LM-SIS is related to the parameter k, where k denotes the root threshold of access policy. It is clear that the size ratio of the shadow is approximately $1/k$ or $9/8k$ while the prime number is p equal to 251 or 257. This is because in the shadow generation procedure, one row of the shadow is generated by k rows of secret image. The shadow size in [9] and [23] is equal to secret image and the shadow size in [22] is $1/k$ of the secret image, which proves the advantages of our scheme in storage. Furthermore, Fig. 10 presents the influence of k on the shadow size. It can be seen that the number of rows of the shadow increases linearly with the number of rows of the secret image and the slope is equal to the size ratio of the shadow.

[1] PyCharm is a Python Integrated Development Environment.

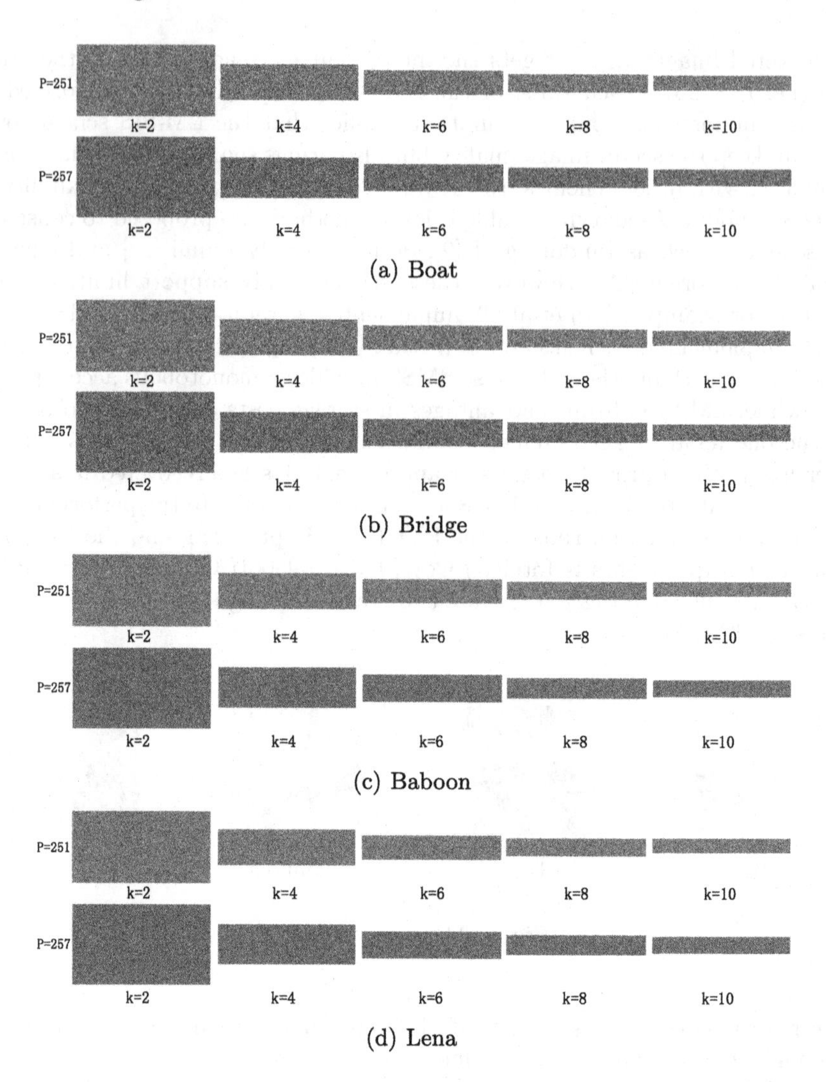

Fig. 9. The shadows

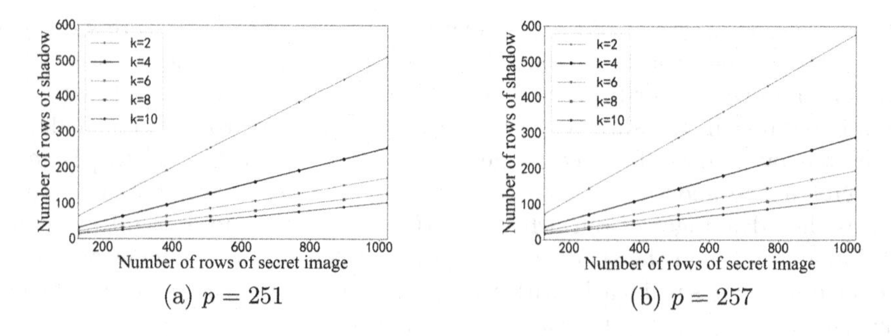

Fig. 10. The size of the shadows

To evaluate the recovered image quality, Fig. 11 shows the recovered images in various cases, the first line are the recovered images over $\mathbb{Z}_{251}$, and the second line are the recovered images over $\mathbb{Z}_{257}$. It can be seen that the recovered images over $\mathbb{Z}_{251}$ are approximately lossless, because only a few pixels larger than 250 are truncated. Meanwhile, the recovered images over $\mathbb{Z}_{257}$ are lossless. In the 251-LM-SIS and 257-LM-SIS schemes, the difference between the recovered images of different schemes is always negligible. In brief, the recovered image loss of 251-LM-SIS is too slight to be perceived visually. Furthermore, for the case $p = 251$, the PSNR and SSIM of such recovered images are given in Fig. 12. Generally speaking, PSNR of 30–40 dB or SSIM over 0.98 indicates that the image quality is good. It can be seen that the PSNR and SSIM of our recovered images are always exceed the reference line (i.e. good visual effect), the PSNR is always greater than 30 dB and the SSIM usually exceeds 0.99. In general, our LM-SIS scheme has high and stable recovered image quality.

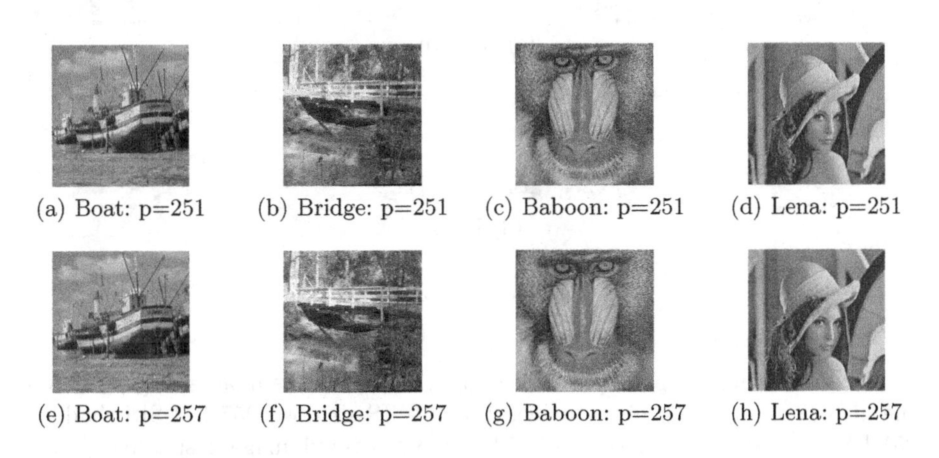

(a) Boat: p=251 (b) Bridge: p=251 (c) Baboon: p=251 (d) Lena: p=251

(e) Boat: p=257 (f) Bridge: p=257 (g) Baboon: p=257 (h) Lena: p=257

Fig. 11. The recovered images

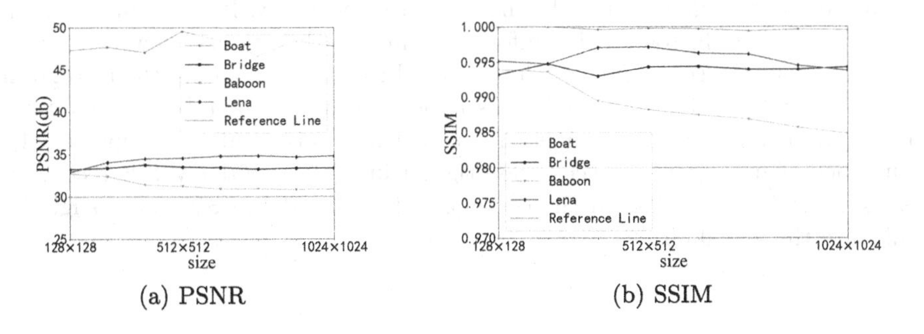

(a) PSNR (b) SSIM

Fig. 12. The PSNR and SSIM of the recovered images

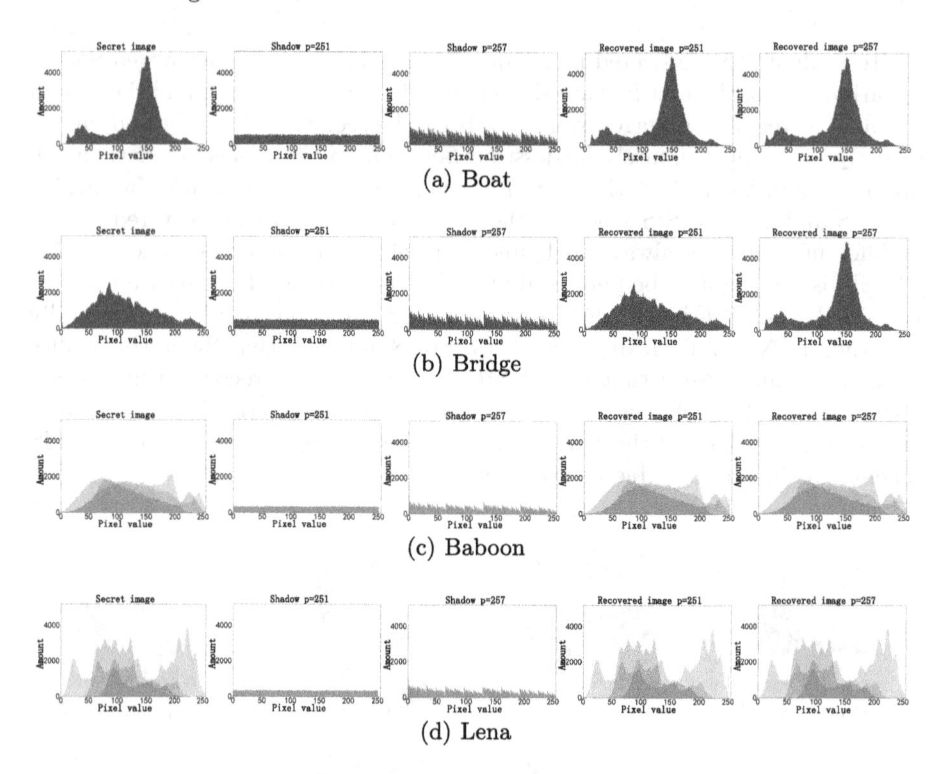

(a) Boat

(b) Bridge

(c) Baboon

(d) Lena

Fig. 13. The histogram

For presenting the pixel probability distribution of the images, Fig. 13 shows the histogram of the original image, 251-LM-SIS shadow, 257-LM-SIS shadow, 251-LM-SIS recovered image and 257-LM-SIS recovered image, respectively. Furthermore, for color images, Fig. 13(c) and Fig. 13(d) present R, G, B pixel distributions by red, green, and blue histogram respectively. In summary, in each case, the shadow histogram is uniform. It implies that the shadow can reveal nothing about the secret image information. In the third column, the histogram of 257-LM-SIS shadow is jagged. This is because we use nine bits to represent a pixel value and save every eight bits as a shadow pixel value. Meanwhile, the difference of histogram between the original image and 251-LM-SIS recovered image is negligible. Thus, the image quality loss of 251-LM-SIS is too slight to influence the visual effect.

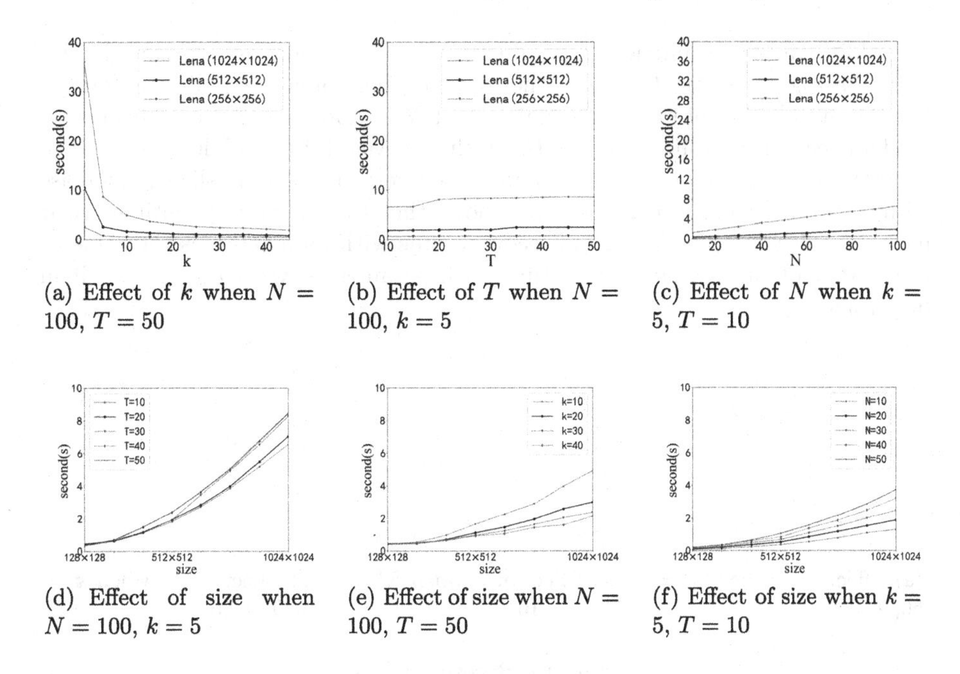

(a) Effect of k when $N = 100$, $T = 50$

(b) Effect of T when $N = 100$, $k = 5$

(c) Effect of N when $k = 5$, $T = 10$

(d) Effect of size when $N = 100$, $k = 5$

(e) Effect of size when $N = 100$, $T = 50$

(f) Effect of size when $k = 5$, $T = 10$

Fig. 14. Shadow generation time

To evaluate the efficiency of the proposed LM-SIS, we simulate the scheme in various cases to get simulation relevant data. The simulation aims to evaluate the influence of the root threshold k, the SMSS generating matrix column number T, and the participant number N on LM-SIS runtime. The test cases of running time are grayscale image Lena, unless specified otherwise. For reliability, the simulation of each case is run 20 times and we take the average runtime as the result shown in Fig. 14. It shows the following conclusions:

1. The shadow generation time is decreasing as k increases.
2. The shadow generation time increases slightly, while T is increasing.
3. The shadow generation time approximately linearly increases with N.
4. The size of the secret image seriously affects shadow generation time. The larger-sized secret image will have a longer shadow generation time with the same size access policy.

A larger k brings a less number of secret matrix units, that is, a secret matrix with more rows can be processed in one operation, so the shadow generation time will be reduced. The increase of T will slightly increase the shadow generation time, this is because the random matrix $\mathbf{R}^*$ has $T-k$ rows, and the preprocessing time of the secret image will increase. N represents the number of participants and the generation time of each shadow is approximately the same, so N has a linear relationship with the shadow generation time. At the same time, the size of the secret image will also affect the shadow generation time. The larger the secret matrix corresponding to the secret image, the more calculations are required.

Additionally, the impact of parameters k, T, $N^{'}$ on recovery time is also shown in Fig. 15, where k denotes the root threshold of access policy, T denotes the column number of the SMSS matrix and N' denotes the number of shadows used for recovery. As shown in Fig. 15(a), the recovered time will decrease while k is increasing. Figure 15(b) implies that the recovery time should slightly increase with T. At the same time, Fig. 15(c) shows that the recovering runtime linear increases with $N^{'}$. The change of recovery time with parameters is the same as the change of shadow generation time with parameters, we won't explain it in detail here.

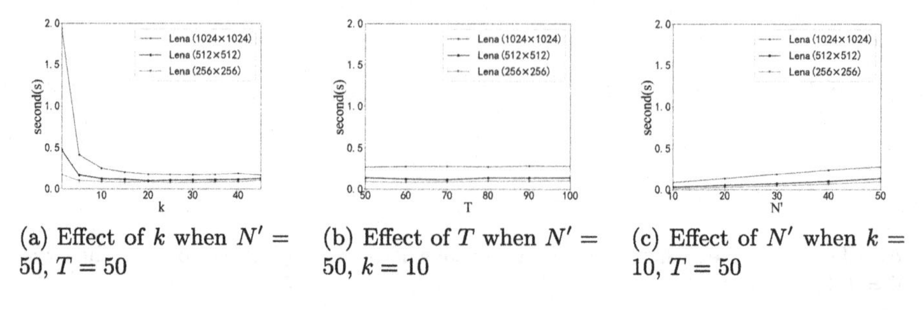

(a) Effect of k when $N' = 50$, $T = 50$
(b) Effect of T when $N' = 50$, $k = 10$
(c) Effect of N' when $k = 10$, $T = 50$

Fig. 15. Recovery time

Finally, Fig. 16 shows the shadow generation time in 251-LM-SIS and 257-LM-SIS affected by parameters k, T, N, where k denotes the root threshold of access policy, T denotes the column number of the SMSS generating matrix and N denotes the row number of SMSS generating matrix. We use the color image Lena as an example to simulate the scheme. The runtime decreases with k and increases with T and N. It can be seen that the shadow generation time is significantly longer than that in Fig. 14. This is because our LM-SIS scheme needs to process the RGB three components of the color image respectively. Furthermore, we can see from Fig. 16 that the runtime difference between 251-LM-SIS and 257-LM-SIS is too slight to be detected. Because converting the eight-bit pixel value of the secret image to nine-bit does not cost a lot of computational overhead, the shadow generation time is mainly concentrated in matrix multiplication.

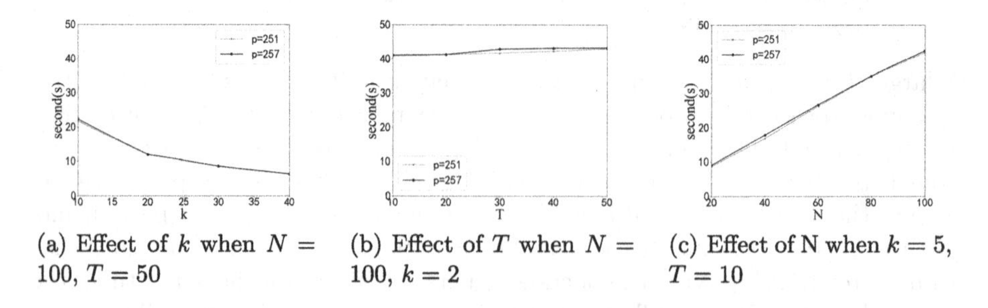

(a) Effect of k when $N = 100$, $T = 50$
(b) Effect of T when $N = 100$, $k = 2$
(c) Effect of N when $k = 5$, $T = 10$

Fig. 16. Shadow generation time of 251-LM-SIS and 257-LM-SIS

7 Conclusions and Future Work

In this paper, we extend the traditional LSSS into the secret matrix and propose an SMSS which is used to share a secret matrix. Furthermore, we present an SMSS based SIS scheme named LM-SIS. Compared with WSIS and ESIS, our scheme provides a more flexible access policy. Thus it can be implemented in more complex scenarios. Meanwhile, the proposed LM-SIS also has advantages in efficiency, storage cost and image quality. The superiority of our scheme can be summarized as follows. Firstly, SMSS can provide a monotonous access policy for LM-SIS and Cauchy matrix improves the efficiency of LM-SIS. Secondly, the shadow size of LM-SIS is smaller than most existing SIS schemes. Finally, both 251-LM-SIS and 257-LM-SIS support high image quality, since the recovered images of such LM-SIS schemes are lossless or approximately lossless. In conclusion, the proposed LM-SIS scheme is flexible, low-load and lossless of image quality. How to design an SIS scheme in which the shadows are meaningful images is our future work.

Acknowledgment. This work was supported by the National Science Foundation of China (No. 61802083, 61862011), the Natural Science Foundation of Guangxi (2018GXNSFBA281164, 2018GXNSFAA138116).

References

1. Li, L., Yuliang, L., Yan, X., Liu, L., Tan, L.: Lossless (k, n)-threshold image secret sharing based on the Chinese remainder theorem without auxiliary encryption. IEEE Access **7**, 75113–75121 (2019)
2. Yan-Xiang, H., Liu, Y.-N.: A progressively essential secret image sharing scheme using hierarchy shadow. J. Inf. Secur. Appl. **47**, 371–376 (2019)
3. Wang, Y., Chen, J., Gong, Q., Yan, X., Sun, Y.: Weighted polynomial-based secret image sharing scheme with lossless recovery. Secur. Commun. Netw. **1–11**, 2021 (2021)
4. Lewko, A.B., Waters, B.: Decentralizing attribute-based encryption. IACR Cryptol. ePrint Arch. **2010**, 351 (2010)
5. Shamir, A.: How to share a secret. Commun. ACM **22**(11), 612–613 (1979)
6. von Maltitz, M., Carle, G.: A performance and resource consumption assessment of secret sharing based secure multiparty computation. In: Garcia-Alfaro, J., Herrera-Joancomartí, J., Livraga, G., Rios, R. (eds.) DPM/CBT -2018. LNCS, vol. 11025, pp. 357–372. Springer, Cham (2018). https://doi.org/10.1007/978-3-030-00305-0_25
7. Yang, C.-N., Yang, Y.-Y.: On the analysis and design of visual cryptography with error correcting capability. IEEE Trans. Circuits Syst. Video Technol. **31**(6), 2465–2479 (2021)
8. Yan, X., Yuliang, L., Yang, C.-N., Zhang, X., Wang, S.: A common method of share authentication in image secret sharing. IEEE Trans. Circuits Syst. Video Technol. **31**(7), 2896–2908 (2021)

9. Yan, X., Liu, X., Yang, C.-N.: An enhanced threshold visual secret sharing based on random grids. J. Real-Time Image Process. **14**(1), 61–73 (2015). https://doi.org/10.1007/s11554-015-0540-4
10. Li, J., Wang, X., Huang, Z., Wang, L., Xiang, Y.: Multi-level multi-secret sharing scheme for decentralized e-voting in cloud computing. J. Parallel Distrib. Comput. **130**, 91–97 (2019)
11. Jia, X., Wang, D., Nie, D., Luo, X., Sun, J.Z.: A new threshold changeable secret sharing scheme based on the Chinese remainder theorem. Inf. Sci. **473**, 13–30 (2019)
12. Beimel, A.: Secure schemes for secret sharing and key distribution. Ph.D. dissertation, Israel Institute of Technology, Technion, Haifa, Israel (1996)
13. Liu, Z., Cao, Z., Wong, D.S.: Efficient generation of linear secret sharing scheme matrices from threshold access trees. IACR Cryptol. ePrint Arch. **2010**, 374 (2010)
14. Wang, J., Huang, C., Xiong, N.N., Wang, J.: Blocked linear secret sharing scheme for scalable attribute based encryption in manageable cloud storage system. Inf. Sci. **424**, 1–26 (2018)
15. Thien, C.-C., Lin, J.-C.: Secret image sharing. Comput. Graph. **26**(5), 765–770 (2002)
16. Zhou, X., Yuliang, L., Yan, X., Wang, Y., Liu, L.: Lossless and efficient polynomial-based secret image sharing with reduced shadow size. Symmetry **10**(7), 249 (2018)
17. Yan, X., Yuliang, L., Liu, L., Wan, S., Ding, W., Liu, H.: Chinese remainder theorem-based secret image sharing for (k, n) threshold. In: Third International Conference Cloud Computing and Security, vol. 10603, pp. 433–440 (2017)
18. Zhou, Z., Yang, C.-N., Cao, Y., Sun, X.: Secret image sharing based on encrypted pixels. IEEE Access **6**, 15021–15025 (2018)
19. Li, P., Yang, C.-N., Chih-Cheng, W., Kong, Q., Ma, Y.: Essential secret image sharing scheme with different importance of shadows. J. Vis. Commun. Image Represent. **24**(7), 1106–1114 (2013)
20. Li, P., Yang, C.-N., Zhou, Z.: Essential secret image sharing scheme with the same size of shadows. Digit. Signal Process. **50**, 51–60 (2016)
21. Li, P., Liu, Z.: An improved essential secret image sharing scheme with smaller shadow size. Int. J. Digit. Crime Forensics **10**(3), 78–94 (2018)
22. Zhen, W., Liu, Y.-N., Wang, D., Yang, C.-N.: An efficient essential secret image sharing scheme using derivative polynomial. Symmetry **11**(1), 69 (2019)
23. Tan, L., Yuliang, L., Yan, X., Liu, L., Li, L.: Weighted secret image sharing for a (k, n) threshold based on the Chinese remainder theorem. IEEE Access **7**, 59278–59286 (2019)
24. Liu, Z., Zhu, G., Ding, F., Kwong, S.: Weighted visual secret sharing for general access structures based on random grids. Signal Process. Image Commun. **92**, 116129 (2021)
25. Yadav, M., Singh, R.: Essential secret image sharing approach with same size of meaningful shares. Multimedia Tools Appl. (1), 1–18 (2021). https://doi.org/10.1007/s11042-021-10625-5
26. Kukreja, S., Kasana, G., Kasana, S.S.: Copyright protection scheme for color images using extended visual cryptography. Comput. Electr. Eng. **91**, 106931 (2021)
27. Nikov, V., Nikova, S.: New monotone span programs from old. IACR Cryptol. ePrint Arch. **2004**, 282 (2004)
28. Drmac, Z., Mezic, I., Mohr, R.: Data driven Koopman spectral analysis in Vandermonde-Cauchy form via the DFT: numerical method and theoretical insights. SIAM J. Sci. Comput. **41**(5), A3118–A3151 (2019)

29. Esfahani, N.N., Stinson, D.R.: On security properties of all-or-nothing transforms. Des. Codes Cryptogr. **89**, 2857–2867 (2021). https://doi.org/10.1007/s10623-021-00958-5
30. Silitonga, A., Jiang, Z., Khan, N., Becker, J.: Reconfigurable module of multi-mode AES cryptographic algorithms for AP SoCs. In: IEEE Nordic Circuits and Systems Conference, pp. 1–7 (2019)

Robust Multi-model Personalized Federated Learning via Model Distillation

Adil Muhammad(✉), Kai Lin, Jian Gao, and Bincai Chen

Dalian University of Technology, Dalian, China
{Muhammadadil2994,22009183}@mail.dlut.edu.cn, {link,china}@dlut.edu.cn

Abstract. Federated Learning(FL) is a popular privacy-preserving machine learning paradigm that enables the creation of a robust centralized model without sacrificing the privacy of clients' data. FL has a wide range of applications, but it does not integrate the idea of independent model design for each client, which is imperative in the areas like healthcare, banking sector, or AI as service (AIaaS), due to the paramount importance of data and heterogeneous nature of tasks. In this work, we propose a Robust Multi-Model FL (RMMFL) framework, an extension to FedMD, which under the same set of assumptions significantly improves the results of each individual model. RMMFL adapted two changes in the FedMD training process: First, a high entropy aggregation method is introduced to soften the output predictions. Second, a weighted ensemble technique is used to weigh the predictions of each client model in line with their performance. Extensive experiments are performed on heterogeneous models using CIFAR/MNIST as benchmark datasets, the simulations results obtained from our study shows that RMMFL exceeds the accuracy by over 5% compare to the baseline method.

Keywords: Federated Learning · Model heterogeneity · Knowledge distillation · Weighted ensemble · Statistical data heterogeneity

1 Introduction

The unprecedented advancement of deep learning approaches stands on artificial neural networks that have provided immense breakthroughs in various AI tasks such as object detection, speech recognition, and text translation, etc. However, the main ingredient for performing these tasks requires vast amounts of training data collected from different users mostly in a privacy-invasive manner. Recently people are growingly more concerned about the privacy of their data. With the formation of data protection regulations like General Data Protection Regulation (GDPR) by the European Union [1] and Health Insurance Portability and Accountability Act (HIPAA) [2] contributed to the development of different Privacy-Preserved Deep Learning approaches such as [3] and Federated Learning(FL) [4]. FL is a distributive machine learning paradigm that facilitates the training of a single global model collaboratively on data silos in a

Y. Lai et al. (Eds.): ICA3PP 2021, LNCS 13157, pp. 432–446, 2022.
https://doi.org/10.1007/978-3-030-95391-1_27

privacy-preserving manner. The main incentive of participating in FL network for individual clients is to leverage the large amounts of shared pool of knowledge from multiple clients. Local model training by individual clients often meets with different constraints including data limitations and poor data quality which results in low-performance inference models.

The prevailing setting of Federated Learning (FL) assumes a network of selected clients (e.g., organizations, institutions, mobile devices) that collectively train a single global model in a secure manner under the orchestration of coordinating server; during the training process clients' data remain local and only model parameters are communicated with the server [5]. However, despite the encouraging merits of FL, technical challenges still inhere [6], especially the presence of system and statistical data heterogeneity which substantially hinder the learning process of FL, makes it more complex and challenging. System heterogeneity refers to the vary of computational power or communication channel (i.e. bandwidth) among clients, range of approaches have been suggested to resolve system heterogeneity such as active sampling [7] and [8]. Statistical heterogeneity attributes to the distinct distribution of data (Non-iid), in a real-world federated network each client institute its own data possibly different from other clients, which, raises the problem of non-iid. Several studies proposed different approaches to address the non-iid problem [9].

In this study, however we focus on another type of heterogeneity called models heterogeneity [10]. In the vanilla FL approach, all clients collaboratively train and agree on a single, global statistical model which is clearly a fair assumption in case of millions in numbers of small edge devices. Although the same assumption can't hold for some areas like digital health, finance and other business sectors where clients are more interested to craft their own unique model, best suits their environmental needs, e.g., in the health sector the privacy of patients data is a topmost priority, institutions like hospitals are not willing to share their patients' information rather they are more open to design their own unique model in a collaborative fashion. To address this problem, [11] proposed FedMD framework, which allows clients to independently design their own models having different architectures. However, the authors didn't consider the performance and contribution of each individual models and used a simple mean aggregation only, which can affect the performance of different models during collaborative learning process particularly in extreme cases of model and data heterogeneity e.g., in learning process the confidence score produce by two or more models to the same data sample may vary, result of heterogeneous architectures and distinct data distributions.

To close the performance gap and stabilize the collaborative learning process, in this paper, we investigate a heterogeneous multi-model federated learning. Under our proposed framework, we introduced a high entropy method for global predictions yielded from the aggregation of multiple models outcome, using a weighted average. The motivation behind increasing the entropy of global predictions is that the class scores produce by a model for single input data contains both correct and incorrect answers in shape of relative probabilities. it is obvious that we are more interested in the correct answer, however, the incorrect answers also contain useful information that resides inside of small probabilities [12], e.g., in a classification task, a label of "Leopard" might provide 0.3 class score to a label of "Jaguar" against 0.01 to a label of "Bear", which means that the model is mistaken but it still sees inherent similarity between the two corresponding labels of Leopard and Jaguar, and that information can be useful. In the same way, models show different performance to the same data sample, some models are more confident than the others depends on their architecture complexity and distinct data distributions. However, the low confident models still provide useful information. Thus, intentionally increasing the entropy of global predictions in each round of collaborative learning phase, provide much more knowledge in the form of soft labels, which results in an increase and stabilize the performance of each model. The contribution of our work is as follows:

☐ First, we propose RMMFL a novel FL framework with a new high entropy aggregation technique that guarantees the stable and robust performance of individual models.
☐ In addition, we use a weighted average ensembles to consider the trust or performance of individual models.
☐ Extensive simulations have been performed on heterogeneous model architectures and benchmark datasets including both iid and non-iid cases. The simulations result shows that proposed framework demonstrates stable and better performance in comparison to the baseline method.

Paper Organization: The rest of this paper is organized as follows: Sect. 2 presents preliminary and related work, Sect. 3 describes in detail the proposed RMMFL framework, Sect. 4 provides experimental setup and comparison between the RMMFL and baseline method, Finally, Sect. 5 presents the concluding remarks.

2 Preliminary and Related Work

This section illustrates the basic concept and related work of Federated learning, Knowledge distillation and Weighted Ensemble

2.1 Federated Learning

Federated Learning (FL) has been introduced and emerged as a promising privacy-preserving collaborative machine learning paradigm [13–15] that enables multiple clients to collectively train a single, global statistical model without sharing their privacy-sensitive data. In canonical FL regime, typically mobile devices collaboratively train a global model by only aggregating model parameters (FedAVG) [5] or gradient information (SGD) [16] under the orchestration of coordinating server without transmitting the training data. Thus, all clients and server receive a common model at the end of learning process.

However, the classical algorithm of FL is not applicable in those areas, when a client require unique personalised learning model [10], several studies focus on personalization in FL. In [17] authors introduced hierarchical clustering for clients as a post-processing step in FL to partition clients into different clusters using bi-partitioning optimal algorithm based on cosine similarity, [18] proposed a new formulation for combination of local and global models, in which each client iteratively train individual local model along with the shared global model. A penalty parameter λ is used for personalization to discourage the clients for being too dissimilar from the mean model, e.g., if λ value increasing the local model will more look like a global model. In contrast, when the value of λ is set to zero pure local model training will occur, in line with this work [19] proposed a new framework named HeteroFL, in which pure heterogeneous clients train local models based on a global model. More recently. [20] and [21] introduced transfer learning in FL for models personalization, their approaches generally use three basic steps: i) First, train a client local model by fine-tuning a pre-trained model over private dataset; ii) second, collaboratively train a common global model over FL; iii) finally, integrate the trained global and local models via transfer learning for a more personalized model. However, all of the aforementioned approaches include both global and local models, and do not direct independent model design. To address the problem of independent model for each client, [11] proposed a FedMD framework based on transfer learning and knowledge distillation, which allows clients to independently design and train their models using private data, before the collaboration phase under orchestration of coordinating server, transfer learning is performed by each individual client model on public dataset until convergence, later fine-tuned on private data.

2.2 Knowledge Distillation

Knowledge distillation is an effective approach for model compression introduced by [12] to encourage a potentially less complex small DNN model to approximate the class probabilities (logits) produced by a large DNN model or ensemble of DNN models. It is a way of transferring the knowledge of a complex cumbersome model through its output predictions into a small model with minimal utility loss. Several studies proposed different methods in conventional knowledge distillation approach to improve the distillation process e.g. [22] used intermediate layer features information as a knowledge along with last layer output predictions to guide the training of a small model.

FedMD [11]and the lately proposed Cronus [23] tried to attain full advantage of knowledge distillation and apply it to FL, they attempted to transfer the knowledge by distilling the model, in which, the logits produced by multiple models for public dataset are averaged per sample. In FedMD, all models are initially pre-trained on public dataset using transfer learning and then iteratively fine-tune on private data for personalization. In contrary, Cronus used jointly the mixture of both public (with soft labels) and private dataset for local model training during collaboration phase.

2.3 Weighted Ensemble

The study of weighted ensemble started in 90's by [24] in classical work. The weighted ensemble takes a weighted average of different models outcome in order to increase the overall performance. The weights allow to show the expected performance or trust of individual model. For classification tasks in neural networks different approaches been proposed such as mixture of experts [25]and boosting [26] used weighted average for multiple models typically in centralized environment to increase the performance.

3 Multi-model Federated Learning

In this section we formally detail our proposed framework for effective multi-model federated learning, in the presence of high entropy aggregation and weighted averaging, we describe RMMFL for clients having heterogeneous model architectures and data distributions.

3.1 Problem Formulation

The population of clients in training process is denoted by n. Each client holds a private labeled dataset D_s, where $D_s := [(x_i^s, y_i^s)]_{i=1}^{n^s}$, the data distribution may or may not be iid potentially depends on client. Moreover, a large public dataset D_p is also available, accessible to all clients, where, $D_p := [(x_i^p, y_i^p)]_{i=1}^{n^p}$. All clients independently design and initialize their learning models M_i without the constraint of being dependent on server or other clients. During the collaborative learning process, a client does not require to share its private data or model parameters with the coordinating server. one communication round of the learning process is performed as follows: First all clients pre-train their models on public dataset D_p until convergence, which is then fine-tuned on individual private data D_s for personalization. After the completion of learning process on both public and private datasets, each client produces predictions (logits) on public dataset and transmit the results to the coordinating server. Server performs the aggregation using Eq. (5) and (2) as described in Algorithm 1 and broadcast high entropy global logit to clients, this process repeats for a fixed number of times. The goal is to accelerate and stable the learning process as well as increase the performance of individual model with a small number of communication rounds.

3.2 High Entropy Aggregation

This part of the section introduces the proposed logit aggregation method called high entropy aggregation (HEA), which intends to increase the entropy of global logits. The definition of entropy for a public logit Y^p is as follows:

$$h_e(Y^p) = -\sum_{i=1}^{n} Y_i^p log Y_i^p. \tag{1}$$

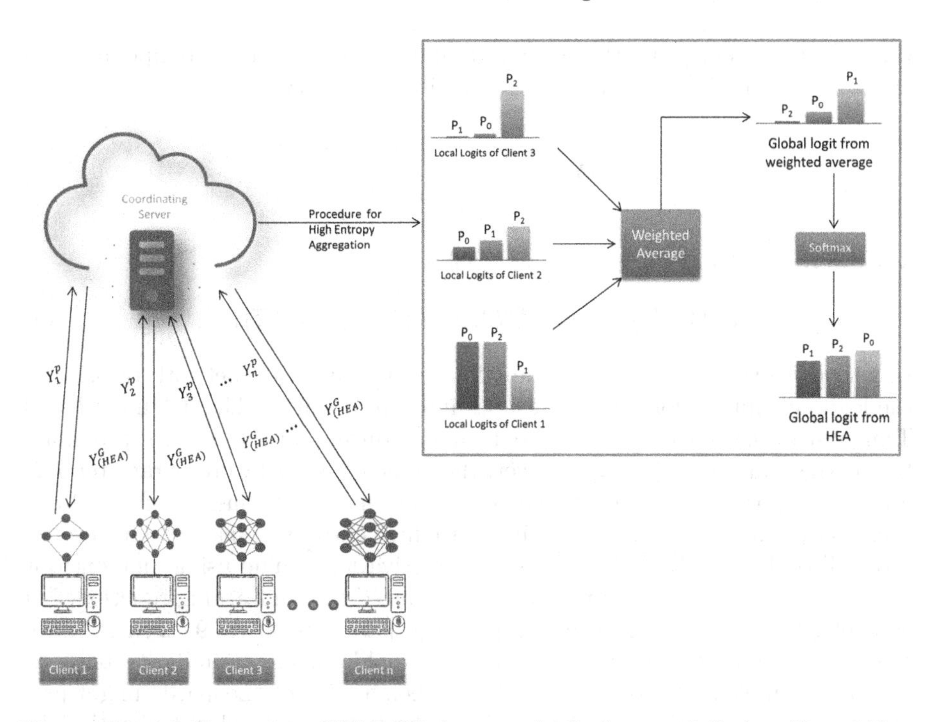

Fig. 1. General illustration of RMMFL framework, the layout of distinct Neural Networks, shows different model architectures for n number of clients. The coordinating server receives logits from n of clients, first performs the weighted average and subsequently increase the entropy of aggregated global logit.

Where h_e donates simple entropy for a logit Y^p. Our motivation behind the proposed aggregation method is that the baseline algorithm considered taking a simple average of multiple uploaded logits from different clients into a single global logit which is susceptible to heterogeneous model architecture and distinct data distribution, the reason for using HEA aggregation method is to stabilize and increase the performance of individual models during the collaborative learning process. The introduction of high entropy in global logits provide more information in form of soft labels to the clients in each communication round. The proposed HEA aggregation method increases the entropy of the global logits produce from multiple uploaded local logits. The procedure is depicted in Fig. 1. In order to increase the global logits entropy yielded from multiple local logits, as an example, we consider the softmax function. Let $Y^G_{(HEA)}$ represents the global logit produce from HEA. The global logit is described as follows:

$$Y^G_{(HEA)} = H_s\left(\frac{1}{N}\sum_{n=1}^{N} Y^p_n|T\right), \tag{2}$$

where $H_s(\cdot|T)$ represents the softmax function with respective temperature T, the softmax function for $h_s(Y^p|T)$ is described as follows:

$$h_s(Y^p|T) = \left(\frac{e^{\frac{y_i^p}{T}}}{\sum_{i=1}^{n} e^{\frac{y_i^p}{T}}} \right). \tag{3}$$

Moreover, the $H_s(Y^p|T)$ is describe as fellows:

$$H_s(Y^p|T) = \{h_s(Y_1^p|T), h_s(Y_2^p|T), ..., h_s(Y_n^p|T)\}. \tag{4}$$

In neural networks, the softmax function at last layer of a network transforms the logits computed for different classes into probabilities. The default value of Temperature parameter T is set to 1 and produce only the relative probabilities learned during training. However, the value of temperature parameter T is directly proportional to the amount of entropy that resides inside softmax function, considering large value of T will induce high entropy in shape of much softer probability distribution over classes. The motivation behind using high entropy is to encourage small class probabilities produced by the learning function. For example, if we set the value of temperature parameter $T = 9$, then it states that the outcome (predictions) in the logits holds inherent similarity between the input sample and corresponding labels. i.e., a label of "Leopard" might provide 0.3 score to a label of "Cheetah" against 0.01 to a label "Elephant". The high entropy will provide further knowledge to all models in form of soft labels during collaboration phase, which will ultimately stabilize the learning process and increase the performance of each model.

3.3 Weighted Averaging

In the proposed weighted averaging method we use a simple but effective technique in order to find optimal weights s_i [27] for each model via grid search, the reason behind our proposed method is that the baseline algorithm used a simple mean function (6) to directly aggregate the predictions produced by multiple models, i.e., $Y_{(SAvg)}^G$ corresponds to the $\frac{1}{N}$ equal weights simple average method, where N is the total number of clients. However, the contribution of clients in the collaborative learning process may not be equal particularly in cases of model and data heterogeneity. The weighted averaging technique for the proposed RMMFL is described as follows:

$$Y_{(WAvg)}^G = \left(\frac{\sum_{i=1}^{n} (s_i)(Y_i^p)}{\sum_{i=1}^{n} (s_i)} \right). \tag{5}$$

3.4 Baseline Approach

We compared the performance of our proposed high entropy aggregation and weighted averaging method with the baseline approach. In the baseline, the coordinating server performs a simple mean aggregation function on uploaded logits

Y_n^p, from n number of clients into a single global logit without using the temperature parameter T. In addition, it used a simple average $\frac{1}{N}$ of equal weights for all models. However, the contribution of models in collaboration learning phase may not be same due to the complexity of their model architectures and distinct data distributions. We named the baseline approach as a simple averaging (SAvg) method and the global logits yielded from the baseline approach is described as follows:

$$Y^G_{(SAvg)} = \left(\frac{1}{N} \sum_{n=1}^{n} Y_n^p \right). \tag{6}$$

Algorithm 1. The RMMFL framework enables multiple robust heterogeneous models in federated learning.

1: **Initialization Phase**
2: **Pre-training:**Each client $n \in [N]$ initialise model $M_i, i = 1, ..., m$ training on public dataset $D_p = [(x_i^p, y_i^p)]_{i=1}^{n^p}$ until convergence
3: After Pre-training, each client randomly select a subset (x_s^i, y_s^i) from its private dataset $D_s = [(x_i^s, y_i^s)]_{i=1}^{n^s}$ and update the model M_i in parallel:
4: **for** each local epoch $e \in \{1, ..., E\}$ **do**
5: Update $w_i \leftarrow$ Local Training (w_i, x_i^s, y_i^s) ▷ w_i is model parameters for M_i
6: **end for**
7: $Y_i^p \leftarrow Predict(w_i, d_p)$ ▷ $d_p \subset D_p$
8: Transmits local logit Y_i^p to a central server
9: **Collaboration Phase**
10: Server aggregates local logits into one global logit $Y^G_{(HEA)}$ via (5) and (2)
11: **for** each communication round $t \in \{1, ..., T\}$ **do**
12: Server randomly samples a public subset $d_p \in D_p$ and distribute to all clients
13: **for** $n \in [N]$ clients **do**
14: **for** each epoch $e \in \{1, ..., E\}$ **do**
15: Digest: $w_i \leftarrow Training(w_i, x_i^p, y_i^p)$ ▷ Train to reach consensus on public dataset
16: **end for**
17: **for** each epoch $e \in \{1, ..., E\}$ **do**
18: Revisit: $w_i \leftarrow Training(w_i, x_i^s, y_i^s)$
19: **end for**
20: $Y_i^p \leftarrow Predict(w_i, d_p)$
21: Transmits local logits Y_i^p to a coordinating server
22: **end for**
23: Server aggregates local logits into one global logit $Y^G_{(HEA)}$ via (5) and (2)
24: **end for**

Algorithm 1 describes our RMMFL framework, which comprises of two learning phases: i) Initialization phase and ii) Collaboration phase. The detailed procedure includes:

☐ In Initialization phase, client n first pre-train its model M_i on public dataset D_p until reach convergence.

☐ After pre-training, clients randomly select a subset of their private data D_s and update their models accordingly.
☐ In Collaboration Phase, Each client make prediction Y_i^p on a randomly sampled subset $d_p \subset D_p$ and upload the local logit to the coordinating server.
☐ Coordinating server perform aggregation on local logits as mentioned in Eq. 5 and 2, later broadcast high entropy global logit to all clients.
☐ In the next round of communication, each client again train its model on the public dataset D_p to reach the consensus on soft-labels (Digest) and then train on its randomly sampled private dataset (Revisit) for personalization D_s.

As described in Algorithm 1 the knowledge learned by one model is transferred to others in the collaboration phase without sharing private data or model parameters. It is pertinent to mention here, that the subset d_p is used to control the cost of communication in each round.

4 Experimental Evaluation

4.1 Experimental Setup

Datasets: In the experiment, we evaluate our proposed framework in two different environmental setups. First, we use MNIST as a public dataset accessible to all clients and a subset of the Federated Extended MNIST (FEMNIST) as a private dataset. In our experiments we considered both IID and non-IID cases, for IID the private dataset for all clients are randomly drawn from FEMNIST. In non-IID case, each client during the training process holds only the letters written by a single writer, while in the testing phase the task for each client is to classify letters written by all writers (Table 1).

Table 1. Summary of public & private datasets.

Collaborative Task	Public	Private
IID	CIFAR10	Subclasses [0,2,20,63,71,82] CIFAR100
Non-IID	CIFAR10	Superclasses [0–5] CIFAR100
IID	MNIST	Letters from [a–f] classes FEMNIST
Non-IID	MNIST	Only letters from one writer FEMNIST

In second environmental setup, we consider CIFAR10 as a public dataset available to all clients and CIFAR100 as a private dataset, which contains 100 subclasses and 20 super-classes, subclasses included as a part of superclasses, e.g., A group of subclasses like leopard, tiger, bear, wolf, and lion etc., all fall under large carnivores super-class. In IID case, each client is required to classify the correct subclasses for test images. The non-IID case is very challenging: in the learning phase, each client has only one subclass data per superclass but clients are required to classify the correct superclass for generic data including all subclasses. For example, during training phase, a client who has seen data only from one subclass, e.g., a bear subclass, is required to classify correctly tiger or other animals from large carnivores superclass in testing phase.

Models and Pre-train Strategy: In both environmental setups, we use 10 clients, each client designs its own unique model differ from other clients in terms of no. of convolutional layers and number of channels see Table 2 and 3 for more details. Firstly, all clients initialize their models and start a pre-training phase on the public dataset D_p until reach convergence. Subsequently, each client fine-tune its model on private data for a small number of iterations. All models typically post accuracy in a range of 98–99% and 71.6–78.2% on MNIST and CIFAR datasets respectively during the pre-training phase.

Table 2. NN models architectures for MNIST/FEMNIST dataset

Name of model	1st layer	2nd layer	3rd layer	Dropout	Pre-train accuracy
NNmodel-0	128	256	None	0.2	98.6%
NNmodel-1	128	384	None	0.2	98.3%
NNmodel-2	128	512	None	0.2	98.8%
NNmodel-3	256	256	None	0.3	98.4%
NNmodel-4	256	512	None	0.4	98.5%
NNmodel-5	64	128	256	0.2	99.0%
NNmodel-6	64	128	192	0.2	98.9%
NNmodel-7	128	192	256	0.2	99.1%
NNmodel-8	128	128	128	0.3	99.0%
NNmodel-9	128	128	192	0.3	99.0%

Table 3. NN models architectures for CIFAR10/CIFAR100 dataset

Name of Model	1st layer	2nd layer	3rd layer	4th layer	Dropout	Pre-train accuracy
NNmodel-0	128	256	None	None	0.2	71.7% ± 1.8
NNmodel-1	128	128	192	None	0.2	72.4% ± 1.4
NNmodel-2	64	64	64	None	0.2	74.3% ± 0.7
NNmodel-3	128	64	64	None	0.3	72.1% ± 2.3
NNmodel-4	64	64	128	None	0.4	71.6% ± 2.9
NNmodel-5	64	128	256	None	0.2	77.3% ± 0.6
NNmodel-6	64	128	192	None	0.2	76.7% ± 1.2
NNmodel-7	128	192	256	None	0.2	75.0% ± 2.0
NNmodel-8	128	128	128	None	0.3	78.2% ± 0.4
NNmodel-9	64	64	64	64	0.2	77.3% ± 0.7

Collaboration Phase: After pre-training of each client model on public dataset and subsequently update the model parameters on private data in parallel for personalization, a collaborative phase starts among clients, during the learning phase each client first make predictions (logits) on a randomly selected subset of public dataset and upload the logits to a coordinating server, server performs a weighted average of uploaded logits into a global logit using Eq. 5 before applying HEA to the global logit as depicted in Fig. 1. Later, the coordinating server broadcast the high entropy global logit to all clients. Afterwards, Each client updates the model by first training on the soft labels yielded via Eq. (2) to approach the consensus on public dataset. Later, fine-tune the model on its own randomly sampled private dataset for personalization. This way, in a cooperative fashion the knowledge learned by one model transfer into other models in the form of soft class probabilities without sharing clients privacy-sensitive data or model parameters with coordinating server. This process of collaboration iterates for a fixed number of communication rounds.

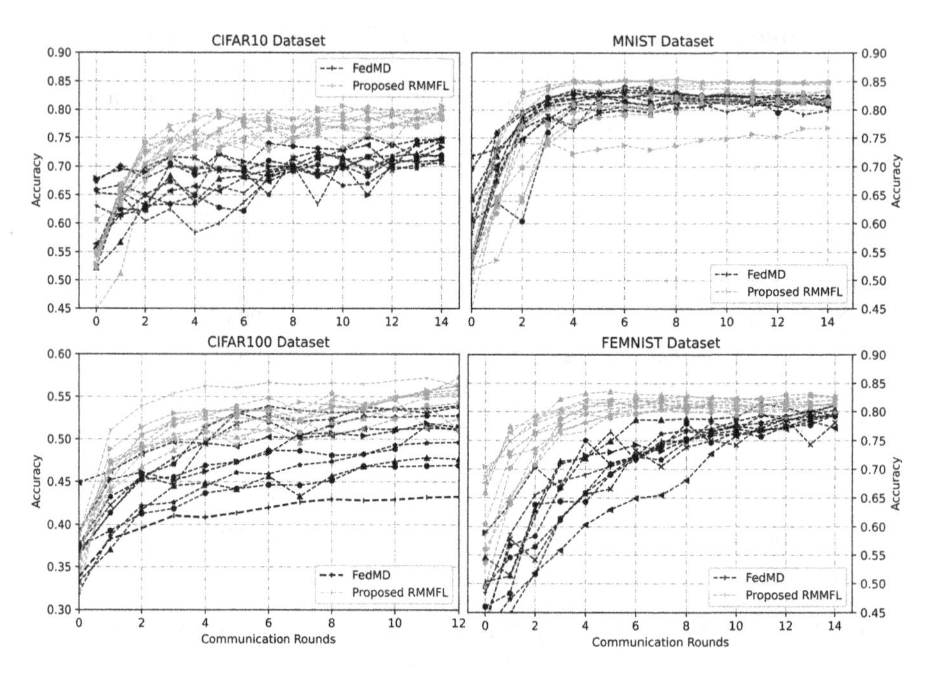

Fig. 2. Shows the performance of each individual model on CIFAR10/100 and MNIST/FEMNIST datasets.

In collaboration learning phase, we used the famous Adam optimizer and initially set the value of learning rate to 1×10^{-3}. Where, for high entropy global logit at server end, we set $T = 9$ the value of temperature parameter in an attempt to soften the probabilities resides in global logit. In each round of communication, a small size of 5000 subset d_p is randomly selected from a public

Table 4. Models performance on CIFAR10/100 IID and non-IID dataset.

Models	Proposed RMMFL		FedMD	
	CIFAR10 IID	CIFAR100 Non-IID	CIFAR10 IID	CIFAR100 Non-IID
NNmodel-0	**79.16** ± 1.83	**55.90**± 0.75	74.10 ± 0.40	43.33 ± 0.30
NNmodel-1	**79.96** ± 0.75	**57.46** ± 0.40	74.71 ± 1.21	52.28 ± 0.22
NNmodel-2	78.88 ± 0.90	**55.10** ± 1.60	76.50 ± 0.84	47.87 ± 0.48
NNmodel-3	**78.33** ± 0.40	53.38 ± 0.67	73.13 ± 1.36	51.65 ± 0.69
NNmodel-4	**78.64** ± 0.27	55.13 ± 0.77	73.48 ± 2.11	51.17 ± 0.25
NNmodel-5	**80.70** ± 0.50	56.16 ± 2.85	75.33 ± 0.75	53.85 ± 0.25
NNmodel-6	78.50 ± 1.50	56.62 ± 1.93	76.66 ± 0.35	52.44 ± 0.56
NNmodel-7	**80.50** ± 1.25	**57.53** ± 0.58	74.73 ± 0.45	45.52 ± 0.98
NNmodel-8	**79.67** ± 1.35	54.24 ± 0.61	72.66 ± 0.74	51.83 ± 0.37
NNmodel-9	79.00 ± 1.28	56.37 ± 0.30	71.27 ± 0.15	53.30 ± 0.20

Table 5. Models performance on MNIST/FEMNIST IID and non-IID dataset.

Models	Proposed RMMFL		FedMD	
	MNIST IID	FEMNIST Non-IID	MNIST IID	FEMNIST Non-IID
NNmodel-0	**84.53** ± 0.10	82.12 ± 0.13	79.47 ± 0.11	81.12 ± 0.12
NNmodel-1	83.25 ± 0.12	82.67 ± 0.18	80.78 ± 0.15	79.45 ± 0.13
NNmodel-2	81.34 ± 0.17	82.54 ± 0.11	82.52 ± 0.18	80.33 ± 0.11
NNmodel-3	**85.08** ± 0.21	79.85 ± 0.22	79.75 ± 0.12	81.04 ± 0.18
NNmodel-4	81.71 ± 0.15	81.45 ± 0.20	82.10 ± 0.12	79.17 ± 0.16
NNmodel-5	76.77 ± 1.32	82.00 ± 0.14	81.39 ± 0.11	79.67 ± 0.12
NNmodel-6	**84.81** ± 0.10	**82.60** ± 0.17	79.70 ± 0.13	77.19 ± 0.12
NNmodel-7	81.17 ± 0.11	81.20 ± 0.13	81.62 ± 0.14	79.11 ± 0.11
NNmodel-8	83.69 ± 0.24	81.25 ± 0.11	81.43 ± 0.14	78.06 ± 0.16
NNmodel-9	**84.88** ± 0.15	82.63 ± 0.15	79.87 ± 0.16	79.35 ± 0.12

dataset for all clients, the reason behind using a small subset of data is an effort to control the communication cost, in an order that it does not scale up with the heterogeneous complexities of client models.

4.2 Performance Analysis

We start out to evaluate the performance of the proposed framework in comparison to baseline method on MNIST/CIFAR benchmark datasets. The accuracy of each individual client model during the collaborative learning process with proposed RMMFL and baseline method are listed in Table 4 and 5 for CIFAR and MNIST datasets respectively. In Fig. 2 it can be easily observed from the learning curves that as the collaborative phase advance, the performance of RMMFL for each individual model improves significantly in a much stable manner. In contrast, the baseline method experienced transient decline after first few rounds of communication. The Table 4 shows that the proposed RMMFL framework distinctively outperform the baseline method on both datasets. RMMFL posted an average accuracy of 79.33% and 55.78% compare to only 74.25% and 50.32% by the baseline method on CIFAR10 and CIFAR100. The experimental results show that the performance of RMMFL exceeds significantly by a margin of over 5.00% on both datasets. Where, the efficacy of RMMFL on MNIST/FEMNIST datasets is just fairly better against the baseline method. However, the proposed method still managed to post accuracy of over 84.00% and 82.00% for some models on MNIST and FEMNIST datasets respectively, see Table 5 for more details.

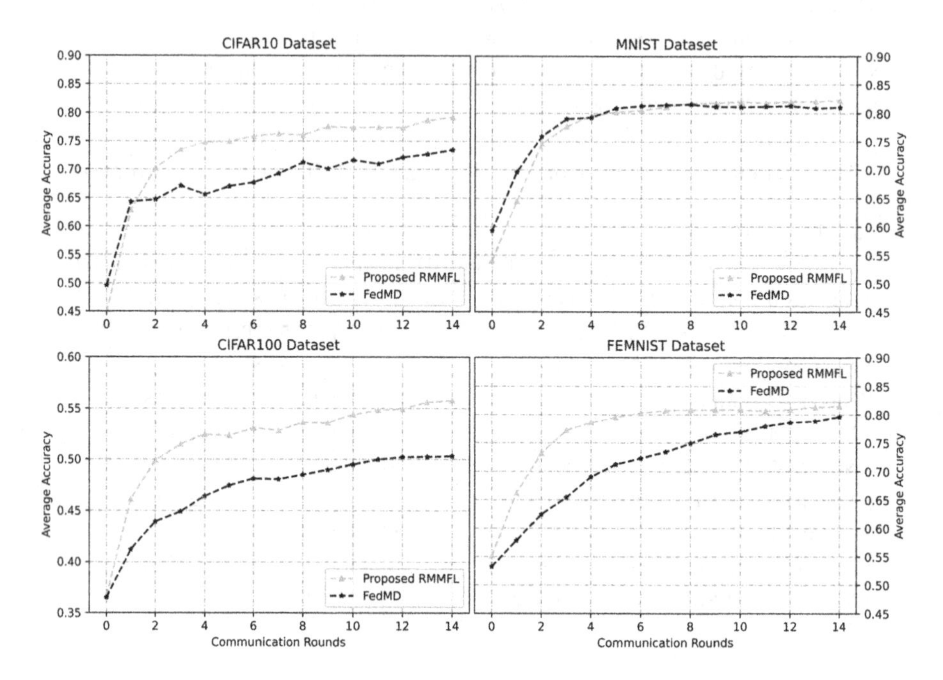

Fig. 3. Shows the average performance of all models during collaboration phase.

We can see from Fig. 2 and 3 that the proposed RMMFL framework shows a much stable performance throughout the whole process on both datasets against the baseline method which validates the approach we adopted for our proposed framework.

5 Conclusion

In this work, we proposed a collaborative framework based on high entropy and weighted aggregation. The fundamental idea was the logit method for RMMFL to accelerate and stabilize the learning process as well as enhance individual model performance. Additionally, we proposed a weighted ensemble method to weigh the predictions of each model according to their performance. The simulations results showed that RMMFL distinctively outperform the baseline method on both iid and non-iid datasets. In future, we will explore more advanced aggregation methods and extend our work to extreme heterogeneous cases to further improve the framework performance. We believe that robust models-based heterogeneous federated learning has a wide scope in healthcare as well as in business applications of deep learning.

References

1. Voigt, P., von dem Bussche, A.: Introduction and 'checklist'. The EU General Data Protection Regulation (GDPR). https://doi.org/10.1007/978-3-319-57959-7
2. Annas, G.J.: HIPAA regulations - a new era of medical-record privacy? N. Engl. J. Med. **348**, 1486 (2003). https://doi.org/10.1056/nejmlim035027. ISSN 0028-4793
3. Shokri, R., Shmatikov, V.: Privacy-preserving deep learning. In: 2015 53rd Annual Allerton Conference on Communication, Control, and Computing, Allerton 2015 (2016). https://doi.org/10.1109/ALLERTON.2015.7447103. ISBN 9781509018239
4. Yang, Q., Liu, Y., Chen, T., Tong, Y.: Federated machine learning: concept and applications. ACM Trans. Intell. Syst. Technol. **10**, 1–9 (2019). https://doi.org/10.1145/3298981, arXiv: 1902.04885. ISSN 21576912
5. McMahan, H.B., Moore, E., Ramage, D., Hampson, S., Arcas, B.A.Y.: Communication-efficient learning of deep networks from decentralized data. arXiv:1602.05629 [cs.LG] (2017)
6. Li, T., Sahu, A.K., Talwalkar, A., Smith, V.: Federated learning: challenges, methods, and future directions. IEEE Signal Process. Mag. **37**(3), 50–60 (2020). https://doi.org/10.1109/MSP.2020.2975749, arXiv:1908.07873. ISSN 15580792
7. Nishio, T., Yonetani, R.: Client selection for federated learning with heterogeneous resources in mobile edge. In: IEEE International Conference on Communications (2019). https://doi.org/10.1109/ICC.2019.8761315, arXiv:1804.08333. ISBN 9781538680889
8. Sahu, A.K., Li, T., Sanjabi, M., Zaheer, M., Talwalkar, A., Smith, V.: On the convergence of federated optimization in heterogeneous networks. arXiv:1812.06127 [cs, stat] (2018)
9. Smith, V., Chiang, C.K., Sanjabi, M., Talwalkar, A.: Federated multi-task learning. In: Advances in Neural Information Processing Systems (2017)
10. Wu, Q., He, K., Chen, X.: Personalized federated learning for intelligent IoT applications: a cloud-edge based framework. IEEE Open J. Comput. Soc. **1**, 35–44 (2020). https://doi.org/10.1109/OJCS.2020.2993259, arXiv:2002.10671. ISSN 15581756
11. Li, D., Wang, J.: FedMD: heterogenous federated learning via model distillation. arXiv:1910.03581 (2019)
12. Hinton, G., Vinyals, O., Dean, J.: Distilling the knowledge in a neural network. arXiv:1503.02531, March 2015
13. Liu, Y., Yuan, X., Xiong, Z., Kang, J., Wang, X., Niyato, D.: Federated learning for 6G communications: challenges, methods, and future directions. CoRR arXiv:2006.02931 (2020)
14. Kairouz, P., et al.: Advances and open problems in federated learning. CoRR arXiv:1912.04977 (2019)

15. Lim, W.Y.B., et al.: Federated learning in mobile edge networks: a comprehensive survey. IEEE Commun. Surv. Tutor. **22**(3), 2031–2063 (2020). https://doi.org/10.1109/COMST.2020.2986024
16. Lin, T., Stich, S.U., Patel, K.K., Jaggi, M.: Don't use large mini-batches, use local SGD. arXiv:1808.07217 [cs.LG] (2020)
17. Sattler, F., Müller, K.-R., Samek, W.: Clustered federated learning: model-agnostic distributed multi-task optimization under privacy constraints. arXiv:1910.01991 [cs.LG] (2019)
18. Hanzely, F., Richtárik, P.: Federated learning of a mixture of global and local models. arXiv:2002.05516 [cs.LG] (2021)
19. Diao, E., Ding, J., Tarokh, V.: HeteroFL: Computation and communication efficient federated learning for heterogeneous clients. arXiv:2010.01264 [cs.LG] (2021)
20. Yang, H., He, H., Zhang, W., Cao, X.: FedSteg: a federated transfer learning framework for secure image steganalysis. IEEE Trans. Netw. Sci. Eng. **8**(2), 1084–1094 (2021). https://doi.org/10.1109/TNSE.2020.2996612
21. Chen, Y., et al.: FedHealth: a federated transfer learning framework for wearable healthcare. arXiv: 1907.09173 [cs.LG] (2021)
22. Romero, A., Ballas, N., Kahou, S.E., Chassang, A., Gatta, C., Bengio, Y.: FitNets: hints for thin deep nets. arXiv:1412.6550, December 2014
23. Chang, H., Shejwalkar, V., Shokri, R., Houmansadr, A.: Cronus: robust and heterogeneous collaborative learning with black-box knowledge transfer. arXiv:1912.11279, December 2019
24. Hashem, S., Schmeiser, B.: Approximating a function and its derivatives using MSE-optimal linear combinations of trained feedforward neural networks. In: Proceedings of the Joint Conference on Neural Networks, pp. 617–620 (1993)
25. Yuksel, S.E., Wilson, J.N., Gader, P.D.: Twenty years of mixture of experts. IEEE Trans. Neural Netw. Learn. Syst. **23**(8), 1177–1193 (2012). https://doi.org/10.1109/TNNLS.2012.2200299
26. Schapire, R.E.: A brief introduction to boosting. In: Proceedings of the 16th International Joint Conference on Artificial Intelligence - Volume 2, IJCAI'99, San Francisco, CA, USA, pp. 1401–1406. Morgan Kaufmann Publishers Inc. (1999)
27. Krogh, A., Vedelsby, J.: Neural network ensembles, cross validation and active learning. In: Proceedings of the 7th International Conference on Neural Information Processing Systems, NIPS'94, Cambridge, MA, USA, pp. 231–238. MIT Press (1994)

V-EPTD: A Verifiable and Efficient Scheme for Privacy-Preserving Truth Discovery

Chang Xu[1(✉)], Hongzhou Rao[1], Liehuang Zhu[1], Chuan Zhang[1], and Kashif Sharif[2]

[1] School of Cyberspace Science and Technology, Beijing Institute of Technology, Beijing, China
{xuchang,rhzinbit,liehuangz,chuanz}@bit.edu.cn
[2] School of Computer Science and Technology, Beijing Institute of Technology, Beijing, China
7620160009@bit.edu.cn

Abstract. Privacy-preserving truth discovery has been researched from many perspectives in the past few years. However, the complex iterative computation and multi-user feature makes it challenging to design a verifiable algorithm for it. In this paper, we propose a novel scheme named V-EPTD that not only protects the privacy information but also verifies the computing in truth discovery. The proposed technique adopts a threshold paillier cryptosystem to solve the multi-user problem so that all parties encrypt the data with the same public key while being unable to decrypt the ciphertext if there are not enough parties. V-EPTD also transforms complex iterative computation into polynomials, uses linear homomorphic hash, and commitment complete verification. The experimentation and analysis show that V-EPTD has good performances for users, verifiers, and the server, both in communication overhead and computation overhead.

Keywords: Verifiable computation · Truth discovery · Privacy-preserving · Homomorphic encryption · Commitment

1 Introduction

Truth discovery is used to extract an answer which is close to the truth from a series of sample data. The deployment and extensive penetration of mobile smart terminals in different technologies have enabled crowdsensing, where people can use redundant resources of their terminals to participate in data sampling and collection, which is then sent to the requesters. Truth discovery can be effectively applied to such data for real information extraction.

Supported by the grant from National Natural Science Foundation of China (No. 61972037).

Y. Lai et al. (Eds.): ICA3PP 2021, LNCS 13157, pp. 447–461, 2022.
https://doi.org/10.1007/978-3-030-95391-1_28

To protect the privacy information during truth discovery in crowdsensing, many works have been presented [16]. However, most of the research in this domain assumes that the server is semi-honest. However, a compromised server can forge data, or return a random result to users to reduce computing costs. To protect users' rights better, a verifiable, as well as privacy-preserving truth discovery protocol is necessary.

Verifiable computing enables a client with a limited computational resource to outsource its computation function F to one or more workers. Several works in this regard have been done, however, the verifiable schemes that have been proposed cannot be applied to truth discovery. The challenge here is dual in nature. On the one hand, the computing functions of truth discovery are complicated, hence a verifiable scheme of a single function cannot verify it. On the other hand, the data sampled in truth discovery is obtained from multiple sources, while the data in existing verifiable schemes is provided by a single source.

To solve the above problems, we propose V-EPTD, which is a verifiable, efficient, and public privacy-preserving computing scheme for truth discovery in crowdsensing. Based on the threshold paillier cryptosystem [4], the server can execute homomorphic computing with multi-source data. In addition, the homomorphic hash is used to verify the result in V-EPTD. It achieves public verification and is efficient for users, the server, and the verifier, which is shown in the evaluation section. We summarize the contributions of this paper below.

- V-EPTD is a privacy-preserving scheme such that both users' sensing data and weight are protected. In the verification phase, the verifier can verify the result without knowing any private information of users. What's more, the data in V-EPTD can be non-integer.
- V-EPTD incurs less computing overhead in the verification phase, which allows a user with limited computing resources to verify the result. It allows anyone to execute the verification phase, i.e., achieving public verification.
- V-EPTD is efficient for all participants so that it has practical application significance. We have done several experiments to support the solutions presented and prove that V-EPTD is an efficient and reliable scheme.

2 Related Works

Truth Discovery: Miao et al. [9] proposed PPTD, the first scheme for privacy-preserving truth discovery. Miao et al. [11] also proposed a privacy-preserving framework that allows a user to submit their sensing result and does not need to participate in iterations. The works of Xu et al. [10] and Zhang et al. [18] have also furthered the truth discovery in a privacy-preserving manner. Xu et al. also proposed a technology named $Double-MaskingwithOne-TimePads$ [14]. It is not only applied in truth discovery, but also applied in [13] in federated learning domains. Zhang et al. [17] designed two schemes in which users are stable and are frequently moving, respectively. To protect the privacy of users, they also explore data perturbation and shift most of the workload to the server, which is the same as [11]. Both [11] and [17] need two non-conspiracy servers.

Verifiable Computing: The functions in verifiable computing having been most researched are matrix inversion, matrix multiplication [5], matrix determinant [19], and polynomial function [6]. For matrix operation verification, perturbation technologies such as data perturbation, permutation mapping [2], and permutation matrix [1] are popular, while for polynomial function verification, researchers prefer to adopt homomorphic encryption, discrete logarithm (DL) problem, and other problems in discrete mathematics.

Verifiable Truth Discovery: To the best of our knowledge, there is only one scheme V-PATD proposed by Xu et al. [15] which is verifiable. It adopts the scheme in [12] to verify the computing of the server and explore differential privacy technique to protect the privacy of the users.

3 Cryptographic Primitives

Several cryptographic protocols have been used in V-EPTD. Here we give a brief detail of these for easier understanding.

3.1 Threshold Paillier Cryptosystem

Damg et al. [4] proposed an additive homomorphic cryptosystem denoted as (U, x)-threshold paillier cryptosystem, where U users participate in the decryption phase while at least x users' private keys are needed to decrypt the ciphertext.

This cryptosystem uses computations of modulo $\mathfrak{n}^{s+1}$, where s is a natural number and $\mathfrak{n}$ is an RSA modulus. Given two primes $\mathfrak{p}$ and $\mathfrak{q}$, we set $\mathfrak{n} = \mathfrak{p}\mathfrak{q}$, and choose a natural number s, so that our plaintext space is $Z_{\mathfrak{n}^s}$. We set the public key as $Tpk = (\mathfrak{g}, \mathfrak{n})$, where $\mathfrak{g} = \mathfrak{n} + 1$. When a user wants to encrypt the plaintext $M \in Z_{\mathfrak{n}}$, it can choose a random integer $r \in Z^*_{\mathfrak{n}^{s+1}}$ and the ciphertext c is computed as $c = \mathbf{Enc}(M) = g^M r^{\mathfrak{n}^s} \mod \mathfrak{n}^{s+1}$.

In this work, we assume a similar (U, x)-threshold paillier cryptosystem, where U private keys $(Tsk_1, ..., Tsk_U)$ are generated and delivered to all users. In decryption phase, each user $\mathcal{U}_i$ calculates its own particular decryption c_i of c with Tsk_i as $c_i = \mathbf{Dec}(M) = c^{2\Delta Tsk_i}$, where $\Delta = U!$. Executing the algorithm *Share Combining* proposed in [4] by at least x users, can get the plaintext M.

Secure Sum Protocol: This protocol is based on the threshold paillier cryptosystem. The inputs are plaintexts of users and the output is the aggregation of these plaintexts. This algorithm is privacy-preserving such that the server and other users do not know the plaintext of the user $\mathcal{U}_i$ after aggregation. The detailed description is given as *Protocol 1* in [9].

3.2 Linear Homomorphic Hash

A linearly homomorphic hash scheme contains three algorithms, i.e., **LHH** = (**H.Gen**, **H.Hash**, **H.Eval**). Note that this construction of **LHH** is collision resistant [3]. The description of these three functions is given below.

- **H.Gen**$(1^k, 1^d)$: Given a security parameter $\boldsymbol{k}$ and a dimension d, this algorithm outputs public parameter **LHH**pp, including parameters of a cyclic group $\mathbb{G}$ that are prime order q, its generator $g \in \mathbb{G}$, and d distinct elements $g_1, ..., g_d \in \mathbb{G}$. For simplicity of presentation, this public parameter will be taken implicitly as the first parameter of **H.Gen** and **H.Hash**.
- **H.Hash**$(\mathbf{x})$: Taking a d-dimensional vector $\mathbf{x}$ as input, this algorithm outputs a linearly homomorphic hash of $\mathbf{x}$ as $h \leftarrow \prod_{i\in[d]} g_i^{\mathbf{x}_i} \in \mathbb{G}$.
- **H.Eval**$(h_1, ..., h_l, \alpha_1, ..., \alpha_l)$: Taking l hashes and l coefficients of linear combination, this algorithm outputs a linear combination of these l hashes as $h^* \leftarrow \prod_{i\in[l]} h_i^{\alpha_i}$.

3.3 Commitment

After a value has been committed by a party, *commitment* does not allow any party to change the value and others cannot know the value before decommitment. The commitment used in this paper is the same as defined in [7]. The formal definition of a non-interactive equivocal commitment is given below.

- **COM.Setup**(1^k): This algorithm generates public parameter **COM**pp, which determines the ciphertext space $\mathcal{C}$ and commitment space $\mathcal{M}$.
- **COM.Commit**(m, r): A message m with randomness r will be committed into a commitment string c by a committer through this algorithm. The commitment string c can be published, while randomness r should be kept secret until de-commitment.
- **COM.Decommit**$(c, m^{'}, r^{'})$: Given a message $m^{'}$ committed with a randomness $r^{'}$, if the commitment string c = **COM.Commit**$(m^{'}, r^{'})$, then output is 1, else output is $\perp$.

4 Truth Discovery

In general, a truth discovery algorithm is composed of weight updating and ground truth updating phases. In the beginning, the ground truths are initialized, and then in each iteration, the weight and truth updates are done, until convergence. The weight update and truth update methods are described below.

Weight Update: In the proposed scheme, we adopt the same method of weight upgrade as given in CRH [8], i.e.,

$$w_i = log(\frac{\sum_{t=1}^{T}\sum_{u=1}^{U} d(\mathbb{V}_i[t] - \mathbb{V}^*[t])}{\sum_{u=1}^{U} d(\mathbb{V}_i[t] - \mathbb{V}^*[t])}), \tag{1}$$

where $d(\cdot)$ is a distance function that measures the differences between users' answers and the estimated ground truth. The estimated ground truth will not change until the next iteration. Distance function $d(\cdot)$ can be designed according to the application scenario [17]. As we discuss continuous data in this work, i.e., a variable can take any value between two given values (for example height or temperature), hence, the distance function becomes $d(\mathbb{V}_i[t] - \mathbb{V}^*[t]) = (\mathbb{V}_i[t] - \mathbb{V}^*[t])^2$.

Truth Update: In this step, a user's weight will not change until the next iteration. The ground truth of a certain task t, can be estimated as $\mathbb{V}^*[t] = \frac{\sum_{i=1}^{U} w_i \cdot \mathbb{V}_i[t]}{\sum_{i=1}^{U} w_i}$, where $\mathbb{V}^*[t]$ represents estimated ground truth for continuous data.

5 System and Threat Models

5.1 System Model

There are two main entities in the V-EPTD system, which are the users $\mathcal{U}$ and the server $\mathcal{S}$. Verifiers can be any one of these.

- Users $\mathcal{U}$: The users send their sensing data to the server in exchange for some reward. However, since they are scattered in space, they do not communicate with each other directly in any way. In contrast, they communicate with each other through the server.
- Server $\mathcal{S}$: It provides a platform where the users can accept or post crowdsensing tasks. It has abundant computation resources and also can communicate with any user.
- Verifiers: The verifiers can be a third party or selected from users. They verify the results returned by the server $\mathcal{S}$ only using the published values from users and the server.

5.2 Threat Model

- Users $\mathcal{U}$: The users are all honest but curious. They execute the protocol honestly and they do not collude with any party. However, they want to know the private information of others.
- Server $\mathcal{S}$: The server is malicious, which means it not only wants to eavesdrop on the private information of users, but it can return a random result to reduce its computing overhead. However, it will not intercept user's messages, which means once there is a message sent to a user $\mathcal{U}_i$, $\mathcal{S}$ will forward the message to $\mathcal{U}_i$ immediately.
- Verifiers: The verifiers are honest but curious. They do not collude with others and execute the verification phase honestly. They are also curious about other parties' private information.

Based on the above system and threat models of the users and server, it can be inferred that in our scheme a user does not need to verify every result returned by other users, however, it needs to doubt that whether the server forged the results during the communication.

6 Proposed Scheme

In this section, we present the details of the V-EPTD scheme.

6.1 Preconditions

To introduce V-EPTD and make its understanding easier, we describe some preconditions here. At first, we assume that there are total T crowdsensing tasks denoted as $\mathcal{T}$, U users denoted as $\mathcal{U}$, and $\mathscr{L}$ iterations during the truth discovery. Without the loss of generality, we set $T \geq \mathscr{L}$, so that we choose T distinct elements $g_1, ..., g_T$ in a linear homomorphic hash cryptosystem. We use (U,x)-threshold paillier cryptosystem, of which the public key is $Tpk = (\mathfrak{g}, \mathfrak{n})$. Note that, in this paper, linear homomorphic hash cryptosystem and threshold paillier cryptosystem use different cyclic groups respectively, while the plaintext space is the same as that of plaintext $M \in Z_E$ and the aggregate value plaintext lies in Z_F such that $F \geqslant U \cdot E$.

In these cryptosystems, we only use integers, while the values in the real world may not be integers. Since we need to verify the result, it is important to control the data accuracy. The solution is to introduce a constant parameter L (of a magnitude of 10). Then a floating-point x can be converted to integer $\hat{x}$ by multiplying L, i.e., $\hat{x} = xL$. We use $\hat{x}$ to denote the converted integer of x. Note that, there are many $\log(x)$ computations, hence, we process these computation as $\widehat{\log(x)} = \log(\widehat{x}/L) * L$.

6.2 Initialization

In this step, some parameters and keys are generated by a Truth Center (TC) which is online in the initialization only and goes offline after the system is started.

In the proposed system, for every user and server $\mathcal{S}$, TC generates the public encryption key $Tpk = (\mathfrak{g}, \mathfrak{n})$ according to [4] and sends the corresponding secret key Tsk_i to the i-th party. In addition, TC will generate public parameters for other cryptosystems which are linear homomorphic hash. The commitment and key agreement algorithm will be adopted according to the specific schemes.

6.3 Privacy-Preserving Truth Discovery

Before starting iterations, users first execute key agreement protocol to generate the pairwise symmetric key with other users participating in the truth discovery. The details of this step are given below.

Step 0: For each user $\mathcal{U}_i, i \in [1, U]$ and each other user $\mathcal{U}_j, j \in [1, U] \setminus i$, $\mathcal{U}_i$ makes a key agreement with $\mathcal{U}_j$ to generate pairwise symmetric keys as $\mathbf{k}_{i,j} \leftarrow \mathbf{KA.Agree}(sk_i, pk_j)$ that are only known by $\mathcal{U}_i$ and $\mathcal{U}_j$.

This step is executed only once before the first iteration, which costs less computing and communication overhead in a dynamic system.

Weight Update: After initialization and pairwise key generation, weight is updated in two steps. Without the loss of generality, we present these steps as the k-th iteration.

Step 1: For each user $\mathcal{U}_i$, $i \in [1, U]$, after they collect their own crowdsensing data, they compute the vector of the distance $\mathbb{DI}_{i_k}$ between their crowdsensing data and estimated ground truths as

$$\mathbb{DI}_{i_k} = ((\mathbb{V}_{i_k}[1] - \mathbb{V}_k^*[1])^2, ..., (\mathbb{V}_{i_k}[T] - \mathbb{V}_k^*[T])^2), \tag{2}$$

where $\mathbb{V}_{i_k}[t]$ represents sensing value for sensing task t by user $\mathcal{U}_i$ in the k-th iteration and $\mathbb{V}_k^*[t]$ represents ground truth for sensing task t in the k-th iteration. Then $\mathcal{U}_i$ aggregates $\mathbb{DI}_{i_k}$ as $Disk_{i_k}$: $Disk_{i_k} = \sum_t^T \mathbb{DI}_{i_k}[t]$, and encrypts $Disk_{i_k}$ and $\log(Disk_{i_k})$ into: $\mathbf{Enc}(\widehat{Disk_{i_k}})$ and $\mathbf{Enc}(\log(\widehat{Disk_{i_k}}))$ respectively.

$\mathcal{U}_i$ also computes $h_{i_k} \leftarrow g_k^{\widehat{Disk_{i_k}}}$, and commitment $c_{i_k} \leftarrow \mathbf{COM.Commit}(h_{i_k}, r_{i_k})$, where g_k represents k-th number. $\mathcal{U}_i$ sends c_{i_k} to the server, while h_{i_k}, r_{i_k} will not be sent to the server until the verification phase. Hence, the users do not need to worry that the server may forge their h_{i_k} by verifying the commitments.

Finally, $\mathcal{U}_i$ stores all values generated in this step, and sends c_{i_k},$\mathbf{Enc}(\widehat{Disk_{i_k}})$, $\mathbf{Enc}(\log(\widehat{Disk_{i_k}}))$ to the server.

Step 2: The server $\mathcal{S}$ receives all (c_{i_k},$\mathbf{Enc}(\widehat{Disk_{i_k}})$, $\mathbf{Enc}(\log(\widehat{Disk_{i_k}}))$) from $\mathcal{U}_i, i \in [1, U]$. Then it aggregates these data as

$$\mathbf{Enc}(\widehat{Sum_k}) = \mathbf{Enc}(\sum_i^U(\widehat{Disk_{i_k}})) = \prod_i^U(\mathbf{Enc}(\widehat{Disk_{i_k}})). \tag{3}$$

Then $\mathcal{S}$ can obtain the plaintext $Sum_k = \widehat{Sum_k}/L$, where $\widehat{Sum_k}$ is the output of *Secure Sum Protocol*, and encrypts it as $\mathbf{Enc}(\log(\widehat{Sum_k})) = \mathbf{Enc}(\log(Sum_k) * L)$. For each user, their own encrypted weight can be computed by $\mathcal{S}$ as

$$\mathbf{Enc}(\widehat{w_{i_k}}) = \mathbf{Enc}(\log(\widehat{Sum_k})) \cdot \mathbf{Enc}(\log(\widehat{Disk_{i_k}}))^{-1}. \tag{4}$$

After getting encrypted weights of U users, $\mathcal{S}$ sends all $\mathbf{Enc}(\widehat{w_{j_k}}), c_{j_k}$, where $j \neq i$ and $j \in [1, U]$ to all users $\mathcal{U}_i, i \in [1, U]$.

Until this point, the update of the weight of k-th iteration is complete. Next, we will introduce the ground truth update.

Ground Truth Update: In the following steps, the ground truth update process is described. Note that, step 3 follows step 2.

Step 3: For each user $\mathcal{U}_i$, $i \in [1, U]$, $\mathcal{U}_i$ decrypts all the encrypted weights it received as $\mathbf{Dec}(\widehat{w_{j_k}})_i$, then encrypts $\mathbf{Dec}(\widehat{w_{j_k}})_i$ into $\mathbf{C\_Dec}$ $(\widehat{w_{j_k}})_i \leftarrow \mathbf{SE.Enc}$ $(\mathbf{Dec}(\widehat{w_{j_k}})_i, \mathbf{k}_{i,j})$, where $j \in [1, U]$ and $j \neq i$, and $\mathbf{SE.Enc}$ (x, k) means symmetric encryption with symmetric key k. Finally, $\mathcal{U}_i$ sends $\mathbf{C\_Dec}(\widehat{w_{j_k}})_i$, where $j \in [1, U]$ and $j \neq i$ to the server $\mathcal{S}$.

Step 4: To help $\mathcal{U}_i$ get its weight plaintext, the server $\mathcal{S}$ sends $\mathbf{C\_Dec}(\widehat{w_{i_k}})_j$, where $i \in [1, U]$ and $i \neq j$ to $\mathcal{U}_i$ after receiving all messages from all users in step 3.

Step 5: In this step, each user $\mathcal{U}_i, i \in [1, U]$ will get their weight by decrypting $\mathbf{C\_Dec}(\widehat{w_{i_k}})_j$ from $\mathcal{U}_j$, where $j \neq i$ and $j \in [1, U]$ to get $\mathbf{Dec}(\widehat{w_{i_k}})_j \leftarrow \mathbf{SE.Dec}(\mathbf{C\_Dec}(\widehat{w_{i_k}})_j, \mathbf{k}_{i,j})$, where $\mathbf{SE.Dec}(x, k)$ means symmetric decryption with symmetric key k. So the plaintext w_{i_k} is available when there are at least n partial decryptions.

Next, $\mathcal{U}_i$ multiplies w_{i_k} by each $\mathbb{V}_{i_k}[t]$ and encrypts them as $\mathbf{Enc}(\widehat{w_{i_k} \cdot \mathbb{V}_{i_k}[t]})$, $t \in [1, T]$, which will be sent to the server $\mathcal{S}$.

$\mathcal{U}_i$ computes $hh_{i_k} \leftarrow \prod_{t=1}^{T} g_t^{\widehat{w_{i_k} \cdot \mathbb{V}_{i_k}[t]}}$, $hw_{i_k} \leftarrow g_k^{\widehat{w_{i_k}}}$, $t \in [1, T]$, commitment $cc_{i_k} \leftarrow \mathbf{COM.Commit}(hh_{i_k}, rr_{i_k})$, and $cw_{i_k} \leftarrow \mathbf{COM.Commit}(hw_{i_k}, rw_{i_k})$. Finally, $\mathcal{U}_i$ stores all messages received and values generated in this step, and sends (cc_{i_k}, cw_{i_k}) to the server $\mathcal{S}$.

Step 6: The server $\mathcal{S}$ now can execute the ground truth update. It first computes the summation of weighted data denoted as Swd and the summation of all users' weighted denoted as Sw by using *secure sum protocol*, i.e., $Swd_{t_k} = \sum_{i=1}^{U} w_{i_k} \cdot \mathbb{V}_{i_k}[t]$ is Swd for t-th task in k-th iteration, and $Sw_k = \sum_{i=1}^{U} w_{i_k}$ is Sw in k-th iteration. Then the ground truth of each task $t \in \mathcal{T}$ in this iteration can be estimated as $\mathbb{V}_k^*[t] = \frac{Swd_{t_k}}{Sw_k}$. Finally, $\mathcal{S}$ sends $\sum_{i=1}^{U} w_{i_k}, cc_{i_k}, i \in [1, U]$ and $\mathbb{V}_{i_k}^*[t], t \in [1, T]$ to each user.

At this point, an iteration of truth discovery is completed. All ground truths will be converged after no more than $\mathscr{L}$ iterations. Then we can start to verify the computation of server $\mathcal{S}$.

6.4 Batch Verification

The details of verification for truth discovery will be described in this phase. We select a user $\mathcal{U}_v$ to be the verifier and assume that there are $\mathscr{L}$ iterations. Before the verification process, each user $\mathcal{U}_i$, where $i \in [1, U]$ and $i \neq v$ needs to upload their own $h_{i_k}, r_{i_k}, hh_{i_k}, rr_{i_k}, hw_{i_k}$, and $rw_{i_k}, k \in [1, \mathscr{L}]$, except $\mathcal{U}_v$. $\mathcal{U}_v$ will de-commit these commitments using **COM.Decommit** algorithm. The process will abort if there is any $\perp$ returned by **COM.Decommit** algorithm.

It is important to note, that any two users can communicate with each other through the server. Since the system adopts a commitment algorithm, the communication will not be forged. This mechanism is based on the verification function of V-EPTD.

Verification of Weight Update: The goal of this verification is to verify the update of the weight computations $\mathbf{Enc}(\widehat{w_i}), i \in [1, U]$.

After receiving the data from other users, $\mathcal{U}_v$ checks whether Eq. 5 holds, where $\widehat{Sum'_k}, k \in [1, \mathscr{L}]$ denotes the sum of all users' *Disk* provided by the server $\mathcal{S}$ in k-th iteration. If it holds, that means $\widehat{Sum'_k}, k \in [1, \mathscr{L}]$ are all correct, and then $\mathcal{U}_v$ publishes $\widehat{Sum'_k}, k \in [1, \mathcal{L}]$ through the server. Then the users can verify whether their own weight is correct by computing $\widehat{w}_i = \log(\widehat{Sum'_k}) - \log(\widehat{Disk_{v_k}})$. If $\widehat{w}_i$ is wrong or Eq. 5 does not hold for any user, that means the server sent forged or incorrect information to the users.

$$\mathbf{H.Hash}((\widehat{Sum'_1}, \cdots, \widehat{Sum'_{\mathcal{L}}})) = \prod_{k=1}^{\mathcal{L}} \prod_{i=1}^{U} h_{i_k} = \prod_{k=1}^{\mathcal{L}} g_k^{\sum_i^U \widehat{Diski_k}} = \prod_{k=1}^{\mathcal{L}} g_k^{\widehat{Sum_k}} \quad (5)$$

Verification of Ground Truth Update: The goal of this verification is to verify the update of the ground truth computation $\mathbb{V}_k^*[t]$, where $k \in [1, \mathcal{L}]$.

$\mathcal{U}_v$ verifies whether all Swd'_{t_k} provided by server $\mathcal{S}$ are correct, where $k \in [1, \mathcal{L}]$ and $t \in [1, T]$. First, $\mathcal{U}_v$ generates $Hswd_k \leftarrow$ $\mathbf{H.Hash}(\widehat{Swd'}_{1_k}, \cdots, \widehat{Swd'}_{T_k})$, $Hhh_k \leftarrow \prod_{i=1}^{U} hh_{i_k}$, and $Hhw_k \leftarrow \prod_{i=1}^{U} hw_{i_k}$, where $k \in [1, \mathcal{L}]$. Next, $\mathcal{U}_v$ checks whether Eq. 6 holds, i.e., all Swd'_{t_k} are correct, otherwise the server $\mathcal{S}$ lied to users. In the same way, all Sw'_{t_k} are correct if $\mathbf{H.Hash}(\widehat{Sw'_1}, \cdots, \widehat{Sw'_{\mathcal{L}}}) = \prod_{k=1}^{\mathcal{L}} Hhw_k$ holds. If Swd'_{t_k}, Sw'_k pass the verification, $\mathcal{U}_v$ can verify each ground truth by checking whether each equation, i.e., $\mathbb{V}_k^*[t] = \frac{\widehat{Swd'}_{t_k}}{\widehat{Sw'}_k}$, in which the verification has failed if any equation does not hold. The verification results can be sent to other users through the server by $\mathcal{U}_v$.

$$\begin{aligned} &\mathbf{H.Eval}(Hswd_1, \cdots, Hswd_{\mathcal{L}}, \alpha_1, \cdots, \alpha_{\mathcal{L}}) \\ &\qquad = \mathbf{H.Eval}(Hhh_1, \cdots, Hhh_{\mathcal{L}}, \alpha_1, \cdots, \alpha_{\mathcal{L}}) \quad (6) \end{aligned}$$

6.5 Discussion

The verification phase can be executed after each iteration, which means the verifier can verify the computing results returned by the server when k-th iteration finishes. Hence, at the end of k-th iteration, users $\mathcal{U}_i, i \in [1, U]$ upload their own $h_{i_k}, r_{i_k}, hh_{i_k}, rr_{i_k}, hw_{i_k}$, and $rw_{i_k}, k \in [1, \mathcal{L}]$ to the server. After receiving above data from the server, the verifier(s) de-commits each commitment. If it is successful in de-committing the commitments, then the verifiers verify the results by check if all $\mathbf{H.Hash}(\widehat{Sum'_k}) = \prod_{i=1}^{U} h_{i_k}$, $Hswd_k = Hhh_k$, $\mathbf{H.Hash}(\widehat{Sw'_1}) = Hhw_k$ hold. If any one of these conditions does not hold, then the verification has failed.

7 Security Analysis

Theorem 1. *Assume there are less than x conspiracy parties and* **COM** *is secure, then V-EPTD can protect input privacy including the sensing data and weight of each user in the whole scheme.*

Proof. In the truth discovery phase, the plaintexts the server knows are $\widehat{Sum_k}$, Swd_{t_k}, Sw_k, and $\mathbb{V}_k^*[t]$, where $t \in T, k \in \mathscr{L}$, which has no weight or sensing data for any user. Moreover, for a user $\mathcal{U}_i$, the weight w_{i_k} and sensing data $\mathbb{V}_{i_k}[t]$ are exchanged between the server and users as ciphertext or are committed into commitments. The commitments can keep the contents secret since there are not enough parameters to de-commit, and there are not more than x conspiracy parties including users and the server. Hence, the ciphertext can not be decrypted, and the plaintexts of $\mathcal{U}_i$'s weight and sensing values will not be disclosed to any parties. In the batch verification phase, the data delivered in the system are $h_{i_k}, r_{i_k}, hh_{i_k}, rr_{i_k}, hw_{i_k}$, and rw_{i_k} for each user. The randomness r_{i_k}, rr_{i_k}, and rw_{i_k} are used to de-commit. As h_{i_k}, hh_{i_k}, and hw_{i_k} are hash values, and since **LHH** is collision-resistant, the plaintext of weight and sensing values are also not available for each party except the owner.

Theorem 2. *Based on the DLP (discrete logarithm problem) and the security of* **COM**, *the aggregation results retain their integrity so that they will pass the verification, if and only if these results are honestly aggregated by the server with an overwhelming probability.*

Proof. The server can break the intergrity of an aggregation result and pass the verification with negligible probability. The detailed proof is similar to *Theorem 1* in [7], so we omit it here.

8 Performance Evaluation

In this section, we evaluate the performance of V-EPTD through two levels of experiments. Both have only one verifier and server in the system. In the first *comparative experiments*, we compare our scheme V-EPTD to V-PATD [15] from two aspects, i.e., computation overhead and communication overhead. Here, we set the number of users and tasks in the range of [10, 50]. In the second large-scale users and tasks experiments, we increase the number of users and tasks both to [100, 300]. The number of iterations for truth discovery is 10. Each simulation was repeated 10 times and the average was taken as the final result.

The security parameter is set as 256 bits. We use a folklore hash commitment scheme which is instantiated with SHA-256 as commitment **COM**. We select ECDH (elliptic curve Diffie-Hellman) with NIST P-256 curve to achieve key agreement. For symmetric encryption, we use AES-OFB mode. In our experiments, we simulate each user, server, and verifier on a system with macOS Catalina on six cores Intel Core i7 2.6 GHz, which utilizes Python ecdsa and Crypto libraries to implement the ECDH and AES computations.

8.1 Comparative Experiments

Computation Overhead: We simulate V-EPTD with two conditions. In the first, the number of users is fixed at 50, and the number of tasks ranges from 10 to 50. Conversely, in the second one, tasks are fixed at 50, and users range from 10 to 50. Compared to V-PATD, the results suggest that V-EPTD has an overall better performance as shown in Fig. 1.

A user's computation overhead is shown in Figs. 1(a) and 1(d), where it remains within 0.2s for both schemes with no significant difference. The data encrypted by a user often is an aggregate value, which causes the computation overhead for a user to be small in V-EPTD. For the server, Figs. 1(b) and 1(e) show that V-EPTD has better performance. The computation overhead of the server in V-EPTD does not exceed 100s, while for V-PATD it exceeds 2500 sec. The results in Figs. 1(c) and 1(f) also show that V-EPTD has a better performance for verifiers. In V-PATD, a verifier assists the server to generate proof of computing frequently, while in V-EPTD, the verifiers can finish the verification with the computation of homomorphic hash.

The overall comparison as shown in Fig. 1 clearly suggests that the time costs for users, server, and verifiers, are better in V-EPTD.

Communication Overhead: In this subsection, we discuss the communication overhead for one iteration. Figure 2 demonstrates the communication overhead of V-PATD and V-EPTD. Similar to the simulation of computation overhead, we set two groups, one fixes users at 50 and changes tasks from 10 to 50, while the other fixes tasks at 50 and changes users from 10 to 50.

The communication requirements of a user are small both in V-PATD and V-EPTD. However, the server (or a verifier) in V-PATD incurs more

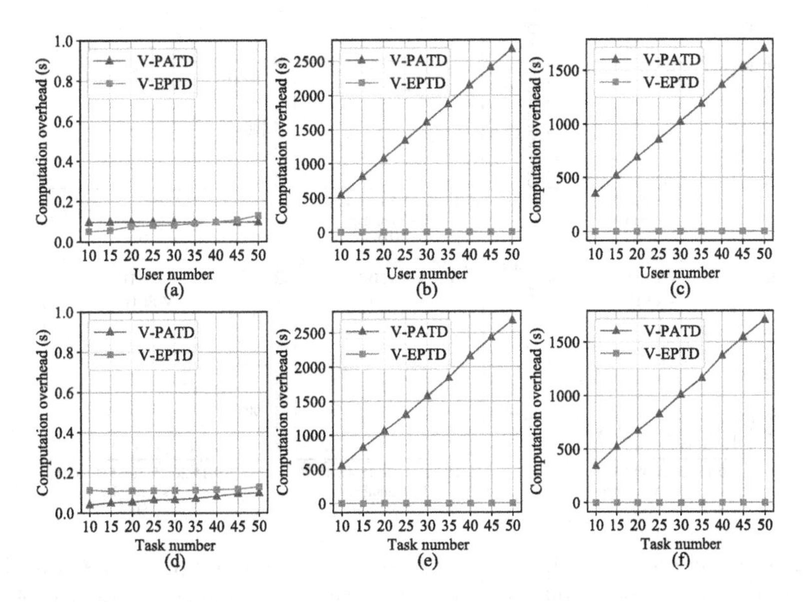

Fig. 1. Computation overhead of a user, server, and verifier: (a), (b), and (c) have a fixed number of tasks as 50. (d), (e), and (f) have fixed number of users as 50. (a) and (d) are the results for users. (b) and (e) are the results for the server. (c) and (f) are the results for a verifier.

Table 1. Communication overhead in KB with 50 users and 50 tasks.

	Initialization				Weight updation				Truth updation			
	V-PATD		V-EPTD		V-PATD		V-EPTD		V-PATD		V-EPTD	
	S	R	S	R	S	R	S	R	S	R	S	R
User	7.716	0	0.031	1.625	0.137	0.04	1.738	1.594	0.004	0	2.414	2.25
Server	0	355.8	76.56	1.656	1061	4.969	1.723	6.41	1269	1296	42.44	47.32
Verifier	0	2.93	0.031	1.625	4.969	1061	1.707	3.316	1296	1269	2.344	5.695

communication overhead than the server (or a verifier) in V-EPTD. The reason is that the server and verifiers in V-PATD need to communicate with each other frequently to complete the verification, while in V-EPTD, the proof is sent after each step so that it reduces the communication frequency and reduces the amount of communication data.

We list the details of communication overhead in Table 1. As there is no division between the verification phase and the truth discovery phase in V-PTAD, hence we also add the communication overhead of verification to other phases for presentation. Here, "S" and "R" refer to the amount of data sent and data received respectively, and the unit is in KB. The communication overhead of server and verifier mainly costs more in weight update and truth update phases.

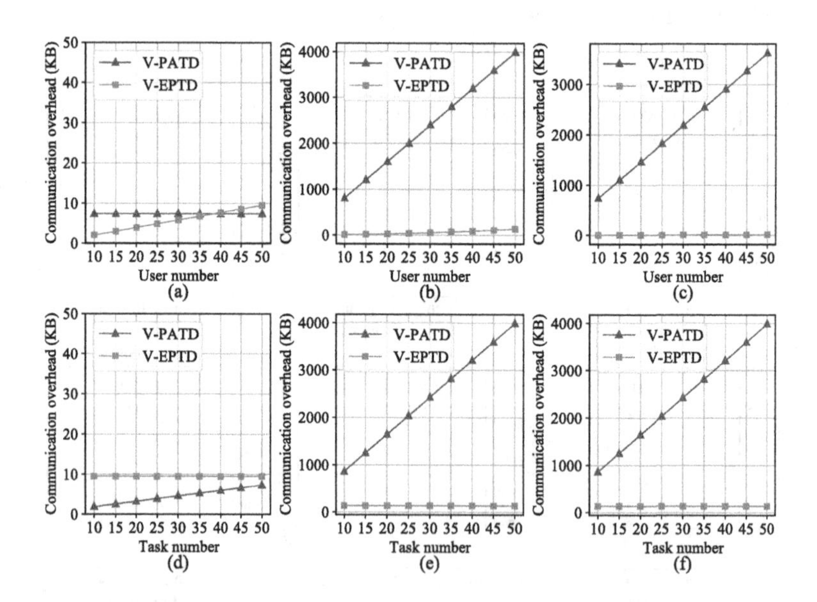

Fig. 2. Communication overhead of a user, server, and verifier: (a), (b), and (c) have a fixed number of tasks as 50. (d), (e), and (f) have fixed number of users as 50. (a) and (d) are the results for users. (b) and (e) are the results for the server. (c) and (f) are the results for a verifier.

8.2 Experiment with Large-scale Users and Tasks

In this experiment, we simulate V-EPTD with a large number of users and tasks. There are two sets of evaluations, where one group fixes users at 300 and changes tasks from 100 to 300, while the other one fixes tasks at 300 and changes users from 100 to 300.

Figures 3(a) and 3(b) demonstrate that even in case of large-scale, the computation overhead is less than 2 s for the proposed scheme. Note that, the cost of verifier shown in Figs. 3(a) and 3(b) is the cost in verification phase, and the verifiers can be chosen from the users. The computation overhead of a user increases with user number, while task number hardly influences it. This feature also exists in communication overhead. Regardless of the number of users, the communication overhead (for that number of users) remains unchanged, as shown in Fig. 3(c). Moreover, as shown in Fig. 3(d), the increase in the number of users has a more obvious impact on the server.

From the above discussion and experimental results, we can conclude that in the case of large-scale users tasks, V-EPTD shows good performance, and hence it can have practical application significance.

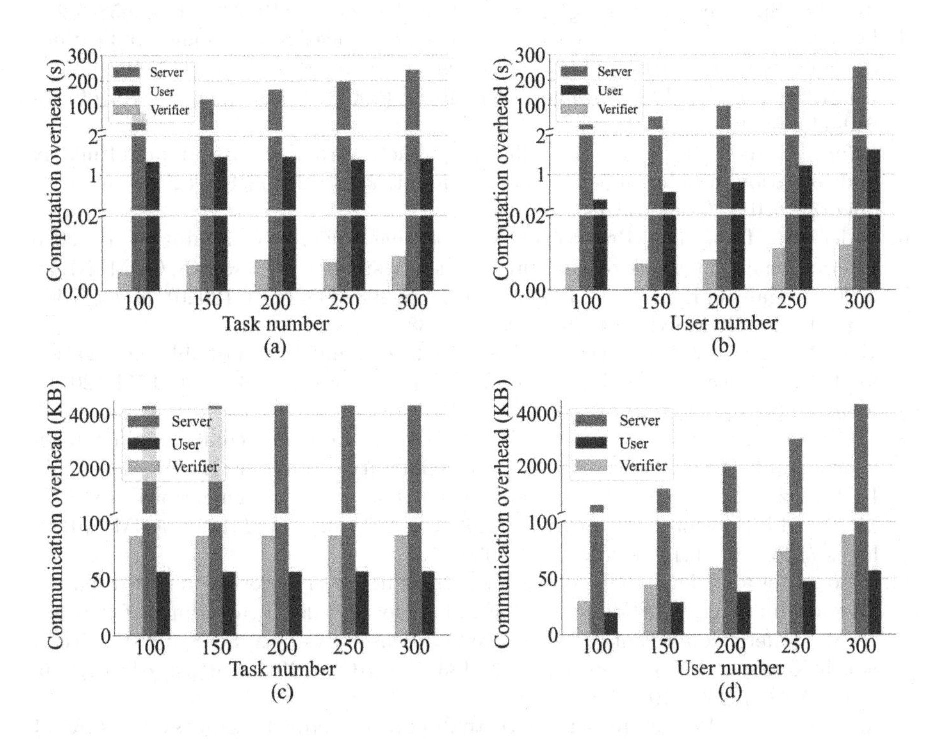

Fig. 3. Computation and communication overhead with large-scale users and tasks: (a) computation overhead of the group with fixed users. (b) computation overhead of the group with fixed tasks. (c) communication overhead of the group with fixed users. (d) communication overhead of the group with fixing tasks.

9 Conclusion

In this paper, we propose a verifiable scheme V-EPTD for privacy-preserving truth discovery. The threshold paillier cryptosystem is adopted to complete privacy-preserving truth discovery. To verify the computing as well as privacy preservation, we adopt a linear homomorphic hash. Moreover, we also show that the server cannot forge the hash values using a commitment guarantee. The experiments show that V-EPTD has a good performance both for computation overhead and communication overhead. In the future, we plan to focus on improving the performance in the scenario of hyper-scale users.

References

1. Matrix mathematics: theory, facts, and formulas with application to linear systems theory, pp. xlii+1139. Princeton University Press (2009)
2. Ajarn, J.J.: Permutations and Combinations. Combinatorial Theory, 2nd edn. (2009)
3. Bellare, M., Goldreich, O., Goldwasser, S.: Incremental cryptography: the case of hashing and signing. In: Desmedt, Y.G. (ed.) CRYPTO 1994. LNCS, vol. 839, pp. 216–233. Springer, Heidelberg (1994). https://doi.org/10.1007/3-540-48658-5_22
4. Damgård, I., Jurik, M.: A generalisation, a simplification and some applications of Paillier's probabilistic public-key system. In: Kim, K. (ed.) PKC 2001. LNCS, vol. 1992, pp. 119–136. Springer, Heidelberg (2001). https://doi.org/10.1007/3-540-44586-2_9
5. Erfan, F., Mala, H.: Secure and efficient publicly verifiable outsourcing of matrix multiplication in online mode. Cluster Comput. **23**(4), 2835–2845 (2020). https://doi.org/10.1007/s10586-020-03049-7
6. Gajera, H., Das, M.L.: Privc: privacy preserving verifiable computation. In: 2020 International Conference on COMmunication Systems & NETworkS, COMSNETS 2020, Bengaluru, India, 7–11 January 2020, pp. 298–305. IEEE (2020). https://doi.org/10.1109/COMSNETS48256.2020.9027488
7. Guo, X., et al.: VeriFL: communication-efficient and fast verifiable aggregation for federated learning. IEEE Trans. Inf. Forensics Secur. **16**, 1736–1751 (2021). https://doi.org/10.1109/TIFS.2020.3043139
8. Li, Q., Li, Y., Gao, J., Zhao, B., Fan, W., Han, J.: Resolving conflicts in heterogeneous data by truth discovery and source reliability estimation. In: Dyreson, C.E., Li, F., Özsu, M.T. (eds.) International Conference on Management of Data, SIGMOD 2014, Snowbird, UT, USA, 22–27 June 2014, pp. 1187–1198. ACM (2014). https://doi.org/10.1145/2588555.2610509
9. Miao, C., et al.: Cloud-enabled privacy-preserving truth discovery in crowd sensing systems. In: Song, J., Abdelzaher, T.F., Mascolo, C. (eds.) Proceedings of the 13th ACM Conference on Embedded Networked Sensor Systems, SenSys 2015, Seoul, South Korea, 1–4 November 2015, pp. 183–196. ACM (2015). https://doi.org/10.1145/2809695.2809719
10. Miao, C., et al.: Privacy-preserving truth discovery in crowd sensing systems. ACM Trans. Sens. Netw. **15**(1), 9:1–9:32 (2019). https://doi.org/10.1145/3277505
11. Miao, C., Su, L., Jiang, W., Li, Y., Tian, M.: A lightweight privacy-preserving truth discovery framework for mobile crowd sensing systems. In: 2017 IEEE Conference on Computer Communications, INFOCOM 2017, Atlanta, GA, USA, 1–4 May 2017, pp. 1–9. IEEE (2017). https://doi.org/10.1109/INFOCOM.2017.8057114

12. Wang, X.A., Choo, K.R., Weng, J., Ma, J.: Comments on "publicly verifiable computation of polynomials over outsourced data with multiple sources". IEEE Trans. Inf. Forensics Secur. **15**, 1586–1588 (2020). https://doi.org/10.1109/TIFS.2019.2936971
13. Xu, G., Li, H., Liu, S., Yang, K., Lin, X.: VerifyNet: secure and verifiable federated learning. IEEE Trans. Inf. Forensics Secur. **15**, 911–926 (2020). https://doi.org/10.1109/TIFS.2019.2929409
14. Xu, G., Li, H., Lu, R.: Practical and privacy-aware truth discovery in mobile crowd sensing systems. In: Lie, D., Mannan, M., Backes, M., Wang, X. (eds.) Proceedings of the 2018 ACM SIGSAC Conference on Computer and Communications Security, CCS 2018, Toronto, ON, Canada, 15–19 October 2018, pp. 2312–2314. ACM (2018). https://doi.org/10.1145/3243734.3278529
15. Xu, G., et al.: Catch you if you deceive me: verifiable and privacy-aware truth discovery in crowdsensing systems. In: Sun, H., Shieh, S., Gu, G., Ateniese, G. (eds.) ASIA CCS'20: The 15th ACM Asia Conference on Computer and Communications Security, Taipei, Taiwan, 5–9 October 2020, pp. 178–192. ACM (2020). https://doi.org/10.1145/3320269.3384720
16. Zhang, C., Xu, C., Zhu, L., Li, Y., Zhang, C., Wu, H.: An efficient and privacy-preserving truth discovery scheme in crowdsensing applications. Comput. Secur. **97**, 101848 (2020). https://doi.org/10.1016/j.cose.2020.101848
17. Zhang, C., Zhu, L., Xu, C., Liu, X., Sharif, K.: Reliable and privacy-preserving truth discovery for mobile crowdsensing systems. IEEE Trans. Dependable Secur. Comput. **18**(3), 1245–1260 (2021). https://doi.org/10.1109/TDSC.2019.2919517
18. Zhang, C., Zhu, L., Xu, C., Ni, J., Huang, C., Shen, X.S.: Efficient and privacy-preserving non-interactive truth discovery for mobile crowdsensing. In: IEEE Global Communications Conference, GLOBECOM 2020, Virtual Event, Taiwan, 7–11 December 2020, pp. 1–6. IEEE (2020). https://doi.org/10.1109/GLOBECOM42002.2020.9322483
19. Zhang, L.F., Safavi-Naini, R.: Protecting data privacy in publicly verifiable delegation of matrix and polynomial functions. Des. Codes Cryptogr. **88**(4), 677–709 (2019). https://doi.org/10.1007/s10623-019-00704-y

FedSP: Federated Speaker Verification with Personal Privacy Preservation

Yangqian Wang[1], Yuanfeng Song[2], Di Jiang[2], Ye Ding[3], Xuan Wang[1], Yang Liu[1], and Qing Liao[1,4](✉)

[1] Harbin Institute of Technology (Shenzhen), Shenzhen 518055, China
19s051041@stu.hit.edu.cn, wangxuan@cs.hitsz.edu.cn, {liu.yang,liaoqing}@hit.edu.cn

[2] AI Group, WeBank Co., Ltd., Shenzhen, China
{yfsong,dijiang}@webank.com

[3] Dongguan University of Technology, Dongguan, China
dingye@dgut.edu.cn

[4] Peng Cheng Laboratory, Shenzhen 518055, China

Abstract. Automatic speaker verification (ASV) has been widely applied in a variety of industrial scenarios. In ASV, the universal background model (UBM) needs to be trained with a large variety of speaker data so that the UBM can learn the speaker-independent distribution of speech features for all speakers. However, the sensitive information contained in raw speech data is important and private for the speaker. According to the recent European Union privacy regulations, it is forbidden to upload private raw speech data to the cloud server. Thus, a new ASV model needs to be proposed to alleviate data scarcity and protect data privacy simultaneously in the industry. In this work, we propose a novel framework named *Federated Speaker Verification with Personal Privacy Preservation*, or FedSP, which enables multiple clients to jointly train a high-quality speaker verification model and provide strict privacy preservation for speaker. For data scarcity, FedSP is based on the federated learning (FL) framework, which keeps raw speech data on each device and jointly trains the UBM to learn the speech features well. For privacy preservation, FedSP provides more strict privacy preservation than traditional basic FL framework by selecting and hiding sensitive information from raw speech data before jointly training the UBM. Experimental results on two pair speech datasets demonstrate that FedSP has superior performances in terms of data-utility and privacy preservation.

Keywords: Speaker verification · Federated learning · Privacy preservation · Sensitive information

1 Introduction

Automatic speaker verification (ASV) aims to verify whether a speech belongs to a specific speaker based on the speaker's known utterances. ASV has become a

Y. Lai et al. (Eds.): ICA3PP 2021, LNCS 13157, pp. 462–478, 2022.
https://doi.org/10.1007/978-3-030-95391-1_29

common verification way in terms of forensics and security. For instance, many commercial smart devices such as mobile phones, AI speakers, and automotive infotainment systems have adopted machine learning-based ASV for unlocking the system or providing a user-specific service. In universal background model (UBM) based ASV, UBM needs to be trained with a large variety of speech data so that the model can learn the speaker-independent feature distribution of all speakers [4,13,22,25]. However, the raw speech data that contains sensitive information is not allowed to upload to the server [3,5,23] according to European Union General Data Protection Regulation (GDPR) [24], which is a privacy preservation regulation. For example, sharing the raw speech data with the server will not only disclose the identity of the speaker but also make the ASV systems vulnerable to spoofing attacks [27,28]. Thus, a new ASV model needs to be proposed to alleviate data scarcity and protecting data privacy simultaneously in the industry.

ASV systems based on federated learning (FL) [9,11,29] can jointly train a high-quality speaker verification model with multiple clients to alleviate the data scarcity problem via repeatedly communicating model parameters, e.g., model weights or certain statistics, between client and server. However, sharing the parameters of ASV in the FL framework is still not secure enough due to it may also disclose user's privacy when using the raw speech data including sensitive information to training model [30]. For example, through model inversion attack [6] the malicious attackers can reconstruct the speaker biometric templates when getting the parameters uploading from the client, thereby leaking the privacy of speaker. Recently, several cryptographic techniques are merged with the FL framework to overcome the above security and privacy issues [1,18]. However, cryptographic based FL frameworks bring heavy overhead and significantly slow down the verification process, which makes it is unsuitable for ASV. Therefore, it is necessary to develop new privacy preservation technologies for FL based speaker verification systems.

In this work, we propose a novel framework named *Federated Speaker Verification with Personal Privacy Preservation*, or FedSP, that can alleviate data scarcity and provide strict privacy preservation at the same time. For data scarcity, FedSP is based on federated learning (FL) framework that can jointly train the UBM well with multiple clients while retaining the raw speech data on each device. For privacy preservation, except for the default privacy preservation provided by the FL framework, FedSP provides more strict privacy preservation by selecting and hiding the sensitive information from the raw speech data before jointly training the UBM. Especially, sharing the parameters of the model during the model training with sensitive information related to the identity of the speaker may disclose the privacy of speaker. Hence, we try to select and hide the sensitive information of raw speech data and then we formulate the selecting process as an NP-hard integer programming problem. Furthermore, a heuristic greedy search algorithm is proposed to obtain a suboptimal solution for the integer programming problem. In summary, the main contributions of this paper as follows:

- FedSP is the first large-scale distributed speaker verification framework based on GMM-UBM, which can simultaneously overcome data scarcity and privacy leakage issues.
- In order to protect the privacy of each client, FedSP tries to select and hide the sensitive information from raw speech data. FedSP formulates the sensitive information selection process as an NP-hard integer programming problem. In this work, we propose a heuristic greedy search algorithm to solve it.
- Experiments on two pair speech datasets validate the effectiveness of FedSP in data-utility and privacy preservation.

The rest of this paper is organized as follows. In Sect. 2, we briefly review the related literature. In Sect. 3, we detail our proposed FedSP framework, followed by experimental results and analyses in Sect. 4. We finally conclude the work in Sect. 5.

2 Related Work

ASV arises unique privacy concerns, because speech data that are used to training the speaker verification model is closely related to the identity of speakers. Several works have been proposed to overcome the privacy challenges in ASV. And these works can be divided into two categories.

For the first category, they mainly focus on corporating cryptographic encryption or salting techniques with existing ASV methods to protect the privacy of speaker. For example, Pathak and Raj et al. [19] merges GMM with Homomorphic encryption (HE) and Secure Multi-Party Computation (SMPC), and summarize how to perform inference and classification on the encrypted GMM-based speaker verification. Manas et al. [20] applied locality sensitive hashing (LSH) transformation to convert and protect speech signals. Yogachandran et al. [21] cooperated randomization technique to propose an i-vector [4] based verification model to verify speakers' voice in the randomized domain. However, all these methods assume the server needs to collect a large variety of speech data to train the universal background model, which is forbidden by GDPR.

For the second category, these methods are based on the FL framework to jointly training the ASV model with local speech data of users by repeatedly communicating the model weights between a server and a group of users. Recently, FL is widely used in many applications such as topic modeling [12], mobile keyboard prediction [15], and visual object detection [17]. However, there only a few work research the problem of using FL framework to preserve the speakers' privacy in ASV. For example, Filip Granqvist et al. [9] using the side information of local speaker, e.g., gender and emotion, to construct an auxiliary model to enriching the speaker embedding network based on the FL framework. However, this method focuses on protecting the privacy of side information rather than the speech data. Hossein Hosseini et al. [11] proposed a framework for training user authentication models named FedUA, which adopts FL framework and random binary embedding to protect the privacy of raw inputs and embedding vectors based on neural networks, respectively. However, FedUA doesn't provide any privacy preservation on the model weights, which make it is vulnerable to model inversion attack. In general, the above works can't simultaneously alleviate data scarcity and protect data privacy for ASV. In this paper, we give an in-depth consideration of how to jointly training ASV model while protecting user's privacy in the FL framework.

3 Framework

As demonstrated in Fig. 1, FedSP is composed of three computational modules: client computation, server merging, and client verification. (1) In the client computation module, we propose a heuristic greedy search algorithm to select and hide the sensitive information from the raw speech data, so that the Baum-Welch statistics [4] leaning in the local client will not contain sensitive information. (2) In the server merging module, the server collects the Baum-Welch statistics of each client and assign updated parameters of UBM to all clients. The Baum-Welch Statistics contain acoustic and phonetic variations in speech data, so merging and using it to update the parameters of UBM can make the UBM fits the acoustic channels of the training data. And the privacy of the speaker will not be disclosed via server and transmission attacks, because the Baum-Welch statistics are calculated based on data that does not contain sensitive information. (3) In the client verification module, each client receives the UBM parameters from the server and derives the hypothesized speaker model based on the local raw speech data contained sensitive information. The sensitive information contained in raw speech data is closely related to the identity of the speaker, so each client can achieve a satisfactory verification performance.

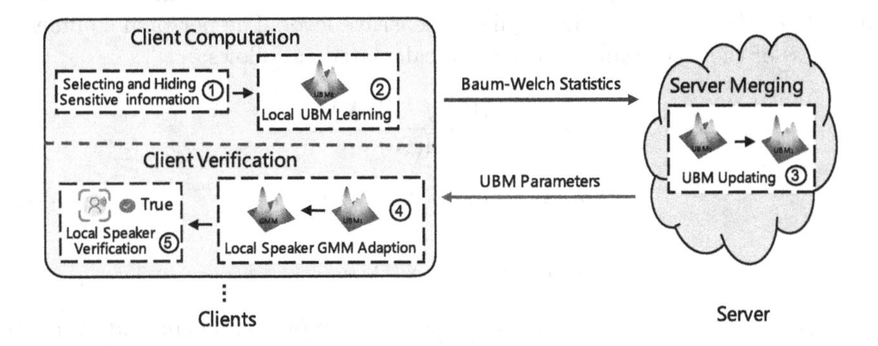

Fig. 1. Federated speaker verification with personal privacy preservation pipeline.

3.1 Client Computation

In this section, we first describe the algorithm for selecting and hiding sensitive information from the raw speech data to protect the privacy of each client. Second, each client executes local UBM learning to get the Baum-Welch statistics without containing sensitive information. At last, clients transmit the local

Baum-Welch statistics without sensitive information to update the parameters of UBM on the server.

Selecting and Hiding Sensitive Information. In this subsection, we present how to select and hide sensitive information (SHS) in detail. In SHS, the sensitive sample is the training vector that closely relates to the sensitive information [5,14,23]. The goal of SHS is to select and hide the sensitive samples from the raw speech data to calculate the Baum-Welch statistics without sensitive information of speaker on local client. Therefore, the speaker's privacy will not be disclosed, when the malicious attackers intercept the Baum-Welch statistics. In GMM-UBM, each training vector of speech data can be divided into different gaussian components in GMM. So we perform sensitive sample selection at the gaussian components level. SHS selects f (i.e., selected fraction) percentage of components containing sensitive information, and then delete the sensitive samples based on that components. In SHS, we first divide the training vectors into different gaussain components according to the posterior probabilities of each training vector. Then we define the *personal confidence score* of each gaussian component and the *distance* between two gaussian components. Finally, the sensitive components selecting process is formulated as an integer programming problem, which is to select a set of gaussian components with maximizing total *personal confident score* and *distance.*

Personal Confidence Score of Gaussian Component. The *personal confidence score* (*pcs*) of a gaussian component is defined as the log-likelihood difference after the vectors belonging to the component are deleted. Therefore, *pcs* can reflect the importance of the component for the speaker. For example, the larger value of *pcs* the component has a higher sensitive level. The personal confidence score pcs_i of the i-th component can be calculated as follows:

$$
\begin{aligned}
pcs_i &= \frac{\Lambda(\boldsymbol{X}) - \Lambda(\bar{\boldsymbol{X}})}{\Lambda(\boldsymbol{X})}, \\
\Lambda(\boldsymbol{X}) &= log(p(\boldsymbol{X}|\lambda_{spk1})), \qquad (1) \\
\Lambda(\bar{\boldsymbol{X}}) &= log(p(\bar{\boldsymbol{X}})|\bar{\lambda}_{spk1})),
\end{aligned}
$$

where $\boldsymbol{X} = \{\boldsymbol{x}_1, ..., \boldsymbol{x}_T\}$ is the raw training vectors of each client and $\bar{\boldsymbol{X}}$ is the training vectors that do not include the vectors belong to the i-th component. And $p(\cdot)$ is the probability density function of GMM. λ_{spk1} and $\bar{\lambda}_{spk1}$ are the GMM model derived from UBM_0 based on $\boldsymbol{X}$ and $\bar{\boldsymbol{X}}$, respectively.

Distance between Gaussian Components. It is not enough to select sensitive components based on *pcs*, because we may select similar components with large *pcs* values. For instance, if the selected components are in the same area of the acoustic space, sensitive components located in other areas can still cause privacy leakage. In order to ensure the diversity of the selected sensitive components, we design a new indicator. In SHS, we recommend ensuring that the distance between all selected components is as large as possible. The *distance* $d_{i,j}$ between two components c_i and c_j is calculated below:

$$d_{i,j} = \sqrt{\sum_{h=1}^{m} (\mu_{i,h} + \mu_{j,h})}, \tag{2}$$

where $\mu_{i,h}$ is the h-th elements of $\boldsymbol{\mu}_i$ and $\boldsymbol{\mu}_i$ is the mean vector of components c_i in λ_{spk1}. $\mu_{j,h}$ is the h-th elements of $\boldsymbol{\mu}_j$ and $\boldsymbol{\mu}_j$ is the mean vector of components c_j in λ_{spk1}.

Optimization Formulation and Solution. As aforementioned, we want to select those gaussian components with maximizing personal confident score and total distance. The objective function of SHS is defined as below:

$$\max \sum_i pcs_i s_i + \alpha \sum_i \sum_{j \neq i} d_{i,j} s_i s_j, \tag{3}$$

where $s_i = 1$ (or 0) indicates that component c_i is selected (or not) and $\sum_i s_i = f * M$, pcs_i denotes the personal confidence score of component c_i and $d_{i,j}$ denotes the distance between component c_i and c_j. α is a trade off parameter and M is the number of gaussian components in GMM. The optimization in Eq. 3 is a typical 0-1 integer programming problem [8,26], which has been shown as a NP-hard problem with $O(2^M)$ search space in exhaustive search. In this work, we propose a greedy component search algorithm (GCS) to solve this problem. Next, we will introduce the GCS in detail.

In GCS, the selected component subset starts from $\boldsymbol{S}_0 = \emptyset$ and adds one component for each step. Supposing that l' is the index of gaussian component selected in the *l-th* step, and component $c_{l'}$ will be added in the component set $\boldsymbol{S}_l$.

$$\boldsymbol{S}_l = \boldsymbol{S}_{l-1} \cup c_{l'}, \tag{4}$$

the components in $\boldsymbol{S}_l = \{c_{i'} | i = 1, ..., l\}$ should meet Eq. (5):

$$\max(\sum_{i=1}^{l} pcs_{i'} + \alpha \sum_{i=1}^{l} \sum_{j \neq i} d_{i',j'}), \tag{5}$$

since $d_{i,j} = d_{j,i}$, $\boldsymbol{S}_l \supset \boldsymbol{S}_{l-1}$ and $\boldsymbol{S}_{l-1} = \{c_{i'} | i = 1, ..., l-1\}$, we can rewrite Eq. (5) as:

$$\max_{l'}((\sum_{i=1}^{l-1} pcs_{i'} + 2\alpha \sum_{i=1}^{l-2} \sum_{j=i+1}^{l-1} d_{i',j'}) + (pcs_{l'} + 2\alpha \sum_{i=1}^{l-1} d_{i',l'})), \tag{6}$$

note that the first part in Eq. (6) is a constant concerning the components selected in the previous $l-1$ steps and the goal becomes to select the component maximizing the second part. So the component $c_{l'}$ selected in l-th step is

$$c_{l'} = \underset{c_i \in \boldsymbol{S}_{M-l-1}}{\arg\max}\ pcs_i + 2\alpha \sum_{j=1}^{l-1} d_{i,j'}, \tag{7}$$

where $\boldsymbol{S}_{M-l-1} = \{c_i | c_i \notin \boldsymbol{S}_{l-1}, i = 1, ..., M\}$.

Selecting and Hiding Sensitive Samples. After getting the component set $\boldsymbol{S}_{f*M}$ that contains sensitive information, we can use it to select sensitive samples. The posterior probability of component c generating the vector $\boldsymbol{x}_t \in \boldsymbol{X}$ is the indicator to select the sensitive samples. The training vector is recognized as a sensitive sample when the posterior probability of the training vector on component c_i is greatest and $c_i \in \boldsymbol{S}_{f*M}$. The sensitive samples will be deleted from the raw training vector $\boldsymbol{X}$ to hide the sensitive information. $\hat{\boldsymbol{X}}$ is the vectors, which delete the sensitive samples from $\boldsymbol{X}$. The posterior probability $r_{c,t}$ of component c generating the training vector $\boldsymbol{x}_t$ is as follows:

$$r_{c,t} = \frac{w_{0,c} \mathcal{N}(\boldsymbol{x}_t | \boldsymbol{\mu}_{0,c}, \boldsymbol{\sigma}_{0,c})}{\sum_{j=1}^{M} w_{0,j} \mathcal{N}(\boldsymbol{x}_t | \boldsymbol{\mu}_{0,j}, \boldsymbol{\sigma}_{0,j})}, \tag{8}$$

where $w_{0,j}, \boldsymbol{\mu}_{0,j}$, and $\boldsymbol{\sigma}_{0,j}$ are the mixture weight, mean vector, and diagonal covariance matrix of j-th component of UBM$_0$, respectively. M is the number of Gaussian components of UBM$_0$. the parameters of UBM$_0$ are collectively represented by the notation $\lambda_0 = \{w_{0,j}, \boldsymbol{\mu}_{0,j}, \boldsymbol{\sigma}_{0,j}\}$, where $j = 1, ..., M$.

The detail of SHS is shown in Algorithm 1.

Algorithm 1. SHS

1: **Input:** raw training vector $\boldsymbol{X}$, pre-trained model UBM$_0$. $\boldsymbol{PCS} = \{pcs_i | i = 1, ..., M\}$, $\boldsymbol{D} = \{d_{i,j} | i \neq j, i, j \in \{1, ..., M\}\}$, and selected fraction f;
2: **Output:** $\hat{\boldsymbol{X}}$;
3: Initially $\boldsymbol{S}_0 = \emptyset$ and $\hat{\boldsymbol{X}} = \emptyset$;
4: **for** $i = 1$ to $f * M$ **do**
5: search for the new component $c_{i'}$ according to Eq.(7);
6: update the other components according to their distance with $c_{i'}$:

$$pcs_j \leftarrow pcs_j + 2\alpha \times d_{i',j}, \quad j \neq i'$$

7: add $c_{i'}$ to the set $\boldsymbol{S}_i$, $\boldsymbol{S}_i = \boldsymbol{S}_{i-1} \cup c_{i'}$;
8: **end for**
9: **for** $t = 1$ to T **do**
10: get the component c', which has the greatest probability of $\boldsymbol{x}_t$ come from it:

$$c' = \underset{c \in \{1,...,M\}}{\arg\max}\ r_{c,t}$$

11: **if** $c' \in \boldsymbol{S}_{f*M}$ **then**
12: $\hat{\boldsymbol{X}} = \hat{\boldsymbol{X}} \cup \boldsymbol{x}_t$;
13: **end if**
14: **end for**

Local UBM Learning. To alleviate the problem of privacy leakage, each local client is the workhorse for UBM learning in FedSP. Especially, FedSP calculates the Baum-Welch statistics on local clients, and that statistics will be uploaded to the server. Thus the Baum-Welch Statistics contain acoustic and phonetic variations in local speech data. At the same time, to prevent the privacy leakage by the Baum-Welch statistics each client using the training vectors processed by SHS, i.e., $\hat{\boldsymbol{X}}$, instead of raw training vector $\boldsymbol{X}$ to do UBM learning. Based on the global UBM_0 received from the server and $\hat{\boldsymbol{X}} = \{\boldsymbol{x}_1, ..., \boldsymbol{x}_{\hat{T}}\}$, we first compute the posterior probability $r_{c,t}$ of $\boldsymbol{x}_t$ on the c-th component of UBM_0 as Eq. (8). Then $r_{c,t}$ is used to calculate the Baum-Welch statistics $\hat{\boldsymbol{r}}_c^i$ and $\hat{\boldsymbol{z}}_c^i$ as follows:

$$\hat{\boldsymbol{r}}_c^i = \sum_{t=1}^{\hat{T}} r_{c,t}[\mathbf{1}], \quad \hat{\boldsymbol{z}}_c^i = \sum_{t=1}^{\hat{T}} r_{c,t} \tag{9}$$

where $[\mathbf{1}]$ is the vector created by filling unit elements and the dimension of it is m. i is the index of client. $\hat{\boldsymbol{r}}_c^i$ and $\hat{\boldsymbol{z}}_c^i$ will be sent to the server for UBM updating. The Baum-Welch statistics do not contain sensitive information, so the privacy of the client will not be disclosed.

3.2 Server Merging: UBM Updating

To alleviate the problem of data scarcity, the server updates the parameters of UBM to model the general acoustic features well. And the UBM updating is based on the Baum-Welch statistics uploaded from local clients. UBM updating on the server needs the help of $\hat{\boldsymbol{r}}_c^i$ and $\hat{\boldsymbol{z}}_c^i$, which are uploaded from the client i. The Baum-Welch statistics are firstly pooled together to form the pooled statistics:

$$\bar{\boldsymbol{r}}_c = \sum_{i=1}^{K} \hat{\boldsymbol{r}}_c^i, \quad \bar{\boldsymbol{z}}_c = \sum_{i=1}^{K} \hat{\boldsymbol{z}}_c^i, \tag{10}$$

where $\hat{\boldsymbol{r}}_c^i$ and $\hat{\boldsymbol{z}}_c^i$ denote the Baum-Welch statistics transmitted from client i with privacy preservation provided by SHS. And K is the number of training clients participating in UBM updating.

Then the server uses the pooled statistics to update the parameters of UBM_0 to get UBM_1, $\lambda_1 = \{w_{1,j}, \boldsymbol{\mu}_{1,j}, \boldsymbol{\sigma}_{1,j}\}$ and $j = 1, ..., M$, according to the following formula:

$$\boldsymbol{\mu}_{1,c} = \frac{\bar{\boldsymbol{z}}_c + \frac{\boldsymbol{\sigma}_{0,c}}{\hat{\sigma}_{\text{UBM}}} \boldsymbol{\mu}_{0,c}}{\bar{\boldsymbol{r}}_c + \frac{\boldsymbol{\sigma}_{0,c}}{\hat{\sigma}_{\text{UBM}}}}, \tag{11}$$

where $\hat{\sigma}_{\text{UBM}}$ represents the covariance prior to indicate that the prior is used for UBM update.

3.3 Client Verification

Local Speaker GMM Adaption. In FedSP, each client derives the speaker's GMM model by adapting the parameters of the UBM_1 based on the raw training vectors $\boldsymbol{X}$. The speaker's GMM model contains sensitive information, which is closely related to the identity of speaker, so each client can verify the identity of a new speech accurately. $\boldsymbol{r}_c$ and $\boldsymbol{z}_c$ calculated based on $\boldsymbol{X}$ according Eq. (9) are the key parameters for adapting the speaker's GMM. Then, $\boldsymbol{r}_c$ and $\boldsymbol{z}_c$ are used to adapt the c-th component of the speaker's GMM, $\lambda_s = \{w_{s,j}, \boldsymbol{\mu}_{s,j}, \boldsymbol{\sigma}_{s,j}\}$ and $j = 1, ..., M$, with the following equation:

$$\boldsymbol{\mu}_{s,c} = \frac{\boldsymbol{z}_c + \frac{\boldsymbol{\sigma}_{1,c}}{\hat{\sigma}_{\text{spk}}}\boldsymbol{\mu}_{1,c}}{\boldsymbol{r}_c + \frac{\boldsymbol{\sigma}_{1,c}}{\hat{\sigma}_{\text{spk}}}}, \tag{12}$$

where $\hat{\sigma}_{\text{spk}}$ indicates the prior for speaker model adaptation.

Local Speaker Verification. Each client can use the global UBM_1 and their GMM model λ_s to compute the log-likelihood ratio to verify the identity of new speeches [22]. Given a new segment of speech $\boldsymbol{X}_{test}$ and the hypothesized speaker S, the task of ASV is to determine if $\boldsymbol{X}_{test}$ is spoken by S. Mathematically, speaker verification can be formulated as:

$$\begin{gathered} \Lambda(\boldsymbol{X}_{test}) = \log p(\boldsymbol{X}_{test}|\lambda_s) - \log p(\boldsymbol{X}_{test}|\lambda_1) \\ \begin{cases} \Lambda(\boldsymbol{X}_{test}) \geq \theta & \boldsymbol{X}_{test} \text{ is from the hypothesized speaker } S \\ \Lambda(\boldsymbol{X}_{test}) < \theta & \boldsymbol{X}_{test} \text{ is not from the hypothesized speaker } S, \end{cases} \end{gathered} \tag{13}$$

3.4 FedSP Workflow

The algorithm of FedSP is presented in Algorithm 2. During the client computation stage, each client first selects and hides sensitive information from the raw speech data according to Algorithm 1. Then each client calculates the Baum-Welch statistics based on the training vectors without sensitive information and uploads the statistics to the server. After that, the server collects the Baum-Welch statistics of each client and assigns update parameters of UBM to all clients. Finally, each client receives the parameters of UBM and uses it to derive the speaker's GMM model based on the raw speech data.

Algorithm 2. FedSP

1: **function** SERVERMERGINGUBMUPDATING( ):
2: initialize UBM_0;
3: **for** each Guassian Component c from 1 to M in UBM_0 **do**
4: $\bar{\boldsymbol{r}}_c = 0$, $\bar{\boldsymbol{z}}_c = 0$;
5: **for** each local speaker i from 1 to K **do**
6: $\hat{r}_c^i, \hat{z}_c^i$ = CLIENTCOMPUTATION(UBM_0,f,c);
7: $\bar{\boldsymbol{r}}_c = \bar{\boldsymbol{r}}_c + \hat{r}_c^i$, $\bar{\boldsymbol{z}}_c = \bar{\boldsymbol{z}}_c + \hat{z}_c^i$;
8: **end for**
9: update UBM from UBM_0 to UBM_1 according Eq.(11);
10: **end for**
11: **return** UBM_1;
12: **end function**
13:
14: **function** CLIENTCOMPUTATION(UBM_0,f,c):
15: get the training speech of speaker i, $\boldsymbol{X}_i$, and processed by SLPP to get $\hat{\boldsymbol{X}}_i$;
16: calculate $\boldsymbol{PCS}$ and $\boldsymbol{D}$ according Eq.(1) and Eq.(2), respectively;
17: $\hat{\boldsymbol{X}}$ =SHS($\boldsymbol{X}$,UBM_0,$\boldsymbol{PCS}$,$\boldsymbol{D}$,f);
18: calculate $\hat{\boldsymbol{r}}_c$ and $\hat{\boldsymbol{z}}_c$ based on $\hat{\boldsymbol{X}}$ just as Eq.(9);
19: **return** $\hat{\boldsymbol{r}}_c$, $\hat{\boldsymbol{z}}_c$;
20: **end function**
21:

4 Experiments

In this section, we first introduce datasets and corresponding settings used in experiments. Second, we introduce the evaluation metrics used in this work. Third, we conduct experiments to verify the effectiveness of FedSP in solving data scarcity and privacy preservation problems. Finally, we state the privacy preservation capability and do the parameter study of FedSP.

4.1 Dataset and Configurations

We conduct our experiments using four real-world speech datasets which include SUD12 [16], ST-AEDS-20180100_1[1], Aishell [2] and TIMIT [7]. The SUD12 dataset contains 61 Chinese speakers and each speaker produces 100 utterances. The ST-AEDS-20180100_1 dataset contains 10 native English speakers and each speaker records about 350 utterances under a silent in-door environment using cellphones. The Aishell dataset contains 400 Chinese speakers under 11 domains such as science & technology, finance, and sport. The TIMIT speech dataset contains 630 speakers of eight major dialects of American English and each speaker reads ten phonetically rich sentences.

The four datasets are divided into two pairs. The first pair of datasets consists of SUD12 and Aishell datasets, which are Chinese speech datasets. In this pair,

[1] http://www.openslr.org/45/.

we use the SUD12 to pre-training the UBM on the server and each client gets one speaker's data of Aishell. The second pair of datasets consists of ST-AEDS-20180100_1 and TIMIT datasets, which are English speech datasets. In this pair, we use the ST-AEDS-20180100_1 to pre-training the UBM on the server and each client gets one speaker's data of TIMIT.

In this experiment, we first use Mel Frequency Cepstral Coef-ficients (MMFCs) method to extract features of speakers from speech dataset. Specially, the dimension of MMFCs is 60 (20 basic + first order + second order) using a 25 ms Hamming window with 10 ms shift. First order and second order are calculated using a 2-frame window. Second, FedSP trains UBM with 256 Gaussian components, i.e., $M = 256$. The value of $\hat{\sigma}_{\mathrm{SPK}} = 0.5$ and $\hat{\sigma}_{\mathrm{UBM}} = 0.07$ in local speaker GMM adapting module and server merging module, respectively. $\alpha = 0.005$ is used to balance the two terms in SHS. Third, $K = 10$, 100 and 200 speakers of different accents are selected randomly from Aishell and TIMIT, respectively, and assigned to different clients. For each client, three sentences were used to jointly training the UBM. At last, to test the performance of the FedSP, we randomly select other 200 speakers that are different from the K speakers used to jointly the UBM.

4.2 Evaluation Metrics

Verification Metric. The metric employed for performance evaluation is the Equal Error Rate (EER). EER is the point that the False Acceptance Rate (FAR) and False Rejection Rate (FRR) become equal. FRR and FAR measure the classification error for target and non-target trials, respectively.

Privacy Metric. The metric employed for measuring the capability of privacy preservation is Kullback-Leibler Distance (KLD) [10]. On the one hand, selecting and hiding the sensitive samples from raw speech data inevitably reduce the amount of data used, thereby increasing EER. Thus using EER to measure the effectiveness of privacy preservation is not rational. On the other hand, FedSP changes the real distribution of the speaker's GMM model by hiding the sensitive samples from the raw training vectors. Therefore, the speaker's real biometric template cannot be reconstructed, when the malicious attackers intercept the Baum-Welch statistics. The smaller the KLD between different speaker's model reconstructed by the server, the smaller the difference between the Baum-Welch statistics calculated by different clients. And the Baum-Welch statistics calculated on local clients contain less sensitive information about the speaker.

4.3 Verification Performance

In FedSP, we jointly train the UBM with multiple clients. In this section, we compare two classical framework learning scenarios in two pair of speech datasets for demonstrating the effectiveness of FedSP in data utility. The details are as follows:

- **Baseline**: This baseline is straightforward. Participants derive their own speaker model based on the pre-trained model without any collaboration.
- **Center**: It is a classical training process for ASV. In this case, the speaker's raw speech data are centrally collected and trained in the server first. Then, the server transmits the parameters of well-trained UBM to the clients and clients derive the speaker's model based on it locally.

The comparative results over two datasets between FedSP and the above method for speaker verification are summarized in Table 1. For the Baseline, we found it is the worst since it cannot handle the data scarcity problem and local clients directly apply the pre-training UBM to derive the speaker's model. For Center and FedSP, they remarkably outperform the Baseline and have lower EER scores as the number of training clients increases. This is because they solve the data scarcity problem by using the users' speech data to jointly train UBM. The performance of FedSP is slightly worse than Center. However Center needs to upload all raw speech data of each client to the central server and FedSP only uploads the Baum-Welch statistics without sensitive information to the central server. Hence, only FedSP can protect the privacy of participants with remain satisfied speaker verification performances.

Table 1. EER scores of three approaches on Aishell and TIMIT datasets.

Method	Baseline	Center			FedSP		
Training Clients K	N/A	10	100	200	10	100	200
Aishell	5.67	2.36	0.57	0.58	2.25	1.5	1.39
TIMIT	7.26	4.84	3.13	2.61	4.88	3.58	3.36

4.4 Effectiveness of SHS in FedSP

The main contribution of this work is that FedSP can protect the privacy of speaker, which is done by hiding the sensitive information from the raw training vectors. The way to show the effectiveness of SHS is to verify whether the Baum-Welch statistics collected from clients will disclose the privacy feature of speakers. When SHS is effective, the Baum-Welch statistics learning by different clients should be more and more alike, and vice versa. And we design the following experiment to verify the effectiveness of SHS:

- **FedSP**$_{INTER}$: In this method, we first use the Baum-Welch statistics that are calculated on the training vectors processed by SHS to reconstruct the GMM model for each client. Then, the KLD is calculated between two different clients' GMM models.
- **FedRS**$_{INTER}$: When the SHS of FedSP is replaced by RS (random sampling), the new framework is named as FedRS. So, in FedSR we randomly delete the same number of vectors as SHS no matter whether the vectors are sensitive samples or not. And the KLD of **FedRS**$_{INTER}$ is calculated as **FedSP**$_{INTER}$.

Figure 2 compares the KLD of the above two methods on Aishell and TIMIT datasets. It can be observed that the KLD of FedSP_{INTER} and FedRS_{INTER} decreases as the selected fraction f increases. This is because that the more sensitive samples are deleted, the fewer sensitive information is included. So the Similarity between the reconstructed GMM models becomes larger. Except for that, the gap between FedSP_{INTER} and FedRS_{INTER} increases, which indicates that SHS is more effective than RS in selecting and hiding sensitive information.

4.5 Privacy Preservation Capability

In this section, we need to define the privacy preservation capability of FedSP and determine when FedSP can provide strict privacy preservation. The stricter the privacy protection, the less sensitive information contained in the Baum-Welch statistics, which are calculated on the training vectors processed by SHS. We say that FedSP can provide strict privacy preservation for speakers when malicious attackers can't intercept sensitive information about the speakers from the Baum-Welch statistics. Except for FedSP_{INTER}, which calculates the KLD of different speakers, we also design a new method named FedSP_{INTRA} to calculate the KLD of the same speaker.

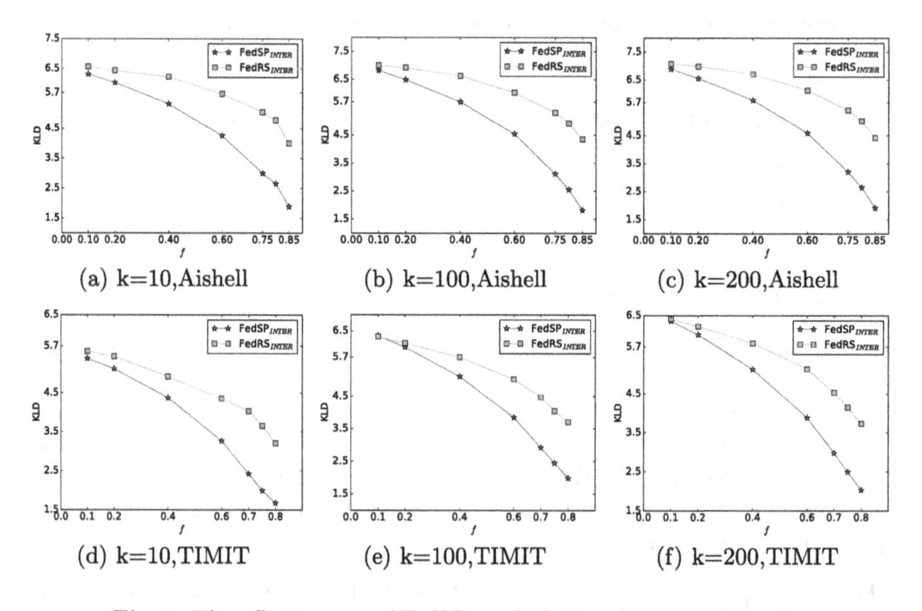

(a) k=10,Aishell (b) k=100,Aishell (c) k=200,Aishell

(d) k=10,TIMIT (e) k=100,TIMIT (f) k=200,TIMIT

Fig. 2. The effectiveness of FedSP on Aishell and TIMIT datasets.

- $\textbf{FedSP}_{INTRA}$: For each client, we calculate the KLD between the real speaker's GMM model get by Eq. (12) and the reconstructed GMM model.

Figure 3 compares FedSP_{INTER} and FedSP_{INTRA} on Aishell and TIMIT datasets. As shown in the Fig. 3, the KLD of FedSP_{INTER} decreases and the

KLD of FedSP$_{INTRA}$ increases as the selected fraction f increase. This is because that as f increase, SHS select and hiding more and more sensitive information for each client. So the sensitive information contained in the statistics which are upload from clients gradually decreases. Besides that, it also can be found that the KLD of FedSP$_{INTER}$ and FedSP$_{INTRA}$ overlap for Aishell and TIMIT dataset, when f is around 0.75. And We state that FedSP can provide strict privacy preservation when the f is greater than or equals to 0.75 for Aishell and TIMIT dataset. In this case, the Baum-Welch statistics of different speakers that are upload to the server are similar and do not contain sensitive information. So the Baum-Welch statistics can not disclose the privacy of speakers.

4.6 Parameter Study

FedSP has two main parameters training clients K and selected fraction f. The verification performance of FedSP is determined by these two parameters together. So in this subsection, we investigate the verification performance of FedSP under different values of these two parameters in Aishell and TIMIT datasets. As shown in Fig. 4, the EER score of FedSP decreases with more training clients and increases with the value of selected fraction increases in both Aishell and TIMIT datasets. In each plot, we show the EER score for Baseline, which is independent of the x-axis. We state that FedSP can provide strict privacy preservation when the f is greater than or equals to 0.75 for Aishell and TIMIT dataset In Sect. 4.5. And in Fig. 4, we also find that $f = 0.75$ for Aisell and TIMIT is a turning point of verification performance. Before the turning point, the EER increases slowly. After the turning point, the EER increases

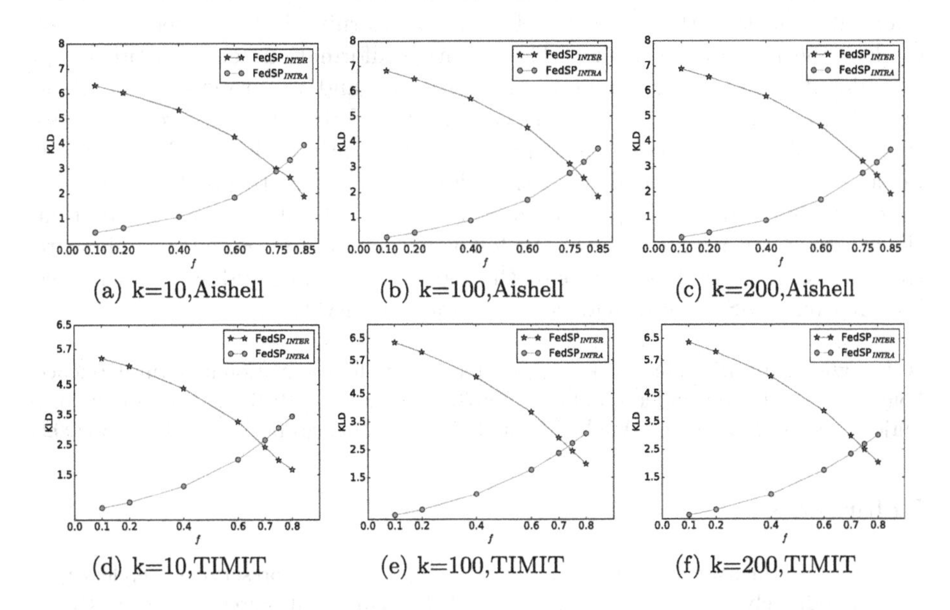

Fig. 3. The privacy preservation capability of FedSP on Aishell and TIMIT datasets.

rapidly. What's more, FedSP still outperforms Baseline when FedSP provides strict privacy preservation. So FedSP can offer an attractive trade-off between data utility and privacy.

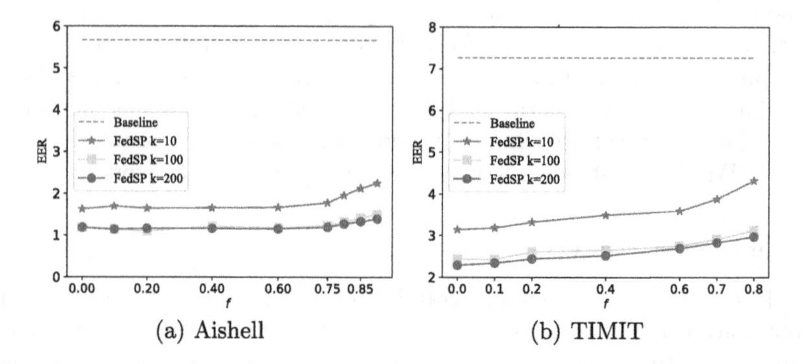

Fig. 4. Verification performance of FedSP with different selection fraction f and training clients K on Aishell and TIMIT datasets.

5 Conclusion

This paper presents a novel framework named FedSP that allows multiple mobile clients to cooperatively train the UBM to solve data scarcity problem while providing strict privacy preservation for participants. In the jointly training procedure, each client executes local learning on the training vectors without sensitive information so that the Baum-Welch statistics learning in local clients will not disclose the privacy of speaker. The sensitive information selection process is formulated as an integer programming problem, and we proposed a heuristic greedy search algorithm to tackling the problem. With the federated architecture in FedSP, a server and a series of clients jointly train a UBM that can model the speaker-independent distribution feature well. To justify our model, we conducted extensive experiments over two speech datasets. We notice that the experimental results support the following points: first, FedSP can alleviate the data scarcity problem; second, through selecting and hiding the sensitive information, FedSP can provide strict privacy preservation.

Acknowledgement. This work is supported in part by the National Natural Science Foundation of China under grant No. 62076079 and No. 61976051, the Guangdong Major Project of Basic and Applied Basic Research under grant No. 2019B030302002.

References

1. Aono, Y., Hayashi, T., Wang, L., Moriai, S., et al.: Privacy-preserving deep learning via additively homomorphic encryption. IEEE Trans. Inf. Forensics Secur. **13**(5), 1333–1345 (2017)

2. Bu, H., Du, J., Na, X., Wu, B., Zheng, H.: Aishell-1: an open-source mandarin speech corpus and a speech recognition baseline. In: 20th Conference of the Oriental Chapter of the International Coordinating Committee on Speech Databases and Speech I/O Systems and Assessment, pp. 1–5 (2017)
3. Chowdhury, M.F.R., Selouani, S.A., O'Shaughnessy, D.: Text-independent distributed speaker identification and verification using GMM-UBM speaker models for mobile communications. In: 10th International Conference on Information Science, Signal Processing and their Applications, pp. 57–60 (2010)
4. Dehak, N., Kenny, P.J., Dehak, R., Dumouchel, P., Ouellet, P.: Front-end factor analysis for speaker verification. IEEE Trans. Audio Speech Lang. Process. **19**(4), 788–798 (2010)
5. Faltlhauser, R., Ruske, G.: Improving speaker recognition using phonetically structured gaussian mixture models. In: Seventh European Conference on Speech Communication and Technology (2001)
6. Fredrikson, M., Jha, S., Ristenpart, T.: Model inversion attacks that exploit confidence information and basic countermeasures. In: Proceedings of the 22nd ACM SIGSAC Conference on Computer and Communications Security, pp. 1322–1333 (2015)
7. Garofolo, J.S.: Timit acoustic phonetic continuous speech corpus. Linguistic Data Consortium (1993)
8. Geng, X., Liu, T.Y., Qin, T., Li, H.: Feature selection for ranking. In: Proceedings of the 30th Annual International ACM SIGIR Conference on Research and Development in Information Retrieval, pp. 407–414 (2007)
9. Granqvist, F., Seigel, M., van Dalen, R.C., Cahill, Á., Shum, S., Paulik, M.: Improving on-device speaker verification using federated learning with privacy. In: Interspeech 2020, 21st Annual Conference of the International Speech Communication Association, Virtual Event, Shanghai, China, 25–29 October 2020, pp. 4328–4332 (2020)
10. Hershey, J.R., Olsen, P.A.: Approximating the kullback leibler divergence between gaussian mixture models. In: IEEE International Conference on Acoustics, Speech and Signal Processing, vol. 4, pp. IV–317 (2007)
11. Hosseini, H., Yun, S., Park, H., Louizos, C., Soriaga, J., Welling, M.: Federated learning of user authentication models. arXiv preprint arXiv:2007.04618 (2020)
12. Jiang, D., et al.: Federated topic modeling. In: Proceedings of the 28th ACM International Conference on Information and Knowledge Management, pp. 1071–1080 (2019)
13. Kumari, T.R.J., Jayanna, H.S.: i-vector-based speaker verification on limited data using fusion techniques. J. Intell. Syst. **29**(1), 565–582 (2020)
14. Lei, Y., Scheffer, N., Ferrer, L., McLaren, M.: A novel scheme for speaker recognition using a phonetically-aware deep neural network. In: IEEE International Conference on Acoustics, Speech and Signal Processing, pp. 1695–1699 (2014)
15. Leroy, D., Coucke, A., Lavril, T., Gisselbrecht, T., Dureau, J.: Federated learning for keyword spotting. In: ICASSP 2019–2019 IEEE International Conference on Acoustics, Speech and Signal Processing (ICASSP), pp. 6341–6345 (2019)
16. Li, L., Wang, D., Zhang, C., Zheng, T.F.: Improving short utterance speaker recognition by modeling speech unit classes. IEEE/ACM Trans. Audio Speech Lang. Process. **24**(6), 1129–1139 (2016)
17. Liu, Y., et al.: FedVision: an online visual object detection platform powered by federated learning. In: Proceedings of the AAAI Conference on Artificial Intelligence, vol. 34, pp. 13172–13179 (2020)

18. Liu, Y., Kang, Y., Xing, C., Chen, T., Yang, Q.: A secure federated transfer learning framework. IEEE Intell. Syst. **35**(4), 70–82 (2020)
19. Pathak, M.A., Raj, B.: Privacy-preserving speaker verification and identification using gaussian mixture models. IEEE Trans. Audio Speech Lang. Process. **21**(2), 397–406 (2012)
20. Pathak, M.A., Raj, B.: Privacy-preserving speaker verification as password matching. In: IEEE International Conference on Acoustics, Speech and Signal Processing, pp. 1849–1852 (2012)
21. Rahulamathavan, Y., Sutharsini, K.R., Ray, I.G., Lu, R., Rajarajan, M.: Privacy-preserving ivector-based speaker verification. IEEE/ACM Trans. Audio Speech Lang. Process. **27**(3), 496–506 (2019)
22. Reynolds, D.A., Quatieri, T.F., Dunn, R.B.: Speaker verification using adapted gaussian mixture models. Digital Signal Process. **10**(1–3), 19–41 (2000)
23. Reynolds, D.A., Rose, R.C.: Robust text-independent speaker identification using gaussian mixture speaker models. IEEE Trans. Speech Audio Process. **3**(1), 72–83 (1995)
24. Voss, W.G.: European union data privacy law reform: general data protection regulation, privacy shield, and the right to delisting. Bus. Lawyer **72**(1), 221–234 (2016)
25. Wang, J., Wang, D., Wu, X., Zheng, T.F., Tejedor, J.: Sequential model adaptation for speaker verification. In: 14th Annual Conference of the International Speech Communication Association, pp. 2460–2464 (2013)
26. Wolsey, L.A., Nemhauser, G.L.: Integer and Combinatorial Optimization, vol. 55. Wiley, Hoboken (1999)
27. Wu, Z., Evans, N., Kinnunen, T., Yamagishi, J., Alegre, F., Li, H.: Spoofing and countermeasures for speaker verification: a survey. Speech Commun. **66**, 130–153 (2015)
28. Wu, Z., et al.: ASVspoof: the automatic speaker verification spoofing and countermeasures challenge. IEEE J. Sel. Top. Signal Process. **11**(4), 588–604 (2017)
29. Yang, Q., Liu, Y., Chen, T., Tong, Y.: Federated machine learning: concept and applications. ACM Trans. Intell. Syst. Technol. **10**(2), 1–19 (2019)
30. Zhu, L., Liu, Z., Han, S.: Deep leakage from gradients. In: Advances in Neural Information Processing Systems, pp. 14774–14784 (2019)

Security Performance Analysis for Cellular Mobile Communication System with Randomly-Located Eavesdroppers

Ziyan Zhu[1], Fei Tong[1,2](✉), and Cheng Chen[1]

[1] Southeast University, Nanjing 211189, Jiangsu, China
ftong@seu.edu.cn
[2] Purple Mountain Laboratories, Nanjing 211111, Jiangsu, China

Abstract. Due to the broadcast characteristics of wireless communication, its security is vulnerable. The security performance of wireless communication has been widely studied. Our research focuses on the security performance analysis of 5G mobile cellular networks. Each network cell is modeled as a regular hexagonal area. For cellular users, we consider their mobile characteristics and a Random Waypoint mobility model is adopted for mobile nodes, with multiple eavesdroppers randomly located in the network. To analyze the security performance between Base Station (BS) and a mobile user, nodal distance distribution is required. We propose an approach to obtain the distribution of the distance between an arbitrary reference node inside or outside the regular hexagon and a mobile node inside the regular hexagon. The distance distribution between the mobile user and the BS located in the center of the regular hexagon is derived, by the above-mentioned approach. Further, we analyze the Signal-to-Noise Ratio of the mobile user and eavesdroppers and the Secrecy Outage Probability in a downlink scenario. The validity of the results is verified by extensive Monte Carlo simulations.

Keywords: RWP · Distance distribution · Mobile node · Secrecy Outage Probability

1 Introduction

Due to open and broadcast nature of wireless communication, eavesdroppers may eavesdrop data from the transmitter. Shannon [1] studied secure transmission of information and established the security system model. Subsequently, Physical Layer Security (PLS), based on Shannon theory has been frequently considered

This work is supported in part by the National Natural Science Foundation of China under Grants 61971131 and 61702452, in part by "Zhishan" Scholars Programs of Southeast University, in part by the Ministry of Educations Key Lab for Computer Network and Information Integration, Southeast University, Nanjing, China, and in part by the Fundamental Research Funds for the Central Universities.

Y. Lai et al. (Eds.): ICA3PP 2021, LNCS 13157, pp. 479–493, 2022.
https://doi.org/10.1007/978-3-030-95391-1_30

in academia since Wyner's seminal work [2]. Secrecy capacity and Secrecy Outage Probability (SOP) are important security targets for PLS performance. A rate at which information can be transmitted secretly from the source to its intended destination is termed an achievable secrecy rate, and secrecy capacity is the maximal achievable secrecy rate. SOP represents the probability that secrecy rate is less than target secrecy rate. Secrecy capacity and SOP are discussed in many relevant security studies.

To satisfy the capacity demands resulting from the massive data growth over the last ten years, posed mainly by video, 5G is proposed as one of the most promising wireless communication technologies, including high carrier frequencies with massive bandwidths, extreme base station and device densities, and unprecedented numbers of antennas [3–5]. These technologies provide services for mobile users with higher data rate and lower latency. To increase the network capacity and reduce pathloss of mmWave in 5G, reducing the coverage radius of cells is an effective approach. Thus, the 5G cellular coverage is smaller than 4G. According to [6,7], the impact of user mobility on 5G small cell networks is enlarged with the decrease of the cell coverage radius. Therefore, it is an important problem to evaluate the impact of user mobility on 5G small cellular networks [8]. Coverage probabilities, outage probability and average bit error rate were analyzed in [8,9]. However, none of them considered security performance in the presence of eavesdroppers.

Making the optimal parameter allocation with the security performance as the optimization goal, and analyzing the security performance of the scenario, are two branches of PLS performance research [10–12]. Analyzing the security performance about different channel model has been considered in many works [13–15]. The distance between the communication nodes is fixed and the channel state information of the scene mentioned above are simulated by Gaussian distribution or Rayleigh distribution.

However, in the actual scenario, we cannot predict the actual location of the eavesdropper and the distance between communication pairs is also not fixed. The channel state information should be modeled based on actual scenarios. The security performance analysis based on distance distribution is frequently considered in academia. The impact of random eavesdroppers' locations on secrecy performance has been investigated in [16–20], where the node distribution was assumed to follow a Poisson Point Process (PPP). However, the PPP model assumes an infinite wireless network with an infinite number of nodes. In addition, for user nodes, they basically are simulated by fixed location, Poisson distribution, or other random geometric distributions. However, none of these tools are suitable for simulating 5G mobile networks, because node position changes from time to time. Therefore, this paper adopts the Random Waypoint mobility model (RWP) [21] to consider the mobility of user nodes to obtain channel state information.

This paper focuses on the node spacial model in finite 5G cellular networks. Signal-to-Noise Ratio (SNR) and SOP, as two vital performance metrics, are analyzed. To this end, we use a simple large-scale fading model to highlight the influence of distance distribution on safety performance indicators. Furthermore, this paper considers distance distribution of mobile user in cellular systems with randomly located eavesdroppers. RWP is adopted for a mobile user. Eavesdroppers are independently and uniformly distributed in the cell, and acting independently to intercept the transmitted message. The contributions of the paper are summarized as follows.

- We analyze SNR at mobile user and eavesdroppers, and SOP of the system.
- In order to get the SNR of mobile user, we propose an approach to obtain the distribution of the distance between an arbitrary reference node inside or outside the regular hexagon and mobile node inside the area.
- In addition, we consider the impact of multiple eavesdroppers, and derive the SNR of the nearest eavesdropper. To this end, we obtain the distribution of k-nearest neighbor distance between BS and eavesdroppers.

The rest of the paper is organized as follows. Section 2 summarizes the related work. Section 3 presents the system model. The approaches to obtaining the distance distributions are detailed in Sect. 4. Further, Sect. 5 derives the SNR and SOP distributions. Performance evaluation is conducted in Sect. 6. Finally, Sect. 7 concludes the paper.

2 Related Wrok

The distance distribution of RWP model was first studied by Bettstetter in [21], which presented a method to derive an approximate formula for the mobile node distribution in a two-dimensional unit square. Inspired by Bettstetter, Hyytia deduced the approximate distribution of the distance between mobile nodes and the center of symmetric geometry in 2-dimensional space, and the distribution of the distance between mobile nodes and the center of n-dimensional unit hypersphere in [22,23]. In [22], the authors used polynomial approximations and central symmetry to obtain the Probability Density Function (PDF), which is difficult to generalize to obtain the distribution of the distance between an arbitrary reference node and mobile nodes. In our paper, we propose a new method to derive the distribution of the distance between an arbitrary reference node inside or outside the regular hexagon and a mobile node inside the area.

Based on the previous researches on RWP, many mobile scenes have been studied by using RWP model. Zhong et al. derived the analytic expressions of the cumulative distribution function and probability density function of the nodal distance between uniformly deployed static nodes and multiple mobile nodes in square and circular areas [24]. The interference statistics in mobile random networks were characterized by incorporating the distance variations of mobile nodes to the channel gain fluctuations [25]. Li et al. studied the received power and the outage probability over Rician fading channel under the mobile wireless network scenarios [26]. The coverage probabilities of small cell and macro cell BSs were derived for all users in 5G small cell networks [8]. The outage probability and average bit error rate were analyzed to quantify the performance over η-μ

fading channels of the mobile system in 5G [9]. However, none of them considered the presence of eavesdroppers.

In order to improve the above researches, this paper focuses on the node spacial model in finite 5G cellular networks. RWP model is used to characterise the mobility of mobile users in the network, and the spacial distribution of a mobile node is derived. The distribution of multiple eavesdroppers is also considered. We use a simple large-scale fading model to highlight the influence of distance distribution on safety performance indicators. Then the SNR of mobile nodes and eavesdroppers and the SOP of the network are analyzed.

3 System Model

The system model considered in this paper is shown in Fig. 1, where BS is located in the center of a regular polygonal cell. RWP is applied for a mobile user and M eavesdroppers are independently and uniformly distributed in the cell. BS communicates with the mobile user in either uplink or downlink mode with a transmission power of P_b. This paper focuses on the downlink security performance analysis. In the downlink mode, the eavesdroppers wiretap the signal that BS sends to the mobile node. All channels are assumed to undergo path loss [16]. Hence, the coefficient modeling the channel between a sender and a receiver can be decomposed as $d^{-\alpha/2}$, where α and d denote the path loss exponent and the distance between a sender and a receiver, respectively.

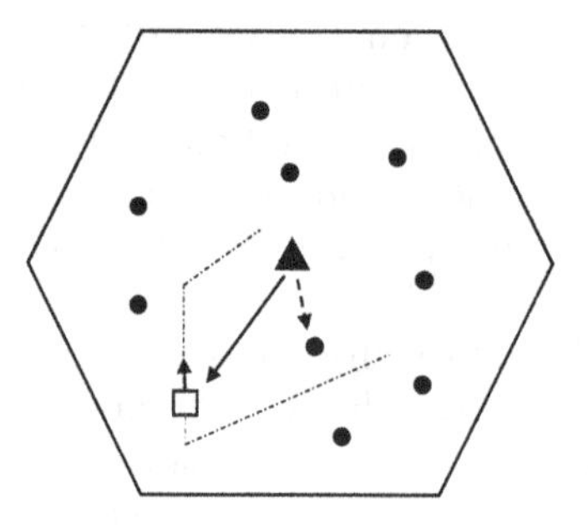

Fig. 1. System model consisting of a BS shown as a triangle, a mobile user shown as a square and multiple eavesdroppers shown as circles underlaying a cell.

3.1 SNR at the Moblie Node and Eavesdroppers

The signal sent from BS received by the mobile user can be given as:

$$S_m = \sqrt{P_b} d_{b,m}^{-\alpha/2} \gamma + z_m \ , \tag{1}$$

where $d_{b,m}$ is the distance between BS and the moblie user, γ is the transmitted information, z_m is the Additive White Gaussian Noise (AWGN) at the moblie user with zero mean and variance of σ^2, and α is path loss exponent. The SNR at the moblie user can be given as:

$$\tau_m = \frac{P_b d_{b,m}^{-\alpha}}{\sigma^2} . \tag{2}$$

Specifically, let τ_m in dB be:

$$SNR_m = 10lg(\tau_m) . \tag{3}$$

The received signal at eavesdropper i can be expressed as:

$$S_{e_i} = \sqrt{P_b} d_{b,e_i}^{-\alpha/2} \gamma + z_{e_i} , \tag{4}$$

where d_{b,e_i} is the distance between BS and eavesdropper i ($i \in 1, 2, ..., M$), z_{e_i} is the AWGN at eavesdropper i with zero mean and variance of σ^2. The SNR at eavesdropper i can be given as:

$$\tau_{e_i} = \frac{P_b d_{b,e_i}^{-\alpha}}{\sigma^2} . \tag{5}$$

Specifically, let τ_{e_i} in dB be:

$$SNR_{e_i} = 10lg(\tau_{e_i}) . \tag{6}$$

3.2 Secrecy Outage Probability

According to [16], the SOP in the downlink mode can be defined as:

$$P_{so} = \Pr(C_s < R_s) , \tag{7}$$

where C_s is the secrecy capacity of the downlink communication, and R_s is a pre-defined target secrecy rate. C_s can be expressed as:

$$C_s = \min_i(\log_2(1 + \tau_m) - \log_2(1 + \tau_{e_i})) . \tag{8}$$

Substitute (8) into (7), the downlink SOP can be reexpressed as:

$$\begin{aligned} P_{so} &= \Pr\left(\min_i(\log_2(1 + \tau_m) - \log_2(1 + \tau_{e_i})) < R_s\right) \\ &= \Pr\left(\min_i\left(\log_2\left(\frac{1 + \tau_m}{1 + \tau_{e_i}}\right)\right) < R_s\right) \\ &= \Pr\left(\min_i\left(\frac{1 + \tau_m}{1 + \tau_{e_i}}\right) < 2^{R_s}\right) . \end{aligned} \tag{9}$$

In order to obtain the above SNR and SOP distributions, we will discuss the relevant distance distribution in the next section.

4 Approaches to Distance Distribution

In mobility management, RWP is a random model that simulates the movement of mobile users and how their position, speed, and acceleration change over time. Due to its simplicity and wide availability, it is one of the most popular mobile models for evaluating mobile ad hoc network performance. Figure 2 shows an example to illustrate the RWP trajectory of the mobile node in a regular hexagon area.

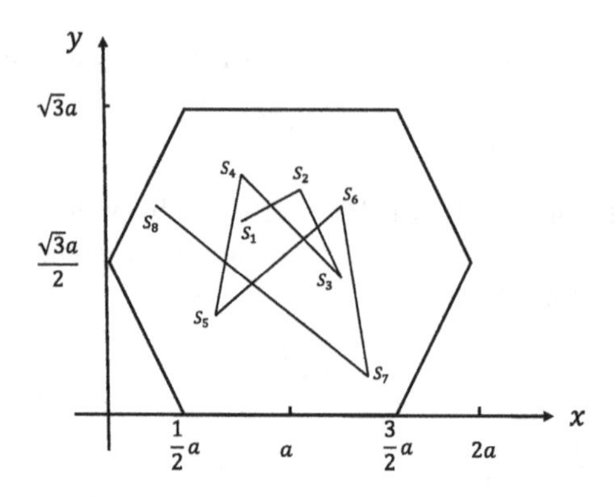

Fig. 2. Mobile node trajectory.

Considering the Cartesian coordinate system established in Fig. 2, below is to show the complete process of RWP in the regular hexagon area: given the simulation time T, randomly and uniformly generate the starting point S_1 of the mobile node. Then the mobile node will select a destination point D_1 and randomly and uniformly select a speed ν from $[\nu_{min}, \nu_{max}]$. The moving time between the destination point D_1 and the starting point S_1 is $\frac{D_1 - S_1}{\nu}$, which completes the 1st-period movement. To simplify the model, we do not consider the effect of pause time t_p and let $t_p = 0$. If there is remaining simulation time, the mobile node immediately selects the next destination point D_2 in the 2nd-period movement, and randomly and uniformly selects a speed from $[\nu_{min}, \nu_{max}]$, then moves from S_2 to D_2, where S_2 is D_1 in the 1st-period movement. The mobile node has a starting point S_k and a destination point D_k in the kth-period movement. Repeat the process until the simulation time runs out.

As shown in Fig. 2, we consider a regular hexagonal cell in a cellular network. In this paper, we investigate the distribution of the distance between an arbitrary reference node denoted as $N_1 = (x_1,\ y_1)$ and a mobile node denoted as $N = (x,\ y)$. The random variable of distance between the moblie node and the reference node is defined as

$$D_m = \sqrt{\Delta x^2 + \Delta y^2}\ , \tag{10}$$

where $\Delta x = x - x_1$, and $\Delta y = y - y_1$. Let ΔX and ΔY denote the random variables of distance between a moblie node and a reference node in x-axis and

y-axis, respectively. The Cumulative Distribution Function (CDF) of D_m denoted as $F_{D_m}(d)$, can be represented as

$$F_{D_m}(d) = \Pr(D_m < d) = \frac{\iint_{\omega} f_{\Delta X}(\Delta x) f_{\Delta Y}(\Delta y) \mathrm{d}\Delta x \mathrm{d}\Delta y}{\iint_{O} f_{\Delta X}(\Delta x) f_{\Delta Y}(\Delta y) \mathrm{d}\Delta x \mathrm{d}\Delta y}, \tag{11}$$

where O denotes the regular hexagon area, and ω denotes the integration area between the circle area defined by $\sqrt{\Delta x^2 + \Delta y^2} < d$ and O, i.e., the shaded part in Fig. 3. $f_{\Delta X}(\Delta x)$ and $f_{\Delta Y}(\Delta y)$ denote the PDFs of ΔX and ΔY, respectively. Let X and Y denote the random variables of moblie node coordinates in x-axis and y-axis, respectively. $f_X(x)$ and $f_Y(y)$ denote the PDFs of X and Y, respectively. Since $f_{\Delta X}(\Delta x)$ and $f_{\Delta Y}(\Delta y)$ are related to $f_X(x)$ and $f_Y(y)$, we below derive $f_{\Delta X}(\Delta x)$, $f_{\Delta Y}(\Delta y)$ by $f_X(x)$ and $f_Y(y)$, respectively. Let the side length of the regular hexagon be a, where $X \in [0, 2a]$ and $Y \in [0, \sqrt{3}a]$.

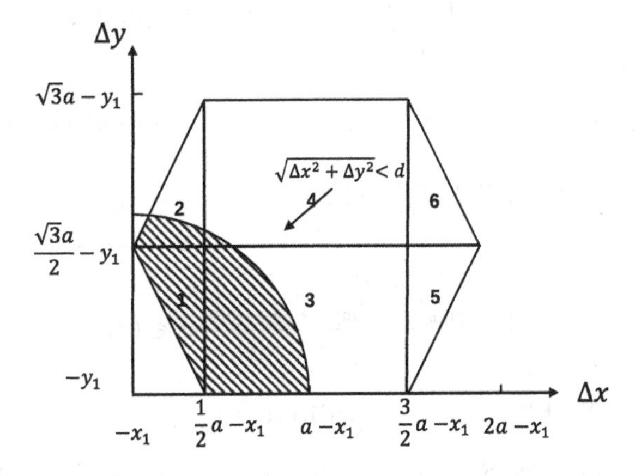

Fig. 3. Integration area.

4.1 PDF of a Moblie Node Coordinates in x-axis

Let the random variables S and D denote the starting and destination coordinates of a movement period in x-axis, respectively. The CDFs $F_S(s)$ and $F_D(d)$ in x-axis can be obtained by the area ratio method. S and D have the same distribution. Therefore,

$$f_S(s) = f_D(d) = \begin{cases} \frac{4s}{3a^2} & s \in [0, \frac{1}{2}a] \\ \frac{2}{3a} & s \in (\frac{1}{2}a, \frac{3}{2}a] \\ \frac{4(2a-s)}{3a^2} & s \in (\frac{3}{2}a, 2a] \end{cases}. \tag{12}$$

In order to derive $f_X(x)$, we first calculate the $F_X(x) = P(X \leq x)$, which denotes the probability that the mobile node is located within $[0, x]$ at an arbitrary instant of time in x-axis. According to [21], we have $P(X \leq x) = \frac{E[L_x]}{E_L}$, where L is the movement distance in each period, L_x is the line segment of L whose coordinate is less than x, $E[L]$ is the mathematical expectation of L in each period, and $E[L_x]$ is the mathematical expectation of L_x.

***E*(*L*):** Let $l(s,d)$ denote the value of the random variable L if $S=s$ and $D=d$. Because of the symmetry of S and D, we restrict the calculation to periods with $s<d$ and then multiply the result by a factor of 2. We have:

$$E(L)=2\int_{s=0}^{2a}\int_{d=s}^{2a} l(s,d)f_S(s)f_D(d)\mathrm{d}d\mathrm{d}s. \tag{13}$$

Finally, we derive $E(L)=\frac{71}{135}a$.

***E*(*L_x*):** Let $l_x(s,d)$ denote the value of the random variable L_x if $S=s$ and $D=d$. $l_x(s,d)$ represents the line segment from s to d whose coordinate is less than x. Similarly, $s<d$ and multiply the result by a factor of 2. We have:

$$E(L_x)=2\int_{s=0}^{2a}\int_{d=s}^{2a} l_x(s,d)f_S(s)f_D(d)\mathrm{d}d\mathrm{d}s\ . \tag{14}$$

***f_X*(*x*):** According to $F_X(x)=P(X\le x)=\frac{E[L_x]}{E_L}$, the PDF $f_X(x)$ of X is given in [27]:

$$f_X(x)=\begin{cases}\frac{12(15a^2x^2-10x^4)}{71a^5} & x\in[0,\frac{1}{2}a]\\ \frac{-105a^2+480ax-240x^2}{142a^3} & x\in(\frac{1}{2}a,\frac{3}{2}a]\\ \frac{-1200a^4+3120a^3x}{71a^5}+\\ \frac{-2700a^2x^2+960ax^3-120x^4}{71a^5} & x\in(\frac{3}{2}a,2a]\end{cases}. \tag{15}$$

4.2 PDF of a Moblie Node Coordinates in *y*-axis

Let the random variables S and D denote the starting and destination points of a movement period in y-axis. We can derive $f_S(s)$ as follows:

$$f_S(s)=\begin{cases}\frac{2(\sqrt{3}a+2s)}{9a^2} & s\in[0,\frac{\sqrt{3}}{2}a]\\ \frac{6\sqrt{3}a-4s}{9a^2} & s\in[\frac{\sqrt{3}}{2}a,\sqrt{3}a]\end{cases}. \tag{16}$$

Similarly, $f_S(s)=f_D(d)$. In order to derive $f_Y(y)$, we first calculate the $F_Y(y)=P(Y\le y)$, which denotes the probability that the mobile node is located within $[0,y]$ at an arbitrary instant of time. Similarly, we have $P(Y\le y)=\frac{E[L_y]}{E_L}$.

***E*(*L*):** Let $l(s,d)$ denote the value of the random variable L if $S=s$ and $D=d$. We have:

$$E(L)=2\int_{s=0}^{\sqrt{3}a}\int_{d=s}^{\sqrt{3}a} l(s,d)f_S(s)f_D(d)\mathrm{d}d\mathrm{d}s\ . \tag{17}$$

According to (17), we derive $E(L)=\frac{41a}{45\sqrt{3}}$.

$\boldsymbol{E(L_y)}$: Let $l_y(s,d)$ denote the value of the random variable L_y if $S = s$ and $D = d$. We have:

$$E(L_y) = 2\int_{s=0}^{\sqrt{3}a}\int_{d=s}^{\sqrt{3}a} l_y(s,d)f_S(s)f_D(d)\mathrm{d}d\mathrm{d}s \ . \tag{18}$$

$\boldsymbol{f_Y(y)}$: According to $F_Y(y) = P(Y \le y) = \frac{E[L_y]}{E_L}$. The PDF $f_Y(y)$ of Y is given in [27]:

$$f_Y(y) = \begin{cases} \frac{20y(27a^3+3\sqrt{3}a^2y-12ay^2-2\sqrt{3}y^3)}{369a^5} & y \in [0, \frac{\sqrt{3}}{2}a] \\ \frac{-360a^4+900\sqrt{3}a^3y-138a^2y^2}{123\sqrt{3}a^5} + & \\ \frac{240\sqrt{3}ay^3-40y^4}{123\sqrt{3}a^5} & y \in (\frac{\sqrt{3}}{2}a, \sqrt{3}a] \end{cases} . \tag{19}$$

4.3 Distance Distribution from an Arbitrary Reference Node to a Mobile Node

According to (11), we need to know $f_{\Delta X}(\Delta x)$ and $f_{\Delta Y}(\Delta y)$ to get $F_{D_m}(d)$. According to (10), we have:

$$\begin{cases} \Delta x = x - x_1 & x \in [0, 2a], \Delta x \in [-x_1, 2a - x_1] \\ \Delta y = y - y_1 & y \in [0, \sqrt{3}a], \Delta y \in [-y_1, \sqrt{3}a - y_1] \end{cases} . \tag{20}$$

Substitute variable x to Δx:

$$\begin{aligned} F_{\Delta X}(\Delta x) &= \Pr(\Delta X \le \Delta x) = \Pr(x - x_1 \le \Delta x) \\ &= \Pr(x \le \Delta x + x_1) = F_X(\Delta x + x_1) \ . \end{aligned} \tag{21}$$

Because $f_{\Delta X}(\Delta x) = {F_{\Delta X}}'(\Delta x)$, it is easy to derive:

$$\begin{aligned} f_{\Delta X}(\Delta x) &= {F_X}'(\Delta x + x_1) = f_X(\Delta x + x_1) \\ f_{\Delta Y}(\Delta y) &= {F_Y}'(\Delta y + y_1) = f_Y(\Delta y + y_1) \ . \end{aligned} \tag{22}$$

Then, $F_{D_m}(d)$ in (11) can be derived by zoning law over the integral region.

4.4 Distance Distributions from BS to Mobile User and Eavesdroppers

BS as the reference node is located in the center of the cell, i.e., $N_1 = (a, \frac{\sqrt{3}a}{2})$. Let $E = (x_e,\ y_e)$ denote an eavesdropper. Let G denote the random variable of the distance between the mobile user and BS, and $F_G(g)$ denote the CDF of G. Let H denote the random variable of the distance between the nearest eavesdropper and BS, and $F_H(h)$ and $f_H(h)$ denote the CDF and PDF of H, respectively.

$\boldsymbol{F_G(g)}$: According to Sect. 4.3, let $(x_1, y_1) = (a, \frac{\sqrt{3}a}{2})$, we can easily obtain $F_G(g)$.

$\boldsymbol{F_H(h)}$**:** According to (8), a minimum secrecy capacity requires a maximum τ_{e_i}. With (5), we need to derive the distribution of H. Similar to the approach adopted in [28], for BS, its distances to M eavesdroppers are ordered as $d_1 \leq d_2 \leq ... \leq d_M$. Let D_k denote the random variable which represents the distance from the k-th nearest neighbor among M eavesdroppers to BS. Then the PDF of D_k is

$$f_{D_k}(d; M) = \frac{M!}{(k-1)!(M-k)!} \times [F(d)]^{k-1}[1-F(d)]^{M-k}f(d) , \tag{23}$$

where $F(d)$ and $f(d)$ are the CDF and PDF of the distance from BS to a randomly-located eavesdropper, respectively. Let $\Delta x = x_e - a$, and $\Delta y = y_e - \frac{\sqrt{3}a}{2}$. Since an eavesdropper is independently and uniformly distributed in the regular polygonal cell, we have:

$$F(d) = \Pr(D < d) = \frac{\iint_{\omega} \mathrm{d}\Delta x \mathrm{d}\Delta y}{\iint_{O} \mathrm{d}\Delta x \mathrm{d}\Delta y} . \tag{24}$$

Then we obtain $F(d)$ and $f(d) = F'(d)$. Let $f_H(h)$ denote the PDF of the distance between the nearest eavesdropper and BS. Therefore, $f_H(h) = f_{D_1}(d; M)$, $F_H(h) = \int_0^h f_H(h)\mathrm{d}h$.

5 Approaches to Obtaining SNR and Secrecy Capacity Distributions

5.1 SNR Distributions of Mobile User and Eavesdroppers

Let Q denote the random variable of SNR at the mobile user, and $F_Q(q)$ denote the CDF of Q. Besides, R denotes the random variable of SNR at the nearest eavesdropper, and $F_R(r)$ denotes the CDF of R. According to (2), we have

$$Q = \frac{P_b G^{-\alpha}}{\sigma^2} . \tag{25}$$

Then, we can derive $F_Q(q)$ as,

$$\begin{aligned} F_Q(q) &= \Pr\left(\frac{P_b G^{-\alpha}}{\sigma^2} < q\right) = \Pr\left(G^{-\alpha} < \frac{\sigma^2 q}{P_b}\right) \\ &= \Pr\left(G > 10^{\frac{\lg \frac{\sigma^2 q}{P_b}}{-\alpha}}\right) = 1 - F_G\left(10^{\frac{\lg \frac{\sigma^2 q}{P_b}}{-\alpha}}\right) . \end{aligned} \tag{26}$$

Similar to $F_Q(q)$, according to (5), we have:

$$F_R(r) = 1 - F_H\left(10^{\frac{\lg \frac{\sigma^2 h}{P_b}}{-\alpha}}\right) . \tag{27}$$

5.2 Secrecy Capacity Distribution

According to (9), let W and V denote the random variables of $1 + \tau_m$ and $\min_i(1 + \tau_{e_i})$, respectively. And let Z denote the random variable of $Z = \frac{W}{V}$. We can derive the CDF $F_Z(z) = P(Z < z)$ of Z:

$$F_Z(z) = \begin{cases} \int_{\frac{w_{min}}{z}}^{v_{max}} \int_{w_{min}}^{zv} f_W(w) f_V(v) \mathrm{d}w \mathrm{d}v & z \in \boldsymbol{z_1} \\ \int_{\frac{w_{min}}{z}}^{\frac{w_{max}}{z}} \int_{w_{min}}^{zv} f_W(w) f_V(v) \mathrm{d}w \mathrm{d}v + & \\ \int_{\frac{w_{max}}{z}}^{v_{max}} \int_{w_{min}}^{w_{max}} f_W(w) f_V(v) \mathrm{d}w \mathrm{d}v & z \in \boldsymbol{z_2} \\ 1 - \int_{zv_{min}}^{w_{max}} \int_{v_{min}}^{\frac{w}{z}} f_W(w) f_V(v) \mathrm{d}v \mathrm{d}w & z \in \boldsymbol{z_3} \end{cases}, \tag{28}$$

where $f_W(w)$ and $f_V(v)$ are PDFs of W and V, respectively. $\boldsymbol{z_1} = \{z|z \in [\frac{w_{min}}{v_{max}}, \frac{w_{max}}{v_{max}}]\}$, $\boldsymbol{z_2} = \{z|z \in (\frac{w_{max}}{v_{max}}, \frac{w_{min}}{v_{min}}]\}$, and $\boldsymbol{z_3} = \{z|z \in (\frac{w_{min}}{v_{min}}, \frac{w_{max}}{v_{min}}]\}$. Let U denote the random variable of $U = P_{so} = \log_2(Z)$,

$$F_U(u) = \Pr(U < u) = \Pr(\log_2 Z < u) = \Pr(Z < 2^u) = F_Z(2^u) , \tag{29}$$

where $F_U(u)$ is the CDF of secrecy capacity. According to (29), we can obtain SOP given different target secrecy rate.

6 Simulation Result

In this section, we verify the distribution derived in the above sections by MATLAB simulation platform, which allows us to simulate the movement characteristics of the mobile node and record its location. We compare the simulation results with the theoretical results.

In the distance distribution simulations, the mobile node moves in a regular hexagonal area with side length of one, and the pause time $t_p = 0$. The velocity is randomly selected from the interval $[0.1s^{-1}, 0.2s^{-1}]$. The position of mobile node at each moment is recorded in the time interval of one second. In each time interval, we obtain the distances between the mobile node and the reference node. After a simulation period of 100, 000 s, the obtained distance samples are applied to estimate the CDF. We use the MATLAB function "ecdf()" to calculate the CDF.

Figures 4 and 5 show the obtained results with different reference nodes. As can be seen from the figures, there is a slight deviation between the simulation results and the analysis results. The authors in [21], pointed out that the two-dimensional movement is composed of two dependent one-dimensional movements. The speed of a node projected along the x-axis is not constant and it is different from the speed along the y-axis. Therefore, an approximation $f_{\Delta X \Delta Y}(\Delta x, \Delta y) \approx f_{\Delta X}(\Delta x) f_{\Delta Y}(\Delta y)$ is adopted within the acceptable range of differences.

In the 5G mobile cellular networks simulations, we assume the noise variance $\sigma^2 = 1$, the transmission-power-to-noise ratio $P_b/\sigma^2 = 10$ dB. The mobile node moves in a regular hexagon with side length of 100 m, and the pause time $t_p = 0$. The velocity is randomly selected from the interval [10 m/s, 15 m/s]. The position of the mobile node at each moment is recorded in the time interval of one second.

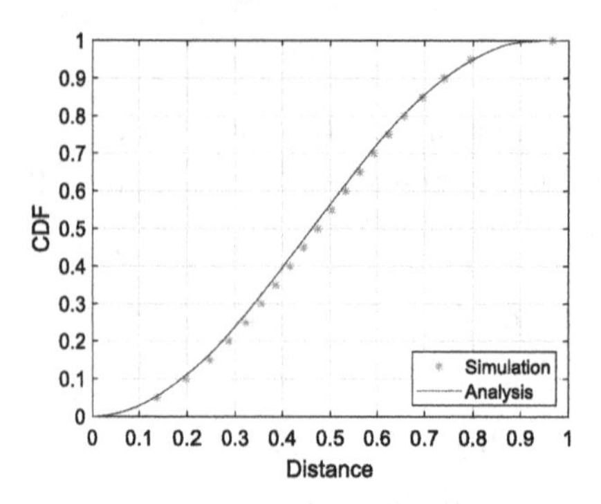

Fig. 4. The CDF of the distance from reference node $(1, \frac{\sqrt{3}}{2})$ to the mobile node.

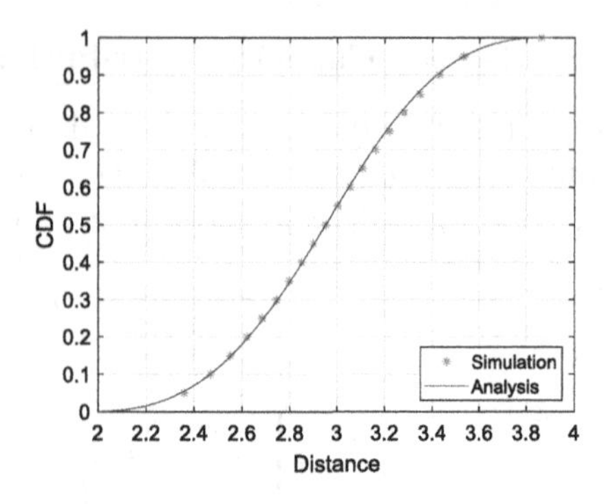

Fig. 5. The CDF of the distance from reference node (3,3) to the mobile node.

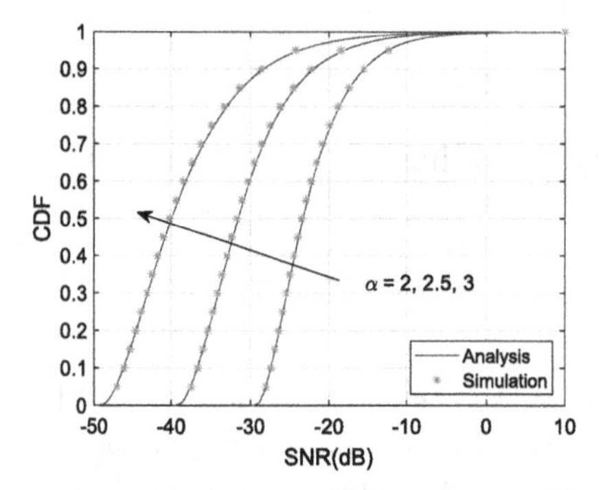

Fig. 6. The CDF of SNR at the mobile node.

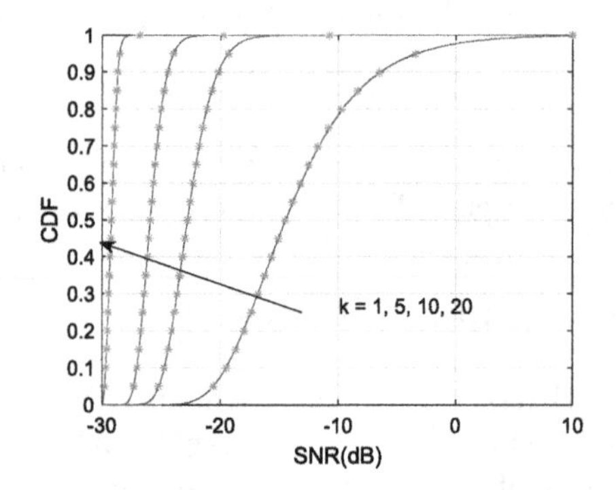

Fig. 7. The CDF of SNR at the eavesdroppers.

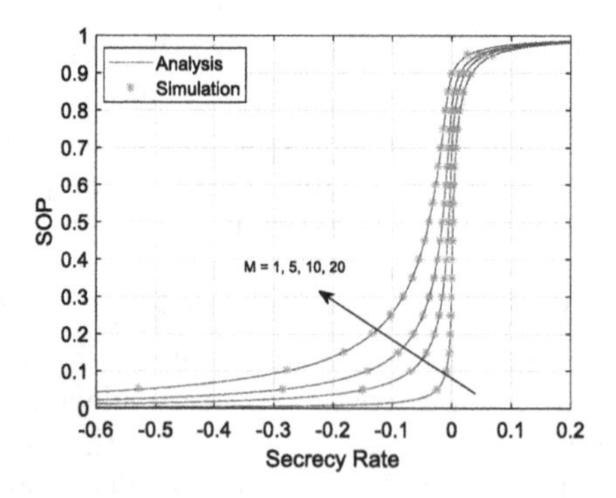

Fig. 8. SOP with $M = 1,\ 5,\ 10,\ 20$, and $\alpha = 2$.

In each time interval, we obtain the distances between the mobile node and BS. After a simulation period of 100, 000 s, the obtained SNR samples are applied to estimate the CDF.

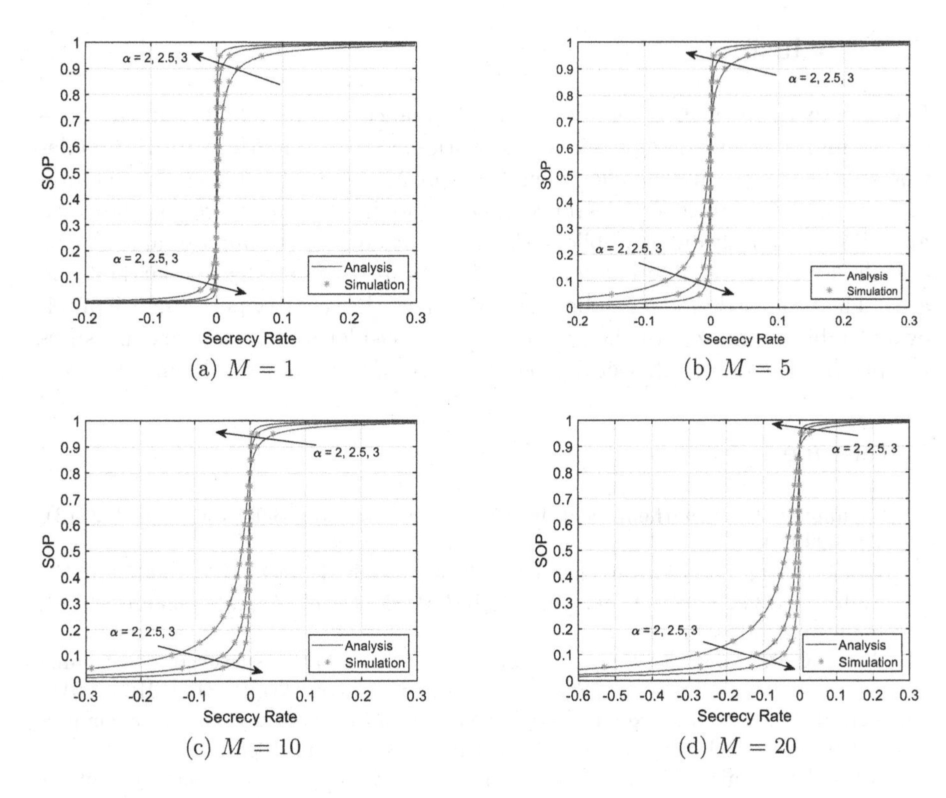

Fig. 9. SOP with M = 1, 5, 10, 20.

Figure 6 shows the CDF of SNR at the mobile node, with the path-loss exponent $\alpha = 2$. As can be seen from the figure, the simulation and analysis results are consistent. In Fig. 6, as α increases, SNR gets smaller. Figure 7 shows the CDF of SNR at the kth-nearest eavesdropper with the path-loss exponent $\alpha = 2$ and $M = 20$. We can find that the farther away the eavesdropper is from the BS, the smaller the SNR at the eavesdropper. For the eavesdropper farthest away from the BS ($k = 20$), the maximum SNR is only -27 dB.

Figure 8 shows the SOP with $\alpha = 2$ and different number of eavesdroppers ($M = 1, 5, 10, 20$) in different target secrecy rate. As shown in Fig. 8, with the increasing number of eavesdroppers, SOP rises sharply. For example, given the target secrecy rate of -0.05, the SOP for $M = 1$ is only 0.03, while for $M = 20$, the SOP reaches 0.45. Figure 9 compares SOP for different target secrecy rate and different α with $M = 1, 5, 10, 20$. From Fig. 9(a) to Fig. 9(d), we can find that the SOP becomes bigger as the number of eavesdroppers increase. According to Fig. 9(d), we can discover that the larger α is, the more difficult it is for eavesdroppers to wiretap signals in the case of a large number of eavesdroppers. However, according to all the above figures, we find that in the mobile scenario, the downlink without security scheme has almost no security, and the risk of eavesdropping is very high.

7 Conclusions

In this paper, we studied the security performance of a mobile user in the presence of multiple eavesdroppers in 5G network scenarios, and Monte Carlo simulation was used to verify the effectiveness of our model. By analyzing the distance distribution from arbitrary reference node to mobile node in the regular hexagon area, SNR distribution at mobile node can be calculated. It can also analyze the downlink SOP with different number of eavesdroppers in the regular hexagon area. The proposed model can be used to analyze the security performance of relevant mobile scenarios. In the future, we will consider adding security measures to optimize the downlink security performance of the mobile cellular network.

References

1. Shannon, C.E.: A mathematical theory of communication. Bell Syst. Tech. J. **27**(3), 379–423 (1948)
2. Wyner, A.D.: The wire-tap channel. Bell Syst. Tech. J. **54**(8), 1355–1387 (1975)
3. Andrews, J.G., et al.: What will 5G be? IEEE J. Sel. Areas Commun. **32**(6), 1065–1082 (2014)
4. Wang, C.-X., Bian, J., Sun, J., Zhang, W., Zhang, M.: A survey of 5G channel measurements and models. IEEE Commun. Surv. Tutor. **20**(4), 3142–3168 (2018)
5. Agiwal, M., Roy, A., Saxena, N.: Next generation 5G wireless networks: a comprehensive survey. IEEE Commun. Surv. Tutor. **18**(3), 1617–1655 (2016)
6. Giust, F., Cominardi, L., Bernardos, C.J.: Distributed mobility management for future 5G networks: overview and analysis of existing approaches. IEEE Commun. Mag. **53**(1), 142–149 (2015)
7. Vasudevan, S., Pupala, R.N., Sivanesan, K.: Dynamic eICIC - a proactive strategy for improving spectral efficiencies of heterogeneous LTE cellular networks by leveraging user mobility and traffic dynamics. IEEE Trans. Wireless Commun. **12**(10), 4956–4969 (2013)
8. Ge, X., Ye, J., Yang, Y., Li, Q.: User mobility evaluation for 5G small cell networks based on individual mobility model. IEEE J. Sel. Areas Commun. **34**(3), 528–541 (2016)
9. Meesa-Ard, E., Pattaramalai, S.: Evaluating the mobility impact on the performance of heterogeneous wireless networks over η - μ fading channels. IEEE Access **9**, 65017–65032 (2021)
10. Dong, L., Han, Z., Petropulu, A.P., Poor, H.V.: Improving wireless physical layer security via cooperating relays. IEEE Trans. Signal Process. **58**(3), 1875–1888 (2010)
11. Laneman, J.N., Tse, D.N.C., Wornell, G.W.: Cooperative diversity in wireless networks: efficient protocols and outage behavior. IEEE Trans. Inf. Theory **50**(12), 3062–3080 (2004)
12. Kapetanovic, D., Zheng, G., Rusek, F.: Physical layer security for massive MIMO: an overview on passive eavesdropping and active attacks. IEEE Commun. Mag. **53**(6), 21–27 (2015)
13. Tang, C., Pan, G., Li, T.: Secrecy outage analysis of underlay cognitive radio unit over nakagami-m fading channels. IEEE Trans. Wireless Commun. **3**(6), 609–612 (2014)

14. Lei, H., Gao, C., Ansari, I.S., Guo, Y., Pan, G., Qaraqe, K.A.: On physical-layer security over SIMO generalized-k fading channels. IEEE Trans. Veh. Technol. **65**(9), 7780–7785 (2016)
15. Lei, H., Gao, C., Guo, Y., Pan, G.: On physical layer security over generalized gamma fading channels. IEEE Commun. Lett. **19**(7), 1257–1260 (2015)
16. Chen, G., Coon, J.P., Di Renzo, M.: Secrecy outage analysis for downlink transmissions in the presence of randomly located eavesdroppers. IEEE Trans. Inf. Forensics Secur. **12**(5), 1195–1206 (2017)
17. Zhou, X., Ganti, R.K., Andrews, J.G., Hjorungnes, A.: On the throughput cost of physical layer security in decentralized wireless networks. IEEE Trans. Wirel. Commun. **10**(8), 2764–2775 (2011)
18. Geraci, G., Singh, S., Andrews, J.G., Yuan, J., Collings, I.B.: Secrecy rates in broadcast channels with confidential messages and external eavesdroppers. IEEE Trans. Wireless Commun. **13**(5), 2931–2943 (2014)
19. Zheng, T.X., Wang, H.M., Yuan, J., Towsley, D., Lee, M.H.: Multi-antenna transmission with artificial noise against randomly distributed eavesdroppers. IEEE Trans. Commun. **63**(11), 4347–4362 (2015)
20. Zheng, T.-X., Wang, H.-M., Yin, Q.: On transmission secrecy outage of a multi-antenna system with randomly located eavesdroppers. IEEE Commun. Lett. **18**(8), 1299–1302 (2014)
21. Bettstetter, C., Resta, G., Santi, P.: The node distribution of the random waypoint mobility model for wireless ad hoc networks. IEEE Trans. Mob. Comput. **2**(3), 257–269 (2003)
22. Hyytiä, E., Lassila, P., Virtamo, J.: Spatial node distribution of the random waypoint mobility model with applications. IEEE Trans. Mob. Comput. **5**(6), 680–694 (2006)
23. Hyytiä, E., Virtamo, J.: Random waypoint model in n-dimensional space. Oper. Res. Lett. **33**(6), 567–571 (2005)
24. Zhong, X., Chen, F., Guan, Q., Ji, F., Hua, Yu.: On the distribution of nodal distances in random wireless ad hoc network with mobile node. Ad Hoc Netw. **97**, 102026 (2020)
25. Gong, Z., Haenggi, M.: Interference and outage in mobile random networks: expectation, distribution, and correlation. IEEE Trans. Mob. Comput. **13**(2), 337–349 (2014)
26. Li, C., Yao, J., Wang, H., Ahmed, U., Shixiao, D.: Effect of mobile wireless on outage and BER performances over rician fading channel. IEEE Access **8**, 91799–91806 (2020)
27. Zhu, Z., Tong, F.: The distance distribution between mobile node and reference node in regular hexagon. arXiv:2111.06553, pp. 1–14 (2021)
28. Tong, F., He, S., Peng, Y., Shi, Z.: A tractable analysis of positioning fundamentals in low-power wide area internet of things. IEEE Trans. Veh. Technol. **68**(7), 7024–7034 (2019)

Short and Distort Manipulations in the Cryptocurrency Market: Case Study, Patterns and Detection

Xun Sun[1], Xi Xiao[1], Wentao Xiao[1], Bin Zhang[2], Guangwu Hu[3](✉), and Tian Wang[4]

[1] Tsinghua Shenzhen International Graduate School, Tsinghua University, Shenzhen, China
{x-sun21,xwt20}@mails.tsinghua.edu.cn, xiaox@sz.tsinghua.edu.cn
[2] Department of New Networks, Peng Cheng National Laboratory, Shenzhen, China
bin.zhang@pcl.ac.cn
[3] School of Computer Science, Shenzhen Institute of Information Technology, Shenzhen, China
hugw@sziit.edu.cn
[4] BNU-UIC Institute of Artificial Intelligence and Future Networks, Beijing Normal University (BNU Zhuhai), Zhuhai, China
tianwang@bnu.edu.cn

Abstract. Recently, with the development of blockchain, cryptocurrencies such as bitcoin are gaining popularity. However, at the same time, there are some problems such as lack of laws and supervision, leading to an endless stream of frauds and market manipulations, which seriously impacts the interests of investors and the development of the economic society. This paper studies Short and Distort, a new manipulation as opposed to Pump and Dump in the cryptocurrency market for the first time. By using the method of case study, Short and Distort is explained. Further, we mine three important patterns with ordinary least squares regression. That is, cryptocurrencies always show price reversal, abnormal trading volume, and widening of bid-ask spreads. Moreover, a model is constructed to detect Short and Distort. Thorough experiments show that our model is superior to the other four methods.

Keywords: Blockchain · Cryptocurrency · Fraud scheme · Short and Distort · Detection model

1 Introduction

Blockchain is a technical solution for collectively maintaining a reliable database in a decentralized manner. Since Satoshi Nakamoto put forward the concept of blockchain in 2008 [1], blockchain has attracted wide attention in the world. Bitcoin, as a cryptocurrency, is the first generation of applications of blockchain. The price of bitcoin has continued to climb, reaching a market capitalization of $1.1 trillion in March, 2021 [2]. However, while blockchain and cryptocurrency are booming, problems such as lack of regulations are gradually exposed in the field, which seriously infringe on the interests of investors [3, 4].

Y. Lai et al. (Eds.): ICA3PP 2021, LNCS 13157, pp. 494–508, 2022.
https://doi.org/10.1007/978-3-030-95391-1_31

Many studies [5, 6] have shown that cryptocurrencies are manipulated by a variety of complex schemes. Pump and Dump (P&D) is a well-known fraud scheme that manipulators spread false or misleading positive information to drive up the price of a stock in order to sell the stock at a higher price. Once the manipulators sell their overvalued shares, the price falls and the innocent investors suffer a loss. Then, is there such a fraud scheme opposite to P&D, which means the manipulators operate to depress the price, rather than driving up the price?

Few researches have studied this kind of opposite fraud scheme in the stock market [7, 8], naming it as Short and Distort (S&D). In the S&D scheme, manipulators taking a short position spread negative rumors to drive down the stock price. When the price is depressed, manipulators buy back to gain profits. The whole process with the P&D process happens to be dual. The details of comparisons between them are shown in Table 1.

Table 1. Comparisons between Pump and Dump (P&D) and Short and Distort (S&D)

Items	Pump and Dump (P&D)	Short and Distort (S&D)
Purpose of manipulation	Drive up prices	Depress prices
Type of spreading information	Positive and misleading	Negative and misleading
Way to profit	Buy low first, then sell high	Sell high first, then buy low
Target of manipulation	Small market capitalization	Medium or large market capitalization
Price trend	Shaped like "Λ"	Shaped like "V"
Regulation in stock market	Strict	Difficult
Familiarity	Well-known	Less well-known

In the stock market, P&D is widely known to people, and it is relatively easier to detect and regulate it than S&D in the stock market. In the past, there have been many cases where manipulators have been punished by the U.S. Securities and Exchange Commission (SEC) because of P&D manipulations. However, S&D also occurs frequently in stock market, but it is difficult to identify and regulate S&D schemes because manipulators operate it so covertly. Thus, cases of SEC Penalties for S&D have been far less than for P&D in the past 20 years [7]. People are also strange to this scheme with few scholars' researches.

What about the emerging cryptocurrency market? Cryptocurrencies are easy to be manipulated due to lack of regulations. With the popularity of cryptocurrencies, plenty of people come into this market, and many scholars pay attention to this area and focus on the fraud schemes, such as P&D in the cryptocurrency market. However, to the best of our knowledge, there are no scholars study S&D schemes in the cryptocurrency market yet. We first study it, trying to warn investors to stay out of the market when the manipulation comes in and provide a method for the supervision of regulators. The main contributions are as follows:

- We study Short and Distort in the cryptocurrency market for the first time. A case study is used to explain its whole process.
- Three important patterns are minded for people to understand the characteristics of Short and Distort.
- We first construct a model to detect Short and Distort in the market. Exhaustive experiments show its superiority. Further, a new dataset of Short and Distort is collected and published.[1]

2 Related Works

In this section, we review the researches on Pump and Dump schemes in the stock and cryptocurrency markets, Short and Distort schemes in the stock market, and some other scams in the cryptocurrency market.

2.1 Pump and Dump Schemes in the Stock and Cryptocurrency Market

Many scholars have studied P&D schemes in the stock market in the early days. In 2006, Aggarwal et al. [9] showed that stocks with low trading volumes were more likely to be manipulated. Massoud et al. [10] found that OTC companies usually carried out stock promotion. Huang et al. [11] showed the manipulations had important effects on market efficiency.

With the rapid growth of cryptocurrencies, P&D schemes in the traditional stock market are also used in the emerging field. Kamps et al. [12] first studied P&D in the cryptocurrency market and proposed a standard to define these activities. Hamrick et al. [13] and Li et al. [14] analyzed P&D events and the new characteristics of P&D. Dhawan et al. [15] studied the reasons for people's participation. Recently, scholars begin to propose some methods to detect P&D. Victor et al. [16], Morgia et al. [17], Mirtaheri et al. [18], Mansourifar et al. [19], and Nilsen et al. [20] quantified and detected P&D scams in the real market. Chen et al. [21] proposed an Apriori algorithm to detect user groups involving in P&D. Xu et al. [22] tried to predict the target tokens of P&D.

2.2 Short and Distort Schemes in the Stock Market

In the traditional stock market, some researchers have studied S&D schemes. Weiner et al. [7] argued that the supervision of S&D should be on the same level as P&D. In 2019, they [8] also found that it was a challenge for companies to pursue claims in S&D. Mitts [23] showed that S&D scams cause a large amount of mispricing.

Activist short selling, in which short-sellers publicly talk down stocks to benefit their short positions, is similar to S&D in concept. Thus, works about activist short selling are also helpful for our research. Zhao [24] argued that companies that are grossly overvalued or uncertain are often the targets. Ogoodniks et al. [25] found activists with a good reputation were more likely to make profits in the short term. Kartapanisde et al. [26] found that activist short selling had a temporary negative impact on the capital market.

[1] https://github.com/DataCodeHub/short-and-distort.

2.3 Other Fraud Schemes in the Cryptocurrency Market

There are also many other fraud schemes in the cryptocurrency market due to lack of regulation. Gandal et al. [5] identified the suspicious trading activities of two "bots", Markus and Willy, on the Mt. Gox exchange. Bartoletti et al. [6] applied data mining techniques to detect Bitcoin addresses related to Ponzi schemes. Nizzoli et al. [27] studied online cryptocurrency manipulations on social media.

3 A Case Study of the Short and Distort Scheme

On the night of March 7, 2018, social media users posted that the Binance exchange had been hacked, and that users' cryptocurrencies in the exchange account had been sold by the hackers via Bitcoin (BTC) or Viacoin (VIA, a small market capitalization cryptocurrency). The hacker's behavior directly led to a sharp rise in the price of VIA in a short time.

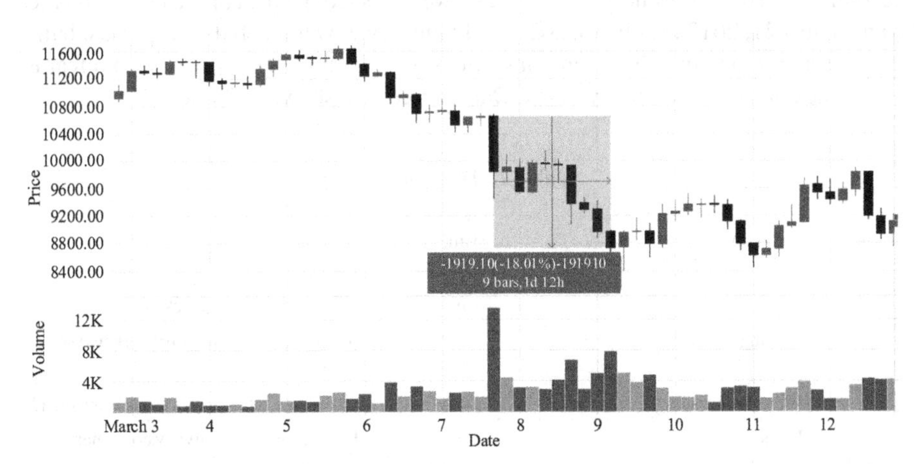

Fig. 1. The candlestick chart of the price and the bar chart of trading volumes around March 8, 2018 [28]. Each candlestick or bar represents 4 h. The top part of figure shows that the price dropped significantly after hacking (at about 20:00 on March 7). As indicated in the blue box in the figure, the price decreased by 18.01% only within 1 day and 12 h (9 bars). The bottom of figure shows that the volume increased sharply at about 20:00 on March 7, peaking over 12K. These two patterns are the signs of S&D scheme as we describe in Sect. 4.

The hacker's action caused panic in the market and uninformed traders joined in the selling activity, which led to prices of cryptocurrencies on all major exchanges fell sharply. Figure 1 shows the price and trading volume of BTC/USD on the Bitstamp exchange around March 8, 2018. It indicates that in just one and a half days after the hacking, the price of bitcoin dropped as much as 18.01%, while the trading volumes increased significantly.

However, hackers didn't end here. They had expected the exchange to stop losses by banning withdrawals. Thus, they just caused bearish sentiment to the market and profited from the orders which they held in advance.

The hacker's behavior in this case shows the characteristics of S&D schemes. First, hackers took short positions in cryptocurrencies on various exchanges in advance. Later, they brought panic to the market by attacking the Binance exchange. In the end, a sharp drop in prices and a surge in trading volumes occurred due to the selling of the uninformed users, and then the hackers were profiting from their short position.

4 Three Patterns of Short and Distorts Schemes

In this section, we conduct a quantitative study through event study methodology and Ordinary Least Squares (OLS) regression analysis to learn some patterns of S&D schemes. First, data collection is introduced.

4.1 Data Collection

According to works of Bouoiyour et al. [29], we collected the data of the bitcoin market from August 21, 2017 to July 30, 2021, including price, volume, bid-ask spreads, bulls, bears, hash, average time between transactions, google trend, gold price, crypto fear & greed index and cryptocurrency index. Data sources are shown in Table 2.

Table 2. Data source.

Variable	Meaning	Time granularity	Source
P	Prices	2 h	https://app.intotheblock.com/
V	Trading volume	2 h	http://data.bitcoinity.org/markets/volume/
S	Bid-ask spreads	1 day	http://data.bitcoinity.org/markets/spread
Bu	Bulls	2 h	https://www.tradingview.com/chart/
Be	Bears	2 h	https://www.tradingview.com/chart/
H	Hash	1 day	https://app.intotheblock.com/coin/BTC
ABT	Average time between transactions	1 day	https://app.intotheblock.com/coin/BTC
GT	Google trend	1 day	https://trends.google.com/trends/explore
GP	Gold price	1 day	https://fred.stlouisfed.org/
CFGI	Crypto fear & greed Index	1 day	https://alternative.me/crypto/fear-and-greed-index/
CRIX	cryptocurrency Index	1 day	http://data.thecrix.de/

Bid-ask spreads are the gap between bid and offer prices. Bulls and Bears represent the prices of the orders of BTCUSDLONGS and BTCUSDSHORTS, which indicates the power of the bulls and the bears, respectively. Hash, which reflects the computational difficulty of bitcoin miners, is used to represent the technical drivers that affect bitcoin's returns. Average time between transactions refers to the time between two bitcoin transactions, which can reflect the activity of the bitcoin market. Google trend denotes the frequency of searching the "bitcoin" keyword in google searches around the world, which can indicate people's interest in bitcoin over a period. Gold price is the dollar price of one ounce of gold. Some scholars believe that bitcoin can replace gold as a hedging tool [30]. Thus, to a certain extent, the returns of bitcoin are related to the gold price. Crypto fear & greed index is used to reflect the panic or greed level of the whole cryptocurrency market. Cryptocurrency Index is used to reflect the returns of the cryptocurrency market portfolio [31]. The bitcoin returns that are different from the market portfolio are considered as abnormal returns, which is necessary for event study methodology in the following section.

Furthermore, we collected 472 "short ideas" (i.e., posts of S&D schemes) about bitcoin on the TradingView website[2]. TradingView is a well-known social network platform for trading and investment in cryptocurrencies. A wide range of investors pay close attention to the opinion posts on the website. Therefore, it is one of the "best ways" for S&D manipulators to spread negative information on TradingView.

In the following, we examine whether any patterns occurred in the bitcoin market around the publication time of these "short ideas".

4.2 Pattern I: Price Reversal

Some scholars have studied that when the stock market experiences S&D manipulation, the stock prices first decline and then reverse [25], as shown in Fig. 2.

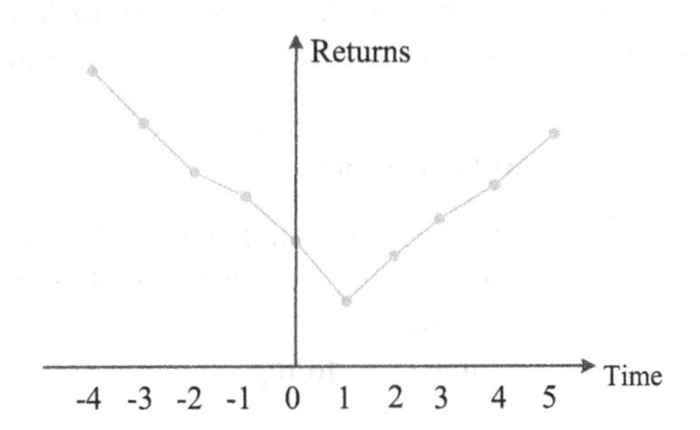

Fig. 2. Diagram of price reversal.

The horizontal axis of Fig. 2 represents the time. "0" refers to the time when S&D manipulation starts or the bad news begin to be published, "1" represents the time of one time step after the publication, "−1" refers to the time of one time step before the publication, and so on. The vertical axis denotes returns.

S&D schemes often exhibit the pattern shown in Fig. 2 [23, 32]. Prices begin to fall before the publication, which may be due to the manipulator making relevant trades in advance or the recent downward tendency of the market (such as "bear market"). In this market, the S&D manipulators observe for a few days and decide to carry out a manipulation. At time step 0, manipulators begin to spread false bad news on the

[2] https://tradingview.com/.

Internet, causing a panic. Then, the uninformed investors sell their assets with prices falling further, and the manipulators buy the assets at lower prices (commonly known as "picking the bottom"). From time step 2 to time step 5, people begin to realize that the news is false, and someone is manipulating the market. Thus, the contrast of the powers of the bulls and the bears gradually reverses the prices towards the true values.

After defining Cumulative Abnormal Returns (*CAR*) in event study methodology [33], we conduct the following OLS regression to test price reversal pattern in the bitcoin market during S&D period.

$$CAR_{10,50} = \beta_0 + \beta_1 CAR_{-20,9} + \beta_2 CAR_{-20,-19} + \beta_3 CBB_{-20,9} + \beta_4 CBB_{-20,-19} + \varepsilon \quad (1)$$

where $CAR_{10,50}$ represents Cumulative Abnormal Returns from 10 × 2 h to 50 × 2 h after the publication of bad news (Note that the time interval is 2 h). $CBB_{-20,9}$ refers to the Cumulative changes of the force of the Bulls and the Bears from 20 × 2 h before and 9 × 2 h after the publication (*C* of *CBB* is short for Cumulative, and *BB* is short for Bulls and Bears). Others follow the same definition. ε is the error term.

The most important coefficient is β_1, which reflects the influence of $CAR_{-20,9}$ on $CAR_{10,50}$. β_1 is expected to be negative because the more negative $CAR_{-20,9}$ is, the more positive $CAR_{10,50}$ is, which means the more prices fall, the more prices reverse in the following days. According to column 2 of Table 3, the value of β_1 is −0.5963, that is, every 1% drop in the returns results in a cumulative reversal of 0.5963% in the following time. T-test shows that the coefficient of $CAR_{-20,9}$ is significant to a degree of 1%, which indicates that the cryptocurrency market with the S&D manipulation is like the stock market. Both of them have the pattern of the extreme price reversal.

4.3 Pattern II: Abnormal Trading Volume

According to the case study in Sect. 3, when the market is manipulated by S&D schemes, the trading volume will surge, significantly higher than that at other times. We test this pattern by the following OLS regression.

First, we define Abnormal Volume (*AV*) to represent the abnormal trading volume of bitcoin:

$$AV = V/EV \quad (2)$$

where *V* refers to volume, and *EV* is expected volume, i.e., the average volume in the estimation window [33].

Then, the following OLS regression is performed:

$$CAV_{-20,9} = \beta_0 + \beta_1 CAR_{-20,9} + \beta_2 CAV_{-20,-19} + \beta_3 CBB_{-20,9} + \beta_4 CBB_{-20,-19} + \varepsilon \quad (3)$$

where $CAV_{-20,9}$ is the cumulative abnormal volume from 20 × 2 h before and 9 × 2 h after the publication of bad news; others have the same meaning as Eq. (1).

According to Sect. 4.3, the negative value of $CAR_{-20,9}$ represents an obvious decline in the returns of bitcoin, which can be used as a sign of S&D schemes. Therefore, the coefficient β_1 reflects the abnormal trading volume caused by S&D schemes. As column 3 in Table 3, when the bitcoin returns fall by 1%, the abnormal trading volume increases by 1.4992%. T-test is significant at the level of 1%, which shows that when the bitcoin market suffers from S&D manipulation, there will be a significant surge in trading volume.

Table 3. Results of OLS regressions of S&D schemes in the bitcoin market.

Independent variables	Dependent variables		Independent variables	Dependent variables
	$CAR_{10,50}$	$CAV_{-20,9}$		$\Delta S2$
$CAR_{-20,9}$	−0.5963*** (−14.062)	1.4992*** (3.407)	$CAR_{-1,1}$	−0.5416*** (−7.913)
$CAR_{-20,-19}$	0.1133 (0.302)	----	H	0.0230 (0.441)
$CAV_{-20,-19}$	----	0.8172*** (51.334)	GP	0.0069 (0.126)
$CBB_{-20,9}$	−0.0019 (−0.590)	0.0333 (0.955)	ABT	0.0262 (0.609)
$CBB_{-20,-19}$	0.0378 (0.769)	−0.5531 (−1.040)	GT	0.0785 (1.572)
β_0	−0.0155	−2.5752	$CFGI$	−0.0398 (−1.134)
			β_0	−0.6070

t statistics in parentheses. $^*p < 0.05$, $^{**}p < 0.01$, $^{***}p < 0.001$.

4.4 Pattern III: Widening of Bid-Ask Spreads

It is indicated in [23] that when informed trading occurs in the stock market, the market makers usually increase the bid-ask spread to cover the potential losses in the future. The manipulator of S&D schemes has already bought or sold certain assets before the publication of bad news, and this constitutes informed trading [23, 34]. Besides, the market makers are only passive receivers of the bad news because they can't control the behavior of the manipulator. The publication of the bad news is bound to bring some uncertainty to the market. Therefore, from the perspective of the market makers, they can only expand bid-ask spreads to make up for some potential losses in the future. In the following, we investigate whether the pattern of widening of bid-ask spreads exists in the cryptocurrency market.

Since we could only find the daily data of some variables such as spreads, we define the dependent and independent variables and test Pattern III with the time granularity of 1 day.

After defining ΔS_2 as the change of the bid-ask spreads of bitcoin from day 0 to day 2 according to the work of Mitts [23], the OLS regression is performed:

$$\Delta S_2 = \beta_0 + \beta_1 CAR_{-1,1} + \beta_2 H + \beta_3 GP + \beta_4 ABT + \beta_5 GT + \beta_6 CFGI + \varepsilon \quad (4)$$

where ΔS_2 represents the change of bid-ask spreads. $CAR_{-1,1}$ refers to the cumulative abnormal returns from the day before to the day after the publication of bad news (Note that the time interval is 1 day). Other variables' meanings can be found in Table 2.

The results of OLS regression are shown in column 4 of Table 3. β_1, the coefficient of $CAR_{-1,1}$, is equal to -0.5416, which indicates that every 1% drop in bitcoin returns increases the bid-ask spreads by 0.5416% two days later. T-test is significant at the 1% level, which strongly proves that the emergence of S&D manipulation in the bitcoin market leads to the widening of bid-ask spread, and shows that S&D manipulation exists in the bitcoin market to a certain extent.

5 Detection of Short and Distort Schemes

In this section, the random forest model is employed to detect S&D schemes in the real market with the patterns obtained in Sect. 3 and Sect. 4, compared with other four detection algorithms.

5.1 Feature Engineering

Section 3 and Sect. 4 study the characteristics of S&D schemes in the cryptocurrency market, showing that markets manipulated by S&D schemes often have the abnormal price and trading volume. Thus, it is possible to judge whether there exist S&D manipulations through the movement of market indicators such as price and volume, which should be also considered in feature engineering.

There are mainly the following considerations in feature selection. First, S&D schemes are phenomena of the cryptocurrency market over a period, rather than at a certain point. Therefore, although the label is marked on a certain day, it is difficult to detect S&D only with the data of that time point. Instead, the market movement characteristics of prices and trading volumes within a time window must be used.

Second, according to the above research, price and trading volume contain much useful information, and thus are important for S&D schemes. It is necessary to fully explore the information contained therein. We focus on processing features by the raw data of price and trading volume, such as average value, maximum, minimum, and standard deviation within the time window. Table 4 shows all the features in the detection model, which are extracted from the time windows.

Table 4. Features used in the detection model.

Feature	Description
P_avg	The average price
P_max	The maximum price
P_min	The minimum price
P_poc	The percentage of price change, defined as the maximum price minus the minimum price divided by the average price
P_std	The standard deviation of the prices
V_avg	The average volume
V_max	The maximum volume
V_min	The minimum volume
V_poc	The percentage of volume change, defined as the maximum volume minus the minimum volume divided by the average volume
V_std	The standard deviation of the volumes

5.2 Detection Model

Sample. The raw data of time series used in the detection model are consistent with regression analysis in Sect. 4. See Sect. 4.1 for details. We set the time window as 20 h (10 time steps). After sliding time window processing and feature engineering, there are 20,046 samples in total, including 3,856 positive cases ($S\&D = 1$) and 16,190 negative cases ($S\&D = 0$).

Model. In the cryptocurrency market, to detect whether S&D appears (`TRUE`) or not (`FALSE`), the detection model is as follows:

$$S\&D = DA(feature1, feature2, \ldots) \tag{5}$$

First, features $(feature1, feature2, \ldots)$ stated in Sect. 5.1 are input into the detection algorithm DA. Then, the algorithm DA outputs the result. A binary variable $S\&D$ is equal to 1 (`TRUE`) when S&D exists in the market, otherwise 0 (`FALSE`).

The detection algorithm DA we employ here is Random Forest (RF), which is a superior ensemble learning algorithm. It uses decision trees as individual learners, and finally outputs the result by integrating the voting of different decision trees. RF is simple and easy to implement with great advantages in dealing with multivariate and overfitting, and usually converges to a lower generalization error with the increase of the number of individual learners.

Since we are the first to study and detect S&D in the cryptocurrency market, there are no previous works to compare. We can only select the other four classical and commonly used algorithms as DA in the model for comparison, which are Naive Bayes Classifier (NBC), Support Vector Machine (SVM), K-Nearest Neighbor (KNN), and Back Propagation (BP).

5.3 Evaluation

Model Evaluation. First, we calculate the confusion matrices, which gives the true and false results of binary classification [35]. According to confusion matrices, some metrics can be further defined to measure the performance of the models, which is commonly used to evaluate the performance of binary classification, as listed in the first column of Table 5.

Table 5. Performance measures.

Metric	Formula	NBC	SVM	KNN	BP	RF
Error	$(FP + FN) / N$	0.4343	0.1519	0.2246	0.1756	0.0821
Accuracy	$(TP + TN) / N$	0.5657	0.8481	0.7754	0.8244	0.9179
FPR	$FP / (TN + FP)$	0.4198	0.4133	0.2125	0.3716	0.2190
Specificity	$TN / (TN + FP)$	0.5802	0.5867	0.7875	0.6284	0.7810
Precision	$TP / (TP + FP)$	0.8499	0.9030	0.9389	0.9083	0.9483
Recall	$TP / (TP + FN)$	0.5622	0.9099	0.7726	0.8708	0.9503
$F1$-score	$2*TP / (N + TP - TN)$	0.7343	0.9065	0.8476	0.8892	0.9493

$N = TP + FN + FP + TN$ represents the total number of test samples, and the deeper the background's color is, the larger the value is. Red indicates the best performance, while the lower Error and *FPR*, and the higher the others, the better.

Error represents the ratio of misclassified samples to the total samples, while Accuracy is just the opposite. Row 2 and row 3 in Table 5 show that RF has the best performance, with an Accuracy of 91.79% and an Error of 8.21%.

Recall (i.e., True Positive Rate (*TPR*) or Sensitivity) is the ratio of the number of samples correctly classified as S&D to the total number of true S&D samples. Recall of RF is the highest (95.03%), followed by SVM (90.99%). BP and KNN are in the middle (lower than 90%), while NBC is the lowest (56.22%).

False Positive Rate (FPR) is the ratio of the number of samples classified as S&D to the total number of true non-S&D samples. FPR of KNN is the lowest (21.25%), followed by RF (21.90%). Both SVM and BP are around 40%, while NBC is the highest (41.98%), indicating that NBC misclassified non-S&D as S&D in many samples. Although FPR of RF is not better than that of KNN, they are similar. What's more, Recall of RF is much higher than that of KNN.

Specificity is the rate of correctly classifying negative cases. In the view of Specificity, KNN is the highest (78.75%), NBC is the lowest (only 58.02%), and the other three models are in the middle (about 60%).

Precision represents how many samples predicted as S&D are true. For Precision, RF is the highest, reaching 94.83%, while the other models are a bit lower than RF.

Moreover, the commonly used evaluation metric is $F1$-score, which is the balance of Recall and Precision [35]. As shown in the last row of Table 5, F1-score of RF is the highest at 94.93%, followed by SVM at 90.65%, and the other three models are all lower than 90%.

AUC is the Area Under Receiver Operating Characteristic Curve (ROC), and the larger values mean the better performances. Figure 3 plots the ROC curves. As can be seen from the legend below in the right, AUC of RF is the largest (0.96), followed by KNN (0.85) and BP (0.83). AUCs of SVM (0.68) and NBC (0.57) are much lower than them.

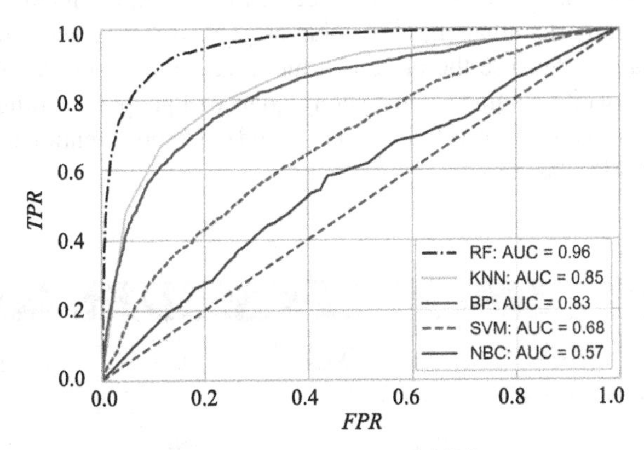

Fig. 3. ROC curves and AUC.

To sum up, as shown in Table 6, from all aspects, RF, the detection algorithm we select, has the best performance compared with the other four methods. BP and SVM also show great performances. KNN follows them, while NBC, as a primary classification algorithm, is not satisfactory at all.

Table 6. Comparisons of performances.

Metric	Rank
Error and accuracy	RF > SVM, BP > KNN > NBC
TPR and *FPR*	RF > SVM > BP, KNN > NBC
Sensitivity and specificity	RF > KNN > BP, SVM > NBC
Precision, recall and *F*1-score	RF > SVM, BP > KNN > NBC
ROC and AUC	RF > KNN, BP > SVM > NBC

Feature Evaluation. Feature Importance Score [35] obtained by rf can be used to evaluate the role of features when detecting S&D in the market. Figure 4 shows the importance scores of 10 features based on the Gini coefficient.

In Fig. 4, the top two features are Minimum Price (*P_min*) and Maximum Price (*P_max*), which suggests that they are more important in detection. This may be because when S&D schemes appear in the market, the prices decline sharply (as described in Pattern I). The minimum and maximum prices catch this tendency, resulting in higher feature importance scores.

In addition, we can also find that in RF detection model, the features of prices (or returns) are more important than those of volumes, such as *P_ avg* and its corresponding *V_ avg*. This may be because the price is a more native variable. People react to the market by quotes, while volume is only a consequence of people's trading behavior as a secondary variable. This suggests that people should pay more attention to price features when studying S&D schemes.

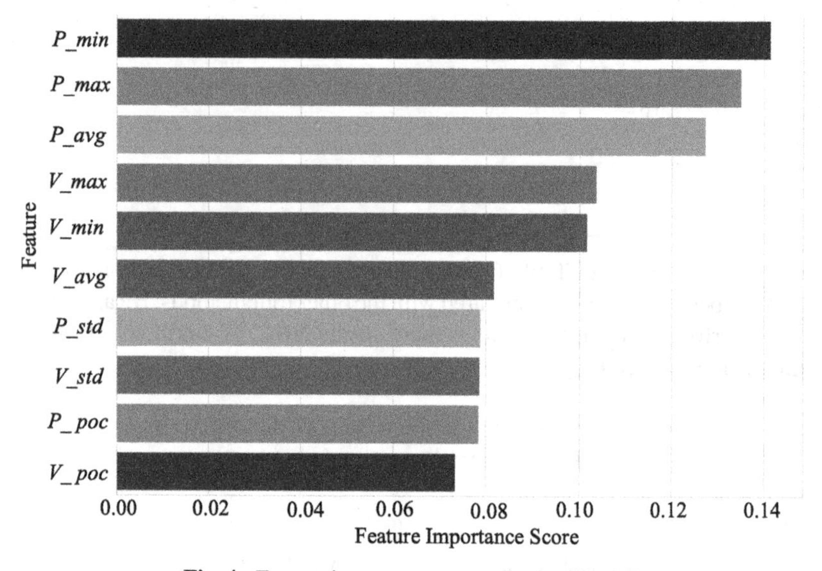

Fig. 4. Feature importance score obtained by RF.

6 Conclusion

This paper studies a new manipulation, Short and Distort in the cryptocurrency market. Through a typical case study, we state the process of the fraud. Using event study methodology and OLS regression, we verify three patterns of Short and Distort, which are price reversal, abnormal trading volume, and widening of bid-ask spreads. Detection of Short and Distort is conducted by the random forest algorithm, which gets the best performance comparing with the other four methods. Our detection model can effectively remind investors away from the market when the fraud scheme comes and provide a method for regulators.

Acknowledgement. This work was supported in part by the National Key Research and Development Program of China (2018YFB1800204), the National Natural Science Foundation of China (61972219), the Research and Development Program of Shenzhen (JCYJ20190813174403598, SGDX20190918101201696), the National Key Research and Development Program of China (2018YFB1800601), the Overseas Research Cooperation Fund of Tsinghua Shenzhen International Graduate School (HW2021013), the Peng Cheng National Laboratory Project (Grant No. PCL2021A02), and Guangdong Major Project of Basic and Applied Basic Research (2019B030302002).

References

1. Nakamoto, S.: Bitcoin: a peer-to-peer electronic cash system. Decentral. Bus. Rev. 21260 (2008)
2. Cryptocurrency Monitoring Website. http://coinmarketcap.com. Accessed 21 Mar 2021
3. Haigh, T., Breitinger, F., Baggili, I.: If I had a million cryptos: cryptowallet application analysis and a trojan proof-of-concept. In: International Conference on Digital Forensics and Cyber Crime (ICDF2C) 2018, LNICST, vol. 259, pp. 45–65. Springer, Cham (2019). https://doi.org/10.1007/978-3-030-05487-8
4. MacRae, J., Franqueira, V.N.: On locky ransomware, al capone and brexit. In: International Conference on Digital Forensics and Cyber Crime (ICDF2C) 2017, LNICST, vol. 216, pp. 33–45. Springer, Cham (2018). https://doi.org/10.1007/978-3-319-73697-6
5. Gandal, N., Hamrick, J.T., Moore, T., Oberman, T.: Price manipulation in the bitcoin ecosystem. J. Monet. Econ. **95**, 86–96 (2018)
6. Bartoletti, M., Pes, B., Serusi, S.: Data mining for detecting bitcoin ponzi schemes. In: 2018 Crypto Valley Conference on Blockchain Technology (CVCBT), pp. 75–84, IEEE, Zug, Switzerland (2018)
7. Weiner, P.M., Weber, R.D., Hsu, K.: The growing menace of "short and distort" campaigns. Thomson Reuters Exp. Anal. (2017)
8. Weiner, P.M., Totino, E.D., Goodman, A.: SEC issues warning to analysts profiting from "short and distort" schemes, opens the door for civil claims. J. Invest. Compl. (2019)
9. Aggarwal, R.K., Wu, G.: Stock market manipulations. J. Bus. **79**(4), 1915–1953 (2006)
10. Massoud, N., Ullah, S., Scholnick, B.: Does it help firms to secretly pay for stock promoters? J. Financ. Stab. **26**, 45–61 (2016)
11. Huang, Y.C., Cheng, Y.J.: Stock manipulation and its effects: pump and dump versus stabilization. Rev. Quant. Financ. Acc. **44**(4), 791–815 (2013). https://doi.org/10.1007/s11156-013-0419-z
12. Kamps, J., Kleinberg, B.: To the moon: defining and detecting cryptocurrency pump-and-dumps. Crime Sci. **7**(1), 1–18 (2018). https://doi.org/10.1186/s40163-018-0093-5
13. Hamrick, J.T., et al.: The economics of cryptocurrency pump and dump schemes. https://papers.ssrn.com/sol3/pa-pers.cfm?abstract_id=3310307 (2018)
14. Li, T., Shin, D., Wang, B.: Cryptocurrency pump-and-dump schemes. https://papers.ssrn.c-om/sol3/papers.cfm?abstract_id=3267041 (2020)
15. Dhawan, A., Putnins, T.J.: A new wolf in town? Pump-and-dump manipulation in cryptocurrency markets. In: Pump-and-Dump Manipulation in Cryptocurrency Markets. https://pape-rs.ssrn.com/sol3/papers.cfm?abstract_id=3670714 (2020)
16. Victor, F., Hagemann, T.: Cryptocurrency pump and dump schemes: quantification and detection. In: International Conference on Data Mining Workshops (ICDMW) 2019, pp. 244–251. IEEE, Beijing, China (2019)

17. La Morgia, M., Mei, A., Sassi, F., Stefa, J.: Pump and dumps in the bitcoin era: real time detection of cryptocurrency market manipulations. In: 29th International Conference on Computer Communications and Networks (ICCCN), pp. 1–9. IEEE, Honolulu, HI, USA (2020)
18. Mirtaheri, M., Abu-El-Haija, S., Morstatter, F., Steeg, G.V., Galstyan, A.: Identifying and analyzing cryptocurrency manipulations in social media. IEEE Trans. Comput. Soc. Syst. **8**(3), 607–617 (2021). https://doi.org/10.1109/TCSS.2021.3059286
19. Mansourifar, H., Chen, L., Shi, W.: Hybrid cryptocurrency pump and dump detection. arXiv 2003.06551, arXiv preprint arXiv:2003.06551 (2020)
20. Nilsen, A.I.: Limelight: real-time detection of pump-and-dump events on cryptocurrency exchanges using deep learning. https://munin.uit.no/bitstream/handle/10037/15733/thesis.pdf?sequence=2&isAllowed=y (2019)
21. Chen, W., Xu, Y., Zheng, Z., Zhou, Y., Yang, J.E., Bian, J.: Detecting "pump & dump schemes" on cryptocurrency market using an improved apriori algorithm. In: 2019IEEE International Conference on Service-Oriented System Engineering (SOSE), pp. 293–2935. IEEE, San Francisco, CA, USA (2019)
22. Xu, J., Livshits, B.: The anatomy of a cryptocurrency pump-and-dump scheme. In: 28th USENIX Security Symposium, pp. 1609–1625. USENIX Association, Santa Clara, CA, USA (2019)
23. Mitts, J.: Short and distort. J. Leg. Stud. **49**(2), 287–334 (2020)
24. Zhao, W.: Activist short-selling. https://www.proquest.com/openview/1559f45788a7fb12-5c3f2282c1e2a875/1?pq-origsite=gscholar&cbl=18750 (2017)
25. Ogorodniks, A., Sirbu, A.: Activist short selling campaigns: informed trading or market manipulation? https://www.sseriga.edu/sites/default/files/2020-05/1Paper_Ogorodniks_S-irbu.pdf (2018)
26. Kartapanis, A.: Activist short-sellers and accounting fraud allegations. https://repositories.lib.utexas.edu/bitstream/handle/2152/74990/KARTAPANIS-DISSERTATION-2019.pdf?-sequence=1&isAllowed=y (2019)
27. Nizzoli, L., Tardelli, S., Avvenuti, M., Cresci, S., Tesconi, M., Ferrara, E.: Charting the landscape of online cryptocurrency manipulation. IEEE Access **8**, 113230–113245 (2020)
28. TradingView Website. https://www.tradingview.com/chart/. Accessed 31 Mar 2021
29. Bouoiyour, J., Selmi, R.: Coronavirus spreads and bitcoin's 2020 rally: is there a link? https://hal.archives-ouvertes.fr/hal-02493309/ (2020)
30. Shahzad, S.J.H., Bouri, E., Roubaud, D., Kristoufek, L., Lucey, B.: Is bitcoin a better safe-haven investment than gold and commodities? Int. Rev. Financ. Anal. **63**, 322–330 (2019)
31. Härdle, W.K., Trimborn, S.: Crix or evaluating blockchain based currencies. https://www.econstor.eu/handle/10419/122006 (2015)
32. Tetlock, P.C.: All the news that's fit to reprint: do investors react to stale information? Rev. Finan. Stud. **24**(5), 1481–1512 (2011)
33. MacKinlay, A.C.: Event studies in economics and finance. J. Econ. Literat. **35**(1), 13–39 (1997)
34. Feng, W., Wang, Y., Zhang, Z.: Informed trading in the bitcoin market. Financ. Res. Lett. **26**, 63–70 (2018)
35. Zhihua, Z.: Machine Learning. Tsinghua University Press, Beijing (2016)

Privacy-Preserving Swarm Learning Based on Homomorphic Encryption

Lijie Chen, Shaojing Fu(✉), Liu Lin, Yuchuan Luo, and Wentao Zhao

College of Computer, National University of Defense Technology, Changsha, China
fushaojing@nudt.edu.cn

Abstract. Swarm learning is a decentralized machine learning method that combines edge computing, blockchain based point-to-point network and coordination while maintaining consistency without the need for a central coordinator for data integration between different medical institutions. Swarm learning solves problems of federated learning that the model parameters are still handled by the central custodian and star structure reduces fault tolerance etc. Although Swarm learning allows fair, transparent and highly regulated shared data analysis on the premise of meeting the privacy law, there will be inference attacks targeting shared model information between nodes, which will lead to privacy disclosure. Therefore, we propose the privacy-preserving Swarm learning scheme based on homomorphic encryption: we use the threshold Paillier cryptosystem to encrypt the shared local model information without affecting the accuracy of the model. In addition, we design a partial decryption algorithm to prevent privacy disclosure caused by aggregation model information. Experiments show that our scheme can guarantee strong privacy and has negligible influence on the accuracy of the final model.

Keywords: Swarm learning · Blockchain · Inference attack · Privacy-preserving · Homomorphic encryption · Partial decryption

1 Introduction

With the development of technology, modern hospitals use electronic medical records instead of paper medical records, which brings great convenience to people's medical treatment. These electronic medical records are also saved in the hospital, resulting in a large number of medical data. These medical data contain the extremely sensitive private information of patients. In addition, medical experimental data and scientific research data are not only related to the privacy of data subjects, industry development, but also related to national security. Therefore, in the research involving medical data, we should consider the increasing divide between what is technically possible and what is allowed because of privacy legislation. Data privacy issues are considered in some research in the medical field [23].

In the research of disease diagnosis method based on artificial intelligence, in addition to considering the rationality of application algorithm, we should

Y. Lai et al. (Eds.): ICA3PP 2021, LNCS 13157, pp. 509–523, 2022.
https://doi.org/10.1007/978-3-030-95391-1_32

also try to make more medical data participate in training. However, from the perspective of law, considering the privacy and confidentiality of medical data, medical data cannot be shared among medical institutions and the reliability of classification obtained by medical institutions using their own medical data for training has certain limitations.

Distributed deep learning can make full use of a large amount of distributed data to improve the accuracy of the training model. Compared with traditional deep learning, distributed deep learning can reduce communication overhead and computation overhead, but it also brings more privacy problems and there has been a lot of research on privacy issues in distributed deep learning in the past few years [2,13,26,27]. Among these existing works, federated learning [8] is the widely adopted system context.

As distributed learning, federated learning is essentially a combination of deep learning and distributed computing, with multiple participants and a parameter server; each participant uses the data stored locally to train a local model separately, and uploads the intermediate gradient obtained from the training to the parameter server. Upon receiving gradients from each participant, the parameter server aggregates these gradients to abtain the global model. All participants can use the updated global model for the next epoch training. This iterative process continues until the specified training time is over or the accuracy of the global model meets the requirements.

In the federated learning framework, although the data involved in training is distributed and stored locally, the privacy of training data will also be leaked. Some researchers have shown that there are inference attacks, which use gradient information to infer important information of training data [9,20]. In addition, the central server in federated learning is vulnerable to attack in an insecure environment. As a distributed ledger, blockchain has features such as unforgeable, traceable, open and transparent, and collective maintenance, which have been applied to many decentralized multi-party settings, such as asset management [24], access control [5], privacy-preserving computation [11] and IoT edge computation [28]. Integrating blockchain into federated learning can effectively mitigate security issues involving the central servers in federated learning.

Swarm learning [22] combines edge of computing and blockchain-based peer-to-peer networks for the integration of medical data between different healthcare facilities. In Swarm learning, participants can safely join the training by registering on blockchain, and smart contracts are used to dynamically select the leader among participants, thus changing the star structure of federated Learning. The chosen leader replaces the role of parameter server in federated learning to collect the local models transmitted by other participants for average aggregation. In Swarm learning, each participant uses federated learning to train the model locally; the experimental results show that the model prediction performance of Swarm learning is better than that of the single node and even if the reduction of training data leads to the decline of Swarm learning's prediction performance, it is still better than that of the single node. Although participants in swarm learning use their local data for training, which ensures the confidentiality of data, there is still a risk of privacy disclosure in Swarm learning: When the participants upload trained locals model to the blockchain, the attacker may launch inference attacks [14,18] and use the local model information to steal data information.

At present, many researchers use differential privacy technology [1,7], secure aggregation [6,25] or homomorphic encryption mechanism [15] to solve the problem of privacy leakage caused by inference attacks against machine learning. In the studies with using differential privacy technology, the trade-off between data privacy and validity should be considered according to the research requirements. Adding too much noise will affect the accuracy of the final model, adding too little noise will weaken the ability to resist inference attack.

Homomorphic encryption has the following properties: the result of addition and multiplication of homomorphic encrypted data after decryption is the same as that obtained by operation directly on plaintext; therefore, homomorphic encryption technology does not affect the accuracy of the model.

We use the threshold Paillier Cryptosystem to encrypt the local model gradient trained by the participants and upload the encrypted local model to the leader. The leader aggregates the encrypted local model to obtain the encrypted aggregation model. It is important to note that we divide the private key used in the threshold Paillier Cryptosystem and allocate it to each participant, no participant has a complete private key in the system. In Swarm learning, not only the local model uploaded is unencrypted, but the aggregation model downloaded by the participants is also unencrypted; therefore, we should also consider that attacks [16,21] on the aggregation model, which can also cause privacy leakage. In our scheme, we design the Partial decryption algorithm of the encrypted aggregation model, which can be used by participants with partial private keys to obtain the decrypted aggregation model locally. The major contributions of this article are as follows:

1. We propose a Privacy-preserving Swarm learning Based on Homomorphic Encryption scheme, which uses homomorphic encryption technology to encrypt the local model parameters obtained by participants' training, which solves the problem of data privacy leakage caused by inference attack against local models in Swarm Learning system.
2. We also designed the partial decryption algorithm of the encrypted aggregation model, which enables the participants to decrypt the encryption aggregation model downloaded from the leader locally when they only have part of the private key, so as to resist attacks against the aggregation model.

The remainder of the paper is organized as follows. We present the related work in Sect. 2 and the preliminary in Sect. 3. Introducing the system architecture of Privacy-preserving Swarm learning in Sect. 4. Privacy analysis is proposed in Sect. 5, followed by experiments in Sect. 6, the conclusion in Sect. 7. Finally, the future work is in Sect. 8.

2 Related Work

In the research related to privacy protection, different encryption technologies are used to enhance the security of system.

[23] proposed a Privacy-Enhanced data aggregation approach using secret sharing. However, this method is targeted at the smart Grid scenario, which includes cloud, edge server and terminal architecture in this paper, and cannot be applied to Swarm learning without a central architecture. In [28], before the data features are submitted to the convolution layer for training, differential privacy technology is used to interfere with the extracted data features. Meanwhile, a new batch normalization technology was proposed to improve the accuracy of the model. In addition, blockchain is used to replace the central server for model aggregation, which solves the problem of single point failure. However, with the increase of the number of training iterations, the added noise is also increasing, which will negatively affect the accuracy of the aggregation model. [17] used the differential privacy technology to design a pre-committed Noise protocol, which can not only add noise to interfere with the model gradient information, but also resist the poisoning and information leakage attacks against the noise protocol. However, because the global model is not protected by noise, if the system is attacked by an attack on the global model, privacy leakage may occur.

[15] pointed out the problem of privacy disclosure in [19]: an honest but curious server may extract the original data from the gradient information, and then [15] used homomorphic encryption technology to encrypt the gradient information. This additive homomorphic property enables the computation across the gradients, which ensures that the private information is not leaked to the honest but curious server, it also ensures that the accuracy of the model is not affected. In [10], the distributed Pailliar encryption was used to encrypt the model parameters of the exchange to protect the privacy of the participants. Combined with zero-knowledge proof technology, the problem of unreliable data transmission of participants is solved. Finally, the system can resist Byzantine attacks and protect privacy at the same time. However, each of [10,15] has a central server, which risks a single point of failure.

Using blockchain instead of the server, Swarm learning can realize completely decentralized machine learning and make full use of medical data distributed in different medical institutions for disease diagnosis or outcome prediction classifiers. However, there are inference attacks against local model and aggregation model in swarm learning, resulting in privacy disclosure of medical data. Therefore, in our work, we mainly use homomorphic encryption technology to encrypt local model and aggregation model to resist inference attacks.

3 Preliminaries

In this section, we briefly introduce the Swarm learning and homomorphic encryption.

3.1 Swarm Learning

Swarm Learning system consists of two parts: Swarm edge nodes and Swarm network (Blockchain); Medical institutions (Swarm edge nodes) that want to participate in the model training need to register through the smart contract of the blockchain, then all registered nodes uniformly download the initial global model from the blockchain and use their local data for local model training. Swarm edge nodes upload the local model parameters of the training to the leader through the Swarm network. The leader is dynamically selected from the Swarm edge nodes by the smart contract of the blockchain. The leader will average the collected local model parameters. If the trained aggregation model meets the requirements, the leader generates a block containing the aggregation model; if not, each Swarm edge nodes will download the aggregation model from the Swarm network to continue the local model training until the aggregation model meets the requirements.

3.2 Homomorphic Cryptographic Scheme

Homomorphic encryption is a powerful cryptographic technique that allows arithmetic circuits to be operated directly on ciphertexts; because of this good nature, people can entrust a third party to process the data without disclosing the information. In homomorphic encryption, the public key is used to encrypt the plaintext and the private key is used to decrypt the ciphertext. Suppose a plaintext m $\in Z_n$, where n is a large positive integer and Z_n is the set of integers module n, we denote the encryption of m as $E_{pk}(m)$. Here we use the threshold variant of Paillier Scheme [3], which satisfies the following properties:

$$E_{pk}(m_1 + m_2) = E_{pk}(m_1)E_{pk}(m_2) \tag{1}$$

where m_1, m_2 are the plaintexts that need to be encrypted.

In this cryptosystem, a participant can encrypt the plaintext m $\in Z_n$ with the public key $pk = (g, n)$ as

$$c = E_{pk}(m) = g^m r^n mod \quad n^2 \tag{2}$$

where r $\in Z_n^*$ (Z_n^* denotes the multiplicative group of invertible elements of Z_n) is selected randomly and privately by this participant. According to Eq. (1) and Eq. (2), the homomorphic properties of this cryptosystem can be described as

$$\begin{aligned} E_{pk}(m_1 + m_2) &= E_{pk}(m_1)E_{pk}(m_2) \\ &= g^{m_1+m_2}(r_1 r_2)^n mod \quad n^2 \end{aligned} \tag{3}$$

where m_1, m_2 are the plaintexts which need to be encrypted, and r_1, $r_2 \in Z_n^*$ are the private randoms.

In this paper, the (p, t) – threshold Paillier cryptosystem is adopted, in which the private key sk is divided (denoted as $sk_1, sk_2, ..., sk_p$) and distributed to p parties. Any single party doesn't have the complete private key. If one party wants to accurately decrypt ciphertext c, it has to cooperate with at least $(t-1)$ other parties. In the decryption step, each party $i \in [1, p]$ can calculate the partial decryption c_i of c with private key sk_i as:

$$c_i = c^{2p! sk_i} \tag{4}$$

Then based on the combining algorithm in [3], at least t partial decryptions can be combined together to get the plaintext m.

4 System Architecture

4.1 System Model

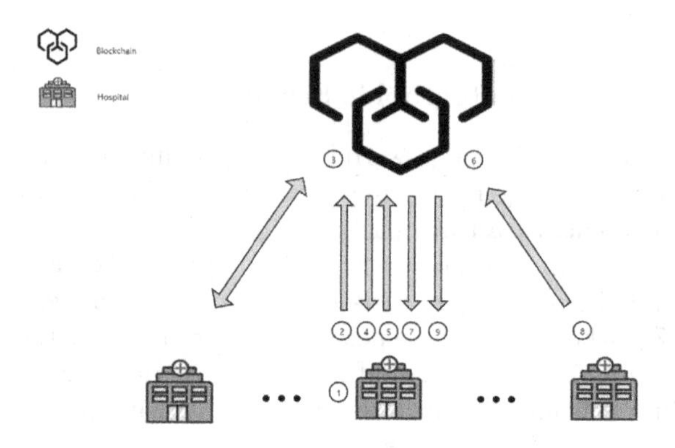

Fig. 1. System model

The system model of Privacy-preserving Swarm learning is shown in Fig. 1. Before the system starts working, we assume a semantically secure (p, t) – threshold Paillier cryptosystem has been given (e.g., established by a trusted key management center). Where p is the total number of all participants participating in the training in the system, and t is the minimum number of participants required to participate in the decryption process. Each participant in the system has the public key $PK = (g, n)$, while the corresponding private key is divided and distributed to each participant, so that each participant has different private key fragments.

As shown in Fig. 1, after the participants register through the blockchain, all participants will perform the following nine steps iteratively:

Step 1: Registered participants download the aggregation model from the blockchain, and use local data to train the model to obtain the local model; then each participant uses the public key to encrypt the local model parameters.

Step 2: Each participant uploads the encrypted local model to the blockchain.

Step 3: The leader aggregates the collected encrypted local models to obtain the encrypted aggregation model (the leader is randomly designated at the beginning).

Step 4: The leader randomly selects $t-2$ participants, who download the encrypted aggregation model from the blockchain; each selected participant uses its private key for partial decryption calculation according to the Eq. (4).

Step 5: The selected $t-2$ participants upload their partial decryptions to the blockchain.

Step 6: The leader also uses its partial private key for partial decryption; the leader sums its partial decryption with the collected $t-2$ partial decryptions of encrypted aggregation model according to the combining algorithm in [3]. So far, the leader has aggregated $t-1$ partial decryptions and obtained sum_{t-1}, and at least one partial decryption is required if the decrypted aggregation model is to be obtained.

Step 7: All participants download the encrypted aggregation model and the sum_{t-1}. Unselected participants can obtain the last partial decryption using their private key, and then sum the sum of sum_{t-1} with the last partial decryption according to the combining algorithm in [3]. The sum is the decrypted aggregation model.

Step 8: After the above steps, the selected $t-2$ participants and the leader have not obtained the decrypted aggregation model, and they still need the last partial decryption. At this time, a participant is randomly selected from the unselected participants, and the participant uploads its partial decryptions to the blockchain.

Step 9: The selected $t-2$ participants and the leader can download the last partial decryption from the blockchain so that they can decrypt the encrypted aggregation model locally; in this way, all participants can use the decrypted aggregation model for the next epoch training.

The above nine steps begin with an initial model, and are then iteratively conducted until the aggregation model meets the model requirements or reaches the size of the epoch.

4.2 System Mechanism

In this section, we will elaborate on the mechanism of the privacy-preserving swarm learning framework. First, we will introduce the *secure sum protocol* in [12], which uses the threshold Paillier cryptosystem, then point out the disadvantages if we directly use this protocol. Second, we will introduce the decryption algorithms of the encrypted aggregation model, which enables the participant with partial private key to decrypt the encrypted aggregation model locally.

4.2.1 *Secure Sum Protocol*

Assuming that there are K participants in the system, denoted as $k = 1, 2, ..., k$. Initially, the system will randomly select a participant as the leader. Let v_k is defined as the local model trained by the k_{th} participant in an iteration.

According to Eq. (2), each participant $k \in K$ encrypts value v_k and sends the ciphertext $E_{pk}(v_k)$ to the leader, the leader calculates $C = E_{pk}(\sum_{k=1}^{K} v_k) = \Pi_{k=1}^{K} E_{pk}(v_k)$ based on Eq. (3); then the leader randomly selects $t-1$ participants and send C to them, each selected participant $k^{'}$ calculates the partial decryption $C_{k'}$ of C based on Eq. (4) and sends $C_{k'}$ to the leader; the leader calculates its partial decryption C_l and then combines C_l with $t-1$ other partial decryptions received from participants according to the combining algorithm in [3] to get the summation$\sum_{k=1}^{K} v_k$; thus, the leader gets the decrypted aggregation model. Other participants can download the decrypted aggregation model directly from the blockchain to continue the training of the next epoch.

According to the above introduction, the participants use homomorphic encryption technology to encrypt the trained local models, which can avoid the risk of privacy disclosure in the process of uploading the local model to the blockchain, and the leader can also use *secure sum protocol* to obtain the decrypted aggregation model. If the aggregation model does not meet the requirements, the participants need to download the decrypted aggregation model from the blockchain to continue local training, but some studies show that the aggregation model itself will also be attacked and cause privacy disclosure. To solve this problem, participants should download undecrypted aggregation models from the blockchain. However, it should be noted that each participant in the system only has part of the private key, and none of the participants has a complete private key, which means that the participants need to cooperate with other participants to obtain the decryption aggregation model. So we design the partial decryption algorithm of the encrypted aggregation model based on the *secure sum protocol*.

4.2.2 *Partial Decryption Algorithm*

The process of partial decryption can be described by Algorithm 1.

In our framework, assuming that there are P participants in the system, denoted as $p = 1, 2, ..., P$. Initially, the system will randomly select a participant as the leader. Let x_p is defined as the local model trained by the p_{th} participant in an iteration. According to Eq. (2), each participant $p \in P$ encrypts value x_p as $E_{pk}(x_p)$ and sends the ciphertext $E_{pk}(x_p)$ to the blockchain, then the leader calculates $C = Agg\ (E_{pk}(x_1), E_{pk}(x_2)..., E_{pk}(x_p))$ based on Eq. (3); the rest is as follows:

Step 1 : After the leader obtains the encrypted aggregation model, the leader randomly selects $t-2$ participants; defining the $t-2$ participants and the leader as a collection named $P_{Selected}$, the remaining participants are defined as a collection named $P_{Unselected}$.

Step 2 : Each selected participant $p_s \in P_{Selected}$ downloads the encrypted aggregation model C and uses its own private key to calculate partial decryption $C_s^{'}$ of C based on Eq. (4); then p_s sends $C_s^{'}$ to the blockchain.

Step 3 : The leader calculates its partial decryption C^l of C and then combines it with $t-2$ other partial decryptions $C_{s1}^{'}, C_{s2}^{'}..., C_{s(t-2)}^{'}$ received from the selected participants to get the partial decryption C_{t-1} of C. Then the leader

Algorithm 1: Partial decryption algorithm of encrypted aggregation model

Input: P encrypted local models $E_{pk}(x_1), E_{pk}(x_2), ..., E_{pk}(x_p)$
Output: Each participant gets the decrypted aggregation model M
1. The leader :
   compute $C = Agg(E_{pk}(x_1), E_{pk}(x_2)...E_{pk}(x_p))$;
   send C to all participant;
2. Each participant $\in P_{Selected}$:
   compute $C_s^{'} = Partially\ Decrypt\ (C)$;
   send $C_s^{'}$ to the leader;
3. The leader :
   compute $C^l = Partially\ Decrypt\ (C)$
   compute $sum_{t-1} = Combine\ (C^l, C_{s1}^{'}...C_{s(t-2)}^{'})$;
4. Each participant in $P_{Unselected}$:
   compute $C_u^{'} = Partially\ Decrypt\ (C)$
   compute $M = Combine\ (\ C_u^{'},\ sum_{t-1})$;
5. One participant $P_u^{'} \in P_{Unselected}$:
   send $C_u^{'}$ of $P_u^{'}$ to all participants in $P_{Selected}$ and the leader;
6. Each participant in $P_{Selected}$ and the leader :
   compute $M = Combine\ (\ C_u^{'},\ sum_{t-1})$;

sends C_{t-1} to the blockchain, all participants in the system download C_{t-1} from blockchain.

Step 4 : Each participant $p_u \in P_{Unselected}$ downloads the encrypted aggregation model C and uses its private key to calculate partial decryption $C_u^{'}$ of C based on Eq. (4), then p_u combines $C_u^{'}$ with C_{t-1} to obtain the decrypted aggregation model M.

Step 5 : For participants in $P_{Selected}$, they randomly select a participant $p_u^{'}$ $\in P_{Unselected}$, and download the partial decryption $C_u^{'}$ of $p_u^{'}$, then they combine $C_u^{'}$ with C_{t-1} locally to obtain the decrypted aggregation model M.

Participants judge whether the decrypted aggregation model meets the requirements. If the model meets the requirements, the leader packages the encrypted aggregation model into blocks, and all participants stop training; If not, all participants continue locally training based on the decrypted aggregation model. The symbol definition is shown in Table 1.

5 Privacy Analysis

In the framework we designed, there are two assumptions: some participants are semi-honest and there is no collusion between the participants. Participants are semi-honest means that the participant will strictly follow the algorithm we designed, but some participants may be malicious and try to infer the original data information by attacking the model information generated in the training process; the second means that all participants in the system will not collude

Table 1. Symbol description

Notation	Definition
x_p	The local model trained by the p-th participant
$E_{pk}(x_p)$	The encrypted local model of the p-th participant
C	The encrypted aggregation model
$P_{Selected}$	The set of selected participants during partial decryption process
$P_{Unselected}$	The set of unselected participants during partial decryption process
C^l	The partial decryption of the leader
$C_s^{'}$	The partial decryption of the selected participant
$C_u^{'}$	The partial decryption of the unselected participant
sum_{t-1}	The sum of $t-1$ partial decryptions
p	The total number of participants in the system
t	The minimum number of participants required in the decryption process
Agg	Model aggregation algorithm

outside the algorithm we designed. These assumptions are reasonable in practice. If participating medical institutions want to obtain correct results, they must strictly abide by the algorithm we designed. In addition, even if some medical institutions cooperate, they will not disclose the privacy of medical data to other medical institutions.

In our scheme, no participant has a complete private key. Therefore, in the process of sharing the encrypted local model, if a malicious attacker launches an attack on the shared model, the attacker cannot decrypt the encrypted local model and thus cannot obtain the information of the local model. In this way, the shared local model will not cause privacy disclosure due to inference attacks.

When the leader obtains the encrypted aggregation model, the leader needs to select $t-2$ participants for partial decryption. Although these $t-2$ participants can obtain the encrypted aggregation model, they do not have a complete private key, both the leader and these $t-2$ participants can only obtain $t-1$ partial decryptions. Therefore, they cannot obtain information from the aggregation model, let alone obtain information from the local model through the encrypted aggregation model. If an attacker launches an attack on the aggregation model in this process, the attacker cannot obtain the information of the aggregation model. In our scheme, the encrypted aggregation model is decrypted locally by the participants, which can avoid the privacy disclosure caused by the attacker's attack on the aggregation model.

In addition to the previous block hash pointer and timestamp, the generated block also stores the encrypted aggregation model. Therefore, the attacker cannot obtain the aggregation model information from the block.

In conclusion, the scheme designed by us can avoid inference attacks against local models and aggregation models. Meanwhile, experiments show that the influence of homomorphic encryption technology used in our scheme on model accuracy can be ignored.

6 Experience

6.1 Experimental Settings

In our experiment, we used Python (version 3.6) and Tensorflow (version 1.4) to implement the Privacy-preserving Swarm learning. We select the popular MNIST dataset [4] for our experiment, which is a very classic dataset in the field of machine learning, with 55,000 training samples, 5, 000 verification samples and 10, 000 test samples. MNIST dataset is divided among different participants so that their data quality vary.

Here we use the convolutional neural network (CNN) to train the model, the CNN consists of five structures: input Layer, Convolutional Layer, Pooling Layer, Fully Connected Network and output Layer. The convolutional neural network layer, pooling layer and full connection layer are collectively called the hidden layer. We set up two hidden layers with 256 neurons in each hidden layer. The number of features in the input sample is 784, and the number of labels in the sample is 10. The weights and bias parameters in Conv layer, Fully Connected layer and Output layer are $w1$ = (784,256) and $b1$ = 256, $w1$ = (256,256) and $b1$ = 256, wo = (256,10) and $b0$ = 10, respectively. The learning rate is 0.1, the number of epochs is 10, and the default number of iterations is 10.

6.2 Experimental Results

In our solution, our goal is to solve the problem of privacy disclosure in Swarm Learning by encrypting the model parameters without affecting the accuracy of the model. We compared our scheme with Swarm Learning in Fig. 2; we set the epoch of federated learning to be 10, and We adopt the popular FedAvg as the FL aggragtion function, which averages the collected local model parameters to derive the global model parameters; each participant has 10 steps per epoch of training. The Fig. 2 shows that before the sixth epoch training, the model accuracy trained by our scheme is slightly lower than that trained by swarm learning, but after the sixth epoch, both the model trained by Swarm learning and the model trained by our scheme begin to plateau with small dips in accuracy, and the difference between the model accuracy of our scheme and that of swarm learning training can be ignored.

From Fig. 2, we can observe that the model improves quickly, but the model accuracy will decrease slightly after the sixth epoch. One reason for this may be that the number of step taken by the training model is not optimal; in our experiment, we fixed the size of epoch to be 10 and changed the size of step, the result is shown in Fig. 3. As we can see, when step = 10 and step = 15, the model accuracy increases faster than when step = 5, but after the 7th epoch, the model accuracy of step = 10 and step = 15 has an obvious downward trend, and the model accuracy of step = 5 is always been an increasing steadily trend. In the end, the three models have little difference in accuracy, but longer training time is needed when step = 5.

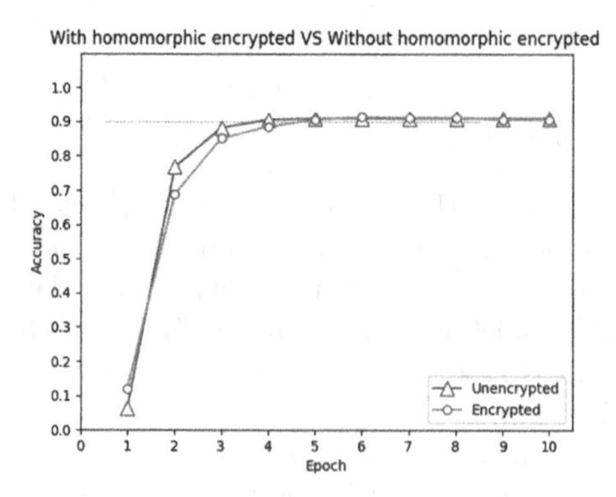

Fig. 2. The model accuracy of our design scheme is compared with that of Swarm learning

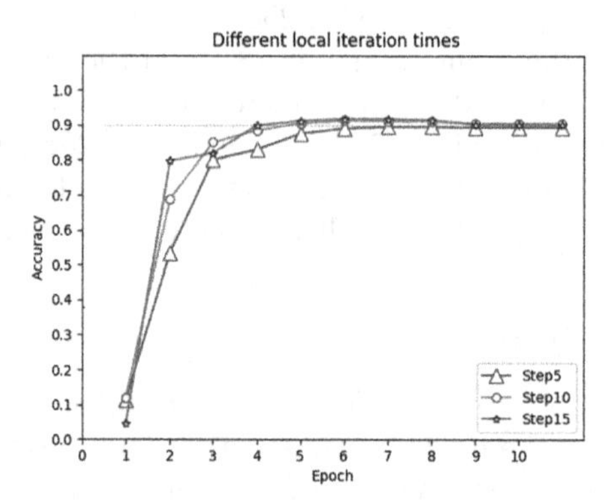

Fig. 3. The model accuracy obtained in the case of different sizes of the step in our scheme

7 Conclusion

In this paper, we mainly focus on the problem of privacy disclosure caused by inference attacks on the model information. In Swarm learning, local models uploaded and aggregated models downloaded by participants are unencrypted, which provides opportunities for malicious attackers to steal data privacy. Therefore, we want to use encryption technology to encrypt the model information, and the encrypted model information will not affect the accuracy of model. The nature of homomorphic encryption technology meets our requirements.

For confidentiality of local model parameters, we employ the Threshold Paillier algorithm to encrypt model parameters; this threshold Paillier splits the private key so that no participant in the system has a complete private key. In our scheme, the participant with part of private key must cooperate with other participants to obtain accurate decrypted model parameters, which makes it impossible for malicious attackers to obtain encrypted local models and model information. In addition, considering that the aggregation model information will also cause privacy disclosure, the leader did not completely decrypt the encrypted aggregation model, and we want to achieve the goal that participants download the undecrypted aggregation model from the blockchain and then decrypt it locally. However, each participant only owns part of the private key. If the encrypted aggregation model is downloaded, the participant cannot decrypt it locally. Therefore based on the *Secure Sum Protocol*, we designed the *Partial Decryption algorithm*. Finally, we implemented our experiment on MNIST, and the experimental results show that our goals have been achieved.

8 Future Work

First, the number of participants is fixed in our scheme, without considering the possibility of offline participants in real life. Because the decryption process requires the cooperation of multiple participants, if a participant is offline, how to deal with part of the private key owned by the participant.

Second, the current focus of our scheme is to protect privacy from an honest but curious adversary. In cases where a malicious adversary wishes to manipulate a global model by poisoning its local data set, additional strategies may be required to verify model quality. For example, we can design the reward and punishment mechanisms in the future to provide rewards to participants who provide good quality local models to encourage participants to better train, and punish participants who provide poor quality local models.

In addition, a model we trained contains more than 100,000 model parameters, and it is time-consuming to encrypt and decrypt data using homomorphic encryption. The size of epoch of participants has an impact on the accuracy of the model and the running time of the whole system. In future experiments, we will consider the tradeoff between model accuracy and time cost.

Acknowledgments. This work is supported by National Nature Science Foundation of China (No. U1811462, No. 62072466, No. 62102430, No. 62102429, No. 62102447), Natural Science Foundation of Hunan Province (No. 2021JJ40688), and the NUDT Grants (No. ZK19-38).

References

1. Abadi, M., Chu, A., Goodfellow, I., Mcmahan, H.B., Zhang, L.: Deep learning with differential privacy. In: The 2016 ACM SIGSAC Conference (2016)
2. Chen, T., Zhong, S.: Privacy-preserving backpropagation neural network learning. IEEE Trans. Neural Networks **20**(10), 1554–1564 (2009)
3. Damgård, I., Jurik, M.: A generalisation, a simplification and some applications of Paillier's probabilistic public-key system. In: Kim, K. (ed.) PKC 2001. LNCS, vol. 1992, pp. 119–136. Springer, Heidelberg (2001). https://doi.org/10.1007/3-540-44586-2_9
4. Deng, L.: The mnist database of handwritten digit images for machine learning research [best of the web]. IEEE Signal Process. Mag. **29**(6), 141–142 (2012)

5. Ding, Y., et al.: Blockchain-based access control mechanism of federated data sharing system. In: 2020 IEEE International Conference on Parallel Distributed Processing with Applications, Big Data Cloud Computing, Sustainable Computing Communications, Social Computing Networking (ISPA/BDCloud/SocialCom/SustainCom), pp. 277–284 (2020)
6. Fereidooni, H., et al.: SAFELearn: secure aggregation for private federated learning. In: 2021 IEEE Security and Privacy Workshops (SPW), pp. 56–62 (2021)
7. Geyer, R.C., Klein, T., Nabi, M.: Differentially private federated learning: a client level perspective. CoRR abs/1712.07557 (2017)
8. Konečný, J., McMahan, H.B., Yu, F.X., Richtárik, P., Suresh, A.T., Bacon, D.: Federated learning: strategies for improving communication efficiency. CoRR abs/1610.05492 (2016)
9. Melis, L., Song, C., De Cristofaro, E., Shmatikov, V.: Inference attacks against collaborative learning (2018)
10. Ma, X., Zhou, Y., Wang, L., Miao, M.: Privacy-preserving byzantine-robust federated learning. Comput. Stand. Interfaces **80**, 103561 (2022)
11. Mei, Q., Xiong, H., Zhao, Y., Yeh, K.H.: Toward blockchain-enabled IOV with edge computing: efficient and privacy-preserving vehicular communication and dynamic updating. In: 2021 IEEE Conference on Dependable and Secure Computing (DSC), pp. 1–8 (2021)
12. Miao, C., et al.: Cloud-enabled privacy-preserving truth discovery in crowd sensing systems. In: Proceedings of the 13th ACM Conference on Embedded Networked Sensor Systems, SenSys 2015, pp. 183–196. Association for Computing Machinery (2015)
13. Mohassel, P., Zhang, Y.: SecureML: a system for scalable privacy-preserving machine learning. In: 2017 IEEE Symposium on Security and Privacy (SP), pp. 19–38 (2017)
14. Nasr, M., Shokri, R., Houmansadr, A.: Comprehensive privacy analysis of deep learning: passive and active white-box inference attacks against centralized and federated learning (2018)
15. Phong, L.T., Aono, Y., Hayashi, T., Wang, L., Moriai, S.: Privacy-preserving deep learning via additively homomorphic encryption. IEEE Trans. Inf. Forensics Secur. **13**(5), 1333–1345 (2018)
16. Salem, A., Zhang, Y., Humbert, M., Fritz, M., Backes, M.: ML-leaks: model and data independent membership inference attacks and defenses on machine learning models. CoRR abs/1806.01246 (2018)
17. Shayan, M., Fung, C., Yoon, C.J.M., Beschastnikh, I.: Biscotti: a blockchain system for private and secure federated learning. IEEE Trans. Parallel Distrib. Syst. **32**(7), 1513–1525 (2021)
18. Shokri, R., Stronati, M., Song, C., Shmatikov, V.: Membership inference attacks against machine learning models. In: 2017 IEEE Symposium on Security and Privacy (SP) (2017)
19. Shokri, R., Shmatikov, V.: Privacy-preserving deep learning. In: Proceedings of the 22nd ACM SIGSAC Conference on Computer and Communications Security, CCS 2015, pp. 1310–1321. Association for Computing Machinery (2015)
20. Shuvo, M.S.R., Alhadidi, D.: Membership inference attacks: analysis and mitigation. In: 2020 IEEE 19th International Conference on Trust, Security and Privacy in Computing and Communications (TrustCom), pp. 1410–1419 (2020)
21. Wang, L., Xu, S., Wang, X., Zhu, Q.: Eavesdrop the composition proportion of training labels in federated learning. CoRR abs/1910.06044 (2019)

22. Warnat-Herresthal, S., Schultze, H., Shastry, K.L., Manamohan, S., Thirumalaisamy, V.P.: Swarm learning for decentralized and confidential clinical machine learning. Nature **594**(7862), 265–270 (2021)
23. Wu, Y., Lu, X., Su, J., Chen, P.: An efficient searchable encryption against keyword guessing attacks for sharable electronic medical records in cloud-based system. J. Med. Syst. **40**(12), 258 (2016)
24. Xenakis, D., Tsiota, A., Koulis, C.T., Xenakis, C., Passas, N.: Contract-less mobile data access beyond 5G: fully-decentralized, high-throughput and anonymous asset trading over the blockchain. IEEE Access **9**, 73963–74016 (2021)
25. Yang, C.S., So, J., He, C., Li, S., Yu, Q., Avestimehr, S.: LightSecAgg: rethinking secure aggregation in federated learning (2021)
26. Yuan, J., Yu, S.: Privacy preserving back-propagation neural network learning made practical with cloud computing. IEEE Trans. Parallel Distrib. Syst. **25**(1), 212–221 (2014)
27. Zhang, Q., Yang, L.T., Chen, Z.: Privacy preserving deep computation model on cloud for big data feature learning. IEEE Trans. Comput. **65**(5), 1351–1362 (2016)
28. Zhao, Y., et al.: Privacy-preserving blockchain-based federated learning for IoT devices. IEEE Internet Things J. **8**(3), 1817–1829 (2021)

Software Systems and Efficient Algorithms

A Modeling and Verification Method of Modbus TCP/IP Protocol

Jie Wang[1,2](✉), Zhichao Chen[1,2], Gang Hou[1,2], Haoyu Gao[1,2], Pengfei Li[1,2], Ao Gao[1,2], and Xintao Wu[1,2]

[1] School of Software Technology, Dalian University of Technology, Dalian, China
wang_jie@dlut.edu.cn

[2] Key Laboratory for Ubiquitous Network and Service Software of Liaoning Province, Liaoning, China

Abstract. With the informatization of industrial control system, industrial communication protocol is facing greater data pressure. At the same time, industrial communication protocols will face more security threats. In this paper, we use the method of transforming STM (State Transition Matrix) model to UPPAAL (a tool for verifying real-time system) model. In order to clearly understand the model and avoid some mistakes in the early stage of modeling, we first establish STM model for Modbus TCP/IP protocol. Finally, it is transformed into UPPAAL model. Five types of attributes are verified by UPPAAL tool, including the verification of unreachable attributes found by STM modeling. These five types of attributes verify the credibility of Modbus TCP/IP protocol. The experimental results show that this method can have a clear understanding of the model in the early stage. After converting STM model into UPPAAL model, more constraints can be found by referring to STM model. Compared with the existing methods, it studies the credibility of the protocol itself. Therefore, the method can find the root of the problem and solve it.

Keywords: Modbus TCP/IP protocol · UPPAAL · STM model · Model transformation

1 Introduction

Industrial communication protocol supports the operation of industrial control system [1] and has the advantages of high credibility and high real-time [2, 3]. However, with the proposal of made in China 2025 plan, it has brought information transformation to the industrial system. Industrial control system will not only greatly improve productivity, but also face great challenges [9]. The development of industrial system informatization means that industrial control system will contact more equipment and process more data, followed by more security threats. Edge computing [6, 7] and big data technology [8] may be applied to industrial control network protocols in the future, and may bring new problems. In this case, the credibility of industrial communication protocol needs more attention. In recent years, the security of industrial control system has become a hot

Y. Lai et al. (Eds.): ICA3PP 2021, LNCS 13157, pp. 527–539, 2022.
https://doi.org/10.1007/978-3-030-95391-1_33

topic, and the credibility detection of industrial control protocol has become a problem that can not be ignored [4, 5].

[10] proposed a detection method based on unsupervised abnormal learning. They developed a multi agent decentralized intrusion detection system based on ant community clustering model. [11] proposed a detection system based on neural network to detect the system at the physical level. [12, 13] proposed a method for in-depth inspection of traffic, and the detection index is abnormal intrusion. [14–16] proposed a solution for firewall. The argument is that only whitelist packets are allowed to pass through the firewall. [17] present an attacker model that makes use of network reconnaissance afforded by the leaked context in conjunction with formal verification and model checking to arbitrarily reason about the underlying topology and reachability of information flow, enabling targeted attacks. [18] employed a deep learning model to learn the structures of protocol frames and propose a fuzzing framework named SeqFuzzer. It successfully detected several security vulnerabilities.

The methods mentioned above are solve the attack of industrial control system. They discuss how to prevent malicious attacks. This paper focuses on the error detection of Modbus TCP/IP protocol itself. We verify the logic of the protocol through experiments. The credibility of the Modbus TCP protocol itself is the basis. If there is a problem with the protocol itself, detecting attacks becomes less important.

This paper presents a method to detect the credibility of industrial control protocol itself. The purpose of this method is to detect whether there are errors in the industrial control protocol. Firstly, the model is established through STM (State Transition Matrix) modeling to have a clearer understanding of Modbus TCP protocol and reduce the errors of early modeling. Then, the model is transformed to UPPAAL (a tool for verifying real-time system) model. The established model can realize the main functions of Modbus TCP protocol. At the same time, we propose five attributes of industrial control protocol. The unreachable attribute is found by STM modeling. Verify the credibility of Modbus TCP protocol by detecting five attributes of Modbus TCP model in UPPAAL.

2 Modeling Tools

2.1 Introduction to UPPAAL

UPPAAL was jointly proposed by Aalborg University and Uppsala University in 1995. It is suitable for systems that can be described as the product of uncertain parallel processes. The research object is modeled by means of timed automata, and then the state space is searched [19].

UPPAAL is a tool for verifying real-time system [20, 21], which is modeled by formal specification language. It has been used to verify routing protocols, train protocols and wireless sensor networks [22–26]. UPPAAL has three functions: system editor, system simulator and verifier. The function of the system editor is to create a system model. The editor is divided into three parts: variable declaration, system template editing and system definition. The system simulator is a confirmation tool. When loading a program, the system simulator will first check the correctness of the program, report an error if it is wrong, and simulate if it is correct. It can check whether there are errors in the possible execution of the model, and can find some errors before verification. The verifier uses computational tree logic to verify, and quickly detects the errors of clock constraints and logic constraints through depth or breadth search.

In terms of verification, UPPAAL provides a BNF syntax for verification, Prop:: = A [] p | E < > p | E [] p | A < > p | P -- > p. E < > p means possible, E < > p is true, which means that if and only if there is a sequence $S_0 \rightarrow S_1 \rightarrow \ldots \rightarrow S_n$ in the conversion system, S_0 is the start state and Sn is p; A [] p means invariant, which is equivalent to not $E_{< >}$ not p; E [] p stands for potentially always. E [] p is true, which means that if and only if there is a sequence $S_0 \rightarrow S_1 \rightarrow \ldots \rightarrow S_i \rightarrow \ldots$, P is valid in all States S_i, and the sequence is infinite or terminates in state (L_n, V_n), for all D: (l_n, V_{n+d}) or satisfies p and Inv(ln), or there is no conversion from (l_n, v_n); A < > p means eventually, which is equivalent to not E [] not p; p -- > q represents lead to, which is equivalent to A [] (p imply A < > q).

2.2 Introduction to STM

STM (State Transition Matrix) is a state machine modeling method based on table structure [27, 28]. The front-end is in table form, and the back-end has strict formal definition, which is used to model the behavior of software system. The table is divided into three types: state, event and processing cell. The state is the state of the system. Event is the precondition for system change. When the event is triggered, the processing cell is activated. The processing cell stores a series of operations performed after the event is triggered. The processing cell has two special cells: ignore cell and unreachable cell, which are expressed as/and respectively ×◦(Table 1).

Table 1. Sketch map of STM

Modle name	S	state1	state2
E		0	1
event1	0	Process 1-1	/
event2	1	×	Process 2-2

STM modeling mode can display the processing mode of all States under different trigger events at the initial stage of modeling. It enables us to have a clearer understanding of the model, especially the understanding of ignoring cells and unreachable cells, and can avoid many errors in the early stage of modeling. In our development process, the later the error is found, the more difficult it is to correct, so it is necessary to use STM modeling.

3 Structure and Function of Modbus TCP/IP Protocol

Modbus TCP/IP protocol is composed of embedding Modbus protocol into the underlying TCP/IP protocol. It uses master-slave communication technology [29]. The master device is generally called Modbus master and the slave device is generally called Modbus slave. Modbus master mainly includes three functions: generating Modbus request messages, sending request messages to Modbus slave, receiving Modbus slave response message and analyzing it. Modbus slave mainly has three functions: receiving Modbus requests messages, storing and parsing request messages, and sending response messages (Fig. 1).

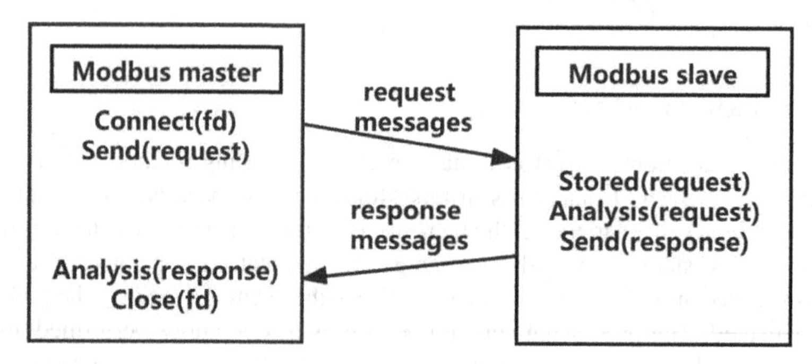

Fig. 1. Schematic diagram of Modbus TCP/IP communication

According to the above analysis of Modbus TCP/IP protocol, this paper proposes four functions of Modbus TCP/IP protocol model:

1. The system has no deadlock;
2. Modbus master can generate request message, process response message and generate processing result;
3. Modbus slave can process the request message, make correct actions after receiving the message and generate the response message;
4. Modbus master and Modbus slave can realize connection and data transmission;

4 Modbus TCP/IP Protocol Model

4.1 Structure Analysis of Modbus TCP/IP Model

Industrial control protocols are generally complex, as is Modbus TCP/IP protocol. The upper layer of Modbus TCP/IP protocol is TCP/IP protocol. This paper does not model it, and finally selects the main function of Modbus TCP/IP protocol as the goal of the model. The model structure takes a Modbus master and a Modbus slave as examples. The key point of Modbus master model implementation is to set the states. According to the main functions described above, three states are selected——MasterWait, MasterExecute and MasterAnalysis. First start from MasterWait, and then enter masterexecute.

MasterExecute is responsible for generating the request message and storing the generated request message, and then enter the analysis states. In this state, Modbusmaster wait for Modbus slave to send the response message. After receiving response message, the Modbus master processes the reply message to judge whether it is an abnormal response, and send different response confirmation according to the response results (Fig. 2).

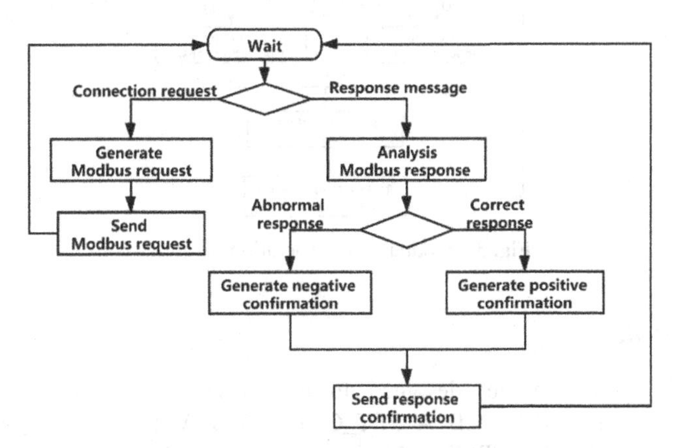

Fig. 2. Modbus master operation diagram

According to the main functions described above, the Modbus slave model selects two states - SlaveWait and SlaveAnalysis. When receiving the request message from the Modbus master, it jumps from the SlaveWait state to the SlaveAnalysis state. The Modbus slave stores the request message, analyzes it after storage, and judges the correctness of the message. If the message is wrong, it generates a Modbus exception response message, Send the abnormal response message to the Modbus master. If the message is correct, judge the function code and select different functions for processing according to the function code. After the message processing is completed, the Modbus slave generates the Modbus response message, and then the Modbus slave sends the response message to the Modbus master. After the sending is completed, it jumps from the SlaveAnalysis state to the SlaveWait state and releases the transaction processing of the Modbus server (Fig. 3).

Modbus PDU is composed of function codes and data. Modbus TCP/IP has a variety of function codes, and there is no essential difference in the implementation of the model. Therefore, it is not necessary to implement each function code in Modbus PDU verification. In this paper, some function codes are selected as cases when we design the model. We finally select function code 01 (reading a coil) and function code 05 (writing a coil). They represent the read-write function of the hardware. They are sufficiently representative.

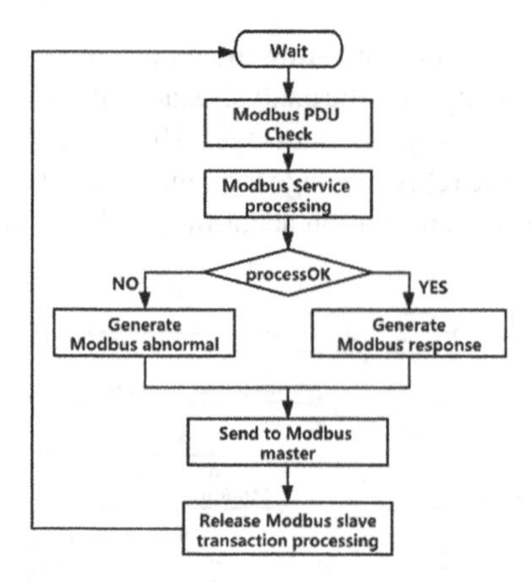

Fig. 3. Modbus slave operation diagram

4.2 STM Model

The Modbus master model designed in this paper selects six event triggers. They are xA_SET_CODE, MB_A_SET_CODE, MB_A_TAG_1, MB_AB_TAG_2, MB_A_TAG_3 and MB_A_TAG_6. Modbus master is in MasterWait at the beginning, and xA_SET_CODE is the startup function of Modbus master. MB_ A_ SET_ Code is set to true. The state of Modbus master jumps to MasterExecute state. In this state, the Modbus master uses the Input() function to input the function code and MB_ A_ TAG_ 1 is set to true (MCC(MB_ A_ CHOOSE_ Code) == 1, 3, 5 and 7 are error and correct function codes 01 and correct and error function codes 05, respectively)Then, in this state, the Modbus master uses the StoreCode() function to store the relevant input data and jump to MasterWait. After Modbus slave sends the response message, MB_ AB_ TAG_ 2 is set to true. In the MasterWait state, the Modbus master jumps to the MasterAnalysis state and MB_ A_ TAG_ 3 set to true. In the MasterAnalysis state, the

Table 2. STM model diagram of modbus master

<table>
<tr><td>Master</td><td>S</td><td>MasterWait</td><td colspan="4">MasterExecute</td><td colspan="4">MasterAnalysis</td></tr>
<tr><td>E</td><td></td><td>0</td><td colspan="4">1</td><td colspan="4">2</td></tr>
<tr><td rowspan="2">xA_SET_CODE</td><td rowspan="2">0</td><td>1</td><td colspan="4">1</td><td colspan="4">2</td></tr>
<tr><td>Start()</td><td colspan="4">xA_SET_CODE =false</td><td colspan="4">xA_SET_CODE =false</td></tr>
<tr><td rowspan="3">MB_A_SET_CODE</td><td rowspan="3">1</td><td rowspan="3">/</td><td>MCC==1</td><td>MCC==3</td><td>MCC==5</td><td>MCC==7</td><td colspan="4" rowspan="3">/</td></tr>
<tr><td>1</td><td>1</td><td>1</td><td>1</td></tr>
<tr><td>Input (1, 1, 3)</td><td>Input (1, 1, 1)</td><td>Input (5, 1, 1)</td><td>Input (5, 3, 1)</td></tr>
<tr><td rowspan="2">MB_A_TAG_1</td><td rowspan="2">2</td><td rowspan="2">/</td><td colspan="4">0</td><td colspan="4" rowspan="2">/</td></tr>
<tr><td colspan="4">StoredCode();</td></tr>
<tr><td rowspan="2">MB_AB_TAG_2</td><td rowspan="2">3</td><td>2</td><td colspan="4" rowspan="2">×</td><td colspan="4" rowspan="2">/</td></tr>
<tr><td>MB_AB_TAG_3=true</td></tr>
<tr><td rowspan="2">MB_A_TAG_3</td><td rowspan="2">4</td><td rowspan="2">/</td><td colspan="4" rowspan="2">/</td><td colspan="4">2</td></tr>
<tr><td colspan="4">AnalysisaB_A()</td></tr>
<tr><td rowspan="3">MB_A_TAG_6</td><td rowspan="3">5</td><td rowspan="3">/</td><td colspan="4" rowspan="3">/</td><td>R!=0</td><td>R!=1</td><td>A>=128</td><td>else</td></tr>
<tr><td>0</td><td>0</td><td>0</td><td>0</td></tr>
<tr><td>C(4)</td><td>C(3)</td><td>C(2)</td><td>C(1)</td></tr>
</table>

Modbus master uses AnalysisB_A() to store the received response message and save MB_ A_ TAG_ 6 set to true. After storage, the Modbus master processes in this state. After processing, the status jumps to MasterWait. The process described above is a cycle of Modbus master model (Table 2).

The Modbus slave model designed in this paper selects four event triggers. They are MB_ AB_ TAG_ 1, MB_ B_ TAG_ CODE_ 1, MB_ B_ TAG_ CODE_ 5 and MB_ AB_ TAG_ 5. Modbus slave is in SlaveWait state at the beginning. When receiving the request message from Modbus master, Modbus slave starts RstoreCode() function. The function of RstoreCode() is to store messages. After the message is stored, the state jumps to SlaveaAnalysis. Modbus slave starts to analyze the message and triggers different events for different function codes. (CB(MB_ PDU_ CODE_ B) == 1 or 3 means that the received function code is 01 function code, and MB_ B_ TAG_ CODE_ 1 is set to true. CB(MB_ PDU_ CODE_ B) == 5 or 7 means that the received function code is 01 function code, then MB_ B_ TAG_ CODE_ 5 is set to true.) After the message processing is completed, the state jumps to the SlaveWait and MB_ AB_ TAG_ 5 is set to true. Finally, Modbus slave sends a response message to Modbus master. The process described above is a cycle of Modbus slave model (Table 3).

Table 3. STM model diagram of modbus slave

<table>
<tr><td>Slave</td><td>S</td><td>SlaveWait</td><td colspan="14">SlaveAnalysis</td></tr>
<tr><td>E</td><td></td><td>0</td><td colspan="14">1</td></tr>
<tr><td rowspan="3">MB_AB_
TAG_1</td><td rowspan="3">0</td><td rowspan="2">1</td><td colspan="2">CB==1</td><td colspan="4">CB==3</td><td colspan="3">CB==5</td><td colspan="3">CB==7</td><td colspan="2">else</td></tr>
<tr><td colspan="2">1</td><td colspan="4">1</td><td colspan="3">1</td><td colspan="3">1</td><td colspan="2">1</td></tr>
<tr><td>RStoreCODE()</td><td colspan="2">MC1=true</td><td colspan="4">MC1=tr
ue</td><td colspan="3">MC5=true</td><td colspan="3">MC5=true</td><td colspan="2">CEE()</td></tr>
<tr><td rowspan="3">MB_B_TAG_
CODE_1</td><td rowspan="3">1</td><td rowspan="3">/</td><td>MB2>=3</td><td colspan="3">MB1+MB2
>=3</td><td colspan="3">M011==0
&&
M012==0</td><td colspan="3">MB1==1
&&
MB2==1</td><td colspan="3">MB1==1
&&
MB2==2</td><td>MB1==2</td></tr>
<tr><td>0</td><td colspan="3">0</td><td colspan="3">0</td><td colspan="3">0</td><td colspan="3">0</td><td>0</td></tr>
<tr><td>C1E1(3)</td><td colspan="3">C1E1(2)</td><td colspan="3">C1E1(4)</td><td colspan="3">C1E2()</td><td colspan="3">C1E2()
MB4=M012</td><td>C1E2()</td></tr>
<tr><td rowspan="3">MB_B_TAG_
CODE_5</td><td rowspan="3">2</td><td rowspan="3">/</td><td colspan="3">MB2==0||
MB2==2^{1201}</td><td colspan="2">MB1>=3</td><td colspan="3">M051==0&&
M052==2^{1201}</td><td colspan="3">MB1==1</td><td colspan="3">MB1==2</td></tr>
<tr><td colspan="3">0</td><td colspan="2">0</td><td colspan="3">0</td><td colspan="3">0</td><td colspan="3">0</td></tr>
<tr><td colspan="3">C5E1(3)</td><td colspan="2">C5E1(2)</td><td colspan="3">C5E1(4)</td><td colspan="3">C5E2()</td><td colspan="3">C5E2()</td></tr>
<tr><td rowspan="2">MB_B_TAG_5</td><td rowspan="2">5</td><td>0</td><td colspan="14" rowspan="2">/</td></tr>
<tr><td>MT2=true</td></tr>
</table>

The STM model completes the functional requirements of Modbus TCP/IP model. Variable names are abbreviated in the above tables. UPPAAL model uses the full name, and the following table is the comparison table of abbreviated names and full names (Table 4).

Table 4. Full name correspondence table

Abbreviated name	Full name	Abbreviated name	Full name
MCC	MB_A_CHOOSE_CODE	R	MB_RSP_MBAP_PROT_A
A	MB_RSP_MBAP_CELL_A	C()	CheckA_B
CB	MB_PDU_CODE_B	MC1	MB_B_TAG_CODE_1
MBi	MB_PDU_DATA_B_i	M0ii	MB_B_DATA_0i_i
CiEi()	CodeiExecutei()	MT2	MB_AB_TAG_2

4.3 Model Transformation

Method of Model Transformation

The transformation from STM model to UPPAAL model mainly includes three parts: the transformation of STM model state, event and processing cell introduced above. The state in the STM model corresponds to the Location in UPPAAL. Events in STM correspond to Guard in edge in UPPAAL. The conditions in the processing cells in STM also correspond to the Guard in the edge in UPPAAL. The function and corresponding variable in the processing cell in STM correspond to the Update in edge in UPPAAL (Fig. 4).

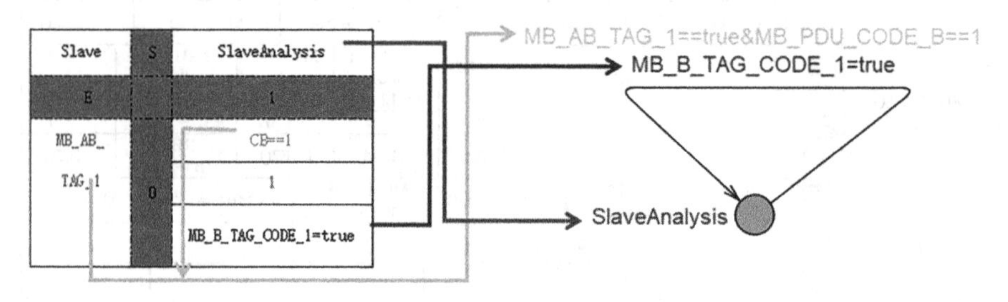

Fig. 4. Model transformation diagram

UPPAAL Modle

See Figs. 5 and 6, Table 5.

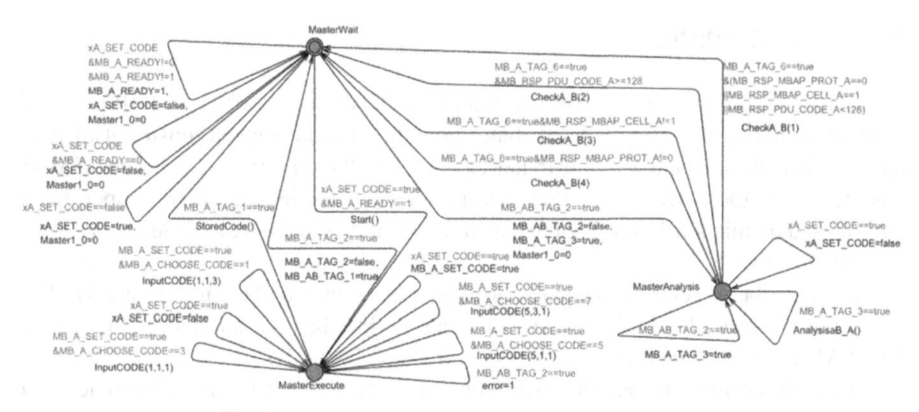

Fig. 5. UPPAAL Model diagram of Modbus master

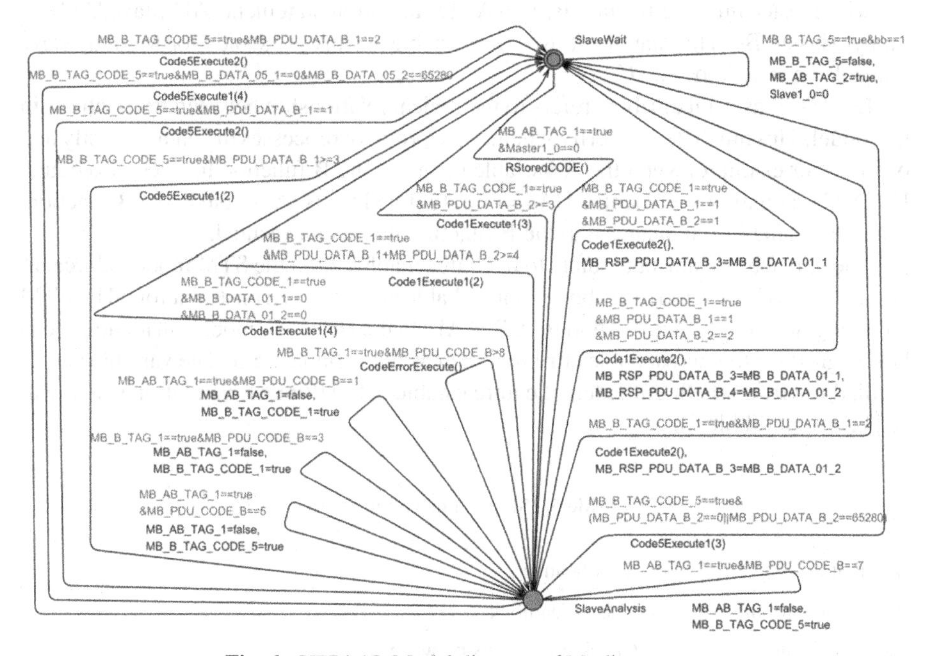

Fig. 6. UPPAAL Model diagram of Modbus master

Table 5. Function table

Function	Significance	Function	Significance
Start()	start function	Input()	Input function code
StoredCode()	Store function code	RStoredCODE()	Restore request message
Code i Execute j()	Analysisa function code i (j: Indicates whether the function code is correct)	AnalysisaB_A()	Analysisa response message
		CheckB_A()	Check message

5 Model Validation

This paper proposes five attribute verification methods for Modbus TCP/IP protocol. They are deadlock attribute, static attribute, dynamic attribute, logical constraint attribute and unreachable attribute. Five attributes can describe the important constraints of Modbus TCP/IP protocol. Deadlock attributes are not described in detail in this paper. Static attributes, dynamic attributes, logical constraint attributes and unreachable attributes are described below.

Static attributes refer to the state correlation between Modbus master and Modbus slave. For example, If Modbus slave is in state A, Modbus master shall be in state B. UPPAAL validation statement: A[]Slave. A imply (Master. B).

Dynamic attribute means that when one of them has a state jump, the other is in a certain state. For example, If the Modbus slave jumps from state A to state C, the Modbus master must be in state B. UPPAAL validation statement: A[] SlaveA_C==1 imply(Master. B). This statement uses the variable slaveA_ C stands for Modbus slave jumps from state a to state C.

Logical constraint attribute refers to the logical relationship between transactions in the model. This attribute can verify whether the protocol processes the data correctly and orderly. For example, when the A variable changes, the B function has been executed. UPPAAL validation statement: A[]A==1 imply (b==1). b is a variable in the B function. Using the value of b indicates that the B function has been executed.

The unreachable attribute refers to the unreachable cell in the STM model. Unreachable cell is has been described above. Unreachable attribute is a constraint found by STM modeling, which is difficult to find for UPPAAL. In the UPPAAL model, an identifier can be designed for the unreachable attribute, such as the variable error. The variable can be verified to prove whether it enters the unreachable cell. UPPAAL validation statement: A[] error==0 (Table 6).

Table 6. Significance of attribute

Attribute	Significance
Deadlock attribute	Check deadlock
Static attribute	Check constraints between states
Dynamic attribute	Check whether the correct jump is performed
Logical constraint attribute	Check whether the corresponding function is executed at each step
Unreachable attribute	Detect whether to enter an unknown location

1. Deadlock attribute
- Deadlock: A [] not deadlock

  Significance: The system has no deadlock

2. Static attributes

- Static1: A[]Slave1.SlaveAnalysis imply (Master1.MasterWait)
  Significance: When the Modbusslave processes the request message, the Modbus master is in the MasterWait.
- Static2: A[] Master1.MasterAnalysis imply(Slave1.SlaveAnalysis)
  Significance: When the Modbus master is in the MasterAnalysis state, the Modbus slave has sent the permission message. But it is in the SlaveAnalysis state incorrectly.

3. Dynamic attributes

- Dynamic1: A[] Master1_0==1 imply(Slave1.SlaveWait)
  Significance: After the Modbus master sends the request message, the Modbus master is in the SlaveAnalysis state incorrectly.
- Dynamic2: A[] Slave1_0==1 imply(Master1.MasterExecute)
  Significance: When Modbus slave sends response message, Modbus master shall be in MasterWait state.

4. Logical constraint attributes

- Logical1: A[]MB_B_TAG_CODE_1==true imply (MB_MBAP_TRAN_B==MB_REQ_MBAP_TRAN_A)
  Significance: MB_ MBAP_ TRAN_ B==MB_ REQ_ MBAP_ TRAN_ A is the internal implementation part of storedcode(). The meaning of this statement is that when the input function code is 01, the request message has been stored when it is processed.
- Logical2: A[] MB_AB_TAG_2==true imply (MB_RSP_PDU_DATA_B_1! = 0)
  Significance: MB_ RSP_ PDU_ DATA_ B_ 1 is the internal implementation part of code I execute j(). The meaning of this statement is that Modbus slave has generated a response message before Modbus master receives the response message.

5. Unreachable attributes

- Unreachable: A[] error==0
  Significance: Modbus master cannot receive and send messages at the same time (Table 7).

Table 7. Significance of attribute

Attribute	Time (s)	Result
Deadlock	0	sat
Static1	0	sat
Static2	0	unsat
Dynamic1	0	sat
Dynamic2	0	unsat
Logical1	0	sat
Logical2	0	sat
Unreachable	0	sat

The experimental results show that static2 and dynamic2 attributes are unsatisfied. The experimental results meet the expected objectives and verify the reliability of Modbus TCP/IP protocol. They also show that the problem of industrial control protocol can be transformed into five types of attributes for verification, and the unsatisfied path can be found by using UPPAAL tool, so as to correct the protocol. UPPAAL can also transform and verify the unreachable attributes found in STM model.

6 Summary

In this paper, we propose our own scheme for the detection of Modbus TCP/IP protocol itself. we use the method of transforming STM model to UPPAAL model. We verify the five attributes of the model. Unreachable attributes are found by STM modeling. The experimental results are satisfactory, which verifies the credibility of Modbus TCP/IP protocol. Therefore, UPPAAL can simulate and verify the industrial control protocol. Using STM modeling method as the basis of early modeling can have a clearer understanding of the model and find more constraints.

Acknowledgments. National Key Research and Development Project (Key Technologies and Applications of Security and Trusted Industrial Control System NO. 2020YFB2009500).

References

1. Galloway, B., Hancke, G.P.: Introduction to industrial control networks. IEEE Commun. Surv. Tutor. **15**(2), 860–880 (2012)
2. Schwartz, M., Mulder, J., Chavez, A.R., et al.: Emerging techniques for field device security. IEEE Secur. Priv. **12**(6), 24–31 (2014)
3. Krotofil, M., Gollmann, D.: Industrial control systems security: what is happening? In: 2013 11th IEEE International Conference on Industrial Informatics (INDIN), pp. 670–675. IEEE (2013)
4. Mo, Y., Kim, T.H.J., Brancik, K., et al.: Cyber–physical security of a smart grid infrastructure. Proc. IEEE **100**(1), 195–209 (2011)
5. Huang, S., Zhou, C.J., Yang, S.H., et al.: Cyber-physical system security for networked industrial processes. Int. J. Autom. Comput. **12**(6), 567–578 (2015)
6. Wang, T., Lu, Y.C., Wang, J.H., et al.: EIHDP: edge-intelligent hierarchical dynamic pricing based on cloud-edge-client collaboration for IoT systems. IEEE Trans. Comput. **70**(8), 1285–1298 (2021)
7. Wang, T., Liu, Y., Zheng, X., et al.: Edge-based communication optimization for distributed federated learning. IEEE Trans. Netw. Sci. Eng. (2021). https://doi.org/10.1109/TNSE.2021.3083263
8. Wu, Y.K., Huang, H.Y., Wu, N.Y., et al.: An incentive-based protection and recovery strategy for secure big data in social networks. Inf. Sci. **508**, 79–91 (2020)
9. Stouffer, K., et al.: Guide to industrial control systems (ICS) security. NIST Spec. Publ. **800**, 82 (2007)
10. Tsang, C., Kwong, S.: Multi-agent intrusion detection system in industrial network using ant colony clustering approach and unsupervised feature extraction. In: 2005 IEEE International Conference on Industrial Technology, pp. 51–56. IEEE (2005)

11. Gao, W., Morris, T., Reaves, B., Richey, D.: On SCADA control system command and response injection and intrusion detection. In: 2010 eCrime Researchers Summit, pp. 1–9. IEEE (2010)
12. Yusheng, W., et al.: IEEE 13th international symposium on autonomous decentralized system (ISADS). IEEE **2017**, 156–162 (2017)
13. Tafto Rodfoss, J.: Comparison of open source network intrusion detection systems (2011)
14. Xu, Y., Yang, Y., Li, T., Ju, J., Wang, Q.: Review on cyber vulnerabilities of communication protocols in industrial control systems. In: 2017 IEEE Conference on Energy Internet and Energy System Integration (EI2), pp. 1–6. IEEE (2017)
15. Khummanee, S., Khumseela, A., Puangpronpitag, S.: Towards a new design of firewall: anomaly elimination and fast verifying of firewall rules. In: The 2013 10th International Joint Conference on Computer Science and Software Engineering (JCSSE), pp. 93–98. IEEE (2013)
16. Cheminod, M., Durante, L., Seno, L., Valenzano, A.: Performance evaluation and modeling of an industrial application-layer firewall. IEEE Trans. Industr. Inf. **14**(5), 2159–2170 (2018)
17. Zhao, H., Li, Z., Wei, H., et al.: SeqFuzzer: an industrial protocol fuzzing framework from a deep learning perspective. In: 2019 12th IEEE Conference on Software Testing, Validation and Verification (ICST), pp. 59–67. IEEE (2019)
18. White, R., Caiazza, G., Jiang, C., et al.: Network reconnaissance and vulnerability excavation of secure DDS systems. In: 2019 IEEE European Symposium on Security and Privacy Workshops (EuroS&PW), pp. 57–66. IEEE (2019)
19. Behrmann, G., David, A., Larsen, K.G.: A tutorial on Uppaal. In: Bernardo, M., Corradini, F. (eds.) Formal Methods for the Design of Real-Time Systems, pp. 200–236. Springer, Heidelberg (2004). https://doi.org/10.1007/978-3-540-30080-9_7
20. Christian, N., Libero, N., Paolo, F.: Sciammarella: modelling and analysis of multi-agent systems using UPPAAL SMC. Int. J. Simulat. Process Model. **13**(1), 73–87 (2018)
21. K, Eun-Young., D, M., H, L.: Probabilistic verification of timing constraints in automotive systems using UPPAAL-SMC. In: Furia, Carlo A., Winter, Kirsten (eds.) Integrated Formal Methods: 14th International Conference, IFM 2018, Maynooth, Ireland, September 5-7, 2018, Proceedings, pp. 236–254. Springer, Cham (2018). https://doi.org/10.1007/978-3-319-98938-9_14
22. Lin, C., Yang, Z., Dai, H., Cui, L., Wang, L., Wu, G.: Minimizing charging delay for directional charging. IEEE/ACM Trans. Netw. **29**(6), 2478–2493 (2021). https://doi.org/10.1109/TNET.2021.3095280
23. Lin, C., Zhou, J.Z., Guo, C.Y., et al.: TSCA: a temporal-spatial real-time charging scheduling algorithm for on-demand architecture in wireless rechargeable sensor networks. IEEE Trans. Mob. Comput. **17**(1), 211–224 (2018)
24. Lin, C., Wang, Z.Y., Deng, J., et al.: mTS: Temporal- and Spatial-Collaborative Charging for Wireless Rechargeable Sensor Networks with Multiple Vehicles, pp. 99–107. IEEE (2018)
25. Lin, C., Shang, Z., Du, W., et al.: CoDoC: A Novel Attack for Wireless Rechargeable Sensor Networks through Denial of Charge, pp. 856–864. IEEE (2019)
26. Lin, C., Zhou, Y., Ma, F., et al.: Minimizing charging delay for directional charging in wireless rechargeable sensor networks. In: IEEE INFOCOM 2019-IEEE Conference on Computer Communications, pp. 1819–1827. IEEE (2019)
27. Modbus, I.: Modbus application protocol specification v1. 1a. North Grafton, Massachusetts (www. modbus. org/specs. php) (2004)
28. Matsumoto, M.: Model checking of state transition matrix. 2nd ITSSV, 2005, pp. 2–11 (2005)
29. Shiraishi, T., Kong, W., Mizushima, Y., et al.: Model checking of software design in state transition matrix. In: SERP 2010: Proceedings of the 2010 International Conference on Software Engineering Research & Practice, Las Vegas, 12–15 July 2010, pp. 507–513 (2010)

Completely Independent Spanning Trees in the Line Graphs of Torus Networks

Qingrong Bian[1,2], Baolei Cheng[1,2(✉)], Jianxi Fan[1], and Zhiyong Pan[1,2]

[1] School of Computer Science and Technology, Soochow University, Suzhou 215006, China
{20195227075,20205227070}@stu.suda.edu.cn, {chengbaolei,jxfan}@suda.edu.cn

[2] Provincial Key Laboratory for Computer Information Processing Technology, Soochow University, Suzhou 215006, China

Abstract. Due to the application in reliable information transmission, parallel transmission and safe distribution of information, and parallel diagnosis algorithm for faulty servers, completely independent spanning trees (CISTs) play important roles in the interconnection networks. So far, researchers have obtained many results on CISTs in many specific interconnection networks, but the results of their line graphs are limited. Some data center networks are constructed based on the line graphs of interconnected networks, such as SWCube, BCDC, AQLCube, etc. Therefore, it is also necessary to study the construction of CISTs in line graphs. A torus network is one of the most popular interconnection networks. The line graph of a torus network is 6-regular, whether there exist 3-CISTs is an open question. In this article, we established the relationship between the completely independent spanning trees in the line graph and the edge division of the original graph. By dividing the edges of the torus network, we can construct three completely independent spanning trees in its line graph in some cases. Some simulation experiments are also implemented to verify the validity.

Keywords: Completely independent spanning trees · Line graph · Interconnection network · Torus network · Algorithm

1 Introduction

Due to the application in reliable information transmission, parallel transmission and safe distribution of information, and parallel diagnosis algorithm for faulty servers [6,8] completely independent spanning trees(CISTs) play important roles in interconnection networks. Therefore, it is very meaningful to construct multiple CISTs in a given interconnection network.

Hasunuma proposed the CIST problem in [8], and he also proposed the following conjecture:

Conjecture 1. Given a $2n$-node-connected interconnection network G with $n \geq 1$, there exist n CISTs in G.

Y. Lai et al. (Eds.): ICA3PP 2021, LNCS 13157, pp. 540–553, 2022.
https://doi.org/10.1007/978-3-030-95391-1_34

For a general graph, it is an NP-hard problem to construct its K completely independent spanning trees, even if $K = 2$ [8].

However, Péterfalvi found a counterexample of it [15]. Thus, the existence and the number of CISTs in a graph is an open and interesting problem, and the construction of CISTs in special networks has received much attention. Constructions of CISTs have been presented in some restricted classes of networks, such as maximal planar networks [8], hypercubes [13], crossed cubes [4], Möbius cubes [14], etc.

The line graph has received considerable attention in recent years. Results have been reported on edge-disjoint Hamilton cycles [12], traceability [17], number of spanning trees [5], structural properties [7], topological indices [16], treewidth [10], etc. The line graphs can be applied to some data center networks by deploying servers at the edge of the original interconnection networks, such as SWCube [11], AQLCube [3], etc.

A torus network is one of the most popular interconnection networks, which can be defined as the Cartesian product of two cycles. Various properties of torus networks have been studied [9,18]. However, few results have been reported on the topic of completely independent spanning trees in their line graphs. In this paper, we propose an algorithm to construct 3-CISTs in the line graph of torus network. By dividing the edges of torus network, 3-CISTs in the line graph of torus network can be constructed in some cases. We mainly obtained the following results:

1. Algorithms to construct CIST-partition in the line graph G of the torus network $T(3, n)(3 \leq n)$ and $T(3m, 3n)(1 \leq m < n)$ are presented.
2. Some simulation results on the line graphs of torus networks to construct 3-CISTs based on python technology are shown.

The rest of this paper is organized as follows. Section 2 provides the terminology and notation used in the paper. In Sect. 3, we present constructions of CIST-partition in the line graph of torus networks in some cases. In the final section, we give the conclusion of the paper.

2 Preliminaries

2.1 Graph Terminology and Notation

An interconnection network can be abstracted as a graph $G(V(G), E(G))$, where $V(G)$ denotes the node set and $E(G)$ denotes the edge set, representing the servers and the links between them, respectively. In this paper, graphs and networks are used interchangeably, and the graphs considered in this paper are undirected, and simple.

Let G be a simple undirected graph. P_1 and P_2 are both the paths between vertices u and v. If P_1 and P_2 have no common edge and no common vertex, then we call P_1 and P_2 are internally disjoint. Let $T_1, T_2, \ldots, T_k$ be spanning trees of G. The paths from u to $v (u \neq v)$ in $T_1, T_2, \ldots, T_k$ are pairwise openly disjoint for any $u, v \in G$, then $T_1, T_2, \ldots, T_k$ are called completely independent spanning trees (CISTs for short) in G.

Let $(V_1, V_2, \ldots, V_k)$ be a partition of the vertex set $V(G)$ and, for $i \neq j$, $B(V_i, V_j, G)$ be a bipartite graph with the edge set $\{uv|uv \in E(G), u \in V_i \text{ and } v \in V_j\}$. If the graph G is clear from the context, we may use $B(V_1, V_2)$ instead of $B(V_1, V_2, G)$. A partition $(V_1, V_2, \ldots, V_k)$ is called a CIST-partition of G if it satisfies the following two conditions: (1) for $i = 1, 2, \ldots, k$, the induced subgraph $G[V_i]$ is connected and (2) for any $i \neq j$, the bipartite graph $B(V_i, V_j)$ has no tree component, that is, every connected component H of $B(V_i, V_j)$ satisfies $|E(H)| \geq |V(H)|$. Araki [1] proved the existence of k CISTs is in fact equivalent to the existence of a CIST-partition. This result plays a key role in our proof.

Theorem 1 [1]. A connected graph G has k completely independent spanning trees if and only if there is a CIST-partition $(V_1, V_2, \ldots, V_k)$ of $V(G)$.

A cycle of length n, where $n \geq 3$, is denoted by C_n. $V(C_n) = \{v_0, v_1, \ldots, v_{n-1}\}$, and $E(C_n) = \{\{v_i, v_{i+1}(\text{mod } n)\}|0 \leq i < n\}$.

Let G and H be two distinct graphs. The Cartesian product $G \times H$ of G and H is defined as follows:
$V(G \times H) = \{(u, v)|u \in V(G), v \in V(H)\}$,
$E(G \times H) = \{\{(u_1, v_1), (u_2, v_2)\}|u_1 = u_2 \text{ and } \{v_1, v_2\} \in E(H), \text{ or } \{u_1, u_2\} \in E(G) \text{ and } v_1 = v_2\}$.

A torus network $T(m, n)$ can be defined as the Cartesian product of two cycles C_m, C_n. When we treat a torus network $C_m \times C_n$, we use symbols u and v for the vertex sets of C_m and C_n, respectively, i.e., $V(C_m) = \{u_0, u_1, \ldots, u_{m-1}\}$ and $V(C_n) = \{v_0, v_1, \ldots, v_{n-1}\}$. A vertex (u_i, v_j) in $C_m \times C_n$ is called the i-row j-column vertex. Also, an edge $\{(u_i, v_j), (u_i, v_{j+1}(\text{mod } n))\}$ (respectively, $\{(u_i, v_j), (u_{i+1}(\text{mod } m), v_j)\}$) in $C_m \times C_n$ is called the i-row edge between the $j, (j+1)(\text{mod } n)$-columns (respectively, the j-column edge between the $i, (i+1)(\text{mod } n)$-rows).

2.2 A Class of Networks—Line Graphs

Given a network G, its *line graph* $LG(G)$ is a graph such that each vertex of $LG(G)$ represents an edge of G and two vertices of $LG(G)$ are adjacent if and only if their corresponding edges share a common endpoint (which are incident) in G. Cheng et al. provided Transformation 1 to demonstrate the construction of a line graph based on an existing network [2].

Transformation 1 [2]. Given a network G, we construct the line graph $LG(G)$ by the following steps:

(1) For every edge started from node x and ended at y in $E(G)$, we add a node $x - y$ to network $LG(G)$, which is referred to as *edge-node*.
(2) For every two adjacent edges (x, y) and (y, z) in G, we connect $x - y$ with $y - z$ in $LG(G)$.

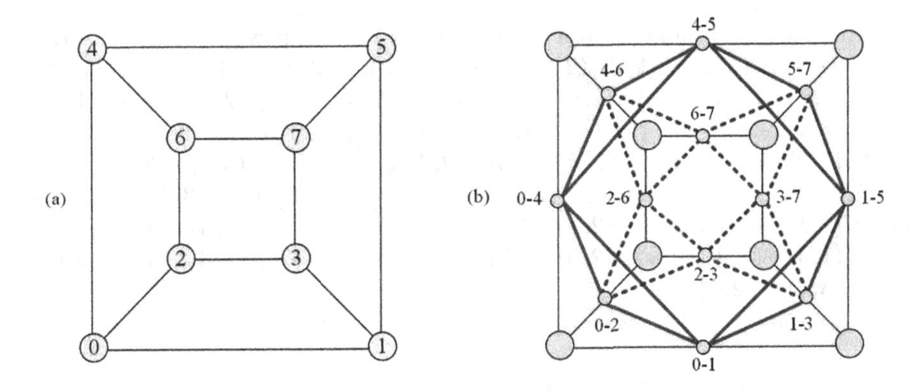

Fig. 1. A network G and its line graph $LG(G)$.

Figure 1 shows the network G and its line graph $LG(G)$. Network $LG(G)$ is derived from network G, where the number of edges in G equals to the number of nodes in $LG(G)$.

Since the vertices in the line graph are the edges in the original graph, the following interesting problem is proposed.

Problem 1. For a torus network, can we divide the edges of it to find the CIST-partition of its line graph.

In the following section, we try to solve this problem by providing a general algorithm for the line graph.

3 CIST-Partition in Line Graphs of Torus Network

3.1 Construction Algorithm of CIST-Partition for Line Graphs of $T(3, n)$ $(n \geq 3)$

According to the definition of line graph $LG(G)$, the vertex of line graph is the edge of the original graph G, so if we want to find the partition of the vertex of $LG(G)$, it is equivalent to find the partition of the edge of G. Now we present an algorithm, called CIST1, to construct the 3-CIST-partition in $LG(T(3, n))(n \geq 3)$. In fact, we divide the edges of torus network as the vertex-partition of $LG(T(3, n))(n \geq 3)$. By Step 1, we divide the edges in $T(3, 3m)(m = n \bmod 3)$. If there is one(or two) column left, we should do some adjustments to the partition in step 2(or 3). Finally, according to step 4 and Transformation 1, the partition of $E(T(3, n))$ is transformed to the CIST-partition in $LG(T(3, n))$.

Example 1. Take $T(3, 7)$ and its line graph, denoted as G for example. Figure 2 shows the construction process of parting edges in torus network. As is shown in Fig. 2(a), edges are divided into three sets after the first steps, in Fig. 2(b), the last edge set is constructed, and adjustment is done on Edge-partition (EP).

The sets obtained after Step 4 is as follow: $\{(1,1)-(1,2), (1,2)-(2,2), (2,2)-(2,3), (2,3)-(3,3), (3,3)-(3,4), (3,4)-(1,4), (1,4)-(1,5), (1,5)-(2,5), (2,5)-(2,6), (2,6)-(2,7), (2,6)-(3,6), (2,7)-(1,7), (1,1)-(3,1), (3,1)-(3,7)\}, \{(2,1)-(2,7), (2,1)-(3,1), (3,1)-(3,2), (3,2)-(1,2), (1,2)-(1,3), (1,3)-(2,3), (2,3)-(2,4), (2,4)-(3,4), (3,4)-(3,5), (3,5)-(1,5), (1,5)-(1,6), (1,6)-(2,6), (1,6)-(1,7), (1,7)-(3,7)\}, \{(1,1)-(1,7), (1,1)-(2,1), (2,1)-(2,2), (2,2)-(3,2), (3,2)-(3,3), (3,3)-(1,3), (1,3)-(1,4), (1,4)-(2,4), (2,4)-(2,5), (2,5)-(3,5), (3,5)-(3,6), (3,6)-(1,6), (3,6)-(3,7), (3,7)-(2,7)\}$.

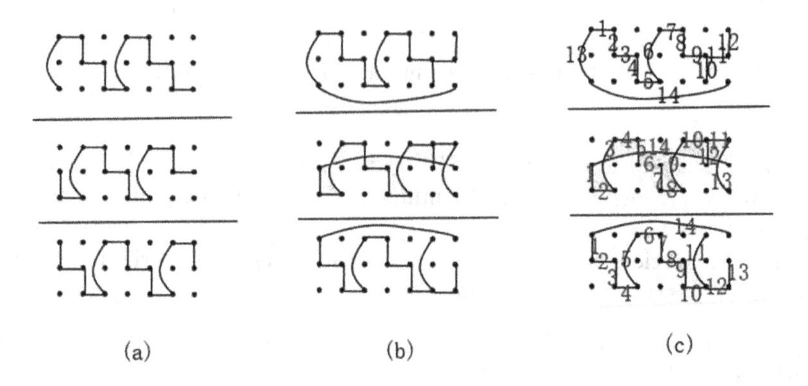

Fig. 2. The edge-partition of $T(3,7)$.

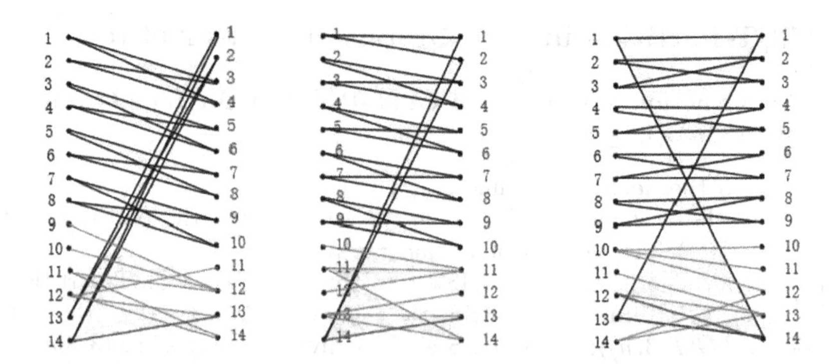

Fig. 3. The CIST-partition in the line graph of $T(3,7)$.

For convenience, we label the vertices in each edge-partition as shown in Fig. 2(c), and use black, red, and purple to represent each partition. Figure 3 shows the sets in G. Obviously, the induced subgraph of each set is connected and any bipartite graph has no tree component. Thus, it is not difficult to verify that the sets above is a CIST-partition of G.

```
Algorithm CIST1
Input: The torus network T(3, n) (n ≥ 3);
Output: A CIST-partition in the line graph of T(3, n);
Begin
Step 1.
1:    for i = 1 to 3 do in parallel
2:       EP[i] = ∅.        /*Initialize partition collection.*/
3:    end for
4:    for i = 1 to 3⌊n/3⌋ step = 3 do in parallel
5:       EP[1] = EP[1] ∪ {(1, i), (1, i+1)}
                  ∪ {(1, i+1), (2, i+1)} ∪ {(2, i+1), (2, i+2)}
                  ∪ {(2, i+2), (3, i+2)} ∪ {(1, i), (3, i)}
                  ∪ {(3, i+2), (3, (i+3)(mod n))}.
6:       EP[2] = EP[1] ∪ {(1, i), (2, i)}
                  ∪ {(2, i), (2, i+1)} ∪ {(2, i+1), (3, i+1)}
                  ∪ {(3, i+1), (3, i+2)} ∪ {(3, i+2), (1, i+2)}
                  ∪ {(1, i+2), (1, (i+3)(mod n))}.
7:       EP[3] = EP[1] ∪ {(2, i), (3, i)}
                  ∪ {(3, i), (3, i+1)} ∪ {(3, i+1), (1, i+1)}
                  ∪ {(1, i+1), (1, i+2)} ∪ {(1, i+2), (2, i+2)}
                  ∪ {(2, i+2), (2, (i+3)(mod n))}.
8:    end for
Step 2.    /*If there is only one column left,
       deal with the extra one column.*/
9: if n mod 3 == 1 do
10:      EP[1] = (EP[1] \ {(3, n-1), (3, 1)})
                  ∪ {(3, n), (3, 1)} ∪ {(2, n-1), (2, n)}
                  ∪ {(2, n), (1, n)}.
11:      EP[2] = (EP[2] \ {(1, n-1), (1, 1)})
                  ∪ {(1, n), (1, 1)} ∪ {(3, n-1), (3, n)}
                  ∪ {(3, n), (2, n)}.
12:      EP[3] = (EP[3] \ {(2, n-1), (2, n)})
                  ∪ {(2, n), (2, 1)} ∪ {(1, n-1), (1, n)}
                  ∪ {(1, n), (3, n)}.
13: end if
Step 3.  /*If there are two columns left, deal with them.*/
14: if n mod 3 == 2 do
15:      EP[1] = EP[1] ∪ {(3, 1), (3, n)} ∪ {(3, n-1), (1, n-1)}
                  ∪ {(1, n-1), (1, n)} ∪ {(1, n), (2, n)}.
16:      EP[2] = EP[2] ∪ {(1, 1), (1, n)} ∪ {(1, n-1), (2, n-1)}
                  ∪ {(2, n-1), (2, n)} ∪ {(3, n), (2, n)}.
17:      EP[3] = EP[3] ∪ {(2, 1), (2, n)} ∪ {(2, n-1), (3, n-1)}
                  ∪ {(3, n-1), (3, n)} ∪ {(3, n), (1, n)}.
18: end if
Step 4.  /*Construct the CIST-partition in
       T(3, n) based on Transformation 1.*/
19: CP = ∅. /*CIST-partition in LG(T(3, n))*/
20: for each set E in EP do
21:      C = ∅.
```

```
22:        for  each edge {(u1,v1),(u2, v2)} in E do
23:            C  =  C ∪ (u1,v1) − (u2, v2).
/* (u1,v1) − (u2, v2) means a node in LG(T(3, n)) */
24:        end for
25:        CP  =  CP ∪ C.
26: end for
27:end
```

3.2 Correctness of the CIST-Partition Obtained by Algorithm CIST1

Now we prove that the set obtained by Algorithm CIST1 is a partition in the line graph G of the torus network $T(3, n)(n \geq 3)$. We present the following theorems.

Theorem 2. *Suppose that EP is the set obtained by Step 2 of Algorithm CIST1, CP is the set obtained by Step 3 of Algorithm CIST1, and G is the line graph of $T(3, n)(n \geq 3)$. EP is a partition of $E(T(3, n))$ and CP is a partition of $V(LG(T(3, n)))$.*

Proof. We have the following cases.

Case 1. $n = 3m(m \in \mathbb{N}^+)$. It is clear that $|EP[j]| = 6m(1 \leq j \leq 3)$, so $\sum_{j=1}^{3} |EP(j)| = 6m \times 3 = 18m = 6n$. By the definition of torus network, $T(3, n)$ has $3n$ vertices, and the degree of each vertex is 4. So $T(3, n)$ has $3n \times 4/2 = 6n$ edges. Thus, $\sum_{j=1}^{3} |EP(j)| = |V(T(3, n))|$. For any $i(1 \leq i \leq m)$, $EP[j] \cap EP[k] = \emptyset(1 \leq j < k \leq 3)$. Thus, EP is a partition of $E(T(3, n))$.

Case 2. $n = 3m + 1(m \in \mathbb{N}^+)$. It is clear that $|EP[j]| = 6m + 2(1 \leq j \leq 3)$ and $EP[j] \cap EP[k] = \emptyset(1 \leq j < k \leq 3)$. $\sum_{j=1}^{3} |EP(j)| = (6m + 2) \times 3 = 6n$. Thus, $\sum_{j=1}^{3} |EP(j)| = |V(T(3, n))|$. Thus, EP is a partition of $E(T(3, n))$.

Case 3. $n = 3m + 2(m \in \mathbb{N}^+)$. The proof is similar to Case 2.

Hence, EP is a partition of $E(T(3, n))$, which implies that CP is a partition of $V(LG(T(3, n)))$ based on Transformation 1. The proof is completed. □

Theorem 3. *Suppose that CP is the set obtained by Step 3 of Algorithm CIST1. The partition CP obtained by Algorithm CIST1 is a CIST-partition in $LG(T(3, n))$.*

Proof. We have the following cases.

Case 1. $n = 3m(m \in \mathbb{N}^+)$. For any $i = 3m'(1 \leq m' \leq m)$, According to Algorithm CIST1, $EP[1] = \cup_{i=1}^{m} \{\{(1, i), (1, i+1)\}, \{(1, i+1), (2, i+1)\}, \{(2, i+1), (2, i+2)\}, \{(2, i+2), (3, i+2)\}, \{(1, i), (3, i)\}, \{(3, i+2), (3, (i+3)(\text{mod } n))\}\}$. We first prove that in the induced subgraph of $G[CP[j]]$ is connected for $1 \leq j \leq 3$. In $CP[1]$, the subset of CP is connected and $(3, i+2)$ is connected to $(3, (i+3)(\text{mod } n))$, so $G[CP[1]]$ is connected. $CP[2]$ and $CP[3]$ are similar to $CP[1]$. Thus, $G[CP[1]]$ is connected after Transformation 1. Then, for each bipartite graph $B(CP[a], CP[b])(1 \leq a < b \leq 3)$, we have the following cases:

Case 1.1. $B(CP[1], CP[2])$ has no tree component. For any $i = 3m'(1 \leq m' \leq m)$, in $B(CP[1], CP[2]), (1, i) - (1, i+1), (1, i+1) - (2, i+1)$ of $CP[1]$ are adjacent to $(1, i+1) - (3, i+1), (1, i+2) - (2, i+2)$ of $CP[2]$, and the number of edges is 4. $(2, i+1) - (2, i+2), (2, i+2) - (3, i+2)$ of $CP[1]$ are adjacent to $(1, i+2) - (2, i+2), (2, i+2) - (2, i+3(\text{mod } n))$ of $CP[2]$. $(1, i) - (3, i), (3, i) - (3, n)$ of $CP[1]$ are adjacent to $(2, i) - (3, i), (3, i) - (3, i+1)$ of $CP[2]$. In this branch $|E| = 4$ and $|V| = 4$, $|E| \geq |V|$. Thus, the bipartite graph $B(CP[1], CP[2])$ has no tree component.

Case 1.2. $B(CP[1], CP[3])$ has no tree component. The proof is similar to Case 1.1.

Case 1.3. $B(CP[2], CP[3])$ has no tree component. The proof is similar to Case 1.1.

Thus, CP obtained by Algorithm CIST1 is a CIST-partition in $LG(T(3, n))$.

Case 2. $n = 3m + 1(m \in \mathbb{N}^+)$. For any $i = 3m'(1 \leq m' \leq m)$, according to Algorithm CIST1, $EP[1] = \cup_{i=1}^{m}\{\{(1, i), (1, i+1)\}, \{(1, i+1), (2, i+1)\}, \{(2, i+1), (2, i+2)\}, \{(2, i+2), (3, i+2)\}, \{(1, i), (3, i)\}, \{(3, n), (3, 1)\}, \{(2, n-1), (2, n)\}, \{(2, n), (1, n)\}\}$. The proof of connectivity is the same as case 1. Then, for each bipartite graph $B(CP[a], CP[b])(1 \leq a < b \leq 3)$, we have the following cases:

Case 2.1. $B(CP[1], CP[2])$ has no tree component. For any $i = 3m'(1 \leq m' \leq m)$, in $B(CP[1], CP[2]), (1, i) - (1, i+1), (1, i+1) - (2, i+1)$ of $CP[1]$ are adjacent to $(1, i+1) - (3, i+1), (1, i+2) - (2, i+2)$ of $CP[2]$, and the number of edges is 4. $(2, i+1) - (2, i+2), (2, i+2) - (3, i+2)$ of $CP[1]$ are adjacent to $(1, i+2) - (2, i+2), (2, i+2) - (2, i+3(\text{mod } n))$ of $CP[2]$. $(1, i) - (3, i), (3, i) - (3, n)$ of $CP[1]$ are adjacent to $(2, i) - (3, i), (3, i) - (3, i+1)$ of $CP[2]$. In the branch $|E| = 4$ and $|V| = 4$, $|E| \geq |V|$. Thus, the bipartite graph $B(CP[1], CP[2])$ has no tree component. In particular, by Step 2, we remove the node $(3, n-1) - (3, n)$ and add $(2, n-1) - (2, n)$ and $(2, n) - (1, n)$ to $CP[1]$. In addition, we remove the node $(2, n-1) - (2, n)$ and add $(1, n-1) - (1, n)$ and $(1, n) - (2, n)$ to $CP[2]$. Thus, we only need to consider whether the subset $\{(2, n-2) - (2, n-1), (2, n-1) - (3, n-1), (2, n-1) - (2, n), (2, n) - (1, n)\}$ of $CP[1]$ and $\{(1, n-1) - (2, n-1)\}, (1, n-1) - (1, n), (1, n) - (3, n), (2, 1) - (2, n)\}$ of $CP[2]$ form tree branches in $B(CP[1], CP[2])$. $(1, n-1) - (2, n-1)$ is adjacent to $(2, n-2) - (2, n-1), (2, n-1) - (3, n-1), (2, n-1) - (2, n)$, and $(2, n-1) - (2, n)$ is adjacent to $(2, 1) - (2, n)$, $(2, 1) - (2, n)$ is adjacent to $(2, n) - (1, n)$, $(2, n) - (1, n)$ is adjacent to $(1, n-1) - (1, n), (1, n) - (3, n)$. In this branch, $|E| = 7$ and $|V| = 8$, fortunately, $(1, n) - (3, n)$ is adjacent to $(3, 1) - (1, n)$. Thus, $\{(2, n-2) - (2, n-1), (2, n-1) - (3, n-1), (2, n-1) - (2, n), (2, n) - (1, n), (3, n) - (3, 1), (1, 1) - (3, 1)\}$ of $CP[1]$ and $\{(1, n-1) - (2, n-1)\}, (1, n-1) - (1, n), (1, n) - (3, n), (2, 1) - (2, n), (2, 1) - (3, 1), (3, 1) - (3, 2)\}$ of $CP[2]$ form a branch. In this branch, $|E| = 12$ and $|V| = 12$, $|E| \geq |V|$. Thus, the bipartite graph $B(CP[1], CP[2])$ has no tree component.

Case 2.2. $B(CP[1], CP[3])$ has no tree component. The proof is similar to Case 2.1.

Case 2.3. $B(CP[2], CP[3])$ has no tree component. The proof is similar to Case 2.1.

Case 3. $n = 3m + 2(m \in \mathbb{N}^+)$. The proof is similar to Case 2.

Hence, CP obtained by Algorithm CIST1 is a CIST-partition in $LG(T(3, n))$.

3.3 Construction Algorithm of CIST-Partition for Line Graphs of $T(3m, 3n)(1 \leq m < n)$

Now we present an algorithm, called CIST2, to construct the CIST-partition in $LG(T(3m, 3n))(1 \leq m < n)$.

```
Algorithm CIST2
Input:  The torus network T(3m, 3n) (1 ≤ m < n);
Output: A CIST-partition in the line graph of T(3m, 3n);
Begin
Step 1.
1:for i = 1 to 3 do in parallel
2:    EP[i]  =  ∅.        /*Initialize partition collection.*/
3:end for
4:for i = 1 to 3m step = 3 do in parallel
5:  for j = 1 to 3n step = 3 do in parallel
6:     EP[1]  =  EP[1] ∪ {(i, j), (i, j+1)}
                   ∪ {(i, j+1), (i+1, j+1)}
                   ∪ {(i+1, j+1), (i+1, j+2)}
                   ∪ {(i+1, j+2), (i+2, j+2)}
                   ∪ {((i+3)(mod 3m), j), (i+2, j)}
                   ∪ {(i+2, j+2), (i+2, (j+3)(mod 3n))}.
7:     EP[2]  =  EP[1] ∪ {(i, j), (i+1, j)}
                   ∪ {(i+1, j), (i+1, j+1)}
                   ∪ {(i+1, j+1), (i+2, j+1)}
                   ∪ {(i+2, j+1), (i+2, j+2)}
                   ∪ {(i+2, j+2), ((i+3)(mod 3m), j+2)}
                   ∪ {(i, j+2), (i, (j+3)(mod 3n))}.
8:     EP[3]  =  EP[1] ∪ {(i+1, j), (i+2, j)}
                   ∪ {(i+2, j), (i+2, j+1)}
                   ∪ {(i+2, j+1), ((i+3)(mod 3m), j+1)}
                   ∪ {(i, j+1), (i, j+2)}
                   ∪ {(i, j+2), (i+1, j+2)}
                   ∪ {(i+1, j+2), (i+1, (j+3)(mod 3n))}.
8:  end for
9:end for
Step 2.  /*Construct the CIST-partition in LG(T(3m, 3n))
```

```
        based on Transformation 1.*/
10: CP = ∅. /*CIST-partition in LG(T(3, n))*/
11: for  each set E in EP do
12:      C = ∅.
13:      for  each edge {(u1,v1),(u2, v2)} in E do
14:          C = C ∪ (u1,v1) − (u2, v2).
/* (u1,v1) − (u2, v2) means a node in LG(T(3m, 3n)) */
15:      end for
16:      CP = CP ∪ C.
17: end for
end
```

3.4 Correctness of the CIST-Partition Obtained by Algorithm CIST2

Now we prove that the set obtained by Algorithm CIST2 is a partition in the line graph G of the torus network $T(3m, 3n)(1 \le m < n)$. We present the following theorems.

Theorem 4. Suppose that EP is the set obtained by Step 1 of Algorithm CIST2, CP is the set obtained by Step 2 of Algorithm CIST2, and G is the line graph of $T(3m, 3n)(1 \le m < n)$. EP is a partition of $E(T(3, n))$ and CP is a partition of $V(LG(T(3, n)))$.

Proof. It is clear that $|EP[k]| = 6mn(1 \le k \le 3)$ and $EP[k] \cap EP[l](1 \le k < l \le 3) = \emptyset$. $\sum_{j=1}^{3} |EP(j)| = 6mn \times 3 = 18mn$. $|V(T(3m, 3n))| = 3m \times 3n \times 4/2 = 18mn$. Thus, $\sum_{j=1}^{3} |EP(j)| = |V(T(3m, 3n))|$. Since $EP[k] \cap EP[l](1 \le k < l \le 3) = \emptyset$. Thus, EP is a partition of $E(T(3m, 3n))$, and CP is a partition of $V(LG(T(3m, 3n)))$.

Theorem 5. Suppose that CP is the set obtained by Step 2 of Algorithm CIST2. The partition CP obtained by Algorithm CIST2 is a CIST-partition in $LG(T(3m, 3n))(1 \le m < n)$.

Proof. By Algorithm CIST2, we obtain $EP[1]$ as follow.

$EP[1] = \cup_{i=1,j=1}^{m,n}\{(i,j),(i,j+1)\},\{(i,j+1),(i+1,j+1)\},\{(i+1,j+1),(i+1,j+2)\},\{(i+1,j+2),(i+2,j+2)\},\{((i+3)(\text{mod } 3m),j),(i+2,j)\},\{(i+2,j+2),(i+2,(j+3)(\text{mod } 3n))\}$.

For any $i = 3m', j = 3n'(1 \le m' \le m, 1 \le n' \le n)$, take $CP[1]$ as an example. The subset of CP is connected, and $(i+2, j+2)$ is connected to $(i+2, (j+3)(\text{mod } 3n))$, $(i+2, j)$ is connected to $((i+3)(\text{mod } 3m), j)$. Thus, $G[CP[1]]$ is connected. $CP[2]$ and $CP[2]$ are similar to $CP[1]$. Thus, $G[CP]$ is connected after Transformation 1. Then, for each bipartite graph $B(CP[a], CP[b])(1 \le a < b \le 3)$, we have the following cases:

Case 1.1. $B(CP[1], CP[2])$ has no tree component. For any $i, j(1 \le i \le m, 1 \le j \le n)$, $\{(i,j)-(i,j+1),(i,j+1)-(i+1,j+1)\}$ of $CP[1]$ and $\{(i+1,j+1)-$

$(i+1, j+2), (i+1, j+1) - (3m, j+1)\}$ of $CP[2]$ form a branch. $\{(i+1, j+1) - (i+1, j+2), (i+1, j+2) - (i+2, j+2)\}$ of $CP[1]$ and $\{(i, j+1) - (i+1, j+2), (i+1, j+2) - (i+1, (j+3) \bmod 3n)\}$ of $CP[1]$ form a branch. $\{(i+2, j) - (i+1, j), (i+2, j) - (i+2, j+1)\}$ of $CP[2]$ and $\{(i, j) - (i, j+1), (i, j) - (3m, j)\}$ of $CP[1]$ form a branch. For each branch, let V and E denote the node set and edge set, respectively. We have $|E|$=4 and $|V|$=4, $|E| \geq |V|$. Thus, the bipartite graph $B(CP[1], CP[2])$ has no tree component.

Case 1.2. $B(CP[1], CP[3])$ has no tree component. The proof is similar to Case 1.

Case 1.3. $B(CP[2], CP[3])$ has no tree component. The proof is similar to Case 1.

Hence, CP obtained by Algorithm CIST2 is a CIST-partition in $LG(T(3m, 3n))$.

3.5 Experiments on CIST-Partition in $LG(T(3m, 3n))(1 \leq m4 < n)$

Here, we adopt python technology to do some simulations in $LG(T(3m, 3n))(1 \leq m5 < n)$. Take $m = 2, n = 3$ as an example. We use the Matplotlib plotting package and Numpy package in Python, where Numpy is used to store each vertex in torus, and draw the edge division obtained by algorithm CIST2 through the Matplotlib plotting package (where the coordinates in row i and column j represent the vertices in row i and column j of torus). The following experimental results were obtained (Fig. 4):

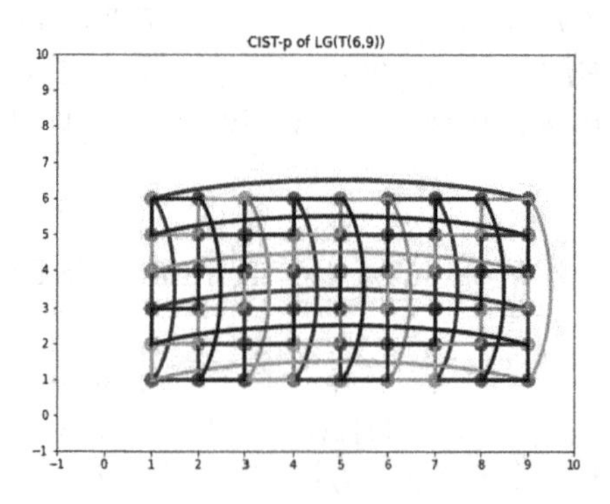

Fig. 4. A CIST-partition in $LG(T(3m, 3n))(1 \leq m < n)$.

From the experimental results, we get EP as follows: $EP[1] = \cup_{i=1,j=1}^{m,n}\{(i, j) - (i, j+1), (i, j+1) - (i+1, j+1), (i+1, j+1) - (i+1, j+2), (i+1, j+2) - (i+2, j+2), ((i+3)(\text{mod } 3m), j) - (i+2, j), (i+2, j+2) - (i+2, (j+3)(\text{mod } 3n))\}$,

$EP[2] = \cup_{i=1,j=1}^{m,n}\{(i, j) - (i+1, j), (i+1, j) - (i+1, j+1), (i+1, j+1) - (i+2, j+1), (i+2, j+1) - (i+2, j+2), (i+2, j+2) - ((i+3)(\text{mod } 3m), j+2)\}$,

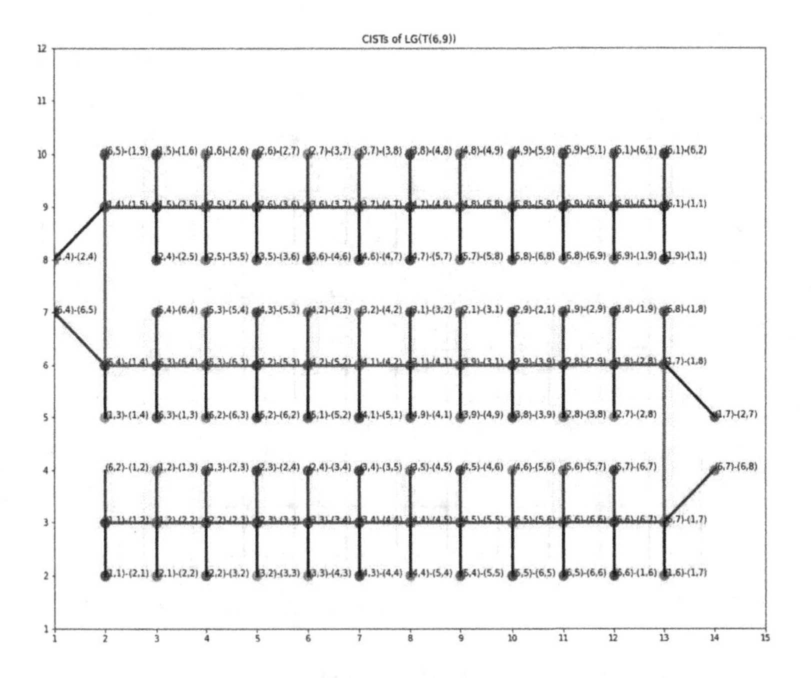

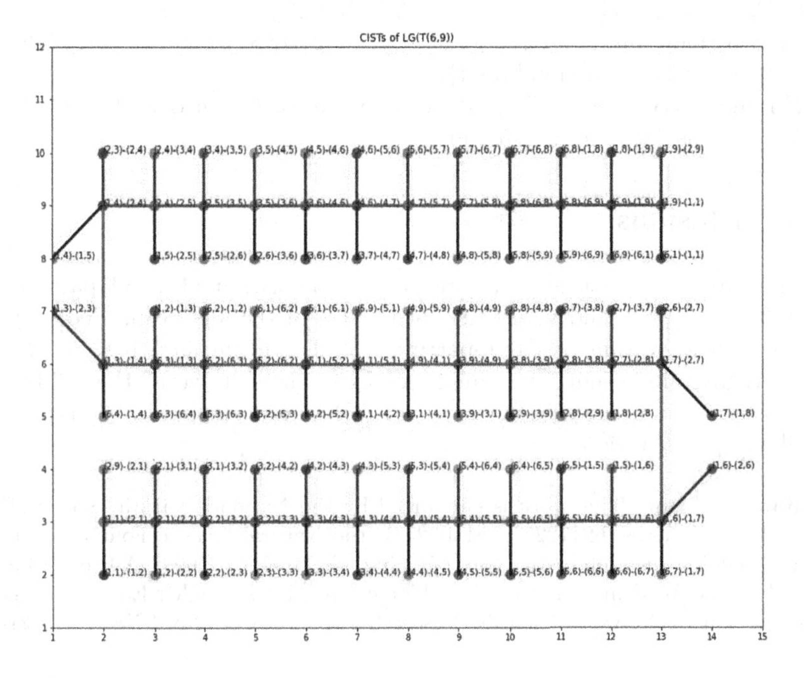

Fig. 5. T_1 and T_2 in CISTs.

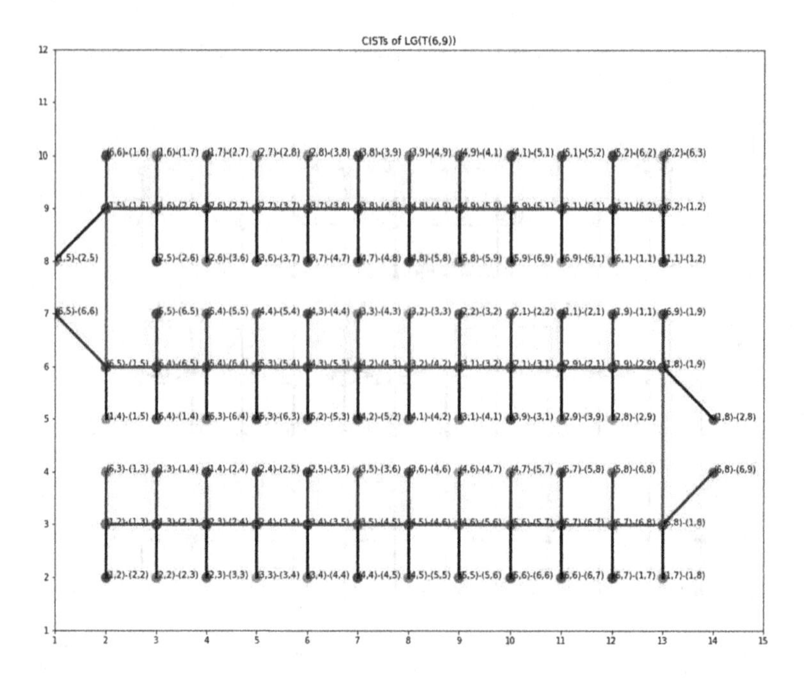

Fig. 6. T_3 in CISTs.

$EP[3] = \cup_{i=1,j=1}^{m,n}\{(i+1,j)-(i+2,j),(i+2,j)-(i+2,j+1),(i+2,j+1)-((i+3)(\text{mod } 3m),j+1),(i,j+1)-(i,j+2))\cup(i,j+2)-(i+1,j+2),(i+1,j+2)-(i+1,(j+3)(\text{mod } 3n))\}$

Through these three CIST-partition, we can easily construct three CISTs of $LG(T(6,9))$, as shown in Fig. 5 and 6.

4 Conclusions

In this paper, we first give a construction algorithm of CIST-partition on $LG(T(3,n))(n \geq 3)$, and verify the correctness of the algorithm. Secondly, the construction algorithm of CIST-partition on $LG(T(3m,3n))(1 \leq m < n)$ is given. We also implement some simulation experiments to verify the validity. For other cases, such as the case where $m = n$, whether there exists CIST-partition is still an open question.

Acknowledgment. This work is supported by the National Natural Science Foundation of China (Nos. 62172291, U1905211), the Natural Science Foundation of the Jiangsu Higher Education Institutions of China (No. 18KJA520009), A Project Funded by the Priority Academic Program Development of Jiangsu Higher Education Institutions, and Future Network Scientific Research Fund Project (No. FNSRFP-2021-YB-39).

References

1. Araki, T.: Dirac's condition for completely independent spanning trees. J. Graph Theory **77**(3), 171–179 (2014)
2. Cheng, B., Fan, J., Li, X., Wang, G., Zhou, J., Han, Y.: Towards the independent spanning trees in the line graphs of interconnection networks. In: Vaidya, J., Li, J. (eds.) ICA3PP 2018. LNCS, vol. 11336, pp. 342–354. Springer, Cham (2018). https://doi.org/10.1007/978-3-030-05057-3_27
3. Chen, G., Cheng, B., Fan, J., Wang, D., Wang, Y.: A secure distribution method for data. Patent No. ZL201910811364.0, 9 July 2021
4. Cheng, B., Wang, D., Fan, J.: Constructing completely independent spanning trees in crossed cubes. Discret. Appl. Math. **219**, 100–109 (2017)
5. Dong, F., Yan, W.: Expression for the number of spanning trees of line graphs of arbitrary connected graphs. J. Graph Theory **85**(1), 74–93 (2017)
6. Hasunuma, T.: Completely independent spanning trees in the underlying graph of a line digraph. Discret. Math. **234**(1), 149–157 (2001)
7. Hasunuma, T.: Structural properties of subdivided-line graphs. J. Discrete Algorithms **31**, 69–86 (2015)
8. Hasunuma, T.: Completely independent spanning trees in maximal planar graphs. In: Goos, G., Hartmanis, J., van Leeuwen, J., Kučera, L. (eds.) WG 2002. LNCS, vol. 2573, pp. 235–245. Springer, Heidelberg (2002). https://doi.org/10.1007/3-540-36379-3_21
9. Hasunuma, T., Morisaka, C.: Completely independent spanning trees in torus networks. Networks **60**(1), 59–69 (2012)
10. Harvey, D.J., Wood, D.R.: Treewidth of the line graph of a complete graph. J. Graph Theory **79**(1), 48–54 (2015)
11. Li, D., Wu, J.: On data center network architectures for interconnecting dual-port servers. IEEE Trans. Comput. **64**(11), 3210–3222 (2015)
12. Li, H., He, W., Yang, W., Bai, Y.: A note on edge-disjoint Hamilton cycles in line graphs. Graphs Comb. **32**, 741–744 (2016)
13. Pai, K.-J., Chang, J.-M.: Constructing two completely independent spanning trees in hypercube-variant networks. Theoret. Comput. Sci. **652**, 28–37 (2016)
14. Pai, K.-J., Chang, R.-S., Chang, J.-M.: A protection routing with secure mechanism in Möbius cubes. J. Parallel Distrib. Comput. **140**, 1–12 (2020)
15. Péterfalvi, F.: Two counterexamples on completely independent spanning trees. Discret. Math. **312**(4), 808–810 (2012)
16. Su, G., Xu, L.: Topological indices of the line graph of subdivision graphs and their Schur-bounds. Appl. Math. Comput. **253**, 395–401 (2015)
17. Tian, T., Xiong, L.: Traceability on 2-connected line graphs. Appl. Math. Comput. **321**, 1339–1351 (2018)
18. Wang, Y., Du, H., Shen, X.: Topological properties and routing algorithm for semi-diagonal torus networks. J. China Univ. Posts Telecommun. **18**(5), 64–70 (2011)

Adjusting OBSS/PD Based on Fuzzy Logic to Improve Throughput of IEEE 802.11ax Network

Chu Xin and Yi-hua Zhu(✉)

School of Computer Science and Technology, Zhejiang University of Technology, Hangzhou 310000, Zhejiang, China
yhzhu@zjut.edu.cn

Abstract. The densely-deployed Wireless Local Area Network (WLAN) exhibit underutilization of radio frequency, low network performance, etc. To enhance frequency spectrum efficiency and improve network performance, the IEEE 802.11ax standard, published in this year, introduces some new technologies, including Orthogonal Frequency Division Multiple Access (OFDMA) and Spatial Reuse (SR). In the SR, Overlapping Basic Service Set (OBSS) Packet Detection (OBSS/PD) is applied to increase parallel transmissions. The standard, however, leaves the unsolved problem of how to set suitable OBSS/PD for the nodes in OBSS. This paper makes efforts to solve this problem and proposes the fuzzy logic based OBSS/PD adjustment (FLOPA) scheme. The proposed FLOPA scheme integrates SR with OFDMA and lets Access Point (AP) maintain the OBSS/PDs for multiple nodes. The FLOPA embeds the fuzzy control system, in which the received signal strength indicator (RSSI), network throughput, and currently-applied OBSS/PD are chosen as input variables and the output variable is used to adjust the OBSS/PD. Simulation results show that the proposed scheme can effectively improve network throughput. Compared with the Carrier Sense Multiple Access with Collision Avoidance (CSMA/CA) scheme in 802.11ax standard, the network throughput can be increased by more than 100%.

Keywords: IEEE 802.11ax standard · Spatial reuse · Carrier sense threshold · Fuzzy control · Orthogonal Frequency Division Multiple Access

1 Introduction

In the Internet of Things (IoT), there are various Wi-Fi devices compatible with IEEE 802.11 standard. IEEE 802.11 standard based Wireless local area networks (WLANs) have been main components of the IoT. According to CISCO's

Supported by National Natural Science Foundation of China (No. 61772470 and 61432015).

Y. Lai et al. (Eds.): ICA3PP 2021, LNCS 13157, pp. 554–570, 2022.
https://doi.org/10.1007/978-3-030-95391-1_35

statistics and prediction, there will be nearly 628 million public Wi-Fi hotspots worldwide by 2023 [1]. As a result, the WLANs tend to be densely deployed. Thus, the coverage of multiple basic service sets (BSSs) share a common area, which forms overlapping basic service sets (OBSS). Then, contention for channel access in the OBSS becomes more serious.

The channel access mechanism adopted by the traditional IEEE 802.11 standard is Carrier Sense Multiple Access with Collision Avoidance (CSMA/CA). The CSMA/CA has exhibited some disadvantages such as low network throughput and insufficient spectrum utilization when it is applied in the densely deployed WLANs.

The IEEE 802.11ax standard [2], which was released in May 2021, aims to improve the performance of densely-deployed WLANs. The standard incorporates several new technologies, including Orthogonal Frequency Division Multiple Access (OFDMA), Multiple Input Multiple Output (MIMO), Target Wake Time (TWT), and Spatial Reuse (SR). The OFDMA divides the whole frequency band into multiple subcarrier sets with each called a Resource Unit (RU) [3]. Access Point (AP) can allocate RUs to multiple mobile stations (STAs) at a time according to channel status and service requirements of the STAs. Thus, an AP can communicate with multiple STAs simultaneously, thereby improving the spectrum efficiency effectively. Meanwhile, OBSS packet detection (OBSS/PD) is adopted in SR, which allows each STA to maintain an OBSS/PD in addition to the default Carrier Sense Threshold (CST) [4] with the traditional CSMA/CA mechanism. The OBSS/PD and the CST are discriminated by the "BSS color" field in the physical layer preamble. Once a STA receives a frame, it immediately checks the BSS color field to see whether it matches its own BSS color. If yes, the received frame is considered as an intra-BSS frame and the STA uses CST to sense the channel state. Otherwise, it is considered an inter-BSS frame, and then the OBSS/PD is used to sense the channel. In general, an OBSS/PD is greater than the default CST at a STA. This increases the concurrent transmission intensity, which helps in improving the spectrum utilization and network throughput.

In the sequel, we refer to a STA or an AP as a node. It should be noticed that, for a node in a BSS, a smaller OBSS/PD, which brings in a high sensitivity, makes a larger sensing area. This may prevent the node from transmitting when a neighboring node belonging to another BSS transmits a frame, which reduces channel utilization. On the other hand, a larger OBSS/PD, which brings in a low sensitivity, makes the node unaware of the ongoing transmission(s), which may cause the node to send data rashly and interfere with data transmissions of other nodes. Therefore, a node faces the challenge of how to set a reasonable OBSS/PD. The IEEE 802.11ax standard specifies the range of OBSS/PD, but does not specify how to set an appropriate OBSS/PD to improve channel efficiency. This paper makes efforts to fill this void in the newly-issued IEEE 802.11ax standard.

Considering we ever succeeded in applying fuzzy logic in Personal Communication System (PCS) Networks [5,6] and fuzzy logic has been widely applied in the Artificial Intelligence (AI)-related applications up to date [7–9], we adopt fuzzy control system in this paper. The main contributions of this paper are as follows:

(1) We propose the fuzzy logic based OBSS/PD adjustment (FLOPA) scheme, which is applicable to the densely-deployed 802.11ax WLANs. The FLOPA takes advantages of the SR and OFDMA, a new feature in IEEE 802.11ax standard. The FLOPA can improve channel efficiency and increase network throughput, in which the AP maintains the OBSS/PDs and CTSs for multiple STAs and initiates parallel transmissions to multiple STAs.
(2) We design a fuzzy control system for the FLOPA. The Received Signal Strength Indicator (RSSI), network throughput, and currently-applied OBSS/PD are chosen as the input of the fuzzy control system. The output of the fuzzy control system is used to adjust the OBSS/PD applied in the future. We use trigonometric function and trapezoidal function as the membership functions and make fuzzy rule for the designed fuzzy control system.
(3) The simulation results show that, compared with the CSMA/CA scheme in 802.11ax standard, the proposed FLOPA scheme can improve the throughput by more than 100%.

The rest of this paper is organized as follows. The related work are surveyed in Sect. 2. The FLOPA scheme is proposed in Sect. 3. The fuzzy control system is elaborated in Sect. 4. The FLOPA scheme is evaluated via simulation in Sect. 5. We conclude the paper in Sect. 6.

2 Related Work

In the densely deployed WLANs, co-channel Interference (CCI) is the main cause for network performance degradation. We have witnessed some newly-published interference-related work. Lee *et al.* [10] propose a link-aware SR scheme, named LSR, that considers the mutual interference and facilitates simultaneous transmissions among OBSSs, and the LSR exploits SR transmission opportunities by selecting the appropriate link for the SR transmission while sufficiently protecting the ongoing transmission. Lanante *et al.* [11] develop an analytical model for IEEE 802.11ax SR and show that the SR gain is tightly linked to the interference range of each BSS in the network.

At present, most of the solutions for relieving the performance degradation resulting from CCI in the OBSS are through adjusting the sensitivity of carrier sensing or controlling the transmit power.

It is feasible to reduce carrier sensing area by adjusting carrier sense threshold, which increases concurrent transmissions. There are many works on dynamic sensitivity control. Afaqui *et al.* [12,13] let a node adaptively adjust its own CST according to the signal reception strength from other nodes. Kulkarni *et al.* [14,15] decrease the CST as packet loss rate increases, which is measured per preset period. Kim *et al.* [16] adjust OBSS/PD based on transmission opportunity to ensure fairness. Lv *et al.* [17] jointly consider frame error rate and transmission opportunity in adjusting CST to improve spatial reuse and ensure fairness. Topal *et al.* [18] adjust CST according to the change in throughput and fairness of the network.

The above works adjust CST without considering whether the receiver succeeds in reception or not. This shortcoming is remedied by some researchers. Selinis *et al.* [19] propose an algorithm that enables nodes to adjust OBSS/PD according to the RSSI from its own BSS and other BSSs. Chau *et al.* [20] propose the adaptive CST adjustment approach that uses the detected hidden terminal and exposed terminal based on the dynamic feedback of nearby transmission. Kim *et al.* [21] set different CSTs based on the combinations of interference sources and receivers, which are obtained through periodic information exchange between APs and STAs.

Most CST adjustment schemes are defaulted to let the nodes use the maximum transmit power, ignoring the distance between AP and STA. This may cause serious interference to the ongoing transmissions. Therefore, Transmit Power Control (TPC) is studied in adjusting CST. Mhatre *et al.* [22] and Yamamoto *et al.* [23] set the pair of transmit power and CST to be inversely proportional, i.e., the product of the transmit power and the CST at each node is set to a constant. Iwai *et al.* [24] propose an algorithm to adjust the pair of transmit power and CST based on AP cooperation, which allows AP to exchange information, including downlink throughput and packet transmission count, with the neighboring APs to ensure throughput and fairness. Ropitault *et al.* [25] use the expected transmission count to dynamically compute the transmit power and CST. Valkanis *et al.* [26,27] adjust the pair of transmit power and CST by taking into account the expected interference at the nodes during concurrent transmission.

In a word, SR technology, which targets at increasing spectrum efficiency, needs to consider the factors other than CST. The FLOPA presented in this paper differs from the existing schemes in that it integrates OFDMA with SR technologies, and uses fuzzy logic to control OBSS/PD.

3 The Proposed FLOPA Scheme

3.1 Network Scenario

The WLAN considered in this paper consisting of m APs, and each AP and its associated STAs form a BSS. Assume all the BSSs operate in the same frequency band, i.e., there exists CCI between BSSs.

Denote m APs by AP_i, $i \in M = \{1, 2, \ldots, m\}$. Let $\Lambda = \{\mathrm{AP}_1, \mathrm{AP}_2, \cdots, \mathrm{AP}_m\}$. The BSS with AP_i is denoted by BSS_i. Let S_i be the set of the STAs in BSS_i, and $\mathrm{STA}_{i,j}$ the j-th STA in S_i. Let $n_i = | S_i |$, which stands for number of the STAs in S_i. Assume the coverage of the m BSSs share a common area, which forms OBSS.

For instance, the OBSS formed by two BSSs is illustrated in Fig. 1, where a circle with solid line represents the sensing range of an AP when the default CST is used, and the one with dotted line represents the sensing range of the AP when OBSS/PD is applied.

It can be seen from the figure, in the case when both AP_1 and AP_2 adopt the default CST, there are 8 STAs in the OBSS, i.e., the common sensing area

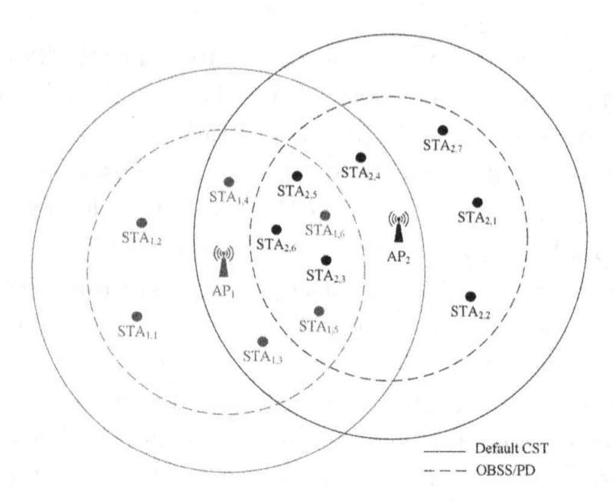

Fig. 1. OBSS WLAN

of AP_1 and AP_2. They are $STA_{1,3}$, $STA_{1,4}$, $STA_{1,5}$, $STA_{1,6}$, $STA_{2,3}$, $STA_{2,4}$, $STA_{2,5}$, $STA_{2,6}$. In the case when both AP_1 and AP_2 adopt OBSS/PD, there are 5 STAs, namely $STA_{1,5}$, $STA_{1,6}$, $STA_{2,3}$, $STA_{2,5}$, $STA_{2,6}$, in the OBSS. Clearly, the default CST has a wider sensing area than the OBSS/PD. Thus, the AP using the default CST has a higher probability of sensing a busy channel, which brings a smaller chance to the AP for gaining the channel and thus the channel utilization may be degraded. For example, at the time when $STA_{2,4}$ is transmitting a packet to AP_2, if AP_1 that adopts the default CST has a packet for $STA_{1,2}$, AP_1 senses a busy channel and then stops contending for the channel. In fact, AP_1 can send packet to $STA_{1,2}$ at this time, because no collision occurs at the receiver. In contrast, when both APs use OBSS/PD, $STA_{2,4}$ is not within the sensing range of AP_1, so AP_1 senses an idle channel and then transmits the packet without deferring.

This example shows that: 1) using the default CST may result in low spectrum utilization and low network throughput; and 2) using an OBSS/PD larger than the default CST, which generates a smaller sensing area, enables multiple pairs of nodes to communicate simultaneously and thus improves spectrum utilization and network throughput.

However, too larger a OBSS/PD leads to interference, which may degrade throughput. Hence, we need to carefully design a scheme for the nodes so that the nodes can set their OBSS/PDs to the values that improve network throughput. Our FLOPA scheme meets this requirement, which adjusts the STAs' OBSS/PDs by fuzzy logic so that network throughput is considerably improved.

3.2 Carrier Sensing Range and Condition of Successful Packet

According to the two-ray path loss model [13], the signal power that $STA_{i,j}$ receives from AP_i can be represented by

$$P_{i,j} = \frac{T_i G_i G_{i,j} h_i^2 h_{i,j}^2}{d_{i,j}^4 \gamma} \tag{1}$$

where T_i is the transmit power of AP_i, G_i is the antenna gain at AP_i, $G_{i,j}$ is the antenna gain at $\mathrm{STA}_{i,j}$, h_i and $h_{i,j}$ represent antenna height of AP_i and $\mathrm{STA}_{i,j}$, respectively, $d_{i,j}$ is the distance between AP_i and $\mathrm{STA}_{i,j}$, and γ is the attenuation coefficient.

Each time an AP intends to initiate a transmission, it needs to contend for the channel by executing CSMA/CA, in which the Clear Channel Assessment (CCA) judges whether the channel is idle or not.

In the case when AP_k interferes with AP_i, the carrier sensing range of AP_i can be expressed as [13]

$$\mathrm{CSR}_i = (\frac{T_k G_k G_i h_k^2 h_i^2}{\mathrm{CST}_i \gamma})^{\frac{1}{4}}. \tag{2}$$

In (2), CST_i is the CST used by AP_i; and T_k, G_k, and h_k are the transmit power, antenna gain, and antenna height at AP_k, respectively. It can be clearly seen from (2) that a larger CST applied at AP_i reduces its carrier sensing range. Usually, a larger CST is adopted when increasing concurrent transmission intensity. However, this may bring in more interference between the communication parties. Therefore, either an AP or a STA needs to carefully choose an appropriate CST.

For $\mathrm{STA}_{i,j}$ to successfully receive a packet, the following two conditions are required [17]:

$$\begin{cases} P_{i,j} \geq R_S, \\ \mathrm{SINR}_{i,j} = \frac{P_{i,j}}{I_{i,j}+N_0} \geq R_{\mathrm{SINR}}, \end{cases} \tag{3}$$

where $P_{i,j}$ stands for the power from AP_i being received at $\mathrm{STA}_{i,j}$, R_S is the minimum sensitivity of successful decoding the received signal, $\mathrm{SINR}_{i,j}$ is the Signal-to-Interference-and-Noise Ratio (SINR) at $\mathrm{STA}_{i,j}$, R_{SINR} is the SINR threshold, N_0 is the noise power, and the interference power at STA_i is $I_{i,j} = \sum_{k \in M \setminus i} P_{k,j}$.

The first condition in (3) desires the received signal strength higher than the receiver sensitivity, while the second one requires the SINR higher than the SINR threshold.

3.3 The FLOPA Scheme

IEEE 802.11ax standard limits transmit power at the nodes when OBSS/PD is in use, which aims to relieve interference with the ongoing transmissions in the network. The relation between transmit power and OBSS/PD is shown in Fig. 2, where T and T_{ref} stand for transmit power and reference power, respectively.

The IEEE 802.11ax standard regulates the relation between transmit power and OBSS/PD as follows:

$$T_{\max} = T_{\mathrm{ref}} - (\mathrm{OBSS/PD} - \mathrm{OBSS/PD}_{\min}), \tag{4}$$

$$\mathrm{OBSS/PD}_{\min} \leq \mathrm{OBSS/PD} \leq \mathrm{OBSS/PD}_{\max}, \tag{5}$$

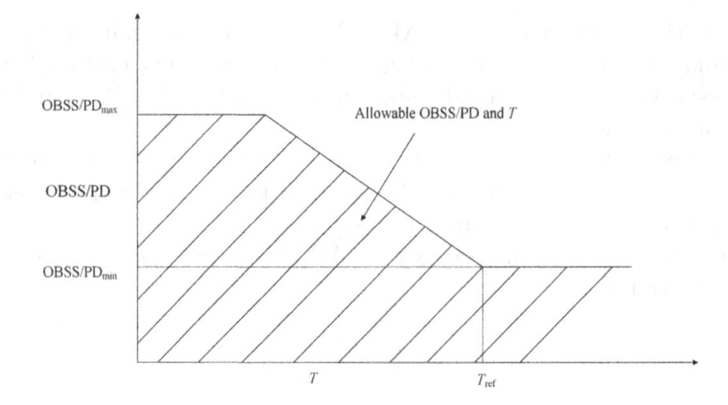

Fig. 2. Relation between OBSS/PD and transmit power [2]

where the subscriptions of "min" and "max" represent the minimum and the maximum, respectively.

In the case of 20 MHz channel bandwidth, the standard sets $\mathrm{OBSS/PD_{min}} = -82$ dBm, $\mathrm{OBSS/PD_{max}} = -62$ dBm, and $T_{\mathrm{ref}} = 21$ dBm.

As mentioned in Sect. 3.1, increase in OBSS/PD can reduce the carrier sensing range, and thus increases concurrent transmission intensity. However, a too high OBSS/PD causes the AP unable to sense the other nodes' transmissions. Therefore, it is very important to set reasonable OBSS/PDs for the nodes.

Unfortunately, the IEEE 802.11ax standard only provides equations (4) and (5) to limit the range of OBSS/PD, but it leaves the unsolved problem of how to set appropriate OBSS/PDs for the nodes.

Nodes are usually with different environment, which causes the nodes to have different probabilities of successfully receiving the data frame transmitted by its AP. With this consideration, our FLOPA lets the AP to maintain multiple OBSS/PD values for different STAs at the same time, and the AP uses the OBSS/PD that corresponds to the STA to which the AP transmits packets. The AP uses the "Information Table", simply referred to as the "InfoTable" below, to record the information relevant to each STA. The InfoTable is shown in Table 1. It contains four fields of "STA-ID", "RSSI", "THR", and "OBSS/PD", which keep the identification of a STA, RSSI at the STA, throughput of the BSS in which the STA stays, and the OBSS/PD for the STA, respectively.

For example, the second tuple in Table 1 means that $\mathrm{STA}_{i,2}$ has $\mathrm{RSSI}_{i,2}$ and uses $\mathrm{OBSS/PD}_{i,2}$, and the BSS that contains $\mathrm{STA}_{i,2}$ has throughput of θ_i. In the table, $\mathrm{RSSI}_{i,j}$ is the received signal strength of $\mathrm{STA}_{i,j}$, θ_i is the throughput of BSS_i, and $\mathrm{OBSS/PD}_{i,j}$ represents the OBSS/PD used when transmitting data to $\mathrm{STA}_{i,j}$. The RSSI of a STA is piggybacked in the packet destined to the AP from the STA.

Figure 3 shows the flow chart of the FLOPA scheme, where N_{RU} is the number of available RUs, and N_{STA} stands for the number of the "eligible STAs". Here, an eligible STA is defined as the one satisfying that: 1) they need to receive

Table 1. STA information table.

STA-ID	RSSI (dBm)	THR (Mbps)	OBSS/PD (dBm)
$STA_{i,1}$	$RSSI_{i,1}$	θ_i	$OBSS/PD_{i,1}$
$STA_{i,2}$	$RSSI_{i,2}$	θ_i	$OBSS/PD_{i,2}$
...	...	...	...

data from the AP; and 2) their OBSS/PDs kept in the InfoTable are higher than the signal energy of inter-BSS frame. Notation N is the number of STAs that need to receive data from the AP. In addition, K is the variable to hold the times of transmissions the AP has conducted, which is used to limit the frequency of adjusting OBSS/PDs for the STAs. That is, once $K = TxTimes$, the fuzzy control system in the FLOPA is applied to adjust the OBSS/PDs, where $TxTimes$ is a preset constant. Moreover, $SuccPkts$ is a variable that holds the number of packets successfully transmitted.

The main steps in the FLOPA scheme are explained as follows:

Step 1. When AP_i has a packet to transmit, it senses the channel status. If a packet is detected, it uses BSS Color technology to recognize the frame as an intra-BSS frame or an inter-BSS frame. For an intra-BSS frame, AP_i extracts the STA's RSSI information carried in the frame and updates the RSSI field in the InfoTable. For an inter-BSS frame, AP_i checks the InfoTable and counts the number of the eligible STAs, i.e., N_{STA}. If N_{STA} is greater than the number of available RUs (i.e., N_{RU}), the AP selects the N_{RU} STAs with the largest OBSS/PDs. If not, AP_i selects N_{STA} STAs. The AP sets its transmit power, which is limited by (4), and then transmits data to the selected STAs via downlink multiple user OFDMA.

Step 2. If AP_i receives an ACK from a STA, which means the STA receives the AP's packet, the AP updates the variable $SuccPkts = SuccPkts+1$. As soon as the AP conducts $TxTimes$ times of transmissions, it updates the InfoTable and calculates the throughput of BSS_i as

$$\theta_i = \frac{SuccPkts \times L}{t}, \tag{6}$$

where L is the length of MAC Protocol Data Units (MPDU), t is the duration from the first transmission to completion of the $TxTimes$-th transmission.

Step 3. AP_i applies the fuzzy control system to adjust the OBSS/PD of the STAs, and updates their OBSS/PDs in the InfoTable. The fuzzy control system will be elaborated in Sect. 4.

Next, we take Fig. 4 as an example to illustrate the usage of different OBSS/PDs in the FLOPA scheme, in which we assume that: 1) AP_1 and AP_2 operate in the same frequency band; 2) AP_1 and AP_2 are each in carrier sensing area of the other when they use the default CST; 3) there are 5 available RUs at AP_1; 4) the InfoTable at AP_1 is given in Table 2; and 5) the transmit power of AP_2 is sensed of -78 dBm by AP_1.

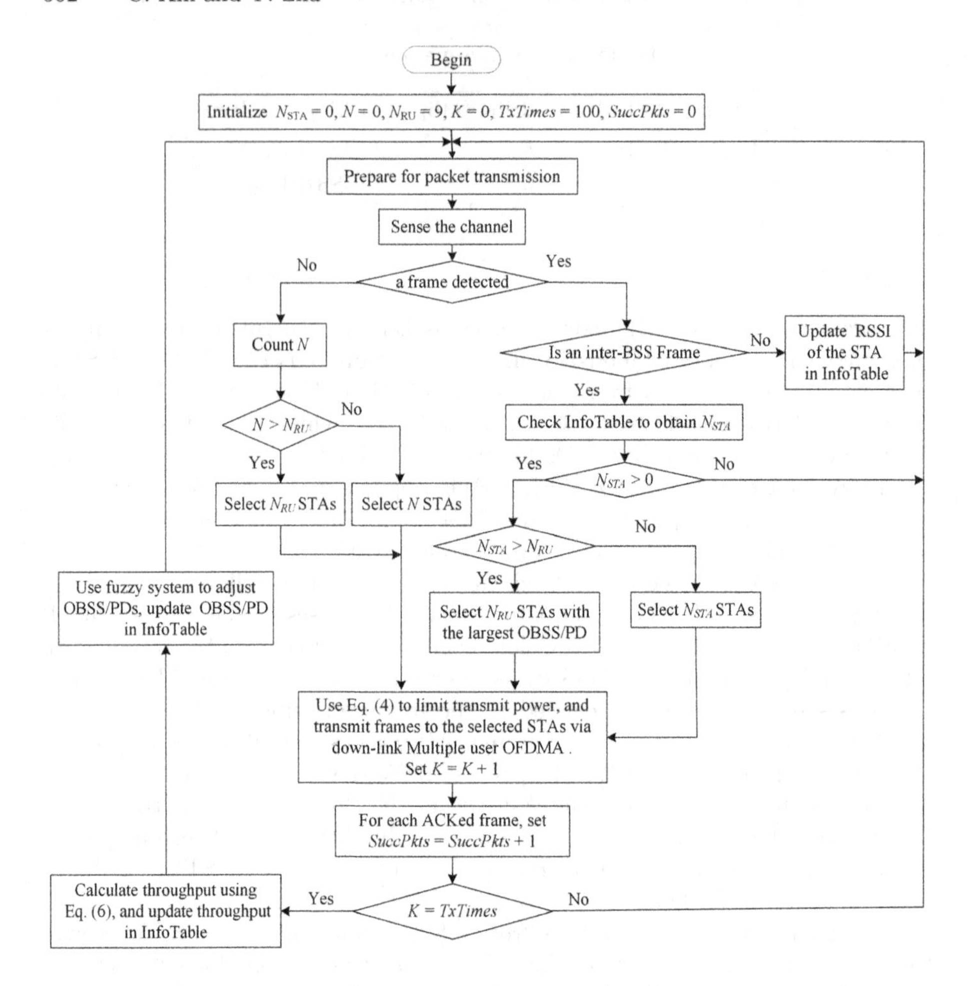

Fig. 3. Flow chart of the FLOPA scheme

Table 2. The STA information table at AP_1.

$STA_{i,j}$	$RSSI_{i,j}$(dBm)	θ_i(Mbps)	$OBSS/PD_{i,j}$(dBm)
$STA_{1,1}$	−66	21	−70
$STA_{1,2}$	−69	21	−75
$STA_{1,3}$	−80	21	−79
$STA_{1,4}$	−76	21	−82

Suppose AP_1 has packets for the associated STAs when AP_2 is transmitting frames to $STA_{2,1}$ and $STA_{2,2}$. Then, AP_1 detects a frame from AP_2 when it senses the channel, and AP_1 recognizes the frame as inter-BSS frame by reading the BSS color field in the frame. Therefore, AP_1 uses OBSS/PD to determine channel status. After checking the InfoTable, AP_1 considers $STA_{1,1}$ and $STA_{1,2}$ as the eligible STAs since their OBSS/PDs are greater than the signal power of AP_2, i.e., -78 dBm. Then, AP_1 transmits data to $STA_{1,1}$ and $STA_{1,2}$, but not to $STA_{1,3}$ and $STA_{1,4}$ due to their OBSS/PDs being no greater than -78 dBm.

Here, we would like to stress that the STAs' OBSS/PDs are dynamically adjusted by the fuzzy control system in the FLOPA scheme, which sets eligible STAs dynamically.

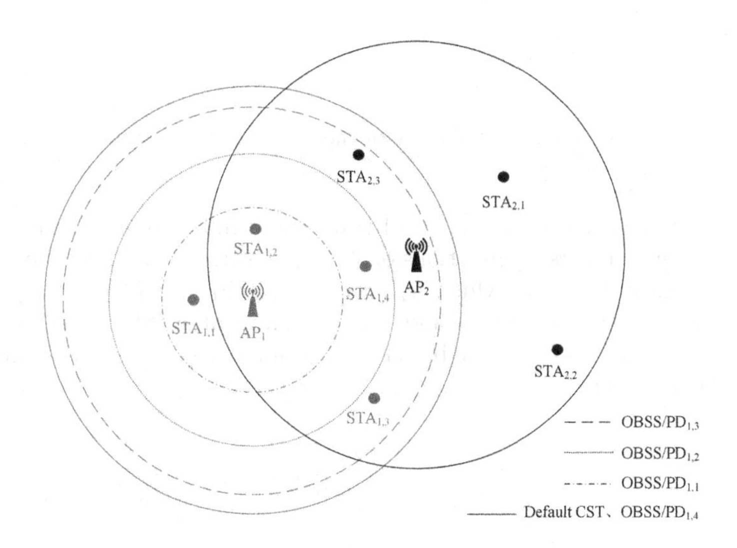

Fig. 4. Usage of different OBSS/PDs

4 Fuzzy Control System

The architecture of the fuzzy control system in the FLOPA scheme is shown in Fig. 5. It has three input variables R, TP, and PD, and one output variable $\triangle PD$. These variables are defined in Table 3.

Table 3. Input and output variables in the fuzzy control system

Variable	Definition
R	RSSI of the signal from the AP
TP	Throughput of BSS
PD	OBSS/PD used for a STA
$\triangle PD$	The change amount in OBSS/PD

As in our previous work [5,6], we adopt the Mamdani fuzzy model [28]. The fuzzy sets of the input variables R and TP are each set to {"Low", "Medium",

"High"}, simply expressed as {L, M, H}; The fuzzy set of the input variable *PD* is set to {"Very Low", "Low", "Medium", "High", "Very High"}, simply as {VL, L, M, H, VH}; and that of the output variable $\triangle PD$ is set to {"Decrease", "Decrease Slightly", "No Change", "Increase Slightly", "Increase"}, simply as {D, DS, NC, IS, I}.

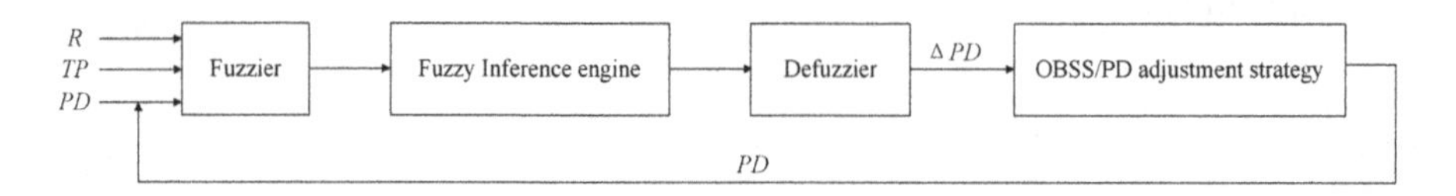

Fig. 5. The architecture of the fuzzy control system

We use trigonometric function and trapezoidal function as the membership functions. The membership functions of the input variables and the output variable are shown in Fig. 6, in which R_L, R_M, R_H, TP_M and TP_H are key points. R_H and R_L are the maximum and the minimum of RSSI, respectively, and $R_M = \frac{1}{2}(R_H - R_L)$. TP_H is set based on the number of STAs associated with the AP, $TP_M = \frac{1}{2}TP_H$.

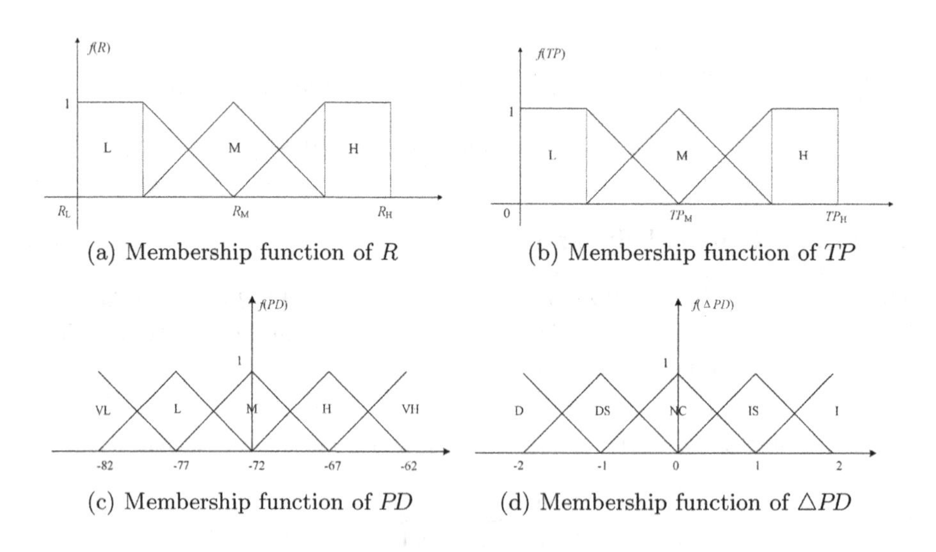

Fig. 6. Membership functions of the fuzzy variables

Generally, if the RSSI at a STA is larger, the STA is more likely to receive packets successfully, and thus a higher OBSS/PD should be set for the STA to increase the chance of concurrent transmissions and vice versa. In addition, if the BSS's throughput and the OBSS/PD of STA is low, the OBSS/PD should be increased to raise the transmission opportunities. If the STA has a high OBSS/PD and a low throughput, the OBSS/PD should be decreased. With the above considerations, we establish the fuzzy rules shown in Table 4.

The maximum-minimum fuzzy inference method is used to obtain the weight of the fuzzy output variable, and after defuzzification the $\triangle PD$ is used to adjust the OBSS/PD according to the following rules:

1) When $-2 \leq \triangle PD \leq -1.5$, $PD = \max\{PD - 2, -82\}$ dBm;
2) When $-1.5 \leq \triangle PD \leq -0.5$, $PD = \max\{PD - 1, -82\}$ dBm;
3) When $-0.5 \leq \triangle PD \leq 0.5$, PD remains unchanged;
4) When $0.5 \leq \triangle PD \leq 1.5$, $PD = \min\{PD + 1, -62\}$ dBm;
5) When $1.5 \leq \triangle PD \leq 2$, $PD = \min\{PD + 2, -62\}$ dBm.

Table 4. Fuzzy rules.

No.	IF			THEN	No.	IF			THEN	No.	IF			THEN
	R	TP	PD	$\triangle PD$		R	TP	PD	$\triangle PD$		R	TP	PD	$\triangle PD$
1	L	L	VL	IS	16	M	L	VL	I	31	H	L	VL	I
2	L	L	L	IS	17	M	L	L	IS	32	H	L	L	I
3	L	L	M	IS	18	M	L	M	IS	33	H	L	M	IS
4	L	L	H	D	19	M	L	H	DS	34	H	L	H	DS
5	L	L	VH	D	20	M	L	VH	D	35	H	L	VH	DS
6	L	M	VL	IS	21	M	M	VL	IS	36	H	M	VL	I
7	L	M	L	IS	22	M	M	L	IS	37	H	M	L	IS
8	L	M	M	NC	23	M	M	M	NC	38	H	M	M	NC
9	L	M	H	DS	24	M	M	H	DS	39	H	M	H	DS
10	L	M	VH	D	25	M	M	VH	DS	40	H	M	VH	DS
11	L	H	VL	IS	26	M	H	VL	IS	41	H	H	VL	IS
12	L	H	L	IS	27	M	H	L	IS	42	H	H	L	IS
13	L	H	M	NC	28	M	H	M	NC	43	H	H	M	NC
14	L	H	H	DS	29	M	H	H	DS	44	H	H	H	NC
15	L	H	VH	D	30	M	H	VH	DS	45	H	H	VH	DS

5 Performance Evaluation

In this section, we evaluate our FLOPA via simulation. The simulation programs are written in MATLAB. We assume that: 1) The m APs operate in the 20 MHz bandwidth channel with 9 available RUs [2]; 2) The number of STAs associated with each AP is 20; and 3) The data packet arrivals at the AP follow Poisson distribution with parameter λ [29–31]. For the sake of ease description, we denote the reciprocal of λ as $w = 1/\lambda$, which stands for the average packet arrival interval.

We use a square with side of 300 meters as the common area of the OBSS formed by the m BSSs. The position of the first AP is fixed in the center of the square, and the other APs are located randomly in the square. In addition, for each AP, its associated STAs are randomly located in coverage of the AP.

The initial OBSS/PDs of APs are set to −82 dBm, and the default transmit power is set to 20 dBm [2]. Moreover, the other simulation parameters are given in Table 5.

Table 5. Main parameters for simulation.

Parameter	Value	Parameter	Value
Bandwidth	20 MHz	DIFS duration	34 μs [4]
Length of data packets	12000 bits [4]	SIFS duration	16 μs [4]
Slot duration	9 μs [4]	BA duration	32 μs [4]
Transmission/Reception gain	0/−2 dB [27]	Minimum contention window	15 [2]
Antenna height	1 m [13]	Maximum contention window	63 [2]
Background noise power	−95 dBm [4]	Allowed OBSS/PD levels	−82 to −62 dBm [2]
Data rate of single RU	8.8 Mbps [2]	Allowed transmit power levels	1 to 20 dBm [2]

We compare the proposed FLOPA scheme with the CSMA/CA scheme in the IEEE 802.11ax standard and the scheme that uses OFDMA with fixed CST, denoted by Fixed-CST, in terms of throughput. We define Percentage of Increased Throughput (PoIT) as $(a - a_1)/a_1$, where a and a_1 represent the throughput of our FLOPA and that of either the Fixed-CST or the CSMA/CA, respectively. The following results are from the average of 10 simulation runs with each mimicking packet delivery for 2 s. In simulation, clock progresses every time slot, and a slot lasts for 9 μs. In summary, each simulation run mimics packet delivery in period of 2×10^6 μs, where simulation clock ticks per 9 μs.

First, fixing $w = 15$, and setting the number of AP, i.e., m, to 5, 10, 15, 20 and 25, respectively, we obtain the simulation results shown in Fig. 7. It can be clearly seen from the figure that: 1) The FLOPA scheme outperforms the CSMA/CA and the Fixed-CST schemes in throughput; and 2) As the number of AP (i.e., m) grows, the throughput difference between the FLOPA and the other compared schemes increases. The reason is that our FLOPA enables the AP to dynamically adjust its sensing area (i.e., OBSS/PD) with the variation in RSSI and throughput so that intensity of parallel transmission is improved, which helps in increasing throughput.

Next, fixing $m = 20$, and letting $w = 10, 15, 20, 25, 30$, we obtain Fig. 8. From the figure, we can observe that 1) Throughput in the FLOPA scheme is always greater than that in the CSMA/CA and the Fixed-CST; and 2) Throughput in the FLOPA scheme fluctuates with variation in packet arrival interval w, which is explained as follows. Our FLOPA scheme applies the fuzzy control system in adjusting OBSS/PD. However, the input variables of the fuzzy control system are not strongly related to the traffic to the AP, causing throughput insensitive to variation in w.

Finally, we plot Fig. 9 to illustrate the process of improving throughput in the FLOPA, where improvement in throughput is measured by PoIT. From Fig. 9, we have the following observations. At the beginning, PoIT is negative, which indicates that the throughput of FLOPA is lower than the Fixed-CST and the CSMA/CA. This is because the fuzzy control system in our FLOPA does not acquires sufficient information to improve throughput. This situation is considerably improved when running time proceeds because the feedbacks from the STAs are received and used in the fuzzy control system.

In summary, the FLOPA scheme can significantly improve the throughput in the OBSS, and it increases throughput nearly 50% more than the Fixed-CST and 110% more than the CSMA/CA.

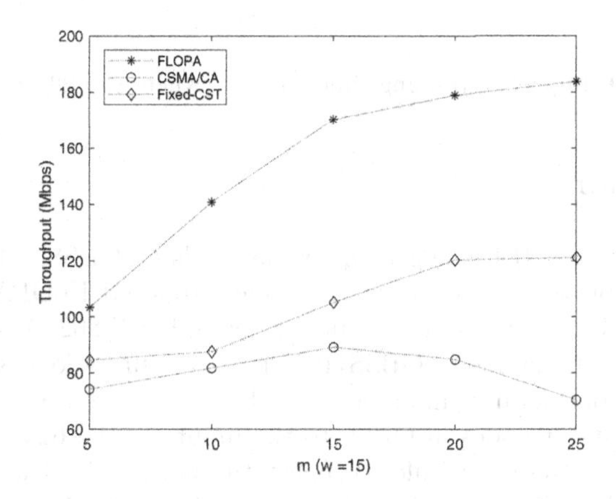

Fig. 7. Impact of m on throughput when $w = 15$

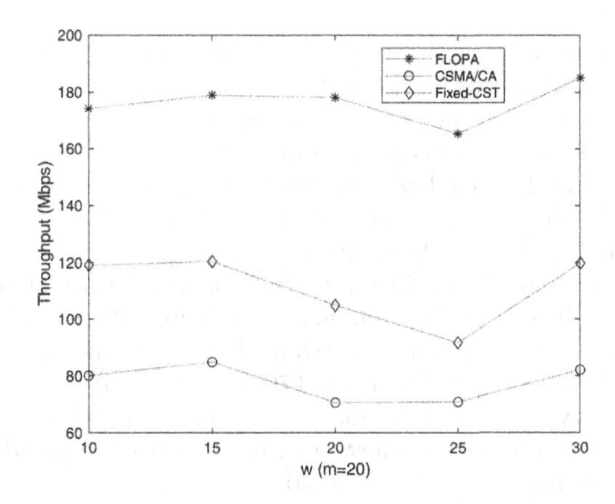

Fig. 8. Impact of w on throughput when $m = 20$

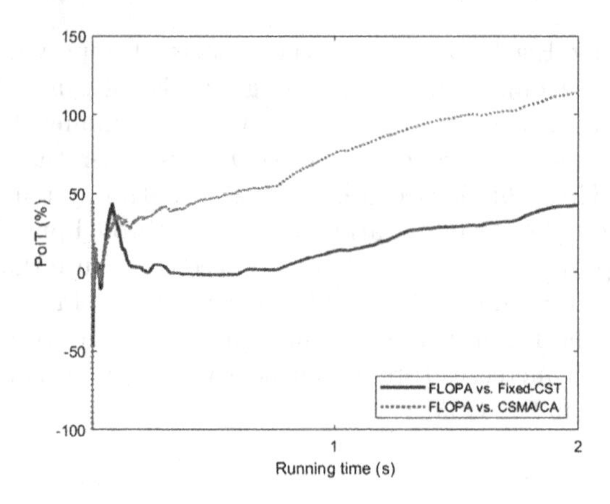

Fig. 9. Impact of running time on PoIT when $w = 20$, $m = 20$

6 Conclusion

Dense deployment of IEEE 802.11 networks leads to OBSS, which makes it a challenge to improve channel utilization. The proposed FLOPA integrates SR with OFDMA, the new technologies introduced in IEEE 802.11ax standard. The FLOPA can adjust the nodes' OBSS/PDs based on fuzzy logic so that network throughput is considerably improved. In the future, we will consider adding packet arrival rate at nodes in OBSS to the inputs in the fuzzy control system and then enhance the fuzzy rule to reflect the newly-added input so that the OBSS/PDs in the nodes can adapt to the variation in traffic to the nodes.

References

1. Cisco Annual Internet Report (2018–2023) (2007). http://physicsweb.org/articles/news/11/6/16/1. Accessed 9 Mar 2020
2. IEEE Standard for Information Technology-Telecommunications and Information Exchange between Systems Local and Metropolitan Area Networks-Specific Requirements Part 11: Wireless LAN Medium Access Control (MAC) and Physical Layer (PHY) Specifications, pp. 1–767, IEEE Std 802.11ax (2021). https://doi.org/10.1109/IEEESTD.2021.9442429
3. Khorov, E., Kiryanov, A., Lyakhov, A., Bianchi, G.: A tutorial on IEEE 802.11ax high efficiency WLANs. IEEE Commun. Surv. Tutor. **21**(1), 197–216 (2019)
4. Wilhelmi, F., Muoz, S.B., Cano, C., Selinis, I., Bellalta, B.: Spatial reuse in IEEE 802.11ax WLANs. Comput. Commun. **170**, 65–83 (2019)
5. Zhu, Y.-H., Pedryez, W.: A fuzzy forwarding pointer location mana gement strategy for personal communication networks. In: 2005 IEEE Networking, Sensing and Control, Tucson, pp. 38–43. IEEE (2005)
6. Zhu, Y.-H., Leung, V.C.M.: A fuzzy distance-based location management scheme for PCS networks. In: 2006 IEEE 63rd Vehicular Technology Conference, Melbourne, pp. 1063–1067. IEEE (2006)

7. Fernandez, A., Herrera, F., Cordon, O., del Jesus, M.J., Marcelloni, F.: Evolutionary fuzzy systems for explainable artificial intelligence: why, when, what for, and where to? IEEE Comput. Intell. Mag. **14**(1), 69–81 (2019)
8. Ahn, S., Couture, S.V., Cuzzocrea, A., et al.: A fuzzy logic based machine learning tool for supporting big data business analytics in complex artificial intelligence environments. In: IEEE International Conference on Fuzzy Systems, New Orleans, pp. 1–6. IEEE (2019)
9. Zhang, W.: From equilibrium-based business intelligence to information conservational quantum-fuzzy cryptography-a cellular transformation of bipolar fuzzy sets to quantum intelligence machinery. IEEE Trans. Fuzzy Syst. **26**(2), 656–669 (2018)
10. Lee, H., Kim, H.-S., Bahk S.: LSR: link-aware spatial reuse in IEEE 802.11ax WLANs. In: 2021 IEEE Wireless Communications and Networking Conference (WCNC), pp. 1–6 (2021). https://doi.org/10.1109/WCNC49053.2021.9417353
11. Lanante, L., Roy, S.: Performance analysis of the IEEE 802.11ax OBSS_PD-based spatial reuse. IEEE/ACM Trans. Netw. (2021). https://doi.org/10.1109/TNET.2021.3117816
12. Afaqui, M.S., Garcia-Villegas, E., Lopez-Aguilera, E., Smith, G., Camps, D.: Evaluation of dynamic sensitivity control algorithm for IEEE 802.11ax. In: IEEE Wireless Communications and Networking Conference, New Orleans, pp. 1060–1065. IEEE (2015)
13. Afaqui, M.S., Garcia-Villegas, E., Lopez-Aguilera, E., Camps-Mur, D.: Dynamic sensitivity control of access points for IEEE 802.11ax. In: IEEE International Conference on Communications, Kuala Lumpur, pp. 1–7. IEEE (2016)
14. Kulkarni, P., Cao, F.: Taming the densification challenge in next generation wireless LANs: an investigation into the use of dynamic sensitivity control. In: IEEE 11th International Conference on Wireless and Mobile Computing, Networking and Communications, Abu Dhabi, pp. 860–867. IEEE (2015)
15. Kulkarni, P., Cao, F.: Dynamic sensitivity control to improve spatial reuse in dense wireless LANs. In: International Conference on Modeling, Analysis and Simulation of Wireless and Mobile Systems, Malta, pp. 323–329. ACM (2016)
16. Kim, Y., Kim, G., Kim, T., Choi, W.: Transmission opportunity-based distributed OBSS/PD determination method in IEEE 802.11ax networks. In: 2020 International Conference on Artificial Intelligence in Information and Communication, Fukuoka, pp. 469–471. IEEE (2020)
17. Lv, Z., Hu, H., Yuan, D., Ran, J.: An adaptive rate and carrier sense threshold algorithm to enhance throughput and fairness for dense WLANs. In: 2017 3rd IEEE International Conference on Computer and Communications, Chengdu, pp. 453–458. IEEE (2017)
18. Topal, O.F., Kurt, G.K., Soysal, A.: Adaptation of carrier sensing threshold to increase throughput in dense 802.11ac wireless networks. In: 2018 Global Information Infrastructure and Networking Symposium, Thessaloniki, pp. 1–6. IEEE (2018)
19. Selinis, I., Katsaros, K., Vahid, S., Tafazolli, S.: Control OBSS/PD sensitivity threshold for IEEE 802.11ax BSS color. In: 2018 IEEE 29th Annual International Symposium on Personal, Indoor and Mobile Radio Communications, Bologna, pp. 1–7. IEEE (2018)
20. Chau, C., Ho, I.W.H., Situ, Z., Liew, S.C., Zhang, J.: Effective static and adaptive carrier sensing for dense wireless CSMA networks. IEEE Trans. Mob. Comput. **16**(2), 355–366 (2017)

21. Kim, S., Yoo, S., Yi, J., Son, Y., Choi, S.: FACT: fine-grained adaptation of carrier sense threshold in IEEE 802.11 WLANs. IEEE Trans. Veh. Technol. **66**(2), 1886–1891 (2017)
22. Mhatre, V.P., Papagiannaki, K., Baccelli, F.: Interference mitigation through power control in high density 802.11 WLANs. In: 26th IEEE International Conference on Computer Communications, Anchorage, pp. 535–543. IEEE (2007)
23. Yamamoto, K., Yang, X., Nishio, T., Morikura, M., Abeysekera, H.: Analysis of inversely proportional carrier sense threshold and transmission power setting. In: 2017 14th IEEE Annual Consumer Communications and Networking Conference, Las Vegas, pp. 13–18. IEEE (2017)
24. Iwai, K., Ohnuma, T., Shigeno, H., Tanaka, Y.: Improving of fairness by dynamic sensitivity control and transmission power control with access point cooperation in dense WLAN. In: 2019 16th IEEE Annual Consumer Communications and Networking Conference, Las Vegas, pp. 1–4. IEEE (2019)
25. Ropitault, T., Golmie, N.: ETP algorithm: increasing spatial reuse in wireless LANs dense environment using ETX. In: 2017 IEEE 28th Annual International Symposium on Personal, Indoor, and Mobile Radio Communications, Montreal, pp. 1–7. IEEE (2017)
26. Valkanis, A., Iossifides, A., Chatzimisios, P.: An interference based dynamic channel access algorithm for dense WLAN deployments. In: 2017 Panhellenic Conference on Electronics and Telecommunications, Xanthi, pp. 1–4. IEEE (2017)
27. Valkanis, A., Iossifides, A., Chatzimisios, P., Angelopoulos, M., Katos, V.: IEEE 802.11ax spatial reuse improvement: an interference-based channel-access algorithm. IEEE Veh. Technol. Mag. **14**(2), 78–84 (2019)
28. Duţu, L.C., Mauris, G., Bolon, P.: A fast and accurate rule-base generation method for mamdani fuzzy systems. IEEE Trans. Fuzzy Syst. **26**(2), 715–733 (2018)
29. Zhu, Y.-H., Leung, V.C.M.: Efficient power management for infrastructure IEEE 802.11 WLANs. IEEE Trans. Fuzzy Syst. **9**(7), 2196–2205 (2010)
30. Nurchis, M., Bellalta, B.: Target wake time: scheduled access in IEEE 802.11ax WLANs. IEEE Wireless Commun. **26**(2), 142–150 (2019)
31. Chen, Q.H., Zhu, Y.-H.: Scheduling channel access based on target wake time mechanism in 802.11ax WLANs. IEEE Trans. Wireless Commun. **20**(3), 1529–1543 (2021)

Two-Stage Evolutionary Algorithm Using Clustering for Multimodal Multi-objective Optimization with Imbalance Convergence and Diversity

Guoqing Li, Wanliang Wang(✉), and Yule Wang

College of Computer Science and Technology, Zhejiang University of Technology, Hangzhou 310023, China
zjutwwl@zjut.edu.cn

Abstract. This paper proposes a two-stage multimodal multi-objective evolutionary algorithm using clustering, termed as TS_MMOEAC. In the first-stage evolution of TS_MMOEAC, the convergence-penalized density (CPD) based k-means clustering is used to divide population into multiple subpopulations. Subsequently, a local archive mechanism is adopted to maintain diverse local Pareto optimal solutions in decision space. In the second-stage evolution, an identical k-means clustering based on distance among individuals in decision space is applied to form multiple subpopulations. In this case, the convergence performance of local Pareto optimal solutions is accelerated. Meanwhile, equivalent Pareto optimal solutions with the imbalance between convergence and diversity are located by a similar local archive method with a larger clearing radius. Experimental results validate the superior performance of TS_MMOEAC, and the proposed TS_MMOEAC is capable of finding equivalent Pareto optimal solutions with the imbalance between convergence and diversity.

Keywords: Two-stage evolutionary algorithm · Clustering · Multimodal multi-objective optimization · Imbalance

1 Introduction

Multiple conflicting objectives are optimized in multi-objective optimization problems (MOPs). In the last two decades, there has been an intense research activity in studying MOPs. Particularly, Pareto-based multi-objective evolutionary algorithms (MOEAs) [1], decomposition-based multi-objective evolutionary algorithms [2], and indicator-based multi-objective evolutionary algorithms [3] are remarkable. The aim of these mentioned MOEAs algorithms and their enhanced versions is to find complete Pareto front (PF) in objective space. However, a special situation for MOPs is overlooked in which multiple Pareto optimal solutions in decision space with the same objective values in objective space. This kind of multi-objective optimization problem is defined as multimodal multi-objective optimization problems (MMOPs) [4].

Y. Lai et al. (Eds.): ICA3PP 2021, LNCS 13157, pp. 571–586, 2022.
https://doi.org/10.1007/978-3-030-95391-1_36

Distinct from previous MOEAs, several proposed MOEAs in recent years have begun to focus on the diversity of Pareto optimal solutions in decision space. For instance, an improving strength Pareto-based evolutionary algorithm (SPEA2) [5], an extended Omni-optimizer procedure (Omni-optimizer) [6], PCA-assisted multi-objective evolutionary algorithm [7]. However, these mentioned MOEAs are not capable of locating equivalent Pareto optimal solutions in decision space, i.e. complete Pareto set (PS). Recently, inspired by popular multi-objective evolutionary algorithms, several multimodal multi-objective evolutionary algorithms (MMEAs) are proposed for solving MMOPs. A framework of decomposition-based MMEAs [8], an indicator-based MMEA [9], and Pareto-based MMEAs [10–12] are proposed one after another. Of these, Pareto-based MMEAs are the most prominent. Niching technologies play an important role in Pareto-based MMEAs. For instance, a double-niched evolutionary algorithm (DNEA) [13], an index-ring topology particle swarm optimization algorithms (MO_Ring_PSO_SCD) [10]. These popular Pareto-based MMEAs can find equivalent Pareto optimal solutions in decision space, where the convergence and diversity of equivalent solutions in decision space are balanced. In other words, the complexity of locating equivalent Pareto optimal solutions in decision space is identical.

However, most popular Pareto-based MMEAs using Pareto-based ranking are ineffective for solving multimodal multi-objective problems with the imbalance between convergence and diversity (MMOP-ICD) in decision space. The main reason is that these solutions, which approximate one or multiple equivalent Pareto optimal solutions, are dominated by an equivalent Pareto optimal solution in evolutionary process. In this case, all equivalent Pareto optimal solutions with the imbalance between convergence and diversity are difficult to be found. To address this issue, a two-stage evolutionary algorithm using clustering (TS_MMOEAC) is developed in this paper for handling MMOP-ICD. The main contributions of this paper are as follows:

1) The proposed TS_MMOEAC is composed of the first-stage evolution and the second-stage evolution. The main purpose of the first-stage evolution using clustering is to find local optimal individuals that gradually approximate equivalent PSs with the imbalance convergence and diversity during evolutionary process.
2) To improve the convergence performance of all individuals, a unique clustering method is involved in second-stage evolution. Particularly, it accelerates the population convergence to one or several equivalent PSs with the imbalance convergence and diversity.

The remainder of this paper is structured as follows: The definition of MMOPs and MMOP-ICD, and popular MMEAs are presented in Sect. 2. The main framework of the proposed TS_MMOEAC is described in Sect. 3. Section 4 validates the superior performance of TS_MMOEAC in comparison to five competing algorithms. Section 5 concludes this paper.

2 Problem Definition and Literature Review

2.1 Problem Definition of MMOP-ICD

There are multiple equivalent Pareto optimal solutions in decision space with the same objective value in objective space for MOPs, and this kind of MOPs is defined as multimodal multi-objective optimization problems (MMOPs). A standard definition of MMOPs is given by Tanabe as follows [14]:

Definition 1. An MMOP is involved in locating all solutions which are equivalent to Pareto optimal solutions.

Definition 2. If $\|f(x_1) - f(x_2)\| < \delta$, two solutions x_1 and x_2 are considered to be equivalent.
where $\|f(x_1) - f(x_2)\|$ is the arbitrary norm between $f(x_1)$ and $f(x_2)$, and δ is a threshold defined by the decision-maker.

In MMOPs, equivalent Pareto optimal solutions have identical convergence and diversity in decision space. Additionally, the complexity of locating equivalent Pareto optimal solutions is uniform. Given the unique scenario that equivalent Pareto optimal solutions do not have identical search complexity. Inspired by imbalanced multi-objective optimization problems [15], the MMOPs with imbalance convergence and diversity (MMOP-ICD) is proposed by Liu [16] and defined as follows:

Definition 3. An MMOP is defined as MMOP-ICD, if one or both of the following conditions are met [16]:

1) The solutions in objective space, which close to one equivalent Pareto optimal solution, are likely to dominate other solutions that approximates another equivalent Pareto optimal solution.
2) The complexity of finding an equivalent Pareto optimal solution in objective space is smaller than that of searching for another equivalent Pareto optimal solution.

2.2 Literature Review

Recently, MMOPs have attracted the attention of researchers. Several MMEAs have been developed for MMOPs. Liang [4] proposed the multimodal multi-objective problems and introduced a niching-based fast and elitist multi-objective genetic algorithms (DN-NSGAII), making a tremendous step forward for MMOPs. An index-based ring topology multimodal multi-objective particle swarm optimization algorithm (MO_Ring_PSO_SCD) is suggested by Yue [10]. In [10], the left and right index-based neighbors of each particle form a ring topology. And local version particle swarm optimization algorithm is adopted to locate equivalent Pareto optimal solutions in decision space. Additionally, several MMOPs and performance metrics are presented in [10, 11]. Inspired by MO_Ring_PSO_SCD, a clustering-based ring topology multimodal multi-objective particle swarm optimization algorithm (MMO-CLRPSO) is proposed by Zhang [12]. In MMO-CLRPSO, the population is divided into multiple subpopulations, and a ring structure is formed among subpopulations. Global version PSO is used within each subpopulation to locate equivalent Pareto optimal solutions in decision space, and local version PSO is applied in all subpopulations to improve the diversity of subpopulations. Subsequently, Lin presented a dual-clustering-based evolutionary algorithm (MMOEA/DC) [17] for MMOPs. In MMOEA/DC, one clustering method is adopted to decision space for maintaining the local PS, and the other clustering method is used to objective space to locate diverse and equivalent Pareto optimal solutions. Additionally, a clearing-based multimodal multi-objective evolutionary optimization using a layer-to-layer strategy [18] is designed for solving MMOPs with local PS.

Several decomposition-based and indicator-based multimodal multi-objective algorithms have also been presented one after another. Tanabe developed a decomposition-based MMEA, called MOEA/D-AD [19], and then proposed a decomposition-based framework to handle MMOPs [8]. Additionally, a niching indicator-based multimodal multi-objective evolutionary algorithm (NIMMO) is also developed by Tanabe [9]. Subsequently, a standard MMOPs benchmark with imbalance convergence and diversity in decision space (MMOP-ICD) is designed [16]. And an evolutionary multimodal multi-objective algorithm via a convergence-penalized density method (CPDEA) is developed for solving MMOP-ICD.

These mentioned MMEAs are capable of finding multiple equivalent PSs with the same diversity and convergence, but most of them cannot find all PSs when dealing with MMOP-ICD since that the diversity and convergence of all equivalent PSs are unbalanced. Therefore, the above-mentioned MMEAs are rarely capable of handling MMOP-ICD. To address this issue, a two-stage multimodal multi-objective evolutionary algorithm using clustering (TS_MMOEAC) is proposed. In the proposed TS_MMOEAC, a two-stage evolutionary strategy is developed to find equivalent Pareto optimal solutions in decision space. The population is divided into multiple subpopulations via the CPD-based k-means clustering method in the first-stage evolution for locating Pareto optimal solutions with high quality, and these high-quality solutions will converge to multiple equivalent PSs with the imbalance in the second-stage evolution. Sustainably, the population is classified into several subpopulations using distance-based k-means clustering for accelerating subpopulations convergence and finding equivalent Pareto optimal solutions with the imbalance in the second-stage evolution. The main framework of TS_MMOEAC is described in Sect. 3.

3 Proposed TS_MMOEAC

3.1 The Main Framework of TS_MMOEAC

In Algorithm 1, the population P with N individuals is initialized in line 1. If the used fitness evaluation number Fes is less than half of the maximum fitness evaluation numbers $MaxFes$, the first-stage evolution (Algorithm 2) is performed in line 3. Otherwise, the second-stage evolution (Algorithm 4) is implemented in line 5. Of these, the local archive mechanism (Algorithm 3) is implemented in both the first-stage and the second-stage evolution. Finally, output the population P. The main framework of TS_MMOEAC is shown in Fig. 1.

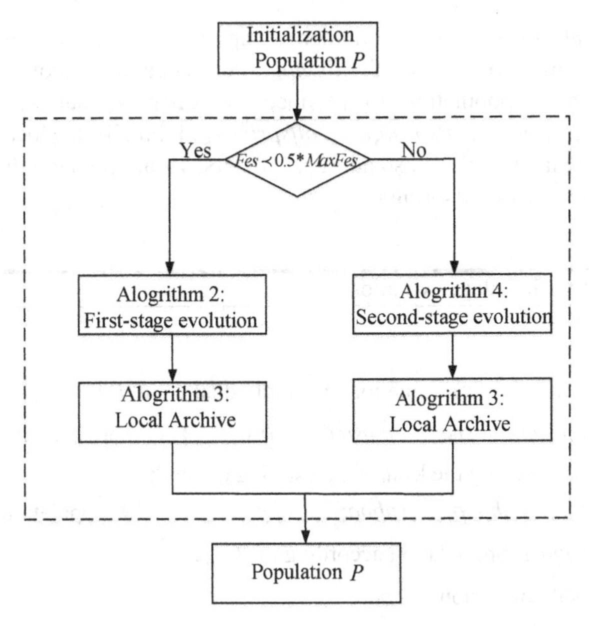

Fig. 1. The main framework of TS_MMOEAC.

Algorithm 1. TS_MMOEAC

Input: N (population size), $MaxFes$ (the maximum fitness evaluation numbers)
Output: P

1. Initialize the population P with N individuals;
2. **while** $Fes < MaxFes$
3. $P \leftarrow$ The first-stage evolution; //Algorithm 2
4. **else**
5. $P \leftarrow$ The second-stage evolution; //Algorithm 4
6. **end**
7. **return** P;

3.2 First-Stage Evolution Strategy

In Algorithm 2, the CPD value of each individual x in the population P is calculated according to the convergence-penalized density method [16] at first. Then, a popular clustering method, k-means clustering, is used to divide the CPD values set into Num clusters $f_{CPD} = \{cluster_1, cluster_2, ..., cluster_i, ..., cluster_{Num}\}$. Then, the population P is divided into Num subpopulations, $P = \{subpop_1, subpop_2, ..., subpop_i, ...subpop_{Num}\}$, according to Num clusters f_{CPD}. The roulette selection operation is utilized to select parents in the first-stage evolution. For each individual $subpop_{i,j}$ in the subpopulation $subpop_i$, the CPD-based roulette selection method is applied to select the parents of $subpop_{i,j}$, and then a reproduction operator is used to generate the offspring $offspring_{i,j}$ of $subpop_{i,j}$. Subsequently, recombining all individuals in each subpopulation to produce the offspring subpopulations $O = \{offspring_1, offspring_2, ..., offspring_i, ..., offspring_{Num}\}$. Finally, the local archive mechanism is used to maintain the outstanding and diverse local optimal individual between the population P and the offspring O.

Algorithm 2 The first-stage evolution

Input P
Output P
1. Calculate the $f_{CPD}(x)$ value for each individual x in P;
2. $f_{CPD} = \{cluster_1, cluster_2, ..., cluster_i, ..., cluster_{Num}\} \leftarrow$ The f_{CPD} is divided into Num clusters using the k-means clustering method;
3. $P = \{subpop_1, subpop_2, ..., subpop_i, ...subpop_{Num}\} \leftarrow$ The population P is divided into Num subpopulation according to f_{CPD};
4. **for** each subpopulation $subpop_i$
5. $offspring_i \leftarrow$ Selecting parents and reproduction $subpop_i$ using roulette selection operation;
6. **end**
7. $O = \{offspring_1, offspring_2, ..., offspring_i, ..., offspring_{Num}\}$;
8. $P = LocalArchive(\{P, O\})$;
9. **return** P

3.3 Local Archive Strategy

For MMOP-ICD, one equivalent Pareto optimal solution is easily located. However, other equivalent solutions are difficult to find since there is an imbalanced diversity and convergence among several unique equivalent solutions. The main reason is that popular MMEAs using dominance relations choose superior solutions and remove poor (local) solutions, but these poor solutions have a greater potential to become equivalent PS. Therefore, popular MMEAs are not capable of locating all equivalent PSs in handling MMOP-ICD. To tackle this issue, inspired by [18, 20], a local archive strategy is presented to maintain diverse local optimal individuals in the first-stage evolution. And these individuals in the local archives have a greater tendency to become equivalent Pareto optimal solutions with imbalanced diversity and convergence in comparison to other equivalent Pareto optimal solutions. The local archive mechanism is described as follows.

First, local Pareto optimal individuals are found in the local archive mechanism. In this step, the mean value of the sum of the distances between all individuals and their Ns nearest neighbors is calculated and denoted as MS. To remove poor local individuals from the population P and the offspring O, the parameter $\theta_r = MS/D$ is viewed as the clearing radius, where D is the decision variable number. For each individual x, any individual x' is within the radius θ_r of the individual x and is dominated by x, and x' is removed. These superior local Pareto optimal solutions are maintained by performing the above operation for each individual x. These poor individuals are removed from the set of the population P and the offspring O, and survived in the archive A. Given this scenario that the number of superior local Pareto optimal individuals is less than the population size N, it is essential to select several superior individuals from A to P. Distinct from the first step, the number of the nearest neighbors is set to $Ns = 0.25 * N_A$, where N_A is the size of the archive A.

Next, the population P is sorted based on the Pareto-dominance ranking, and the population $P = \{F_1, F_2, ..., F_m\}$ is divided into m layers. A sharing method is applied to obtain diverse global and local solutions in both decision space and objective space. For each layer F_i, $i = 1, 2, ..., m$, if the individual numbers in the F_i layer are more than N_F, where $N_F = 0.5 * N$ is the maximum individual numbers in each layer, several individuals with poor sharing values in the F_i layer are removed. The sharing value of each individual $F_{i,q}$, $i = 1, 2, .., M$, $q = 1, 2, ..., |F_i|$ in F_i is calculated by Eq. (1).

$$f_{DS}(F_{i,q}) = \sum_{k=1}^{K} sh_{dec}(F_{i,q}, F_{i,k}) + sh_{obj}(F_{i,q}, F_{i,k}) \tag{1}$$

where $F_{i,k}$ is the k-th nearest neighbors of $F_{i,q}$, $sh_{dec}(F_{i,q}, F_{i,k})$ and $sh_{obj}(F_{i,q}, F_{i,k})$ are the sharing values of the individual $F_{i,q}$ in decision space and objective space, respectively. The $sh_{dec}(F_{i,q}, F_{i,k})$ and $sh_{obj}(F_{i,q}, F_{i,k})$ are calculated as follows [20].

$$\begin{aligned} sh_{dec}(F_{i,q}, F_{i,k}) &= \max\{0, 1 - d_{dec}(q, k)/\sigma_{dec}\}, \\ sh_{obj}(F_{i,q}, F_{i,k}) &= \max\{0, 1 - d_{obj}(q, k)/\sigma_{obj}\}. \end{aligned} \tag{2}$$

where $d_{dec}(q, k)$ and $d_{obj}(q, k)$ are the Euclidean distance between $F_{i,q}$ and $F_{i,k}$ in both decision space and objective space, respectively. σ_{dec} and σ_{obj} are the niche radius in both decision space and objective space, respectively.

f_{DS} evaluates the density of the individuals in F_i between decision space and objective space, simultaneously. The larger f_{DS}, the worse the diversity. Removing the individual with the largest f_{DS} value from F_i to maintain the diverse individuals in both decision space and objective space. In the sharing method, the individual number in each layer $F_i, i = 1, 2, ..., q$ is no more than N_F. Add each layer F_i to P' until the number of P' is greater than N. The pseudo-code for the local archive mechanism is presented in Algorithm 3.

Algorithm 3 *LocalArchive*(P)

Input P
Output P

1. $MS \leftarrow$ the mean value of the sum of the distances between all individuals and its Ns nearest neighbors is calculated;
2. $\theta_r = MS / D$, $A = \varnothing$, $P' = \varnothing$;
3. **for** each individual x in P
4.   **if** anyone individual x' in P is within the radius θ_r of x and is dominated by x
5. $P = \{P \setminus x'\}, A = \{A \cup x'\}$;
6.   **end**
7. **end**
8. **if** $|P| < N$
9. $P \leftarrow$ select several superior individuals from A to P;
10. **else**
11. $P = \{F_1, F_2, ..., F_q\} \leftarrow sort(P)$;
12.   **for** each F_i in P
13. $f_{DS} \leftarrow$ The sharing value of each individual $F_{i,q}$ in F_i is calculated;
14.   **while** $|F_i| > N_F$
15.   Remove the individual with the largest f_{DS} value from F_i;
16.   **end**
17. $P' = \{P' \cup F_i\}$;
18.   **if** $|P'| \geq N$
19.   **break;**
20.   **end**
21.   **end**
22. **end**
23. $P = P'$;
24. **return** P;

3.4 Second-Stage Evolution Strategy

When the computing resources used by the first-stage evolution are greater than half of the total computing resources (0.5 * *MaxFes*), the second-stage evolution of the proposed TS-MMOEAC is performed and shown in Algorithm 4. The second-stage evolution is similar to the first stage, and there are three main distinctions. The first is that the clustering method in the second-stage evolution is slightly different. Instead of the CPD-based k-means clustering, the distance-based k-means clustering is used to divide the population into multiple subpopulations. The individuals nearby are clustered into a subpopulation in decision space, and the distance-based k-means clustering used in the second-stage evolution accelerates the convergence of subpopulations. The second distinction is that the tournament selection method for each subpopulation is used to choose the parents based on f_{DS} values in the second-stage evolution. The main reason is that each subpopulation is more likely to approach these solutions with the smallest f_{DS} values and locate diverse Pareto optimal solutions. The third difference is that the θ_r value is larger for dominating poor local Pareto optimal solutions and maintaining global Pareto optimal solutions in P, where θ_r is equal to MS in the local archive in Algorithm 3. The main purpose is to preserve elite local optimal solutions and move them towards equivalent Pareto optimal solutions with imbalanced diversity and convergence.

Algorithm 4 The second-stage evolution

Input P
Output P
1. $P = \{subpop_1, subpop_2, ..., subpop_i, ...subpop_{Num}\} \leftarrow$ The population P is divided into Num subpopulations according to the Euclidean distance.
2. **for** each subpopulation $subpop_i$
3. $offspring_i \leftarrow$ Selecting parents and reproduction $subpop_i$ using tournament selection operation;
4. **end**
5. $O = \{offspring_1, offspring_2, ..., offspring_i, ..., offspring_{Num}\}$;
6. $P = \{P, O\}, P = LocalArchive(P)$;
7. **return** P

---

4 Experimental Results and Discussion

4.1 Experimental Setup

To validate the performance of the proposed TS_MMOEAC for solving MMOP-ICD, several state-of-the-art algorithms are selected as the competing algorithms of TS_MMOEAC, including DNEA [13], MO_Ring_PSO_SCD [10], TriMOEATAR [21], MMOEA/DC [17], and CPDEA [16]. Among them, MO_Ring_PSO_SCD, TriMOEATAR, MMOEA/DC, and CPDEA are popular MMEAs that are published in IEEE Transactions on Evolutionary Computation in recent years. The proposed TS_MMOEAC and its competing algorithms are implemented in the standard MMOP-ICD benchmark. The MMOP-ICD benchmark consists of 4 types of imbalanced distance minimization problems, and it involves 12 problems, including IDMP-M2-T1, IDMP-M2-T2, IDMP-M2-T3, IDMP-M2-T4, IDMP-M3-T1, IDMP-M3-T2, IDMP-M3-T3, IDMP-M3-T4, IDMP-M4-T1, IDMP-M4-T2, IDMP-M4-T3, and IDMP-M4-T4. Please refer to the literature [16] for further information on MMOP-ICD.

Three performance metrics, IGDX [10], IGDF [11], and IGDM [16], are applied to test the superior performance of TS_MMOEAC. The IGDX implies the distance between obtained PS and true PS in decision space. IGDF measures the distance between obtained PF and true PF in objective space. IGDM indicates the diversity and the convergence performance of obtained Pareto optimal solutions in both decision space and objective space, simultaneously. The smaller IGDX, IGDF, and IGDM are desired.

4.2 Experimental Result

In this experiment, the population size and the fitness evaluations are set to $30 * 2^{D-1}$ and $9000 * 2^{D-1}$, respectively, where D is the decision variable number. The parameters involved in all competing algorithms are maintained with the original papers. In the proposed TS_MMOEAC, these parameters, *Num*, *Ns*, and K, are involved. *Ns* and K is set to 50 and 5 according to [18, 20], respectively. Additionally, the parameter *Num* is discussed in Sect. 4.3. To guarantee a fair experimental comparison between TS_MMOEAC and its competing algorithms, the experiment is performed 30 times independently. The IGDX, IGDF, and IGDM performance metrics of TS_MMOEAC and its compared algorithms are present in Tables 1, 2, and 3, respectively. Additionally, the Wilcoxon rank-sum test with a significant level $\alpha = 0.05$ is used to show a significant difference between the proposed TS_MMOEAC and its competing algorithms, where the symbol "+", "−", and "∼" indicates that the proposed TS_MMOEAC is significantly worse than, better than, and similar to its competing algorithms.

It is observed from Table 1 that TS_MMOEAC performs the best result on 10 out of 12 test problems, including IDMP-M2-T1, IDMP-M2-T2, IDMP-M2-T3, IDMP-M2-T4, IDMP-M3-T1, IDMP-M3-T2, IDMP-M3-T3, IDMP-M4-T2, IDMP-M4-T3, and IDMP-M4-T4. And TS_MMOEAC is superior to DNEA, MO_Ring_PSO_SCD, and TriMOEATAR in all problems. The IGDX values of TS_MMOEAC are also significantly superior to MMOEA/DC in the benchmark problems except for IDMP-M4-T1. TS_MMOEAC performs slightly worse than CPDEA in IDMP-M3-T4 and IDMP-M4-T1, but the performance of TS_MMOEAC is superior to CPDEA. The Pareto optimal solutions obtained by DNEA, MMOEA/DC, CPDEA, and TS_MMOEAC on two problems in decision space are clearly shown in Fig. 2 and Fig. 3, respectively. It is clear that DNEA and MMOEA/DC only find one or multiple Pareto optimal solutions, but cannot locate all equivalent solutions in the decision space. CPDEA is similar to TS_MMOEAC in that they are capable of finding all equivalent solutions in decision space. Therefore, we can infer that the proposed TS_MMOEAC is superior to its competing algorithms in decision space, and it is capable of locating equivalent solutions with imbalance convergence and diversity in decision space.

It can be seen from Table 2 that the IGDF performance of DNEA outperforms the proposed TS_MMOEAC and other competing algorithms in objective space. The main reason is that DNEA is originated from MOEAs which use a niching technique to search for Pareto optimal solutions in both decision space and objective space. But the IGDF value of TS_MMOEAC is superior to MO_Ring_PSO_SCD and TriMOEATAR in all problems and outperforms MMMOEA/DC and CPDEA in most cases. It is implied that the convergence of Pareto optimal solutions obtained by TS_MMOEAC in objective space is better than other competing algorithms except for DNEA.

The performance of the proposed TS_MMOEAC in both objective space and decision space is evaluated by IGDM, simultaneously. Since the IGDF values of TS_MMOEAC

Table 1. The IGDX performance of TS_MMOEAC and its competing algorithms

Problem	DNEA	MO_Ring_ PSO_SCD	Tri MOEATAR	MMOEA/DC	CPDEA	TS_ MMOEAC
IDMP-M2-T1	6.06E−01 (2.05E−01)−	2.21E−01 (3.03E−01)−	6.30E−01 (1.66E−01)−	6.92E−02 (2.05E−01)−	2.02E−03 (1.04E−04)−	**1.60E−03 (1.12E−04)**
IDMP-M2-T2	5.61E−01 (2.55E−01)−	6.74E−01 (5.15E−04)−	6.52E−01 (1.18E−01)−	9.17E−02 (2.32E−01)−	2.09E−03 (1.11E−04)−	**1.70E−03 (7.88E−05)**
IDMP-M2-T3	3.42E−01 (3.40E−01)−	1.45E−01 (2.02E−01)−	5.10E−01 (2.79E−01)−	1.14E−01 (2.55E−01)−	2.94E−03 (3.65E−04)−	**2.04E−03 (6.41E−05)**
IDMP-M2-T4	6.51E−01 (1.23E−01)−	1.30E+00 (3.12E−03)−	6.73E−01 (6.09E−05)−	4.26E−01 (3.28E−01)−	1.96E−03 (8.26E−05)−	**1.61E−03 (1.25E−04)**
IDMP-M3-T1	6.65E−01 (3.03E−01)−	1.26E−01 (1.87E−01)−	8.61E−01 (2.04E−01)−	3.75E−02 (7.40E−02)−	1.13E−02 (1.92E−04)−	**1.07E−02 (2.82E−04)**
IDMP-M3-T2	5.98E−01 (3.01E−01)−	8.34E−02 (1.05E−01)−	8.09E−01 (2.29E−01)−	5.36E−02 (9.19E−02)−	1.13E−02 (1.63E−04)−	**1.07E−02 (5.36E−04)**
IDMP-M3-T3	5.17E−01 (2.60E−01)−	1.67E−02 (1.93E−03)−	6.63E−01 (2.12E−01)−	1.04E−01 (1.54E−01)−	1.22E−02 (2.31E−04)−	**1.06E−02 (1.80E−04)**
IDMP-M3-T4	7.55E−01 (2.76E−01)−	5.29E−02 (8.66E−02)−	8.75E−01 (1.99E−01)−	2.64E−01 (2.49E−01)−	**1.12E−02 (1.59E−04)+**	3.50E−02 (7.37E−02)
IDMP-M4-T1	1.17E+00 (3.88E−02)−	5.98E−01 (3.90E−01)−	1.19E+00 (8.38E−04)−	1.54E−01 (2.11E−01)+	**8.64E−03 (8.03E−05)+**	4.32E−01 (3.58E−01)
IDMP-M4-T2	1.03E+00 (2.03E−01)−	1.26E−01 (1.31E−01)−	1.06E+00 (1.68E−01)−	1.55E−01 (2.31E−01)−	8.71E−03 (1.40E−04)−	**8.07E−03 (3.46E−04)**
IDMP-M4-T3	8.28E−01 (3.25E−01)−	1.51E−02 (1.33E−03)−	9.16E−01 (2.26E−01)−	6.67E−02 (1.40E−01)−	1.85E−02 (4.86E−02)−	**8.52E−03 (1.02E−04)**
IDMP-M4-T4	1.06E+00 (2.34E−01)−	1.74E−02 (3.42E−03)−	1.04E+00 (2.29E−01)−	1.82E−01 (1.71E−01)−	3.53E−02 (8.11E−02)−	**1.72E−02 (4.86E−02)**
±/~	0/12/0	0/12/0	0/12/0	1/11/0	2/10/0	

Table 2. The IGDF performance of TS_MMOEAC and its competing algorithms

Problem	DNEA	MO_Ring _PSO_SCD	Tri MOEATAR	MMOEA/DC	CPDEA	TS_ MMOEAC
IDMP-M2-T1	**1.23e−3 (1.55e−5)+**	2.55E−03 (2.19E−04)−	1.76e−3 (1.04e−4) −	1.59e−3 (5.26e−4)−	1.55e−3 (9.70e−5) −	1.33e−3 (6.78e−5)
IDMP-M2-T2	**1.24e−3 (1.42e−5)+**	2.15E−03 (2.28E−04)−	1.78e−3 (1.69e−4) −	1.42e−3 (7.23e−5)−	1.47e−3 (4.59e−5)−	1.28e−3 (5.37e−5)
IDMP-M2-T3	**1.29e−3 (5.24e−5)~**	2.20E−03 (2.17E−04)−	1.14e−2 (1.98e−2) −	1.62e−3 (9.28e−5)−	1.56e−3 (5.33e−5)−	1.30e−3 (3.15e−5)
IDMP-M2-T4	**1.23e-3 (1.16e-5)+**	2.12E+02 (5.13E−02)−	1.74e−3 (1.75e−4)−	1.47e−3 (8.68e−5)−	1.51e−3 (6.07e−5)−	1.33e−3 (1.10e−4)
IDMP-M3-T1	**6.15e−3 (2.18e−4)+**	1.00E−02 (5.63E−04)−	1.31e−2 (6.65e−4)−	8.17e−3 (4.44e−4)−	6.75e−3 (1.54e−4)−	6.61e−3 (2.42e−4)
IDMP-M3-T2	**6.14e−3 (2.00e−4)+**	8.34E−03 (2.82E−04)−	1.30e−2 (7.49e−4)−	7.82e−3 (3.04e−4)−	6.70e−3 (1.70e−4)~	6.66e−3 (2.23e−4)

(continued)

Table 2. (*continued*)

Problem	DNEA	MO_Ring _PSO_SCD	Tri MOEATAR	MMOEA/DC	CPDEA	TS_ MMOEAC
IDMP-M3-T3	**6.35e−3 (2.58e−4)+**	8.76E−03 (4.12E−04)−	1.28e−2 (8.02e−4)−	8.58e−3 (5.07e−4)−	7.21e−3 (1.95e−4)−	6.79e−3 (1.87e−4)
IDMP-M3-T4	**6.13e−3 (2.21e−4)+**	8.41E−03 (3.35E−04)−	1.31e−2 (7.28e−4)−	7.53e−3 (7.85e−4)−	6.76e−3 (1.63e−4)−	6.60e−3 (2.06e−4)
IDMP-M4-T1	**5.78e−3 (2.02e−4)+**	1.71E−02 (8.22E−04)−	2.64e−2 (1.44e−3)−	7.47e−3 (3.55e−4)−	6.62e−3 (2.14e−4)−	6.20e−3 (4.17e−4)
IDMP-M4-T2	**5.79e−3 (2.13e−4)+**	1.07E−02 (4.12E−04)−	2.62e−2 (1.51e−3)−	6.99e−3 (3.13e−4)−	6.31e−3 (2.33e−4)~	6.28e−3 (2.52e−4)
IDMP-M4-T3	**6.00e−3 (2.96e−4)+**	1.08E−02 (3.70E−04)−	2.74e−2 (7.57e−3)−	8.20e−3 (3.15e−4)−	6.66e−3 (1.73e−4)−	6.47e−3 (2.02e−4)
IDMP-M4-T4	**5.86e−3 (2.29e−4)+**	1.08E−02 (4.82E−04)−	2.62e−2 (1.32e−3)−	6.96e−3 (4.07e−4)−	6.23e−3 (2.49e−4)~	6.25e−3 (2.40e−4)
±/~	11/0/1	0/12/0	0/12/0	0/12/0	0/9/3	

Table 3. The IGDM performance of TS_MMOEAC and its competing algorithms

Problem	DNEA	MO_Ring_ PSO_SCD	Tri MOEATAR	MMOEA /DC	CPDEA	TS_ MMOEAC
IDMP-M2-T1	4.54E−01 (1.50E−01)−	2.33E−01 (1.97E−01)−	4.80E−01 (9.14E−02)−	6.40E−02 (1.50E−01)−	1.44E−02 (7.82E−04)−	**1.14E−02 (6.13E−04)**
IDMP-M2-T2	4.21E−01 (1.86E−01)−	5.05E−01 (5.69E−04)−	4.94E−01 (5.35E−02)−	7.96E−02 (1.69E−01)−	1.42E−02 (6.68E−04)−	**1.18E−02 (5.49E−04)**
IDMP-M2-T3	2.77E−01 (2.32E−01)−	1.75E−01 (1.42E−01)−	4.69E−01 (9.21E−02)−	9.70E−02 (1.85E−01)−	1.75E−02 (2.30E−03)−	**1.21E−02 (3.38E−04)**
IDMP-M2-T4	4.86E−01 (8.94E−02)−	1.00E+00 (0.00E+00)−	5.04E−01 (4.37E−04)−	3.24E−01 (2.40E−01)−	1.39E−02 (6.23E−04)−	**1.16E−02 (5.64E−04)**
IDMP-M3-T1	5.59E−01 (1.78E−01)−	2.45E−01 (1.23E−01)−	7.02E−01 (9.87E−02)−	1.18E−01 (6.53E−02)−	8.08E−02 (1.59E−03)−	**7.66E−02 (1.81E−03)**
IDMP-M3-T2	5.20E−01 (1.95E−01)−	2.09E−01 (8.08E−02)−	6.83E−01 (1.29E−01)−	1.30E−01 (8.13E−02)−	8.02E−02 (1.24E−03)−	**7.64E−02 (2.88E−03)**
IDMP-M3-T3	4.82E−01 (1.61E−01)−	1.16E−01 (1.49E−02)−	6.31E−01 (1.10E−01)−	1.78E−01 (1.32E−01)−	8.40E−02 (1.66E−03)−	**7.35E−02 (1.24E−03)**
IDMP-M3-T4	6.20E−01 (1.77E−01)−	1.73E−01 (7.37E−02)−	7.13E−01 (1.03E−01)−	2.96E−01 (1.83E−01)−	**8.03E−02 (1.05E−03)+**	9.95E−02 (6.78E−02)
IDMP-M4-T1	7.57E−01 (5.31E−05)−	5.62E−01 (1.52E−01)−	7.83E−01 (1.82E−03)−	1.86E−01 (1.53E−01)~	**6.32E−02 (7.71E−04)+**	3.59E−01 (2.21E−01)
IDMP-M4-T2	7.13E−01 (9.00E−02)−	2.52E−01 (1.02E−01)−	7.59E−01 (5.60E−02)−	1.84E−01 (1.65E−01)−	6.21E−02 (1.12E−03)−	**5.88E−02 (1.39E−03)**
IDMP-M4-T3	6.38E−01 (1.68E−01)−	1.06E−01 (9.69E−03)−	7.38E−01 (7.45E−02)−	1.22E−01 (1.08E−01)−	7.29E−02 (4.10E−02)−	**5.76E−02 (6.93E−04)**
IDMP-M4-T4	7.27E−01 (1.30E−01)−	1.28E−01 (2.46E−02)−	7.58E−01 (1.14E−01)−	2.29E−01 (1.46E−01)−	8.47E−02 (6.95E−02)−	**6.72E−02 (4.19E−02)**
±/~	0/12/0	0/12/0	0/12/0	0/11/1	2/10/0	

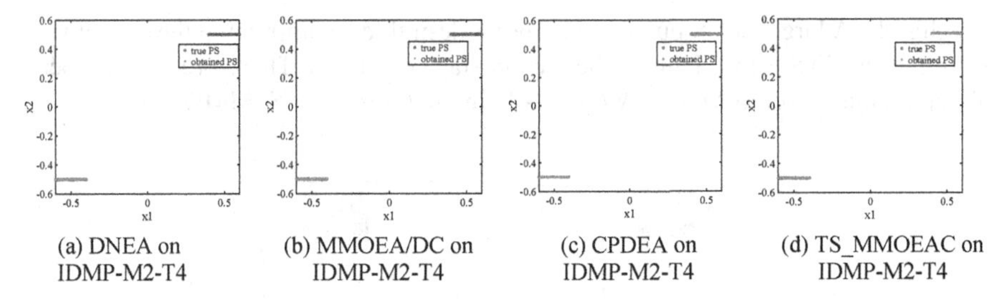

(a) DNEA on IDMP-M2-T4 (b) MMOEA/DC on IDMP-M2-T4 (c) CPDEA on IDMP-M2-T4 (d) TS_MMOEAC on IDMP-M2-T4

Fig. 2. Obtained PS on IDMP-M2-T4 by four algorithms.

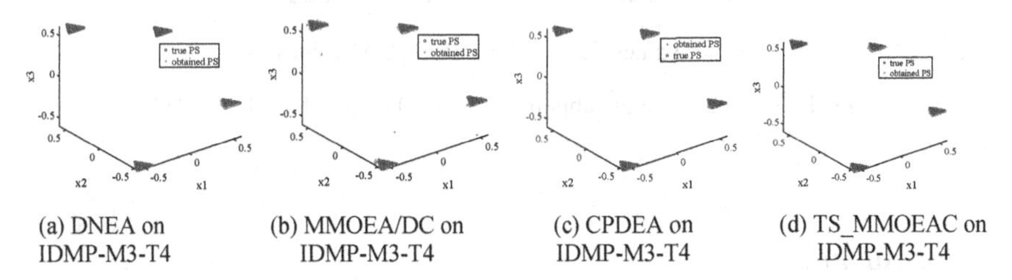

(a) DNEA on IDMP-M3-T4 (b) MMOEA/DC on IDMP-M3-T4 (c) CPDEA on IDMP-M3-T4 (d) TS_MMOEAC on IDMP-M3-T4

Fig. 3. Obtained PS on IDMP-M3-T4 by four algorithms.

and its competing algorithms are approximate in objective space, and the IGDM metric is similar to IGDX for six algorithms in Table 3. It is clear that TS_MMOEAC also obtains 10 best values, and CPDEA gains 2 best results in 12 test problems in Table 3. Additionally, the IGDM value of the proposed TS_MMOEAC is significantly better than five competing algorithms.

From the above analysis, it is deduced that the proposed TS_MMOEAC is capable of handling such multimodal multi-objective optimization problems with imbalance convergence and diversity in decision space and is superior to currently popular MMEAs.

4.3 The Influence of Subpopulation Numbers

An experiment, TS_MMOEAC with different subpopulation numbers, is designed to investigate the influence of subpopulation numbers on the performance of TS_MMOEAC. In this experiment, the subpopulation numbers are set to $Num = N/5(D-1)$, $Num = N/10(D-1)$, $Num = N/15(D-1)$, and $Num = N/20(D-1)$, respectively, where N and D are population size and decision variable numbers, respectively. The IGDX and IGDM performance of TS_MMOEAC are shown in Fig. 4. In Fig. 4, the numbers on the horizontal axis denote the following problems: 1 = IDMP-M2-T1, 2 = IDMP-M2-T2, 3 = IDMP-M2-T3, 4 = IDMP-M2-T4, 5 = IDMP-M3-T1, 6 = IDMP-M3-T2, 7 = IDMP-M3-T3, 8 = IDMP-M3-T4, 9 = IDMP-M4-T1, 10 = IDMP-M4-T2, 11 = IDMP-M4-T3 and 12 = IDMP-M4-T4.

It is observed that the IGDX and IGDF performance of TS_MMOEAC is similar on 12 problems except for IDMP-M3-T4, IDMP-M4-T1, and IDMP-M4-T2. It implies that the influence of the subpopulation numbers on the performance of TS_MMOEAC is negligible. A larger subpopulation number is desirable for improving the convergence performance of TS_MMOEAC in the second-stage evolution. Therefore, the number of subpopulations is set to $Num = N/5(D-1)$ in the proposed TS_MMOEAC.

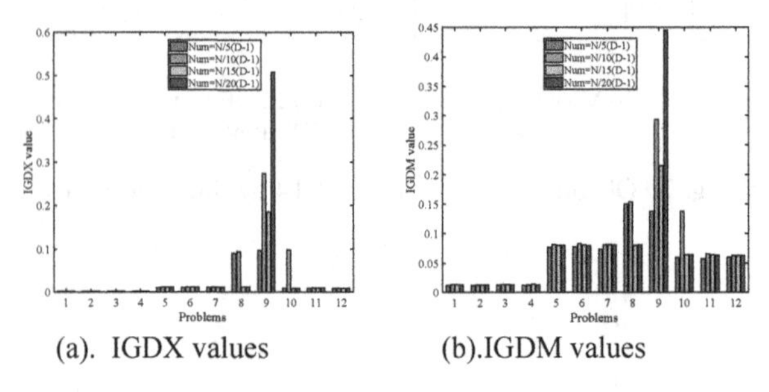

(a). IGDX values (b).IGDM values

Fig. 4. The influence of subpopulation numbers for TS_MMOEAC.

5 Conclusion

In this paper, we suggest a two-stage multimodal multi-objective evolutionary algorithm using clustering (TS_MMOEAC) for solving the multimodal multiobjective problem with the imbalance between convergence and diversity. The first-stage evolution using CPD-based clustering in TS_MMOEAC is to locate diverse local optimal solutions in decision space. Because these superior local optimal solutions have the potential to become equivalent Pareto optimal solutions with the imbalance between convergence and diversity in comparison to other equivalent Pareto solutions in the second-stage evolution. The second-stage evolution uses distance-based clustering to find all equivalent Pareto optimal solutions. The imbalance Pareto optimal solutions are more likely to be found in all subpopulations. Additionally, the convergence of the Pareto optimal solutions is remarkably boosted in the second-stage evolution. The experimental results validate the effectiveness of the two-stage evolutionary algorithm for solving the imbalanced MMOPs.

The two-stage evolutionary algorithm is capable of locating equivalent Pareto solutions with the imbalance between convergence and diversity, and our future work is to develop the proposed TS_MMOEAC to address real-world MMOPs.

Acknowledgments. This work is supported by the National Natural Science Foundation of China (No. 61873240).

References

1. Deb, K., Pratap, A., Agarwal, S., Meyarivan, T.: A fast and elitist multiobjective genetic algorithm: NSGAII. IEEE Trans. Evol. Comput. **6**(2), 182–197 (2002)

2. Zhang, Q., Li, H.: MOEA/D: a multiobjective evolutionary algorithm based on decomposition. IEEE Trans. Evol. Comput. **11**(6), 712–731 (2007)
3. Zitzler, E., Künzli, S.: Indicator-based selection in multiobjective search. In: Yao, X., et al. (eds.) PPSN 2004. LNCS, vol. 3242, pp. 832–842. Springer, Heidelberg (2004). https://doi.org/10.1007/978-3-540-30217-9_84
4. Liang, J., Yue, C.T., Qu, B.Y.: Multimodal multi-objective optimization: a preliminary study. In: Yao, X., Deb, K. (eds.) IEEE Congress on Evolutionary Computation, , vol. 1, pp. 2454–2461. IEEE (2016)
5. Kim, M., Hiroyasu, T., Miki, M., Watanabe, S.: SPEA2+: improving the performance of the strength pareto evolutionary algorithm 2. In: Yao, X., et al. (eds.) PPSN 2004. LNCS, vol. 3242, pp. 742–751. Springer, Heidelberg (2004). https://doi.org/10.1007/978-3-540-30217-9_75
6. Deb, K., Tiwari, S.: Omni-optimizer: a generic evolutionary algorithm for single and multi-objective optimization. Eur. J. Oper. Res. **185**(3), 1062–1087 (2008)
7. Zhou, A.M., Zhang, Q.F., Jin, Y.C.: Approximating the set of pareto-optimal solutions in both the decision and objective spaces by an estimation of distribution algorithm. IEEE Trans. Evol. Comput. **13**(5), 1167–1189 (2009)
8. Tanabe, R., Ishibuchi, H.: A framework to handle multimodal multiobjective optimization in decomposition-based evolutionary algorithms. IEEE Trans. Evol. Comput. **24**(4), 720–734 (2020)
9. Tanabe, R., Ishibuchi, H.: A niching indicator-based multi-modal many-objective optimizer. Swarm Evol. Comput. **49**, 134–146 (2019)
10. Yue, C.T., Qu, B.Y., Liang, J.: A multiobjective particle swarm optimizer using ring topology for solving multimodal multiobjective problems. IEEE Trans. Evol. Comput. **22**(5), 805–817 (2017)
11. Li, G., et al.: A SHADE-based multimodal multi-objective evolutionary algorithm with fitness sharing. Appl. Intell. **51**(12), 8720–8752 (2021). https://doi.org/10.1007/s10489-021-02299-1
12. Zhang, W.Z., Li, G.Q., Zhang, W.W., Liang, J., Yen, G.G.: A cluster based PSO with leader updating mechanism and ring-topology for multimodal multi-objective optimization. Swarm Evolut. Comput. **50**, 100569 (2019)
13. Liu, Y., Ishibuchi, H., Nojima, Y., Masuyama, N., Shang, K.: A double-niched evolutionary algorithm and its behavior on polygon-based problems. In: Auger, A., Fonseca, C.M., Lourenço, N., Machado, P., Paquete, L., Whitley, D. (eds.) PPSN 2018. LNCS, vol. 11101, pp. 262–273. Springer, Cham (2018). https://doi.org/10.1007/978-3-319-99253-2_21
14. Tanabe, R., Ishibuchi, H.: A review of evolutionary multimodal multiobjective optimization. IEEE Trans. Evol. Comput. **24**(1), 193–200 (2019)
15. Liu, H.L., Chen, L., Deb, K., Goodman, E.D.: Investigating the effect of imbalance between convergence and diversity in evolutionary multiobjective algorithms. IEEE Trans. Evol. Comput. **21**(3), 408–425 (2017)
16. Liu, Y.P., Ishibuchi, H., Yen, G.G., Nojima, Y., Masuyama, M.: Handling imbalance between convergence and diversity in the decision space in evolutionary multimodal multiobjective optimization. IEEE Trans. Evol. Comput. **24**(3), 551–565 (2020)
17. Lin, Q.Z., Lin, W., Zhu, Z.X., Gong, M.G., Coello, C.A.C.: Multimodal multi-objective evolutionary optimization with dual clustering in decision and objective spaces. IEEE Trans. Evol. Comput. **25**(1), 130–144 (2021)
18. Wang, W.L., Li, G.L., Wang, Y.L., et al.: Clearing-based multimodal multi-objective evolutionary optimization with layer-to-layer strategy. Swarm Evol. Comput. (2021). https://doi.org/10.1016/j.swevo.2021.100976

19. Tanabe, R., Ishibuchi, H.: A Decomposition-based evolutionary algorithm for multi-modal multi-objective optimization. In: Auger, A., Fonseca, C.M., Lourenço, N., Machado, P., Paquete, L., Whitley, D. (eds.) PPSN 2018. LNCS, vol. 11101, pp. 249–261. Springer, Cham (2018). https://doi.org/10.1007/978-3-319-99253-2_20
20. Liu, Y.P., Ishibuchi, H., Nojima, Y., Masuyama, N., Han, Y.Y.: Searching for local Pareto optimal solutions: a case study on Polygon-based problems. In: Zhang, M.J., Tan, K.C. (eds.) 2019 IEEE Congress on Evolutionary Computation (CEC 2019), pp. 896–903. IEEE (2019)
21. Liu, Y.P., Yen, G.G., Gong, D.W.: A multimodal multiobjective evolutionary algorithm using two-archive and recombination strategies. IEEE Trans. Evol. Comput. **23**(4), 660–674 (2019)

SLA: A Cache Algorithm for SSD-SMR Storage System with Minimum RMWs

Xuda Zheng[1], Chi Zhang[1], Keqiang Duan[1], Weiguo Wu[1](✉), and Jie Yan[2]

[1] Xi'an Jiaotong University, Xi'an, China
wgwu@mail.xjtu.edu.cn
[2] Hikvision (Beijing) Co., Ltd., Beijing, China

Abstract. To satisfy the low-cost and massive data storage requirements imposed by data growth, Shingled Magnetic Recording (SMR) disks are extensively employed in the area of high-density storage. The primary drawback of SMR disks is the write amplification issue caused by sequential write constraints, which is becoming more prominent in the field of non-cold archives. Although a hybrid storage system comprised of Solid State Disks (SSDs) and SMRs may alleviate the aforementioned issue, current SSD cache replacement algorithms are still limited to the management method of Least Recently Used (LRU). The LRU queue, on the other hand, is ineffective in reducing the triggers of Read-Modify-Write (RMW), which is a critical factor for the performance of SMR disks. In this paper, we propose a new SMR Locality-Aware (SLA) algorithm based on a band-based management method. SLA adopts the Dual Locality Compare (DLC) strategy to solve the hit rate reduction problem caused by the traditional band-based management method, as well as the Relatively Clean Band First (RCBF) strategy to further minimize the number of RMW operations. Experiments indicate that, compared to the MSOT method, the SLA algorithm can maintain a similar hit rate as the LRU, while reducing the number of RMWs by 77.2% and the SMR disk write time by 95.1%.

Keywords: Shingled Magnetic Recording · Replacement algorithm · Hybrid storage system · Spatial locality

1 Introduction

Shingled Magnetic Recording (SMR) increases disk capacity with minimal manufacturing changes. SMR disks remove gaps between tracks by writing tracks in an overlapping way to produce a higher areal density than disks utilizing Conventional Magnetic Recording (CMR) technology [12]. Due to its large capacity and low cost [5], SMR disks are widely utilized in data centers for cold archive

This work is supported by the National Key Research and Development Program (2017YFB1001701), the National Science Foundation of China (61972311), Shandong Provincial Natural Science Foundation (ZR2019LZH007).

Y. Lai et al. (Eds.): ICA3PP 2021, LNCS 13157, pp. 587–601, 2022.
https://doi.org/10.1007/978-3-030-95391-1_37

storage, and low-frequency storage. However, as the requirement for low-cost storage grows, the issue of how to convert SMR disks to standard storage has become increasingly pressing.

The three types of SMR interface implementations are Host Managed, Drive Managed, and Host Aware, which correspond to the three types of disk devices, HM-SMR, DM-SMR, and HA-SMR, respectively. The DM-SMR disk has a Persistent Buffer (PB) that caches non-sequential write requests and manages sequential write constraints internally, enabling both sequential and random write operations. The Shingled Translation Layer (STL) cleans up the PB periodically [13], which blocks the IO requests and causes a great jitter in application performance. In this paper, DM-SMR disks are used in our algorithms and experiments, and we shall refer to DM-SMR disk as SMR disk in the following section.

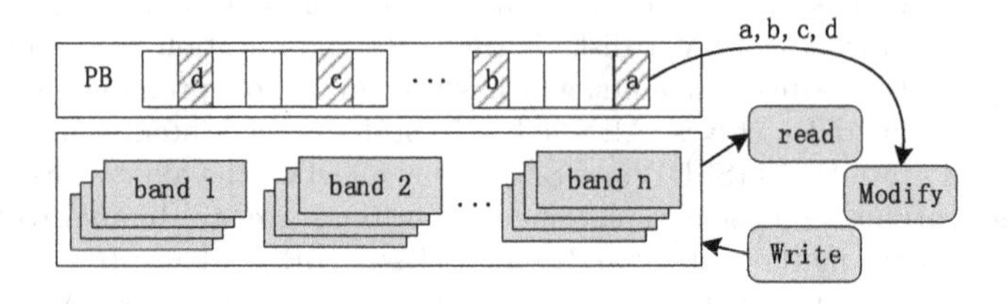

Fig. 1. The process of RMW operation

The SMR disk can only conduct sequential writes, random write data must be stored in the PB to enable in-place update write operations. The Read-Modify-Write (RMW) operation is introduced to write back the buffered data to the band region, When the PB is almost full. As illustrated in Fig. 1, the RMW operation first loads the whole band to which the evicted block belongs into the memory, then all blocks belonging to the band in the PB will be modified, and finally, the entire band will be written back into the SMR band area. While RMW operation solves the sequential write constraints, it also causes significant write amplification and performance fluctuations [4].

A hybrid storage system comprised of SSD and SMR disks relieves the aforementioned issues. However, traditional cache replacement algorithms, such as LRU, do not take the intrinsic characteristics of SMR disks into consideration, resulting in significant write amplification. The trade-off between write amplification and hit rate is the subject of several later cache replacement algorithms for SMR drives [4,6]. Sun et al. [1], on the other hand, utilize Pearson Correlation Coefficients (PCC) to quantify the relativity between the factors and I/O time and demonstrate that the number of RMWs is the key factor affecting the performance of SMR disks. This paper is also based on the above observations.

In this paper, we present the **S**MR **L**ocality-**A**ware (SLA) cache replacement algorithm, which is designed specifically for SSD-SMR hybrid storage system and focuses on how to minimize the number of RMWs. The cache block described in the following refers to the SSD cache block since the algorithm is SSD-oriented. Compared to the MOST algorithm, the SLA algorithm can reduce the number of RMWs by 77.2% and the disk write time by 95.12% while maintaining a hit rate similar to that of the LRU. The main contributions of this paper are as follows.

1) The spatial locality of evicted blocks and the eviction sequence of clean/dirty blocks are proposed as two key factors that influence the frequency of RMWs. Following that, we conducted theoretical analysis and comparison experiments to demonstrate the effectiveness of these two factors, as well as to address the side effects brought about by them.
2) It is presented that increasing spatial locality, i.e. employing a band-based management method, is crucial to minimizing RMWs. We investigate the footprint and recency of the band-based management system to figure out why it has such a poor hit rate. In this paper, The term "footprint" refers to the number of cache blocks per unit band, while "recency" refers to the number of recently accessed blocks per unit band. Following that, a band-based management method based on dual locality (spatial and temporal locality) is suggested to achieve a similar hit rate to LRU.
3) In the band-based management method, we estimate the costs of clean and dirty blocks and propose an Relatively Clean Band First (RCBF) eviction mechanism. The RCBF strategy can evict the cache blocks at the lowest cost, decreasing the number of RMWs even further.

The rest of this paper is organized in the following manner. Section 2 describes the related work. Section 3 presents the motivation for our work. The architecture of the SLA algorithm is described in Sect. 4. The SLA algorithm is evaluated in a simulated environment and on a real disk in Sect. 5. Finally, Sect. 6 concludes this paper with a summary of the SLA algorithm.

2 Related Work

SMR Device. Early discussions on SMR disks focused on the working principle [19,24], STL design [26,28], database [5,9,25,27], and file system [3,10]. Through the skylight approach, Aghayev et al. [2] found the main features of two drive-managed SMR disks, including PB size, band size, and so on, after commercial usage of SMR disks. Skylight also serves as a theoretical foundation for further SMR Disk Research. Ma et al. [7] employs a two-level buffer cache architecture, which keeps hot data in the filter buffer and sends cold data to the SMR disk. K-Framed Reclamation (KFR) [23] reduces performance recovery time by breaking down the reclamation process into more fine-grained KFramed reclamation processes. Ma et al. [11] manages the SMR disks by replacing the PB with a built-in NAND flash to achieve faster cleaning.

DM-SMR Based Hybrid Storage. In a hybrid storage system composed of SSDs and DM-SMRs, current research mainly focuses on SSD cache replacement algorithms. The major cache replacement algorithms based on SSD and

DM-SMR hybrid storage systems can be classified into two categories: LRU-like and Band-based. One of the fundamental components of the LRU-like algorithm is the LRU queue. Since the LRU does not consider the inherent characteristics of SMR disks, it is not beneficial to reduce the number of RMWs. The band-based algorithm can gather the eviction cache blocks as much as possible, thereby reducing the number of RMWs at the source. Currently existing LRU-like algorithms include Partially Open Region for Eviction (PORE) [4] and SMR-Aware Co-design (SAC) [1], and Band-based algorithms include MOST [6]. For the MOST policy, When the cache block needs to be evicted, the band with the **most** cache blocks will be selected.

The PORE algorithm organizes the cache queue using LRU, which limits the LBA range of evicted dirty blocks, lowering the rate of write amplification. However, PORE does not distinguish between clean blocks and dirty blocks. The SAC algorithm extends the PORE algorithm by separating clean and dirty blocks into two LRU queues for management, but its fundamental algorithm remains LRU and does not depart from conventional algorithm restrictions. The MOST algorithm tends to minimize write amplification and for the first time presents a band-based management method. All cache blocks belonging to the same band are grouped together, and the MOST policy evicts the band with the most cache blocks. The degree of aggregation of evicted cache blocks, i.e., spatial locality, can be maximized using the band-based management. However, the MOST algorithm ignores the effect of cache hit rate, which is the fundamental purpose of cache.

HA-SMR Based Hybrid Storage. Current research focuses on the recognition and restructuring of I/O streams in a hybrid storage system made of SSDs and HA-SMRs. ZoneTier [15] is a hybrid storage system that employs SSDs and HA-SMRs to effectively manage all non-sequential writes using a zone-based storage tiering and caching co-design technique. Xie et al. [16] takes use of HA-SMR drives' inherent host-awareness to filter both sequential and innocuous non-sequential writes out of SSD and write them straight to HA-SMR disk. Liu et al. [20] aims to improve the performance of a HA-SMR drive by rearranging out-of-order writes belonging to the same zone and using the SSD cache to absorb update writes and small random writes.

3 Motivation

The data cached in PB is written back in the form of RMW due to the overlapping of the internal tracks of the SMR disk. Correspondingly, the performance fluctuations of SMR disk is caused by RMW. As a consequence, to guarantee the efficiency of the cache, the replacement algorithm must be adaptively adjusted according to the specific storage medium characteristics, i.e. RMW. Therefore, the motivation of this paper is how to effectively minimize the number of RMWs.

The reduction of the number of RMWs is the key factor in optimizing SMR disk performance. The degree of spatial locality of evicted blocks and the sequence of evicted clean/dirty blocks are two major factors that influence

RMWs. We think that the better the locality of the evicted block, that is, the more concentrated the distribution, the fewer RMWs will be triggered. Similarly, since clean blocks do not generate write-back operation, it do not trigger RMW. The remainder of this section will demonstrate the effectiveness of these two factors.

Algorithm 1. LRU_SBSC eviction operation

Input: The missed block

Output: The number of blocks to evict

1: **if** $cache\_full == TREU$ **then**
2: $evict\_block = pop\_buf(LRU \rightarrow tail)$
3: $band\_num = calculate\_band\_number(evict\_block)$
4: **while** $cur\_evict\_num < max\_evict\_num$ **do**
5: delete_from_LRU(evict_block)
6: write_to_smr(evict_block)
7: **if** $band\_num \rightarrow next \ \ != NULL$ **then**
8: $evict\_block = band\_num \rightarrow next$
9: **end if**
10: **end while**
11: **end if**
12: **return** cur_evict_num

We design the LRU_SBSC (Same Band, Same Cleaning) algorithm to verify the impact of the spatial locality of evicted blocks. The algorithm organizes the queue in an LRU manner. When the queue is full, the last block in the queue is chosen for eviction, and all blocks in the same band are evicted at the same time. To demonstrate the influence of the evicted block's spatial locality on the performance of the SMR disk, we count the size of the SPL (Spatial Locality) indicator. Equation 1 depicts the $SPL_{average}$ calculation formula. *band_blocks* represents cache blocks for a single band that are evicted in a cycle (a cycle is defined as the process of evicting a certain number of cache blocks), and *band count* represents the number of bands to which these cache blocks belong.

$$SPL_{average} = \frac{\sum_{k=1}^{n} band\_blocks_k}{band\ count} \tag{1}$$

We conducted experiments based on *src* trace(detailed information in Tab. 1) to get realistic $SPL_{average}$ and RMWs values. The abscissa indicates the number of evicted cycles, while the ordinate reflects the $SPL_{average}$ value, as illustrated in Fig. 2(a). The value of $SPL_{average}$ varies dynamically in relation to the granularity of the eviction. The granularity is 64, 256, 1024, and 4096 correspondingly, with each unit representing a 4 KB physical block. The experimental findings are presented in Fig. 2(b), where the number of RMWs reduces as $SPL_{average}$ increases, but increases when $SPL_{average}$ is too high. In Sect. 4.3, we will introduce how to avoid side effects caused by excessive $SPL_{average}$.

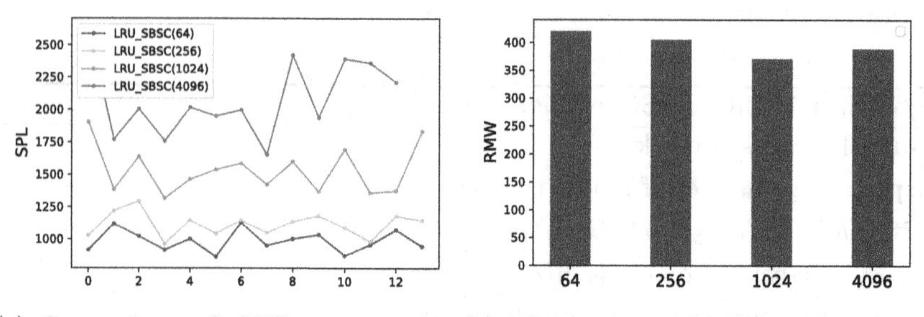

(a) Comparison of $SPL_{average}$ under different eviction granulates (b) The number of RMWs under different $SPL_{average}$

Fig. 2. The effect of $SPL_{average}$ on the number of RMWs

The second factor is the sequence of evicted clean and dirty blocks. The effect of this factor on RMWs is self-evident, since clean blocks do not need to be written back, and no RMW is triggered. However, merely prioritizing the eviction of clean blocks [14] is not a suggested replacement algorithm, since it will have a detrimental impact on the hit rate of the cache. In Sect. 4.4, we will discuss how to circumvent this issue.

4 Design and Algorithm

We developed the SLA algorithm taking into account the substantial effect of the aforementioned two factors on RMW. Section 4.1 primarily introduces three key components of the SLA algorithm. Sections 4.2, 4.3, and 4.4 detail the specific functions of the three components.

4.1 Overview

The architecture of the SLA algorithm is depicted in Fig. 3. The algorithm is made up of three key components: Band-based Management (BM), Dual Locality Compare (DLC), and Relatively Clean Band First (RCBF). We classify cache blocks based on three criteria: physical location, recent access, and clean/dirty data. The data blocks of the same color indicate that they belong to the same band, **F** represents the number of data blocks in the band, **R** represents the data that has been accessed recently, and **C** represents clean data.

The BM component determines its band number by calculating the offset of the cache blocks [1], and it manages and evicts cache blocks belonging to the same band in a consistent manner. The DLC component augments the eviction mechanism with the temporal locality of the cache block and the spatial locality of the band, ensuring the band-based management method's hit rate. To obtain the final eviction sequence, the RCBF component sorts the bands filtered by DLC according to the weight of the clean data block.

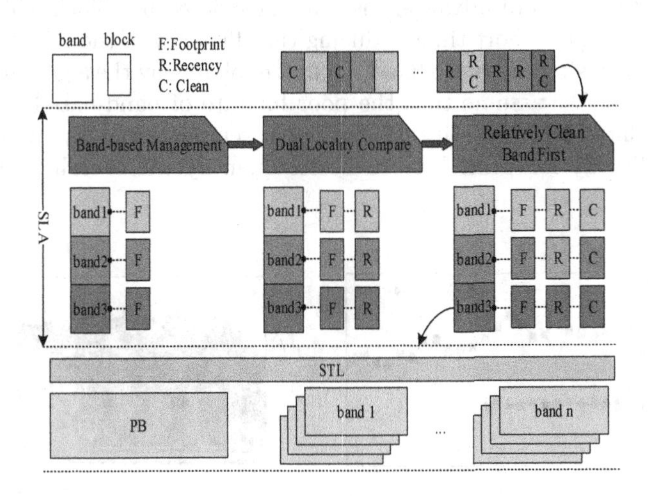

Fig. 3. SLA algorithm architecture

4.2 Band-Based Management

For a long time, the temporal locality of LRU and its variants has been the primary consideration in system cache management, while the spatial locality of cache blocks has been largely ignored [8]. In this paper, we argue that spatial locality is just as important as temporal locality. The reason for this is due to the following two factors. First, compared to the first-level cache, the temporal locality of SSDs is degraded as a second-level cache [22]. Second, SMR disks show unique spatial locality characteristics as a result of the sequential write constraints. Considering the above factors, the SLA algorithm proposed in this paper adopts a more spatially localized band-based management method.

We organize and manage cache blocks in the form of bands, and evict them in units of bands, to enhance the spatial locality of evicted cache blocks. The MOST algorithm has a similar management method, but it is solely dependent on the number of cache blocks in the band, and the band with the most cache blocks is evicted first. This eviction mechanism ensures that the cache block's spatial locality is optimum, but it also significantly lowers the hit rate. Our SLA algorithm employs the Dual Locality Compare and Relatively Clean Band First (RCBF) strategies to minimize RMW while maintaining a high hit rate.

4.3 Dual Locality Compare

Under the band-based management method, all blocks in the cache are statistically sorted according to the bands. When a band is evicted, the band with the most blocks is chosen first. Although this basic method ensures that each evicted block has the greatest spatial locality, it has two apparent flaws. The first disadvantage is that it does not account for the degree of cold and hot, resulting in a relatively low hit rate of the algorithm while lowering overall system performance. The second disadvantage is that the evicted hot block will be re-read into the memory in a short time, reducing the efficiency of the RMW operation.

Before addressing these two flaws, we first explain why the algorithm has such a poor hit rate. We assume that the poor hit rate of band-based management method is related to its locality. We represent temporal locality by recency and spatial locality by footprint here to better quantify the overall locality of the band.

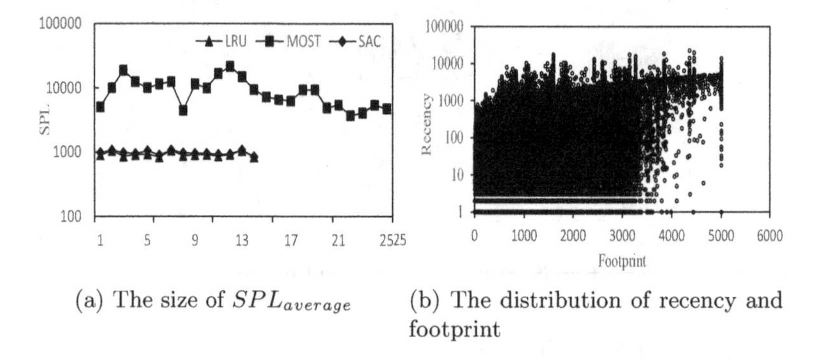

(a) The size of $SPL_{average}$ (b) The distribution of recency and footprint

Fig. 4. Characteristics of traditional band-based algorithms

To obtain an authentic locality, we also conducted experiments based on *src* trace. The footprint and recency of all bands are first counted, and the results are presented in Fig. 4(b). It can be observed that when a band's footprint is high, its recency is comparatively concentrated and high, while when the footprint of a band is low, its recency is more dispersed and low. It should be noted that evicting the band with the most cache blocks first almost always results in the eviction of more hot data blocks. This is why the band-based management method has a poor hit rate.

The $SPL_{average}$ value of the three algorithms, MOST, LRU, and SAC, is then calculated. The number of blocks in a single band is 5000, which must be stated in advance. The reason for this is that DM-SMR drives are made up of bands ranging in size from 15 MB to 40 MB [2], in this paper, we assume the band is 20 MB [1]. The $SPL_{average}$ values of the LRU and SAC algorithms are both about 1000, as shown in Fig. 4(a), while the MOST algorithm is about 10000, which is double the number of blocks per unit band. The $SPL_{average}$ value of MOST algorithm also demonstrates that the same band has been evicted many times in a unit cycle.

$$W_{TPL}(b) = \sum F(cur\_time - reference\_time) \quad (2)$$

$$W_{SPL}\left(b\right) = SPL_{band}\left(b\right) \tag{3}$$

Through the above analysis of the spatial and temporal locality of bands, it is observed that the influence of temporal locality and spatial locality on the hit rate of band-based algorithms is comprehensive. To quantify this influencing factor, we express the weight of temporal locality and spatial locality by $W_{TPL}\left(b\right)$ and $W_{SPL}\left(b\right)$, as shown in Eq. 2, 3. The temporal locality calculation Equation of a single data block [17] is $F\left(x\right) = \left(\frac{1}{p}\right)^{\alpha x}$ ($p \geq 2$, $0 \leq \alpha \leq 1$), and $SPL_{band}\left(b\right)$ represents the total number of blocks in the entire band. The complexity will grow as we count the access time and computation weight of each block. In this case, we borrow from *Calculus* and consider several blocks accessed at the approximate time as the same computation, or we use the number of recent accesses instead.

$$W_{band}\left(b\right) = \alpha W_{SPL}\left(b\right) - \beta W_{TPL}\left(b\right) + \gamma W \tag{4}$$

The overall weight of the band is expressed as $W_{band}\left(b\right)$. We may deduce from the above explanation that the $W_{band}\left(b\right)$ is inversely proportional to temporal locality and directly proportional to spatial locality. Because the locality value is a linear accumulation, the final calculation formula of $W_{band}\left(b\right)$ is given in Eq. 4. W is the value linked to clean blocks and dirty blocks, as discussed in the next section. The coefficients of these three terms are represented by α, β, and γ, which are all uniformly set to 1.

4.4 Relatively Clean Band First

The cost of clean blocks and dirty blocks in conventional HDDs is not much different. However, because of the write amplification of SMR disks, the cost of evicted dirty blocks is considerably greater than that of clean blocks. Taking this into consideration, we recalculated the eviction costs of clean and dirty blocks, represented by $Cost_{clean}$ and $Cost_{dirty}$, respectively. As shown in the Eq. 5 and Eq. 6, $Cost_{access\_miss}$ and $Cost_{write\_back}$ indicate the time needed to access the missed block and the time required to write the cache block back to disk, respectively. $P\left(access\right)$ represents the probability of the block being re-access.

$$Cost_{clean} = Cost_{access\_miss} * P(access) \tag{5}$$

$$Cost_{dirty} = Cost_{write\_back} + Cost_{access\_miss} * P(access) \tag{6}$$

The eviction mechanism in SLA algorithm takes the band as the unit, so we use $Cost_{band}$ to represent the cost of evicting a band. The calculation formula is as shown in Eq. 7, where N_{clean} and N_{dirty} represent the number of clean and dirty data blocks in the band, and N represents the sum of the two. So the above formula can be simplified to the Eq. 8.

$$Cost_{band} = Cost_{clean} * N_{clean} + Cost_{dirty} * N_{dirty} \tag{7}$$

$$Cost_{band} = Cost_{write\_back} * N_{dirty} + Cost_{access\_miss} * N * P(access) \quad (8)$$

The candidate eviction band has a comparable N value in the band-based management method, so it may be considered as a fixed value. Similarly, $P(access)$ can be regarded as a fixed value in a fixed eviction algorithm and trace. As a result, the eviction cost of a band is proportional to the number of dirty blocks evicted, i.e., $Cost_{band} \propto N_{dirty}$.

Given that the number of blocks in the candidate band is about equal, the more clean blocks there are, the fewer dirty blocks there are. As a result, we recommend using the Relatively Clean Band First (RCBF) eviction strategy. The number of clean blocks in the band is taken as a positive factor by RCBF and added to the computation of $W_{band}(b)$, which is the value of W in Eq. 4.

5 Evaluation

This section will conduct a comprehensive evaluation of our proposed SLA algorithm. Section 5.1 introduces the experimental settings. We conduct performance comparison experiments between SLA and other typical caching algorithms in Sects. 5.2 and 5.3 respectively. Section 5.2 verifies the performance of the hit rate, RMWs, and other parameters through the emulator. Then the experiments to evaluate the performance of the SLA algorithm on a real disk are presented in Sect. 5.3.

5.1 Experimental Setup

The experiments are carried out in the Ubuntu 20.04.1 version based on the kernel 5.11.0-34-generic environment, with an AMD EPYC 7302 16-Core CPU @ 3.0 GHz and 16 G DRAM. All experiments are based on the Seagate 8TB SMR drive model ST0008AS0002 [18]. We set the block size to 4 KB and artificially decrease the effective SSD capacity, i.e. the size of the actual cache, to 256 K blocks [4,13]. The experiments use five real-world enterprise I/O traces released by Microsoft Research in Cambridge [21], which represent five different write request ratio traces from 20% to 90%. The specific information is shown in Table 1. Since this algorithm is oriented to standard storage, write-only traces are not used for experiments.

To demonstrate the efficacy of the SLA algorithm, we also compared it to three other algorithms: LRU, MOST, and SAC. LRU and MOST represent the traditional LRU-like algorithm and Band-based algorithm, respectively. SAC stands for the state-of-the-art cache replacement algorithm in the hybrid storage system comprised of SSDs and DM-SMRs. When the SSD cache space is Insufficient, the LRU algorithm chooses the least recently used blocks for eviction, the MOST method chooses the band with the most cache blocks, and the SAC algorithm chooses the least costly blocks for eviction.

Table 1. Real-world traces details

Trace	Total requests (*10^6)	Write percent	Written LBA range (GB)	Accessed LBA range (GB)
src	14	0.832	3.8	3.93
mds	2.9	0.704	3.58	3.73
stg	6	0.682	7.29	7.58
web	9.6	0.464	7.11	8.35
usr	12.8	0.279	6.42	6.92

5.2 Emulation Experiments

Due to the fact that we cannot get the RMW information from the real disk, we performed experiments in a emulator environment on five different traces. SLA achieves a significant advantage compared with its competitors, As illustrated in Fig. 5, we first evaluate the number of RMWs and hit rates produced by various algorithms across five traces. The number of RMWs triggered by the SLA algorithm has reduced to various degrees when compared to LRU-like algorithms such as LRU and SAC, with an average reduction of 91.8% and 72.6%, respectively. When compared to the MOST algorithm, it decreased by 86.6%, 60.7%, and 38.8%, respectively, for *src*,*usr*, and *web*. While *mds* and *stg* do not trigger RMWs, the number of RMWs triggered by the five traces decreased by an average of 77.2%.

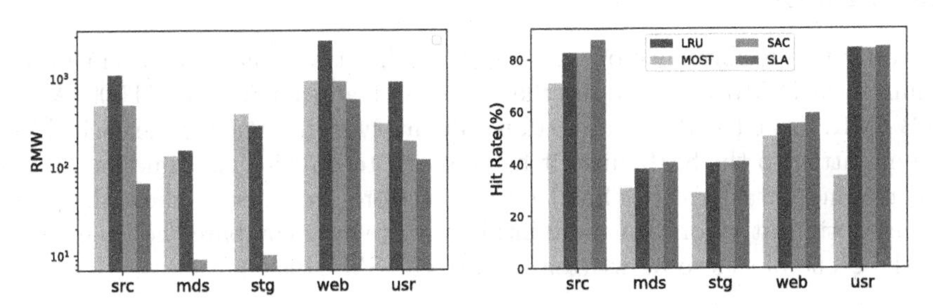

(a) Comparison of the number of RMWs under different traces

(b) Comparison of hit rates under different traces

Fig. 5. The performance of the four algorithms under different traces

The SLA algorithm maintains the same hit rate as the LRU-like algorithm, which compensates for the poor hit rate of band-based management methods. The hit rate of SLA demonstrates that the reduction in RMW is not achieved by evicting clean blocks at the cost of hit rate, but rather as a consequence of the combination of spatial and temporal locality.

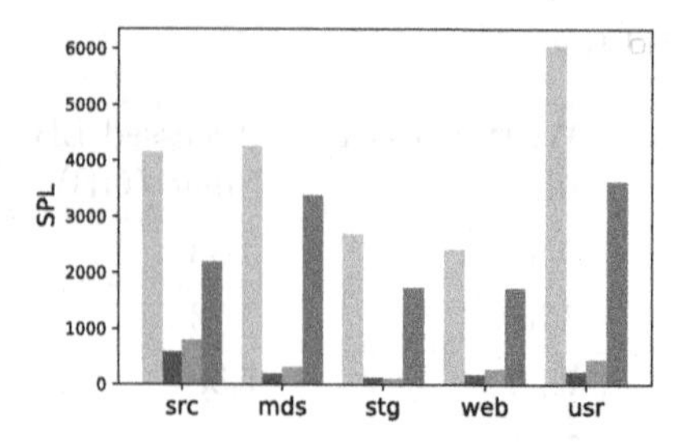

(a) SPL size comparison under different read/write ratios

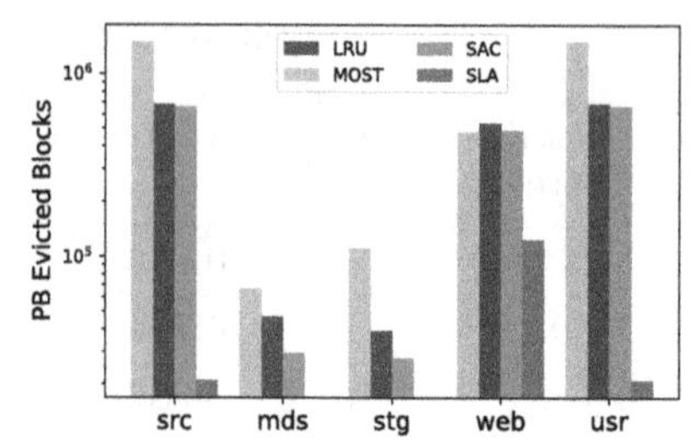

(b) Comparison of PB evicted blocks under different read/write ratios

Fig. 6. The performance of the four algorithms under different read/write ratios

Further analysis is shown in Fig. 6, the number of PB write-back blocks of SLA algorithm is greatly reduced compared to LRU-like algorithms. Compared to MOST algorithm under src, usr, and web, it is reduced by 98.6%, 92.1%, and 74.2%, respectively. Because there is no RMWs is triggered in mds and stg, the number of PB write-back blocks is 0. In comparison to the MOST algorithm, the $SPL_{average}$ value can be slightly adjusted, demonstrating that the SLA algorithm reduces the probability of the eviction of hot data. Simultaneously, the huge gap in $SPL_{average}$ shows that the spatial locality of band-based management methods, such as SLA and MOST, is far superior to that of LRU-like management methods.

5.3 Real Disk

To verify the performance of SLA and the other three cache replacement algorithms in real SMR disks, we run the above five traces on Seagate ST0008AS0002 8TB SMR. Since the RMW operation only involves the time for the cache blocks to be written to the SMR disk, it has nothing to do with the time for the SSD and reading data from the SMR disk. Therefore, in order to show the performance optimization of SLA more intuitively, we only compare the time it takes for the cache blocks to be written back to the SMR disk.

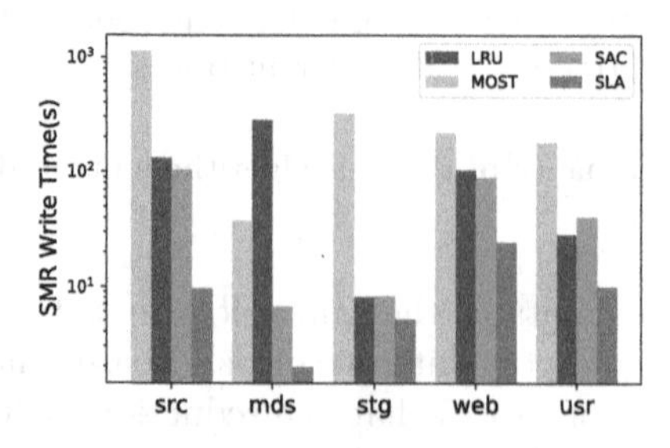

Fig. 7. Disk write time of four algorithms under real SMR disk

The experimental results are shown in Fig. 7. Compared with LRU-like algorithms, the SMR disk write time of the SLA algorithm is greatly reduced. Furthermore, compared with the MOST algorithm, the SLA algorithm reduces under five different traces by 99.1%, 94.8%, 98.4%, 88.8%, and 94.5%, respectively. The disk write time of the five traces decreased by an average of 95.12%.

6 Conclusion

To alleviate the RMWs issue on SMR disks, we propose a new SMR Locality-Aware (SLA) algorithm based on a band-based management method, which optimizes the two key factors affecting RMWs by using DLC and RCBF strategies. Compared with algorithms such as LRU, MOST, and SAC, SLA triggers the least RMWs while maintaining a high hit rate. It is worth mentioning that if the load cannot offer adequate locality for the SLA algorithm, or if the locality is insufficient, the hit rate of SLA will be significantly reduced (the same for LRU-like). To make things worse, since the RCBF strategy is locality-dependent, this issue will be amplified, which is also our next optimization goal.

References

1. Sun, D., Chai, Y.: SAC: a co-design cache algorithm for emerging SMR-based high-density disks. In: Proceedings of the Twenty-Fifth International Conference on Architectural Support for Programming Languages and Operating Systems, pp. 1047–1061. ACM, New York (2020)
2. Aghayev, A., Shafaei, M., Desnoyers, P.: Skylight-a window on shingled disk operation. ACM Trans. Storage **11**(4), 1–28 (2015)
3. Chen, S., Yang, M., Chang, Y., Wu, C.: Enabling file-oriented fast secure deletion on shingled magnetic recording drives. In: 2019 56th ACM/IEEE Design Automation Conference, pp. 1–6. IEEE (2019)
4. Wang, C., Wang, D., Chai, Y., Wang, C., Sun, D.: Larger cheaper but faster: SSD-SMR hybrid storage boosted by a new SMR-oriented cache framework. In: Proceedings 33rd International Conference Massive Storage System Technology (2017)
5. Yao, T., Tan, Z., Wan, J., Huang, P., Zhang, Y., et al.: SEALDB: an efficient LSM-tree based KV store on SMR drives with sets and dynamic bands. IEEE Trans. Parallel Distrib. Syst. **30**(11), 2595–2607 (2019)
6. Xiao, W., Dong, H., Ma, L., Liu Z., Zhang, Q.: HS-BAS: a hybrid storage system based on band awareness of shingled write disk. In: 2016 IEEE 34th International Conference on Computer Design, pp. 64–71. IEEE (2016)
7. Ma, C., Shen, Z., Wang, Y., Shao, Z.: Alleviating hot data write back effect for shingled magnetic recording storage systems. IEEE Trans. Comput. Aided Des. Integr. Circuits Syst. **38**(12), 2243–2254(2018)
8. Jiang, S., Ding, X., Chen, F., Tan, E., Zhang, X.: DULO: an effective buffer cache management scheme to exploit both temporal and spatial locality. In: Proceedings of the 4th Conference on USENIX Conference on File and Storage Technologies, pp. 1–8. USENIX Association (2005)

9. Liang, Y., Chen, T., Chi, C., Wei, H., Shih, W.: Enabling a B+-tree-based data management scheme for key-value store over SMR-based SSHD. In: 2020 57th ACM/IEEE Design Automation Conference, pp. 1–6. IEEE (2020)
10. Aghayev, A., Ts'o, T., Gibson, G.: Evolving Ext4 for shingled disks. In: 15th USENIX Conference on File and Storage Technologies, pp. 105–120. USENIX Association (2017)
11. Ma, C., Shen, Z., Han, L., et al.: FC: built-in flash cache with fast cleaning for SMR storage systems. J. Syst. Architect. **98**, 214–220 (2019)
12. Gibson, G., Ganger, G.: Principles of operation for shingled disk devices. In: Proceedings 3rd USENIX Workshop Hot Topics in Storage and File Systems, pp. 1–5. USENIX Association (2011)
13. Xie, X., Yang, T., Li, Q., et al.: Duchy: achieving both SSD durability and controllable SMR cleaning overhead in hybrid storage systems. In: Proceedings of the 47th International Conference on Parallel Processing, pp. 1–9. ACM (2018)
14. Park, S., Jung, D., Kang, J., et al.: CFLRU: a replacement algorithm for flash memory. In: Proceedings of the 2006 International Conference on Compilers, Architecture and Synthesis for Embedded Systems, pp. 234–241. ACM (2006)
15. Xie, X., Xiao, L., Du, D.H.: ZoneTier: a zone-based storage tiering and caching co-design to integrate SSDS with SMR drives. ACM Trans. Storage **15**(3), 1–25 (2019)
16. Xie, X., Xiao, L., Ge, X., et al.: SMRC: an endurable SSD cache for host-aware shingled magnetic recording drives. IEEE Access **6**, 20916–20928 (2018)
17. Lee, D., Choi, J., Kim, J H., et al.: On the existence of a spectrum of policies that subsumes the least recently used (LRU) and least frequently used (LFU) policies. In: Proceedings of the 1999 ACM SIGMETRICS International Conference on Measurement and Modeling of Computer Systems, pp. 134–143. ACM (1999)
18. Seagate 2016. http://www.seagate.com/wwwcontent/product-content/hdd-fam/seagate-archive-hdd/enus/docs/100757960h.pdf. Accessed 22 Sept 2021
19. Cassuto, Y., Sanvido, M., et al.: Indirection systems for shingled-recording disk drives. In: Proceedings of the 2010 IEEE 26th Symposium on Mass Storage Systems and Technologies, pp. 1–14. IEEE (2010)
20. Liu, W., Zeng, L., Feng, D., et al.: ROCO: using a solid state drive cache to improve the performance of a host-aware shingled magnetic recording drive. J. Comput. Sci. Technol. **34**(1), 61–76 (2019)
21. Narayanan, D., Donnelly, A., Rowstron, A.: Write off-loading: practical power management for enterprise storage. ACM Trans. Storage (TOS) **4**(3), 1–23 (2008)
22. Ye, F., Chen, J., Fang, X., et al.: A regional popularity-aware cache replacement algorithm to improve the performance and lifetime of SSD-based disk cache. In: 2015 IEEE International Conference on Networking, Architecture and Storage, pp. 45–53. IEEE (2015)
23. Ma, C., Wang, Y., Shen, Z., Shao, Z.: KFR: optimal cache management with K-framed reclamation for drive-managed SMR disks. In: 2020 57th ACM/IEEE Design Automation Conference, pp. 1–6. IEEE (2020)
24. Feldman, T., Gibson, G.: Shingled magnetic recording: areal density increase requires new data management. login Mag. USENIX SAGE **38**(3), 22–30 (2013)
25. Yao, T., Wan, J., Huang, P., et al.: GearDB: a GC-free key-value store on HM-SMR drives with gear compaction. In: 17th USENIX Conference on File and Storage Technologies, pp. 159–171. USENIX Association (2019)
26. He, W., Du, D.H.: Novel address mappings for shingled write disks. In: 6th USENIX Workshop on Hot Topics in Storage and File Systems, pp. 1–5. USENIX Association (2014)

27. Chen, S., Liang, Y., Yang, M.: KVSTL: an application support to LSM-tree based key-value store via shingled translation layer data management. IEEE Trans. Comput. (2021)
28. Yang, T., Wu, H., Huang, P., et al.: A shingle-aware persistent cache management scheme for DM-SMR disks. In: 2017 IEEE International Conference on Computer Design, pp. 81–88. IEEE (2017)

Trajectory Similarity Search with Multi-level Semantics

Jianbing Zheng[1], Shuai Wang[1], Cheqing Jin[1](✉), Ming Gao[1], Aoying Zhou[1], and Liang Ni[2]

[1] East China Normal University, Shanghai, China
{jbzheng,51215903058}@stu.ecnu.edu.cn,
{cqjin,mgao,ayzhou}@dase.ecnu.edu.cn
[2] Nanjing University of Science and Technology, Nanjing, China

Abstract. With the widespread popularity of intelligent mobile devices, massive trajectory data have been captured by mobile devices. Although trajectory similarity search has been studied for a long time, most existing work merely considers spatial and temporal features or single-level semantic features, thus insufficient to support complex scenarios. Firstly, we define multi-level semantics trajectory to support flexible queries for more scenarios. Secondly, we present a new "spatial + multi-level semantic" trajectory similarity query, and then propose a framework to find k most similar ones from a trajectory database efficiently. Finally, to hasten query processing, we build a multi-layer inverted index for trajectories, design 4 light-weight pruning rules, and propose an adaptive updating method. The thorough experimental results show that our approach works efficiently in extensive and flexible scenarios.

Keywords: Multi-level semantics · Inverted index · Trajectory similar query

1 Introduction

With the popularization and development of ubiquitous computing and positioning technology, users' location information is frequently collected from daily life by mobile intelligent devices. The massive trajectory data not only reflect a person's daily behavior, but also indicate the activity pattern of a user group or even the whole city. Therefore, trajectory analysis is involved in many fields, such as precision marketing, statistical analysis and policy making.

Similarity search is a typical issue in trajectory data management. Given a query trajectory, the goal is to find one or more trajectories close to it. Currently, most of the existing works that mainly focus on spatial and temporal features of trajectory, are lack of effective utilization of trajectories' semantics. In fact, with the popularity of intelligent devices and applications, more and more data beyond space and time are collected and converted into semantic information. Trajectory similarity search that integrates semantics is more meaningful in real

Y. Lai et al. (Eds.): ICA3PP 2021, LNCS 13157, pp. 602–619, 2022.
https://doi.org/10.1007/978-3-030-95391-1_38

life, which therefore supports more scenarios [1]. Moreover, since the semantic attributes of trajectory points are hierarchical, it can be defined as a specific attribute name, or an attribute level above the specific attribute name. In most cases, although semantic attributes may not be identical at the current level, they may be similar at higher levels. Therefore, comparing the hierarchical semantic similarity between trajectories has wide application, such as transportation modes and trip purpose. Take Fig. 1 as an example (the X-axis and Y-axis are longitude and latitude respectively), although the four trajectories are far away and not similar for spatial dimension, they are similar in transportation mode and trip purpose for semantic dimension.

Transportation Modes Classification and Planning. Multi-level semantic trajectories can be queried and applied in a variety of ways, which cannot be realized simultaneously by single-level semantics. For example, the four trajectories represent four students going to school, where Tr_1 and Tr_3 take the bus, Tr_4 takes the subway, and Tr_2 drives to school. If someone wants to choose a favorite mode from all possible public transportation modes to school, the high-level semantics, "Public Transportation", can be used to query all the trajectories in transportation mode. Moreover, integrating multi-level semantics, the trajectories with the same destination in the same region can be found as an alternate path. For example, Tr_1 and Tr_4 are close in space, and both public transport. When there are few buses, students on Tr_1 (bus) can choose Tr_4 (subway) to go to school in the same area through the query of high-level semantics, "Public Transportation", and space.

Trip Purpose Classification and Planning. With the combination of low and high level semantic tags, more personalized services can be provided. For example, when travelling to Beijing, we not only want to visit the Palace Museum and the Great Wall (low-level semantic tag), but also taste Beijing's special food and theater (high-level semantic tag), which cannot be achieved simultaneously with single-level semantics. In this example, with the high-level semantic tag ("Special Food"), multiple trajectories passing through different specialty restaurants will be found for people to choose.

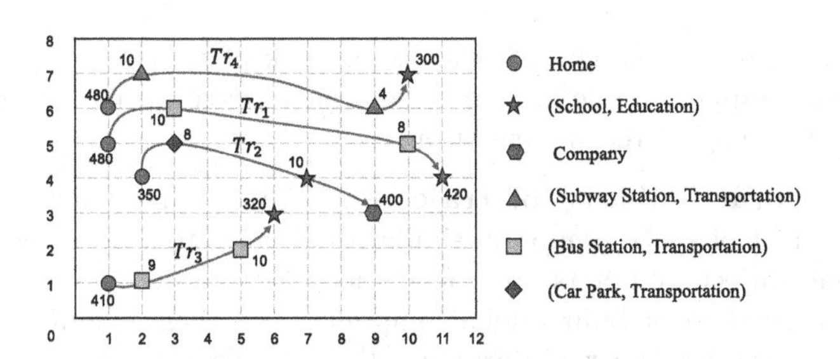

Fig. 1. An example of multi-level semantic trajectories.

However, there are several challenges in trajectory data analysis. Since massive trajectory data are collected by mobile devices every day, it is challenging to convert discrete trajectory points into stay points and then attach accurate semantic tags. In addition, previous studies only considered single-level semantics, and Jaccard Index was usually used to evaluate semantic similarity of trajectories. However, since Jaccard Index no longer works for multi-level semantics, how to define an appropriate similarity evaluation method is also challenging. Furthermore, processing large-scale trajectory data raises another challenge. The conventional methods, which compare trajectories one by one, is quite expensive. Therefore, it is critical to design efficient pruning methods and feasible indexes for optimization.

In this paper, we propose a similar trajectory search framework integrating hierarchical semantic features. We first mine stay points from raw trajectories, then choose appropriate multi-level semantic tags for each stay point to generate semantic trajectories. To evaluate the difference between a pair of trajectories, we define an efficient trajectory distance based on time, space and semantics. Meanwhile, we introduce spatial and semantic indexes, and propose several lightweight pruning rules to optimize query processing. To sum up, we make the following contributions in this paper:

- We introduce a new way to evaluate trajectory distance based on spatio-temporal and hierarchical semantic features.
- We present an efficient query processing method, which relies on indexes and four pruning rules.
- We verify the performance of our proposed method on two different real datasets.

The rest of our paper is organized as follows. Section 2 reviews related work in recent years. Section 3 defines some important concepts. Section 4 introduces our novel framework in detail. Section 5 verifies the performance of our proposed methods through experiments. Section 6 summarizes the paper briefly.

2 Related Work

Trajectory similarity research, a foundational task of trajectory data management, plays an important role in many fields, such as traffic management, urban planning and intelligent recommendation.

Spatio-Temporal Similarity of Trajectory. Early studies mainly focused on the measurement of spatio-temporal similarity of trajectory [2,3]. To deal with the case that trajectory points are not aligned, Ta et al. proposed a similarity measurement method of bidirectional mapping, and then generated a feature code for each trajectory for rapid pruning [4]. In [5], Shang et al. combined time and space linearly to calculate trajectory similarity, and proposed a two-stage algorithm to support parallel retrieval of spatial and temporal dimensions, which improves query efficiency.

Semantic Similarity of Trajectory. With the emergence of the blending of location and text, since more and more trajectories carry text information, more researches focus on trajectory similarity search with semantics [6,7]. In [8], Zheng et al. proposed fuzzy keyword query for semantic trajectory. Given a set of query keywords, by calculating the pairwise semantic edit distance, top-k trajectories with the smallest distance are returned.

Spatial and Semantics Similarity of Trajectory. He work that combines space and semantics has also been studied in recent years [9,10]. [11] proposed a top-k spatial keyword activity trajectory query method, aiming to find a set of trajectories that both geographically close and semantically meet the query requirement. Compared with the existing fuzzy keyword query, this query can find more similar trajectories. In [12], Chen et al. proposed a divide-and-conquer algorithm to deduce the boundary of spatial similarity and textual similarity between two trajectories, which realizes trajectory pruning without calculating the exact value of trajectory similarity and improves the query efficiency.

Although the trajectory similarity search based on spatio-temporal dimensions is intuitive, how to reflect the semantic attribute is challenging. Also, it is relatively unitary to compute trajectory similarity only from semantic dimension. Therefore, the combination of space and semantics is more effective and suitable for more scenarios. But most existing researches only consider single level semantics, without in-depth consideration of the hierarchy of semantics. And the existing method of calculating trajectory distance based on single-level semantics is not suitable for multi-level semantics. Consequently, we define a new way to compute the distance between trajectories, based on which an efficient method by integrating several pruning rules is given.

Index of Trajectory. In addition, to improve query efficiency, it is necessary to build indexes in both spatial and semantic dimensions. [13] described the trajectory representation and storage, and lists the indexing methods for spatial text trajectory data: quadtree, R-tree, grid index and Z-order curve. Space-efficient index representations and processing frameworks are crucial for trajectory data, and the majority of trajectory search solutions [14] rely on an R-tree [15], which store all points from the raw trajectories. Since trajectory datasets such as T-drive [16] often contain millions of points, the R-tree must manage an enormous number of maximum bounding rectangles (MBR), which have prohibitive memory cost in practice. Simpler Grid-index solutions are sometimes more appropriate in such scenarios [17]. Therefore, we combine grid index and inverted index to establish spatial index and semantic index.

3 Problem Definition

In this section, we define the query formally, along with some core concepts.

Definition 1. Trajectory (Tr). A trajectory Tr is a series of n points ordered by time, $Tr = (p_1, p_2, ..., p_n)$, where $p_i = (t_i, loc_i)$ means that the moving object is located at loc_i at time t_i.

Definition 2. Stay Point (*SP*). A stay point $s = (loc, len, c)$ represents that a moving object has stayed at *loc* for *len* time, and c is a label to represent the corresponding semantics.

As aforementioned in Sect. 1, each stay point may have concrete semantics, e.g., restaurant, cinema and university. The semantics can be structured hierarchically. For example, through the POI information of the map and manual correction, we define the multi-level semantics of Tsinghua University as (educational institution, school, university, Tsinghua University). Furthermore, some special multi-level semantics may be customized for different persons, such as home and working area, as illustrated in Fig. 2.

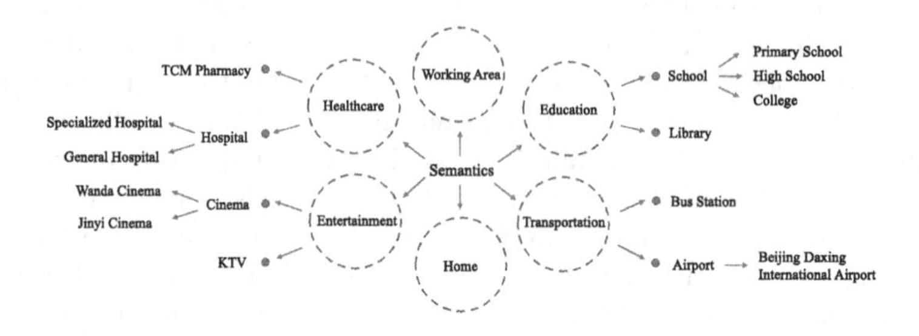

Fig. 2. An example of semantics.

Definition 3. Semantic Trajectory (*ST*). A semantic trajectory *ST* is a series of chronological ordered stay points of a moving object, $ST = (s_1, s_2, ..., s_m)$.

To compute the distance between a pair of semantic trajectories A and B (Spatial Semantic distance, S^2D), we integrate both spatial distance (SpD) and semantic distance (SeD) at the same time, as shown below.

$$S^2D(A, B) = \beta \cdot SpD(A, B) + (1 - \beta) \cdot SeD(A, B) \tag{1}$$

where $\beta \in [0, 1]$ leverages the importance of each feature. We set β a greater value if more attention is paid to space, and vice versa. The spatial distance, $SpD(A, B)$, is computed as normalized Euclidean distance.

$$SpD(A, B) = \begin{cases} \sum_{i=1}^{|A|} \|A.loc_i, B.loc_i\|, & |A| \leq |B| \\ MAX\_DIST, & |A| > |B| \end{cases} \tag{2}$$

$$\|A.loc_i, B.loc_i\| = \frac{dis(A.loc_i, B.loc_i) - minDis}{maxDis - minDis + 1} \tag{3}$$

where $dis(A.loc_i, B.loc_i)$ represents the Euclidean distance between point $A.loc_i$ and $B.loc_i$, $maxDis$ and $minDis$ are the maximum and minimum values of the distance between track A and B.

According to Eq. (2), if the length of B is smaller than that of A, we simply return MAX_DIST, a specific BIG value. Otherwise, we compare the first $|A|$ pairs of points. Note that $SpD(A, B) = SpD(B, A)$ only if $|A| = |B|$. If $|A| \neq |B|$, $SpD(A, B) \neq SpD(B, A)$. In comparison, the semantic distance, SeD, is a bit more complex, since the two labels may be identical at any level, or totally different at all levels. Hence, the semantic difference between two labels, $sd(c_1, c_2)$, and the overall semantic distance between two semantic trajectories, $SeD(A, B)$, are defined below.

$$sd(c_1, c_2) = \begin{cases} 0, & c_1 = c_2 \\ 1 - \alpha^g, & \text{identical in the upper } g\text{-th level} \\ 1, & \text{different at all levels} \end{cases} \tag{4}$$

$$SeD(A, B) = \begin{cases} \sum_{i=1}^{|A|} sd(A.c_i, B.c_i), & |A| \leq |B| \\ MAX\_DIST, & |A| > |B| \end{cases} \tag{5}$$

where $\alpha \in (0, 1)$ in Eq. (4) reflects the decay rate. $\forall c_1, c_2, sd(c_1, c_2) \in [0, 1]$. We use α to control the decay rate of semantic distance if two labels are identical at the g-th upper level, i.e., a smaller α value means higher decay rate.

Definition 4. Spatial Semantic Similar query (S^2Sim *query*)**.** Given a set of semantic trajectories $\Phi = \{ST_1, ST_2, ..., ST_g\}$, a query trajectory q, an adjustment coefficient β and an integer k, $S^2Sim(\Phi, q, \beta, k)$ query returns a subset Φ', such that (1) $|\Phi'| = k$, and (2) for any $ST_i \in \Phi'$ and $ST_j \in \Phi \setminus \Phi'$, $S^2D(ST_i, q) \leq S^2D(ST_j, q)$.

Take Fig. 1 as an example, suppose ST_1 is a query trajectory, $k = 1$ and $\alpha = 0.7$. Table 1 illustrates S^2D values between ST_1 and others in different β. If we consider space and semantics equally ($\beta = 0.5$), we return ST_4 because $S^2D(ST_1, ST_4) = 0.5 \times 8.74 + 0.5 \times 0.6 = 4.67$ is the minimal one. Similarly, if we only focus on semantics ($\beta = 0$) or space ($\beta = 1$) respectively, the query result is ST_3 or ST_2 accordingly.

Table 1. S^2D between ST_1 and other three trajectories in different β.

ST	SpD	SeD	S^2D			$wSpD$	$wSeD$	S^2D		
			$\beta = 0$	$\beta = 0.5$	$\beta = 1$			$\beta = 0$	$\beta = 0.5$	$\beta = 1$
ST_2	7.81	2.3	2.3	5.06	**7.81**	1.857	0.536	0.536	1.197	1.857
ST_3	20.02	0	**0**	10.01	20.02	4.504	0	**0**	2.252	4.504
ST_4	8.74	0.6	0.6	**4.67**	8.74	1.833	0.005	0.005	**0.919**	**1.833**

$$wSpD(A,B) = \begin{cases} \sum_{i=1}^{|A|} w_i \cdot \|A.loc_i, B.loc_i\|, & |A| \leq |B| \\ MAX\_DIST, & |A| > |B| \end{cases} \tag{6}$$

$$wSeD(A,B) = \begin{cases} \sum_{i=1}^{|A|} w_i \cdot sd(A.c_i, B.c_i), & |A| \leq |B| \\ MAX\_DIST, & |A| > |B| \end{cases} \tag{7}$$

As a moving object may stay at a stay point for different time length, the stay time length also acts as an important factor, which results in a weighted version. Equation (6) and (7) depict the weighted version of distance, where the i-th weight is $w_i = \frac{\min(A.len_i, B.len_i)}{\sum_{j=1}^{\min(|A|,|B|)} \min(A.len_j, B.len_j)}$.

The right part of Table 1 shows the weighted distance, where the stay time length of each stay point comes from in Fig. 1. If we consider space and semantics equally ($\beta = 0.5$), we return ST_4 because $S^2D(ST_1, ST_4) = 0.5 \times 1.833 + 0.5 \times 0.005 = 0.919$ is the minimal one. Similarly, if we focus on semantics ($\beta = 0$) or space ($\beta = 1$) respectively, the query result is ST_3 or ST_4 accordingly.

4 Framework

In this section, we detail our framework, shown in Fig. 3. Our framework contains three parts: semantic trajectory generation, index construction and query processing. In the first part, stay points are generated and then tagged with concrete semantics. In index construction part, we build spatial index and semantic index to quickly retrieve trajectories. Furthermore, we propose several pruning rules to quickly update the candidate set. Finally, we return top-k similar trajectories.

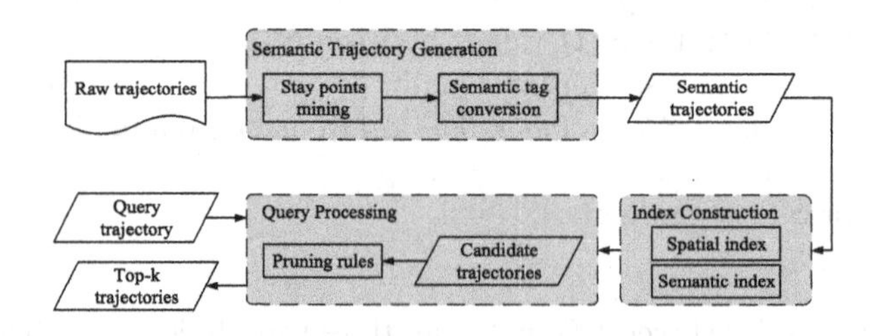

Fig. 3. Processing framework.

4.1 Semantic Trajectory Generation

To generate semantic trajectory, we mine stay points from raw trajectories, and then set appropriate labels. Algorithm 1 describes the steps to generate semantic trajectory. The first point is initialized as the left endpoint of a segment (line 1). If a segment $(p_l, ..., p_{r-1})$ satisfies the conditions $||loc_r, loc_l|| > \delta_{dis}$ and $t_r - t_l \geq \delta_t$, it means $||loc_{r-1}, loc_l|| \leq \delta_{dis}$ and we treat the mean of this segment as a stay point and put it into st (lines 3–6). Otherwise, if $t_r - t_l < \delta_t$, we ignore this segment and update the left endpoint (line 7). After traversing all points, if $l < |Tr|$, it means $||loc_{|Tr|}, loc_l|| \leq \delta_{dis}$. And if $t_{|Tr|} - t_l \geq \delta_t$, we also treat the mean of the segment $(p_l, ..., p_{|Tr|})$ as a stay point and put it into st (lines 10–12).

After getting stay points, we select the nearest point of interest (POI) as semantic label. Finally, the raw trajectory Tr is transformed into a semantic trajectory st.

Algorithm 1. GenST

Input: Tr: raw trajectory, δ_{dis}: distance threshold, δ_t: temporal threshold;
Output: st: semantic trajectory;

1: $l \leftarrow 1$, $st \leftarrow \emptyset$;
2: **for** $r = 2$ to $|Tr|$ **do**
3:   **if** $||loc_r, loc_l|| > \delta_{dis}$ **then**
4:     **if** $t_r - t_l \geq \delta_t$ **then**
5:       $s.loc \leftarrow$ the center of points $p_l, ..., p_{r-1}$, $s.len \leftarrow t_r - t_l$, $st.add(s)$;
6:     **end if**
7:     $l \leftarrow r$;
8:   **end if**
9: **end for**
10: **if** $l < |Tr|$ **and** $t_{|Tr|} - t_l \geq \delta_t$ **then**
11:   $s.loc \leftarrow$ the center of points $p_l, ..., p_{|Tr|}$, $s.len \leftarrow t_{|Tr|} - t_l$, $st.add(s)$;
12: **end if**
13: **return** st;

4.2 Query Processing

The most straightforward solution to find similar trajectories is to scan the whole trajectory database, and compute the distance to the query trajectory one by one, which is, however, inefficient for large-scale dataset. To improve query efficiency, it is necessary to build indexes in both spatial and semantic dimensions, and propose efficient light-weight pruning rules, which is the basic idea of this paper.

Index Construction. We have combined grid index and inverted index to establish spatial index and semantic index. According to the proposed trajectory similarity formula, we calculate the spatial and semantic distance point by point. Therefore, we put the i-th point of each trajectory together to construct hierarchical spatial and semantic index. The final number of levels is determined by the longest trajectory in the dataset.

For spatial dimension, we use grid index to handle trajectories, where the whole space is divided into multiple basic cells with same size, and each trajectory is mapped to the corresponding cell. Let a and b denote two arbitrary points in the space, function $mindis(cell(a), cell(b))$ computes the minimal distance between two cells containing a and b respectively. Note that $mindis(cell(a), cell(b)) = 0$ if they are at the neighbor cell. For semantic dimension, we build an inverted index for semantic information. And each node is represented as (c, trs), where c refers to the label and trs is a set of trajectories. Since the label is structured hierarchically, function $upper(c_i, g)$ gets the label in the g-th upper level. For example, *upper* (bus station, 1) = transportation, since transportation is the highest level of bus station, as shown in Fig. 2.

Index Maintenance. When a new trajectory is added to the database, it is unnecessary to rebuild the existing index, but only to update it incrementally. Spatial index and semantic index have the same essential structure, which are both multilevel inverted index.

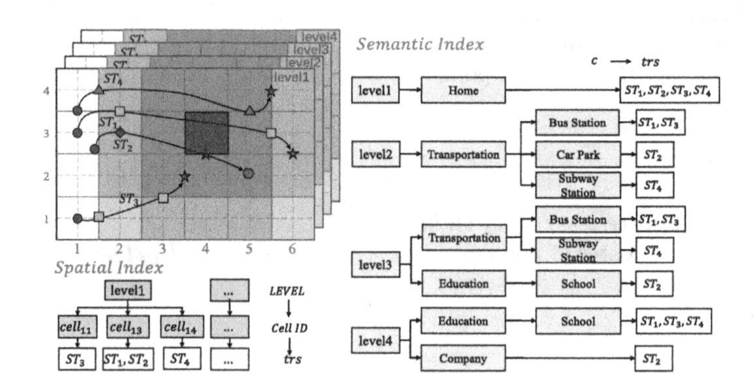

Fig. 4. An example of spatial and semantic indexes.

Figure 4 illustrates the spatial and semantic indexes for the trajectories in Fig. 1. The maximum number of four trajectories is 4, so the spatial index and semantic index have 4 levels respectively. The left part is spatial index, where the space is divided into 24 cells, each with a size of 2 km × 2 km. The minimal distances between $cell_{43}$ and $cell_{42}$, $cell_{23}$ and $cell_{21}$ (calling *mindis* function) are 0, 2 and $2\sqrt{2}$, respectively. The right part is semantic index, and each level is a tree-based structure, which satisfies the hierarchy of semantics. One trajectory may fall on many nodes, and one node will contain many trajectories. So, we can easily get trajectories using one semantic label. At level 2, ST_1 and ST_3 will be quickly extracted via "Bus Station".

Pruning Rules. Besides indexing, it is necessary to devise light-weight pruning rules to hasten query processing. In other words, given partial information, we are capable of judging whether a trajectory is a candidate or not. The main idea is that the distance between two trajectories can be accumulated point by point. Consequently, if the partial distance based on one or several stay point(s) is evaluated exceeding a given threshold, the whole trajectory can be filtered safely.

Pruning Rule 1 (Spatial Pruning). Given an upper bound τ, an adjustment parameter β, a query semantic trajectory q, and an arbitrary semantic trajectory r, $S^2D(q,r) > \tau$ if there exist positions $i_1, i_2, ..., i_d$, such that

$$\sum_{j=1}^{d} mindis(cell(q.loc_{i_j}), cell(r.loc_{i_j})) > \frac{\tau}{\beta} \tag{8}$$

Proof. According to Eq. (2), we have:

$$SpD(q,r) \geq \sum_{j=1}^{d} \| q.loc_{i_j}, r.loc_{i_j} \| \geq \sum_{j=1}^{d} mindis(cell(q.loc_{i_j}), cell(r.loc_{i_j})) > \frac{\tau}{\beta}$$

Integrating with Eq. (1), we have $S^2D(q,r) \geq \beta \cdot SpD(q,r) > \tau$. □

Pruning Rule 2 (Semantic Pruning). Given an upper bound τ, an adjustment parameter β, a query semantic trajectory q, and an arbitrary semantic trajectory r, $S^2D(q,r) > \tau$ if there exist positions $i_1, i_2, ..., i_d$, such that

$$\sum_{j=1}^{d} sd(q.c_{i_j}, r.c_{i_j}) > \frac{\tau}{1-\beta} \tag{9}$$

Proof. According to Eq. (5), we have:

$$SeD(q,r) \geq \sum_{j=1}^{d} sd(q.c_{i_j}, r.c_{i_j}) > \frac{\tau}{1-\beta}$$

Integrating with Eq. (1), we have $S^2D(q,r) \geq (1-\beta) \cdot SeD(q,r) > \tau$. □

Framework. We propose a query framework which is **T**op-**k** similar trajectories query based on **I**ndexes and **P**runing (TKIP). In Algorithm 2, we get spatial and semantic candidate sets according to Algorithm 3 and 4 at first, then we get top-k similar trajectories from the candidate set. Among them, the selection of spatial candidate set needs to calculate the nearest spatial distance. In order to improve the calculation efficiency, we generate cell distance coordinate pair set *DcSet* in advance and arrange them in ascending order of distance. Specifically, the *DcSet* contains two elements, the cell distance value *dist*, and its corresponding

Algorithm 2. TKIP

Input: ISP: spatial index, ISE: semantic index, q: query trajectory, $DcSet$, β, k;
Output: Ψ: top-k similar trajectories;

1: $\tau \leftarrow \infty$, $\Psi \leftarrow \emptyset$, $g \leftarrow 0$, $step \leftarrow 0$, $\sigma \leftarrow 0$, $lb_1 \leftarrow 0$, $lb_2 \leftarrow 0$;
2: **while** $lb_1 \cdot \beta + lb_2 \cdot (1-\beta) \leq \tau$ **do**
3: **if** β=1 **then**
4: $S \leftarrow GetSpCandidate(q, ISP, step, DcSet)$, $step \leftarrow step + 1$;
5: **else if** β=0 **then**
6: $S \leftarrow GetSeCandidate(q, ISE, g)$, $g \leftarrow g + 1$;
7: **else**
8: $S_1 \leftarrow GetSpCandidate(q, ISP, step, DcSet)$;
9: $S_2 \leftarrow GetSeCandidate(q, ISE, g)$;
10: $S \leftarrow S_1 \cup S_2$, $step \leftarrow step + 1$, $\sigma \leftarrow \sigma + (1-\beta)$, $g \leftarrow g + \lfloor \sigma \rfloor$;
11: **end if**
12: $S \leftarrow$ remove trajectories in S which length less than $|q|$;
13: **if** $S \neq \emptyset$ **then**
14: $\Psi \leftarrow k$ nearest trajectories in $\Psi \cup S$, $\tau \leftarrow \max(\{S^2D(q,r)|r \in \Psi\})$;
15: **end if**
16: $lb_1 \leftarrow DcSet_{step}.dis$, $lb_2 \leftarrow$semantic distance of the g-th level;
17: **end while**
18: **return** Ψ;

coordinate pair $CoorSet$. And the $CoorSet$ contains the cell distance between X-axis and Y-axis formed as $(\Delta x, \Delta y)$.

Algorithm 2 details the steps to process TKIP. We first initialize upper bound τ, candidate set Ψ, the level of multi-level semantics g, current distance *step*, adjusting parameter σ, spatial lower bound lb_1 and semantic lower bound lb_2 (line 1). For each iteration, we use the pruning rules to determine whether to terminate the algorithm (line 2). If not, we should determine the value of parameter β at first. If $\beta = 0$ or 1, then we only need to get a candidate set of one dimension, spatial or semantic, and update the corresponding parameters *step* or g (lines 3–6). Otherwise, both the spatial and semantic candidate sets are generated to get the final candidate set, and the parameter *step*, σ and g are updated accordingly (lines 7–10). Furthermore, we remove the trajectories in S whose trajectory length is less than $|q|$ (line 12). If candidate set S is not empty, we keep the k nearest trajectories in Ψ and update threshold τ (lines 13–15). At the end of each iteration, we update the lower bounds lb_1 and lb_2 (line 16). Finally, Ψ only contains the top k similar trajectories.

Algorithm 3 details the steps to get the spatial candidate set of similar trajectories. We first initialize spatial candidate set S_1 (line 1). For i from 1 to $|q|$, we get the cell id of the i-th point in q and all the cell id in the i-th level of ISP. According to the *DcSet*, we get the set of coordinate pairs corresponding to the current step, generate the corresponding candidate cells based on the query cell and put them in set DS. Next we take the intersection of cells which unmarked in *DcSet* and HCI as the candidate set and put them in S_1 (lines 2–7). Finally, we mark the trajectories in S_1 in the index to prevent re-access (line 8). The

Algorithm 3. GetSpCandidate

Input: q: query trajectory, ISP: spatial index, $step$: current distance step, $DcSet$;
Output: S_1: spatial trajectory candidate set;
1: $S_1 \leftarrow \emptyset$;
2: **for** i from 1 to $|q|$ **do**
3: $h \leftarrow cell(q.loc_i)$, $HCI \leftarrow$ get all the cell id in the i-th level of ISP;
4: $A \leftarrow DcSet_{step}.CoorSet$, $DS \leftarrow \{h.x + A.\Delta x, h.y + A.\Delta y\}$;
5: $mTr \leftarrow$ unmarked trajectories in $DS \cap HCI$;
6: $S_1 \leftarrow S_1 \cup mTr$;
7: **end for**
8: Mark the trajectories in S_1 in the index to prevent re-access;
9: **return** S_1;

Algorithm 4. GetSeCandidate

Input: q: query trajectory, ISE: semantic index, g;
Output: S_2: semantic trajectory candidate set;
1: $S_2 \leftarrow \emptyset$;
2: **for** i from 1 to $|q|$ **do**
3: $r \leftarrow upper(q.c_i, g)$, $R \leftarrow$ Get i-th level of ISE;
4: $Tr \leftarrow$ trajectories in the g-th level from R which contians r and unmarked;
5: $S_2 \leftarrow S_2 \cup Tr$;
6: **end for**
7: Mark the trajectories in S_2 in the index to prevent re-access;
8: **return** S_2;

semantic candidate set is obtained in a manner similar to the spatial candidate set, as detailed in Algorithm 4.

When updating the candidate set, we use a priority queue to maintain the candidate set and its complexity is $O(\log k)$. Therefore, the time complexity is $O(L \cdot |S| \cdot (|q| + \log k))$, where $O(L)$ is the number of iterations and $O(|S| \cdot (|q| + \log k))$ is the cost of updating the candidate set.

4.3 Weighted Query Processing

Our proposed framework (Algorithm 2) still works for weighted query version (wTKIP), except the pruning rules should be modified accordingly. The complexity of this weighted version (wTKIP) is the same as the original version (TKIP). To simplify the formula and proof, let $\lambda_i = \frac{\min(q.len_i, r.len_i)}{q.length}$, where $q.length = \sum_{j=1}^{|q|} q.len_j$ means the total stay time length of q. According to Eq. (6) and (7), the weight w_i is computed as $\frac{\min(q.len_i, r.len_i)}{\sum_{j=1}^{\min(|q|,|r|)} \min(q.len_j, r.len_j)}$, where $\sum_{j=1}^{\min(|q|,|r|)} \min(q.len_j, r.len_j)$ is within $[\min(q.len_i, r.len_i), q.length]$. Thus, w_i is within $[\lambda_i, 1]$.

Pruning Rule 3 (Weighted Spatial Pruning). Given an upper bound τ, an adjustment parameter β, a query semantic trajectory q, and an arbitrary semantic trajectory r, $S^2D(q,r) > \tau$ if there exist positions $i_1, ..., i_d$, such that

$$\sum_{j=1}^{d} \lambda_{i_j} \cdot mindis(cell(q.loc_{i_j}), cell(r.loc_{i_j})) > \frac{\tau}{\beta} \tag{10}$$

Proof. According to Eq. (6), we have:

$$\begin{aligned} SpD(q,r) &\geq \sum_{j=1}^{d} w_{i_j} \cdot \parallel q.loc_{i_j}, r.loc_{i_j} \parallel \\ &\geq \sum_{j=1}^{d} w_{i_j} \cdot mindis(cell(q.loc_{i_j}), cell(r.loc_{i_j})) \\ &\geq \sum_{j=1}^{d} \lambda_{i_j} \cdot mindis(cell(q.loc_{i_j}), cell(r.loc_{i_j})) > \frac{\tau}{\beta} \end{aligned}$$

Integrating with Eq. (1), we have $S^2D(q,r) \geq \beta \cdot SpD(q,r) > \tau$. □

Pruning Rule 4 (Weighted Semantic Pruning). Given an upper bound τ, an adjustment parameter β, a query semantic trajectory q, and an arbitrary semantic trajectory r, $S^2D(q,r) > \tau$ if there exist positions $i_1, ..., i_d$, such that

$$\sum_{j=1}^{d} \lambda_{i_j} \cdot sd(q.c_{i_j}, r.c_{i_j}) > \frac{\tau}{1-\beta} \tag{11}$$

Proof. According to Eq. (7), we have:

$$SeD(q,r) \geq \sum_{j=1}^{d} w_{i_j} \cdot sd(q.c_{i_j}, r.c_{i_j}) \geq \sum_{j=1}^{d} \lambda_{i_j} \cdot sd(q.c_{i_j}, r.c_{i_j}) > \frac{\tau}{1-\beta}$$

Integrating with Eq. (1), we have $S^2D(q,r) \geq (1-\beta) \cdot SeD(q,r) > \tau$. □

5 Experiments

In this section, we evaluate our proposed framework upon two real datasets. One dataset named NJ is the base-station access logs of 5,000 users in Nanjing from Oct. 10 to Oct. 31, 2020. Another dataset named SZ is the base-station access logs of 8,000 users in Suzhou from Mar. 10 to Mar. 31, 2021. The sizes of the two datasets are about 3.4 GB and 6.17 GB respectively. In addition, we use a real file with 982,200 POIs containing real location and the corresponding multi-level semantic information. All codes were written in C++. The experiments run on a PC with Windows 10 Pro, Intel(R) Core(TM) i7-8550U CPU @1.80 GHz, and 16 GB of RAM.

We transform the raw trajectories into semantic trajectories by setting $\delta_{dis} = 1\,\text{km}$ and $\delta_t = 900\,\text{s}$, then we can get 90,423 and 154,096 semantic trajectories respectively. As mentioned above, since no existing work considers spatial dimension and hierarchical semantics at the same time, we use LINEAR and weighted LINEAR (wLINEAR) as the baseline methods, which computes the distance between the query trajectory and other trajectories one by one. LINEAR directly traverses the trajectory set, calculates the distance using the trajectory distance formula S^2D defined in this paper, and returns the closest one. The difference between wLINEAR and LINEAR is that wLINEAR calculates the distance using the weighted trajectory distance formula S^2D.

We build spatial index and semantic index in this paper. Figure 5 illustrates the construction time and memory consumption of index under different factors and datasets.

As shown in Fig. 5(a) and (d), with the increment of dataset size, the index construction time and memory consumption also rise accordingly. Figure 5(b) and (e) show that with the increase of grid cell size, the construction time of spatial index changes little, and the memory consumption tends to decrease due to the reduction of the number of cells that need to be placed in memory.

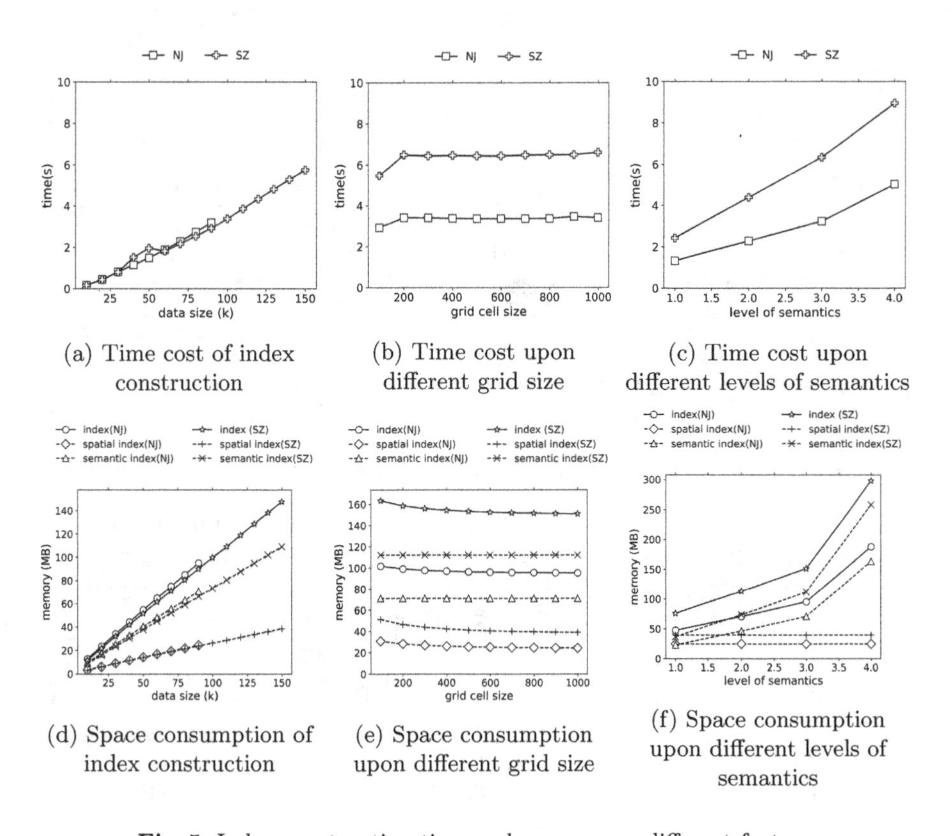

(a) Time cost of index construction

(b) Time cost upon different grid size

(c) Time cost upon different levels of semantics

(d) Space consumption of index construction

(e) Space consumption upon different grid size

(f) Space consumption upon different levels of semantics

Fig. 5. Index construction time and memory on different factors.

Figure 5(c) and (f) illustrate that with the increment of semantic level, the construction time and memory consumption of semantic index will increase significantly. Moreover, the level of semantics will lead to different depth and number of nodes in semantic index. In conclusion, through the analysis of the above experiments, it is found that the index construction time will not increase significantly with the increase of semantic level or the number of grid cells, and basically remains unchanged. This shows that our index structure is reasonable and efficient.

5.1 Performance Comparison

We report the performance of our framework by comparing with the baseline methods. We randomly select one query trajectory, and use our algorithms and baseline methods to process queries in different dataset size and record these query times.

Figure 6(a) reports the performance comparison. We observe that TKIP and wTKIP are much faster than LINEAR and wLINEAR in different cases.

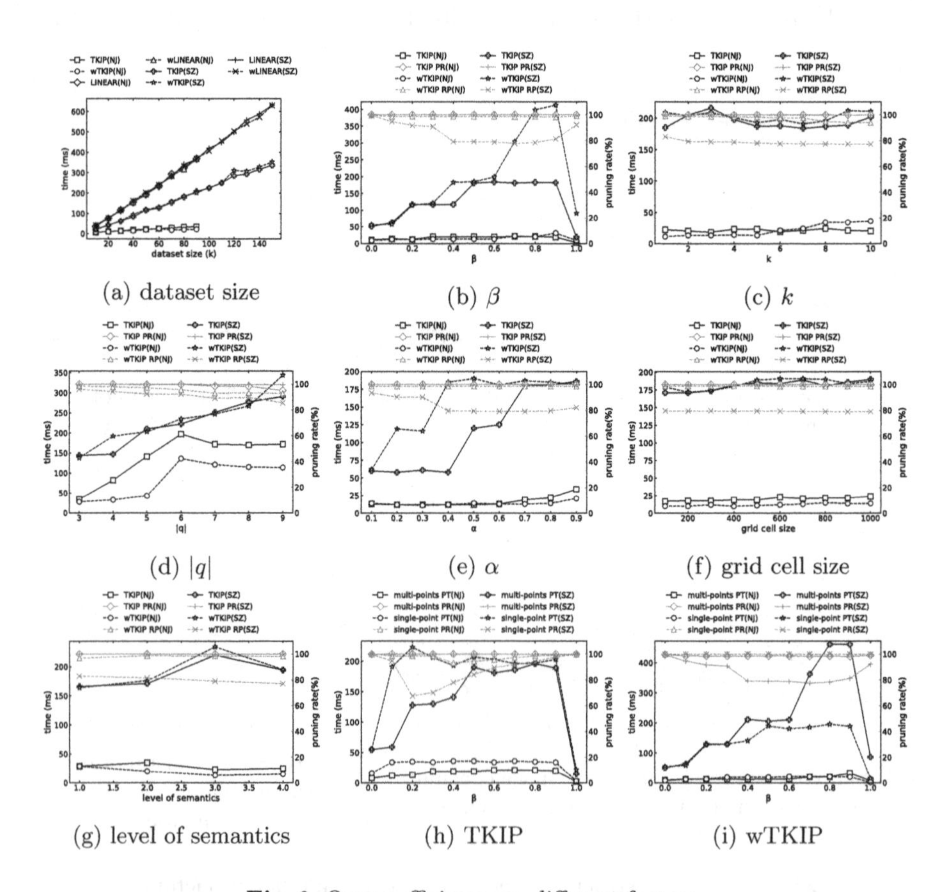

(a) dataset size (b) β (c) k

(d) $|q|$ (e) α (f) grid cell size

(g) level of semantics (h) TKIP (i) wTKIP

Fig. 6. Query efficiency on different factors.

However, due to different data structures of SZ and NJ, the results presented by the two experimental datasets are not completely consistent. There is also slight difference in query performance between different query methods. When the NJ dataset size is 90k, TKIP and wTKIP only take about less than 50 ms, while LINEAR and wLINEAR take about 500 ms.

5.2 Parameter Analysis

Then, we investigate the impact of the parameters. We randomly select one query trajectory to process queries under different parameters, and record query time and pruning rate (PR).

Figure 6(b) shows that the influence of β on query efficiency and pruning rate under different experimental data. Query efficiency is highest when β=0 or 1, because only one distance, spatial or semantic distance, is concerned at this time, thus greatly reducing the query time. In other cases, the higher the β value, the lower the query efficiency. This is because spatial features will receive more attention when β increases, and the query efficiency of spatial indexes is lower than that of semantic indexes, because the pruning rules of semantic indexes are simpler than those of spatial indexes. Figure 6(c) shows that with the increase of k, the query efficiency decreases. The reason is that the greater the k value is, the greater the upper bound τ will be, which may reduce the pruning efficiency. Figure 6(d) shows that with the increase of $|q|$, the query efficiency may be reduced. Because the calculation time of the trajectory distance is a factor that affects the efficiency, which is proportional to the length of the trajectory.

In summary, we find that parameter settings do affect query time and pruning rate. Moreover, the query time is not only determined by the pruning rate, but also other factors, such as the time consumed by merging trajectories.

5.3 Analysis of Indexes and Pruning Rules

We first analyze the impact of different α and index structures on query efficiency, and then compare pruning rules, finally verify the effectiveness.

Figure 6(e) shows the query efficiency on different α. The result shows that a smaller α may reduce the query time. And the weighted query method is more deeply affected by α than the unweighted query method. Moreover, when $\alpha = 0.4$, there will be an inflection point when query time increases. In addition, Fig. 6(f) and (g) illustrate the query efficiency under different grid cell size and semantic level, respectively. We find that with the increase of grid cell size, query efficiency decreases, but the effect is insignificant. Due to the different data distribution of the two experimental datasets, the efficiency of weighted query based on the experimental data of SZ is generally low, mainly due to the low pruning rate. Moreover, the level of semantics has little influence on the query efficiency, whether weighted or unweighted.

Our pruning rules determine whether the partial distance based on one or several stay point(s) exceeds a threshold. Single point pruning only considers one point, while multi-points pruning considers as many as points as possible.

To compare the efficiency of single point and multi-points pruning, we record the query time and pruning rate under two pruning rules. Figure 6(h) and (i) illustrate the comparison of TKIP and wTKIP respectively. Figure 6(h) shows the query efficiency of multi-point pruning is higher than single-point pruning under TKIP. However, under wTKIP, as shown in Fig. 6(i), the weight of query method has a great influence on pruning methods.

6 Conclusion

In this paper, we study similarity search, a typical issue in trajectory data management. Unlike most existing works that only focus on spatio-temporal and single-level semantic features, we integrate multi-level semantic features and propose a framework to find top-k similar trajectories efficiently. We first build spatial and semantic indexes to quickly retrieve trajectories. To improve query efficiency, we also propose several light-weight pruning rules to filter invalid trajectories and update the candidate set continuously and an adaptive updating method of candidate set based on β value. In practice, parameter β is used to adjust spatial and semantic attention, so as to adapt to more scenarios. The experiments based on real datasets show that our approach is not only efficient, but also suitable for flexible scenarios.

In future, we intend to improve our index structure, and propose more pruning rules to filter out more illegal trajectories in time. Furthermore, we also intend to analyze the scenarios with different β value.

Acknowledgments. This work is partially supported by National Science Foundation of China (U1911203, 61877018 and U1811264).

References

1. Tao, Y., Papadias, D.: The MV3R-tree: a spatio-temporal access method for timestamp and interval queries. In: VLDB, pp. 431–440 (2001)
2. Wang, S., Ferhatosmanoglu, H.: PPQ-trajectory: spatio-temporal quantization for qerying in large trajectory repositories. Proc. VLDB Endow. **14**(2), 215–227 (2020)
3. Shang, Z., Li, G., Bao, Z.: DITA: distributed in-memory trajectory analytics. In: Proceedings of the 2018 International Conference on Management of Data, pp. 725–740 (2018)
4. Ta, N., Li, G., Xie, Y., Li, C., Hao, S., Feng, J.: Signature-based trajectory similarity join. IEEE TKDE **29**(4), 870–883 (2017)
5. Shang, S., Chen, L., Wei, Z., Jensen, C.S., Zheng, K., Kalnis, P.: Parallel trajectory similarity joins in spatial networks. VLDB J. **27**(3), 395–420 (2018)
6. Dai, Y., Shao, J., Wei, C., Zhang, D., Shen, H.T.: Personalized semantic trajectory privacy preservation through trajectory reconstruction. World Wide Web **21**(4), 875–914 (2018)
7. Kontarinis, A., Zeitouni, K., Marinica, C., Vodislav, D., Kotzinos, D.: Towards a semantic indoor trajectory model. In: EDBT/ICDT Workshops (2019)
8. Zheng, B., Yuan, N.J., Zheng, K., Xie, X., Sadiq, S., Zhou, X.: Approximate keyword search in semantic trajectory database. In: ICDE, pp. 975–986 (2015)

9. Zhao, K., Chen, L., Cong, G.: Topic exploration in spatio-temporal document collections. In: Proceedings of the 2016 International Conference on Management of Data, pp. 985–998 (2016)
10. Belesiotis, A., Skoutas, D., Efstathiades, C., Kaffes, V., Pfoser, D.: Spatio-textual user matching and clustering based on set similarity joins. VLDB J. **27**(3), 297–320 (2018). https://doi.org/10.1007/s00778-018-0498-5
11. Zheng, K., Zheng, B., Xu, J., Liu, G., Liu, A., Li, Z.: Popularity-aware spatial keyword search on activity trajectories. World Wide Web **20**(4), 749–773 (2016). https://doi.org/10.1007/s11280-016-0414-0
12. Chen, L., Shang, S., Jensen, C.S., Yao, B., Kalnis, P.: Parallel semantic trajectory similarity join. In: ICDE, pp. 997–1008 (2020)
13. Wang, S., Bao, Z., Shane Culpepper, J., Cong, G.: A survey on trajectory data management, analytics, and learning. ACM Comput. Surv. (CSUR) **54**(2), 1–36 (2021)
14. Chen, Z., Shen, H.T., Zhou, X., Zheng, Y., Xie, X.: Searching trajectories by locations-an efficiency study. In: SIGMOD, pp. 255–266 (2010)
15. Guttman, A.: R-trees: a dynamic index structure for spatial searching. ACM SIGMOD Rec. **14**(2), 45–47 (1984)
16. Yuan, J.: T-drive: driving directions based on taxi trajectories. In: GIS, pp. 99–108 (2010)
17. Zheng, K., Shang, S., Yuan, N.J., Yang, Y.: Towards efficient search for activity trajectories. In: ICD, pp. 230–241 (2013)

Nonnegative Matrix Factorization Framework for Disease-Related CircRNA Prediction

Cheng Yang[1], Li Peng[1](✉), Wei Liu[2], Xiangzheng Fu[3], and Ni Li[3]

[1] College of Computer Science and Engineering, Hunan University of Science and Technology, Xiangtan, Hunan, China
plpeng@hnu.edu.cn

[2] College of Information Engineering, Xiangtan University, Xiangtan, Hunan, China

[3] College of Information Science and Engineering, Hunan University, Changsha, Hunan, China

Abstract. Nowadays, more and more scholars and related studies have shown that circular RNA (circRNA) is related to the occurrence and development of human diseases. It is necessary to use computational approach to predict potential and unknown associations between circRNA and disease, which will save time and money in developing drugs for treating disease. In this paper, we propose a new computational model called robust nonnegative matrix factorization framework (RNMF), which learns a robust association discriminative representation by using the similarity information structure of network space and $\ell_{2,1}$-norm loss function. Specifically, we use integrated circRNA similarity network and integrated disease similarity network to update known association matrix from the perspective of matrix multiplication at first. Then, the problems of noise and outliers of elements in the updated association matrix are well addressed by the $\ell_{2,1}$-norm loss function. Finally, the computational model is expressed as an optimization problem of objective function solved by iterative algorithm. Experimental results show that our model can achieve better performance than state-of-the-art methods in several aspects.

Keywords: CircRNA-disease associations · Nonnegative Matrix Factorization · Computational prediction model

1 Introduction

Circular RNAs (circRNAs) are a new class of non-coding molecules with covalently closed loop structures, high stability, lacked of 5′-end caps and 3′-end tails, which play an important role in gene regulation [1, 2]. There is increasing evidence that circRNA plays an important role in a variety of diseases related to human, including neurological diseases, prion diseases, osteoarthritis and diabetes [3]. Moreover, circRNA has been abnormally expressed in several types of cancer, including colorectal cancer, hepatocellular carcinoma. Based on the function of circRNA in cancer, circRNA can be used as a biosignature for diagnosis or tumor, and it will hopefully provide new therapeutic targets for cancer therapy [4]. Due to the artificial biological experiments are time-consuming and laborious to uncover potential circRNA associated with disease. It is necessary to

Y. Lai et al. (Eds.): ICA3PP 2021, LNCS 13157, pp. 620–631, 2022.
https://doi.org/10.1007/978-3-030-95391-1_39

use computational prediction model to predict potential and unknown circRNA-disease associations, which will save time and money in the development of drugs and diseases treatment. Here are some algorithm frameworks for predicting unknown associations, such as collaborative filtering algorithms [5–7]. However, in this paper, we use our proposed RNMF framework to speculate about potential circRNA-disease associations. Finally, to evaluate the performance of our computational prediction framework, we compare our framework with three existing computing methods including KATZHCDA [8], iCircDA-MF [9], LLCDC [10]. We performed a 5-fold cross validation (5-CV) of the verified circRNA-disease associations. The results show that the AUC value obtained by our model is higher than other methods, which demonstrates that the performance of our method is better than other competitive methods.

2 Related Work

With the rapid development of science and technology, high-throughput sequencing analysis of circRNA molecules revealed that circRNA showed tissue and developmental specific expression. Most importantly, it played a key role in a variety of cellular processes. And specific circRNAs can act as microRNA sponges and regulate gene transcription [11]. In recent years, many computational methods have been proposed to predict the potential association between circRNA and diseases, which can be used to provide effective candidate disease-causing circRNAs for diseases. For example, there are the following calculation methods.

Wei et al. [9] proposed method called iCircDA-MF. In this method, the reconstructed circRNA-disease association matrix was generated by using the interaction spectra of similar circRNAs and similar diseases, and then the association score between circRNA and disease was calculated by nonnegative matrix factorization.

Li et al. [12] proposed the method called SIMCCDA, using the known circRNA-disease association, the integrated circRNA similarity matrix, and the integrated disease similarity matrix to construct the model and complete the accelerated induction matrix on the model, and then to calculate the score matrix.

Ping et al. [13] proposed a method which only relies on topological information from known LnRNA-disease association networks to identify potential disease-related LncRNA.

Xiao et al. [14] proposed an integrated framework called ICDA-CMG, which utilized the known circRNA association with disease, disease similarity and circRNA similarity to obtain the most promising candidate circRNA related to disease through matrix complete graph learning model.

Nonnegative matrix factorization and inductive matrix completion belong to the same class of methods, which capture linear features between circRNAs and diseases. But, known associations between circRNA and disease are too few, which cannot simulate such a complex association by traditional nonnegative matrix model.

To tackle this problem, we used the reconstructed disease network information matrix and the reconstructed circRNA network information matrix by completing matrix multiplication on the known association matrix to increase the potential association information. In other words, using matrix multiplication to update the information that inside the association matrix inevitably bring outliers and noise. Therefore, how to select valid information from the updated association matrix is very critical to the model. The loss function based on $\ell_{2,1}$-norm is robust to outliers in data points and the $\ell_{2,1}$-norm regularization effective selects feature on all data points in matrix [24]. loss function based on $\ell_{2,1}$-norm and regularization based on $\ell_{2,1}$-norm also are used to doing image representation, semi-supervised multi-view learning [15, 16]. Inspired by their success, we used loss function with $\ell_{2,1}$-norm to well address the noise and outliers in the model, which makes the model more robust and efficient to select the correlation information of circRNA-disease association. Therefore, the RNMF can be summarized as follows: Firstly, we obtained the integrated similarity matrix of circRNAs and diseases respectively through calculation method. And then, we use matrix multiplication to update the association matrix from the perspectives of circRNA and disease, respectively. Finally, we infer the underlying circRNA-disease associations by using RNMF.

3 Materials and Methods

In this study, we propose a robust nonnegative matrix factorization framework to enhance traditional matrix factorization methods for predicting circRNA-disease association. Firstly, we present a target formula for circRNA-disease association prediction. Next, we give the technical details that calculating the similarity network for disease and circRNA respectively, association matrix updating, $\ell_{2,1}$-norm based loss function, and solution of problem formula.

3.1 Data Sources

This section describes the preparation of data used in the RNMF framework.

CircRNA-Disease Associations
We retrieve known circRNA-disease associations from CircR2Disease database (https://bioinfo.snnu.edu.cn/CircR2Disease/), which contain 739 verified circRNA-disease association with 661 circRNAs and 100 diseases. We removed redundant data and limit their association with humans. Finally, we obtained 650 associations among 585 circRNAs and 88 diseases, whose interaction sparsity is about 0.0126. In this paper, the dataset of circRNAs is represented by $C = \{c_1, c_2, c_3, \cdots c_m\}$ and the dataset of diseases is represented by $D = \{d_1, d_2, d_3, \cdots d_n\}$, where m and n represent the numbers of circRNAs and diseases, respectively. We defined the association matrix $A_{(i,j)} \in R^{(m \times n)}$ to represent circRNA-disease assocaitions. If the circRNA c_i has interaction with disease d_j in the resultant dataset, the value of $A(i,j)$ is 1; otherwise, it is 0.

Disease-Disease Similarities
We downloaded the ontological features of the disease from https://disease-ontology.org. Then, according to Wang [17] method, we use the "dosim" function in DOSE software (R package) [18] to calculate the semantic similarity $S_d\left(d_i, d_j\right)$ between disease d_i and d_j.

Gaussian Interaction Profile Kernel Similarity for Diseases and circRNAs
According to the hypothesis that similar diseases tend to be related to functionally similar

circRNA and vice versa [17]. According to the way of calculating Gaussian interaction profile (GIP) [19], we can calculate the GIP kernel similarities of circRNAs and diseases by association matrix A. This method is widely used to calculate the similarity measure matrix of noncoding RNA and disease. The GIP kernel similarity $CGK_{(c_i,c_j)}$ between circRNAs c_i and circRNAs c_j can be measured based on Eqs. (1) and (2).

$$CGK_{(c_i,c_j)} = \exp\left(-u\left\|A(c_i) - A\left(c_j\right)\right\|^2\right) \tag{1}$$

$$u = \frac{1}{\frac{1}{N_c}\sum_{i=1}^{N_c}\|A(c_i)\|^2} \tag{2}$$

Where $A(c_i)$ represent the i th rows of association A, $A\left(c_j\right)$ represent the j th rows of association A, u represents the regularization parameter controlling the kernel bandwidth. N_c represents the total number of circRNAs.

At the same way, the GIP kernel similarity $DGK\left(d_i, d_j\right)$ between disease d_i and disease d_j can be calculated by Eqs. (3) and (4).

$$DGK_{(d_i,d_j)} = \exp\left(-u\left\|A(d_i) - A\left(d_j\right)\right\|^2\right) \tag{3}$$

$$u = \frac{1}{\frac{1}{N_d}\sum_{i=1}^{N_d}\|A(d_i)\|^2} \tag{4}$$

Where $A(d_i)$, $A\left(d_j\right)$ represent the i th and j th columns of the association matrix A, u is the regularization parameter and N_d the total number of diseases.

CircRNA-CircRNA Similarities

The functional similarity between the two circRNAs is based on the assumption that circRNA in disease groups sharing similarity semantic are often functionally similar [20]. Then, the functional similarity measure of $FS\left(c_i, c_j\right)$ between c_i and c_j can be expressed as:

$$FS\left(c_i, c_j\right) = \frac{\sum_{d \in DT_i} SD\left(d, DT_j\right) + \sum_{d \in DT_j} SD(d, DT_i)}{|DT_i| + \left|DT_j\right|} \tag{5}$$

Especially, $SD = (S_d + DGK)/2$, and DT_i, DT_j were two disease sets associated with c_i and c_j respectively. $|DT_i|$ and $\left|DT_j\right|$ represent the number of diseases in the two disease sets associated with c_i and c_j, respectively. Suppose that the disease set is defined as $DT_n = \{d_{n1}, d_{n2}, d_{n3}, \cdots, d_{nk}\}$. Then, the similarity score $SD(d, DT_n)$ of the disease set corresponding to disease d is defined as:

$$SD(d, DT_n) = \max_{1 \le x \le k}(SD(d, DT_x)) \tag{6}$$

Where k represents the number of diseases in disease set DT_n, and x represents a certain integer value from 1 to k.

3.2 Data Incorporation

The incorporated similarities of circRNAs ($RCS_{(c_i,c_j)}$) and diseases ($RDS_{(d_i,d_j)}$) were computed as:

$$RCS_{(c_i,c_j)} = \begin{cases} FS(c_i, c_j) & \text{if } c_i \text{ and } c_j \text{ has functional similarity} \\ CGK(c_i, c_j) & \text{Otherwise} \end{cases} \tag{7}$$

$$RDS_{(d_i,d_j)} = \begin{cases} S_d(d_i, d_j) & \text{if } d_i \text{ and } d_j \text{ has semantic similarity} \\ DGK(d_i, d_j) & \text{Otherwise} \end{cases} \tag{8}$$

Inspired by [21], we use matrix multiplication to update the association matrix, it can formulae as:

$$A' = Max(RCS * A, A * RDS) \tag{9}$$

There, A' is the new circRNA-disease association matrix.

3.3 Nonnegative Matrix Factorization Framework

Nonnegative matrix factorization is widely used in various data processing fields because of its good recovery mechanism of missing information. It decomposes matrix $A' \in \mathbb{R}^{(m \times n)}$ into two nonnegative low rank submatrixes $C \in \mathbb{R}^{(m \times k)}$ and $D \in \mathbb{R}^{(n \times k)}$, where $k < \min(m, n)$ is the dimension of the characteristic subspace, and then the outer products of these two nonnegative low rank submatrixes can well approximate the original matrix A', that is, $A' \cong CD^T$ [22–24]. We use matrix multiplication to update the adjacency matrix in order to capture the potential interaction characteristics, which will inevitably bring noise and outliers to limiting the prediction performance of the model. To address this problem, we use $\ell_{2,1}$-norm to constrain loss function in order to reduce the impact of outliers and noise on the performance of the model.

$\ell_{2,1}$-norm is a continuous and smooth loss function, and its mathematical meaning is similar to Frobenius norm. The smoother the loss function, the better the robustness of the algorithm. Then, inspired by [24], the initial sparse circRNA-disease prediction problem can be defined as:

$$\min_{C,D} \left\| A' - CD^T \right\|_{2,1} \quad s.t. \;\; C \geq 0, D \geq 0 \tag{10}$$

Based on [24, 25], the formula (10) can be expressed as:

$$\min_{C,D} \|A' - CD\|_{2,1}^{T} = \min_{C,D} Tr\left[\left(A' - CD^T\right)^T B\left(A' - CD^T\right)\right] \tag{11}$$

Where $B \in \mathbb{R}^{m \times m}$ is a diagonal matrix and its diagonal element is $B_{ii} = \frac{1}{\|(A' - CD^T)_i\|_2}$. $Tr(\cdot)$ represents trace function.

According to [26], the formula of (11) may not make full use of the local structure inherent in circRNA and disease network similarity feature space. Therefore, we add graph regularization constraints to solve this problem. In graph regularization theory, the relative position of information data points remains unchanged in the feature space of circRNA and disease. In other words, after the dimensionality reduction of the original data, it still maintains its internal local geometric information in the low dimensional space. We formulate it as:

$$\begin{aligned} \tfrac{1}{2}\textstyle\sum_{i,j=1}^{m} \left\|C(c_i) - C(c_j)\right\|^2 RCS_{(c_i,c_j)} = Tr\left(C^T G_c C\right) \\ \tfrac{1}{2}\textstyle\sum_{e,b=1}^{n} \|C(c_e) - C(c_b)\|^2 RDS_{(c_e,c_b)} = Tr\left(D^T G_D D\right) \end{aligned} \tag{12}$$

Where $G_C = I_C - RCS$, $G_D = I_D - RDS$ represent Laplacian matrices in circRNA and disease feature similarity space, respectively. I_C and I_D are two diagonal matrices, and the row sum of RCS and RDS characteristic matrices are taken as the diagonal elements of I_C and I_D, respectively. So, the formula (11) can be expressed as:

$$\begin{aligned} &\min_{C,D} Tr\left[\left(A - CD^T\right)^T B\left(A - CD^T\right)\right] + \beta\left(Tr\left(C^T G_c C\right) + Tr(D^T G_D D)\right) \\ &s.t.\ C \geq 0,\ D \geq 0 \end{aligned} \tag{13}$$

Where β represents the regularization term coefficient. Then, we use Lagrange multipliers to solve the optimization problem in Eq. (13), the Lagrange function L_f is as follows:

$$\begin{aligned} L_f = Tr(A'^T BA) = -2Tr(A'^T BCD^T) + Tr(DC^T BCD^T) \\ +\beta(Tr(C^T G_c C) + Tr(D^T G_D D)) + Tr(\Psi C^T) + Tr(\Phi D^T) \end{aligned} \tag{14}$$

Then, we got low rank submatrixes C, D according to KKT condition [27] and iteration termination condition. Finally, the predicting result of A^* can be defined as:

$$A^* = CD^T \tag{15}$$

In addition, each element of the prediction matrix A^* represents the predicting score of the disease associated with the corresponding circRNA. And the flowchart of RNMF is shown in Fig. 1.

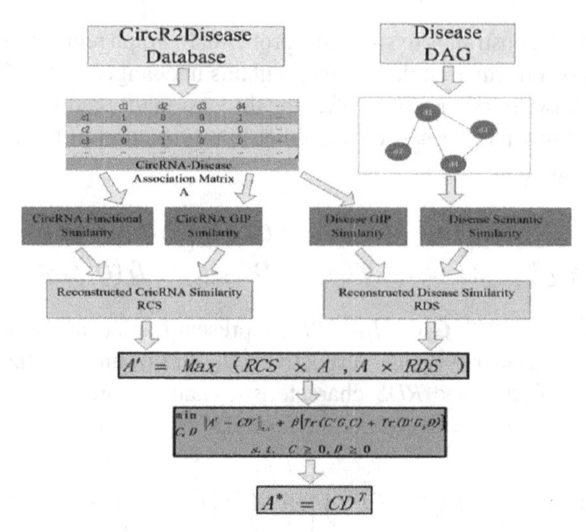

Fig. 1. The framework of RNMF for predicting potential circRNA-disease associations

Algorithm 1 details the iterative alternating algorithm of RNMF, line 3 means taking the maximum value of the corresponding element between matrix $(RCS * A)$ and matrix $(A * RDS)$. Line 5 represents the elements on diagonal matrix B are normal L_2 norm of column vectors in $(A' - CD^T)$ matrix.

Algorithm 1 : RNMF

Input: The original association matrix $A \in R^{m \times n}$,similarity matrices $RCS \in R^{m \times m}$ and $RDS \in R^{n \times n}$,subspace dimensionality k, regularization coefficient β.

Output: Predicted association matrix A^*.

1. randomly initialize two nonnegative matrices $C \in R^{m \times k}$ and $D \in R^{n \times k}$;
2. calculate the diagonal matrices I_D , I_C for RDS and RCS matrices, respectively;
3. reconstruct association matrix $A^{'} = Max(RCS * A, A * RDS)$;
4. **repeat**
5. calculate diagonal matrix B by following rules:

$$B = \begin{bmatrix} \frac{1}{\left\|(A'-CD^T)_1\right\|_2} & & \\ & \ddots & \\ & & \frac{1}{\left\|(A'-CD^T)_n\right\|_2} \end{bmatrix}$$

6. update C and D by the following rules:

$$C_{ik} \leftarrow C_{ik} \left(\frac{BA'D + \beta(RCS)C}{BCD^T D + \beta I_C C} \right)_{ik}$$

$$D_{jk} \leftarrow D_{jk} \left(\frac{(A')^T BC + \beta(RDS)D}{DC^T BC + \beta I_D D} \right)_{jk}$$

7. **until convergence or reaching upper limit of the iteration**;
8. calculate the predicted association matrix:$A^* = C * D^T$;
9. **return A***;

---

4 Experiments and Results

In this section, we first introduce the evaluation indicators and baseline methods, and then prove the effectiveness of our proposed method by comparing the experimental results with the baseline methods. Finally, we discuss parameter design.

4.1 Evaluation Indicators

We perform 5-fold cross validation to evaluate the performance of our model. In 5-CV, the known circRNA-disease association was regarded as a positive sample and divided it into five equal parts, then, randomly disassociate one subset as the test set, and the remaining four subsets as the training set and negative samples are unknown associations between circRNA and disease. In this model, the GIP similarity between circRNA and disease depended on known associations. Therefore, circRNA similarity and disease similarity matrices should be recalculated in each repeat experiment.

The framework predictive performance was evaluated by plotting the true positive rate (TPR) and false positive rate (FPR) at different threshold points using receiver operating characteristic (ROC) curves. TPR is the ratio of correctly identified positive samples to all positive samples, and FPR is the ratio of incorrectly identified negative samples to all negative samples and it can be defined as:

$$TPR = \frac{TP}{TP + FN} \tag{16}$$

$$FPR = \frac{FP}{FP + TN} \tag{17}$$

Among them, TP is true positive, FN is false negative, FP is false positive, TN is true negative. The value of AUC was calculated as the corresponding area under the ROC curve. When the value of AUC is greater than 0.5, the prediction performance of the framework is better with the increase of AUC value. In order to highlight the superior performance of our framework. Several evaluation metrics are also adopted to measure the performances of our prediction framework, i.e., the area under precision recall curve (AUPR), accuracy (ACC), F-measure (F1), precision and recall.

4.2 Baseline Methods

To demonstrate the validity of our proposed method, we compared it with the following baseline methods, which were based on machine learning or matrix factorization:

- **KATZHCDA** [8]: KATZHCDA built a heterogeneous network by using the similarity matrix of circRNA and disease, and then used KATZ method to predict circRNA-disease association.
- **iCircDA-MF** [9]: iCircDA-MF applied non-negative matrix factorization to predict the interaction on the association matrix which reconstructed by circRNA and disease similarity neighbors.
- **LLCDC** [10]: LLCDC uses LLC method [28] to encode association matrix to reconstruct the similarity matrix of circRNA and disease respectively, and then uses label propagation method to obtain correlation prediction on the similarity matrix.

4.3 Performance Comparison

For parameters in the RNMF framework, we set $\beta = 0.002$ and $k = 60$. Figure 2 shows the corresponding ROC curve of our method and baseline approaches. KATZHCDA, iCircDA-MF, LLCDC achieve reliable performance with AUC values of 0.9303, 0.8503 and 0.8617, respectively, and our method has improved with an AUC value of 0.9621. These comparisons show that our approach can achieve better performance at baseline.

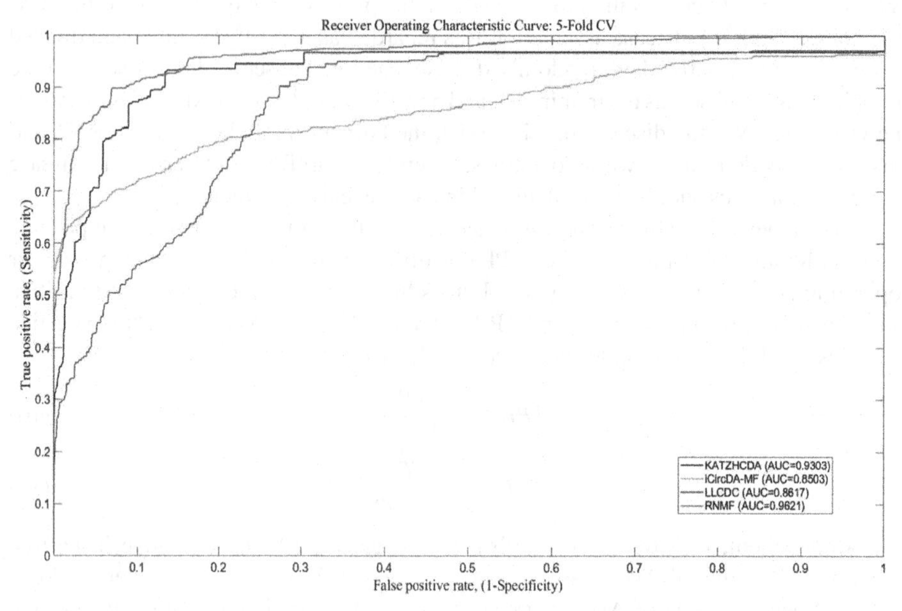

Fig. 2. Performance comparison in terms of AUC and ROC curve

4.4 Impact of Parameters

1) Impact of β: The parameter β controls the weight of regularization loss in RNMF. On the one hand, if β is too small, the model may run the risk of overfitting. On the other hand, if β is too large, the model will rely heavily on regularization losses. Then the sensitivity of β were studied under the condition of K = 60. Figure 3 shows experimental results with different values of β. We can see the best performance of model in standard validation data set is achieved when $\beta = 0.002$.
2) Impact of k: Parameter k controls the number of circRNA and disease latent expression factors. Too little amount of k may lead to incomplete expression of potential impact factors of circRNA and disease, while too much amount of k may lead to over-expression of circRNA and disease, leading to over-fitting of the framework. Figure 4 shows experimental results with different values of β, which is conducted under condition of $\beta = 0.002$ in standard validation data set. As can be seen from Fig. 4, the value of AUC increases with the increase of k value. When k value is 60,

AUC value reaches the maximum value, and then when k is greater than 60, AUC value begins to decrease. So, we set $k = 60$ in our framework.

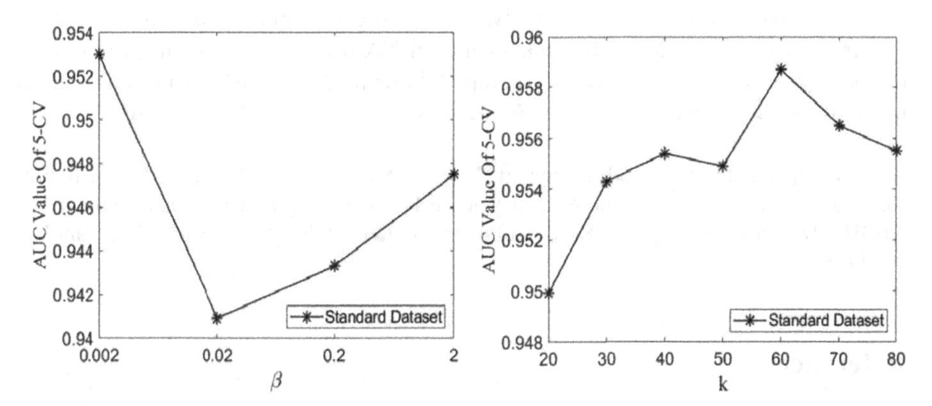

Fig. 3. Impact of parameter β

Fig. 4. Impact of parameter k

In the performance comparison experiments, all associations are split into five subsets for 5-CV. We set $\beta = 0.002, k = 60$ for RNMF framework because these parameters can lead to the best performance. Similarly, Models KATZHCDA, LLCDC, and CircDA-MF all adopt the optimal parameters in 5-CV. To avoid the bias of data split, the performances are measured by averaging the results of 10 independent runs. As shown in Table 1, RNMF can produce better results than the other three baseline model.

Table 1. Performances of different models for 5-CV

Methods	AUC	AUPR	F1	Accuracy	Precision	Recall
RNMF	**0.9594**	**0.5102**	**0.5648**	**0.9902**	**0.6388**	**0.5062**
KATZHCDA	0.9278	0.3965	0.3653	0.9841	0.3676	0.3631
LLCDC	0.8753	0.2792	0.3223	0.9828	0.3216	0.3231
iCircDA-MF	0.8614	0.2050	0.1835	0.9376	0.1099	0.3154

5 Conclusion

Traditional biological experiments are expensive, time-consuming and laborious, resulting in a very limited number of disease associations with circRNA verified by experiment. Using computational models to identify disease-related circRNA can provide a better understanding of disease pathogenesis at the molecular level, meanwhile, which can save time and money for biological experiments and help identify biomarkers of disease and design of drug. In this study, we proposed a robust nonnegative matrix framework for predicting circRNA-disease association, which named RNMF. Firstly, we integrated circRNA-disease associations from CircR2Disease set. Second, we calculated the circRNA function similarity according to the association matrix and disease semantic similarity, then, we exploit the matrix multiplication operation to update the association matrix from circRNA and disease respect by integrated circRNA similarity and integrated disease similarity. Finally, we used robust nonnegative matrix factorization framework to predict all the unknown circRNA-disease associations. To evaluate the performance of our method, we compared our method with four more advanced methods. 5-CV results show that our framework is superior to other methods.

Acknowledgements. This work is supported by the National Natural Science Foundation of China (Grant No. 61902125), the Natural Science Foundation of Hunan Province (Grant No. 2019JJ50187), and the Scientific Research Project of Hunan Education Department (Grant No. 19C1788).

References

1. He, J., Xie, Q., Xu, H., Li, J., Li, Y.: Circular RNAs and cancer. Cancer Lett. **396**, 138–144 (2017)
2. Vo, J.N., Cieslik, M., Zhang, Y., Shukla, S., Xiao, L., Zhang, Y., et al.: The landscape of circular RNA in cancer. Cell **176**(4), 869–81.e13 (2019)
3. Jiao, S., Wu, S., Huang, S., Liu, M., Gao, B.: Advances in the identification of circular RNAs and research into circRNAs in human diseases. Front. Genet. **12**(387) (2021)
4. Zhang, H., Jiang, L.-H., Sun, D.-W., Hou, J.-C., Ji, Z.-L.: CircRNA: a novel type of biomarker for cancer. Breast Cancer **25**(1), 1–7 (2017). https://doi.org/10.1007/s12282-017-0793-9
5. Chen, X., Liang, W., Xu, J., Wang, C., Li, K.C., Qiu, M.: An efficient service recommendation algorithm for cyber-physical-social systems. IEEE Trans. Netw. Sci. Eng. 1–1 (2021)
6. Lei, X., Fang, Z., Guo, L.: Predicting circRNA–disease associations based on improved collaboration filtering recommendation system with multiple data. Front. Genet. 10(897) (2019)
7. Liang, W., Xiao, L., Zhang, K., Tang, M., He, D., Li, K.C.: Data fusion approach for collaborative anomaly intrusion detection in blockchain-based systems. IEEE Internet Things J. 1–1 (2021)
8. Fan, C., Lei, X., Wu, F.-X.: Prediction of CircRNA-disease associations using KATZ model based on heterogeneous networks. Int. J. Biol. Sci. **14**(14), 1950–1959 (2018)
9. Wei, H., Liu, B.: iCircDA-MF: identification of circRNA-disease associations based on matrix factorization. Brief. Bioinform. **21**(4), 1356–1367 (2019)
10. Ge, E., Yang, Y., Gang, M., Fan, C., Zhao, Q.: Predicting human disease-associated circRNAs based on locality-constrained linear coding. Genomics **112**(2), 1335–1342 (2020)
11. Meng, X., Li, X., Zhang, P., Wang, J., Zhou, Y., Chen, M.: Circular RNA: an emerging key player in RNA world. Brief. Bioinform. **18**(4), 547–557 (2016)
12. Li, M., Liu, M., Bin, Y., Xia, J.: Prediction of circRNA-disease associations based on inductive matrix completion. BMC Med. Genomics **13**(5), 42 (2020)
13. Ping, P., Wang, L., Kuang, L., Ye, S., Iqbal, M.F.B., Pei, T.: A novel method for LncRNA-disease association prediction based on an lncRNA-disease association network. IEEE/ACM Trans. Comput. Biol. Bioinf. **16**(2), 688–693 (2019)
14. Xiao, Q., Zhong, J., Tang, X., Luo, J.: iCDA-CMG: identifying circRNA-disease associations by federating multi-similarity fusion and collective matrix completion. Mol. Genet. Genomics **296**(1), 223–233 (2020). https://doi.org/10.1007/s00438-020-01741-2

15. Li, Z., Tang, J., He, X.: Robust structured nonnegative matrix factorization for image representation. IEEE Trans. Neural Netw. Learn. Syst. **29**(5), 1947–1960 (2018)
16. Liang, N., Yang, Z., Li, Z., Xie, S., Sun, W.: Semi-supervised multi-view learning by using label propagation based non-negative matrix factorization. Knowl. Based Syst. **228**, 107244 (2021)
17. Wang, D., Wang, J., Lu, M., Song, F., Cui, Q.: Inferring the human microRNA functional similarity and functional network based on microRNA-associated diseases. Bioinformatics **26**(13), 1644–1650 (2010)
18. Yu, G., Wang, L.-G., Yan, G.-R., He, Q.-Y.: DOSE: an R/Bioconductor package for disease ontology semantic and enrichment analysis. Bioinformatics **31**(4), 608–609 (2014)
19. van Laarhoven, T., Nabuurs, S.B., Marchiori, E.: Gaussian interaction profile kernels for predicting drug–target interaction. Bioinformatics **27**(21), 3036–3043 (2011)
20. Chen, X., Clarence Yan, C., Luo, C., Ji, W., Zhang, Y., Dai, Q.: Constructing lncRNA functional similarity network based on lncRNA-disease associations and disease semantic similarity. Sci. Rep. **5**(1), 11338 (2015)
21. Gu, C., Liao, B., Li, X., Li, K.: Network consistency projection for human miRNA-disease associations inference. Sci. Rep. **6**(1), 36054 (2016)
22. Lee, D.D., Seung, H.S.: Learning the parts of objects by non-negative matrix factorization. Nature **401**(6755), 788–791 (1999)
23. Lee, D.D., Seung, H.S.: Algorithms for non-negative matrix factorization. In: Proceedings of the 13th International Conference on Neural Information Processing Systems, pp. 535–541. MIT Press, Denver (2000)
24. Nie, F., Huang, H., Cai, X., Ding, C.: Efficient and robust feature selection via joint norms minimization. In: Proceedings of the 23rd International Conference on Neural Information Processing Systems – vol. 2, pp. 1813–1821. Curran Associates Inc., Vancouver, British Columbia, Canada (2010)
25. Long, X., Xiong, J., Chen, L.: Robust automated graph regularized discriminative non-negative matrix factorization. Multim. Tools Appl. **80**(10), 14867–14886 (2021). https://doi.org/10.1007/s11042-020-10410-w
26. Cai, D., He, X., Han, J., Huang, T.S.: Graph regularized nonnegative matrix factorization for data representation. IEEE Trans. Pattern Anal. Mach. Intell. **33**(8), 1548–1560 (2011)
27. Facchinei, F., Kanzow, C., Sagratella, S.: Solving quasi-variational inequalities via their KKT conditions. Math. Program. **144**(1–2), 369–412 (2013). https://doi.org/10.1007/s10107-013-0637-0
28. Wang, J., Yang, J., Yu, K., Lv, F., Huang, T., Gong, Y.: Locality-constrained Linear coding for image classification. In: 2010 IEEE Computer Society Conference on Computer Vision and Pattern Recognition, pp. 3360–3367 (2010)

A Fast Authentication and Key Agreement Protocol Based on Time-Sensitive Token for Mobile Edge Computing

Zisang Xu[1], Wei Liang[2(✉)], Jin Wang[1], Jianbo Xu[2], and Li-Dan Kuang[1]

[1] Changsha University of Science and Technology, Changsha, China
{xzsszx111,jinwang}@csust.edu.cn
[2] Hunan University of Science and Technology, Xiangtan, Hunan, China
weiliang99@hnu.edu.cn, jbxu@hnust.edu.cn

Abstract. Compared with cloud computing, Mobile Edge Computing (MEC) can sink some services and functions located in cloud servers to edge nodes to reduce network latency and provide real-time services. However, MEC not only inherits the security issues in cloud computing, but also faces new security risks. To ensure the security and privacy of messages transmitted in the channel, a secure Authentication and Key Agreement (AKA) protocol is essential. However, the existing AKA protocols are not lightweight enough or require a cloud server or a trusted third party to participate in the authentication process. Therefore, this paper designs a fast AKA protocol based on time-sensitive token for MEC. With this protocol, the terminal node can achieve fast authentication through the applied token, and the authentication process only requires a related edge node to participate. Simulation results based on ProVerif and informal security analysis show that our protocol can resist various common attacks. The comparison with related protocols shows that our protocol only needs to spend very little computational and communicational cost to authenticate a terminal node with a token.

Keywords: Authentication · Cryptography · Key agreement · Mobile edge computing · Security

1 Introduction

With the widespread application of Cyber-Physical Systems (CPS) and Internet of Things (IoT), there is a greater demand for storage and computing resources of local terminals [1,11]. Cloud computing enables local terminals with limited

This work is supported in part by the National Natural Science Foundation of China under Grants 61872138, in part by the Research Foundation of Education Bureau of Hunan Province, China (Grant 2020C0080 and Grant 19C0031), and in part by the Natural Science Foundation of Hunan Province under Grant 2020JJ5603.

Y. Lai et al. (Eds.): ICA3PP 2021, LNCS 13157, pp. 632–647, 2022.
https://doi.org/10.1007/978-3-030-95391-1_40

resources to outsource massive data storage and costly computing tasks to cloud servers with abundant computing resources. However, achieving low latency and providing real-time services are two key challenges of cloud computing [18]. Therefore, researchers are increasingly focusing on Mobile Edge Computing (MEC), which can effectively reduce network latency and provide real-time services.

A typical MEC architecture usually consists of three layers: edge device layer, edge computing layer and core facility layer [19]. Terminal Nodes (TNs) at the edge device layer are usually resource-constrained devices, responsible for collecting raw data and interacting with edge nodes. The edge computing layer is composed of several Edge Nodes (ENs). Edge nodes usually have certain computing, storage, and communication resources, and are mainly responsible for receiving, processing, and forwarding the raw data from the edge device layer. The core facility layer is usually a cloud server, which has abundant computing resources [15].

Although MEC greatly improves the performance problems of cloud computing, it also brings new security challenges. For example, a large number of edge nodes are usually considered to be in a semi-trusted or malicious environment, which makes them a target for adversaries. Therefore, to ensure data privacy and security in MEC, a secure and reliable Authentication and Key Agreement (AKA) protocol is essential. However, the AKA protocol proposed for MEC on the one hand needs to be lightweight enough to be deployed in resource-constrained TNs. On the other hand, some protocols still require the participation of a cloud server or a trusted third partie when authenticating the TN. In this article, we propose a token-based AKA protocol for MEC, which has the following characteristics:

- When each TN is authenticated for the first time, it can apply for a token with a life cycle. In the subsequent authentication, the lifetime of the token will be consumed.
- When a TN has a token, it only needs to spend a small computational cost and only needs to verify the validity of the token by EN, which can effectively improve the authentication efficiency.
- Our protocol can negotiate three different session keys between the TN and the cloud server, the EN and the cloud server, and the TN and EN. The TN can use different keys according to the attributes of the transmitted message.

The rest of this paper is organized as follows. Section 2 surveys related works. Section 3 introduces system models. Section 4 describes the proposed protocol in detail. The security and correctness of our protocol are presented in Sect. 5. The performance comparison between our protocol and related protocols is in Sect. 6, and the final section summarizes the whole paper.

2 Related Works

Li et al. [9] summarized the difference and relationship between cloud computing and edge computing, and analyzed the security weaknesses in edge computing and related applications. To solve these security weaknesses, different researchers have proposed many solutions. Tsai and Lo [17] proposed an authentication scheme based on identity and trusted third parties for mobile cloud computing. However, He et al. [4] pointed out that the computational and communicational costs of Tsai and Lo's scheme [17] are too high. They improved Tsai and Lo's scheme [17] and proposed a lightweight AKA protocol for smart grids based on Elliptic Curve Cryptography (ECC). In the same year, Odelu et al. [14] pointed out that the scheme of He et al. [4] could not resist the ephemeral key leakage attack and could not guarantee the privacy of the smart meter certificate. Therefore, they proposed an improved AKA scheme for smart grid. In 2018, Jiang et al. [6] pointed out that Tsai and Lo's [17] scheme cannot achieve mutual authentication and cannot resist impersonation attack. However, they did not give a clear solution, but gave some suggestions to enable relevant researchers to avoid this problem in future work.

More recently, Kaur et al. [7] proposed an AKA protocol for MEC, which only uses ECC and hash function operations, and also uses discrete logarithm problems, computational Diffie-Hellman, and timestamps to defend against various attacks. The scheme proposed by Kumar et al. [8] uses hybrid cryptography to implement mutual authentication and key agreement in smart energy networks, and also ensures the anonymity of identity. But their scheme makes the neighborhood area network gateway have to manage and maintain a large number of symmetric keys for multiple home area network gateways. Based on Physically Uncloneable Functions (PUF), Gope and Sikdar [3] proposed an AKA protocol for smart grid, and Gope et al. [2] proposed an AKA protocol for industrial wireless sensor networks. However, devices based on field-programmable gate array are more suitable for using PUF [16]. Therefore, the application environment of these protocols may be restricted. Jia et al. [5] proposed an AKA scheme for MEC that can guarantee user anonymity and untraceability. In their scheme, the MEC server does not store any user's private information, and there is no need for a trusted third party during the authentication process. To ensure the anonymity of users in MEC, Li et al. [10] proposed a new authentication architecture, and based on this architecture, proposed an AKA scheme for MEC. Based on the blockchain [12,13], Wang et al. [18] proposed an AKA protocol for smart grid edge computing infrastructure, which implements conditional traceability and revocability through the smart contract technology of the blockchain, and does not require a trusted third party. In general, most of the existing MEC protocols are not lightweight enough, and a trusted third party or registration center is required to participate in the authentication process.

3 System Models

3.1 Network Model

There are three nodes in our network model, namely TN, EN, and Cloud Server Node (CSN), as shown in Fig. 1. TNs are usually mobile and are resource-constrained devices. ENs usually have certain computing and storage resources.

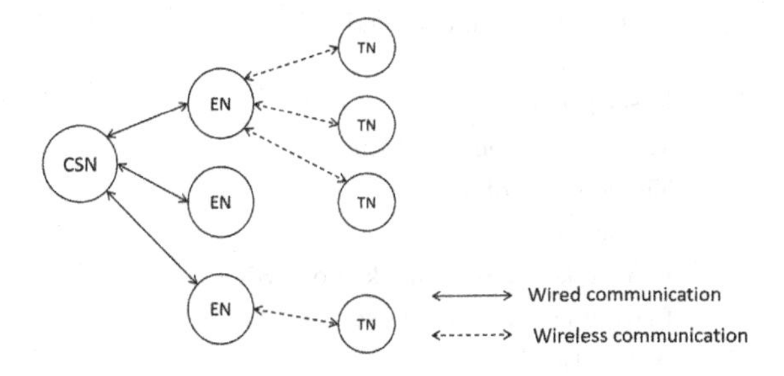

Fig. 1. The network model used in our protocol.

The CSN is usually rich in various resources. The EN and the TN usually use wireless communication, and the EN and the CSN usually use wired communication. Each CSN can manage several ENs, and each EN can also manage several TNs.

3.2 Threat Model

The threat models used in this protocol are listed below. (1) The CSN are considered as a trusted node. (2) An adversary can intercept exchanged messages over unsecured channels. In addition, he/she can inject new messages into the network and replace, replay, or modify previously intercepted messages. (3) An adversary can capture any number of TNs or ENs, and once a TN or EN is captured, the adversary can extract relevant secret parameters from the TN's or EN's memory.

4 Proposed Protocol

Table 1 lists the definitions of the symbols used in our protocol. The proposed protocol has five phases. The initialization phase is executed by the System Administrator (SA), and the TN registration phase and the EN registration phase are executed in a secure environment. When a $\mathrm{TN_i}$ does not have a valid token, the authentication without token phase is executed, otherwise, the authentication with token phase is executed. After a $\mathrm{TN_i}$ performs the authentication without token phase, the session key TK_1 between the $\mathrm{TN_i}$ and the CSN and the session key TK_2 between the $\mathrm{EN_j}$ and the CSN can be obtained. After a $\mathrm{TN_i}$ performs the authentication with token phase, the session key SK between the $\mathrm{TN_i}$ and the $\mathrm{EN_j}$ can be obtained. For $\mathrm{TN_i}$, it can decide whether to use TK_1 or SK according to the attributes of the transmitted data. For example, for extremely sensitive and private data, the $\mathrm{TN_i}$ should use TK_1 to prevent $\mathrm{EN_j}$ from obtaining these data. For general data, SK can be used to enable $\mathrm{EN_j}$ to process the data, or send it to the CSN after integration, which can reduce the communication and computation load of the CSN. The details of each phase are as follows.

Table 1. Symbols used in our protocol.

Symbols	Description
TID_i	Temporary identity of the TN_i
ID_i	The identity of TN_i
IDE_j	The identity of EN_j
s, P_{pub}	Private key and public key of CSN
W_i, S_i	Private key and public key of TN_i
AT	Authentication token
Atr	Attributes of the token applied for by TN
RL_i	The remaining usable life of the token
T_1 to T_6	Timestamp
T_{new}	The timestamp when the latest information was received
ΔT	Maximum communication transmission delay
E_k, D_k	Symmetric encryption and decryption algorithm
TK_1, TK_2	Symmetric key
SK	Session key
$\oplus$	Bitwise XOR operation
(a, b)	Concatenation of data a and data b

4.1 Initialization Phase

In this phase, the SA needs to perform the following steps.

Step I1: Generates a cyclic additive group G_1 of order p, a cyclic multiplicative group G_2 of order p, a generator Q of G_1, and $e : G_1 \times G_1 \rightarrow G_2$ is a bilinear map.

Step I2: Picks a private key $s \in Z_p$, and computes the corresponding public key through $P_{pub} = sQ$.

Step I3: Publishes parameters $\{p, G_1, G_2, Q, e, P_{pub}, h(.), E_k, D_k\}$, where $h(.) : \{0,1\}^* \rightarrow \{0,1\}^l$ is a hash function.

4.2 EN Registration Phase

Before entering the system, each EN needs to complete the registration with the CSN through this phase. The process at this phase is as follows.

Step ER1: The CSN generates a random number a_j and KE_j, a unique identity IDE_j, computes $A_j = a_jQ$ and sends parameters a_j, A_j, KE_j, and IDE_j to the corresponding EN_j through a secure channel.

Step ER2: The $\mathrm{EN_j}$ stores tuple $(IDE_j,\ a_j,\ A_j,\ KE_j)$ in its memory. The CSN stores tuple $(IDE_j,\ KE_j)$ in its memory.

4.3 TN Registration Phase

Due to the limited storage space of the EN, it is impossible for EN to store all authentication parameters related to the TN. Therefore, each TN needs to register with the CSN and store these authentication parameters in the CSN. The process at this phase is as follows.

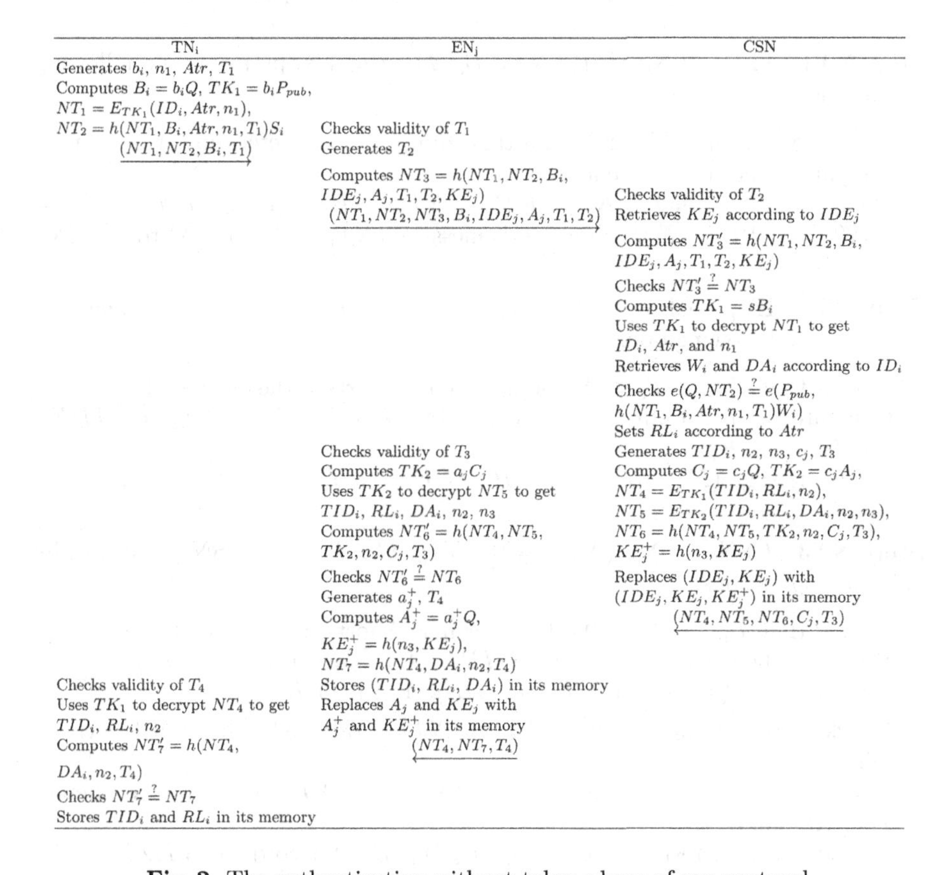

Fig. 2. The authentication without token phase of our protocol.

Step TR1: The CSN generates a random number r, a unique identity ID_i, computes $W_i = h(ID_i)$, $S_i = sW_i$, $DA_i = h(ID_i, r, S_i)$ and sends parameters ID_i, S_i, W_i, and DA_i to $\mathrm{TN_i}$ through a secure channel.

Step TR2: The $\mathrm{TN_i}$ stores tuple $(ID_i,\ S_i,\ W_i,\ DA_i)$ in its memory. The CSN stores tuple $(ID_i,\ W_i,\ DA_i)$ in its memory.

4.4 Authentication Without Token Phase

When a TN_i does not have a valid token, it needs to perform this phase to complete mutual authentication with the CSN and apply for a new token. In this phase, TN_i will send a parameter Atr about the attributes of the token, which contains information about how long the token needs to be used. When the CSN receives this parameter, it will set the remaining usable life RL_i of the token according to Atr. In addition, the parameters used to verify the token will be sent by the CSN to the corresponding EN_j. Figure 2 shows the authentication without token phase in our protocol. The details of this phase are as follows:

Step NT1: $TN_i \rightarrow EN_j$: (NT_1, NT_2, B_i, T_1), the TN_i performs the following operations:

- Generates a random b_i and n_1, a timestamp T_1, and determines the value of the parameter Atr as needed.
- Computes $B_i = b_iQ$, $TK_1 = b_iP_{pub}$, $NT_1 = E_{TK_1}(ID_i, Atr, n_1)$, $NT_2 = h(NT_1, B_i, Atr, n_1, T_1)S_i$, and sends message ($NT_1$, NT_2, B_i, T_1) to the EN_j.

Step NT2: $EN_j \rightarrow$CSN: (NT_1, NT_2, NT_3, B_i, IDE_j, A_j, T_1, T_2). The EN_j performs the following operations:

- Checks that $T_{new} - T_1 < \Delta T$ holds or not. Aborts if the check fails.
- Generates a timestamp T_2, and computes $NT_3 = h(NT_1,\ NT_2,\ B_i,\ IDE_j,\ A_j,\ T_1,\ T_2,\ KE_j)$.
- Sends message (NT_1, NT_2, NT_3, B_i, IDE_j, A_j, T_1, T_2) to the CSN.

Step NT3: CSN$\rightarrow$ EN_j: (NT_4, NT_5, NT_6, C_j, T_3), the CSN performs the following operations:

- Checks that $T_{new} - T_2 < \Delta T$ holds or not. Aborts if the check fails.
- Retrieves the corresponding tuple (IDE_j, KE_j) according to IDE_j.
- Computes $NT_3' = h(NT_1, NT_2, B_i, IDE_j, A_j, T_1, T_2, KE_j)$.
- Checks whether the condition $NT_3' \stackrel{?}{=} NT_3$ is satisfied. Aborts if the check fails.
- Computes $TK_1 = sB_i$, and uses TK_1 to decrypt NT_1 to get the parameters ID_i, Atr, and n_1.
- Retrieves the corresponding tuple (ID_i, W_i, DA_i) according to ID_i.
- Checks whether the condition $e(Q, NT_2) \stackrel{?}{=} e(P_{pub}, h(NT_1, B_i, Atr, n_1, T_1)W_i)$ is satisfied. Aborts if the check fails.
- According to Atr, sets an appropriate remaining usable life RL_i for the token.
- Generates a unique TID_i, a random n_2, n_3, and c_j, a timestamp T_3, and computes $C_j = c_jQ$, $TK_2 = c_jA_j$, $NT_4 = E_{TK_1}(TID_i, RL_i, n_2)$, $NT_5 = E_{TK_2}(TID_i, RL_i, DA_i, n_2, n_3)$, $NT_6 = h(NT_4, NT_5, TK_2, n_2, C_j, T_3)$, $KE_j^+ = h(n_3, KE_j)$.
- Replaces (IDE_j, KE_j) with (IDE_j, KE_j, KE_j^+) in its memory.
- Sends message (NT_4, NT_5, NT_6, C_j, T_3) to EN_j.

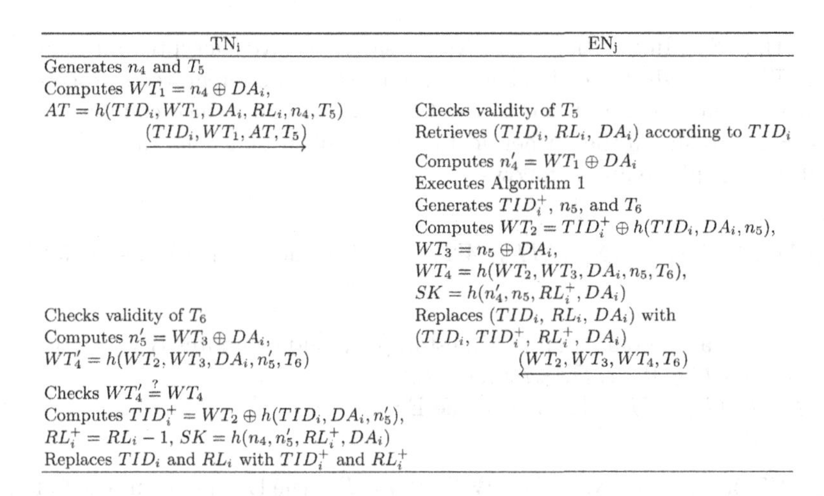

Fig. 3. The authentication with token phase of our protocol.

Step NT4: EN_j → TN_i: (NT_4, NT_7, T_4), the EN_j performs the following operations:

- Checks that $T_{new} - T_3 < \Delta T$ holds or not. Aborts if the check fails.
- Computes $TK_2 = a_j C_j$, and uses TK_2 to decrypt NT_5 to get the parameters TID_i, RL_i, DA_i, n_2, and n_3.
- Computes $NT_6' = h(NT_4, NT_5, TK_2, n_2, C_j, T_3)$, and checks whether the condition $NT_6' \stackrel{?}{=} NT_6$ is satisfied. Aborts if the check fails.
- Generates a random a_j^+, a timestamp T_4, and computes $A_j^+ = a_j^+ Q$, $KE_j^+ = h(n_3, KE_j)$, $NT_7 = h(NT_4, DA_i, n_2, T_4)$.
- Stores the tuple (TID_i, RL_i, DA_i) and replaces A_j and KE_j with A_j^+ and KE_j^+ in its memory.
- Sends message (NT_4, NT_7, T_4) to TN_i.

Step NT5: The TN_i performs the following operations:

- Checks that $T_{new} - T_4 < \Delta T$ holds or not. Aborts if the check fails.
- Uses TK_1 to decrypt NT_4 to get the parameters TID_i, RL_i, and n_2.
- Computes $NT_7' = h(NT_4, DA_i, n_2, T_4)$.
- Checks whether the condition $NT_7' \stackrel{?}{=} NT_7$ is satisfied. Aborts if the check fails.
- Stores the parameters TID_i and RL_i in its memory.

4.5 Authentication with Token Phase

In this phase, when the TN_i has a valid token, it only needs to complete mutual authentication and key agreement with the EN_j, which can effectively reduce the load of the CSN. In addition, the EN_j can safely receive encrypted messages sent by the TN_i through the session key SK, so the EN_j can integrate or process these messages before sending them to the CSN, which can reduce communicational costs. Figure 3 shows the authentication with token phase in our protocol. The details of this phase are as follows:

Step WT1: $TN_i \rightarrow EN_j$: (TID_i, WT_1, AT, T_5), the TN_i performs the following operations:

- Generates a random n_4 and a timestamp T_5, and computes $WT_1 = n_4 \oplus DA_i$, $AT = h(TID_i, WT_1, DA_i, RL_i, n_4, T_5)$.
- Sends (TID_i, WT_1, AT, T_5) to the EN_j.

Step WT2: $EN_j \rightarrow TN_i$: (WT_2, WT_3, WT_4, T_6), the EN_j performs the following operations:

- Checks that $T_{new} - T_5 < \Delta T$ holds or not. Aborts if the check fails.
- Retrieves the corresponding tuple (TID_i, RL_i, DA_i) according to TID_i.
- Computes $n_4' = WT_1 \oplus DA_i$, and executes Algorithm 1.
- Generates a new unique TID_i^+, a random n_5, a timestamp T_6, computes $WT_2 = TID_i^+ \oplus h(TID_i, DA_i, n_5)$, $WT_3 = n_5 \oplus DA_i$, $WT_4 = h(WT_2, WT_3, DA_i, n_5, T_6)$, $SK = h(n_4', n_5, RL_i^+, DA_i)$.
- Replaces tuple (TID_i, RL_i, DA_i) with tuple $(TID_i, TID_i^+, RL_i^+, DA_i)$ in its memory.
- Sends message (WT_2, WT_3, WT_4, T_6) to the TN_i.

Step WT3: The TN_i performs the following operations:

- Checks that $T_{new} - T_6 < \Delta T$ holds or not. Aborts if the check fails.
- Computes $n_5' = WT_3 \oplus DA_i$, $WT_4' = h(WT_2, WT_3, DA_i, n_5', T_6)$, and checks whether the condition $WT_4' \stackrel{?}{=} WT_4$ is satisfied. Aborts if the check fails.
- Computes $TID_i^+ = WT_2 \oplus h(TID_i, DA_i, n_5')$, $RL_i^+ = RL_i - 1$, $SK = h(n_4, n_5', RL_i^+, DA_i)$.
- Replaces TID_i and RL_i with TID_i^+ and RL_i^+ in its memory.

In order to prevent jamming/desynchronization attacks, we designed Algorithm 1, and the details of Algorithm 1 are as follows. Note that for each TN_i, there is a $flag\_sku_i$ in the corresponding EN_j. When the TN_i and the EN_j complete the mutual authentication and generate a SK, $flag\_sku_i$ will be set to 0. When the TN_i uses the SK to communicate with the EN_j, the EN_j sets $flag\_sku_i$ to 1, otherwise $flag\_sku_i$ remains at 0.

5 Security Analysis

5.1 Simulation Based on ProVerif

ProVerif is a widely used automated security verification tool that can verify whether each parameter in the AKA protocol will be leaked and whether the AKA protocol process is executed correctly [20]. Therefore, we use ProVerif in this section to prove the security of our protocol. Figures 4 and 5 respectively shows the simulation codes of the authentication without token phase and the authentication with token phase in our protocol. Figure 6a and 6b respectively show the simulation results of the authentication without token phase and the authentication with token phase of our protocol. According to Fig. 6a and 6b, it can be found that the simulation results of ProVerif prove that in our protocol, ID_i, TK_1, TK_2, SK, DA_i, and RL_i cannot be obtained by the adversary, and the protocol is executed in order.

Algorithm 1. Algorithm used to ensure the consistency of RL_i of the communication parties

1: **if** $RL_i == 0$ **then**
2: The token is invalid, and abort this session;
3: **else**
4: Compute $AT' = h(TID_i, WT_1, DA_i, RL_i, n'_4, T_5)$;
5: **if** $AT' == AT$ **then**
6: $RL_i^+ = RL_i - 1$;
7: Continue authentication without token phase;
8: **else**
9: Compute $AT' = h(TID_i, WT_1, DA_i, RL_i + 1, n'_4, T_5)$;
10: **if** $AT' == AT$ && $flag\_sku_i$==0 **then**
11: $RL_i^+ = RL_i$;
12: Continue authentication without token phase;
13: **else if** $AT' == AT$ && $flag\_sku_i$==1 **then**
14: Warn that the TN may be compromised and abort this session;
15: **else**
16: Abort this session;
17: **end if**
18: **end if**
19: **end if**

5.2 Informal Security Analysis

Eavesdropping Attack. In our protocol, the adversary needs to get TK_1, TK_2, or SK to get valuable information. However, the adversary cannot obtain a_j, b_i, s, c_j, n_3, n_4, RL_i, and DA_i through eavesdropping attacks, so he/she cannot calculate TK_1, TK_2, or SK.

TN's Anonymity and Untraceability. Only NT_1 contains ID_i, but the adversary cannot obtain TK_1, so ID_i cannot be obtained. In addition, there is no fixed parameter in each message exchanged. As for TID_i, when the authentication fails, the TID_i will remain unchanged, but as long as the authentication is successful, the TID_i will be updated. If the adversary interferes with TN repeatedly and causes authentication failure, the TN can easily detect the abnormality and issue an alarm. Therefore, the adversary cannot track a certain TN through any parameter.

Impersonation Attack. There is a verification parameter in every message exchanged in our protocol, which are NT_2, NT_3, NT_6, NT_7, AT, and WT_4. The adversary needs to be able to create a valid verification parameter to pass the verification. However, the adversary cannot obtain n_1 to n_5, S_i, TK_2, DA_i, KE_j or RL_i, so our protocol can resist impersonation attacks.

Capture Attack. According to the threat model, the adversary has the ability to capture any number of TN or EN. At this time, it is necessary to ensure that the adversary cannot create any new legal TN or EN. If the adversary wants to create a new legal $\mathrm{TN_i}$, he/she needs to create a legal ID_i and S_i. If the adversary

```
free c:channel.

type QP.
const q: QP.

fun pm(QP,bitstring):QP.
equation forall x:bitstring,y:bitstring;
pm(pm(q, x), y) = pm(pm(q, y), x).
fun e(QP,QP):QP.
equation forall x:QP,y:QP,s:bitstring;
e(pm(y,s),x)=e(pm(x,s),y).

fun senc(bitstring, QP): bitstring.
reduc forall m: bitstring, k: QP;
sdec(senc(m,k),k) = m.
fun con(bitstring,bitstring):bitstring.
reduc forall m: bitstring, n: bitstring;
lde(con(m,n)) = m.
reduc forall m: bitstring, n: bitstring;
rde(con(m,n)) = n.
fun h(bitstring):bitstring.
fun QP_to_BS(QP):bitstring [typeConverter].
fun BS_to_QP(bitstring):QP [typeConverter].

event acc1().
event acc2().
event acc3().
event acc4().

query secret IDi.
query secret TK1.
query secret TK2.
query event (acc4())==¿event (acc3()).
query event (acc3())==¿event (acc2()).
query event (acc2())==¿event (acc1()).

let TN(IDi:bitstring,Si:QP,
DAi:bitstring,Ppub:QP,Q:QP) =
new bi:bitstring;
new n1:bitstring;
new Atr:bitstring;
new T1:bitstring;
let Bi=pm(Q,bi) in
let TK1=pm(Ppub,bi) in
let NT1=senc(con(IDi,con(Atr,n1)),TK1) in
let NT2=pm(Si,h(con(NT1,con(QP_to_BS(Bi),
con(Atr,con(n1,T1)))))) in
out (c,(NT1,NT2,Bi,T1));
in (c,(NT4:bitstring,NT7:bitstring,
T4:bitstring));
let tn2=rde(rde(sdec(NT4,TK1))) in
let tNT7=h(con(NT4,con(DAi,
con(tn2,T4)))) in
if (tNT7=NT7) then event acc4().

let EN(IDEj:bitstring,aj:bitstring,
Aj:QP,KEj:bitstring) =
in (c,(NT1:bitstring,NT2:QP,
Bi:QP,T1:bitstring));
new T2:bitstring;
let NT3=h(con(NT1,con(QP_to_BS(NT2),
con(QP_to_BS(Bi),con(IDEj,con(QP_to_BS(Aj),
con(T1,con(T2,KEj)))))))) in
out (c,(NT1,NT2,NT3,Bi,IDEj,Aj,T1,T2));
in (c,(NT4:bitstring,NT5:bitstring,
NT6:bitstring,Cj:QP,T3:bitstring));
let TK2=pm(Cj,aj) in
let tDAi=lde(rde(rde(sdec(NT5,TK2)))) in
let tn2=lde(rde(rde(rde(sdec(NT5,
TK2))))) in
let tn3=rde(rde(rde(rde(sdec(NT5,
TK2))))) in
let tNT6=h(con(NT4,con(NT5,
con(QP_to_BS(TK2),con(tn2,
con(QP_to_BS(Cj),T3)))))) in
if (tNT6=NT6) then event acc3();
new naj:bitstring;
new T4:bitstring;
let nKEj=h(con(tn3,KEj)) in
let NT7=h(con(NT4,con(tDAi,
con(tn2,T4)))) in 0.

let CSN(KEj:bitstring,s:bitstring,
Wi:QP,DAi:bitstring,Ppub:QP,Q:QP)=
in (c,(NT1:bitstring,NT2:QP,NT3:bitstring,
Bi:QP,IDEj:bitstring,Aj:QP,T1:bitstring,
T2:bitstring));
let tNT3=h(con(NT1,con(QP_to_BS(NT2),
con(QP_to_BS(Bi),con(IDEj,con(QP_to_BS(Aj),
con(T1,con(T2,KEj)))))))) in
if (tNT3=NT3) then event acc1();
let TK1=pm(Bi,s) in
let IDi=lde(sdec(NT1,TK1)) in
let Atr=lde(rde(sdec(NT1,TK1))) in
let tn1=rde(rde(sdec(NT1,TK1))) in
if(e(NT2,Q)=e(pm(Wi,h(con(NT1,
con(QP_to_BS(Bi),con(Atr,con(tn1,
T1)))))),Ppub)) then event acc2();
new RLi:bitstring;
new TIDi:bitstring;
new n2:bitstring;
new n3:bitstring;
new cj:bitstring;
new T3:bitstring;
let Cj=pm(Q,cj) in
let TK2=pm(Aj,cj) in
let NT4=senc(con(TIDi,con(RLi,n2)),TK1) in
let NT5=senc(con(TIDi,con(RLi,con(DAi,
con(n2,n3)))),TK2) in
let NT6=h(con(NT4,con(NT5,
con(QP_to_BS(TK2),con(n2,
con(QP_to_BS(Cj),T3)))))) in
let nKEj=h(con(n3,KEj)) in
out (c,(NT4,NT5,NT6,Cj,T3)).

process
new IDi:bitstring;
new DAi:bitstring;
new IDEj:bitstring;
new KEj:bitstring;
new s:bitstring;
new Wi:QP;
new Q:QP;
new aj:bitstring;
let Ppub=pm(Q,s) in
let Aj=pm(Q,aj) in
let Si=pm(Wi,s) in
(!TN(IDi,Si,DAi,Ppub,Q))|(!EN(IDEj,aj,
Aj,KEj))|(!CSN(KEj,s,Wi,DAi,Ppub,Q)))
```

Fig. 4. The simulation code of the authentication without token phase of our protocol.

wants to create a new legal $\mathrm{EN_j}$, he/she needs to create a legal IDE_j and KE_j. Obviously, the adversary cannot get the legal ID_i and IDE_j from all known ID_i and IDE_j. In addition, due to elliptic curve discrete logarithm problem, no matter how many S_i is obtained by the adversary, s cannot be calculated in polynomial time. KE_j will be updated after each successful authentication, and it is protected by $h()$. Therefore, the adversary cannot create a new legal TN or EN by capture attacks.

Replay Attack. In our protocol, every transmitted message contains a timestamp. In addition, the verification parameters in each message, namely NT_2, NT_3, NT_6, NT_7, AT, and WT_4, contain the information of the timestamp. According to the analysis of the impersonation attack part, the adversary cannot create new legal verification parameters, which makes the timestamp immutable. Therefore, our protocol can resist replay attacks.

Forward and Backward Secrecy. In our protocol, there are 3 different symmetric keys, namely TK_1, TK_2, and SK. Since TK_1 and TK_2 are both generated based on Diffie-Hellman key exchange protocol, so their forward and backward secrecy can be guaranteed. For $SK = h(n_4, n_5, RL_i^+, DA_i)$, the n_4 and n_5 are

```
free c:channel.
fun xor(bitstring,bitstring):bitstring.
equation forall m:bitstring,n:bitstring;
xor(xor(m,n),n)=m.
fun minus(bitstring):bitstring.
fun con(bitstring,bitstring):bitstring.
fun h(bitstring):bitstring.

event acc1().
event acc2().
query secret SK.
query secret RLi.
query secret DAi.
query event (acc2())==¿event (acc1()).

let TN(TIDi:bitstring,RLi:bitstring,
DAi:bitstring) =
new n4:bitstring;
new T5:bitstring;
let WT1=xor(n4,DAi) in
let AT=h(con(TIDi,con(WT1,con(DAi,
con(RLi,con(n4,T5)))))) in
out (c,(TIDi,WT1,AT,T5));
in (c,(WT2:bitstring,WT3:bitstring,
WT4:bitstring,T6:bitstring));
let tn5=xor(WT3,DAi) in
let tWT4=h(con(WT2,con(WT3,con(DAi,
con(tn5,T6))))) in
if (tWT4=WT4) then event acc2();
let nTID=xor(WT2,h(con(TIDi,
con(DAi,tn5)))) in
let nRLi=minus(RLi) in
let SK=h(con(n4,con(tn5,
con(nRLi,DAi)))) in 0.

let EN(TIDi:bitstring,RLi:bitstring,
DAi:bitstring) =
in (c,(TIDi:bitstring,WT1:bitstring,
AT:bitstring,T5:bitstring));
let tn4=xor(WT1,DAi) in
let tAT=h(con(TIDi,con(WT1,con(DAi,
con(minus(RLi),con(tn4,T5)))))) in
if (tAT=AT) then event acc1();
let nRLi=minus(RLi) in
new nTIDi:bitstring;
new n5:bitstring;
new T6:bitstring;
let WT2=xor(nTIDi,h(con(TIDi,
con(DAi,n5)))) in
let WT3=xor(n5,DAi) in
let WT4=h(con(WT2,con(WT3,
con(DAi,con(n5,T6))))) in
let SK=h(con(tn4,con(n5,
con(nRLi,DAi)))) in
out (c,(WT2,WT3,WT4,T6)).

process
new TIDi:bitstring;
new DAi:bitstring;
new RLi:bitstring;
(!TN(TIDi,RLi,DAi))|
(!EN(TIDi,RLi,DAi))
```

Fig. 5. The simulation code of the authentication with token phase of our protocol.

random numbers, and the RL_i^+ is never directly sent in the channel. Therefore, the forward and backward secrecy of SK can also be guarantee.

6 Performance Analysis

In this section, we compare our protocol with the protocol proposed by Jia et al. [5], Li et al. [10], and Wang et al. [18] in terms of computational cost. We refer to the authentication without token phase as P1, and the authentication with token phase as P2.

We use symbols t_e, t_h, t_{sym}, t_{pm}, and t_{bp} to represent the computation time required to implement one modulo exponential operation, one general hash function operation, one symmetric encryption or decryption, one point multiplication operation on ECC, and one bilinear pairing operation respectively. According to [21], we summarize the required computation time to perform different operations, which shown in Table 2. As for communicational cost, we use $|G|$, $|G_T|$, Z_p, $|T|$, $|ID|$ to represent the size of the element in G_1, G_2, Z_p, timestamp, and the identity of TN or EN, which is 1024, 1024, 160, 64, 64 bits.

Table 3 shows the computational and communicational cost comparison between our protocol and related protocols. It can be found from Table 3 that

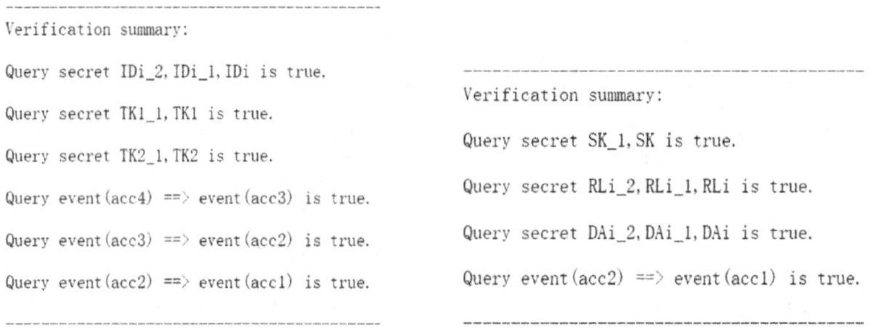

```
--------------------------------------------
Verification summary:

Query secret IDi_2,IDi_1,IDi is true.

Query secret TK1_1,TK1 is true.

Query secret TK2_1,TK2 is true.

Query event(acc4) ==> event(acc3) is true.

Query event(acc3) ==> event(acc2) is true.

Query event(acc2) ==> event(acc1) is true.

--------------------------------------------
```

(a) The authentication without token phase of our protocol.

```
--------------------------------------------
Verification summary:

Query secret SK_1,SK is true.

Query secret RLi_2,RLi_1,RLi is true.

Query secret DAi_2,DAi_1,DAi is true.

Query event(acc2) ==> event(acc1) is true.

--------------------------------------------
```

(b) The authentication with token phase of our protocol.

Fig. 6. The simulation results of different phase in our protocol.

Table 2. Computation time required for different operations.

Symbols	Execute by TN	Execute by EN or CSN
t_e	2.249 ms	0.339 ms
t_h	0.056 ms	0.007 ms
t_{sym}	0.224 ms	0.028 ms
t_{pm}	13.405 ms	2.165 ms
t_{pa}	0.081 ms	0.013 ms
t_{bp}	32.713 ms	5.427 ms

P2 greatly reduces the communicational and computational costs. For P1, the computational cost of the TN is the lowest, while the computational cost of EN and CSN and the total communicational cost are higher. However, compared with the TN, both EN and CSN have more resources. On the other hand, our protocol does not always implement P1, but more implements P2. For example, if our protocol executes P1 once and gets a token lifetime of 2, that is, the TN can execute P2 twice, then the average computational cost for each authentication of EN and CSN is $(16.413 + 0.035 + 0.035)/3 = 4.828$ms, and the total communicational cost is $(7808 + 992 + 992)/3 = 3264$ bits. Therefore, our protocol can greatly reduce communicational and computational cost, thereby improving authentication efficiency and reducing network latency.

Table 3. The comparison of the computational and communicational cost between our protocol and related protocols.

Protocols	Computational of TN/user	Computational of EN and CSN	Communicational cost
Jia et al. [5]	$4t_{pm} + t_e + 5t_h$	$t_{bp} + 5t_{pm} + 3t_{pa} + 5t_h$	$4G + 2T + 2Z_q + ID$
	= 56.149 ms	= 16.326 ms	= 4608 bits
Li et al. [10]	$6t_{pm} + 4t_{pa} + 5t_h$	$t_{bp} + 4t_{pm} + 3t_{pa} + 2t_h$	$3G + 4Z_q$
	= 81.034 ms	= 14.14 ms	= 3712 bits
Wang et al. [18]	$4t_{pm} + t_{pa} + 5t_h$	$4t_{pm} + t_{pa} + 6t_h$	$3G + 2Z_q + 2T$
	= 53.981 ms	= 8.715 ms	= 3712 bits
P1	$3t_{pm} + 2t_h + 2t_{sym}$	$t_{bp} + 5t_{pm} + 7t_h + 4t_{sym}$	$6G + 5T + 8Z_q + ID$
	= 40.775 ms	= 16.413 ms	= 7808 bits
P2	$4t_h$ = 0.224 ms	$5t_h$ = 0.035 ms	$2T + 5Z_q + ID$ = 992 bits

7 Conclusion

This paper designs a fast AKA protocol based on time-sensitive token for MEC. In this protocol, each TN can apply for a token with a life cycle. When a TN owns a token, the communicational and computational costs required for authentication can be greatly reduced, and the authentication process only requires the participation of the relevant EN instead of a trusted third party. The simulation results of ProVerif and informal security analysis prove that our protocol can resist various common attacks. The comparison with related protocols shows that although TN needs to spend some communicational and computational costs to perform the authentication without token phase to apply for a token, once it gets the token, it can perform the authentication with token phase with low computational and communicational costs. If multiple authentications are performed, it is easy to find that our protocol can greatly reduce computational and communicational costs. In future work, we consider combining blockchain to implement a cross-EN fast authentication protocol.

References

1. Chen, X., Liang, W., Xu, J., Wang, C., Li, K.C., Qiu, M.: An efficient service recommendation algorithm for cyber-physical-social systems. IEEE Trans. Netw. Sci. Eng. 1 (2021). https://doi.org/10.1109/TNSE.2021.3092204
2. Gope, P., Das, A.K., Kumar, N., Cheng, Y.: Lightweight and physically secure anonymous mutual authentication protocol for real-time data access in industrial wireless sensor networks. IEEE Trans. Ind. Inform. **15**(9), 4957–4968 (2019). https://doi.org/10.1109/TII.2019.2895030
3. Gope, P., Sikdar, B.: Privacy-aware authenticated key agreement scheme for secure smart grid communication. IEEE Trans. Smart Grid **10**(4), 3953–3962 (2019). https://doi.org/10.1109/TSG.2018.2844403
4. He, D., Wang, H., Khan, M.K., Wang, L.: Lightweight anonymous key distribution scheme for smart grid using elliptic curve cryptography. IET Commun. **10**(14), 1795–1802 (2016). https://doi.org/10.1049/iet-com.2016.0091
5. Jia, X., He, D., Kumar, N., Choo, K.K.R.: A provably secure and efficient identity-based anonymous authentication scheme for mobile edge computing. IEEE Syst. J. **14**(1), 560–571 (2020). https://doi.org/10.1109/JSYST.2019.2896064
6. Jiang, Q., Ma, J., Wei, F.: On the security of a privacy-aware authentication scheme for distributed mobile cloud computing services. IEEE Syst. J. **12**(2), 2039–2042 (2018). https://doi.org/10.1109/JSYST.2016.2574719
7. Kaur, K., Garg, S., Kaddoum, G., Guizani, M., Jayakody, D.N.K.: A lightweight and privacy-preserving authentication protocol for mobile edge computing. In: 2019 IEEE Global Communications Conference (GLOBECOM), pp. 1–6 (2019). https://doi.org/10.1109/GLOBECOM38437.2019.9013856
8. Kumar, P., Gurtov, A., Sain, M., Martin, A., Ha, P.H.: Lightweight authentication and key agreement for smart metering in smart energy networks. IEEE Trans. Smart Grid **10**(4), 4349–4359 (2019). https://doi.org/10.1109/TSG.2018.2857558
9. Li, X., Chen, T., Cheng, Q., Ma, S., Ma, J.: Smart applications in edge computing: overview on authentication and data security. IEEE Internet Things J. **8**(6), 4063–4080 (2021). https://doi.org/10.1109/JIOT.2020.3019297
10. Li, Y., Cheng, Q., Liu, X., Li, X.: A secure anonymous identity-based scheme in new authentication architecture for mobile edge computing. IEEE Syst. J. **15**(1), 935–946 (2021). https://doi.org/10.1109/JSYST.2020.2979006
11. Liang, W., Ning, Z., Xie, S., Hu, Y., Lu, S., Zhang, D.: Secure fusion approach for the internet of things in smart autonomous multi-robot systems. Inf. Sci. **579**, 468–482 (2021). https://doi.org/10.1016/j.ins.2021.08.035, https://www.sciencedirect.com/science/article/pii/S0020025521008355
12. Liang, W., Xiao, L., Zhang, K., Tang, M., He, D., Li, K.C.: Data fusion approach for collaborative anomaly intrusion detection in blockchain-based systems. IEEE Internet Things J. 1 (2021). https://doi.org/10.1109/JIOT.2021.3053842
13. Liang, W., Zhang, D., Lei, X., Tang, M., Li, K.C., Zomaya, A.Y.: Circuit copyright blockchain: blockchain-based homomorphic encryption for IP circuit protection. IEEE Trans. Emerg. Topics Comput. **9**(3), 1410–1420 (2021). https://doi.org/10.1109/TETC.2020.2993032
14. Odelu, V., Das, A.K., Wazid, M., Conti, M.: Provably secure authenticated key agreement scheme for smart grid. IEEE Trans. Smart Grid **9**(3), 1900–1910 (2018). https://doi.org/10.1109/TSG.2016.2602282

15. Qiu, T., Chi, J., Zhou, X., Ning, Z., Atiqzzzaman, M., Wu, D.O.: Edge computing in industrial internet of things: architecture, advances and challenges. IEEE Commun. Surv. Tutorials **22**(4), 2462–2488 (2020). https://doi.org/10.1109/COMST.2020.3009103
16. Trimberger, S.M., Moore, J.J.: FPGA security: motivations, features, and applications. Proc. IEEE **102**(8), 1248–1265 (2014). https://doi.org/10.1109/JPROC.2014.2331672
17. Tsai, J.L., Lo, N.W.: A privacy-aware authentication scheme for distributed mobile cloud computing services. IEEE Syst. J. **9**(3), 805–815 (2015). https://doi.org/10.1109/JSYST.2014.2322973
18. Wang, J., Wu, L., Choo, K.K.R., He, D.: Blockchain-based anonymous authentication with key management for smart grid edge computing infrastructure. IEEE Trans. Ind. Inform. **16**(3), 1984–1992 (2020). https://doi.org/10.1109/TII.2019.2936278
19. Xiao, Y., Jia, Y., Liu, C., Cheng, X., Yu, J., Lv, W.: Edge computing security: state of the art and challenges. Proc. IEEE **107**(8), 1608–1631 (2019). https://doi.org/10.1109/JPROC.2019.2918437
20. Xu, Z., Li, X., Xu, J., Liang, W., Choo, K.K.R.: A secure and computationally efficient authentication and key agreement scheme for internet of vehicles. Comput. Electr. Eng. **95**, 107409 (2021). https://doi.org/10.1016/j.compeleceng.2021.107409
21. Xu, Z., Liang, W., Li, K.C., Xu, J., Jin, H.: A blockchain-based roadside unit-assisted authentication and key agreement protocol for internet of vehicles. J. Parallel Distrib. Comput. **149**, 29–39 (2021). https://doi.org/10.1016/j.jpdc.2020.11.003

Ferproof: A Constant Cost Range Proof Suitable for Floating-Point Numbers

Yicong Li, Kuanjiu Zhou[✉], Lin Xu, Meiying Wang, and Nan Liu

School of Software, Dalian University of Technology, Dalian 116024, China
zhoukj@dlut.edu.cn

Abstract. The decentralization of the blockchain has led to the leakage of privacy data at the transaction layer, causing information security issues. The zero-knowledge range proof can confidentially verify that the transaction amount belongs to a legal positive integer range without revealing the transaction, effectively solving the privacy leakage of the blockchain transaction layer. The currently known blockchain range proof scheme is unable to process floating-point numbers, which limits the application fields of range proof. Moreover, the existing solutions can be further optimized in terms of proof speed, verification speed, and computational cost. We propose Ferproof, a constant cost range proof with floating-point number adaptation. Ferproof improves the zero-knowledge protocol based on Bulletproofs to optimize the proof structure, and a Lagrangian inner product vector generation method is issued to make the witness generation time constant and the commitment is constructed according to the floating-point number range relationship to implement floating-point range proof. Ferproof only relies on the discrete logarithm assumption, and the third-party credibility is not required. The communication cost and time complexity of Ferproof are constant. According to the experimental results, compared with the most advanced known range proof scheme, Ferproof's proof speed is increased by 40.0%, and the verification speed is increased by 29.8%.

Keywords: Blockchain · Privacy protection · Zero-knowledge proof · Range proof · Vector inner product commitment

1 Introduction

As a global distributed database technology, blockchain has the inherent characteristics of decentralization and global node consensus, which can effectively solve the third-party trust entrustment and reduce credit costs. However, the openness and collective maintenance of on-chain data causes any individual node to verify the validity of other nodes' transactions on the distributed ledger at any time, which poses a threat to user privacy data leakage. Although the UTXO

Supported by organization Key Research and Development Program Project of Ministry of Science and Technology (No. 2019YFD1101104).

Y. Lai et al. (Eds.): ICA3PP 2021, LNCS 13157, pp. 648–667, 2022.
https://doi.org/10.1007/978-3-030-95391-1_41

model used by virtual currencies such as Bitcoin [1] must provide script signatures for things scripts to verify the validity at the time of transaction [2], the amount of UTXO transactions is still transparent. Moreover, when a user transaction request is uploaded to the chain and submitted to the operating system for verification, the user's private data will be leaked to the system in the form of calculation parameters, the confidentiality of the data on the chain is weakened. Hence, based on ensuring the decentralization of data, how to strengthen the confidentiality and security of transaction information on the chain is a typical problem of the blockchain [3]. Zero-knowledge proof [4] is a cryptographic theory that solves data confidentiality verification, and it is widely used to solve the problem of blockchain privacy protection. Range proof is an important specific application of zero-knowledge proof theory, which essentially proves that the submitted promise value is within a specified secret data interval or belongs to a member of the public set without revealing the promise value content.

Currently, among the known range proof schemes for confidential blockchain transactions, when a problem scale is n bits, it proves that the time complexity is optimal as $O(n)$. When the scope of the problem is too large, blockchain platforms with limited computing on the chain still require high computing costs. Although it is known that some constant schemes of proof and verification time are implemented, that is, the time cost is independent of the problem scale, and these schemes do not consider the constancy of evidence generation time. Hence, their overall computational cost remains linear and there is still much room for optimizing the computational speed. In addition, the currently known range proof schemes can only generate proofs for integers, and range proof schemes for floating-point numbers have not been investigated in-depth.

2 Related Work

2.1 Confidential Transactions and Non-interactive Proof

To address the issue of on-chain privacy and security, Maxwell et al. proposed the concept of confidential transaction (CT) [5], which is to hide the transaction amount within a quantitative range by constructing a transaction commitment [6]. To achieve confidential transactions, Maxwell et al. proposed the Borromean ring signature technique [7], and the Borromean ring signature was then applied in Monero [8,9] in combination with the Pedersen promise [10]. Using zero-knowledge proof theory, the Ethereum [11] platform designed a non-interactive zero-knowledge proving (NIZK) tool for smart contract programs to provide the user verification privacy protection for complex smart contract programs [12]. However, NIZK is limited in on-chain applications because the limited computing power of smart contracts and the high cost of blockchain communication make NIZK not suitable for verification through smart contracts [13].

zk-SNARK [14,15] was issued as a concise non-interactive zero-knowledge proof scheme. The fast verification time and constant evidence size of zk-SNARK make it suitable for verification by smart contracts [16,17]. zk-SNARK is also used in the Zcash [18] zero-knowledge confidential trading platform.

zk-SNARK effectively satisfies the concept of confidential transactions using number-theoretic assumptions, but zk-SNARK requires the third-party Trusted Setup to host the public reference characters (CRS) required for the operation, which means that a zk-SNARK based proof system can be trusted only if the third-party trust mechanism is fully trusted, this conflicts with the decentralized nature of blockchain. To address the problem of credible mechanism, Wahby et al. [19] proposed the Hyrax protocol for zk-SNARK so that the logic circuit verification operation relies only on the discrete logarithm assumption without a credible mechanism. Srinath Setty et al. [20] also proposed a new construction based on zk-SNARK that does not require the third-party trustworthiness. Ben-Sasson et al. proposed zk-STARK [21] using HASH function collisions, which focuses on replacing the number-theoretic assumptions of zk-SNARK by symmetric encryption through HASH function collisions to eliminate the credible mechanism. However, the evidence generated by zk-STARK is large in length and not concise enough in generation. In 2019, a more new construction forms of zk-SNARK were issued, such as Sonic [22] by Maller et al. and Plonk [23] by Ariel Gabizon et al.

2.2 Range Proof

Range proof is a research hotspot of zero-knowledge proof technology. Lipmaa et al. [24] used Lagrange's theorem on the sum of four squares that any non-negative integer can be composed of four squares to achieve the range proof for non-negative integers. Groth et al. [25] performed a square number reduction of Lipmaa's work to achieve a non-negative integer range proof based on triple sums of squares. However, the security of [24,25] relies on the strong RSA assumption, which leads to a large commitment and is not suitable for on-chain verification.

In 2018, Benedikt Bunz et al. [26] proposed Bulletproofs short evidence range proof, which utilizes vector inner product commitments based on the polynomial commitment scheme proposed by Jonathan Bootle et al. [27] to reduce the evidence size and time complexity to logarithmic order without relying on the third-party trustworthy mechanisms. Cong Deng et al. [28] proposed Cuproof based on Bulletproofs to construct inner product vectors of constant dimension using Lagrange's theorem on sums of four squares to achieve flexible integer interval range proofs with constant proof time and verification time. However, Cuproof implements flexible range proofs by performing the aggregation of two proofs, thus increasing the time overhead.

3 Contribution

We propose Ferproof, a novel constant cost range proof suitable for floating-point numbers, which achieves constant witness generation time, proof time, and verification time with faster computing time. In addition, Ferproof supports range proofs for integers and floating-point numbers with arbitrary flexible range intervals.

The innovation of Ferproof is reflected in: (1) Ferproof applies an improved zero-knowledge protocol structure by proposing a novel form of commitment construction that allows the proof system to perform polynomial commitment verification on multivariate quadratic arithmetic expressions without splitting the expressions to perform aggregated proofs. In the case of the same problem scale, the commitment dimension generated by Ferproof is only half that of [28]. (2) We propose a inner product vector generation method, using the maximum square split algorithm to generate the inner product vector for the original problem of Ferproof, optimizes the time complexity of the relational inner product vector search solution of the problem down to $O(1)$, makes the witness generation time constant, and discretizes the vector dimension to shrink the average computational cost. (3) We propose a floating-point range relation expression for Ferproof and transform it into a polynomial commitment with a new protocol structure, and applies the inner product vector generation method to optimize the inner product vector generation time to achieve a general efficient range proof with constant computational cost.

We provided a complete implementation of Ferproof and compared Ferproof with other advanced known range proof schemes. The experimental results show that the comprehensive computing performance of Ferproof is more outstanding than other schemes.

4 Preliminaries

Before we present Ferproof, we first review the definition of symbols and related underlying commitment tools.

4.1 Symbol Definition

Table 1 shows the specific symbol definitions. Moreover, in the symbolic representation of vectors, bold lowercase letters are used to represent vectors, for example, $\mathbf{a} = (a_1, ..., a_n) \in \mathbb{Z}_p^n$; capitalized bold letters represent the matrix, for example, $\mathbf{A} \in \mathbb{Z}_p^{n\times m}$ represent matrix with n rows and m columns. In addition, $\langle \mathbf{a},\mathbf{b} \rangle = \sum_{i=1}^{n} a_i \cdot b_i$ represents the inner product of vectors $\mathbf{a}$ and $\mathbf{b}$; $\mathbf{a} \circ \mathbf{b} = (a_1 \cdot b_1, ..., a_n \cdot b_n)$ represents the dot product of vectors $\mathbf{a}$ and $\mathbf{b}$; $c \cdot \mathbf{b} = (c \cdot b_1, ..., c \cdot b_n)$ represents the product of constant c and vector $\mathbf{b}$.

Further, we also extend the above symbolic definition to the vector representation of group elements. Specifically, let $\mathbf{g} = (g_1, ..., g_n)$ represents the generator vector of the cyclic group vector $\mathbb{G}^n$; The binding commitment for vector $\mathbf{a}$ can be expressed as $C = \mathbf{g}^{\mathbf{a}} = \prod_{i=1}^{n} g_i^{a_i}$.

4.2 Commitment

Definition 1 (commitment): A non-interactive commitment scheme consists of a set of probabilistic polynomial time (PPT) algorithms $(SETUP, COM)$. The $SETUP$ algorithm generates the common parameter pp and security parameter λ for the commitment scheme. The COM algorithm defines the mapping relationship among the message space $\mathrm{M_{pp}}$, the random number space $\mathrm{R_{pp}}$ and the commitment space $\mathrm{C_{pp}}$: $\mathrm{M_{pp}} \times \mathrm{R_{pp}} \rightarrow \mathrm{C_{pp}}$. For the message $x \in \mathrm{M_{pp}}$, the commitment algorithm generates random numbers $r \stackrel{\$}{\longleftarrow} \mathrm{R_{pp}}$ uniformly and randomly, and calculates commitment $\boldsymbol{com} = COM_{\mathrm{pp}}(x, r)$.

Definition 2 (Homomorphic Commitments): The homomorphism of commitment means that in the space of commutative groups $\mathrm{M_{pp}}$, $\mathrm{R_{pp}}$ and $\mathrm{C_{pp}}$, for any $x_1, x_2 \in \mathrm{M_{pp}}$, $r_1, r_2 \in \mathrm{R_{pp}}$, the following relationship exists: $COM\left(x_1, r_1\right) + COM\left(x_2, r_2\right) = COM\left(x_1 + x_2, r_1 + r_2\right)$.

Definition 3 (Hiding Commitment): Hiding commitment means that the following relationship holds for all PPT adversaries:

$$\left| P \left[b = b' \middle| \begin{array}{c} \mathrm{pp} \leftarrow SETUP\left(1^{\lambda}\right), \\ \left(x_0, x_1\right) \in \mathrm{M_{pp}}^2 \leftarrow A\left(\mathrm{pp}\right), \\ b \stackrel{\$}{\leftarrow} \{0,1\}, r \stackrel{\$}{\leftarrow} R_{\mathrm{pp}}, \\ \boldsymbol{com} = COM\left(x_b, r\right), \\ b' \leftarrow A\left(\mathrm{pp}, \boldsymbol{com}\right) \end{array} \right] - \frac{1}{2} \right| \leq \eta\left(\lambda\right),$$

where of the $\eta(\lambda)$ represents a negligible function. When the probability of the $\eta(\lambda)$ of the above relationship approaches zero, the commitment is hidden, that is, the content of the commitment is indistinguishable in calculation.

Definition 4 (Binding Commitment): Binding commitment means that the following relationship holds for all PPT adversaries:

$$P \left[\begin{array}{c} COM(x_0, r_0) = COM(x_1, r_1) \\ \wedge x_0 \neq x_1 \end{array} \middle| \begin{array}{c} \mathrm{pp} \leftarrow SETUP(1^{\lambda}), \\ x_0, x_1, r_0, r_1 \leftarrow A(\mathrm{pp}) \end{array} \right] \leq \eta\left(\lambda\right).$$

When the probability of the $\eta(\lambda)$ of the above relationship approaches zero, the commitment is binding, that is, the content of the commitment is non-repudiation.

Definition 5 (Pedersen Commitment): Pedersen commitment aims to hide the secret value x under the bound message space $\mathrm{M_{pp}}$. The commitment structure is defined as follows:

$$\begin{array}{l} \mathrm{M_{pp}}, \mathrm{R_{pp}} \in \mathbb{R}_{\mathrm{p}}, \mathrm{C_{pp}} \in \mathbb{G} \\ SETUP : g, h \stackrel{\$}{\leftarrow} \mathbb{G} \\ COM(x, r) = (g^x h^r) \end{array} .$$

Definition 6 (Pedersen Vector Commitment): Pedersen vector commitment aims to hide the secret vector $\mathbf{x}$ under the bound message space $\mathrm{M_{pp}}$. The construction form of commitment is defined as follows:

$$\begin{array}{l} \mathrm{M_{pp}} \in \mathbb{Z}_p^n, \mathrm{R_{pp}} \in \mathbb{Z}_p, \mathrm{C_{pp}} \in \mathbb{G} \\ SETUP : \mathbf{g} = \left(g_1, \ldots g_n\right), h \leftarrow \mathbb{G} \\ COM\left(\mathbf{x}^n, r\right) = \left(h^r \mathbf{g}^{\mathbf{x}}\right) = h^r \prod_i g_i^{x_i} \in \mathbb{G} \end{array} .$$

Definition 7 (Polynomial Commitment): Assuming that the prover **P** has polynomials $t(X) = \sum_{i=0}^{n} t_i X^i$, the polynomial commitment aims to hide and prove that the prover **P** master polynomials $t(X)$. On the polynomial commitment of $t(X)$, firstly, the Pedersen commitment $(t_0, ..., t_n)$ is constructed for the polynomial coefficients $C_i = g^{t_i} h^{r_i} \in \mathbb{G}$. Then, the verifier **V** selects the random number x and sends it to the prover **P**. The prover **P** calculates the random term $\sum_{i=0}^{n} r_i x_i$ and $t(x)$ according to x and sends it to the verifier **V**. The verifier **V** verifies whether the following equation is true:

$$g^{t(x)} h^{\sum_{i=0}^{n} r_i x^i} = \prod_{i=0}^{n} C_i^{x^i} \in \mathbb{G}.$$

According to the conclusion of Fiat-Shamir heuristic [29,30], the proof system can use HASH function to generate random number x at the prover, and change the commitment construction process to non-interactive.

Definition 8 (Vector Inner Product Commitment): The vector inner product commitment aims to hide and prove that the prover **P** grasps the vectors $\mathbf{a_1}$ and $\mathbf{a_2}$ to establish $\langle \mathbf{a_2}, \mathbf{a_2} \rangle = c$. The construction form of the commitment is defined as follows:

$$\begin{aligned}
&\mathrm{M_{pp}} \in \mathbb{Z}_p^n, \mathrm{C_{pp}} \in \mathbb{G} \\
&SETUP : \mathbf{g} = (g_1, ...g_n), \mathbf{h} = (h_1, ...h_n), u \xleftarrow{\$} \mathbb{G} \\
&COM(\mathbf{a_1}^n, \mathbf{a_2}^n, \langle \mathbf{a_1}, \mathbf{a_2} \rangle) \\
&= \left(\mathbf{g}^{\mathbf{a_1}} \cdot \mathbf{h}^{\mathbf{a_2}} u^{\langle \mathbf{a_1}, \mathbf{a_2} \rangle}\right) = \prod_i g_i^{a_{1i}} \cdot \prod_i g_i^{a_{2i}} u^{\langle \mathbf{a_1}, \mathbf{a_2} \rangle} \in \mathbb{G}
\end{aligned}$$

5 General Efficient Range Proof with Constant Computational Cost

First, in Sect. 5.1, we propose the improved zero-knowledge protocol for Ferproof to optimize the operation performance. In Sect. 5.2, an inner product vector generation method is issued to solve the Lagrange square sum search with constant computational cost and the inner product vector dimension compression for Ferproof respectively. In Sect. 5.3, We show the specific technical details of Ferproof by applying an improved zero-knowledge protocol and inner product vector generation method.

5.1 Improved Zero-Knowledge Proof Protocol

Based on the idea of vector inner product commitment, polynomial commitment and Lagrange's square sum theorem, we construct a novel protocol structure and verification scheme and issues an improved zero-knowledge protocol.

Table 1. Summary of notations

Notation	Description
$\mathbf{P}$	A prover
$\mathbf{V}$	A verifier
$\mathbb{G}$	A cyclic group with prime order
$\mathbb{Z}_p$	An integer ring with module p
$\mathbb{G}^n$	Vector spaces with dimension n on $\mathbb{G}$
$\mathbb{Z}_p^n$	Vector spaces with dimension n on $\mathbb{Z}_p$
$\mathbb{Z}_p^*$	$\mathbb{Z}_p \backslash \{0\}$
g, h	The point generated by two cyclic groups $\mathbb{G}$
$\mathbf{g}, \mathbf{h}$	The vector with dimension 4 formed by g and h
$\mathbf{v_1}$, $\mathbf{v_2}$	A unequal vector groups about v
v	A private message
V	Vector inner product commitment about v
$\widehat{V}$	The differential commitment about v
$\mathbf{a}, \mathbf{b}$	Lagrangian inner product vectors of v
a_i, b_i	The elements of $\mathbf{a}$ and $\mathbf{b}$
$\langle \mathbf{a}, \mathbf{b} \rangle$	The inner product of vectors $\mathbf{a}$ and $\mathbf{b}$
A, B	The Pedersen vector commitment constructed for vector $\mathbf{a}$ and $\mathbf{b}$
$\mathbf{s_L}$, $\mathbf{s_R}$	The vector blinding factors selectedon the integer ring
S	The Pedersen vector commitment constructed for $\mathbf{s_L}$ and $\mathbf{s_R}$
α, ρ, φ, μ, τ_1, τ_2, x	The random value selected in the integer ring with module p
T_1, T_2	The Pedersen commitment constructed for τ_1, τ_2
z	A non-zero random number selected on the integer ring p
X	A unknown quantity
$l(X)$, $r(X)$	Polynomial constructed from $\mathbf{a}, \mathbf{b}$ and X
$l(x)$, $r(x)$	Substituting x into the polynomials $l(X)$ and $r(X)$
t	The inner product of $l(x)$ and $r(x)$
t_0, t_1, t_2	One-variable quadratic polynomial coefficients
m	A proof value
d_l, d_u	The upper and lower bounds of m interval respectively
f_m, f_{dl}, f_{du}	The decimal places of m, d_l, d_u respectively
f_{max}	The maximum value in f_m, f_{dl}, f_{du}
w	The f_{max} power of 10

Our protocol is divided into two parts: The first part is to propose a differential commitment structure to construct vector inner product commitment and differential commitment about multivariate quadratic arithmetic expression instead of Pedersen vector commitment which directly constructs the expression value to promise the validity and immutability of the calculation of secret value v without revealing it. The second part, the vector inner product commitment about $v \geq 0$ is constructed by using Lagrange's square sum theorem and transformed into polynomial commitment for promising the non-negativity of secret value without revealing v. Finally, a new commitment verification scheme is proposed at the verifier end of the protocol.

Firstly, the protocol is integer oriented. The purpose of the protocol is to convince the verifier $\mathbf{V}$ that the prover $\mathbf{P}$ holds a non-negative integer v without revealing its value. The prover **P** splits the integer v into a group of unequal vectors so that the inner product of this group of vectors is equal to v and the splitting process is random and not unique. Then, the vector inner product commitment about the vector group is generated to bind the vector splitting results. At the same time, the vector group and random value about v are used to generate the difference commitment. The dimension of the difference commitment is the same as that of the vector inner product commitment, the only difference is the committed vector term. The vector inner product commitment vector term is bound to v, while the differential commitment vector term is bound to random values. The specific process is defined as follows:

$$v = \langle \mathbf{v_1}, \mathbf{v_2} \rangle \wedge \mathbf{v_1} \neq \mathbf{v_2} \tag{1}$$

$$V = h^v \cdot \mathbf{g}^{\mathbf{v_1}} \cdot \mathbf{h}^{\mathbf{v_2}} \wedge h \in \mathbb{G}_p \wedge \mathbf{g}, \mathbf{h}, V \in \mathbb{G}_p^4 \tag{2}$$

$$\varphi \xleftarrow{\$} \mathbb{Z}_p \tag{3}$$

$$\widehat{V} = g^{\varphi} \cdot \mathbf{g}^{\mathbf{v_1}} \cdot \mathbf{h}^{\mathbf{v_2}} \wedge g \in \mathbb{G}_p \wedge \mathbf{g}, \mathbf{h}, V \in \mathbb{G}_p^4. \tag{4}$$

$\widehat{V}$ represents the differential commitment about v, and the construction form of the differential commitment is similar to that of Pedersen vector commitment. If the vector elements of $\mathbf{v_1}$ and $\mathbf{v_2}$ are formalized as expression elements about v, any multivariate quadratic expression about v can be hidden and bound through the construction of vector inner product commitment and differential commitment, without constructing expression commitment about v through Pedersen commitment of v.

After completing the construction of the first part, it is proved that the system constructs the vector inner product commitment about $v \geq 0$. Specifically, according to the Lagrange square sum theorem, an equivalent relationship can be constructed for v, as shown in expression (5):

$$\exists \{a_i | i \in \{1,2,3,4\} \wedge a_i \in \mathbb{Z}\}, \sum_{i=1}^{4} a_i^2 = v \Leftrightarrow v \geq 0, \tag{5}$$

where a_i is expressed as four integers respectively. If and only if there are four integers so that the sum of squares is equal to v, $v \geq 0$ holds. Although reference [25] states that when $v-1$ is an integer multiple of 4, any v can be expressed as the sum of three square numbers. However, the verification protocol of vector inner product commitment (definition 8 presentation) can be executed recursively only when the vector dimension is an integer power of 2. Hence, the Lagrange square sum dimension used in this section is still 4.

According to the Lagrange square sum representation of v, construct the inner product about v in the form of expression (6) (7):

$$a_1^2 + a_2^2 + a_3^2 + a_4^2 = v \tag{6}$$

$$\begin{cases} \mathbf{a} = (a_1, a_2, a_3, a_4) \\ \langle \mathbf{a}, \mathbf{a} \rangle = v \end{cases}, \tag{7}$$

where $\mathbf{a}$ represents the Lagrangian inner product vector of v, so $\mathbf{a}$ is different from $\mathbf{v_1}$ and $\mathbf{v_2}$. For different v, the dimension of vector $\mathbf{a}$ is not the same. For example, when v itself is square, the dimension of its inner product vector $\mathbf{a}$ is only 1. So, we will introduce the vector dimension differentiation algorithm in Sect. 3.2 to make the proof system take only the smallest dimension vector $\mathbf{a}$ that can represent v.

After calculating $\mathbf{a}$, in order not to disclose the value of v, the proof system cannot directly send $\mathbf{a}$ to the verifier $\mathbf{V}$. Hence, the prover $\mathbf{P}$ needs to construct a Pedersen vector commitment about vector $\mathbf{a}$.

$$\alpha \xleftarrow{\$} \mathbb{Z}_p \tag{8}$$

$$A = h^{\alpha} \cdot \mathbf{g}^{\mathbf{a}} \cdot \mathbf{h}^{\mathbf{a}} \in \mathbb{G}_p \tag{9}$$

According to the polynomial commitment, the prover $\mathbf{P}$ constructs a set of univariate first-order polynomials $l(X)$ and $r(X)$ about the vector $\mathbf{a}$, and calculates the inner product polynomial $t(X)$ about $l(X)$ and $r(X)$. Since the prover $\mathbf{P}$ cannot directly send the polynomials $l(X)$ and $r(X)$ to the verifier $\mathbf{V}$, the proving system introduces a set of vector blinding factors to blind the polynomial, and constructs Pedersen vector commitment about the blinding factors. The specific process is defined as follows:

$$\mathbf{s_L}, \mathbf{s_R} \xleftarrow{\$} \mathbb{Z}_p^4 \tag{10}$$

$$z \xleftarrow{\$} \mathbb{Z}_p^* \tag{11}$$

$$\begin{cases} l(X) = \mathbf{a} \cdot z + \mathbf{s_L} \cdot X \in \mathbb{Z}_p[X] \\ r(X) = \mathbf{a} \cdot z + \mathbf{s_R} \cdot X \in \mathbb{Z}_p[X] \end{cases} \tag{12}$$

$$\begin{aligned} t(X) &= \langle l(X), r(X) \rangle \\ &= \langle \mathbf{a}, \mathbf{a} \rangle \cdot z^2 \\ &+ (z \cdot \langle \mathbf{a}, \mathbf{s_L} \rangle + z \cdot \langle \mathbf{a}, \mathbf{s_R} \rangle) \cdot X + \langle \mathbf{s_L}, \mathbf{s_R} \rangle \cdot X^2 \\ &= t_0 + t_1 \cdot X + t_2 \cdot X^2 \in \mathbb{Z}_p[X] \end{aligned} \tag{13}$$

$$\rho \xleftarrow{\$} \mathbb{Z}_p \tag{14}$$

$$S = h^{\rho} \mathbf{g}^{\mathbf{s_L}} \mathbf{h}^{\mathbf{s_R}} \in \mathbb{G}_p \tag{15}$$

$$\mathrm{P} \rightarrow \mathrm{V} : A, S. \tag{16}$$

Since $l(X)$ and $r(X)$ are subtly designed, if and only if $l(X)$ and $r(X)$ are calculated correctly, The numerical expression of $t(X)$ constant term t_0 can be expressed as formula (17):

$$t_0 = \langle \mathbf{a}, \mathbf{a} \rangle \cdot\ z^2 = v \cdot\ z^2. \tag{17}$$

According to the polynomial commitment, the prover $\mathbf{P}$ constructs the Pedersen commitment on the coefficient of non-zero term, and generates the coefficient random number polynomial and the committed random number polynomial. Then, the non-zero random number x is generated on the integer ring to calculate the numerical results of polynomials $l(x)$ and $r(x)$. So far, the prover **P** completes the calculation of all evidence and sends the evidence set to the verifier. The specific process is defined as follows:

$$\tau_1, \tau_2 \xleftarrow{\$} \mathbb{Z}_p \tag{18}$$

$$T_i = g^{\tau_i} h^{t_i} \in \mathbb{G}_p \quad i = \{1, 2\} \tag{19}$$

$$\mathbf{P} \rightarrow \mathbf{V} : T_1, T_2 \tag{20}$$

$$\mathbf{V} : x \xleftarrow{\$} \mathbb{Z}_p^* \tag{21}$$

$$\tau_x = \tau_1 \cdot x + \tau_2 \cdot x + \varphi \cdot z^2 \in \mathbb{Z}_p \tag{22}$$

$$\mu = \alpha \cdot z + \rho \cdot x \qquad \in \mathbb{Z}_p \tag{23}$$

$$\begin{cases} \mathbf{l} = l(x) = \mathbf{a} \cdot z + \mathbf{s}_L \cdot x \\ \mathbf{r} = r(x) = \mathbf{a} \cdot z + \mathbf{s}_R \cdot x \end{cases} \tag{24}$$

$$t = \langle \mathbf{l}, \mathbf{r} \rangle \in \mathbb{Z}_p \tag{25}$$

$$\mathbf{P} \rightarrow \mathbf{V} : \tau_x, \mu, \mathbf{l}, \mathbf{r}, t. \tag{26}$$

The verifier **V** obtains the evidence set and constructs the verification equation. If and only if the verification of equation (26) (27) is true, the verification is successful, that is, the verifier **V** believes that the prover **P** has a non-negative integer v; Otherwise, the verification fails.

$$\hat{V}^{Z^2} \cdot g^{\tau_x} h^t = V^{Z^2} \cdot T_1^x \cdot T_2^{x^2} \tag{27}$$

$$h^{\mu} \cdot \mathbf{g}^{\mathbf{l}} \cdot \mathbf{h}^{\mathbf{r}} = A^z \cdot S^x \tag{28}$$

The above zero-knowledge proof scheme is an interactive scheme. To transform the scheme into a non-interactive proof scheme, the proof system uses the Fiat-Shamir heuristic, that is, the random number is generated by the local HASH function instead of the verifier to send the random number interactively to the prover for better application in the blockchain.

5.2 Inner Product Vector Generation Method

Relying on the zero-knowledge protocol proposed in Sect. 5.1, it can not fully realize the constancy of proof time and calculation cost in the calculation under the actual chain. Because it is very difficult to solve the Lagrange inner product vector **a** of v in expression (6). The traditional Lagrange square sum algorithm for v needs to use the enumeration method to traverse the numerical domain $[1, \sqrt{v}]$, so its time complexity is $O(n^2)$. When the number of bits of v is too large, the time cost of calculating Lagrange square sum by enumeration method will be very high.

In this section, a maximum square splitting algorithm is proposed to solve the Lagrange square sum search problem in a constant time. Specifically, for a given non-negative integer v, calculate the square number a less than and closest to v, that is, solve the Lagrange square sum inner product vector **a**. let $v = v_0$ and make the element $a_i \in \{a_i|min(v_i - a_i) \wedge \sqrt{a_i} \in \mathbb{N}^*\}$ of vector **a**. Then execute formula (28) and iterate until $v = 0$.

$$v_{i+1} = v_i - a_i \quad i \in \{0, ..., n\}, \tag{29}$$

where n represents the number of iterations, that is, the Lagrangian inner product vector dimension. The a_i solved in each iteration constitutes the Lagrangian inner product vector $\mathbf{a} = (a_1, ..., a_n)$ with respect to v. Because the number of iterations is not fixed, it is impossible to completely limit the dimension of the inner product vector **a** to 4 by only 4 iterations.

Algorithm 1. Maximum square splitting algorithm

Require: v
Ensure: **a**, **b**

```
1: tp_list ← []
2: tp_v ← v
3: ns_dic ← {inseparable dictionary}
4: for i = 0 to 3 do /*Solve for the inner product vector a*/
5:     tp_list [i] ← int (sqrt (tp_v))
6:     if power (int (sqrt (tp_v))) == tp_v then
7:         a← tp_list
8:         return a, b
9:     end if
10:    if i == 3 then /*Determine whether the dimension of A reaches 4*/
11:        a← tp_list
12:        tp_v = tp_v − power (sqrt_tp_v)
13:        if tp_vinns_dic then
14:            b← ns_dic [tp_v]
15:            return a, b
16:        end if
17:        for j = 0 to 3 do/*Solve for the inner product vector b*/
18:            tp_list [i] ← int (sqrt (tp_v))
19:            if power (int (sqrt (tp_v))) == tp_v then
20:                b← tp_list
21:                return a, b
22:            end if
23:            tp_v = tp_v − power (int (sqrt (tp_v)))
24:        end for
25:    end if
26:    tp_v = tp_v − power (int (sqrt (tp_v)))
27: end for
```

If the above iterative process is recorded as a function $x_{n+1} = f(x_n) \wedge f(x) = \sqrt{x}$ and let $x_0 = v_0$, $f(x)$ is second-order convergence. When $f(x)$ converges four times, v_3 will be much less difficult to solve than v_0. When the number of iterations is greater than 4, the maximum square splitting algorithm will solve v_3 separately to limit the vector dimension. When the number of iterations is greater than 4, v_3 is solved again. Specifically, an inseparable dictionary is constructed for v_3. The key of the inseparable dictionary mainly stores the value of v that cannot be solved in 4 iterations, and the value of the dictionary stores the solution vector corresponding to each v. v_3 first finds the corresponding solution vector according to the index key of the inseparable dictionary, if the query is successful, it will directly return the dictionary value; if the query fails, continue to use the maximum square division to solve. The inseparable dictionary can be expanded according to the size of v. Finally, the Lagrange inner product vector about v_3 is expressed as $\mathbf{b} = (b_1, ..., b_n)$, and the dimension of $\mathbf{b}$ should be the same as $\mathbf{a}$, If not, add element 0. So far, the Lagrangian inner product relation of v is shown as (29).

$$v = \langle \mathbf{a}, \mathbf{a} \rangle + \langle \mathbf{b}, \mathbf{b} \rangle = \sum_{i=1}^{n} {a_i}^2 + \sum_{j=1}^{n} {b_j}^2 \quad (30)$$

The complete maximum square splitting algorithm is shown as Algorithm 1.

Algorithm 2. Vector dimension differentiation algorithm

Require: a, b
Ensure: a, b

```
1: vec_len ← 0
2: if b == ∅ then /*Let's see if vector b is empty*/
3:     if len(a) == 3 then
4:         vec_len ← 4
5:     else
6:         vec_len ← len(a)
7:     end if
8: else
9:     vec_len ← 4
10: end if
11: for i = 0 to 4 do
12:     a.remove()/*Delete the last element of vector a*/
13:     b.remove()/*Delete the last element of vector b*/
14: end for
15: return a, b
```

To further reduce the average vector dimension and compress the average operation time, the proof system differentiates the dimensions of inner product vectors $\mathbf{a}$ and $\mathbf{b}$ by deleting the vector 0 element, while ensuring that the dimensions of $\mathbf{a}$ and $\mathbf{b}$ are the same, the dimensions n of a and b are compressed and differentiated from $n = 4$ to $n \in \{1, 2, 4\}$. Among them, since the verification process of vector inner product commitment is executed by making the dimension return to $n = n/2$, the dimension after $\mathbf{a}$ and $\mathbf{b}$ compression cannot be 3. The specific process of vector dimension differentiation algorithm is shown in Algorithm 2.

In Algorithm 2, when $\mathbf{b} \neq \varnothing$, there is $a_i \neq 0 \quad i \in \{1, 2, 3, 4\}$, that is the dimensions $\mathbf{a}$ and $\mathbf{b}$ are incompressible.

5.3 Specific Description of Ferproof Scheme

This section describes the overall implementation process and core technical details of Ferproof scheme.

Specifically, for any floating-point number, the flexible non-negative range interval relationship can be expressed as shown in formula (30) and formula (31):

$$m \in [d_l, d_u] \wedge d_l > 0 \wedge d_u > 0 \tag{31}$$

$$m - d_l > 0 \wedge d_u - m > 0 \wedge d_l > 0 \wedge d_u > 0, \tag{32}$$

where m represents the proof value of floating-point number, d_l and d_u represent the upper and lower bounds of floating-point number interval respectively. For non-negative range interval, $d_u - m < 0$ is true if and only if $m - d_l > 0$, on the contrary, $d_u - m > 0$ is true when $m - d_l < 0$. Hence, the above range relationship can be transformed into expression (32):

$$\begin{aligned} &m \in [d_l, d_u] \wedge d_l > 0 \wedge d_u > 0 \Leftrightarrow \\ &(m - d_l)(d_u - m) > 0 \wedge d_l > 0 \wedge d_u > 0. \end{aligned} \tag{33}$$

For the range relation of floating-point numbers, take the maximum decimal places of the upper and lower bounds d_l and d_u of the interval and the proof value m as the idempotent of 10, and multiply it by the interval boundary value and the proof value at the same time, the range relation is still valid.

$$w \in \left\{10^{f_{\max}} \left| f_{\max} = \max\left(f_m, f_{dl}, f_{du}\right) \right.\right\} \tag{34}$$

$$\begin{aligned} &m \in [d_l, d_u] \wedge d_l > 0 \wedge d_u > 0 \Leftrightarrow \\ &(w \cdot m - w \cdot d_l)(w \cdot d_u - w \cdot m) > 0 \wedge d_l > 0 \wedge d_u > 0 \Leftrightarrow \\ &w^2 \cdot \left(m \cdot d_u - m^2 - d_l \cdot d_u + d_l \cdot m\right) > 0 \wedge d_l > 0 \wedge d_u > 0 \end{aligned} \tag{35}$$

If $f_{max} = 0$, expression (34) can be expressed as an integer range relationship, At this time, $w = 1$. The proof system switches between floating-point number and integer range proof by calculating f_{max}.

The prover $\mathbf{P}$ uses the range relation to construct vector inner product commitment and difference commitment. The construction method is defined as follows:

$$\begin{cases} \mathbf{v_1} = (w \cdot m, -w \cdot m, -w \cdot d_l, w \cdot d_l) \\ \mathbf{v_2} = (w \cdot d_u, w \cdot m, w \cdot d_u, w \cdot m) \end{cases} \tag{36}$$

$$v = \langle \mathbf{v_1}, \mathbf{v_2} \rangle \tag{37}$$

$$V = h^v \mathbf{g}^{\mathbf{v_1}} \mathbf{h}^{\mathbf{v_2}} \wedge \mathbf{g}, \mathbf{h}, V \in \mathbb{G}_p^4 \tag{38}$$

$$\widehat{V} = g^{\varphi} \cdot \mathbf{g}^{\mathbf{v_1}} \mathbf{h}^{\mathbf{v_2}} \wedge \varphi \xleftarrow{\$} \mathbb{Z}_p \wedge \mathbf{g}, \mathbf{h}, \widehat{V} \in \mathbb{G}_p^4 \tag{39}$$

$$\mathbf{P} \rightarrow \mathbf{V} : V, \widehat{V}. \tag{40}$$

The proof system divides the relationship into two groups of vectors $\mathbf{v_1}$ and $\mathbf{v_2}$ constructed by $w \cdot m$, $w \cdot a$ and $w \cdot b$, and constructs the vector inner product commitment V and Pedersen vector commitment $\widehat{V}$ about $\mathbf{v_1}$ and $\mathbf{v_2}$ as the difference commitment.

The range relation is transformed into the Lagrange square sum, which can be expressed as (40):

$$w^2(m - d_l)(d_u - m) = \langle \mathbf{a}, \mathbf{a} \rangle + \langle \mathbf{b}, \mathbf{b} \rangle = \sum_{i=1}^{n} {a_i}^2 + \sum_{j=1}^{n} {b_{ii}}^2 \quad n \in \{1, 2, 4\}. \tag{41}$$

The proof system uses the inner product vector generation method to obtain the Lagrange inner product vector $\mathbf{a} = (a_1, ..., a_n)$ and $\mathbf{b} = (b_1, ..., b_n)$ with respect to the range relationship of floating-point numbers and dimensional differentiation.

Then, the prover $\mathbf{P}$ constructs the Pedersen vector commitment of inner product vectors $\mathbf{a}$ and $\mathbf{b}$, and selects the blinding factor to construct the corresponding commitment. The specific process is defined as follows:

$$\alpha \xleftarrow{\$} \mathbb{G}_p \tag{42}$$

$$A = h^{\alpha} \mathbf{g}^{\mathbf{a}} \mathbf{h}^{\mathbf{a}} \in \mathbb{G}_p \tag{43}$$

$$\varepsilon \xleftarrow{\$} \mathbb{G}_p \tag{44}$$

$$B = h^{\varepsilon} \mathbf{g}^{\mathbf{b}} \mathbf{h}^{\mathbf{b}} \in \mathbb{G}_p \tag{45}$$

$$\rho \xleftarrow{\$} \mathbb{G}_p \tag{46}$$

$$\mathbf{s}_L, \mathbf{s}_R \xleftarrow{\$} \mathbb{Z}_p^n \tag{47}$$

$$S = h^{\rho} \mathbf{g}^{\mathbf{s}_L} \mathbf{h}^{\mathbf{s}_R} \in \mathbb{G}_p \tag{48}$$

$$\mathbf{P} \rightarrow \mathbf{V} : A, B, S. \tag{49}$$

Since the vectors $\mathbf{a}$ and $\mathbf{b}$ undergo dimensional differentiation, the dimensions of blind vectors $\mathbf{s}_L$ and $\mathbf{s}_R$ are variable.

After completing the commitment construction of $\mathbf{a}$ and $\mathbf{b}$, according to the polynomial commitment construction method, the prover $\mathbf{P}$ constructs polynomials $l(x)$ and $r(x)$ and calculates $\mathbf{l}$, $\mathbf{r}$ and t. The construction and calculation process is defined as follows:

$$\mathbf{l} = l(x) = \mathbf{a} \cdot z + \mathbf{b} \cdot z + \mathbf{s}_L \cdot x \in \mathbb{Z}_p^n \tag{50}$$

$$\mathbf{r} = r(x) = \mathbf{a} \cdot z + \mathbf{b} \cdot z + \mathbf{s}_R \cdot x \in \mathbb{Z}_p^n \tag{51}$$

$$\begin{aligned} t &= \langle \mathbf{l}, \mathbf{r} \rangle \\ &= (\langle \mathbf{a}, \mathbf{a} \rangle + \langle \mathbf{b}, \mathbf{b} \rangle + 2 \cdot \langle \mathbf{a}, \mathbf{b} \rangle) \cdot z^2 \\ &+ (\langle \mathbf{a} \cdot z + \mathbf{b} \cdot z, \mathbf{s}_L \rangle + \langle \mathbf{a} \cdot z + \mathbf{b} \cdot z, \mathbf{s}_R \rangle) \cdot x + \langle \mathbf{s}_L, \mathbf{s}_R \rangle \cdot x^2 \\ &= (\langle \mathbf{a}, \mathbf{a} \rangle + \langle \mathbf{b}, \mathbf{b} \rangle) \cdot z^2 + 2 \cdot \langle \mathbf{a} \cdot z, \mathbf{b} \cdot z \rangle \\ &+ (\langle z \cdot (\mathbf{a} + \mathbf{b}), \mathbf{s}_L \rangle + \langle z \cdot (\mathbf{a} + \mathbf{b}), \mathbf{s}_R \rangle) \cdot x + \langle \mathbf{s}_L, \mathbf{s}_R \rangle \cdot x^2 \in \mathbb{Z}_p \end{aligned} \tag{52}$$

$$\begin{aligned} t_0 &= (\langle \mathbf{a}, \mathbf{a} \rangle + \langle \mathbf{b}, \mathbf{b} \rangle) \cdot z^2 + 2 \cdot \langle \mathbf{a} \cdot z, \mathbf{b} \cdot z \rangle \\ &= v \cdot z^2 + 2 \cdot \langle \mathbf{a} \cdot z, \mathbf{b} \cdot z \rangle . \end{aligned} \tag{53}$$

According to the Fiat-Shamir heuristic, the prover **P** can generate polynomial random number z by using the HASH function $H(A, B, S)$ promising A, B and S to replace the way of generating random number at the verifier for the purpose of non-interactive proof.

Subsequently, the prover **P** calculate:

$$\lambda, \tau_1, \tau_2 \xleftarrow{\$} \mathbb{Z}_p \tag{54}$$

$$T_i = g^{\tau_i} h^{t_i} \in \mathbb{G}_p \quad i = \{1, 2\} \tag{55}$$

$$\tau_x = \tau_1 \cdot x + \tau_2 \cdot x^2 - z^2 \cdot \phi + 2z^2 \cdot \lambda \in \mathbb{G}_p \tag{56}$$

$$\mu = \alpha \cdot z + \varepsilon \cdot z + \rho \cdot x \in \mathbb{Z}_p \tag{57}$$

$$K = g^{\lambda} \cdot h^{\langle a, b \rangle} \in \mathbb{G}_p \tag{58}$$

$$\boldsymbol{P} \rightarrow \boldsymbol{V} : T_1, T_2, \mu, \tau_x, K, \mathbf{l}, \mathbf{r}, t. \tag{59}$$

Among them, after completing the construction of T_i, τ_x and μ, it is proved that the system constructs Pedersen commitment K about $\langle \mathbf{a}, \mathbf{b} \rangle$ is to ensure the zero-knowledge of **a** and **b** at the verification end. The construction form is shown in expression (56).

The prover **P** sends the evidence set to the verifier **V**. The verifier **V** verifies whether equation (53) (54) is true:

$$\hat{V}^{Z^2} \cdot g^{\tau_x} h^t = V^{Z^2} \cdot K^{2 \cdot z^2} \cdot T_1^x \cdot T_2^{x^2} \tag{60}$$

$$h^{\mu} \cdot \mathbf{g}^{\mathbf{l}} \cdot \mathbf{h}^{\mathbf{r}} = A^z \cdot B^z \cdot S^x \in \mathbb{G}. \tag{61}$$

If and only if the verification of equation (53) (54) is valid, the verification is successful; Otherwise, it will fail.

Through the concepts of vector inner product commitment and polynomial commitment, it can be inferred that Ferproof is complete, that is, for any honest prover **P**, **P** forges evidence to make the probability of the verifier **V** passing the verification can be minimal to negligible; Ferproof is correct, that is, for any verifier **V**, the probability of failure in verification can be minimal to negligible when **P** provide the effective evidence; Ferproof has zero-knowledge, that is, the verifier **V** cannot obtain any additional knowledge about m, d_l and d_u in the verification process.

6 Experimental Results and Analysis

6.1 Theoretical Effect of Scheme

Table 2 shows that the communication cost comparison of different schemes. n represents the number of binary digits of the problem being processed. Since the Bulletproofs and Shellproof construct polynomials according to binary, the number of cyclic group elements used is related to the number of binary digits. While Ferproof and Cuproof construct fixed dimensional inner product vectors according to Lagrange's theorem, therefore, their communication cost is constant. The number of cyclic group elements used by Ferproof and Cuproof is 2. Ferproof constructs differential commitment and uses the maximum square splitting algorithm to generate inner product vectors, hence its random number value on the integer ring is 2 more than Bulletproofs and Shellproof.

Table 2. Communication cost comparison

Scheme	$\mathbb{G}_p$	$\mathbb{Z}_p$	$\mathbb{G}_q$	$\mathbb{Z}_q$
Bulletproofs	$2\log(n)+4$	5	0	0
Shellproof	$2\log(n)+4$	5	0	0
Cuproof	2	4	2	3
Ferproof	**2**	**7**	**0**	**0**

Table 3 shows that the theoretical operation time complexity of different schemes. The overall operation time complexity of Ferproof is the lowest and the overall calculation cost is constant due to leveraging the maximum square splitting algorithm to obtain the inner product vector under a constant number of iterations.

Table 3. Computational time complexity comparison

Scheme	Proof and verification time	Witness generation time	Total time
Bulletproofs	$O(4log(n))$	$O(n)$	$O(n)$
Shellproof	$O(2log(n))$	$O(n)$	$O(n)$
Cuproof	$O(1)$	$O(n)$	$O(n)$
Ferproof	$\boldsymbol{O(1)}$	$\boldsymbol{O(1)}$	$\boldsymbol{O(1)}$

6.2 Scheme Experimental Performance

To evaluate the performance of Ferproof and ensure the same experimental conditions, we adopt Python language to implement Ferproof and several known state-of-the-art schemes respectively. Secp256k12 elliptic curve is used as cyclic group in the experiments. 400 sets of 0–2 integer power range samples, 400 sets of flexible integer range samples, 400 sets of flexible floating-point numbers and integer mixed range samples with the problem scale increasing from 8 bit to 128 bit are used as test samples respectively. The CPU model used in the experimental operation is AMD 4000, the running memory is 16G, and the experimental program is run by single thread. We uploaded all the implemented code to github, the github link is https://github.com/liyicong763/Ferproof_for_Python.

Table 4. Performance and proof size of 4 schemes

Method	Problem scale	Gate	Proof size	Prove time	Verify time
Bulletproofs	8 bit	8	491	265.06	115.02
	16 bit	16	553	471.10	165.02
	32 bit	32	620	968.70	326.07
	64 bit	64	691	1844.41	578.13
	128 bit	128	759	3482.50	1094.23
Average:	–	–	623	1406.35	455.69
Shellproof	8 bit	8	490	375.59	136.03
	16 bit	16	556	408.07	138.04
	32 bit	32	622	758.23	229.03
	64 bit	64	688	1437.23	416.09
	128 bit	128	754	2980.12	891.83
Average:	–	–	**622**	1191.85	362.20
Cuproof	8 bit	6	2068	364.08	117.03
	16 bit	6	2078	365.66	116.02
	32 bit	6	2088	364.08	116.12
	64 bit	6	2099	369.08	114.02
	128 bit	6	2119	363.09	114.02
Average:	–	–	2090	365.20	115.44
Ferproof	8 bit	6	740	220.04	80.01
	16 bit	6	739	219.04	82.01
	32 bit	6	738	222.03	80.01
	64 bit	6	740	216.13	82.02
	128 bit	6	738	219.11	81.01
Average:	–	–	739	**219.27**	**81.01**

Table 4 and Fig. 1 show the performance of the different range proof schemes under different problem sizes respectively. Since Bulletproofs and Shellproof use a binary single bit to construct a polynomial commitment, the dimensions of the commitment, proof sizes, and computational time increased stepwise with the expansion of the number of binary bits in the problem. Since Cuproof uses the Lagrangian Sum of Squares theorem to treat the sample problem as a constant-dimensional vector, the length of the evidence and the calculation time are irrelevant to the number of problem bits, but Cuproof needs to disassemble the sample problem into two proof tasks and aggregate them, so increased time overhead. Ferproof uses an improved protocol structure to construct evidence without disassembling the sample problem, and further reduce the computational cost.

The experimental results show that the Ferproof proposed in this paper is the best in terms of computational time, the average proof time dropped to 219.27 ms, the average verification time dropped to 81.01 ms, and the average evidence size reached 739 bytes. Figure 1 further visualizes the computational time of different schemes. Ferproof has achieved impressive computational speed and stability. The computational time of Ferproof is constant and optimal.

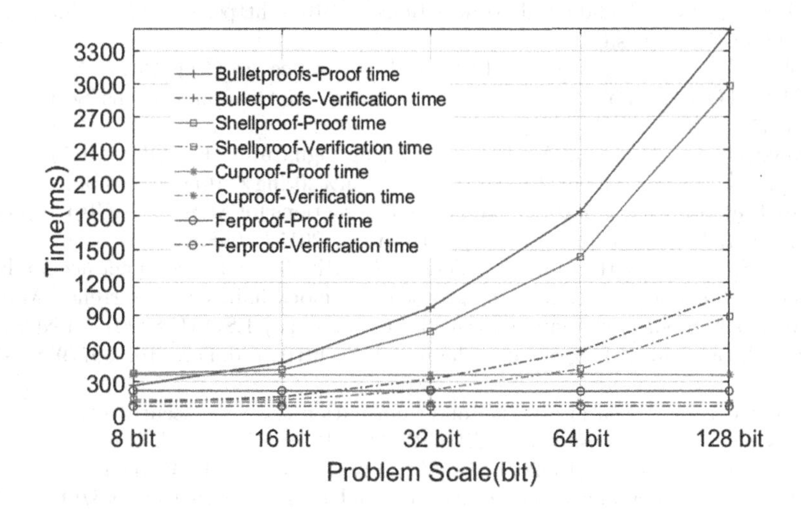

Fig. 1. Comparison of computational time

7 Conclusion

Considering the high cost of witness generation time and the limitation of processing data types in the existing range proof schemes based on vector inner product commitments, we propose the range proof scheme with constant computational cost and suitable for floating-point numbers, called Ferproof. To implement Ferproof, we propose the novel commitment structure for constructing a zero-knowledge proof protocol based on Lagrange inner product vector, and inner product vector generation method with constant computational cost. Through experiments, it is proved that Ferproof has great advantages in terms of witness generation time and adaptation of floating-point numbers. Compared with existing methods, its prove time and verify time have also been reduced by 40.0% and 29.8%, respectively. In our future research, the optimization of protocol structure will be considered to reduce the selection of random numbers of integer rings. In addition, the implementation of aggregation proof by Ferproof will also be the focus of the next research.

References

1. Nakamoto, S.: Bitcoin: a peer-to-peer electronic cash system[EB/OL]. SSRN Electron. J. (2008). https://bitcoin.org/bitcoin.pdf

2. Bonneau, J., Miller, A., Clark, J., Narayanan, A., Kroll, J.A., et al.: Research perspectives and challenges for bitcoin and cryptocurrencies, pp. 104–121. IEEE (2015)
3. Liu, M., Chen, Z., Shi, Y., Tang, L., Cao, D.: Research progress of blockchain in data security. Chin. J. Comput. **44**(1), 1–27 (2021)
4. Goldreich, O., Oren, Y.: Definitions and properties of zero-knowledge proof systems. J. Cryptol. **7**(1), 1–32 (1994). https://doi.org/10.1007/BF00195207
5. Maxwell, G.: Confidential transactions (2016). https://people.xiph.org/greg/confidential_values.txt
6. Blum, M.: Coin flipping by telephone. In: Advances in Cryptology: A Report on CRYPTO 81, CRYPTO 81, IEEE Workshop on Communications Security, pp. 11–15 (1981)
7. Maxwell, G., Poelstra, A.: Borromean ring signatures [EB/OL] (2015). http://diyhpl.us/bryan/papers2/bitcoin/Borromean%20ring%20signatures.pdf
8. Noether, S.: Ring signature confidential transactions for Monero. IACR Cryptology ePrint Achieve (2015). https://eprint.iacr.org/2015/1_098.pdf
9. Sun, S.-F., Au, M.H., Liu, J.K., Yuen, T.H.: RingCT 2.0: a compact accumulator-based (linkable ring signature) protocol for blockchain cryptocurrency Monero. In: Foley, S.N., Gollmann, D., Snekkenes, E. (eds.) ESORICS 2017. LNCS, vol. 10493, pp. 456–474. Springer, Cham (2017). https://doi.org/10.1007/978-3-319-66399-9_25
10. Pedersen, T.P.: Non-interactive and information-theoretic secure verifiable secret sharing. In: Feigenbaum, J. (ed.) CRYPTO 1991. LNCS, vol. 576, pp. 129–140. Springer, Heidelberg (1992). https://doi.org/10.1007/3-540-46766-1_9
11. Wood, G.: Ethereum: a secure decentralised transaction ledger [EB/OL] (2014). http://gavwood.com/paper.pdf
12. McCorry, P., Shahandashti, S.F., Hao, F.: A smart contract for boardroom voting with maximum voter privacy. In: Kiayias, A. (ed.) FC 2017. LNCS, vol. 10322, pp. 357–375. Springer, Cham (2017). https://doi.org/10.1007/978-3-319-70972-7_20
13. Campanelli, M., Gennaro, R., Goldfeder, S., Nizzardo, L.: Zero-knowledge contingent payments revisited: attacks and payments for services. In: Conference on Computer and Communications Security. ACM (2017)
14. Parno, B., Howell, J., Gentry, C., Raykova, M.: Pinocchio: nearly practical verifiable computation. In: Proceedings of the 34th IEEE Symposium on Security and Privacy, pp. 238–252. IEEE (2013)
15. Ben-Sasson, E., Chiesa, A., Genkin, D., Tromer, E., Virza, M.: SNARKs for C: verifying program executions succinctly and in zero knowledge. In: Canetti, R., Garay, J.A. (eds.) CRYPTO 2013. LNCS, vol. 8043, pp. 90–108. Springer, Heidelberg (2013). https://doi.org/10.1007/978-3-642-40084-1_6
16. Parno, B., Howell, J., Gentry, C., Raykova, M.: Pinocchio: nearly practical verifiable computation. Commun. ACM **59**, 103–112 (2016)
17. Groth, J.: On the size of pairing-based non-interactive arguments. In: Fischlin, M., Coron, J.-S. (eds.) EUROCRYPT 2016. LNCS, vol. 9666, pp. 305–326. Springer, Heidelberg (2016). https://doi.org/10.1007/978-3-662-49896-5_11
18. Peterson, P.: Zcash-transaction linkability [EB/OL], 25 January 2017. https://z.cash/_blog/transaction-linkability.html
19. Wahby, R.S., Tzialla, I., Shelat, A., Thaler, J., Walfish, M.: Doubly-efficient zkSNARKs without trusted setup. In: Proceedings of 2018 IEEE Symposium on Security and Privacy (SP), pp. 2375–1207. IEEE (2018)
20. Setty, S., Angel, S., Lee, J.: Verifiable state machines: proofs that untrusted services operate correctly. ACM SIGOPS Oper. Syst. Rev. **54**(1), 40–46 (2020)

21. Ben-Sasson, E., Bentov, I., Horesh, Y., Riabzev, M.: Scalable, transparent, and post-quantum secure computational integrity. IACR Cryptology ePrint Archive, 2018:46 (2018)
22. Maller, M., Bowe, S., Kohlweiss, M., Meiklejohn, S.: Sonic: zero-knowledge snarks from linear-size universal and updatable structured reference strings. Cryptology ePrint Archive (2019)
23. Gabizon, A., Williamson, Z.J., Ciobotaru, O.: PLONK: permutations over lagrange-bases for Oecumenical noninteractive arguments of knowledge. Cryptology ePrint Archive, 2019:953 (2019)
24. Lipmaa, H.: On Diophantine complexity and statistical zero-knowledge arguments. In: Laih, C.-S. (ed.) ASIACRYPT 2003. LNCS, vol. 2894, pp. 398–415. Springer, Heidelberg (2003). https://doi.org/10.1007/978-3-540-40061-5_26
25. Groth, J.: Non-interactive zero-knowledge arguments for voting. In: Ioannidis, J., Keromytis, A., Yung, M. (eds.) ACNS 2005. LNCS, vol. 3531, pp. 467–482. Springer, Heidelberg (2005). https://doi.org/10.1007/11496137_32
26. Bünz, B., Bootle, J., Boneh, D., Poelstra, A., Wuille, P., Maxwell, G.: Bulletproofs: short proofs for confidential transactions and more. In: 2018 IEEE Symposium on Security and Privacy (SP), pp. 315–334. IEEE (2018)
27. Bootle, J., Cerulli, A., Chaidos, P., Groth, J., Petit, C.: Efficient zero-knowledge arguments for arithmetic circuits in the discrete log setting. In: Fischlin, M., Coron, J.-S. (eds.) EUROCRYPT 2016. LNCS, vol. 9666, pp. 327–357. Springer, Heidelberg (2016). https://doi.org/10.1007/978-3-662-49896-5_12
28. Deng, C., Tang, X., et al.: Cuproof: a novel range proof with constant size. IACR Cryptology e Print Achieve (2021)
29. Feige, U., Fiat, A., Shamir, A.: Zero-knowledge proofs of identity. J. Cryptol. **1**(2), 77–94 (1988). https://doi.org/10.1007/BF02351717
30. Rivest, R.L., Shamir, A., Adleman, L.: A method for obtaining digital signatures and public-key cryptosystems. Commun. ACM **21**(2), 120–126 (1978)

NEPG: Partitioning Large-Scale Power-Law Graphs

Jiaqi Si, Xinbiao Gan(✉), Hao Bai, Dezun Dong, and Zhengbin Pang

College of Computer, National University of Defense Technology, Changsha, China
xinbiaogan@nudt.edu.cn

Abstract. We propose Neighbor Expansion on power-law graph(NEPG), a distributed graph partitioning method based on a specific power-law graph that offers both good scalability and high partitioning quality. NEPG is based on a heuristic method, Neighbor Expansion, which constructs the different partitions and greedily expands from vertices selected randomly. NEPG improves the partitioning quality by selecting the vertices according to the properties of the power-law graph. We put forward theoretical proof that NEPG can reach the higher upper bound in partitioning quality. The empirical evaluation demonstrates that compared with the state-of-the-art distributed graph partitioning algorithms, NEPG significantly improved partitioning quality while reducing the graph construction time. The performance evaluation demonstrates that the time efficiency of the proposed method outperforms the existing algorithms.

Keywords: Graph partitioning · Distributed graph processing · Power-law graph · Heuristic algorithm · Experimental evaluation

1 Introduction

In the framework of graph data processing that has emerged in the past decade, such as Pregel [12], Giraph [2], GraphX [7], X-Stream [17], GPS [11], Gemini [24], GraM [22], ScaleGraph [19], LA3 [1], PowerLyra [3], PGX.D [9], PowerGraph [6], Pregel+ [16], and Trinity [18], graph partitioning algorithm plays an essential role in analyzing large-scale, real-world graph data. All of these systems mentioned keep a large graph distributed onto multiple machines and use communication between machines to query and exchange part of the information of the whole graph [5]. Therefore, the amount of communication in the whole information exchanging process depends on how the graph information is stored in the machine. In other words, the lower the storage of graph information is, the lower is the volume and the communication time in processing graph data.

While vertex partitioning is often used in traditional graph partitioning problems, recent research has shown that edge partitioning is more effective in reducing the communication volume of query processing in large-scale real-world graphs. In reality, most of the large-scale real-world graphs have the skewed-degree distribution [10], which means that most vertices have fewer neighbors,

Y. Lai et al. (Eds.): ICA3PP 2021, LNCS 13157, pp. 668–690, 2022.
https://doi.org/10.1007/978-3-030-95391-1_42

whereas a few have more neighbors. Under this condition, vertex partitioning will lead to a serious imbalance in the workload of graph data processing. On the other hand, there are more edges than vertices in a graph, and edge partitioning can greatly reduce the amount of data stored. Since the edge partitioning problem is NP-hard, we try to use a heuristic algorithm to solve the edge partitioning problems for large-scale graphs.

The research of edge partitioning algorithms needs to pay attention to the following core problems: quality, elapsed time, and scalability. Today, the data graph of all kinds of large companies is getting bigger and bigger. Take the real-world graphs as an example, the social graphs of Facebook and Wechat have over one trillion vertices [4]. The real-world graphs have the power-law distribution, which brings a challenge for us to scale the traditional edge partitioning algorithm to exceedingly large-scale graphs.

Unfortunately, the existing edge partitioning algorithms are not satisfactory when dealing with trillions of edge graphs. Current Hash-based graph partitioning methods, such as 1D hash, 2D hash, DBH [23], Hybrid Hash [3], usually do not produce high partition quality because of the randomness of hash operations and bring the problems of load imbalance. Based on random allocation, Hybrid Ginger [3] proposed a method of iterative low-degree vertex selection for optimization, which restrictively improves performance and load balancing. On the other hand, the high-quality partitioning method suitable for small-scale graphs cannot be scaled to large-scale graphs. Even in the state-of-the-art high-quality distributed methods, each method has its shortcomings. Because it's a limited way to improve performance by adding machines, Sheep [13] cannot scale to the trillion-edge graph. HDRF, the most advanced streaming algorithm, consumes little memory but can not yield the same high partition quality as the best in-memory algorithms.

In this paper, we propose Neighbor Expansion on power-law graph (NEPG), a distributed graph partitioning method based on the specific power-law graph (Kronecker graph) that can provide good scalability and high partitioning quality. NEPG is based on a greedy heuristic algorithm named parallel greedy expansion. The key idea of the algorithm is that the whole process starts from multiple selected vertices under conditional restrictions in parallel and greedily expands each edge set in the case of ensuring the minimum local vertex cuts. NEPG expands greedily in parallel based on selecting the initial vertices according to the properties of the graph, which theoretically improves the quality lower bound of the heuristic algorithm.

The key contributions of this paper are as follows:

(1) **We propose Neighbor Expansion on power-law graph (NEPG)**, a distributed graph partitioning algorithm. NEPG outperforms the state-of-the-art algorithms.

(2) **We introduce seed vertex selection optimization**, an effective and fast method for vertex selection, which is helpful to provide high-quality partitions. We propose the method according to the characteristics and properties of the kronecker graph.

(3) **We provide the theoretical upper and lower bounds of partition quality**, which is a theoretical gap ignored by other papers when focusing on experiments.

(4) **The experimental evaluation manifests that NEPG is competitive enough compared with the advanced graph partitioning algorithm**. Experiments on the graphs with up to 4 billion edges generated by the RMAT generator demonstrate that NEPG outperforms the state-of-the-art algorithms.

2 Background

2.1 Notation and Problem Definition

We assume $G(V,E)$ be an undirected and unweighted graph that consists of vertices, V, and edges, E. The goal of edge partitioning is to divide E into k $(k > 1)$, $k \in N$ different partitions $P = p_1, \ldots, p_k$, and $p_i \cap p_j = \emptyset$, $i \neq j$. We use $e_{u,v}$ to represent an edge connecting vertices, v and u. Figure 1 illustrates edge partitioning and vertex partitioning with $k = 3$ partitions on an undirected graph. When an edge (u, v) is assigned to a partition p_i, both u and v are covered by p_i. For a partition p_i, denote $V(p_i) = \{x|(x, y) \in p_i\}$ as the set of vertices in p_i. When a vertex v is covered by a partition, which means that the v is replicated on that partition and it may be replicated on the other partition. Now, we try to minimize the replication factor $RF(p_1, ..., p_k) = \frac{1}{|V|}\sum_{i=1}^{k}|V(p_i)|$. By minimizing the replication factor, the amount of communication overhead between distributed computing nodes can be minimized.

In edge partitioning, the edges are assigned to partitions and the vertices are divided. Bourse et al. have analyzed the difference between both cut types, proving that vertex cuts are smaller than edge cuts on power-law graphs. Here, we review the existing edge partitioning methods.

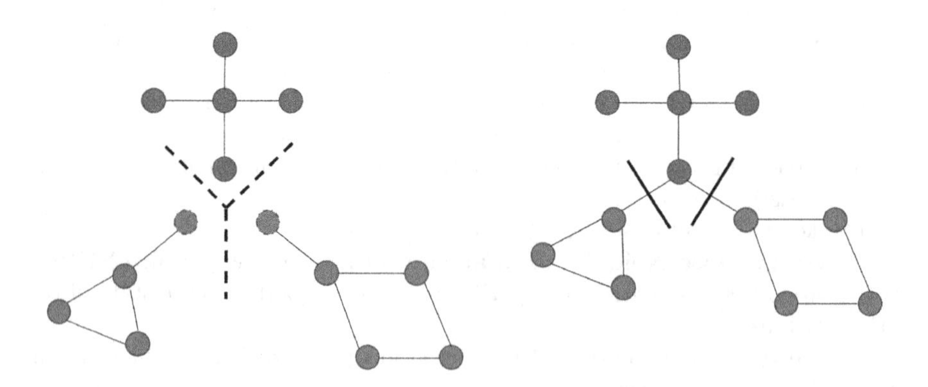

Fig. 1. Edge partition vs vertex partition

2.2 Edge Partitioning

One of the mainstream methods now is based on the random hash. Among the random hash, the most simple and direct method is 1D-hash partitioning, where the edge is randomly assigned to a one-dimensional partitioning space. Corresponding to this is another method, 2D-hash partition, in which the edges are assigned to the two-dimensional partition space by hashing adjacent vertices respectively. The latest hash-based method, Hybrid Hashing [3], and DBH [23] utilize the degree of vertices, where edges are randomly assigned so that high-degree vertices are divided into more partitions. Hybrid Ginger is a heuristic algorithm that is used to iteratively optimize allocation after hash partitioning to improve partition quality, which mitigates the poor locality and high interference among threads. Random allocation in hash operations leads to lower partition quality, but at the same time, it can scale to large-scale graphs because of its lightweight property.

Recently, the methods applied to large-scale graphs are also based on streaming methods, such as FENNEL-based [20] edge partitioning, HDRF. Before the algorithm starts, the whole graphs is represented as the edge stream. Streaming algorithms pass through the edge stream, only process one (or a small number of) edge(s) at a time. It does not need to know the information of the whole graph and consumes less memory, but it also brings the problem of low partition quality. These methods are based on the hypothetical sequence, which further reduces the partition quality.

Sheep [13] is a more advanced distributed edge processing method, which partitions the tree by converting the graph into an elimination tree in parallel. But Sheep [13] does not provide a theoretical guarantee of partitioning quality.

Most related works are two methods: DNE [8] and HEP [15]. DNE is an improved method to scale NE to the large-scale graphs. NE is the state-of-the-art greedy algorithm based on the expansion of the edge set. Because the entire graph is deployed on the main memory on a single machine, the scalability of NE is limited. We listed some limitations of DNE. (1) The algorithm needs to load the complete graph into memory. Although this prevents DNE from being used to partition graphs that exceed the memory capacity of the available compute node. (2) DNE needs to acquire the information of which partition the edge has been assigned to, which leads to high memory and run-time overhead. HEP is a hybrid edge partitioning system that combines DNE, the state-of-the-art in-memory partitioning with HDRF, the state-of-the-art streaming partitioning. It balances the need for memory and quality. But the quality can not meet our requirements.

3 Neighbor Expansion

In this section, we first explain the main ideas of the greedy expansion approach, which forms the basis of our method. Next, we explore the technical implementation details of scaling this method to a large-scale graph and give our approach.

3.1 Edge Partitioning Based on Expansion

We first introduced the process of an edge set for partitioning. And we review the greedy heuristics for efficient expansion.

For $G(V, E)$, let X be an edge set ($M \subset E$). We refer to the vertex set $B(M) := \{v | v \in V(M) \wedge \exists\, e_{v,u} \in E \backslash M\}$ as boundary of M. Before the expansion of M, let's firstly select $v \in B(M)$, then $M \leftarrow M \cup \{e_{v,u} | e_{v,u} \in E \backslash M\}$. If M is empty, v will be selected at random from $V(E \backslash M)$.

We first select the initial vertex as the starting point of the expansion process, which are boundary vertices. At the same time, the core edge set is created to store the partitioned edges. Then the boundary set begins to expand, from which the edges with the lowest score are selected to join the core set. The expanded boundary set will be removed from E. When a partition is built, the algorithm repeats the above steps. When E is empty, the whole algorithm terminates. There are two principles to be followed in the whole expansion process: (1) When selecting vertices into the boundary set, try to minimize the increase of vertex replications. (2) During expansion, try to allocate additional edges which do not increase the replication vertices anymore. Figure 2 manifests how these principles are used in the actual expansion process.

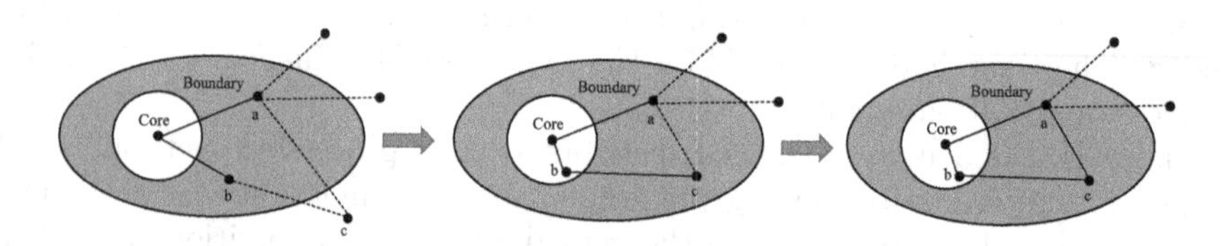

Fig. 2. (1) Vertex b has fewer external neighbors than vertex a, which is the better choice. (2) When we allocate $e_{a,c}$, we do not increase vertex replication.

3.2 Vertex Selection from Boundary

We clarify the criteria for selecting vertices from the boundary. Let $p_i(t)$ be the set of partitions allocated until t time, where t is the number of rounds iterated in the expansion process. In this paper, we use $V(p_i(t))$ to represent the number of p_i's vertex replications at t time.

We define $D_{rest}(v)$ to represent the increase of the vertex replications of the vertex v from time t to $t+1$. There is the following equation.

$$|V(p_i(t+1))| - |V(p_i(t))| = D_{rest}(v) \tag{1}$$

which is equal to the v's degree in the remaining edges. Thus, we use v_{min} to represent the new vertex that minimizes the increase of the vertex replications:

$$V_{\min} := \underset{x \in B}{\arg\min} \mathrm{D}_{\mathrm{rest}}(x). \tag{2}$$

For example, $e_{a,c}$ is additionally allocated during the expansion. The allocation of $e_{a,c}$ does not increase the number of the vertex replications because one of the edges to a and one of the edges to c are already allocated to the same partition.

3.3 Challenges to Distributed Algorithm

The core idea of the proposed method is to implement the expansion in each partition in parallel, which is simple but may encounter various challenges when scaling to exceedingly large-scale graphs. The challenges are summarized as follows:

Memory. The memory size limits the scalability of the partitioning algorithm. The parallel expansion algorithm needs to load the entire graph from the beginning, which will be deployed in distributed memory for scalability and efficiency. The existing methods use CSR to store the information of the graph, but the assigned vertex information is not deleted from the neighbor node's adjacency list, which will result in a certain waste of memory.

Concurrent Allocation. Concurrency issues appear during the parallel expansion when multiple expanded parts simultaneously try to allocate a vertex or edge. When replicating the element among the distributed process, the global synchronization of the allocated is necessary among the distributed processes to maintain the consistency of global information.

Selection of Initial Vertices. When the distance between the initial vertices selected by the parallel process is too close or unreasonable, the allocation process will bring more resource competition, which affects the partition quality. The selection of vertices needs to be combined with the properties of the graph. When a large graph is stored in a distributed system, the neighbor vertex information with a large number of hops needs to be obtained by communication between distributed processes, which brings some challenges to selecting suitable initial vertices.

4 Our Distributed Approach

In this section, we will propose our approach NEPG. The approach we proposed is mainly used for the pre-processing of distributed graph application algorithms (such as breadth-first algorithm and single-source shortest path algorithm). We should try our best to match the number of partitions generated by the method with the number required by the distributed graph processing system. In general, each partition is assigned to a machine. Therefore, the whole process is to divide the input large graph into $|P|$ parts and then deploy the $|P|$ partitions to the distributed machines. Since $|P|$-way edge partitioning is an NP-hard problem, we try to use the heuristic method to partition the graph.

We give a brief explanation of the memory aspect. The entire input graph needs to be stored in the main memories on the multiple machines during the expansion. Another point is that the selection of the vertex from the boundary is managed in a single machine. The whole process is divided into two distributed processes: *expansion process* and *allocation process*. The expansion processes manage the vertex selection and expansion of the boundary. The allocation processes manage the input graph in the distributed way and manage the allocation of vertices and edges.

4.1 Basic NE Algorithm

Before introducing our method, we briefly introduce the basic NE algorithm and analyze it. First, NE defines the core set and the boundary set. The boundary set starts to expand, and then select the appropriate vertices into the core set (the set for storing partitioned vertices). Before the expansion, a seed vertex is selected. The seed vertices are placed in the core set represented by C. The method places all neighboring vertices of each seed vertex in a boundary set denoted by B_i. Edges that connect vertices between or within C and B_i are assigned to the current partition p_i. In the expansion step, the vertex form B_i is selected that has the lowest D_{rest}, the lowest number of neighbors that are neither in B_i nor and C. NE moved the vertex from B_i to C. For every vertex v in B_i, the algorithm calculates the external degree of v. Finally, edges between v and vertices in B_i and C are assigned to the current partition p_i and the algorithm removed the edges from the graph. Then, the next expansion step is performed again to determe the vertex in B_i with the lowest D_{rest} and move it to C. Once a partition reaches its upper limit, the remaining edges will overflow to the next partition. When the partition is built, all of the edges will be removed from the graph and the algorithm will restart from the seed vertex. The algorithm stops after the whole graph is partitioned.

4.2 Distributed NE

DNE is an improved method to scale NE to large-scale graphs. Unlike NE, DNE starts from multiple random vertices and greedily expands each edge set in parallel. The expansion process mentioned above is consistent with the NE algorithm. DNE made improvements during the allocation process, so we briefly show how to make these improvements. The overall idea of the distribution process is as follows: If the vertex v and u are involved in $V(p(t))$, then allocation of $e_{u,v}$ will never increase replications. Based on this fact, DNE optimizes vertex allocation and improves partition quality. DNE mainly solves two problems: the management of concurrent allocation and the two-hop neighbor search. The key idea of the algorithm is to first allocate the one-hop-neighbor edges and then allocate two-hop-neighbor edges that meet the above condition. When the one-hop-neighbor allocation finished, DNE executed the synchronization to detect two-hop neighbors on the other expansion processes to make the allocation information consistent among the processes. Specifically, the whole process consists of 4 phases as follows.

(1) **Expansion and One-Hop-Neighbor Allocation.** For each vertex and its associated partition $< v, p >$, v's one-hop-neighbor edge $e_{u,v}$ and vertex u are allocated to p in parallel. And $< e_{u,v}, p >$ is assigned to the set of one-hop-neighbor edges and $< u, p >$ is assigned to the boundary set. DNE solves the conflict among different processes to allocate the same edge by taking the CAS performance. For example in Fig. 3, vertex 1 is the initial vertex of the expansion process. Then, $e_{1,2}$ and $e_{1,3}$ are allocated in blue lines which are black before. And vertex 2 and vertex 3 become new boundaries in the next iterations. In iteration 4, when allocating $e_{3,4}$, there is a conflict between the two allocation processes. In this case, unless the memory of the allocated process exceeds the upper limit, the edge with the allocation conflict is given priority to the process with a small sequence number.

(2) **Synchronization of Vertex's Allocation Information.** The algorithm synchronizes vertices that are newly allocated in the previous neighbor allocation. In this way, all replicated vertices among the different processes will be assigned the same allocation ids. The local information of the new boundaries in the process will be sent to the replications on the other machines. The allocation ids information is inserted in the boundaries vertices, which is needed when assigning two-hop-neighbors. In the example, the allocation information of vertex 2 and 3 need to be synchronized among the processes with the replications of vertex 2 and 3 at iteration.

(3) **Two-Hop-Neighbor Allocation.** As we mentioned before, two-hop neighbors of the selected vertex may include the edges that will not increase the vertex replication. For a set of triangular edges, if both u and v have been already allocated to the same partitions p, and the two-hop-neighbor edge $e_{u,v}$ will be allocated to p, too. The two-hop-neighbor edges will be inserted in the edge set. Allocation information is synchronized among processes and then will be sent to the expansion processes. In the example, $e_{1,2}$ and $e_{1,3}$ are allocated in blue lines. Then, $e_{2,3}$ is allocated to the same partition at iteration 3 as well.

(4) **Calculation of Local D_{rest} of the New Boundaries.** After the three stages: Expansion and One-hop-neighbor Allocation, Synchronization of Vertex's Allocation Information and Two-hop-neighbor Allocation, the algorithm will calculate the score of D_{rest} in each process for every vertex in the boundary set. In the final stage, All local scores will be exchanged among the processes and aggregated to calculate the overall D_{rest} score, which will be helpful for us to compare the replication factor (RF) later. We briefly summarize the shortcomings of DNE. (1) DNE eagerly keeps eyes on which partition the edges have been assigned to and which mirror node the neighbor information of the vertices belong to. (2) DNE still doesn't resolve the problem that the seed vertices are randomly selected, which will result in the lower bound instability of the whole partitioning quality.

Limitations of DNE. We briefly summarize the limitations of DNE. (1) To prevent NE from processing graphs that exceed the memory of its available compute nodes, DNE loaded the complete graph into memory. (2) The vertices initially chosen by DNE are selected randomly, which doesn't combine the properties and characteristics of the whole graph.

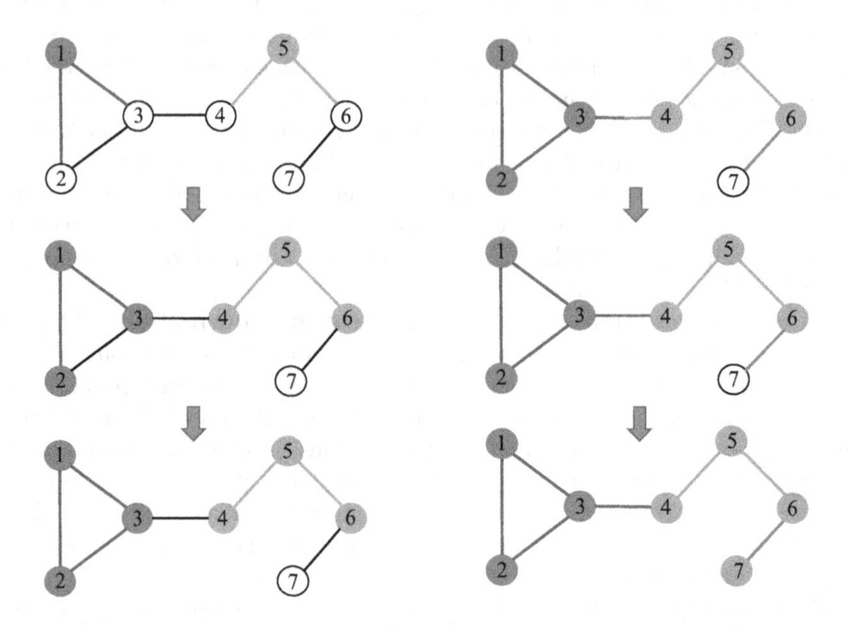

Fig. 3. Overview of DNE (Color figure online)

4.3 NEPG

Seed Vertex Selection. NEPG is based on a specially optimized algorithm for the power-law graph. Graph500 is an international large-scale graph computing competition, which uses the RMAT generator to generate Kronecker graphs. The Kronecker graph model is based on a recursive construction. The model is described in terms of the Kronecker product of matrices. We make clear the definition of the kronecker product. Given two matrices $A = [a_i, a_j](i < n, j < m)$ and B of sizes $n \times m$ and $n' \times m'$, the Kronecker product matrix C of matrices A and B is given by

$$C = A \otimes B = \begin{pmatrix} a_{1,1}B & a_{1,2}B & \cdots & a_{1,m}B \\ a_{2,1}B & a_{2,2}B & \cdots & a_{2,m}B \\ \vdots & \vdots & \ddots & \vdots \\ a_{n,1}B & a_{n,2}B & \cdots & a_{n,m}B \end{pmatrix} \in R^{nn' \times mm'} \tag{3}$$

The generator iteratively performs the kronecker product from an initial matrix $\begin{pmatrix} a & b \\ c & d \end{pmatrix}$ $(a = 0.57, b = c = 0.19, d = 0.05)$. The graph generated follows the densification power-law (DPL). The probability p_{uv} of an edge (u,v) in k-th Kronecker power $P = P_k$ can be calculated in $O(k)$ time as follows:

$$p_{uv} = \prod_{i=0}^{k-1} P_1 \left[|\frac{\mathrm{u}-1}{N_1^i}|(\mathrm{modN}_1) + 1, |\frac{\mathrm{v}-1}{N_1^i}|(\mathrm{modN}_1) + 1 \right] \tag{4}$$

The generated graph has a degree distribution and diameter similar to that of a large realistic graph. Most vertices have small degrees and few vertices have high degrees. To avoid a large number of conflicts during the greedy expansion of the vertex set, we select the vertices with extreme high-degree as the seed vertices. According to the value of the initial matrix of RMAT as a reference, we select vertices in the first half of the edge set to ensure a distance of 2 hops or more when selecting different initial vertices, which reduces the conflict during the expansion of the edge set and improves the partition quality. Figure 4 manifests an example of our selection of initial seed vertices. We omit the drawing of extra edges and use color to express the state of vertices. We use red for high-degree vertices while black for low-degree vertices. Assuming that the average degree in the graph is 16, we set a degree threshold of $\tau = 100$, which means all vertices with a degree of 100 or more are considered to be high-degree (In the example, u, v and w are high-degree vertices). When u is selected initially as the seed vertex, we consider the distance from u to the next high vertex. The distance between v and u is a one-hop distance, while the distance between w and u is a two-hop distance. To avoid premature conflicts between high-degree vertices in the expansion phase, we give priority to high-degree vertices with larger distances. In the example, we choose u and w as the seed vertices.

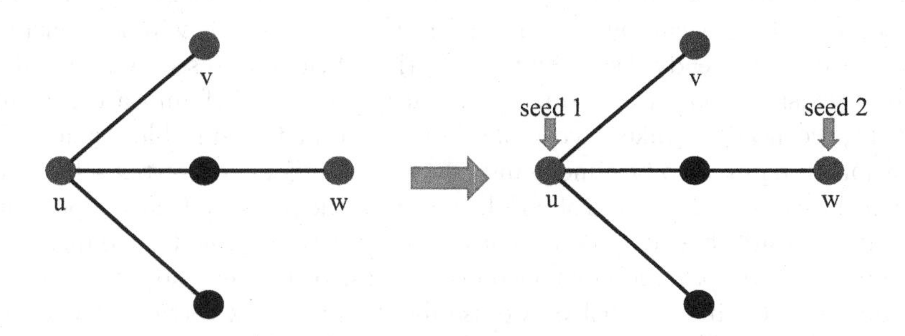

Fig. 4. Seed vertex selection

In the graph generated by graph500, the vertices are arranged in order by IDs. We give priority to selecting seed vertices in the vertices set with low ID. When one seed vertex u is selected and then the next seed vertex v is selected, we traverse the adjacency matrix of vertex u. If the distance between v and u is two hops or more, we choose v as the next seed vertex, otherwise we do the same for the next high-degree vertex. Before determining the next seed vertex, we need to traverse the adjacency matrix of the previously selected seed vertices, which will bring some communication overhead but will improve the partition quality to a greater extent. On the other hand, we control the number of all selected initial vertices to make them more close to the statistical law of the kronecker graph (power-law graph) in distribution. In graph500, the four parameters a, b, c, d of the initial matrix are 0.57, 0.19, 0.19 and 0.19. So we try our best to make the proportion of the selected vertex ID in the vertex order set close to $0.57 : 0.19 : 0.19 : 0.05$. With the increase of the scale of the generated graph, the degree distribution of vertices should be closer to the theoretical value. The improvement effect of this operation is not significant when the scale of the graph is small, and the quality will be improved with the increase of the scale of the graph.

In NEPG, we exploit this property as follows: We choose the high-degree vertices far enough apart as the initial vertices of the core set, which reduces the conflict overhead and partition quality loss. Compared with randomly selecting initial vertices, our method can ensure a higher lower bound of quality. By controlling the number distribution of the initial seed vertices, we try to improve the partition quality to approach the theoretical upper bound. In the process of expansion, NEPG is close to the original expression of NE algorithm.

Further Optimization. When there are no more vertices in B_i that can be moved to C, we call a function to find a suitable vertex outside in B_i that can be moved to C which is neither a high-degree vertex nor a vertex in C. When the above happens, the following situations occur: (1) When the seed vertices are selected and the boundary sets begin to expand. (2) When the whole graph is being partitioned, the components are not connected to each other, and some of them have finished expansion. (3) When there are only high-degree vertices in B_i and no more vertices can be moved to C. In the first case, we will not consider it because we have already discussed the selection of the initial seed vertices above. Since other situations where suitable vertices can not be found are extremely frequent, we need to take a cost operation to find the suitable vertices. The operation is required to be timely and effective. Randomized vertex selection, as done in the expansion process of NE, brings with the problem that the possibility of hitting the suitable vertices becomes lower over time while the boundary sets become bigger. As partition expansion continues, vertex selection becomes more and more challenging. In NEPG, we use flag bits to mark vertices that once be considered inappropriate, and no longer consider these vertices in the subsequent expansion process. The operation has been used since initialization to ensure that partition quality does not decrease with expansion.

5 Parallel Expansion Rate Control in Iteration

In this section, we will introduce the control operation of the rate of parallel expansion in each iteration. Although we have taken some operations such as optimizing the initial vertex selection, marking unsuitable vertices with flag bits, the communication among the allocation processes will encounter a bottleneck when the number of iterations gradually increases. The main reason for this phenomenon is that there are many barrier operations when communications between the allocation occur. The number of iteration rounds depends on the number of partitions $|P|$ and the number of edges allocated in each partition per iteration n. Suppose the $G = (V, E)$, and the number of iteration rounds is equal to $|E|/(n \cdot |P|)$. The more edges we allocate in each iteration, the less time it takes to process the whole graph. But at the same time, the partition quality decreases. We need to find a balance between partition quality and graph processing time, which is the reason we introduced the expansion factor $\lambda (0 < \lambda < 1)$. Suppose that when the boundary set expands, the outermost vertices number is B_{new}. In the expansion, $\lambda \cdot |B_{new}|$ vertices are expanded for allocating new edges. We control the speed of expansion by adjusting the size of λ. As λ increases, the speed of expansion increases, resulting in a reduction in the overall execution time. Too large λ will lead to the great communication burden and partition quality loss, and too small λ will lead to the slow program execution speed.

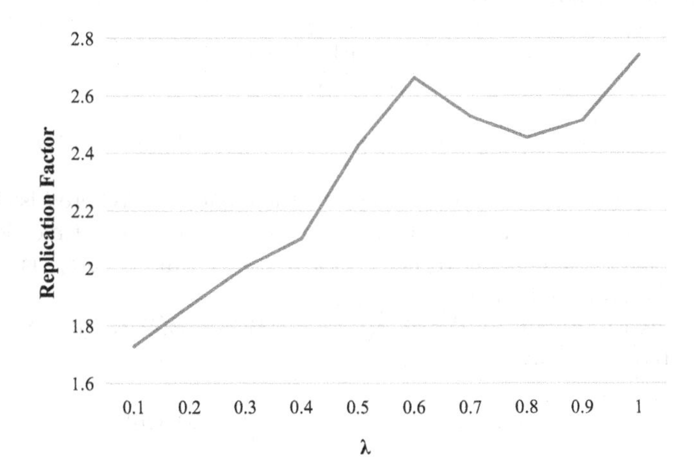

Fig. 5. The relationship between λ and RF

We try different λ to explore the internal relationship between the size of the graph and the λ. We also did experiments to explore the relationship between the replication factor and λ (which also determines the number of iterations) on 16 partitions. Figure 5 manifests that the effect of λ on replication factor is not significant. According to the experiment, the number of iterations linearly decreases as the increase of λ. If $\lambda = 1.\ 0$, which means all boundary vertices will be chosen at every iteration, it consumes a lot of computing resources in a short time, and can not guarantee the quality of partitions. In reality, the number of iterations is much smaller than the hops of the graph radius. With the exponential increase of λ, the replication factor will increase as a whole. When the λ is close to 1, the processing quality of the whole graph data is exceedingly poor. According to the experiment, we infer that the value of λ is proportional to edge factor. We find similar characteristics in 4, 8, 32, and 64 partitions. By comparing the experimental results, We choose $\lambda = 0.\ 0625$ in reality to maximize quality and efficiency.

6 Theoretical Demonstration

In this section, We will give the lower and upper limits of the partition quality of NEPG, and then give the efficiency analysis of NEPG. We give the partition quality (by comparing the replication factors) for the multi-parallel expansion of NEPG. The upper bound of partitioning quality of the graph muti-expansion has been given in [8], but there is no proper scaling and we give a better theoretical derivation.

Theorem 1. *(Upper Bound). For an undirected graph $G(V, E)$, the replication factor is obtained by computing all the data of edge partitions. We finally get the theoretical upper bound of RF.*

$$\mathrm{RF} \leq \frac{|\mathrm{E}| + |V|}{|V|}$$

where $|E|$ represents the number of edges, $|V|$ represents the number of vertices.

Proof. We will construct a function to prove the inequality. Where t is the round of the iteration; $E_{rest}(t)$ is the set of non-allocated edges at t time; $V_{rest}(t)$ is the set of vertices that not have been allocated at t time. $V(p(t))$ is the set of vertices in the allocated edges.

The key is to prove that $\Delta\varphi = \varphi(t{+}1){-}\varphi(t) < 0, \varphi(t) \leq \varphi(0)$. So, $\Delta\varphi$ can be represented as follow:

$$\Delta\varphi = \Delta|E_{rest}| + \Delta|V_{rest}| + \sum_{p \in P} \Delta|V(p)|$$

When we select a vertex x_p from the boundary set B_p to the core set in parallel at t time. The reduction of E_{rest} is divided into two parts: x_p's one-hop-neighbor edges n_p^{one} and x_p's two-hop-neighbor edges n_p^{two} added to $p(t)$. The number of edges n_p^{one} depends on $\lambda \cdot |B_{new}|$ mentioned before. When p's expansion has finished already, n_p^{one} and n_p^{two} are equal to zero. We represent $\Delta|E_{rest}|$ as follows.

$$\Delta|Erest| = -\sum_{p \in P} (n_p^{one} + n_p^{two})$$

$\Delta|V_{rest}|$ is also a monotone decreasing function. We introduce auxiliary functions $\psi\ (p)$ to represent $\Delta|V_{rest}|$,

$$\Delta|\mathrm{Vrest}| \leq -\sum_{p \in P} \psi(\mathrm{p}), \psi(\mathrm{p}) = \begin{cases} 0 \text{ (p ends at t or earlier)} \\ 1 \text{ (Other)} \end{cases}$$

which means if the partition is terminated at t time or earlier, there is no vertex will be allocated. In the two-hop neighbor allocation phase, no extra vertices will be allocated. On the other hand, if the partition is still in the expansion phase, the number of x_p's one-hop-neighbor that will be allocated is greater than 1.

Next, let's discuss the change of the $\Delta|V(p)|$, $|V(p(t))|$ is a monotonously increasing function over t. It is not difficult to deduce the inequality: $\Delta|\mathrm{V(p)}| \leq \mathrm{n_p^{one}} + 1$, where means that $\Delta|V(p)|$ is less than the number of x_p's allocated neighbor and itself.

Finally, we integrate the inequality derived above and get the time change of our constructed function $\Delta\phi(t)$ from time t to $t+1$ as follows:

$$\begin{aligned} \Delta\phi &= \Delta|E_{rest}| + \Delta|\mathrm{V}_{rest}| + \sum_{p \in P} \Delta|V(p)| \\ &\leq - \sum_{\mathrm{p \in P}} \{(\mathrm{n_p^{one}} + \mathrm{n_p^{two}}) + \psi(\mathrm{p}) - (\mathrm{n_p^{one}} + 1)\} \\ &= - \sum_{\mathrm{p \in P}} (\mathrm{n_p^{two}} + \psi(\mathrm{p}) - 1) \leq 0 \end{aligned}$$

Because $\Delta\phi \leq 0$ for all t, and $\phi(t) \leq \phi(0)$. Therefore the theorem 1 is proved.

$$RF = \sum_{p \in P} \frac{|V(p)|}{|V|} = \frac{\phi(t)}{|V|} \leq \frac{\phi(0)}{|V|} = \frac{|\mathrm{E}| + |\mathrm{V}|}{|\mathrm{V}|}$$

Theorem 2. *(Lower Bound) As for the undirected graph $G(V, E)$, we refer to the operation of initial vertex optimization in NEPG get the theoretical lower bound of RF.*

$$\mathrm{RF} \geq \frac{|\mathrm{V}| + |\mathrm{P}| - 1}{|\mathrm{V}|}$$

where $|E|$ represents the number of edges, $|V|$ represents the number of vertices.

Proof. First of all, we first analyze the degree distribution of large real graphs. Figure 6 manifests the degree distribution of graph500. As we expected, the actual situation is consistent with the theory, and the degree distribution of graph500 is consistent with that of the power-law graph. The degree distribution image of the power-law graph will show a heavy tail. Most vertices have small degrees, and a small number of vertices have large degrees. According to the vertex selection method in NEPG, the algorithm starts with vertices with high degrees (theoretically the most influential vertex). And we adjusted the distance between the high-degree vertices to avoid the premature occurrence of conflict. When discussing the lower bound of the replication factor, we assume the most extreme case: all the partitions end up fighting for the last vertex, or there is only one vertex at the junction of each partition. Imagine a situation where there are a total of P partitions, and in the last partition, that partition has only one mirror node in all the other $P-1$ partitions. Then there is the following equation:

$$\mathrm{RF} = \frac{1}{|\mathrm{V}|} \sum_{i=1}^{k} |V(p_i)| = \frac{|V| + |P| - 1}{|V|}$$

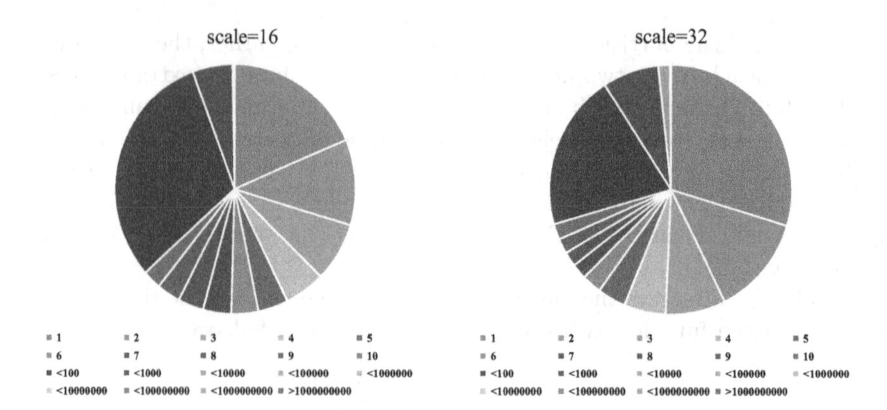

Fig. 6. Degree distribution of Graph500

where in addition to the vertices in the whole graph, it includes the mirror nodes on the remaining $P-1$ partition. When the situation tends to a more general scene, there is much more than one node competing between partitions. Figure 7 manifests an example, where the figure on the left is extreme: the boundary between all partitions has only one vertex, while the figure on the right manifests the situation that more than one vertex. As more nodes compete, the increment on the left side of the above equation exceeds $|P|-1$. Even if we use the optimization of vertex selection, it is ideal to have only one vertex copy between partitions. In the actual experiment, the setting value of edge factor in graph500 is generally 16. There is still a lot of work to be done for us to reach the lower limit of quality. Thus, Theorem 2 is proved.

DNE is a state-of-the-art partitioning algorithm. In [8], the theoretical upper bound of DNE is compared with other distributed edge partitioning methods (Random, Grid and DBH). In the case of power-law graphs, DNE provides a better upper bound than the existing distributed methods. In this paper, we introduce the optimization method of initial seed vertex selection, and a give better quality upper bound and lower bound.

Theorem 3. *(Efficiency) We analyze the time complexity of the NEPG. Excluding extreme cases, the input graph is evenly distributed on each machine; and the workload in each machine is evenly assigned to its computing units. The worst-case time complexity of running the program on each unit is* $O\left(\frac{d|E|(|P|+d)}{n|P|}+d^2\right)$, *where* $|E|$, $|P|$ *and* d *are the number of the edges, partitions, and maximum degree.*

Proof. Initialization has a complexity of $O(d^2)$, as each high-degree vertex's two-hop-neighbor is visited during the vertex selection phase. In the expansion phase, the number of messages in B_{new} is $O(|E|/P)$ in the worst case. When computing on each unit, the time complexity is $|B_{new}|/n = O(|E|/n|P|)$. For each vertex u,

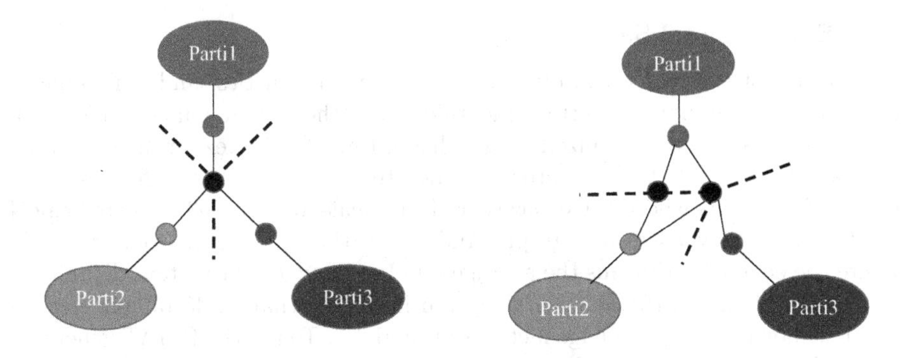

Fig. 7. When partitioning, the most extreme case is that the boundary between every two partitions in all partitions is a vertex. In contrast, it is more common to have more vertices within the boundary.

$O(d)$ neighbors are processed. When detecting the intersection of two one-hop-neighbor (searching for two-hop-neighbor), the time complexity is $O|P|$. And when assigning two-hop-neighbor vertices, the time complexity is $O|d|$. For each vertex u, at most $O|d|$ neighbors are processed. Therefore, the time complexity of the expansion process is $O\left(\frac{|E|}{n|P|}\right) \times O(d) \times O(|P| + \mathrm{d}) = O\left(\frac{d|E|(|P|+d)}{n|P|}\right)$, and the total time complexity of the algorithm is $O\left(\frac{d|E|(|P|+d)}{n|P|} + d^2\right)$.

7 Empirical Evaluation

In this section, we discuss the empirical analysis of quality, scalability, and efficiency. Our conclusions are as follows.

Highest Quality. In the power-law graphs of different scales of graph500, NEPG can produce higher quality partitions than other distributed in-memory algorithms or streaming algorithms.

Good Scalability. Compared with other state-of-the-art high-quality distributed in-memory algorithms, NEPG can use less than 1/10 of the memory at run time. But the usage of memory is still much more than the streaming algorithm. Because when processing graph data, streaming algorithm passes through the edge stream, only looking at one (or a small number of) edge(s) at a time. Streaming algorithm does not need to know the information of the whole graph, but deals with the edges that traversed like a stream of water. Streaming algorithms consume little memory, but do not yield the same high-quality partitioning as the state-of-the-art in-memory algorithm.

Comparable Efficiency. The running time of NEPG is comparable to the state-of-the-art high-quality distributed methods. The time complexity of the NEPG is a little higher than that of DNE, but achieves better partition quality.

7.1 Experimental Setup

The graphs of the real world often have their particularities, and it is difficult for us to get common properties and rules from them. Our paper is based on the partition strategy optimized by graph500, thus, further experiments will be conducted using RMAT generator to generate a graph with real features. We use RMAT graphs whose vertex sizes are from Scale20 to Scale30, where ScaleN is referred to 2^N vertices in a graph. And we set the edge factor from 2^4 to 2^{10}, where the edge factor means the average number of edges per vertex (default set to 2^4). For example, the Facebook's graph has more than 1. 45 billion vertices and 1 trillion edges, which edge factor is around 2^{10}. To use the RMAT generator to simulate such a large graph, we need to set the scale ≥ 31 and the edge factor ≥ 16.

7.2 Benchmark Partitioning Algorithm

We compare NEPG with 4 distributed partitioning methods. NE is a basic partition method based on vertex greedy expansion. DNE is the state-of-the-art high-quality distributed edge partitioning method based on NE. Its greatest advantage is being able to scale across large clusters of machines and can process exceedingly large-scale graphs. In some cases, DNE cannot always maintain partition balance. ParMETIS is the standard multi-level vertex partitioning. It is used as a benchmark partition algorithm by other papers because of its good quality of graph partition. It implements a multi-level partitioning approach with coarsening and refinement phases and can get better replication factors. HEP is a hybrid partition method of in-memory algorithm and streaming algorithm, which uses the HDRF (the state-of-the-art high-quality streaming algorithm) for high-degree vertices and NE++ (optimized NE version for the hybrid algorithm) for low-degree vertices. HEP uses CSR (the data representation structure), which avoids a lot of random access and cache misses caused when performing the neighborhood expansion steps. There are dozens of partitioning methods, and we cannot compare them with our method. Other partitioning methods may have advantages in terms of replication factor, running time, scalability or efficiency. In papers [8,15], DNE and HEP have been compared with the previous state-of-the-art algorithms and have shown some advantages. Hence, we argue that we made a reasonable and scientific choice of benchmark partitioning algorithms. Compared to DNE, HEP has less memory consumption, better replication factor, and better partition balance control. But DNE has better partition quality. Compared with the above algorithms, NEPG has a better replication factor and partition balance. Because NEPG operates the optimizing functions according to the properties of the input graph, the partition quality is closer to the theoretical optimal value.

7.3 Performance Metrics

We measured the replication factor, run time (including the construction of the graph and preprocessing time), and scalability.

7.4 Experiments

We evaluate the performance of Tianhe-CC in TianHe prototype system [21] and use a cluster of 128 nodes. We perform experiments with k = 4, 16, 32, 64 partitions. We perform HEP by setting the value of τ (the threshold of distinguishing high-degree vertices from low-degree vertices): 100, 10, and 1. For each graph, we repeat the experiment five times and record the mean values of the standard deviation.

7.5 Quality Evaluation

The quality of partitioning is evaluated utilizing the replication factor in Sect. 7.6. For each graph, we repeat the experiment five times to get the mean value of the replication factor. Figure 8 manifests the replication factor of the Kronecker graphs generated by the RMAT generator on partitions from 4 to 64.

For the scale of the experimental graphs, we set the scale from 16 to 20, and the edge factor is set to 16.

Overall, NEPG outperforms the other methods in large-scale graphs. When we use the HEP method, setting the corresponding τ to 1, 10, and 100. The smaller the number of partitions, the smaller the gap between NEPG and other methods. When the scale of the graph is large enough, other partitioning methods are no longer competitive with NEPG.

Moreover, the partitioning quality of NEPG is stable. DNE will affect the partition quality according to the quality of the selected seed vertices. But NEPG has good and stable processing quality, thanks to the optimal selection of initial vertices. In large real-world graphs, partition quality may be unpredictable because of the different characteristics of the graph. In this paper, only the graphs generated by RMAT generator are used as the experimental input graph. In the actual situation, there is a positive correlation between the replication factor and the edge factor. And the graph will become more complicated as the edge factor increases. It is a great challenge for us to find the "center" of the graph and to build enough stable and high-quality partitions. Under the condition of the same edge factor, the replication factor of the graph with a difference of tens times in size has no significant change, which indicates that the size of the replication factor depends more on the edge factor. The challenge of getting high-quality partitions comes more from the complexity of the graph.

There had been related works to discuss the disadvantages of each partitioning algorithm before we proposed the NEPG. To summarize, we classify all the partitioning algorithms into two categories: in-memory algorithm and streaming algorithm. Because the streaming algorithm does not grasp the overall information of the graph, the partition quality obtained is often not ideal. Then we can divide the memory algorithms into three categories. The first categories are based on the hashing, such as 2D-Random, Hybrid Ginger, and Spinner, which can not obtain stable high partition quality because of the randomness. The second category is converting the partitioning problem into other similar problems, such as tree partitioning (Sheep [13]), which are the "golden method" under

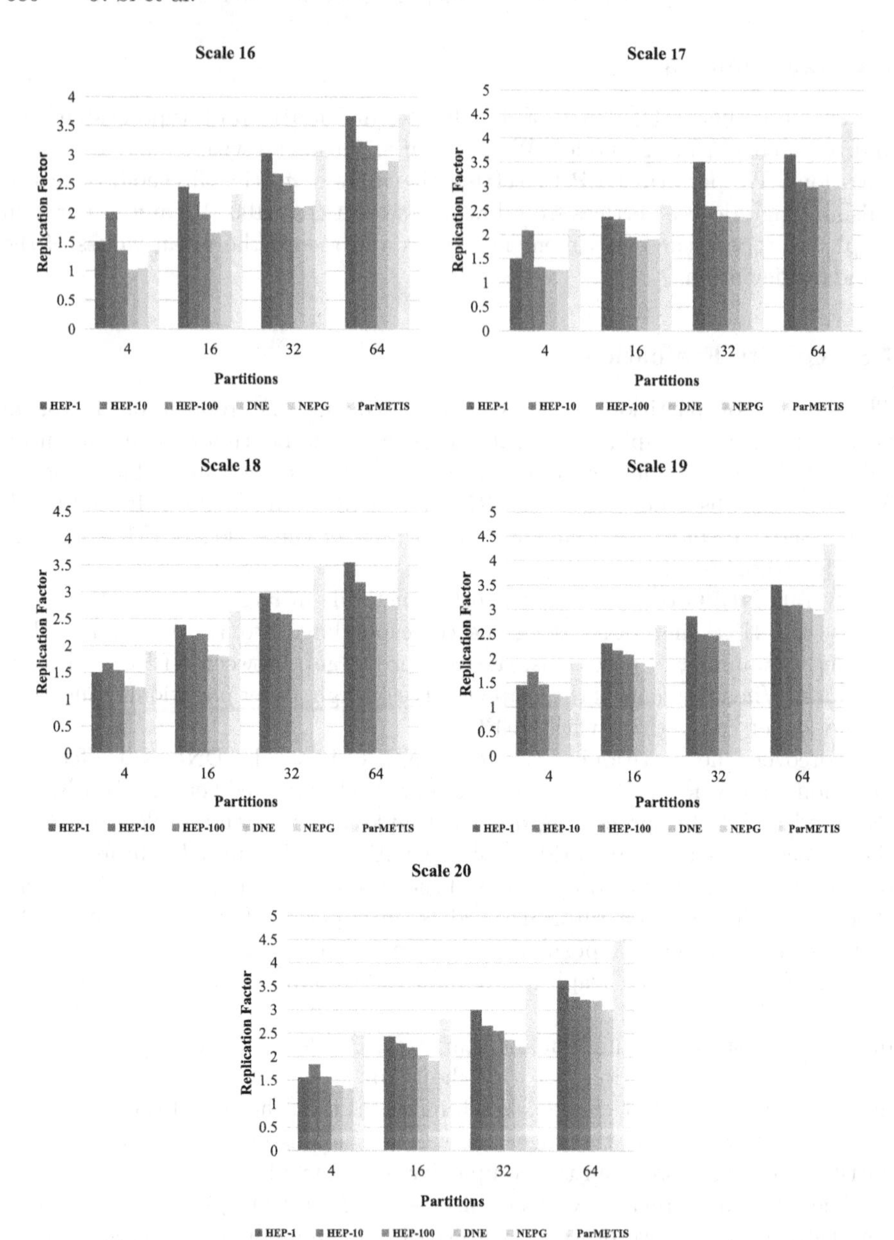

Fig. 8. Replication factor of kronecker graphs

special conditions. The final category is heuristic algorithms, such as DNE and ParMETIS. Relying only on heuristic algorithms can only improve the quality of partitions as a whole, but can not guarantee stability. Because of the complexity of exceedingly large-scale graphs, the conflict will occur earlier sometimes during parallel expansion, resulting in exceedingly poor partition quality. HEP is a mixture of DNE and another streaming algorithm HDRF, so it has the disadvantages of both streaming algorithms and heuristics algorithms. On the other hand, NEPG makes initial vertex optimization based on a heuristic algorithm, which can provide high-quality partition (consistent with the upper bound of replication factor derived above).

7.6 Performance Evaluation

First of all, let's clear the following conclusions. (1) The choice of partitioner depends on the scale and complexity of the graph. For extremely short processing, the hash-based algorithms are the best option. The hash-based algorithms are efficient and scalable since they only include light-weight hash calculations and local refinements. The hash-based algorithms that are often used now are Random, Oblivious [6], Hybrid Ginger [3], and Spinner [14]. (2) For long-running graph processing jobs, like multiple subsequent runs of Breadth First Search (BFS) and Single Source Shortest Path (SSSP), it is undoubtedly a better choice that will be beneficial to perform high-quality graph partitioning. (3) When all algorithms are competitive in partition processing, vertex balance also becomes an important factor in determining whether it is a good partition quality. For small-scale graphs, the partition quality gap processed by each algorithm is small. For large-scale graphs, partitioning in data with higher complexity brings a great challenge to the scalability and effectiveness of the methods.

Memory Consumption. Memory consumption is an important indicator to measure the scalability of an algorithm. We measure the instantaneous maximum of memory from graph construction to the end of the algorithm and represent it as M_{max}. The whole process of measuring memory is done on 64 machines.

Both DNE, NEPG, and HEP are massively parallel algorithms based on NE, which have advantages over other partitioning algorithms in memory saving. DNE's memory savings benefit from two aspects: vertex-cut and no more memory-consuming data structure. In general, the replication of vertices is more efficient than that of edges. Compared with other partitioning algorithms that use hash mapping and extra arrays, DNE uses CSR to store data from core sets, which saves more memory. Based on the optimization of DNE, HEP uses three more measures to optimize memory consumption. First, HEP uses the streaming algorithm for high-degree vertices, which does not need to load the graph into memory. Moreover, HEP saves additional memory by pruning the adjacency list of high-degree vertices and by avoiding auxiliary data structures to track whether edges have been allocated. Compared with DNE, NEPG adds the operations such as pre-processing of initial vertex selection and adding auxiliary markers, which increases lightweight memory consumption but achieves higher partition quality.

Elapsed Time. We will evaluate the efficiency with the elapsed time. And the elapsed time will be equal to the graph partitioning time plus the time of running the graph500 program, not including the time to load the input graph. Figure 9 manifests the harmonic_mean_TEPS of processing the graph with the same edge factors into 16 partitions with different scale, which is an indicator of the number of instantaneous processing edges in graph500. As can be seen from the Fig. 9, NEPG has the minimum elapsed time.

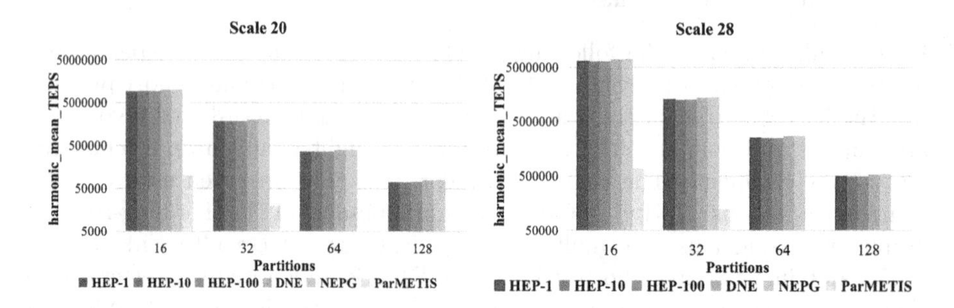

Fig. 9. Performance comparison

Scalability. Figure 10 manifests the scalability of NEPG to the large-scale graph. We perform experiments on large servers. We fix the number of vertices per machine as 2^{22} and change the number of nodes (= the number of partitions). We process Scale24 on 4 nodes, Scale26 on 16 nodes, Scale28 on 64 nodes, Scale29 on 128 nodes. With the increase of nodes, the elapsed time will increase proportionally. When the node running the partition algorithm becomes larger and larger, the load imbalance will occur in the expansion process. Under the overall situation, the increase of the number of nodes will bring higher communication.

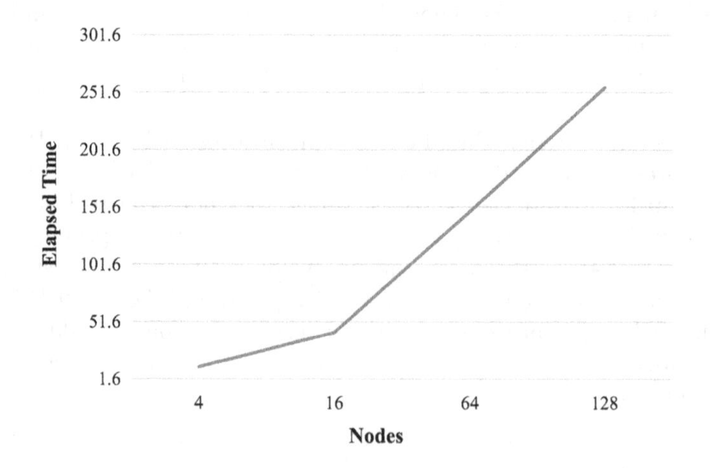

Fig. 10. Scalability

8 Conclusion

In this paper, we presented NEPG, a distributed edge partitioning algorithm based on greedy expansion, which steadily provides high-quality partitions of Kronecker Graph with the utilization of seed vertex selection optimization. And we provide the theoretical upper bound and lower bound as well. In future work, we aim to reduce memory consumption by improving the data storage structure. Beyond that, we plan to expand our work to more complicated graphs, such as hypergraphs and dynamic graphs.

Acknowledgement. This work is supported in part by National Key R&D Program of China (Grant No. 2018YFB0204300), Excellent Youth Foundation of Hunan Province (Dezun Dong), National Postdoctoral Program for Innovative Talents (Grant No. BX20190091).

References

1. Ahmad, Y., et al.: LA3: a scalable link-and locality-aware linear algebra-based graph analytics system. Proc. VLDB Endowment **11**(8), 920–933 (2018)
2. Arulraj, J.: Apache Giraph (2018)
3. Chen, R., Shi, J., Chen, Y., Zang, B., Guan, H., Chen, H.: PowerLyra: differentiated graph computation and partitioning on skewed graphs. ACM Trans. Parallel Comput. (TOPC) **5**(3), 1–39 (2019)
4. Ching, A., Edunov, S., Kabiljo, M., Logothetis, D., Muthukrishnan, S.: One trillion edges: graph processing at Facebook-scale. Proc. VLDB Endowment **8**(12), 1804–1815 (2015)
5. Gan, X.: Customizing graph500 for Tianhe pre-exacale system. arXiv preprint arXiv:2102.01254 (2021)
6. Gonzalez, J.E., Low, Y., Gu, H., Bickson, D., Guestrin, C.: PowerGraph: distributed graph-parallel computation on natural graphs. In: 10th USENIX Symposium on Operating Systems Design and Implementation (OSDI 12), pp. 17–30 (2012)
7. Gonzalez, J.E., Xin, R.S., Dave, A., Crankshaw, D., Franklin, M.J., Stoica, I.: GraphX: graph processing in a distributed dataflow framework. In: 11th USENIX Symposium on Operating Systems Design and Implementation (OSDI 14), pp. 599–613 (2014)
8. Hanai, M., Suzumura, T., Tan, W.J., Liu, E., Theodoropoulos, G., Cai, W.: Distributed edge partitioning for trillion-edge graphs. arXiv preprint arXiv:1908.05855 (2019)
9. Khayyat, Z., Awara, K., Alonazi, A., Jamjoom, H., Williams, D., Kalnis, P.: Mizan: a system for dynamic load balancing in large-scale graph processing. In: Proceedings of the 8th ACM European Conference on Computer Systems, pp. 169–182 (2013)
10. Leskovec, J., Chakrabarti, D., Kleinberg, J., Faloutsos, C., Ghahramani, Z.: Kronecker graphs: an approach to modeling networks. J. Mach. Learn. Res. **11**(2), 985–1042 (2010)
11. Low, Y., Gonzalez, J., Kyrola, A., Bickson, D., Guestrin, C., Hellerstein, J.M.: Distributed GraphLab: a framework for machine learning in the cloud. arXiv preprint arXiv:1204.6078 (2012)

12. Malewicz, G., et al.: Pregel: a system for large-scale graph processing. In: Proceedings of the 2010 ACM SIGMOD International Conference on Management of Data, pp. 135–146 (2010)
13. Margo, D., Seltzer, M.: A scalable distributed graph partitioner. Proc. VLDB Endowment **8**(12), 1478–1489 (2015)
14. Martella, C., Logothetis, D., Loukas, A., Siganos, G.: Spinner: scalable graph partitioning in the cloud. In: 2017 IEEE 33rd International Conference on Data Engineering (ICDE), pp. 1083–1094. IEEE (2017)
15. Mayer, R., Jacobsen, H.-A.: Hybrid edge partitioner: partitioning large power-law graphs under memory constraints. In: Proceedings of the 2021 International Conference on Management of Data, pp. 1289–1302 (2021)
16. Roy, A., Mihailovic, I., Zwaenepoel, W.: X-stream: edge-centric graph processing using streaming partitions. In: Proceedings of the Twenty-Fourth ACM Symposium on Operating Systems Principles, pp. 472–488 (2013)
17. Salihoglu, S., Widom, J.: Optimizing graph algorithms on pregel-like systems (2014)
18. Shao, B., Wang, H., Li, Y.: Trinity: a distributed graph engine on a memory cloud. In: Proceedings of the 2013 ACM SIGMOD International Conference on Management of Data, pp. 505–516 (2013)
19. Suzumura, T., Ueno, K.: ScaleGraph: a high-performance library for billion-scale graph analytics. In: 2015 IEEE International Conference on Big Data (Big Data), pp. 76–84. IEEE (2015)
20. Tsourakakis, C., Gkantsidis, C., Radunovic, B., Vojnovic, M.: FENNEL: streaming graph partitioning for massive scale graphs. In: Proceedings of the 7th ACM International Conference on Web Search and Data Mining, pp. 333–342 (2014)
21. Wang, R., et al.: Brief introduction of TianHe exascale prototype system. Tsinghua Sci. Technol. **26**(3), 361–369 (2020)
22. Wu, M., et al.: Gram: scaling graph computation to the trillions. In: Proceedings of the Sixth ACM Symposium on Cloud Computing, pp. 408–421 (2015)
23. Xie, C., Yan, L., Li, W.-J., Zhang, Z.: Distributed power-law graph computing: theoretical and empirical analysis. In: NIPS, vol. 27, pp. 1673–1681 (2014)
24. Zhu, X., Chen, W., Zheng, W., Ma, X.: Gemini: a computation-centric distributed graph processing system. In: 12th USENIX Symposium on Operating Systems Design and Implementation (OSDI 16), pp. 301–316 (2016)

Accelerating DCNNs via Cooperative Weight/Activation Compression

Yuhao Zhang, Xikun Jiang, Xinyu Wang, Yudong Pan, Pusen Dong, Bin Sun, Zhaoyan Shen, and Zhiping Jia(✉)

School of Computer Science and Technology, Shandong University, QingDao, China
yu_hao@mail.sdu.edu.cn, {shenzhaoyan,jzp}@sdu.edu.cn

Abstract. Deep convolutional neural networks (DCNNs) have achieved great success in various applications. Nevertheless, training and deploying such DCNNs models require a huge amount of computation and storage resources. Weight pruning has emerged as an effective compression technique for DCNNs to reduce the consumption of hardware resources. However, existing weight pruning schemes neglect to simultaneously consider accuracy, compression ratio and hardware-efficiency in the design space, which inevitably leads to performance degradation. To overcome these limitations, in this work, we propose a cooperative weight/activation compression approach. Initially, we observe spatially insensitive consistency, namely the insensitive weight values that have no impact on accuracy distributed at the same spatial position in a mass of channels. Based on the key observation, we propose a cluster-based weight pattern pruning technique, which converges the channels with spatially insensitive consistency into a cluster and prunes them into the weight pattern with a uniform shape. In addition, there are many sparsity rows in the activation matrix due to the effect of the activation function–Rectifying Linear Unit (ReLU). We study the sparsity rows and find that rows with larger sparsity degree are insensitive to accuracy. Hence, we further propose a sparsity-row-based activation removal technique, which directly eliminates insensitive rows of activation. The two proposed techniques allow the hardware to execute at excellent parallel granularity and achieve better compression ratio with negligible accuracy loss. Experiment results on several popular DCNNs models show that our scheme reduces the computation by 63% on average.

Keywords: Deep convolutional neural networks (DCNNs) · Weight · Activation · Clustering · Compression

1 Introduction

Deep learning, especially deep convolutional neural networks (DCNNs), has been gradually applied to various domains, ranging from image classification [1], object detection [2] to speech recognition [3]. This is mainly due to their ability to achieve unprecedented accuracy. The high accuracy of DCNNs comes with

Y. Lai et al. (Eds.): ICA3PP 2021, LNCS 13157, pp. 691–706, 2022.
https://doi.org/10.1007/978-3-030-95391-1_43

hundreds of millions of learnable parameters. Training and executing a modern DCNNs algorithm with heavy parameters usually desires a significant amount of computation and storage resources [4,5]. For instance, VGGNet [6], a typical DCNNs model that achieves a 7.3% top-5 error rate for image classification with ImageNet dataset [7], has a model size of about 548M and requires 1.5×10^{10} multiply-and-accumulate (MAC) operations per image to perform classification. This makes it difficult to deploy DCNNs on embedded and mobile devices with limited computation and storage resources [8–10].

Weight pruning has been emerged as an effective compression approach to reduce the requirement of hardware resources for DCNNs [11–15]. It leverages the redundancy of weight parameters and prunes these weights to zeros, resulting in sparse DCNNs models. A number of existing studies have been proposed various pruning schemes to compress DCNNs. For example, Deep Compression [12] converts smaller weights, regarding as unimportant or insensitive values, to zero, followed by a retraining process to fine tune the residual weights for maintaining the original accuracy. Since the insensitive weights are randomly distributed, thereby noticeably posing the irregular sparsity in DCNNs. Although the above non-structured weight pruning method can achieve excellent compression ratio and accuracy, the irregular sparsity prevents hardware from fully utilizing their parallel potential to improve performance.

To avoid the drawbacks of irregular sparsity, the structured pruning schemes [13,14,16] have been exploited to enhance hardware-friendly. They prune the whole channels [13] or even filters [16] to obtain considerable compression ratio and hardware-efficiency but result in relatively high accuracy drop. To address the disadvantages derived from both the non-structured and structured pruning schemes, semi-structured pruning has been proposed to achieve both high accuracy and high hardware efficiency. For instance, pattern-based weight pruning method [17] for DCNNs assigns a pattern for each filter and trains the pattern-based weights for recovering accuracy. Nevertheless, it sacrifices the compression ratio. The existing pruning schemes ignore to simultaneously consider accuracy, compression ratio and hardware-efficiency in the design space. The resulting performance degradation of DCNNs becomes a critical problem.

To solve the above problem, in this paper, we propose a cooperative weight/activation compression approach for DCNNs. Initially, we fully analyze the spatial distribution of weights for several typical DCNNs models and observe that spatially insensitive consistency exists in numerous channels of filters. To be specific, insensitive weight values that have no impact on accuracy are located in the same spatial position for abundant channels. Hence, we propose a cluster-based weight pattern pruning technique to utilize the above characteristic of distribution, resulting in regular sparsity in DCNNs. Firstly, we adopt a clustering algorithm to cluster the weight channels with spatially insensitive consistency in the same layer. Secondly, for the channels in a cluster, the insensitive weight values at the same position are pruned, thereby remaining the uniform shape of channels. The cluster-based weight pattern pruning technique not only maintains the advantages of hardware friendliness and high accuracy of semi-structured pruning, but also improves the compression ratio of DCNNs by clustering method.

In addition, most existing neural network libraries, such as cuDNN for GPU and openBLAS for CPU, express the convolution operation as general matrix multiplication (GEMM) [18]. It unrolls the activation feature maps and filters of DCNNs into large matrices by image to column (im2col) transformation. Since each convolution operation in DCNNs is commonly followed by an activation function called a Rectifying Linear Unit (ReLU) that returns zero for negative inputs and yields the input itself for the positive ones, there are many sparsity rows in the unrolled activation matrices. We carefully study them and observe that rows with larger sparsity degree are insensitive to accuracy. Hence, we further propose a sparsity-row-based activation removal technique, which dynamically eliminates insensitive rows of activation matrices to improve performance.

In summary, this paper aims to comprehensively consider accuracy, compression ratio and hardware-efficiency of DCNNs to further improve the performance. To this end, we thoroughly analyze the distribution characteristics and sensitivity of weight and activation. Based on the analysis, we propose a cooperative weight/activation compression approach for DCNNs. The approach consists two techniques: 1) a cluster-based weight pattern pruning technique; 2) a sparsity-row-based activation removal technique. The two proposed techniques allow the hardware to execute at highly parallel granularity and achieve better compression ratio with negligible accuracy loss. The main contributions of this work are as follows:

- We carefully analyze the spatial distribution characteristics of weights and observe that insensitive weights distribute at the same position in numerous channels. Hence, we propose a cluster-based weight pattern pruning technique, which adopts the clustering algorithm to cluster weight channels with spatially insensitive consistency and prunes them into the weight pattern with a uniform shape.
- We thoroughly study the sparsity rows of activation matrix and find that rows with larger sparsity have no impact on accuracy. Therefore, we propose a sparsity-row-based activation removal technique, which eliminates insensitive rows of activation.
- Evaluation results on several popular DCNNs models show that the proposed cooperative weight/activation approach achieves 63% reduced computation on average.

The rest of this paper is organized as follows: Sect. 2 introduces the background and motivation of our study. Section 3 presents the proposed cooperative weight/activation compression approach in detail. Section 4 describes evaluation results for the proposed approach. Finally, Sect. 5 concludes this paper.

2 Background and Motivation

2.1 DCNNs Basics

DCNNs learn levels of representation and abstraction to make sense of data (e.g., image, text, sound) via various layers connected to each other. Figure 1 shows an overview of DCNNs which mainly consists of convolutional layers (CONV), max

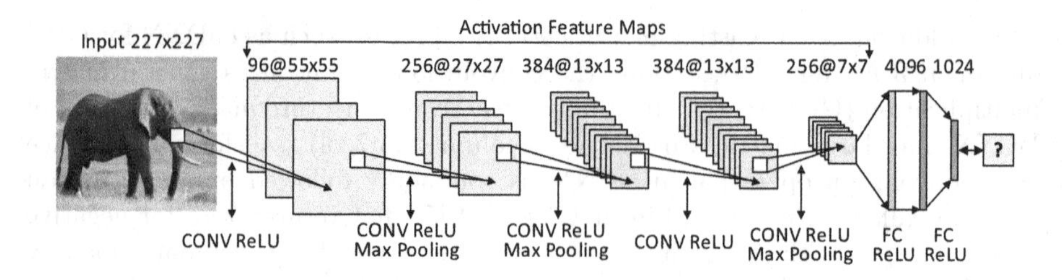

Fig. 1. An overview of DCNNs model.

pooling layers (MP) and fully connected layers (FC). The CONV layers extract features information by filters on its input (activation) feature maps. They can be defined as:

$$C_O(w,x,y) = \sum_{n=0}^{N-1} \sum_{\alpha=0}^{K_x-1} \sum_{\beta=0}^{K_y-1} i_n(x+\alpha, y+\beta) \times f^w(n,\alpha,\beta). \tag{1}$$

Where $C_O(w,x,y)$ represents an output neuron. It can be seen that the CONV layers need to do $N \times K_x \times K_y$ MAC operations to get a simple output neuron. The MP layers perform a subsampling operation on the extracted features to reduce the data dimensions and mitigate overfitting issues. The FC layers are used to make the final inference results. The computation in FC can also be defined by Eq. (1), where N is the length of each kernel and K_x and K_y are equal to one.

To make DCNNs learn the complex functional mapping between inputs and outputs, each CONV and FC layers always follow by activation functions. ReLU is the most popular activation function in the typical DCNNs, such as AlexNet, VGGNet, and ResNet. The ReLU function can be defined as:

$$ReLU(C_O) = \begin{cases} 0, & C_O <= 0 \\ C_O, & C_O > 0. \end{cases} \tag{2}$$

It returns zero for negative values and the values themselves for the positive ones. In CONV and FC layers, there are a larger number of negative outputs, indicating that a large fraction of ReLU outputs are zero. Hence, the output feature maps of this layer (that is, the input feature map of next layer) have abundant zero activation values.

2.2 DCNNs Pruning

In DCNNs, performing such intensive MAC operations with hundreds of millions of weights at CONV and FC layers requires huge amount of computation and storage resources. Weight pruning is a main stream model compression technique that reduces the redundancy in the number of weights, requiring significantly less hardware resources. As shown in Fig. 2, three main approaches of weight pruning are 1) non-structured pruning; 2) structured pruning; and 3) semi-structured pruning. Note that the weights marked gray are pruned.

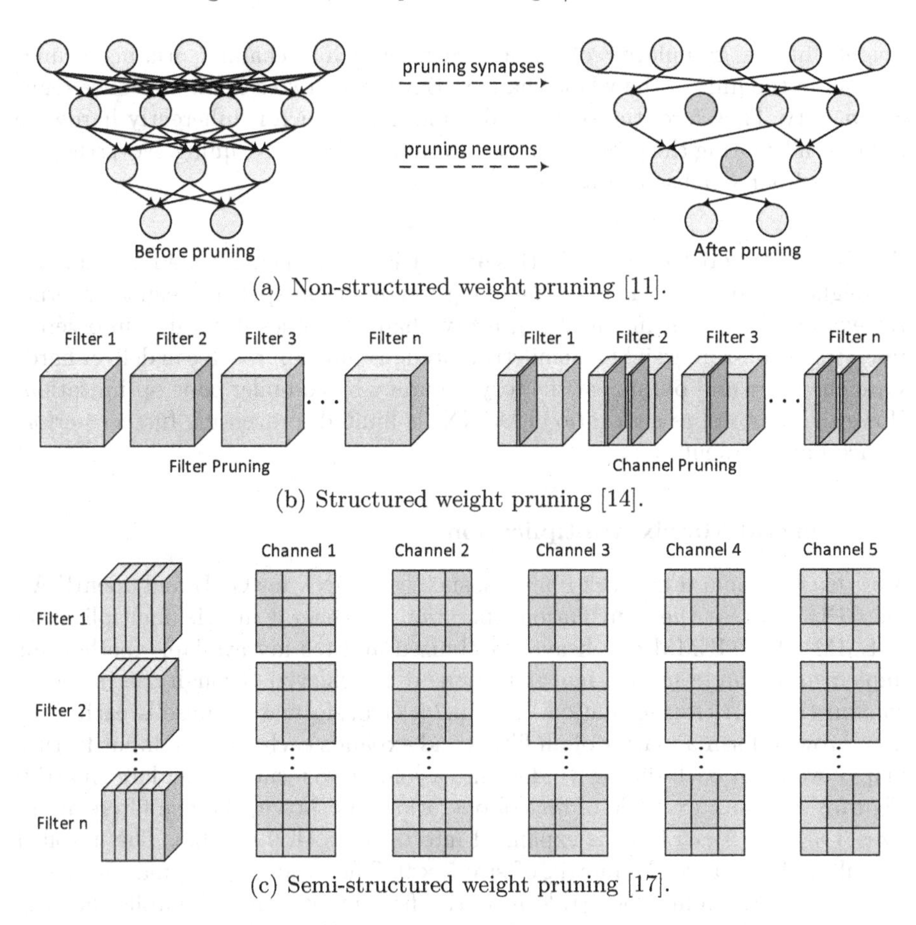

(a) Non-structured weight pruning [11].

(b) Structured weight pruning [14].

(c) Semi-structured weight pruning [17].

Fig. 2. Weight pruning schemes for DCNNs models.

Non-structured Pruning. This approach prunes the weights that are below a small threshold. Such weights can be converted to zero because their contribution to final output is negligible. However, the small weights are randomly distributed, thereby resulting in irregular sparsity. The irregular sparsity incurs several challenges for memory access and computation [17]: 1) the poor data locality require the indices for the compressed weight representation, which causes low memory performance; 2) it introduces heavy control flow instructions, which degrades instruction-level parallelism; 3) it poses poor thread-level parallelism due to the load imbalance. Hence, although the non-structured weight pruning method can result in a high compression ratio without accuracy loss, it does not translate into the performance improvement due to the hardware unfriendly.

Structured Pruning. The approach has two representative schemes: filter pruning and channel pruning, as shown in Fig. 2 (b). Filter pruning removes weights in the granularity of the entire filter, while channel pruning prunes weights in the unit of the whole channel. Both these two schemes produce regular sparsity. Therefore, the structured pruning approach is inherently hardware friendly, alleviating the problems posed by non-structured pruning. Nevertheless, it suffers from serious accuracy loss.

Semi-structured Pruning. In this approach, for each channel, a fixed number of weights are pruned, and the remaining weights form specific weight patterns. An example is shown in Fig. 2 (c), every channel prunes 4 weights into zeros, reserving 5 non-zero weights. Semi-structured pruning approach can deliver hardware efficiency and obtain satisfactory accuracy by compiler code optimization. However, the compression ratio of DCNNs is limited, preventing further performance improvements.

2.3 General Matrix Multiplication

Most existing neural network libraries, such as cuDNN for GPU and openBLAS for CPU, express the convolution operation as general matrix multiplication (GEMM) [18]. GEMM is a heavily optimized and readily available method for implementing large matrix multiplication. It is achieved through the image to column (im2col) transformation. The im2col transformation unrolls each filter into a row to form a matrix of all filters and expands each block of input feature maps covered by each sliding window into a column to form another large matrix. Figure 3 shows an example of im2col operation. As shown, the five filters are of sizes $3 \times 3 \times 1$. Each filter is expanded into a row with 9 weights. The unrolled five filters form a weight matrix of size 5×9. The input feature map is of size $4 \times 4 \times 1$. Assuming the stride is 1, the first input window unrolls into the first column of the activation matrix, and so on. Hence, the input feature map unrolls an activation matrix of size 9×4. The entire convolution operation is performed by executing one single dot product over the two large matrices using an efficient GEMM routine.

2.4 Motivation

Previous weight pruning works end up missing the opportunity to simultaneously consider accuracy, compression ratio and hardware-efficiency of DCNNs in the design space for further performance improvement. Therefore, we are motivated to propose a compression approach that satisfies the above three metrics to accelerate DCNNs.

Initially, we fully analyze the spatial distribution of weights by conducting some preliminary experiments on several DCNNs models and observe an interesting phenomenon. We observe that the position of smaller weight values between a large number of channels is close to or exactly the same. The smaller weight values, which have been demonstrated to have no impact on accuracy [11], are termed as insensitive weight values. This means spatially insensitive consistency

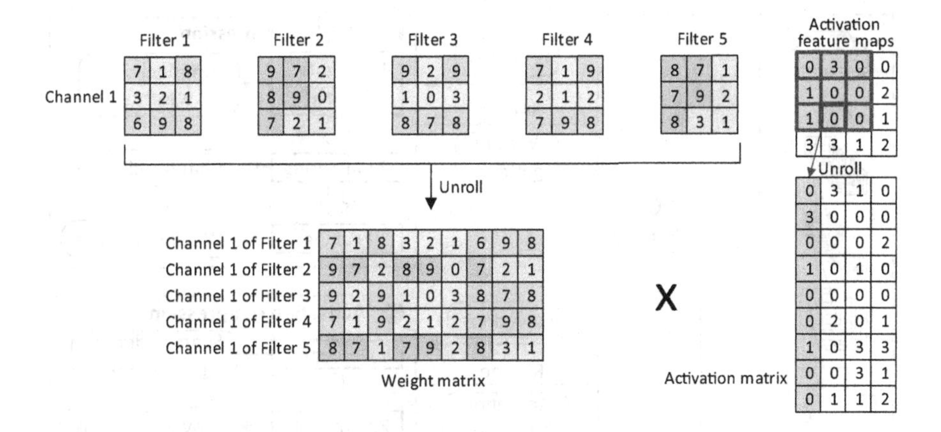

Fig. 3. An example of im2col operation.

exists in numerous channels of filters. To be specific, the insensitive weight values are distributed at the same spatial position in these channels. As shown in Fig. 3, the channel 1 of filter 1, 3, and 4 has above spatial distribution characteristic. Pruning the insensitive weight values can generate regular sparsity in a large quantity of channels, while the reserving sensitive weight values form weight patterns with uniform shapes.

Furthermore, we thoroughly study the activation matrix characteristics after im2col transformation. We find that the rows of activation matrix with different sparsity degree have different impact on the DCNNs accuracy. Specifically, the rows with larger sparsity degree do not affect the accuracy and they are termed as insensitive rows in activation matrix. In Fig. 3, rows 2, 3 and 5 of the activation matrix are insensitive. Removing the insensitive rows directly can significantly reduce the amount of computation without accuracy loss.

The sensitivity analysis for weight and activation values offers us a new opportunity to improve the performance of DCNNs. Therefore, we propose a cooperative weight/activation compression approach. The approach consists of a cluster-based weight pattern pruning technique to prune the weight channels with insensitive consistency into the same shape and a sparsity-row-based activation removal technique to eliminate the insensitive rows.

3 The Cooperative Weight/Activation Compression

In this section, we first present the workflow overview of the proposed cooperative weight/activation compression approach, then separately introduce the cluster-based weight pattern pruning technique and sparsity-row-based activation removal technique in detail.

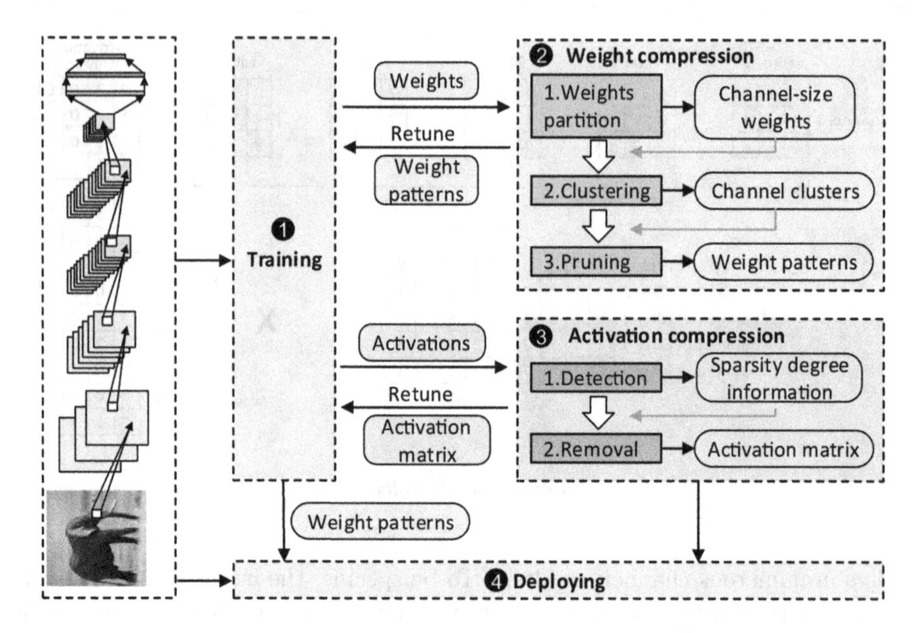

Fig. 4. An overview of workflow for the weight/activation compression approach.

3.1 An Overview of Workflow

The cooperative weight/activation compression approach workflow can be depicted in Fig. 4. It consists of four steps: Firstly, we train the DCNNs based on datasets to obtain a basic model with weight parameters(❶). Secondly, we use cluster-based weight pattern pruning technique to compress weights in the well-trained model(❷). The produced weight patterns have specified shapes. After weight compression, accuracy is significantly impacted. Hence, retraining is critical to recover the accuracy by retuning the values of the reserved weight patterns. During retraining, only the values of the weight patterns change, their specific shapes do not. Thirdly, we propose a sparsity-row-based activation removal technique for activation compression(❸). It eliminates the rows with larger sparsity degree in the activation matrix and is applied dynamically in the forward propagation of DCNNs. Finally, the lightweight DCNNs model with well-trained weight patterns is deployed on the hardware platform for inference(❹). At inference run time, we also use the sparsity-row-based activation removal technique to further improve performance. Algorithm 1 formally shows the workflow of the cooperative weight/activation compression approach.

3.2 The Cluster-Based Weight Pattern Pruning

The cluster-based weight pattern pruning technique aims to significantly prune insensitive weight values and makes the reserved sensitive weight values show several specific shapes in channel granularity. As ❷ shown in Fig. 4, its implementation goes through three phases: 1) the weights of the pre-trained DCNNs

Algorithm 1. The Workflow of Cooperative Weight/Activation Compression

1: ***Training process:***
2: Pre-training ($DCNNs, data$);
3: **for** each layer in DCNNs **do**
4: Cluster-based weight pattern pruning ($weights$);
5: **end for**
6: Retraining;
7: **for** forward propagation in DCNNs **do**
8: **for** each layer in DCNNs **do**
9: Sparsity-row-based activation removal ($activation\ matrix$);
10: **end for**
11: **end for**
12: Retraining;
13: **return** Weight patterns;
14: ***Inference process:***
15: **for** forward propagation in DCNNs **do**
16: **for** each layer in DCNNs **do**
17: Sparsity-row-based activation removal ($activation\ matrix$);
18: **end for**
19: **end for**
20: **return** Inference results

model is carved into weight groups of channel size, which we called weight channels; 2) we adopt a clustering algorithm to cluster the weight channels, thereby getting several clusters; 3) pruning the weight channels that belong to a cluster into weight patterns with uniform shape.

To enable the cluster-based weight pattern pruning technique, two challenges must be addressed as follows: 1) how to appropriately define the clustering standard of weights channels; 2) how to accurately determine the shape of weight patterns in different channel clusters. To address the first challenge, we propose that weight channels belong to a cluster if the position of sensitive weights in channels is close to or exactly same. To this end, we assign a value that represents sensitivity degree to each weight in channel according to its absolute value. The greater the absolute value of the weights is, the greater the value of sensitivity degree is. Then, we multiply the sensitivity values of the weights at the corresponding positions in the two weight channels and add the products to obtain the similarity value. The similarity value is used to measure the similarity degree of the two weight channels. The larger the similarity value is, the more consistent the position of the sensitive values of the two weight channels is, that is, the more similar the two weight channels are. To handle the second challenge, we adopt the method of preserving the average larger weight. Specifically, for the weight channels in a cluster, we calculate the average absolute weight value of the same position. The weights located in the position with the higher average value are retained, which form a uniform shape.

Figure 5 shows an example of the cluster-based weight pattern pruning technique. As shown, the channel 1 of filter 1, 3, and 4 belong to a cluster, they are

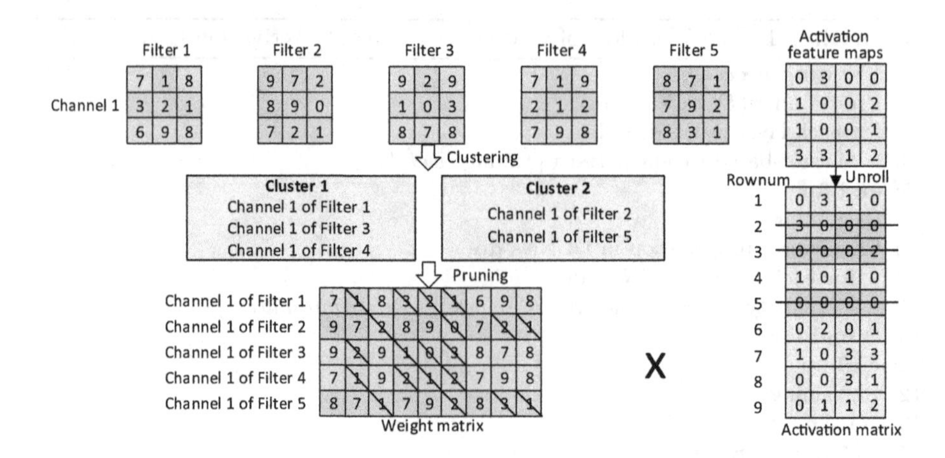

Fig. 5. An example of the weight/activation compression approach.

Algorithm 2. The Cluster-based Weight Pattern Pruning Technique

---

Input: Cluster number M, Channel set *channel_set*, Centroid set *centroid_set*, *weights*;
Output: Weight patterns;
1: ***Weight partition:***
2: **for** *weights* in each layer **do**
3: Weight partition (*weights*)
4: **end for**
5: **return** Weight channels;
6: ***Clustering:***
7: Initialize *centroid_set* as random set with M;
8: **while** *centroid_set* updated **do**
9: **for** each channel in *channel_set* **do**
10: **for** each each centroid in *centroid_set* **do**
11: Distance=sum $(channel_{sensitivity} \cdot centroid_{sensitivity})$;
12: **if** Distance< Min_distance **then**
13: *channel_set* [channel] $\subset$ *centroid_set*[centroid];
14: **end if**
15: **end for**
16: **end for**
17: **for** i from 1 to M **do**
18: *centroid_set*[i]=mean($\sum_{channel \in i} Channels$)
19: **end for**
20: **end while**
21: **return** Channel clusters;
22: ***Pruning:***
23: **for** each channels for a cluster **do**
24: Computing average weights;
25: Pruning the smaller average weight;
26: **end for**
27: **return** Weight patterns

Algorithm 3. The Sparsity-row-based Activation Removal Technique

1: **for** each layer in DCNNs **do**
2:    **for** each rows in activation matrix **do**
3:       Sparsity degree statistical function (*weights*);
4:       **if** Sparsity degree > Threshold **then**
5:          Removing the rows;
6:       **end if**
7:    **end for**
8: **end for**

pruned into weight patterns with a uniform shape. While the channel 1 of filter 2 and 5 belong to another cluster, they are pruned into a pattern of the same shape. Algorithm 2 presents the process of the weight compression technique.

3.3 The Sparsity-row-based Activation Removal

The sparsity-row-based activation removal technique can eliminate the rows of activation matrix that have larger sparsity degree. It is consisted of two steps, as ❸ shown in Fig. 4. First, we design a detector to detect the numbers of zero for each row in the activation matrix after im2col transformation. It allows us to obtain the sparsity information of rows. Then we compare all the sparsity degree of rows to the predefined threshold. The corresponding rows will be eliminated if the sparsity degree is larger than the threshold. This can be applied to all the activation matrices of each layer for activation compression. An example as shown in Fig. 5, the sparsity degree of rows 2, 3 and 5 respectively are 3/4, 3/4 and 1 in the activation matrix. They are greater than the threshold 2/4, thereby removing directly. Algorithm 3 shows the process of sparsity-row-based activation removal technique.

4 Evaluation

In this section, we show the experimental setup and present the results and analysis of the proposed approach.

4.1 Experimental Setup

We conduct the experiments on four typical DCNNs models (i.e. LeNet5, AlexNet, VGG16, ResNet18). They are trained based on MNIST [19] and CIFAR-10 [20] datasets using PyTorch framework. We first train the original models as baseline, and then respectively implement the cluster-based weight pattern pruning technique (Cluster-WP-Prune), sparsity-row-based activation removal technique (ACT-Remove), the cooperative weight/activation compression method (ACT-WP) and random non-structured weight pruning (Ran-Prune). We show the comparison results in terms of accuracy, parameters compression ratio and the reduction of multiply-and-accumulation (MAC) operations.

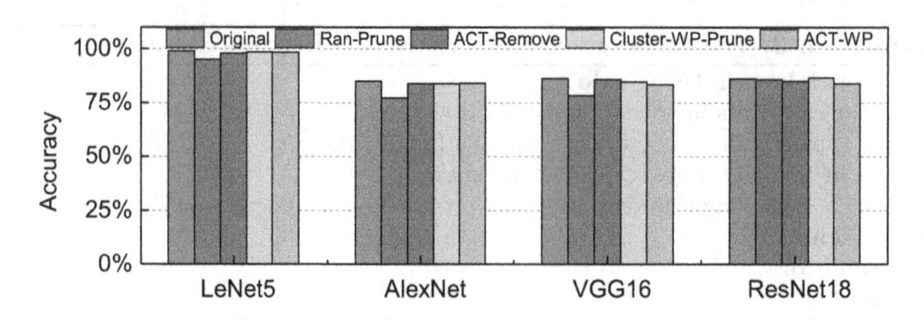

Fig. 6. Accuracy.

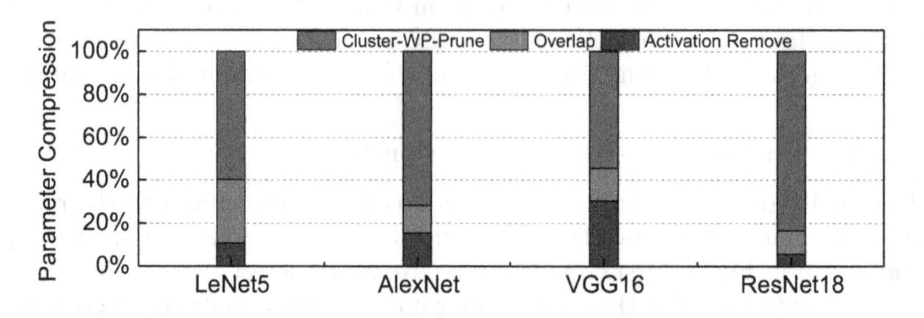

Fig. 7. Compression ratio.

4.2 Experimental Results

Accuracy and Compression Ratio. Figure 6 shows the top-1 accuracy of four models using these methods. It can be seen that cluster-based weight pattern pruning gains better accuracy than random-unstructured weight pruning at the same pruning ratio. The accuracy improvement mainly comes from the exploration of inner relations between different filter channels. Meanwhile, compared with the original model, the accuracy loss of our methods is negligible with greatly parameters compression. Figure 7 shows the proportion of parameter reduction brought by weight pruning and activation removal in convolution layers. There is an overlap part in reduction, where both weight pruning and activation removal can eliminate redundant parameters. It can be seen that the cluster-based weight pattern pruning method reduces more parameters than the activation removal technique.

Computation Reduction. We count the number of MAC operations for DCNNs after using our approach and show the normalized results in Fig. 8. As shown, both cluster-based weight pattern pruning and sparsity-row-based activation removal can reduce MAC operations. Our weight pruning technique shows better results than activation removal, which brings 50% reduction of MAC on average. The combination of the two techniques contributes more to computation reduction. The cooperative weight/activation compression brings

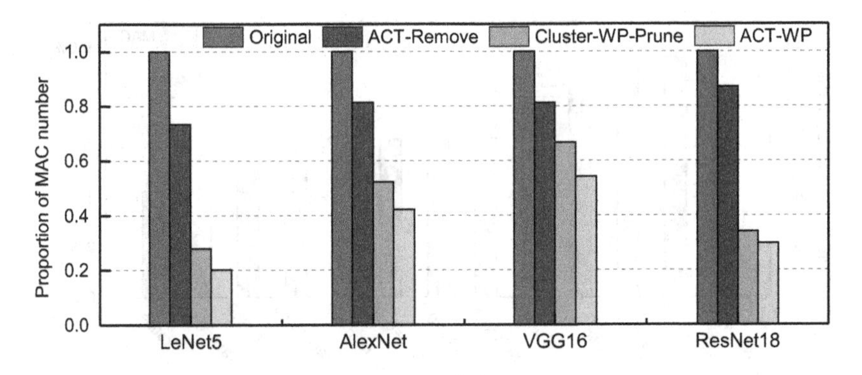

Fig. 8. The reduction of MAC operations.

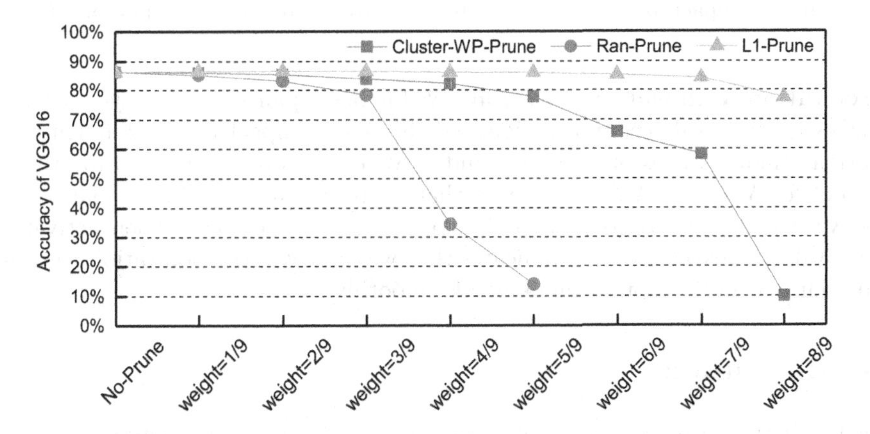

Fig. 9. The comparison of three weight pruning methods on VGG16.

about 10% reduction than only using cluster-based weight pattern pruning. It has only 37% of the MAC operations of original DCNNs on average. In addition, our cooperative approach is more efficient for DCNNs with deeper layers. It removes 70% MAC computation in ResNet-18, better than AlexNet and VGG16.

Weight Pruning Comparison. In this part, we compare our cluster-based weight pattern pruning technique with normal L1-non-structured [11] and random-non-structured weight pruning methods on VGG16. We set different pruning ratios for convolution layers and fixed ratio for linear layers to learn how accuracy varies. The results are shown in Fig. 9. For cluster-based weight pattern pruning, accuracy decreases slightly before pruning almost half of the weights. While exceeding 5/9 percent of pruning ratio, accuracy drops down suddenly. At the same pruning ratio, our cluster-based weight pattern pruning obtains higher accuracy than random-non-structured weight pruning. Although L1-non-structured weight pruning gains the best accuracy among three pruning methods, it is not hardware friendly, resulting in impeding acceleration.

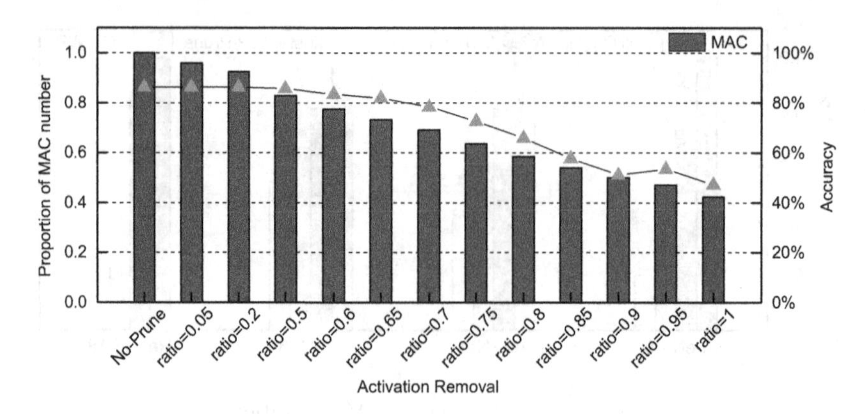

Fig. 10. The impact of the sparsity-row-based activation removal technique on VGG16.

Activation Removal Exploration. We further explore the sparsity-row-based activation removal technique. Figure 10 shows the impact of the different activation removal ratios on accuracy and MAC operations normalized to original model for VGG16. We can observe that there is almost a linear relationship between activation ratio and MAC operations (or accuracy) when activation removal ratio exceeds 0.5. It indicates that we can set activation ratio according to accuracy and computation demands smoothly.

5 Conclusion

In this paper, we propose a cooperative weight/activation compression approach, which consists of a cluster-based weight pattern pruning technique and a sparsity-row-based activation removal technique. The cluster-based weight pattern pruning technique prunes weight channels in a cluster into weight patterns with a uniform shape. The sparsity-row-based activation removal technique dynamically eliminates insensitive rows of activation matrices. The two proposed techniques achieve better compression ratio with negligible accuracy loss and allow the hardware to execute with high parallelism. Evaluation results show that the compression approach reduces the computation by 63% on average.

Acknowledgements. This research is supported by the grants from the National Science Foundation for Young Scientists of China (Grant No.61902218), the National Natural Science Foundation of China (Grant No. 92064008).

References

1. Krizhevsky, A., Sutskever, I., Hinton, G.E.: ImageNet classification with deep convolutional neural networks. In: Advances in Neural Information Processing Systems (NeurIPS) (2012)

2. Redmon, J., Divvala, S., Girshick, R., Farhadi, A.: You only look once: unified, real-time object detection. In: IEEE Conference on Computer Vision and Pattern Recognition (CVPR) (2016)
3. Amodei, D., et al.: Deep speech 2: end-to-end speech recognition in English and Mandarin. In: The International Conference on Machine Learning (ICML) (2016)
4. Jouppi, N.P., et al.: In-datacenter performance analysis of a tensor processing unit. In: Proceedings of the 44th Annual International Symposium on Computer Architecture (ISCA) (2017)
5. Akhlaghi, V., Yazdanbakhsh, A., Samadi, K., Gupta, R.K., Esmaeilzadeh, H.: SnaPEA: predictive early activation for reducing computation in deep convolutional neural networks. In: ACM/IEEE 45th Annual International Symposium on Computer Architecture (ISCA) (2018)
6. Simonyan, K., Zisserman, A.: Very deep convolutional networks for large-scale image recognition. In: International Conference on Learning Representations (ICLR) (2015)
7. Deng, J., Dong, W., Socher, R., Li, L.-J., Li, K., Fei-Fei, L.: ImageNet: a large-scale hierarchical image database. In: IEEE Conference on Computer Vision and Pattern Recognition (CVPR) (2009)
8. Wu, Y., Wang, Z., Shi, Y., Hu, J.: Enabling on-device CNN training by self-supervised instance filtering and error map pruning. IEEE Trans. Comput. Aided Des. Integr. Circ. Syst. **39**(11) (2020)
9. Bateni, S., Liu, C.: NeuOS: a latency-predictable multi-dimensional optimization framework for DNN-driven autonomous systems. In: 2020 USENIX Annual Technical Conference (USENIX ATC) (2020)
10. Ma, X., et al.: PCONV: the missing but desirable sparsity in DNN weight pruning for real-time execution on mobile devices. In: Proceedings of the AAAI Conference on Artificial Intelligence (AAAI) (2020)
11. Han, S., Pool, J., Tran, J., Dally, W.: Learning both weights and connections for efficient neural network. In: NeurIPS (2015)
12. Han, S., Mao, H., Dally, W.J.: Deep compression: compressing deep neural networks with pruning, trained quantization and Huffman coding. arXiv preprint arXiv:1510.00149 (2015)
13. He, Y., Zhang, X., Sun, J.: Channel pruning for accelerating very deep neural networks. In: Proceedings of the IEEE International Conference on Computer Vision (ICCV) (2017)
14. Wen, W., Wu, C., Wang, Y., Chen, Y., Li, H.: Learning structured sparsity in deep neural networks. In: Advances in Neural Information Processing Systems (NeurIPS) (2016)
15. Yazdani, R., Riera, M., Arnau, J.-M., González, A.: The dark side of DNN pruning. In: 2018 ACM/IEEE 45th Annual International Symposium on Computer Architecture (ISCA) (2018)
16. Li, H., Kadav, A., Durdanovic, I., Samet, H., Graf, H.P.: Pruning filters for efficient convnets. arXiv preprint arXiv:1608.08710 (2016)
17. Niu, W.: PatDNN: achieving real-time DNN execution on mobile devices with pattern-based weight pruning. In: Proceedings of the Twenty-Fifth International Conference on Architectural Support for Programming Languages and Operating Systems (ASPLOS) (2020)
18. Mogers, N., Radu, V., Li, L., Turner, J., O'Boyle, M., Dubach, C.: Automatic generation of specialized direct convolutions for mobile GPUs. In: Proceedings of the 13th Annual Workshop on General Purpose Processing using Graphics Processing Unit (GPGPU) (2020)

19. LeCun, Y., Bottou, L., Bengio, Y., Haffner, P.: Gradient-based learning applied to document recognition. Proc. IEEE **86**(11), 2278–2324 (1998)
20. Krizhevsky, A., Hinton, G.: Learning multiple layers of features from tiny images. Technical report, Citeseer (2009)

PFA: Performance and Fairness-Aware LLC Partitioning Method

Donghua Li[1], Lin Wang[1], Tianyuan Huang[1], Xiaomin Zhu[2], and Shichao Geng[1(✉)]

[1] Shandong Normal University, Jinan, China
[2] Shandong Institute of Big Data, Jinan, China
zhuxiaomin@sdibd.cn

Abstract. In server for cluster, the number of running programs is increasing. The benefit of consolidating multiple programs is good for server utilization, but it also leads to the program's performance degradation. Severe performance degradation can result in significant losses. Therefore, it is essential to divide the shared resource to support consolidation. Our experiment showed that even some shared resources such as CPU cores and memory had been divided, the performance of programs still drop down significantly compared to the program running alone, then we found out the primary reason was the contention for LLC. In this paper, we proposed the LLC partitioning method to improve the performance for consolidation programs. We classify the LLC usage type of the program by analyzing the LLC behavior, then allocate reasonable LLC ways according to the LLC usage type. Meanwhile, we monitor the program's performance in real-time and allocate the LLC ways dynamically. The experiment found that compared with the default LLC allocation method, our method reduced the performance loss by an average of 6.73% and improved the fairness by 0.03. Compared with the CPA method, our method reduced the performance loss by an average of 4.86%.

Keywords: Co-scheduling · LLC partitioning · LLC management · Program's performance

1 Introduction

To make full use of the server resources, consolidating multiple programs is the common method. In previous works, in order to improve the performance of the program, it was common to run a high-priority program with multiple low-priority programs together [1]. This method sacrifices the performance of the low-priority programs to guarantee the performance of the high-priority programs. However, in many cases, we believe that the co-running programs are equally important, rather than just guaranteeing the performance of high-priority programs.

D. Li and L. Wang—Contributed equally to this work.

Y. Lai et al. (Eds.): ICA3PP 2021, LNCS 13157, pp. 707–721, 2022.
https://doi.org/10.1007/978-3-030-95391-1_44

Consolidating multiple programs can improve server utilization. Meanwhile, contention for shared resources can also lead to unpredictable performance degradation [2]. To improve the performance of programs, container technology is often used to isolate shared resources to reduce resource contention between programs. As is shown in Fig. 1, in order to reduce the contention of shared resources, we divided the CPU cores and memory by Cgroups [3]. Different programs were divided into separate groups to reduce the CPU and memory contention. But the last level cache is still one of the important shared resources of the program. We tested the baseline performance of programs after dividing the core and memory resources and compared it with the performance when the programs consolidated together, as is shown in Fig. 2. Compared to the programs executing alone, milc's performance dropped by 5.2%, facesim's performance dropped by 5.8%, and Omnetpp's performance dropped by 32%. Therefore, we turn our attention to the problem of LLC contention for shared resources between programs.

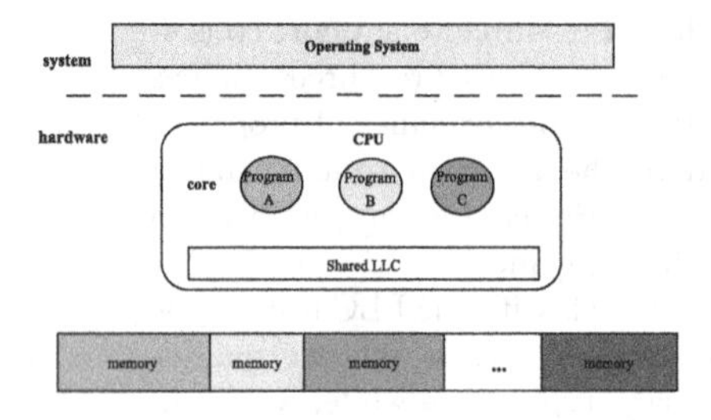

Fig. 1. Multiple programs compete for LLC.

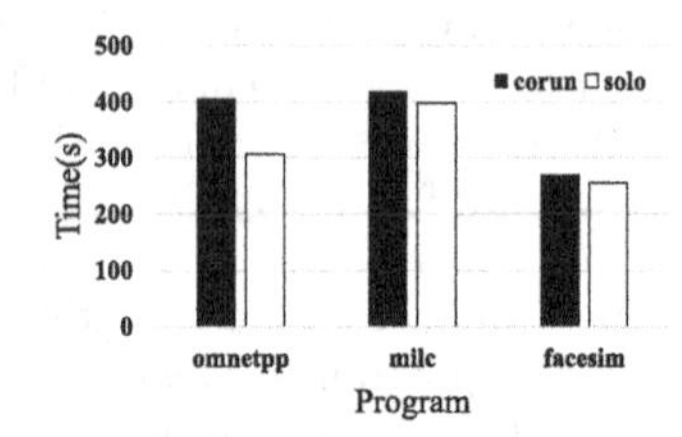

Fig. 2. Performance comparison of programs running alone and programs running together.

Intel provides Resource Director Technology (RDT) [4] for monitoring and managing the LLC resource. However, to make reasonable use of this technology on a server, it is necessary to have the information of the exact LLC space required by the program. The amount of LLC space occupied by a program depends on the behavior of the program and its co-runner. If the program is allocated less LLC space than it actually needs, it will suffer a large number of LLC misses. Meanwhile, a big LLC allocation will lead to poor cache utilization, which will result in insufficient LLC occupied by other programs. Our work aims to use RDT to mitigate the impact of inter-program cache contention and improve the performance of each program.

In this paper, an LLC partitioning method is proposed: PFA, which can improve the performance of the consolidated programs. PFA is implemented by: 1) precisely analyzing the LLC behavior of the programs and classifying different types of LLC usage according to the LLC behavior of the programs; 2) guiding LLC partitioning for multiple co-located programs based on the type of LLC usage; 3) dynamic adjusting the LLC resources at runtime to improve the performance and fairness for multiple co-located programs. Our approach achieves maximum utilization of LLC resources while improving the performance of programs.

The rest of this article is organized as follows: Sect. 2 gives the motivation, Sect. 3 discusses the method, and Sect. 4 presents the experimental methodology, Sect. 5 evaluates the experimental results, Sect. 6 discusses the related work, and we summarize our work and propose future research in Sect. 7.

2 Motivation

Commonly multi-core processors provide the shared last-level cache (LLC), which can speed up the processor's load and store. When multiple programs compete for the shared LLC, it is likely to result in the program's performance degradation due to un-reasonable LLC allocation. For example, if a program does not perform better due to more LLC space, it will cause other consolidated programs to perform worse due to poor LLC. Recently Intel provided LLC resource monitoring and partitioning technology RDT for the Xeon processors, RDT can avoid LLC contention by assigning non-overlapping CLOS (Classes of Service) to each program. It can effectively mitigate performance degradation due to shared LLC resources.

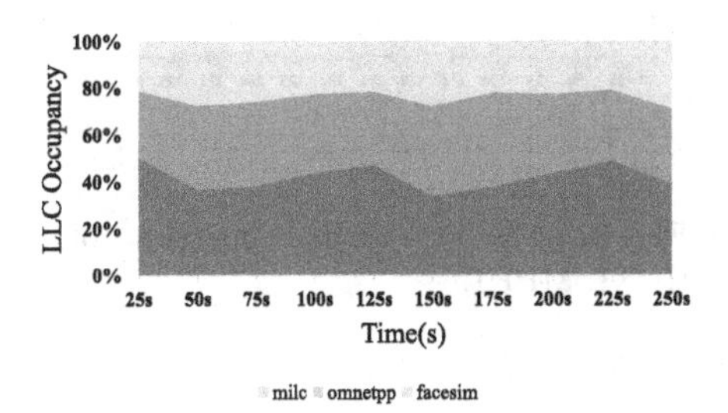

Fig. 3. Cache occupancy when omnetpp, milc and facesim are running together.

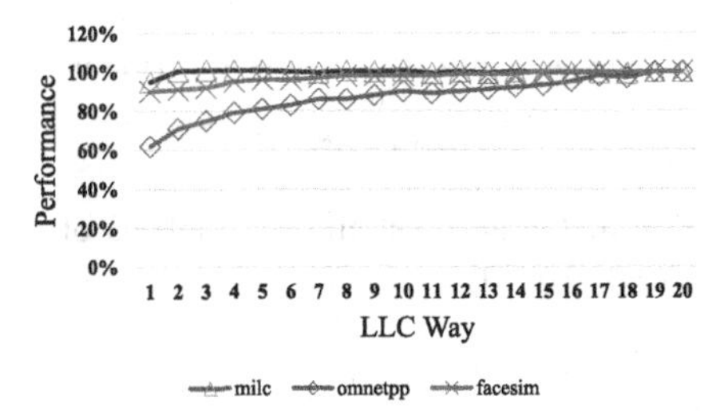

Fig. 4. Performance of programs omnetpp, milc, facesim with different LLC space.

We have realized the LLC monitor and LLC partition for co-running programs using CMT (Cache Monitoring Technology) and CAT (Cache Allocation Technology). The server used in the experiment has 20 ways of LLC with a total size of 15 MB, and the capacity of each LLC way is 0.75 MB. As is shown in Fig. 3, When the programs omnetpp, facesim and milc are running together, it can be seen that facesim and omnetpp take up about 20% of the LLC size separately, while milc takes up almost half of the LLC size.

Then we tested the performance of the programs with different LLC spaces. In this test, different programs are isolated by Cgroups on different cores. As shown in Fig. 4, omnetpp is the most sensitive program to LLC, and the performance of this program is improving with the LLC space increases. When allocating 4-way LLC for the program, it takes 20% more time to complete than allocating the entire LLC (baseline). Milc has achieved the optimal performance using only 2-way LLC. However, when multiple programs run together, milc takes up half of the LLC, which causes less LLC space of the LLC-sensitive program omnetpp. As a result, omnetpp's performance drops down significantly.

Therefore, we need to profile the LLC usage characteristics of the programs and allocate reasonable LLC resources according to the LLC usage characteristics. This method can mitigate unreasonable resource allocation due to resource contention and improve the performance of co-running programs. However, different programs have different sensitivity to LLC, and it is a challenge to identify the characteristics of the LLC usage and how much space should be allocated for the program. Moreover, the performance of the program, as well as the utilization of the whole system, should be considered.

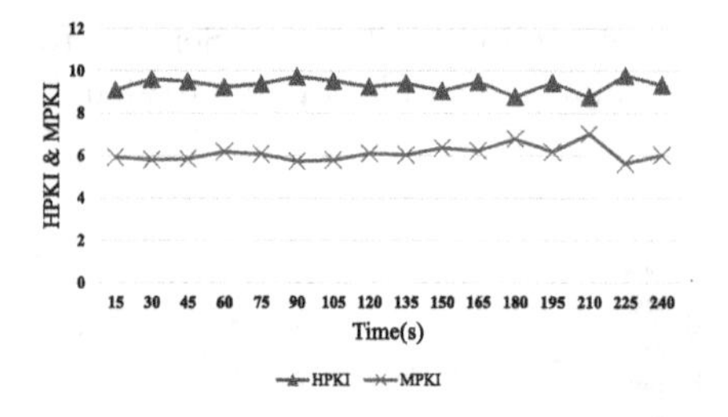

Fig. 5. Omnetpp's runtime metrics of HPKI, MPKI.

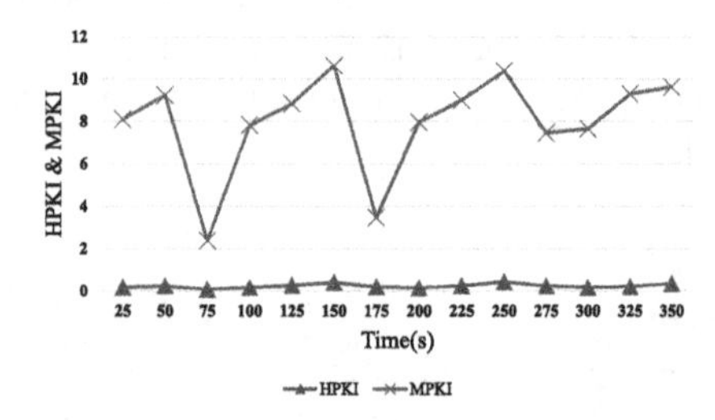

Fig. 6. Milc's runtime metrics of HPKI, MPKI.

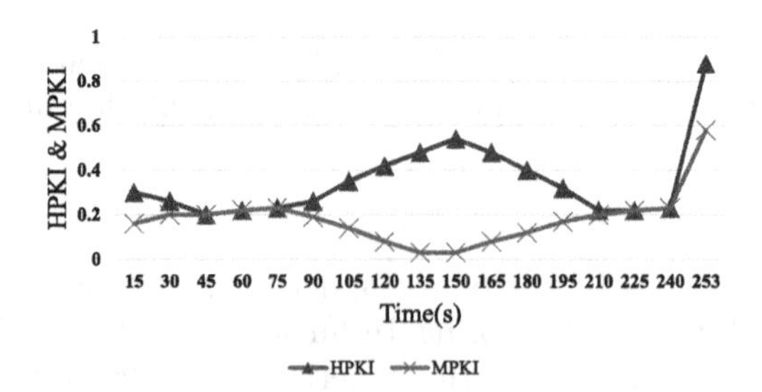

Fig. 7. Facesim's runtime metrics of HPKI, MPKI.

3 Method

3.1 Overview

In this work, a performance and fairness-aware LLC partitioning method (PFA) is proposed. The PFA method consists of three functional modules, as is shown in Algorithm 1. In Module 1, PFA calculates the program's LLC performance metrics of HPKI, MPKI. In Module 2, FPA classifies the program's LLC behavior and then predicts the program's LLC demand based on the LLC usage type. To improve the performance of programs and avoid severe performance degradation, in module 3, PFA monitors the performance of co-running programs in real-time and dynamically adjusts LLC resources based on the performance metric *slack*. The details of PFA will be described in Sect. 3.2.

Algorithm 1. The PFA function.

1: **Module 1: Obtain the LLC usage characteristics of the program**
2: For all programs do: Read instruction, mem_load_uops_retired.l3_hit, mem_load_uops_retired.l3_miss events
3: Compute metrics HPKI, MPKI
4: **Module 2: Allocate LLC space based on LLC usage characteristics**
5: Measure HPKI, MPKI
6: **if** LLC-sensitive program
7:     allocate 8 ways LLC
8: **else if** LLC-insensitive program
9:     allocate 2 ways LLC
10: **else if** LLC-aggressive program
11:     allocate 2 ways LLC
12: **else**
13:     allocate 6 ways LLC
14: **Module 3: Adjust LLC space according to program's performance**
15: **for** each program **do**
16:     slack = $(T_{corun} - T_{solo})/T_{solo}$
17: **end**
18: **if** slack <0.1 && LLC way >2 **then**
19:     downsize(); //decrease LLC space
20: **else if** slack >0.3 **then**
21:     upsize(); //increase LLC space
22: **else** all program's slack >0.3
23:     reduce the number of programs
24: **end**

3.2 Detail

Program Characterization. In order to analyze the LLC behavior of programs, PFA uses HPKI (LLC hits per kilo-instruction) and MPKI (LLC misses per kilo-instruction) as the evaluation metrics, and each program is executed alone using all LLC. Figures 5, 6 and 7 respectively describe the runtime indicators HPKI and MPKI when omnetpp, milc, and facesim are executed separately. The HPKI and MPKI of omnetpp are large during the runtime, and the MPKI is smaller compared to the HPKI, but the minimum value is almost equal to 6. The HPKI of milc is particularly small, almost equal to 0, and the maximum value does not exceed 0.5, but the MPKI is large, with a maximum value equal to about 11. The HPKI and MPKI of facesim are small and are less than 1, fluctuating between 0.1 and 0.5. By combining with Fig. 4, we found that the HPKI of the cache-sensitive omnetpp is much larger than that of the cache-insensitive facesim, while milc, which has a very low cache utilization, has a very small HPKI and a large MPKI.

Table 1. Classification of program LLC usage types.

LLC usage type	HPKI	MPKI	Program
LLC-aggressive	<0.5	>1.5	milc streamcluster
LLC-sensitive	>1	>0.5	mcf omnetpp xalancbmk canneal GemsFDTD
LLC-insensitive	≤ 1	≤ 1	bwave zeusmp lbm libquantum named povray sjeng tonto blackscholes bodytrack facesim ferret fluidanimate freqmine swaption vips x264 raytrace
LLC-medium	else	else	sphinx3 leslie3d x264 zeusmp

We further studied SPEC CPU2006 [5] and Parsec3.0 [6] and found some commonalities. The HPKI and MPKI values of LLC-sensitive programs are large, the HPKI and MPKI values of LLC-insensitive programs are small, the HPKI of LLC-underutilized programs is small, but the MPKI is large, and the HPKI and MPKI of LLC weakly-sensitive programs do not belong to the above three cases. By dividing the HPKI and MPKI thresholds, we classified the LLC usage types of programs, and defined programs with HPKI >1 and MPKI >0.5 as LLC-sensitive programs, as is shown in Table 1. Programs with HPKI ≤ 1 and MPKI ≤ 1 are defined as LLC-insensitive type. Programs with HPKI <0.5 and MPKI >1.5 are defined as LLC aggression type. Previous methods [7,8] used miss ratio (the number of misses for a given number of instructions) to indicate the LLC sensitivity of a program. But this is not accurate, just like milc, which has a large MPKI, but the performance does not increase with the LLC space. We used two metric methods, HPKI and MPKI, to effectively identify this class of programs and defined programs that do not fall within the above three value ranges as LLC-medium program.

Resource Allocation. How to allocate LLC resources appropriately is a challenge in managing LLC. If a program is allocated less LLC space than it actually needs, it will suffer from abundant LLC misses; and a relatively larger LLC allocation will result in poor LLC utilization since it is not possible to estimate the LLC requirements of the program in advance. The LLC usage type of the program provides help in estimating the LLC requirements of the program.

LLC-aggressive programs have a small HPKI, and a large MPKI, which indicates the LLC reusability of such programs is poor. Occupying a large LLC space does little help for improving the program's performance. For such programs, allocating 10% of the LLC space, i.e., 2-way LLC, can almost achieve the baseline performance (for the case of 15M LLC space and CLOS equal to 20). LLC-sensitive programs will gradually improve the program's performance as the LLC space increases. When the allocated LLC space is reduced from 8-way to 2-way, its performance decreases by 21.2% on average, and some programs even decrease by 45%. Allocating 40% of the LLC space (8-way) can achieve 85% of the baseline performance. For such programs, we allocate 8-way LLC space. LLC-insensitive programs are insensitive to LLC capacity, and it's HPKI and MPKI are so small that the performance degradation is less than 1% when the allocated LLC space is reduced from 14 to 2. Therefore, we allocate 2-way LLC for such programs. For LLC-medium programs, allocating 6-way LLC can achieve 90% baseline performance, which is less sensitive than LLC-sensitive programs but sensitive than LLC-insensitive programs. For this kind of program, 30% of the LLC space is allocated, i.e., 6-way LLC. The initial allocation of LLC is based on the sensitivity of the program to LLC. PFA considers resource utilization while ensuring the performance of the program.

Resource Adjustment. Since there has other shared resources contention, such as memory bandwidth contention between programs, which can affect the performance of programs running together, the initial allocation of LLC may not be quite perfect. In order to further ensure the program's performance and fairness, it is necessary to adjust LLC resources in real-time. PFA dynamically adjusts the LLC space of consolidated programs according to the *slack* metric (Eq. 1). T_{corun} represents the runtime of the program when multiple programs are running together. T_{solo} represents the runtime of the program when it is executed individually. We use the *slack* metric to represent the performance degradation of the program.

$$slack = \frac{T_{corun} - T_{solo}}{T_{solo}} \tag{1}$$

To improve the performance of the programs while considering fairness, PFA has handled the possible cases accordingly. The specific steps are as follows:

Step1: If there is a program with *slack* >0.3, it means the program has serious performance violation due to lack of LLC space, and the LLC space of the program needs to be increased. We take the following measures. First, PFA monitors whether there are any unallocated LLC resources, and if it is, the adapter increases the LLC space occupied by the current program and re-executes step1. If it is not, the LLC space occupied by other programs with *slack*

<0.3 and occupying space greater than 2-way is reduced, and the reduced space is allocated to the program with *slack* >0.3. If the *slack* metrics of the program is still less than 0.3 and the LLC way is greater than 2 after reducing the LLC space, LLC adapter continues to reduce the LLC space occupied by the program and increases the LLC space of the program with *slack* >0.3, until the program's *slack* metric less than 0.3. If the *slack* of the program after reducing the LLC space is greater than 0.3, it means that the operation has seriously affected the performance of the program, and the LLC modification operation in the previous step is restored.

Step2: If the *slack* metrics of programs are all less than 0.3, PFA first determines whether there is still unused LLC space. If it is, allocate the LLC space to programs with large *slack*.

Step3: If there are two programs whose *slack* metrics is greater than 0.3, LLC adapter allocates the free LLC way to the program according to the priority order of LLC-sensitive, LLC-medium, LLC-insensitive, and LLC-aggressive. If the *slack* metrics is still greater than 0.3 after resource adjustment, the number of programs should be reduced. PFA ensures as much performance and fairness as possible for each program.

4 Experimental Methodology

4.1 Experiment Platform

In our experiment, we evaluated PFA on a server with the processor of Intel Xeon E5-1650 and with the LLC capacity of 15M. The server run with the OS of Ubuntu 18.04, and the kernel is 4.15.0. Table 2 has more detailed parameters of the server.

Table 2. Experimental testbed configuration.

Component	Description
Processor	Intel(R) Xeon(R) CPU E5-1650 v4 @ 3.60 GHz, 6 physical cores
L1 cache	Private, 32 KB, L1D/L1I
L2 cache	Private, 256K
LLC	Shared, 15M, 20ways, set-associative
OS	Ubuntu 18.04 with kernel 4.15
Memory	62 GB, DDR4

4.2 Design and Composition

As is shown in Fig. 8, PFA consists of three parts: 1) LLC behavior monitor; 2) LLC feature classifier; 3) LLC adapter. The LLC behavior monitor uses Perf hardware counters to collect runtime characteristics such as instructions, LLC hits and LLC misses. LLC feature classifier generates LLC allocation schemes based on runtime characteristics, and LLC adapter adjusts LLC space of the program in real-time.

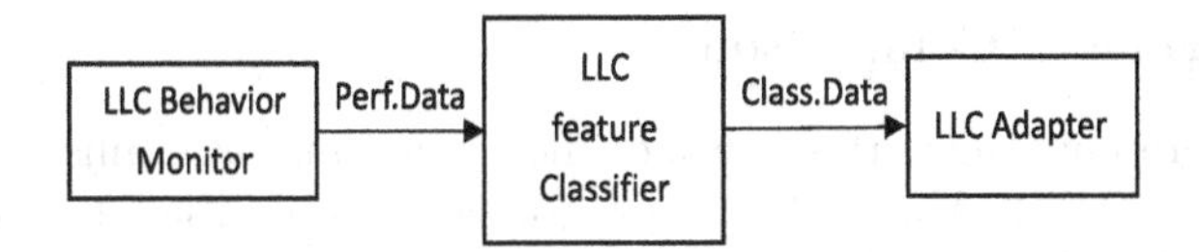

Fig. 8. The main composition structure of PFA.

4.3 Resource Isolation

To mitigate performance interference between programs, we utilize the following resource isolation mechanisms: 1) Core/thread isolation: Use Linux OS Cgroups to bind different programs to different physical cores to reduce the interference caused by thread contention. 2) Memory isolation: Use the Linux operating system's Cgroups to divide different programs into different memory spaces. 3) LLC isolation: Different types of programs are separated under different CLOS to mitigate the impact of LLC contention, as is shown in Fig. 9.

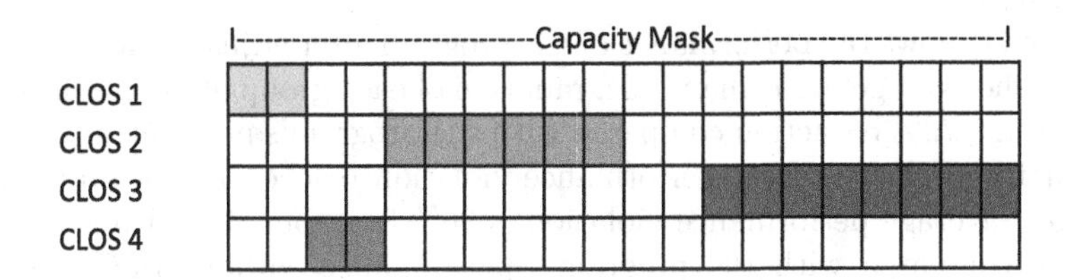

Fig. 9. Different programs isolated in different CLOS.

4.4 Index Calculation Method

We use Linux Perf hardware counters to obtain the following events: instructions, mem_load_uops_retired.l3_hit, mem_load_uops_retired.l3_miss.

Mem_load_uops_retired.l3_hit: Counts retired load uops which data sources were data hits in the last-level (L3) cache. Mem_load_uops_retired.l3_miss: Count misses in last-level (L3) cache. The API provided by pqos library can monitor the cache usage size of the program in real-time.

5 Evaluation

5.1 Benchmark

To address the challenges and opportunities of rapid change, we selected 27 programs from Parsec [6], and SPEC CPU2006 [5] with execution time constraints. SPEC CPU2006 uses the "ref" input set, and PARSEC uses the "native" input set. In all experiments, each program is isolated in a separate core and uses the memory space required for a separate runtime.

5.2 Performance Comparison

Figure 10 represents the performance of the co-running programs when the LLC is un-managed (UM). About 60% of programs get more than 10% performance degradation than baseline. It is particularly severe for LLC-sensitive programs, the LLC-sensitive programs get 44% performance degradation on average, and some program combinations get 80% performance degradation. On average, the LLC-medium programs get 13% performance degradation, and LLC-aggressive programs get 14% performance degradation. There are 7 groups of programs with performance loss greater than 30%. Specifically, in group1 and group5, the LLC-aggressive program milc takes up much more cache space than it actually needs, resulting in a 25% and 32% performance degradation of the LLC-sensitive program omnetpp due to LLC contention. Similarly, in group2, the performance of the LLC-sensitive program canneal degrades by 42% compared to baseline. It becomes very important to identify the LLC-aggressive programs and regulate the LLC occupation of such programs. Therefore, to implement multiple programs running together, the performance loss due to cache contention needs to be mitigated.

Figure 11 shows the performance of the program after adjustment by the PFA method. The average performance degradation of each group of programs is 17%, which has a 6.73% reduction compared with the program's performance without LLC partition. The smallest performance violation reaches 4% for group1, and the largest average performance violation is 36% for group 4, which gets a 13% reduction compared with the program's performance without LLC partition. The performance violations of the programs in Group 9 are all less than 5%, which is almost equal to the baseline performance of the program. In group1, the LLC-aggressive program milc has achieved a partial performance improvement, rather than a performance decrease after the LLC space is reduced. This is due to the fact that milc has very little LLC hit rate, and the improvement in LLC hits all come from the co-running program omnetpp. This part of LLC hits reduces the pressure on memory bandwidth, thereby reducing the average queue time for accessing memory. The reduction in queuing time reduces the latency of milc accessing memory, so the performance of milc is improved. Similarly, PFA improved the performance of the LLC-sensitive program in groups 2, 5, 7, and 10 by an average of 13.6% compared to UM. In the UM case, the performance of LLC-sensitive programs in group3 group6 and group8 decrease by 48%, 25% and 59%; respectively, it is because LLC-medium programs with medium LLC utilization take up much LLC space. In PFA, we slightly reduce the LLC space occupied by LLC-medium and allocate it to LLC-sensitive programs. After tuning, the performance of LLC-sensitive programs improved by an average of 21%, while the performance of LLC-medium programs decreased by an average of less than 1%. For group9, PFA identifies the LLC-aggressive program milc, reduces the cache space occupied by the milc, and allocates it to the relatively severe performance violation of sphinx3. In group4, the bandwidth of the other two programs is underutilized due to heavy bandwidth contention by program bwave. As a result, even if each program is assigned the proper amount

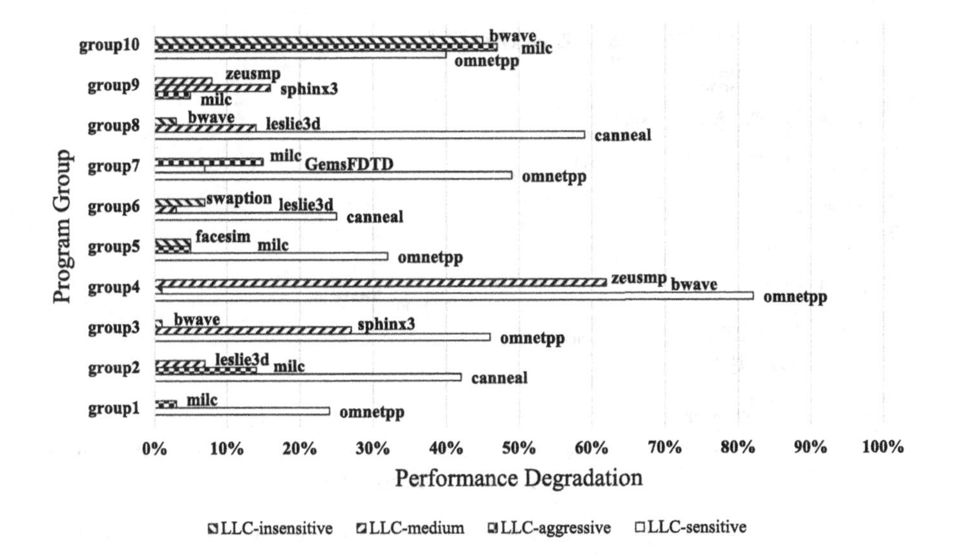

Fig. 10. Program's performance in the UM case.

of cache space, the benefits from cache are reduced due to the long queueing time for accessing memory. This issue will be further investigated in our future work. Through experiments, we found that only two groups of programs have a performance loss of more than 30%, and the PFA method guaranteed 70% of the baseline performance of eight groups of programs.

We also compared our method to Lucia's CPA approach [9], which focused on reducing the turnaround time of multiple programs running together while ignoring the performance violations of each program. With the CPA approach, the average performance of the program decreased by 22%, which reduced the performance loss by 2% compared to UM, but the number of groups with performance loss greater than 30% is still five groups, as shown in Fig. 12. The reason is that the idea of CPA partial sharing, although it can improve LLC utilization, can destroy the performance of the program. In the cases of groups 4 and 7, the LLC-medium program shared some of the space of the LLC-sensitive program, and multiple LLC-sensitive programs shared the same cache space. This arrangement does not maximize the performance of the program.

5.3 Fairness Comparison

We focus on the performance loss of each program running on a server rather than just optimizing the performance of one of the programs. Therefore, we also consider the issue of performance fairness among programs. In previous work, Kenzo et al. [10] defined the unfairness of programs using Eq. 2, μ represent the average degree of consolidated program's performance degradation and σ represent the standard deviation of program's performance degradation.

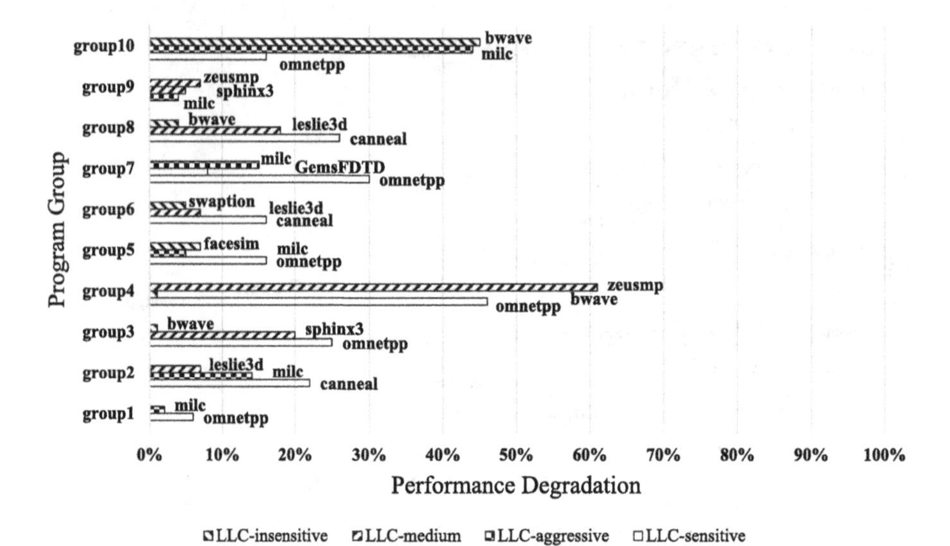

Fig. 11. Program's performance in the PFA case.

$$unfairness = \frac{\mu}{\sigma} \quad (2)$$

As is shown in Fig. 13, our method performed better in almost every group. Compared to UM, the unfairness of the program in group2 is reduced by 0.4. Compared to the CPA method, the unfairness of the program in group5 is reduced by nearly 0.4. In group 10, the unfairness of our method is not good. The reason is that PFA sufficiently improves the performance of omnetpp by an average of 17% compared to UM and CPA.

6 Related Work

Together with other shared resources, LLC also plays an important role in the program's performance. This paper summarizes the related work of LLC partition for the program's performance. Cache partitioning is divided into two main categories based on the partitioning motive. The first class of partitioning methods performs cache partitioning based on resource usage characteristics [9,11,12]. CoPart [11] dynamically analyzes programs characteristics and coordinates the allocation of LLC and memory bandwidth to improve the fairness of the consolidated programs. Lucia et al. [9] dynamically analyze the LLC behavior of programs, found the relationship between LLC behavior and IPC, and proposed a phase-aware CPA approach. It identifies and limits the LLC space occupied by Squanderer-type programs, improves LLC utilization using partial sharing, and reduces program turnaround time. Guillaume et al. [12] precisely model the behavior of programs on an LLC-partitioned platform, answering the question of how much core and LLC space should be allocated to each program

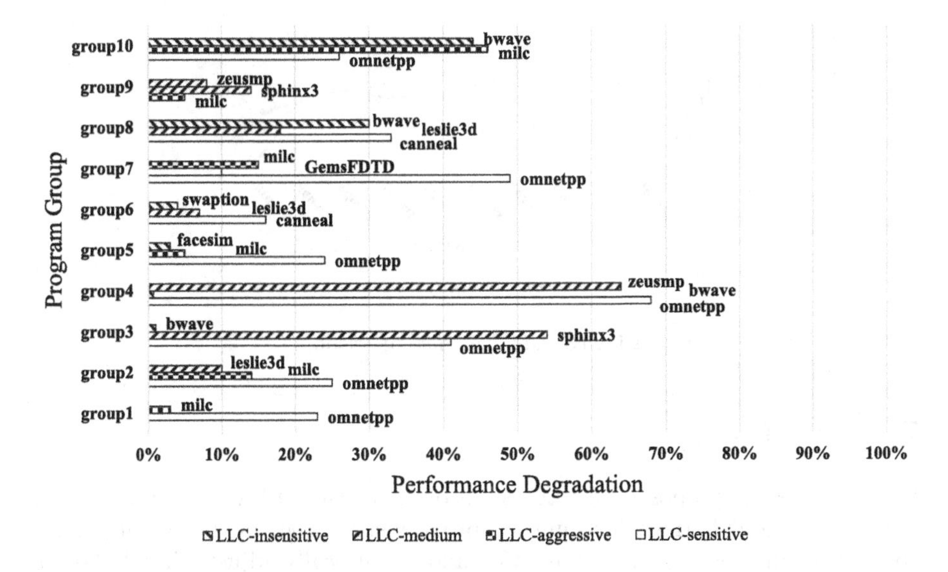

Fig. 12. Program's performance in the CPA case.

to maximize platform efficiency and determining which combinations of programs benefit most from LLC partitioning. The second class of cache partitioning methods is based on program performance [1,13–15]. Common performance metrics are throughput rate, fairness and response time. Chen et al. [13] propose BIG-C, a container-based resource management framework. It implements latency-critical programs and batch programs to run together, which improves resource utilization while ensuring the performance of latency-critical programs. Xiang et al. [14] explore partial sharing of LLC between programs and propose the Dynamic LLC Allocation and Partial Sharing (DCAPS) framework. This framework allows some degree of LLC overlap and contention, which can achieve higher LLC utilization without causing performance degradation. Heracles [15] supports batch and latency-critical task consolidation using hardware and software isolation mechanisms. Real-time monitoring and offline analysis are used to detect sources of interference and to prevent interference through four isolation mechanisms, but it can only guarantee the performance of one latency-sensitive program on one server. Konstantinos et al. [1] propose a dynamic LLC partitioning scheme, DICER, which adjusts LLC allocations according to the needs of high-priority programs and allocates free LLC resources to low-priority programs. This scheme brings the performance of high-priority programs close to that of standalone execution and improves system utilization.

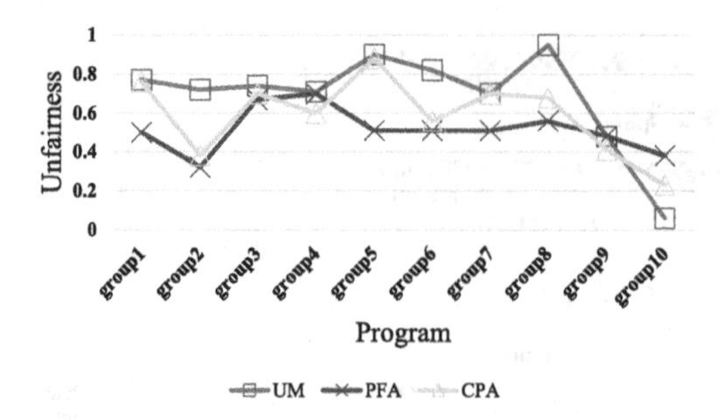

Fig. 13. Unfairness comparison.

7 Conclusion

In this paper, we proposed an LLC partitioning method PFA that improves the performance and fairness for multiple programs. It predicts the required LLC space through program characterization and dynamically adjusts the LLC occupancy of each program based on performance metrics. In this way, PFA achieves high performance for each consolidated program. We characterized nearly 30 programs, and PFA improved the program's performance by 6.73% and improved the program's fairness by 0.03 compared with UM case. It enables more than 80% of programs to achieve 70% of their baseline performance. The average performance of the programs was improved by 4.86% compared to the CPA approach.

Memory bandwidth is also one of the important shared resources in multiple programs running together. Memory bandwidth and LLC occupancy are closely related and interact with each other. Programs can access memory faster through LLC, but lower memory bandwidth reduces the benefits of LLC. How to reconcile LLC space with memory bandwidth will be further investigated in our future work.

References

1. Nikas, K., Papadopoulou, N., Giantsidi, D., Karakostas, V., Goumas, G., Koziris, N.: DICER: diligent cache partitioning for efficient workload consolidation. In: Proceedings of the 48th International Conference on Parallel Processing, pp. 1–10 (2019)
2. Sfakianakis, Y., Kozanitis, C., Kozyrakis, C., Bilas, A.: QuMan: profile-based improvement of cluster utilization. ACM Trans. Archit. Code Optim. (TACO) **15**(3), 1–25 (2018)
3. CGroups (2021). https://www.kernel.org/doc/Documentation/cgroup-v1/cgroups.txt
4. Intel(R) Resource Director Technology (2021). https://github.com/intel/intel-cmt-cat
5. SPECCPU2006 (2021). https://www.spec.org/cpu2006/
6. PARSEC (2021). https://parsec.cs.princeton.edu/

7. Tang, L., Mars, J., Soffa, M.L.: Contentiousness vs. sensitivity: improving contention aware runtime systems on multicore architectures. In: Proceedings of the 1st International Workshop on Adaptive Self-Tuning Computing Systems for the Exaflop Era, pp. 12–21 (2011)
8. Qureshi, M.K.: Adaptive spill-receive for robust high-performance caching in CMPs. In: 2009 IEEE 15th International Symposium on High Performance Computer Architecture, pp. 45–54. IEEE (2009)
9. Pons, L., Sahuquillo, J., Selfa, V., Petit, S., Pons, J.: Phase-aware cache partitioning to target both turnaround time and system performance. IEEE Trans. Parallel Distrib. Syst. **31**(11), 2556–2568 (2020)
10. Selfa, V., Sahuquillo, J., Eeckhout, L., Petit, S., Gomez, M.E.: Application clustering policies to address system fairness with intel's cache allocation technology. In: 2017 26th International Conference on Parallel Architectures and Compilation Techniques (PACT), pp. 194–205. IEEE (2017)
11. Park, J., Park, S., Baek, W.: CoPart: coordinated partitioning of last-level cache and memory bandwidth for fairness-aware workload consolidation on commodity servers. In: Proceedings of the Fourteenth EuroSys Conference 2019, pp. 1–16 (2019)
12. Aupy, G., Benoit, A., Goglin, B., Pottier, L., Robert, Y.: Co-scheduling HPC workloads on cache-partitioned CMP platforms. Int. J. High Perform. Comput. Appl. **33**(6), 1221–1239 (2019)
13. Chen, W., Rao, J., Zhou, X.: Preemptive, low latency datacenter scheduling via lightweight virtualization. In: 2017 USENIX Annual Technical Conference (USENIXATC 17), pp. 251–263 (2017)
14. Xiang, Y., Wang, X., Huang, Z., Wang, Z., Luo, Y., Wang, Z.: DCAPS: dynamic cache allocation with partial sharing. In: Proceedings of the Thirteenth EuroSys Conference, pp. 1–15 (2018)
15. Lo, D., Cheng, L., Govindaraju, R., Ranganathan, P., Kozyrakis, C.: Heracles: improving resource efficiency at scale. In: Proceedings of the 42nd Annual International Symposium on Computer Architecture, pp. 450–462 (2015)

SGP: A Parallel Computing Framework for Supporting Distributed Structural Graph Clustering

Xiufeng Xia, Peng Fang, Yunzhe An, Rui Zhu(✉), and Chuanyu Zong

School of Computer Science, Shenyang Aerospace University, Shenyang, China
{xiufengxia,anyunzhe,zhurui,zongcy}@sau.edu.cn

Abstract. Structural graph clustering is an important problem in the domain of graph data management. Given a large graph G, structural graph clustering is to assign vertices to clusters where vertices in the same cluster are densely connected to each other and vertices in different clusters are loosely connected to each other. Due to its importance, many algorithms have been proposed to study this problem. However, no effort focuses on the distributed graph environment. In this paper, we propose a parallel computing framework named SGP (short for Statistics-based Graph Partition) to support large graph clustering under distributed environment. We first use historical clustering information to partition graph into a group of clusters. Based on the partition result, we can properly assign vertexes to different nodes based on connection relationship among vertex. When a clustering request is submitted, we can use properties leading by the partition for efficiently clustering. Finally, we conduct extensive performance studies on large real and synthetic graphs, which demonstrate that our new approach could efficiently support large graph clustering under distributed environment.

Keywords: Structural graph · Cluster · Distributed · Core-vertex

1 Introduction

This paper studies the problem of distributed graph clustering [1,2]. Given a very large graph G, graph clustering is partitioning G into a group of sub-graphs. In each cluster G_i, there is a dense set of edges among vertices in G_i. Given another cluster G_j, there are few edges among vertices belonging to different clusters.

Structural Graph clustering has many applications, ranging from community detection, pattern recognition, as well as biological network and outlier detection. For example, graph clustering could be used in the domain of recommender system, which recommends property goods for users based on the relationship among different users. Take an example of Fig. 1. $\{u_1, u_2, \ldots, u_8\}$ are 8 users. We find that the relationship among $\{u_1, u_2, \ldots, u_4\}$ is close, they are clustered into the same cluster G_1. Similarity, the relationship among $\{u_5, u_6, \ldots, u_8\}$ is close, they are clustered into another cluster G_2. When recommending goods to users, since u_1 and u_2 have bought the computer "ThinkPad T490s", the recommender system can recommend "ThinkPad T490s" to the

Y. Lai et al. (Eds.): ICA3PP 2021, LNCS 13157, pp. 722–736, 2022.
https://doi.org/10.1007/978-3-030-95391-1_45

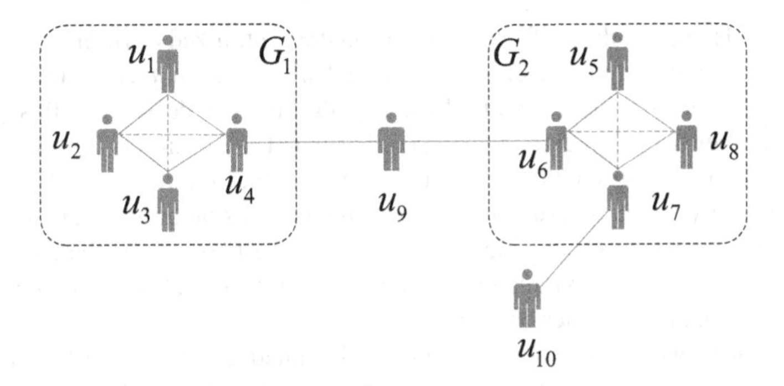

Fig. 1. Structural graph clustering

users u_3 and u_4 respectively. Since the relationship among users in G_1 and G_2 is loose, the goods "ThinkPad T490s" will not be recommended to users in G_2.

Due to the importance of graph clustering, many researchers have studied this problem, and many definitions are proposed, such as modularity-based, graph partitioning-based, density-based, and so on. However, as is discussed in [5], the above definitions cannot distinguish outliers and bridges. Here, we call a vertex as an outlier if it does not belong to any cluster, and connect with no more than two clusters. In addition, we call a vertex as a hub-vertex if it connects with at least two clusters. In order to overcome this problem, structural graph clustering was proposed to [5]. It evaluates the similarity among different vertexes based on the $\sigma(u, v)$. Specially, let G be a graph that is partitioned into $\{G_1, G_2, \ldots, G_m\}$. We regard vertices u and v are structure-similar to each other if the in-equality $\sigma(u, v) \leq \epsilon$ is satisfied. Here, $\sigma(u, v)$ refers to the structure-similarity between u and v, ϵ is an user-defined threshold. In addition, we call a vertex $v \in G_i$ as a core-vertex if there are no less than μ neighbors in G_i that are structure-similar to v. Thirdly, for each vertex in G that does not belong to $\{G_1, G_2, \ldots, G_m\}$, they are hub-vertices or outlier-vertices. Back to the example of Fig. 1. u_9's neighbours belong to G_1 and G_2 respectively, we call it as a hub-vertex. u_{10}'s neighbours only belong to G_2, we call it as an outlier-vertex. Here, given a vertex u, we call v as a *neighbour* of u if there is an edge between them.

To the best of our knowledge, there are mainly three efforts studying this problem, which are SCAN, SCAN++, and pSCAN. The key of these efforts is using the key of DBSCAN for partitioning vertexes into a group of clusters. The SCAN algorithm repeatedly processes vertexes that are not assigned to any cluster. In particularly, if a vertex is evaluated as a core-vertex, a new cluster is generated. The drawback of this algorithm is it has to compute the structural similarity for every pair of adjacent vertices in G. Compared with the SCAN algorithm, SCAN++ avoids computing structural similarity between two-hop-away vertexes. However, its performance is also sensitive to the scale of graph. The reason is that it also has to spend highly computational cost in computing structural similarity between every two adjacent vertexes. pSCAN further reduces the cost of computing structural similarity among vertexes. The authors propose many techniques to speed up structure-similar computation.

Challenges. However, these efforts all focus on the centralized environment. When the graph scale is large, a signal machine cannot load all vertexes and edges to the memory. Thus, an efficient distributed graph clustering algorithm is desired. In this paper, we study the problem of graph clustering under distributed environment. In order to achieve this goal, the following challenges should be overcome. Firstly, how to partition the graph. Intuitively, given a vertex and its neighbours, if its neighbours are contained in many servers, we have to spend highly commutation cost. Secondly, the cluster request may be based on different parameters, i.e., different μ and σ. How to efficiently to find the clustering result is another challenge.

In this paper, we propose a novel framework named SGP(short for Statistics-based Graph Partition). It first finds two parameters based on the historical clustering requests, and partition graph into a group of sub-graphs based on the clustering result. Next, we construct a group of cluster-servers, and a hub-server. Specially, we assign vertexes in the same cluster into a signal cluster-server as much as possible. In particularly, if a vertex has out-neighbours, i.e., a vertex in a cluster has neighbours in other clusters, we assign its reference to the hub-server. Intuitively, when clustering, even a vertex's neighbours are distributed at other clusters, we only need to access one server, i.e., the hub-server. In this way, the highly communication cost could be reduced as much as possible. Via deeply studying the natural properties of the clustering results, when new clustering requests are submitted, we only need to access a small number of vertexes. Last of all, we maintain the similarity among vertexes based on the clustering results. In this way, we could avoid the similarity computation when other cluster requests are submitted.

Contributions. In summary, the contributions of this paper are as follows.

- A Statistics-based Preprocessing Algorithm: We first execute pScan algorithm to partition the whole graph into a group of clusters based on the historical clustering requests. Based on the clustering result, we can obtain the prior knowledge of relationships among vertexes. More important, we can compute the similarity among different vertexes. In this way, when other clustering requests are submitted, we can avoid similarity computation.
- A Greedy based Partition Algorithm: Based on the clustering result, we propose a greedy algorithm to assign vertexes to different servers based on both the clusters' scale and vertexes' out-neighbours. Based on the assigning result, we can guarantee that vertexes in the same cluster are assigned to, as few as, servers. In this way, the high communication cost could be reduced as much as possible.
- A Parallel Clustering Algorithm: We propose two observations. Based on these observations, we propose a parallel computing algorithm. When other clustering requests are submitted, we can use the current partition result to answer requests in many cases. Even if not, we only need to access a small number of vertexes other than accessing all vertexes in the graph to support the clustering requests. In this way, the highly computational cost could be effectively reduced.

The rest of this paper is organized as follows. Section 2 is the related work and the problem definition. Section 3 explains the framework SGP. Section 4 reports our experimental results. Section 5 concludes this paper.

2 Preliminary

In this section, we first review some important existing results of graph clustering. Then, Sect. 2.2 describes the problem definition.

2.1 Related Works

Graph clustering plays an important role in the domain of social networks, communications networks [14–16] and so on. Many researchers have fully studied this types of problem. To the best of our knowledges, existing methods include modularity-based methods, graph partitioning [2], and structural graph clustering methods. In this section, we mainly focus on the problem of structural graph clustering.

The Centralized-Based Graph Clustering. The current graph clustering research mainly involves three types of methods, namely SCAN, SCAN++ and pSCAN [1,3]. Scan is a non-overlapping cluster discovery algorithm improved from the density-based clustering algorithm. During the processing, it can find bridge servers and outliers in the graph. For vertices of unassigned clusters, check whether the server is a core vertex, and a new cluster is generated if it is a core vertex. But the disadvantage of the scan algorithm is that it needs to calculate every pair of adjacent vertices in the graph. Although the SCAN++ algorithm does not need to calculate each pair of adjacent vertices, its calculation cost is proportional to the ratio of the graph; when the graph is large enough, the cost of calculating the structural similarity of the adjacent vertices will increase greatly. pSCAN further optimizes the computational cost of clustering.

Besides exact algorithms, some approximation techniques are also proposed to support graph clustering. Their key idea is using Scan algorithm to obtain approximate clustering results. The main limitation of this kind of methods is that the obtained clustering result is an approximate result. Although it improves the efficiency of structural graph clustering, it reduces the accuracy of the result.

The Parallelizing-Based Graph Clustering. Some researchers study the problem of Parallelizing Scan [5–8]. From all efforts, anySCAN is the first framework that could support parallel graph clustering over shared memory architectures and multi-core CPUs. The key behind it is processing vertices in blocks, and then acquired results are merged into an underlying cluster structure consisting of the so-called super-nodes for building clusters. Consequently, anySCAN uniquely is a both interactive and work-efficient parallel algorithm. Takahashi et al. proposed another framework named SCAN-XP. As is discussed in [8], it is a parallel version of SCAN, and it could efficiently work under Intel Xeon Phi Co-processors. However, since SCAN-XP requires all structural similarities to be calculated, compared with anySCAN, it performs worse than that of anySCAN.

The Dynamic Structural Graph Clustering. Besides graph clustering over static graph, some researchers study the problem of graph clustering over dynamic structural graph. Among all this type of algorithms [6,9], the algorithm DENGRAPH is used for detecting communities over large and dynamic social networks. danySCAN is another framework, which processes updates in a bulk mode. Compared with other efforts, it can process dynamic weighted graphs.

2.2 Problem Definition

Let $G(V, E)$ be a graph, V be the vertex set, and V be the edge set. $|V|$ refers to the number of vertexes of G. $|E|$ refers to the number of edges in G. For each vertex $v \in V$, if another vertex $u \in V$ is connected with v, we call u as a neighbour of v. Assuming the vertex u has $\mathsf{d}(u)$ neighbours, the degree of u is $\mathsf{d}(u)$. Back to the example of Fig. 1. The degree of vertex u_4, i,.e., $\mathsf{d}(u_4)$ equals to 4. In addition, let $N[u]$ be the *closed neighborhoods* of the vertex u. It contains u and all neighbours of u. $|N[u]|$ refers to the number of elements in $N[u]$, which equals to $\mathsf{d}(u)$+1. In Fig. 1, $N[u]$ of vertex u_6 is 5. In the following, we will formally explain the concept of similarity.

Definition 1 ***Similarity.*** *Given two vertex u and v in G, the similarity between two vertices, i.e., denoted as $\sigma(u, v)$, refers to the number of common vertices in $D[u]$ and $D[v]$ normalized by the geometric mean of their cardinalities. It is computed based on Eq. 1.*

$$\delta(u, v) = \frac{|D[u] \cap D[v]|}{|D[u] \cup D[v]|} \tag{1}$$

Back to the example in Fig. 1. $N[u_4] = \{u_1, u_2, u_3, u_4, u_9\}$ $N[u_3] = \{u_1, u_2, u_3, u_4\}$. The common vertexes of u_2,u_5 is $\{u_1, u_2, u_3, u_4\}$, That is $\delta(u_3, u_4) = |\{u_1, u_2, u_3, u_4\}|$. Similarly, $N[u_1] = \{u_1, u_2, u_3, u_4\}$, $N[u_4] = \{u_1, u_2, u_3, u_4, u_9\}$, then the common vertexes of u_1, u_4 is $\{u_1, u_2, u_3, u_4\}$. Therefore, $\delta(u_1, u_4)$ equals to $\{u_1, u_2, u_3, u_4\}$. We find that the more common vertices of two vertexes have, the greater the their structural similarity. Let $N_\sigma[u]$ be a subset of u's neighbours. For each element u' in $N_\sigma[u]$, $\delta(u, v)$ must be no less than σ. As is shown in Fig. 1, $N_{0.8}[u_2] = \{u_1, u_2, u_3, u_4\}$.

Definition 2 ***Core-vertex.*** *Let σ and μ be two threshold, i,e., $0 \leq \sigma \leq 1$ and $\mu \geq 2$. For each vertex $u \in \mu$, only when $N_\sigma[u] \geq \mu$ is satisfied, the vertex u is regarded as a core-vertex. Otherwise, the vertex u is regarded as a non-core vertex.*

Note that, let v be a core-vertex of the clustering G_i. For any other vertex v', if $\delta(v, v') \geq \sigma$, v' is also regarded as an element of G_i. In other words, a vertex is regarded as an element of G_i if it is a core-vertex of G_i, or the similarity between it and another core-vertex in G_i is larger than σ. After explaining the concept of *core-vertex* and *non-core-vertex*, we now explain the concept of *outlier-vertex* and *bridge-vertex*. Let G be a graph, μ and σ be two parameters, and $\{C_1, C_2, \ldots, C_m\}$ be a group of clusters under μ and σ. For each vertex $u \in G$, if it does not belong to any cluster, and only connect with at more one cluster, we regard it as an outlier-vertex. If it does not belong to any cluster, but connect with at least one cluster, we regard it as a bridge-vertex. Last of this section, we formally discuss the problem definition.

Definition 3 ***Problem Definition.*** *Let σ and μ be two threshold, i,e., $0 \leq \sigma \leq 1$ and $\mu \geq 2$. Given a graph $G(V, E)$, graph clustering is to partition G into a group of clusters $\{C_1, C_2, \ldots, C_m\}$, and a group of bridge-vertexes and outlier-vertexes. For each vertex $v \in C_i$, it is regarded as a core-vertex or the similarity between v and a core-vertex $v' \in C_i$ is larger than σ.*

As is shown in Fig. 1, assuming the parameters σ and μ are set to 0.8 and 4, the graph G could generate two clusters, i.e., G_1 and G_2. In addition, G_1 is $\{u_1, u_2, u_3, u_4\}$, and G_2 is $\{u_5, u_6, u_7, u_8\}$. u_{10} is the outlier,u_9 is the brigde, vertices that are associated with only one of the vertices are called neighbor points.

3 The Framework SGP

In this section, we discuss the framework *SGP* . First of this section is the framework overview. Next, we discuss the partition algorithm. Last of this section is the clustering algorithm under different parameters.

3.1 The Framework Overview

Let $Q\{q_1, q_2 \ldots, q_m\}$ be a set of historical clustering requests under parameters $\langle\sigma_1, \mu_1\rangle, \langle\sigma_2, \mu_2\rangle, \ldots, \langle\sigma_m, \mu_m\rangle$. We first compute the median of σ based on $\{\sigma_1, \sigma_2 \ldots, \sigma_m\}$, and the median of μ based on $\{\mu_1, \mu_2 \ldots, \mu_m\}$ respectively. Next, we execute pScan algorithm to partition the graph G into a group of clusters $\{G_1, G_2 \ldots, G_m\}$(also finds all bridge-vertexes and outlier-vertexes) based on μ and σ.

Let the capacity of each server E_i be C. We assign $\{G_1, G_2 \ldots, G_m\}$ to servers $\{E_1, E_2 \ldots, E_n\}$. The assignment rule is if $|G_i| + |G_j| \leq C$, they can not be assigned to the same server. Otherwise, they can be assigned to the same server. In addition, if a cluster G_i satisfies that $|G_i| \geq C$, we should repeatedly partition vertexes in G_i under other parameters until the sum of vertexes and edges in each cluster is smaller than C. In addition, in order to reduce the communication cost, we construct a few hub-servers. They contain three types of vertexes: *hub-vertexes*, *outlier-vertexes* and *m-vertexes*. Here, we call a vertex as a *m-vertex* if it is a member of a cluster G_i and connected with vertexes in the other clusters. We will discuss the partition algorithm in Sect. 3.2.

When a clustering request $q\langle\sigma_q, \mu_q\rangle$ is submitted, we compare $\langle\sigma_q, \mu_q\rangle$ with $\langle\sigma, \mu\rangle$. Based on the comparison result, we can divide vertexes in each original cluster into three-types: *cluster-vertexes*, *uncluster-vertexes*, *c-vertexes*. Here, we call a vertex as a cluster-vertex if it must be belonged to its original cluster, and we need not to find a new cluster for it. We call a vertex as an uncluster-vertexes if it must not be belonged to its original cluster, and we need to find a new cluster for it. We call a vertex as a *c-vertex* if we should further evaluate whether it is belonged to a new cluster or the original cluster. We will discuss the algorithm in Sect. 3.3.

Take an example in Fig. 2. Given a graph $G(E, V)$, we totally find three clusters, i.e., C_1, C_2 and C_3 based on the parameters 0.4 and 6. They are assigned to 3 servers respectively. In addition, we use a hub-server to maintain all *hub-vertexes*, *outlier-vertexes*

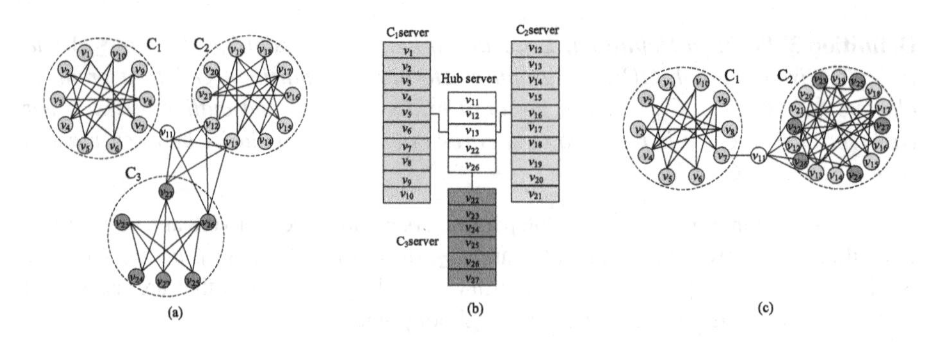

Fig. 2. The framework overview

and *m-vertexes*. When a clustering request $q\langle 0.5, 8\rangle$ is submitted, since $0.5 \geq 0.4$ and $8 \geq 6$, we can assert that, for each vertex $v \in C_1$, it is not belonged to any other clusters, i.e., $\{C_2\}$ under $q\langle 0.5, 8\rangle$. In addition, all outlier-vertex under $\langle 0.4, 6\rangle$ are still outlier-vertexes.

3.2 The Partition Algorithm

This section discusses the partition algorithm. As is discussed in Sect. 3.1, we first partition the whole graph G into a group of clusters $\{G_1, G_2 \ldots, G_m\}$(also find all bridge-vertexes and outlier-vertexes) based on μ and σ. For each cluster G_i, if $|G_i| \geq C$, we use $\mu \times 1.1$ and $\sigma \times 1.1$ as new parameters, and repeat pScan algorithm to further partition G_i until the size of each cluster is smaller than C. Here, C refers to the maximize size of a signal server. Note that, after partitioning, we maintain the similarity among edges, which helps us avoid the similarity computation.

In order to reduce communication cost, we construct a summary graph SG based on the partition result. We regard each cluster as a vertex. Accordingly, SG contains vertexes $\{g_1, g_2, \ldots, g_m\}$. Here, g_i corresponds to the cluster G_i. In addition, given two clusters G_i and G_j, assuming there is w_{ij} edges among vertexes in G_i and G_j, the weight of edge e_{ij} is set to w_{ij}. Next, we sort edges in SG based on these edges' weights. Thirdly, we use the key of greedy to construct hub-servers. Specially, let e_{max} be the edge with maximal weight, and it is connecting G_i and G_j. If $|G_i| + |G_j| \leq C$, we assign G_i and G_j into the same server C_1. Otherwise, they are assigned into two servers C_1 and C_2. In addition, the reference of *m-vertexes* in G_i and G_j are assigned into the hub-server.

From then on, we repeat the above operations. Specially, let e_{sec} be the edge with secondly maximal weight, and it is connecting G_i' and G_j'. We try to assign them to non-empty servers. If no non-empty server can contain them, we assign them into new servers. Next, we use the similar manner to update hub-server. In particular, if the hub-server turns to the full, we should assign them to a new hub-server. In addition, if G_i' is assigned to the server C_u, we need not to maintain *m-vertexes* for G_i' and other clusters in C_u. In this case, we remove the corresponding edges from SG. When all clusters are assigned to a server, the algorithm is terminated.

Algorithm 1: The Partition Algorithm

```
Input: The Graph G, Historical Query Set Q
Output: The Server Set C
1  σ ←getMedian(Q.σ);
2  μ ←getMedian(Q.μ);
3  Cluster 𝒢 ← pScan(G, σ, μ);
4  for i from 1 to m do
5      if |G_i| ≥ C then
6          𝒫 ← 𝒫∪pScan(P_i);
7  for i from 1 to |V| do
8      for j from 1 to V[i].n do
9          V_ij.s ← comSim(V_ij, V_i);
10         if same(V_ij, V_i)=true then
11             V_ij.c ← true;
12         else
13             V_ij.c ← flase;
14 SG ←construction(𝒫);
15 Weight W ← sort(SG);
16 C_1 ← G_1;
17 Hub-Server H ← ∅;
18 for i from 2 to |W| do
19     for j from i to |C| do
20         if |C_j| + |G'_i| ≤ C then
21             C_j ← C_j ∪ G'_i;
22             bMerge ← ture;
23     H ← update();
24 return;
```

As is shown in Algorithm 1, we first partition the whole graph G into a group of clusters $\{G_1, G_2, \ldots, G_m\}$ (line 1 to line 6). Next, we maintain the similarity among vertex, and label whether they are *m-vertex*es or not (line 7 to line 13). Thirdly, we construct a summary graph SG (line 14–15). Last of all, Next, we use the key of greedy to assign clusters to suitable servers. Our goal is to guarantee that the reference of *m-vertex*es are contained in, as few as, hub-servers. Intuitively, if most *m-vertex*es are contained in the same server, the communication cost could be reduced a lot.

3.3 The Clustering Algorithm Under Distributed Environment

Before discussing the algorithm, we first propose two observations. We use them for efficiently clustering under distributed graph. For simplify discussing our algorithm, we assume that all clusters are generated under the same parameters μ and σ. In addition, we assume that each server only contains one cluster. However, we want to highlight that our proposed algorithm could work under other cases.

Observation 1. *Let G be a graph that is partitioned into clusters $\{G_1, G_2, \ldots G_m\}$ under parameters μ and σ. Once a clustering request $q\langle\sigma_q, \mu_q\rangle$ is submitted, if $\sigma \geq \sigma_q$ and $\mu \geq \mu_q$, for each two vertexes v_i and v_j in G_u, they are also belonged to the same cluster.*

Observation 1 implies once a new clustering request is submitted, if $\sigma \geq \sigma_q$ and $\mu \geq \mu_q$, we can assert that all core-vertexes in each cluster under σ and μ are all core-vertexes under σ_q and μ_q. Under this case, we only need to execute the following operations. Firstly, we should check whether non-core vertexes are turning to core-vertexes. Secondly, we should check whether some clusters are merged into a new cluster. Thirdly, we should check whether some outlier-vertexes could be assigned into a cluster.

Algorithm 2: The Local Clustering Algorithm

```
  Input: The cluster sub-graph CG, query q
  Output: The Cluster Set S
1 if σ_q ≤ σ ∧ μ_q ≤ μ then
2     Check Set M ← getCheckVectex(CG);
3     for i from 1 to |M| do
4         if M[i].c = CORE∧checkCore(M[i])=ture then
5             S ← S ∪ M[j];
6 if σ_q ≥ σ ∧ μ_q ≥ μ then
7     sort(CG);
8     for i from 1 to |CG| do
9         pScan(CG[i]);
10 else
11    repeat Line 6-9;
12    S ←findMNewCoreV();
13 send(S);
14 return;
```

Observation 2. *Let G be a graph that is partitioned into clusters $\{G_1, G_2, \ldots G_m\}$ under parameters μ and σ. Once a clustering request $q\langle\sigma_q, \mu_q\rangle$ is submitted, if $\sigma \leq \sigma_q$ and $\mu \leq \mu_q$, for each two vertexes $v_i \in G_u$ and $v_j \in G_v$, they are not belonged to the same cluster.*

Observation 2 implies once a new clustering request is submitted, if $\sigma \leq \sigma_q$ and $\mu \leq \mu_q$, we can assert that, for each two vertexes $v_i \in G_u$ and $v_j \in G_v$, they are not belonged to the same cluster. Assuming G_u is contained in the server C_i, and G_v is contained in the server C_j, when clustering vertexes in C_i, we need not to access vertexes in C_j. Obviously, it helps us reduce the costly communication cost a lot. In this case, given a cluster G_i, we sort vertexes in G_i based on their neighbour amount.

Next, we access each vertex in G_i, evaluate whether it is a core-vertex, and construct clutters based on vertexes in G_i. Since the algorithm is similar with PScan, we skip the details for saving space. Under other cases, the operations is similar with that of discussed before. One difference is we should check whether core vertexes are turning to non-core-vertex.

We now formally discuss the clustering algorithm. It contains two steps, which are *local-clustering* and *global-clustering*. In the first step, each server should execute the clustering in a *parallel* way based on the observations discussed before. As shown in Algorithm 2, if $\sigma \geq \sigma_q$ and $\mu \geq \mu_q$, we execute pScan in each local server. In this case, no communication cost is spent. We want to highlight that since we have re-computed the similarity among every two vertexes, we could use the key of greedy to speed up the pScan. If $\sigma \leq \sigma_q$ and $\mu \leq \mu_q$, we only access non-core vertexes, evaluate whether they turn to core vertexes. Furthermore, if existing non-cores vertexes turn cores vertexes and they are m-vertexes, we should send them to the hub-servers. Under other cases, we execute pScan in each local server, and send a few vertexes to the hub-server, i.e., similar with the manner under case two.

In the global clustering step, we mainly check whether some clusters under local-clusters could be merged together. Specially, when the hub-server receives all the m-vertexes, we check whether existing vertexes whose similarity is larger than the parameter σ_q. If existing, their corresponding clusters could be merged together. Since the algorithm is simple, we skip the details for saving space.

4 Experiment

In this section, we conduct extensive experiments to demonstrate the efficiency of SGP framework. The experiments are based on both real datasets and synthetic datasets. In the following, we first explain the datasets used in our experiments and the settings of our experiments, and then report our findings.

4.1 Experiment Settings

Data Sets. In total, ten datasets are used in our experiments, including two real datasets, namely AMAZON, and LIVEJOURNAL, and 8 synthetic datasets. LIVEJOURNAL is a free *on-line* blogging community where users declare friendship among each other. It contains 400727 vertexes and 3200440 edges respectively. AMAZON is a free *on-line* blogging community where users declare friendship among each other. It contains 3997962 vertexes and 34681189 edges. Besides real data set, another eight synthetic datasets are also used. They are used for evaluating the stability of our proposed algorithm.

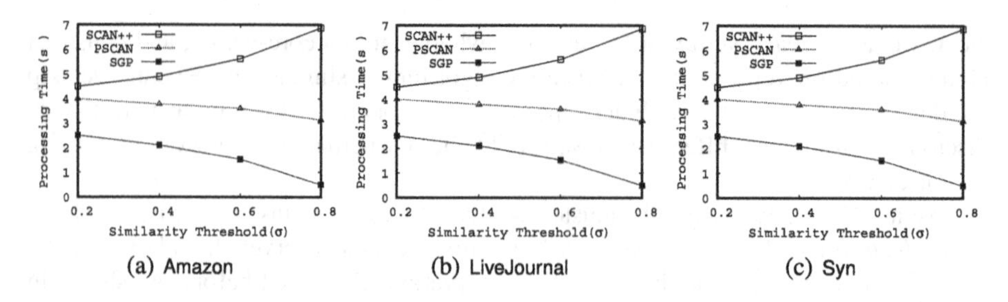

(a) Amazon (b) LiveJournal (c) Syn

Fig. 3. Running time comparison of different algorithms under different parameter σ

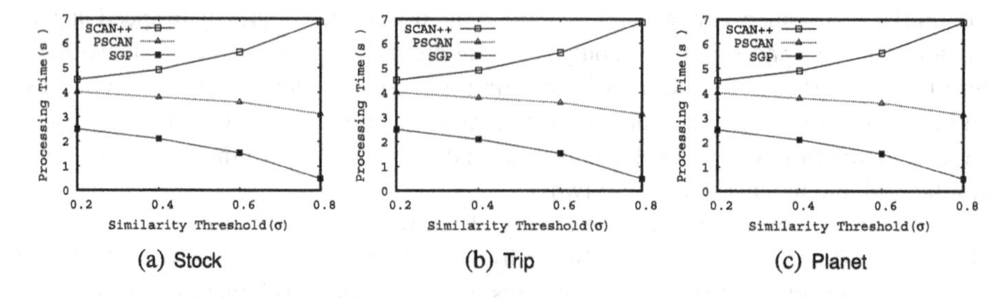

(a) Stock (b) Trip (c) Planet

Fig. 4. Communication cost comparison of different algorithms under different parameter σ

Parameter Setting. In our study, we consider three parameters, i.e., the parameter σ, μ, and the capacity of servers, i.e., denoted as C. Here, the parameter σ ranges from 0.2 to 0.8 with defaulted value 0.5; the parameter μ ranges from 4 to 16 with defaulted value 10. Lastly, the capacity of server ranges from 1 to 5 with defaulted value 3.

Performance Metrics and Experiment Environment. The *computational cost*, *communication cost* are employed as the main performance metrics. The total computational cost is evaluated by the time used to process a signal clustering request. The communication cost is evaluated by the amount of accessing vertexes in the other servers. In addition to SGP framework, we implement pScan, k-scan++, two state-of-the-art approaches under distributed environment. Under these two algorithms, vertexes are randomly assigned to different servers. All the algorithms are implemented with C++, and all the experiments are conducted on a CPU i7 with 32 GB memory, running Microsoft Windows 10. The total number of servers is 10.

4.2 Comparison Between SGP and Existing Algorithms

In this section, we compare the performance of SGP with its competitors. The running time of different algorithms is reported in Fig. 3, corresponding to AMAZON, LIVE-JOURNAL, and SYN, the three datasets used in our experiments. We firstly report the running time of all the algorithms under different σ values in Figs. 3(a)–3(c). Among all the σ values evaluated, we can observe that SGP outperforms existing algorithms consistently, for all three datasets. Obviously, SGP on average consumes only 39% of

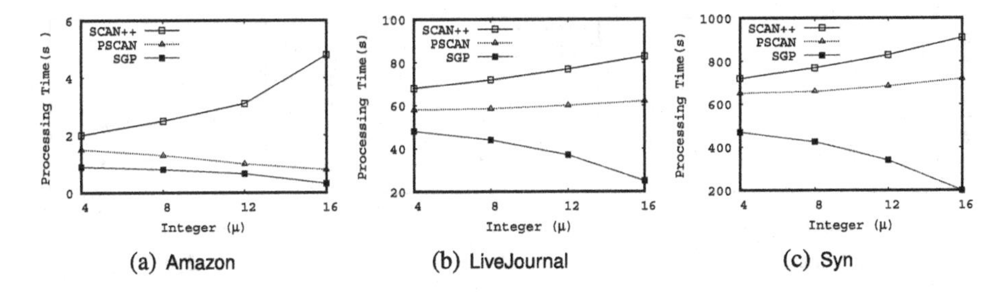

Fig. 5. Running time comparison of different algorithms under different parameter μ

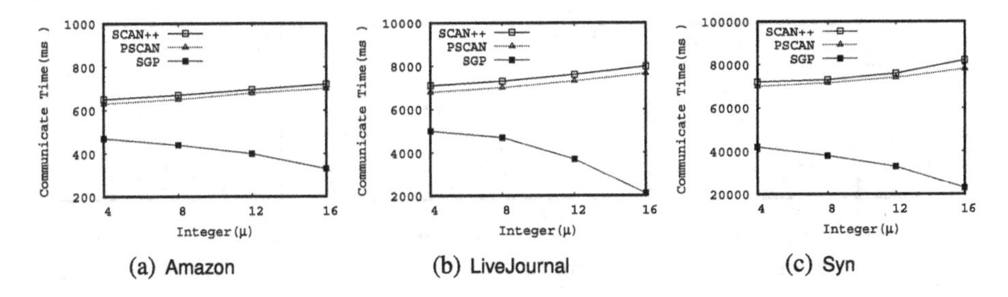

Fig. 6. Communication cost comparison of different algorithms under different parameter μ

scan++'s and 66% of pScan's running time respectively. The significant improvement lies on the fact that SGP has pre-compute the similarity among vertexes. Thus, we need not to spend high cost in computing similarity among vertexes. Another reason is, in many cases, we only need to access a few non-core vertexes. Institutively, since the number of vertexes we should access it small, the overall computational cost is low.

Next, We report the communication cost of all the algorithms under different σ values in Fig. 4(a)–4(c). Among all the σ values evaluated, we can observe that SGP outperforms existing algorithms consistently, for all three datasets. For example, SGP on average consumes only 22% of scan++'s and 25% of pScan's running time respectively. The significant improvement lies on the fact that SGP has the ability of partitioning vertexes to different servers based the collection relationship among different vertexes. Given two vertexes v_i and v_j, if the collection relationship between them is high, they may be assigned into the same servers with a relatively high probability. Thus, when processing newly submitted clustering requests, only a few number of vertexes have to access vertexes in the other servers. Another reason is, based on the partition result, many vertexes are needed not to be accessed. Accordingly, their corresponding neighbours located at other servers are avoided accessing. Thus, the communication cost of SGP is lower than the other algorithms.

Thirdly, we compare the performance of SGP with algorithms scan++ and pScan under differernt μ. The running time of different algorithms is reported in the same three datasets used in our experiments. We firstly report the running time of all the algorithms under different μ values in Fig. 5(a)–5(c). Among all the μ values evaluated, we can observe that when the parameter μ values increases, the running time of the

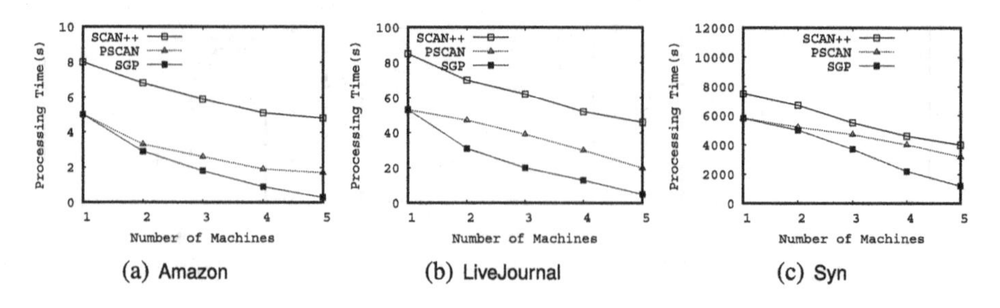

(a) Amazon (b) LiveJournal (c) Syn

Fig. 7. Running time comparison of different algorithms under different sever amount

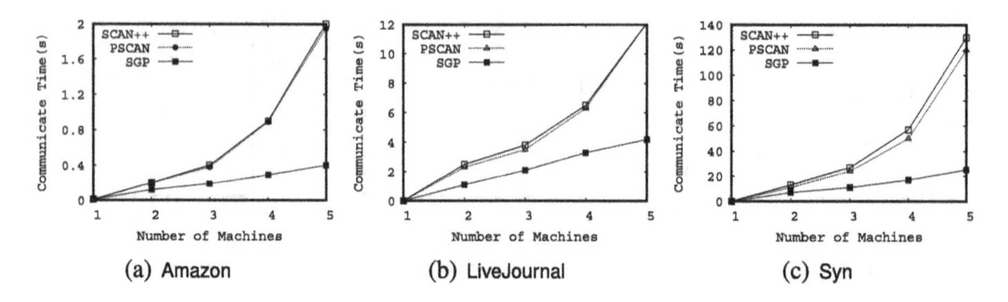

(a) Amazon (b) LiveJournal (c) Syn

Fig. 8. Communication cost comparison of different algorithms under different sever amount

SGP is only 45% of the scan++ and 54% of the pScan. However, when the amount of data increases, the scan++ and pScan require longer calculation time, while the SGP is different. The reason behind it is the SGP does not need to calculate all the core vertexes.

Next, We report the communication cost of all the algorithms under different μ values in Fig. 6(a)–6(c). Among all the μ values evaluated, we can observe that SGP outperforms existing algorithms consistently, for all three datasets. When the amount of data increases, the data distributed on each server requires longer communication time for clustering calculations. The SGP algorithm only calculates the data on some servers, so the communication time is only 44% of scan++ and 46% of pScan, it can be seen that SGP can save a lot of communication cost.

Fifthly, we compare the performance of SGP with its competitors under different server capacity. The running time of different algorithms is reported in Fig. 7, corresponding to AMAZON, LIVEJOURNAL, and SYN, the three datasets used in our experiments. We firstly report the running time of all the algorithms under different server values in Fig. 7(a)–7(c). Among all the server values evaluated, we can observe that SGP outperforms existing algorithms consistently, for all three datasets. For the three algorithms in this article, increasing the number of servers will significantly increase the running time. However, the SGP algorithm only uses scan++' 62% and pScan' 76% of the running time. It is mainly because that scan++ and pScan algorithms have to spend highly cost in computing similarity among vertexes. In addition, the clustering under these two algorithm has to access all vertexes. However, SGP only needs to access a small number of vertexes.

Last of this section, we report the communication cost of all the algorithms under different server values in Fig. 8(a)–8(c). Among all the server values evaluated, we can observe that SGP outperforms existing algorithms consistently, for all three datasets. For example, When the number of servers increases, it takes more communication time, but the SGP algorithm only uses 28% of scan++ and 20% of pScan the communication time. Obviously, the SGP algorithm is better for processing distributed graph clustering than the other two algorithms. Besides the reasons discussed above, the more the server amount, the higher the communication cost. However, the cost of SGP increases slowest of all. The reason behind it our proposed algorithm has the ability of assigning closely connected vertexes to, as few as, servers. Thus, our propose framework is not sensitive to the number of servers.

5 Conclusion

Structural graph clustering is an important problem in the domain of graph data management. In order to support large graph clustering under distributed environment, we propose a novel framework named SGP. It first pre-clusters vertexes in the graph based on the historical clustering requests. The benefit is we can evaluate the connection relationship among vertexes, and properly assign vertexes into suitable servers. In order to support new clustering requests, we propose a novel algorithm. It uses properties leading by the partition for reducing computational and communication cost. Extensive experiments demonstrate the superior performance of SGP.

Acknowledgment. This paper is partly supported by the National Natural Science Foundation for Young Scientists of China (61702344, 61701322), the Natural Science Foundation of Liaoning Province under Grant No. 2019-ZD-0224.

References

1. Xu, X., Yuruk, N., Feng, Z., et al.: SCAN: a structural clustering algorithm for networks. In: Proceedings of the 13th ACM SIGKDD International Conference on Knowledge Discovery and Data Mining, San Jose, 12–15 August 2007, pp. 824–833. ACM, New York (2007)
2. Chang, L., Yu, J.X., Qin, L., Lin, X., Liu, C., Liang, W.: Efficiently computing k-edge connected components via graph decomposition. In: Proceedings of SIGMOD 2013 (2013)
3. Clauset, A., Newman, M.E.J., Moore, C.: Finding community structure in very large networks. Phys. Rev. E **70**, 066111 (2004)
4. Huang, X., Cheng, H., Qin, L., Tian, W., Yu, J.X.: Querying k-truss community in large and dynamic graphs. In: Proceedings of SIGMOD 2014 (2014)
5. Shiokawa, H., Fujiwara, Y., Onizuka, M.: SCAN++: efficient algorithm for finding clusters, hubs and outliers on large-scale graphs. PVLDB **8**(11), 1178–1189 (2015)
6. Wang, L., Xiao, Y., Shao, B., Wang, H.: How to partition a billion server graph. In: Proceedings of ICDE 2014 (2014)
7. Zhao, W., Martha, V.S., Xu, X.: PSCAN: a parallel structural clustering algorithm for big networks in MapReduce. In: Proceedings of the 27th IEEE International Conference on Advanced Information Networking and Applications, Barcelona, 25–28 March 2013, pp. 862–869. IEEE Computer Society, Washington (2013)

8. Ding, C.H.Q., He, X., Zha, H., Gu, M., Simon, H.D.: A min-max cut algorithm for graph partitioning and data clustering. In: Proceedings of ICDM 2001 (2001)
9. Shi, J., Malik, J.: Normalized cuts and image segmentation. IEEE Trans. Pattern Anal. Mach. Intell. **22**(8), 888–905 (2000)
10. Holme, P., Kim, B.J.: Growing scale-free networks with tunable clustering. Phys. Rev. E **65**(2), 026107 (2002)
11. Gibbons, A.: Algorithmic Graph Theory. Cambridge University Press, Cambridge (1985)
12. Zhang, Y., Parthasarathy, S.: Extracting analyzing and visualizing triangle k-core motifs within networks. In: Proceedings of ICDE 2012 (2012)
13. Xin, R.S., Gonzalez, J.E., Franklin, M.J., et al.: GraphX: a resilient distributed graph system on Spark. In: Proceedings of the 1st International Workshop on Graph Data Management Experiences and Systems, New York, 23 June 2013, vol. 2. ACM, New York (2013)
14. Wang, T., Yucheng, L., Wang, J., Dai, H.-N., Zheng, X., Jia, W.: EIHDP: edge-intelligent hierarchical dynamic pricing based on cloud-edge-client collaboration for IoT systems. IEEE Trans. Comput. **70**(8), 1285–1298 (2021)
15. Wang, T., Liu, Y., Zheng, X., Dai, H.-N., Jia, W., Xie, M.: Edge-based communication optimization for distributed federated learning. IEEE Trans. Netw. Sci. Eng. (2021). https://doi.org/10.1109/TNSE.2021.3083263
16. Youke, W., Huang, H., Ningyun, W., Wang, Y., Bhuiyan, Md.Z.A., Wang, T.: An incentive-based protection and recovery strategy for secure big data in social networks. Inf. Sci. **508**, 79–91 (2020)

LIDUSA – A Learned Index Structure for Dynamical Uneven Spatial Data

Zejian Zhang, Yan Wang(✉), and Shunzhi Zhu

Xiamen University of Technology, Xiamen 361000, Fujian, China
wangyan@xmut.edu.cn

Abstract. Based on the problem that existing learned indexes are difficult to adjust dynamically with data changes, a learned index Structure for dynamical uneven spatial data (LIDUSA) is presented in our paper. To handle with the problem of poor KNN query performance on sparse regions, LIDUSA could dynamically adjust data layout by merging and splitting corresponding grid cells, and relearn mapping function of this region to make data points stored in adjacent sparse grid cell also stored in neighboring disk pages. It combines the advantage of tree-shaped indexes, which could be adjusted dynamically, and that of learned indexes. In this paper, extensive experiments are conducted on real-world dataset and synthetic datasets. From experiment results, it could be seen that LIDUSA is twice as fast as other existing indexes in the scenario of KNN query, which will greatly extend the applicable scope of learned indexes.

Keywords: LIDUSA · Spatial index · Learned index

1 Introduction

With the development of artificial intelligence (AI) technology, machine learning is also used in DBMS for improving efficiency of data access. For a long time, tree-shaped index technologies have been thought as common methods for improving performance of database query, which includes B-tree (for one-dimensional data) and R-tree (for multi-dimensional data). The ideas of these methods are to filter irrelevant regions from a tree's top to bottom, layer by layer. Then description of each region, which corresponds to a node in a tree, should be stored. As height of the tree increases, so does overheads of querying and maintenance. With the amount of data grows and the dimension of spatial datasets becomes larger, the storage overheads of tree-shaped indexes become bigger, which even exceed the size of data which is needed to index.

Therefore, another way to reduce the tree's height is inspired. One could see that if the height of a tree is small, the layers to be processed in a query are also small. Then, the performance of queries would be improved. One idea is to construct a function for mapping coordinates of data points to locations of pages where they are stored. Such indexes are called learned indexes. When a spatial query is issued, according to the inputted coordinate, the DBMS system could get location of corresponding page by

Y. Lai et al. (Eds.): ICA3PP 2021, LNCS 13157, pp. 737–753, 2022.
https://doi.org/10.1007/978-3-030-95391-1_46

using a mapping function. So, it is thought as a kind of special R-tree with only one layer, which is more efficient than that with multiple layers.

The key of this method lies in improving accuracy of the underlying mapping function with a prediction model. However, in the scenario of spatial datasets, the coordinates of data points are usually multi-dimensional, but data locations – always expressed with disk page number – are one-dimensional. Then, the aim of prediction model is to map a multi-dimensional data point to one-dimension. Given that there exist strict restrictions on space and time for DBMS, the prediction model used here should be compact, which would be not so accurate. Then, it is necessary to introduce intermediate variables to reduce mapping errors. The first step is to map a multi-dimensional coordinate to a one-dimensional representation. Then the one-dimensional logical representation could be mapped to a storage location. Such methods are described in LISA [1], ZM [10], etc.

In the scenario of multi-dimensional datasets, the influence of each dimension is equal for distance-based queries, but how data is sorted is mainly determined by only one dimension, maybe a hybrid dimension. How to choose the best dimension for sorting data is hard. Especially, under the real-world circumstances, the distributions of data points are changing all the time. Is the best dimension in initial stage always the best at all times? Obviously, the answer is "NO". So, the sorting method, or mapping function, in a learned index should also automatically adjust with change of data distributions.

The existing learned index methods do not solve this problem well. For example, LISA divides the whole space into a series of grid cells, and sorts data points with two steps. In the first step, data points in different grid cells are sorted according to their corresponding grid cells' id, which heavily depends on organization method of grid cells. However, LISA does not present a dynamical adjusting method of grid cells. Then, two points which are close in space according to their mapped one-dimensional representations, may be stored far away from each other. This will seriously hurt performances of spatial queries on such index, especially for KNN (K-Nearest Neighbors) queries.

Let's take Fig. 1 as an example. Figure 1(a) shows a distribution map of starting positions, represented as black points, of taxi orders in Chengdu, a mega-city in southwest of China. At the bottom of Fig. 1(a), a blue point represents the center of a KNN query. To clearly exhibit the drawbacks of existing learned indexes for KNN queries, we enlarge the grid cells involved in the query, which are shown in Fig. 1(b). The dotted lines in blue are boundaries of grid cells, and a semi-circle in black represents the query scope. The grid cells involved in this query are marked as a black rectangle. The ids of these grid cells are marked in red. It is assumed that the grid cells are sorted based on horizontal axis firstly, and vertical axis secondly. Then the ids of grid cells in the black rectangle are far away, such as 12, 24, 36, …, and 108. So, the storage positions of data points in these grid cells are also far away. During query processing, data points that satisfy the condition are obtained by expanding search radius gradually. To read all the data points in the black semi-circle, it is needed to scan pages from the 12th to the 108th grid cells. However, if these data are stored in less grid cells, the query overhead is drastically reduced.

In order to solve the above problem, in this paper, a Learned Index structure for Dynamical Uneven Spatial dAta (**LIDUSA** for short) is proposed. When a grid cell is sparse, it should be merged to the adjacent grid cell. Otherwise, it will be split into two

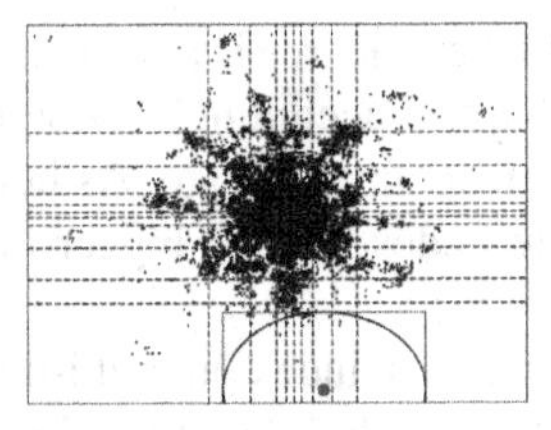

(a)Distribution map of taxi orders

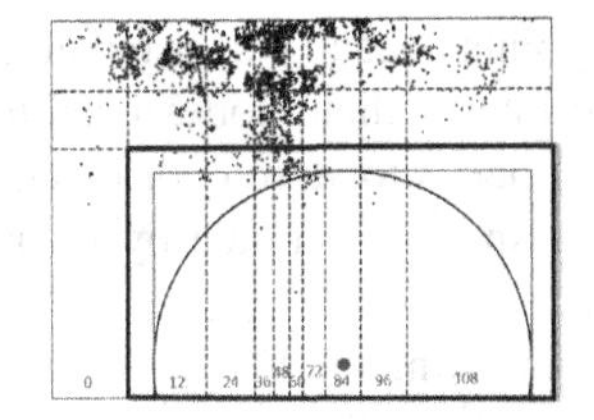

(b) Enlarged view of the KNN query area

Fig. 1. An Example of a KNN query on existing learned indexes used for an uneven dataset.

grid cells, where there store equal number of points. Then, the system needs to recalculate one-dimensional representations of data points in the merged grid cell, adjusts storage location of those points to make spatially adjacent data stored also on adjacent pages. Due to the number of points in a sparse grid cell is small, the data adjustment overhead is also small. Then, with dynamic insertions and deletions of data points, LIDUSA could automatically adjust grid cells to ensure low IO cost.

Our contributions in this paper are summarized as follows.

1. We propose a new learned index structure (LIDUSA), which can adjust scope of sparse or dense grid cells dynamically, to improve the applicability of learned indexes.(Section 3)
2. We design efficient algorithms to process range query or KNN query using LIDUSA. (Section 4)
3. We conduct extensive performance evaluations with real-world and synthetic datasets. The results demonstrate that LIDUSA outperforms alternative methods in terms of storage consumption and IO cost for range and KNN queries. (Section 5)

In addition, Sect. 2 reviews the related work. Section 6 concludes this paper and points out future directions.

2 Related Work

Up to now, lots of index methods about spatial datasets have been proposed, which could be categorized into three parts. One is to index data points with space partitioning, another is to index data points with point partitioning, the last is to index data points after mapping multi-dimensional coordinates into one-dimensional representations. Typical methods of the first category are QuadTree [3], Octree [4] and KD-tree [2], which are respectively oriented to the space of two dimensions, three dimensions and k dimensions. A typical method of the second category is R-tree [5], on which many variants have been invented [6–8]. Typical methods of the third category include the use of Hilbert space filling curve [9] or Z-order curve [10] to map multi-dimensional coordinates to one dimensional.

The problem of the first category is that the amount of data points in each sub-space deviates greatly, then the index tree would be unbalanced and query performance on sub-space consisted of more data points would also be poor. The second category could ensure that data points in each part is balanced, but when the tree is adjusted, the ranges corresponding to intermediate nodes need to be changed accordingly. Moreover, each intermediate node needs to store coordinates of corresponding range. As height of the tree increases, the space consumed by intermediate nodes may be bigger than that of data to be indexed.

Among these indexing methods, the idea of learning an index on multi-dimensional space is a great breakthrough. However, the difficulty lies in that obtaining a learned index needs to sort the original data points firstly. As is well known, multi-dimensional data has multiple sorting methods. The intuitive idea is to use methods in the third category to map coordinates of data points in a multi-dimensional space to a one-dimensional space, and then build a learned index on those mapped one-dimensional representations. There is a big challenge to design a mapping function with no error, in meanwhile, it is pointed out that a simple multi-dimensional CDF could not be used as a good mapping function under this circumstance. That would cause a phenomenon that, even if one-dimensional representations of two data points are close, they could also be far apart in a multi-dimensional space. So, the ideas of methods in the third category are to design accurate mapping functions, then based on those mapped one-dimensional representations, a robust and applicable multi-dimensional index could be learned. To learn these mapping functions, machine learning technologies could be used.

SageDB [11] is known as a pioneer for introducing a machine learning method into DBMS. Its idea is to use distributions of data points and queries for learning the dimension for splitting and corresponding granularity for grid cell generation. In that paper, a locality sensitive hashing algorithm is used for searching data points. However, SageDB did not present an adjustment strategy for adapting to distribution changes.

Z-ordering model (ZM) is used in [16] for learning an index on spatial datasets. In that paper, Z-order curve [12] is used to map multi-dimensional coordinates to one-dimensional values, on which a recursive model index (RMI) is built. The query performance of the learned ZM index is better than R-tree. Moreover, the space cost is much smaller than R-tree.

As a variant of grid index, Flood [14] is an expansion of the multi-dimensional index field. Its innovation lies in the use of machine learning methods to optimize data layout. For data points in a multi-dimensional space, Flood firstly calculates importance of each dimension, then sorts grid cells in that space according to importance of dimensions. In Flood, importance of each dimension depends on three main factors: frequency of filters on that dimension, relevance between that dimension and other dimensions, and relevance of that dimension to query loads. Based on synthetic datasets and query loads, Flood learns a cost model by using random forest regression [15]. Then, it iteratively selects a dimension for sorting data points, and uses the cost model to choose optimal layout among multiple candidate layouts.

LISA [1] also uses the idea of splitting original multi-dimensional space into grid cells. Firstly, based on the Lebesgue measure [13], it calculates one-dimensional representation of data points in each grid cell. Then, the order of each grid cell is added to its one-dimensional representation. This method focuses on training a piece-wise linear regression model for predicting id of the page where a corresponding queried data point is stored. In the field of bioinformatics, LISA applied the idea of learning indexing to

DNA sequence search [18]. Similarly, Sapling [19] used the idea of learning indexing to accelerate genome lookup. In addition, the idea of learning index [20] is able to speed up the sorting algorithm by quickly mapping elements to the approximate positions.

Among those above methods, we could see that, a common problem for obtaining learned indexes lies in that they ignore dynamic changes of data distribution. This will do harm to query performance greatly, especially, in the scenario of k-nearest neighbors (KNN) query. The purpose of KNN query is to return k nearest data points of a query point. In the process of distance calculation, contribution of each dimension is equal. The existing methods of learned index often sort data points based on the mapped one-dimensional representations, which focus on the most important dimension. If the results of a KNN query cross several grid cells, the dimension importances concerned in the mapping function are inconsistent with that embodied in data distribution near the query point. Thus, large extra page scans are inevitable. Although, in traditional R-tree index field, there are a lot of researches on improving performance of KNN queries. However, in previously mentioned methods, such as SageDB and ZM, the KNN query is not focused enough. In ML-index [17] and LISA, the support for KNN queries are discussed. But, experimental results show that, when a query point falls in a sparse region, the performance of KNN query is much worse than that in Rtree. Therefore, in this paper, we proposes LIDUSA for adjusting grid cells dynamically to improve performance of KNN queries.

3 Architecture

In LIDUSA, recorded information is divided into three parts. The first part is about grid cells (shown in top-left part of Fig. 2), which includes range of grid cells, the number of points in each grid cell, and the number of inserted or deleted points in each grid cell since the last adjustment. With these information, corresponding grid cells could be obtained for a range query or a KNN query, and automatic adjustments of grid cells could be processed. The second part is a mapping function between original multi-dimensional coordinates of data points and one-dimensional representations (shown in upper right part of Fig. 2). With this function, data points could be sorted based on the mapped one-dimensional representations. Thus, according to the number of data points that could be stored in a page, the data points in a grid cell are stored orderly in a series of pages. In other words, with this function and inputted multi-dimensional coordinates, the corresponding page could be found easily. In order to improve the accuracy of mapping function, piece-wise functions are generally used, where points correspond to a piece are called a segment. The third part stores correspondence between a segment and its corresponding pages. To reduce query cost, points of a segment are stored closely, i.e., stored in consecutive pages Then, we could store page numbers, in ascending order, of a segment with a linked list. In other words, the array formed by all the linked lists is the third part (shown in lower part of Fig. 2).

Then, the procedure to construct LIDUSA also consists of three steps. The first is grid cell generation. In the original space V shown in Fig. 3(a), each dimension is divided into several parts to generate initial grid cells shown in Fig. 3(b). Secondly, according to number of points in a grid cell, current grid cell and its adjacent grid

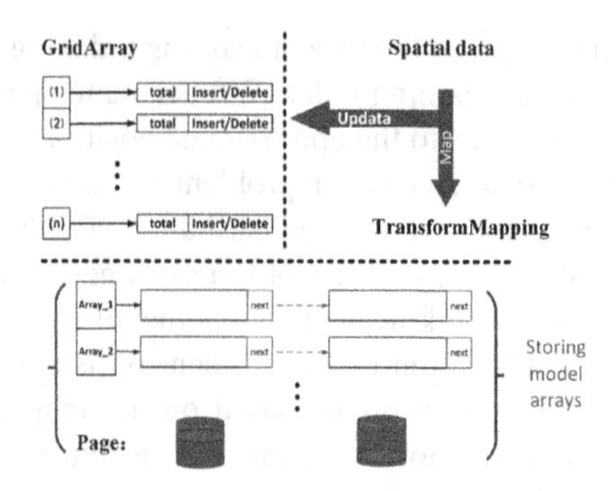

Fig. 2. LIDUSA stores information

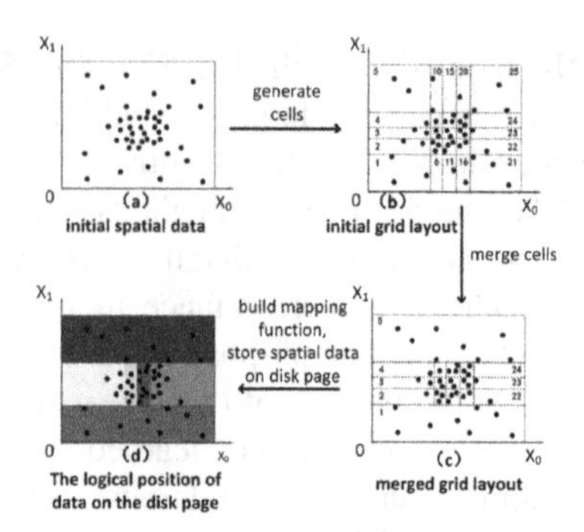

Fig. 3. Flow Diagram of LIDUSA's Initialization

cells are adjusted. As too many points are deleted from one grid cell, the grid cell and its adjacent may be merged into a new grid cell automatically. Otherwise, the corresponding grid cell may be split into two. As shown in Fig. 3(c), the top part and bottom part are merged grid cells. Finally, a piece-wise mapping function is established to map multi-dimensional coordinates to one-dimensional representations. And data points are stored sequentially according to the mapped one-dimensional representations. In Fig. 3(d), one color represents points correspond to a piece in mapping function, i.e. points in one segment.

3.1 Preliminaries

Table 1. Notations

Symbol	Definition
D	The dimensionality of data coordicates
V	the whole space with$V = [0, X_0) \times \cdots \times [0, X_{d-1})$
T	every axis into the same number of parts
C_i	The ith grid cell
Ω	The maximum number of keys stored in a page
ω	Threshold to control when to merge a grid cell
λ	Threshold to control when to split a grid cell
α	Threshold for judging sparse grids

The symbols involved in this paper are listed in Table 1. Among them, d is the dimension of a spatial dataset, V is the data space to be studied, and the dimensions of V are consistent with dimensions of data points. When initializing grid cells in space V, T is the number of regions divided in each dimension, C_i represents the ith divided grid cell. Because data dimensions are different for different spatial datasets, the maximum number of data points that could be stored in a disk page is also different. In this paper, we uniformly use Ω to represent the maximum number of data points that could be stored in a disk page. The last three parameters are hyper-parameters, ω is a threshold for determining whether a grid cell is to be merged with its adjacent cell, λ is for determining whether a grid cell is to be split, and α is used for judging whether a grid cell is sparse compared with other grid cells along the same dimension. See Sect. 5.1 for descriptions of experimental settings of these super-parameters, such as α, ω, and λ.

3.2 Initialization of Grid Cells

In data space V, each dimension is partitioned along the x_i axis to make data points distributed in each grid cell as evenly as possible. The whole space V can be represented by the union of a series of disjoint grid cells. During this process, the order of dimensions to partition is determined by sparseness of each dimension. Firstly, the sparsest dimension is selected to partition, and data points should be evenly distributed in each parts along this dimension. Then, according to its sparseness, each dimension is chosen sequentially from sparse to dense for partitioning. The partitioned results of dimensions are shown in Fig. 3(b). In order to simplify the partition procedure, it is assumed that each dimension is partitioned into T parts. Therefore, the whole space V is partitioned into $T * T$ grid cells in total.

Because each grid cell is a hyper-cube, C_i could be represented by points correspond to the left-lower corner and right-upper corner, $(l_0,\ldots,l_{d-1})$ and $(u_0,\ldots,u_{d-1})$, so C_i could be described as $[l_0,u_0) \times \cdots \times [l_{d-1},u_{d-1})$. Then, the merging and splitting of a grid cell could be easily realized by only modifying the coordinates of its left-lower and right-upper corners.

The details are shown in Algorithm 1. The input consists of three parts: a range of space V, an array of coordinates of data points *arrPoint* and number of parts partitioned along each dimension T. The output is an array of grid cells *arrCell*, with each element includes id and range of the corresponding grid cell. Firstly, based on *arrSortedDim*, the function *get_range_on_dim* is called repeatedly to partition along each dimension, and return T partition points of corresponding dimension in a multi-dimensional array *arrRange* (Line 2–4). And, all grid cells are looped over to find if the width of current grid cell along i-th dimension exceeds α times the average width of grid cells along that dimension. If so, the grid cell is judged as sparse, and function *merge_cell* is called to merge it with its adjacent grid cell (Line 7–13). The details of function *merge_cell* are described in Sect. 3.4.

Algorithm 1 init_cells

Input: *V*, ***arrPoint***, ***T***
Output: ***arrCell***
1: *arrSortedDim* = sort_dim(*V*, *arrPoint*) *//to sort all dimensions by sparsity*
2: **for** *i* =0 **to** len(*arrSortedDim*)-1
3: *arrRange*[*arrSortedDim*[*i*]] = get_range_on_dim(*arrSortedDim*[i], *arrPoint*, *V*, *T*)
4: **end for**
5: arrCell = get_cell(*arrRange*); *//to obtain grid cells coordinates and sort grid cells*
6: arrAvgWidthOnDim = get_avg_width(arrCell); *//to obtain the average width of grid cells in each dimension*
7: **for each** *cell* **in** *arrCell*
8: **for** i =0 **to** len(*arrSortedDim*)-1
9: **if** get_width_on_dim(*cell*, *i*) > α * arrAvgWidthOnDim[i]
10: *arrCell* = merge_cell(*cell*, *arrCell*)
11: **end if**
12: **end for**
13: **end for each**
14: **return arrCell**

3.3 Mapping Function

Here, multi-dimensional coordinates of data points are mapped to one-dimensional representations, which are closely related to pages they are stored. In order to reduce the complexity, the procedure is divided into two steps. The one is to sort the data points within their corresponding grid cell to get one-dimensional representations within a grid cell. The other is to incorporate the order of grid cells into these one-dimensional representations obtained above. Then all the data points are divided into several segments based on their one-dimensional coordinates, with each segment corresponding to a piece. The learning procedure is to get a piece-wise linear function for mapping one-dimensional representations to their segment numbers.

The coordinate mapping within a grid cell adopts a Lebesgue measure. Assuming that data point x is located in grid cell C_i, the coordinate of x is $(x_0, \ldots, x_{d-1})$, and the range of C_i is $[\theta_{j0}^{(0)}, \theta_{j0+1}^{(0)}) \times \cdots \times [\theta_{jd-1}^{(d-1)}, \theta_{jd-1+1}^{(d-1)})$, then x's coordinate within the grid cell is $\mathcal{M}(x) = \frac{\mu(H_i)}{\mu(C_i)}$, where $H_i = \left[\theta_{j0}^{(0)}, x_0\right) \times \cdots \times \left[\theta_{jd-1}^{(d-1)}, x_{d-1}\right)$, and μ is the Lebesgue measure. Next, we add id of the grid cell into the one-dimensional representation, then the final representation is $\mathrm{O(x)} = \mathrm{M(x)} + \mathrm{i}$, where i is the id of grid cell C_i.

The function to learn, called $\mathcal{S}(x)$, is composed of k monotonic piece-wise linear functions S_k. That is to say, $\mathcal{S}(x) = \mathcal{F}_k(x)$. The input of function *S* is *O(x)*, and output is the segment number that x belongs to. The relationship between values of *O* and a segment id is a many-to-one relationship.

Each segment corresponds to a linked list, where each element have two items, *pageAddress* and *mappingRange*. The former stores the address of corresponding disk pages, and the latter stores the range of one-dimensional representations *O(x)* of data points stored in corresponding disk pages.

3.4 Merging and Splitting of Grid Cells

With the continuous insertion and deletion of data points, the number of data points in a grid cell will change dynamically. Then, LIDUSA needs to record the changes of data points in each grid cell real-timely to automatically adjust grid cells.

Merging of Grid Cells: The procedure of grid cell merging is shown in Algorithm 2. The input comprises of three parts, the grid cell to be merged (*cell*), an array of all grid cells (*arrCell*), and the number of partitions along each dimension (*T*). The output is an array of all the merged grid cells (*arrMergedCell*). The condition for grid merging is that total number of data points in these two grid cells, *neighborCell* and *cell*, is less than $\omega*\Omega$ (Line 4).

Algorithm 2 merge_cell

Input: ***cell, arrCell:T***
Output: ***arrMergedCell:***
1: *arrPoint* = get_point(*cell*)
2: **while** len(*arrPoint*) <= ω*Ω
3: *neighborCell* = get_neighbor_cell(*cell, arrCell, T*) *//to find the neighboring grid cell*
4: **if** get_point_num(*cell*) + get_point_num(*neighborCell*) <= ω*Ω
5: *arrPoint* = union_points(*cell, neighborCell*) *//to collect data points*
6: del_page_address(*neighborCell*) *//to delete page of neighborCell from segment*
7: del_mapping_range(*neighborCell*) *//to delete map range of neighborCell from segment*
8: *arrMergedCell* = update_merged_cell(*arrCell, cell, neighborCell*)
9: **end if**
10: **end while**
11: ins_points(*arrPoint*) */to insert data points into the corresponding segment*
12: *seg* = get_segment(*cell*) *//to find segment where cell is located*
13: *sk* = retrain_sp(*seg*, get_data(*seg*), *arrPoint*) *//to retrain the mapping function of this segment*
14: update_Sk(*sk, seg*)
15: **return** *arrMergedCell*

Data Insertion: The implementation of function *ins_points* is shown in Algorithm 3. The input is an array of points to be inserted.

Algorithm 3 ins_points

Input: ***arrPoint***

1: **for each** *p* **in** *arrPoint*
2: *s* = get_seg_no(*p*) *//to calculate the segment number where p is located*
3: *page* = get_last_page_from_page_list(*s*) *//to find last page of p's corresponding page*
4: **if** get_num_of_points(*page*) < Ω *//when page is not full*
5: append_point(*page*, *p*)
6: **else**
7: *arrAppendedPoint* = sort(union(get_point_from_page(*page*), *p*)) *//to union and sort data points of page and p*
8: write_1st_half(*page*, *arrAppendedPoint*) *//to save first half of arrAppendedPoint to page*
9: *new_page* = extend_page(*s*); *//to extend a new page*
10: write_2nd_half(*new_page*, *arrAppendedPoint*) *//to save the second half part of arrAppendedPoint to new_page*
11: adjust_page_range(*s*, *page*, *new_page*) *//to adjust 1-D range of two pages in segment s*
12: **end if**
13: **end for each**

Splitting of Grid Cells: The inputted parameters are similar to Algorithm 2. The output is an array of all grid cells in space *V* after splitting. To avoid non-convex grid cells after repeatedly splitting and merging, only the previously merged grids could be split. At first, corresponding pages stored in LIDUSA are also adjusted, which means deleting data points in *arrNewCellPoint* from pages corresponding to *cell*, and inserting them into pages corresponding to *newCell* (Line 4–8). Based on the segment where *newCell* is located, mapping function S_k is retrained and updated (Line 9–11). Finally, all the grid cells that have been split are returned (Line 12).

Algorithm 4 split_cell

Input: ***cell*, *arrCell*, *T***;
Output: ***arrSplitCell***

1: *arrPoint* = get_points(*cell*) *// to store all data points in cell*
2: *newCell* = get_split_cell(*cell*, *T*) *//to get the range of new grid cell to split*
3: *arrNewCellPoint* = filter_cell_points(*arrPoint*, *newCell*) *//to choose newCell's corresponding points from arrPoint*
4: del_page_address(newCell)
5: del_mapping_range(newCell)
6: *arrSplitCell* = update_split_cell(*arrCell*, *cell*, *newCell*)
7: del_points(*cell*, *arrNewCellPoint*)
8: ins_points(*newCell*, *arrNewCellPoint*)
9: *seg* = get_segment(*newCell*)
10: *sk* = retrain_sp(seg, get_data(*seg*), *arrNewCellPoint*)
11: update_Sk(*sk*, *seg*)
12: **return** *arrSplitCell*

Data Deletion: The details of function *del_points* are shown in Algorithm 5. The input is an array of points to be deleted.

Algorithm 5 del_points

```
Input: arrPoint
1: for each p in arrPoint
2:   s = get_seg_no(p)  //to calculate the segment number where p is located
3:   page = get_located_page(s, p)  //to find p's corresponding page
4:   arrRemainedPoint = minus(get_points_from_page(page), p)  //to get the remaining
points of page, excluding p
5:   if len(arrRemainedPoint) > 0
6:     del_point_from_page(page, p)
7:     adjust_page_range(s, new_array(page))  //to adjust 1-D range of page in segment
s
8:   else
9:     remove_page_from_seg(s)  //to remove the page from segment s
10:  end if
11: end for each
```

4 Query Processing with LIDUSA

4.1 Range Query

The procedure of the range query is described in Algorithm 6. There is an input parameter *qc,* which denotes a query hyper-cube, and an output parameter *arrPage,* which denotes an array of pages containing data points within *qc*. Based on the one-dimensional representation, *arrCell* is looped over to obtain the upper bound and lower bound of each grid cell, which are stored in *o_lower* and *o_upper* respectively. Next, the corresponding segment numbers *s_lower* and *s_upper* could be found according to *o_lower* and *o_upper* (Line 3–4).

Algorithm 6 range_query

```
Input: qc
Output: arrPage
1: arrCell = intersect_with_cells(qc)  //to find all grid cells intersecting qc
2: for each cell in arrCell:
3:     o_lower, o_upper = get_bounded_ovalue(cell)
4:     s_lower = get_seg_by_mvalue(o_lower), s_upper = ger_seg_by_mvalue(o_upper)
5:     for s = s_lower to s_upper:
6:        arrSegPage = get_all_pages_from_seg(s)
7:        for each page in arrSegPage :
8:           if page_intersect_with(m_lower, m_upper):
9:              arrPage = append(arrPage, page)
10:          end if
11:       end for each
12:    end for
13: end for each
14: return arrPage
```

4.2 KNN Query

Given a query point, the purpose of KNN query is to find its K nearest neighboring data points. While performing a KNN query, a multi-dimensional hyper-cube centered on the query point is generated, and a range query is performed in this hyper-cube. Then, the found data points within the corresponding inscribed hyper-sphere are sorted, and top K of them are returned. If the number of such points is less than K, the side of that hyper-cube is increased and search process is repeated again. The details are shown in Algorithm 7. There are three inputs, one is query point q in space V, the second is the number of data points K to be returned, and the third is the dimension d of V. The output is an array of points, named *arrPoint,* consisting of K nearest neighboring data points of q.

Algorithm 7 KNN_query

```
Input: q, K, d
Output: arrPoint
1:  sideLen = LR(x)  //to predict the initial side length
2:  while len(arrCandidate) < K
3:    qc = build_hypercube(q, sideLen)  //to construct a range query
4:    arrPage = range_query(qc)  //to run the range query and get all pages
5:    for each page in arrPage:
6:      for each point in get_points(page):
7:        if dist(point, q) < sideLen
8:          arrCandidate= append(arrCandidate, point)
9:        end if
10:     end for each
11:   end for each
12:   sideLen = increment_length(sideLen, len(arrCandidate), K, d)  //to expand side length
13: end while
14: arrPoint = filter_KNN(arrCandidate, q, K)  //to sort points by distance
15: return arrPoint
```

5 Experiment

5.1 Experiment Settings

Datasets. We use a real-world dataset and synthetic datasets for experiments. In order to conduct a comparison between them, we scale horizontal and vertical axis both range from 0 to 500.

Didi. This dataset is from Didi Chuxing and contains all start points and end points of taxi orders in Chengdu city on November 1, 2016. These points are shown within latitude and longitude axes in Fig. 4. To train a prediction model, we randomly divides it into two parts, using a ratio of 1:1. One is for training, named **init-Didi**, the other is for inserting data points, name **ins-Didi**. In **init-Didi**, the number of data points is 209,423, which range from 103.5998 °E to 104.483069 °E and 30.32164 °N to 31.01437 °N, while the coordinates of data points in **ins-Didi** range from 103.25635 °E to 104.704311 °E and 29.51723 °N to 31.33911 °N. In addition, it is also necessary to test the impact of data deletions, a dataset named **del-Didi** is formed by randomly choosing half of data points from **ins-Didi**. The geographic range of **del-Didi** is from 103.25635 °E to 104.54141 °E and 30.17778 °N to 31.13284 °N.

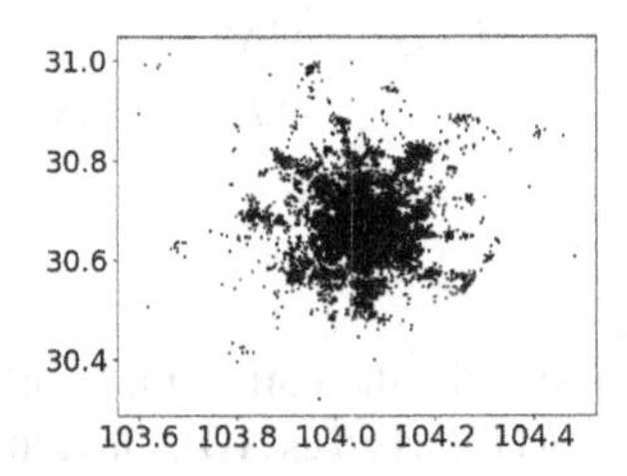

Fig. 4. Geography Distribution Map of samples from Didi

Synthetic Data Sets. Based on normal distribution, with mean value taken as 250 and variance set as 30, 60 and 100, three two-dimensional data sets, named as **Normal_30**, **Normal_60** and **Normal_100** are randomly generated, where there are 500,000 two-dimensional points respectively. In all these synthetic datasets, if it is closer to the center of space, data points will be denser, which is quite different from the uneven distribution of Didi. Similarly, each dataset is divided into three parts, one is for training, the other two are for testing the impact of data insertions and deletions. The generating process is the same as Didi, which is omitted here for the sake of space limit.

Evaluation Metrics. Here, we use three metrics to evaluate the effectiveness of LIDUSA: size, IO and IO ratio. Size indicates the disk storage space that an index consumes. IO indicates the average number of pages to be loaded during a query. We also use the metric IO ratio to indicate the ratio of an index's cost to R-tree's cost.

Parameter Settings. The parameter settings for experiments are shown in Table 2. Disk page size (PS) represents the storage space size of each disk page. MBR size (MS) represents the space size required for each data point. Page address size (AS) represents the space size occupied by the page address. α is a width ratio of a grid cell on a specified dimension to average width of all grid cells on that dimension. When $\alpha = 4$, the model performance is the best; ω is computed as the ratio of total number of data points in a grid cell to be merged to the maximum number of data points that could be accommodated in one page. When $\omega = 0.5$, the performance of system is the best; λ is a ratio of total number of data points in a grid cell to be split to the maximum number of data points that could be accommodated on one page.When $\lambda = 1.5$, the performance of system is the best. (Note that, results shown in this section are the average of 5 trials).

5.2 Didi Dataset

Memory Consumption. In Fig. 5, memory space consumed by three indexes – such as LIDUSA, LISA and R-tree – on Didi are reported. These experimental results correspond to three stages: when importing of init-DiDi has been finished (marked as Init);

Table 2. Parameter settings

Parameter	Setting
Disk page size (PS)	4096 bytes
MBR size (MS)	$8 \times 2 \times d = 16\,d$ bytes
Page address size (AS)	4 bytes
Ω	PS/(MS + AS)
T	240

when inserting of ins-Didi has been finished after Init (marked as AI); when deleting of Del-Didi has been finished after AI (marked as AD). Obviously, LISA and LIDUSA consume much less memory space than R-tree. In addition, compared with LISA, although LIDUSA needs to store information about merging and splitting of grid cells, it does not have a significant increase in memory consumption, which is only 6% bigger than that of LISA.

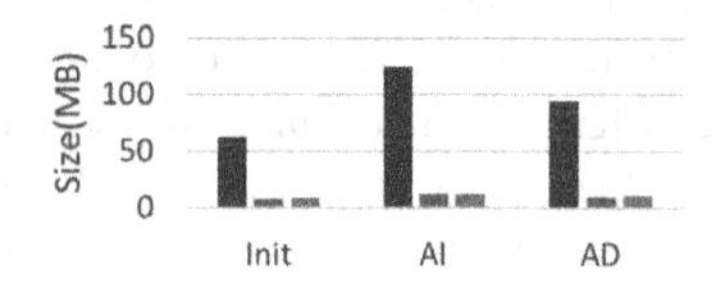

Fig. 5. Memory consumption on Didi (MB)

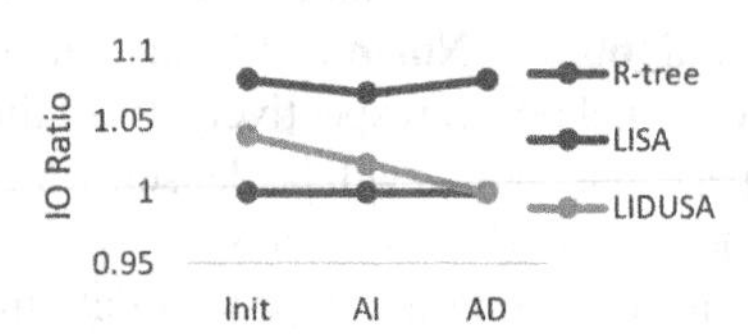

Fig. 6. Performance comparison of range queries on Didi

Range Query. In Fig. 6, for stages of Init, AI and AD, effects of range queries performed on R-tree, LISA and LIDUSA are respectively reported. The unit of vertical axis is IO Ratio, which represents the ratio of IO cost of candidate indexes to that of R-tree. It could be seen that in a scenario of range query, the performance of these three differs little. Among them, R-tree is slightly better than LIDUSA, and LISA is the worst.

KNN Query. In experiments of KNN queries, we also test query performance of LIDUSA, LISA and R-tree. The experimental results are shown in Fig. 7(a)–(c). The horizontal axis represents K-value of KNN queries, and the vertical axis represents average IO consumption. As could be seen from that figure, IO cost of LIDUSA is about 75% of R-tree and only 50% less of LISA. Although, as the value of K increases, IO costs of all these indexes also increase. And, LIDUSA also performs the best as k-value increases in all these cases, and R-tree follows, LISA is the worst. The reason is that when K is larger, the query hyper-cube should expand on both axises, but mapping function of LISA could only focuses on one primary axis. So, it results in higher error rate. On the contrary, merging of sparse grid cells is considered in LIDUSA, where direction of grid cell merging is orthogonal to that of grid cell sorting. Then, if grid cells are sorted based on vertical axis firstly, the sparse grid cell will be merged to its horizontal neighbors. In this way, search in both directions could be taken into account, query cost is significantly reduced. The low efficiency of R-tree lies in that it needs to filter corresponding regions layer by layer. During this process, the number of pages to be read is also larger than that of LIDUSA.

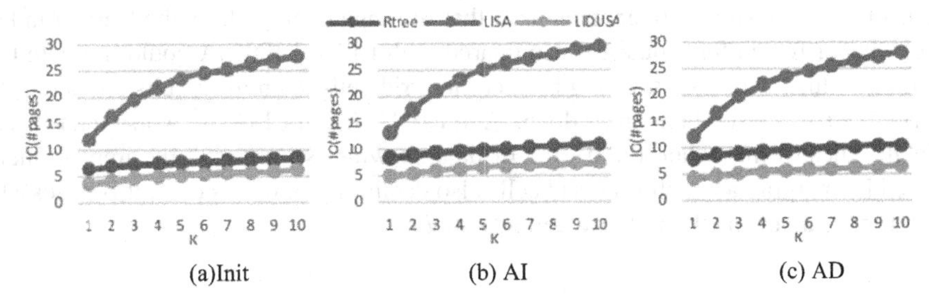

(a)Init (b) AI (c) AD

Fig. 7. Comparison of KNN query performance on DiDi

5.3 Synthetic Datasets

Memory Consumption. Memory space consumed by R-tree, LISA, and LIDUS are shown in Fig. 8. Similar to Didi, LISA and LIDUSA consume much less memory than R-tree. Also, compared with LISA, LIDUSA does not have a significant increase in memory consumption, although it stores more information about merging and splitting of grid cells.

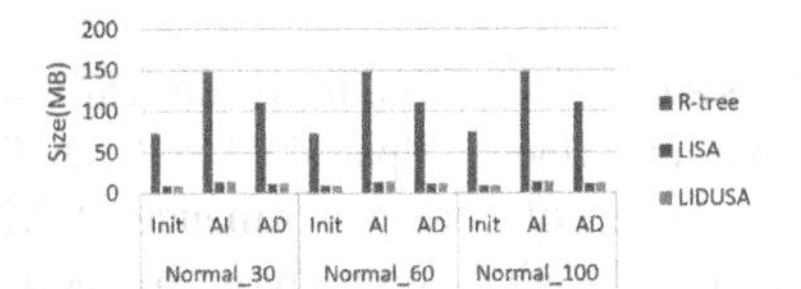

Fig. 8. Memory consumption on the Synthetic Dataset (MB)

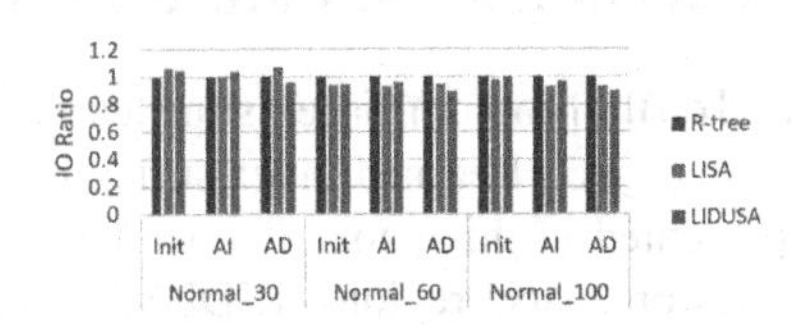

Fig. 9. Performance comparison of range queries on the Synthetic Dataset

Range Query. In Fig. 9, on Normal_30, Normal_60 and Normal_90, IO costs of range queries on R-tree, LISA and LIDUSA are shown. As could be seen from this figure, in other two datasets, except Normal_30, both LISA and LIDUSA are better than R-tree. This is due to the reason that data distributions of the latter two are more balanced than Normal_30. Then, by adjusting mapping function, the learned indexes could mitigate the harm of data skewing. In addition, in stage of AD, LIDUSA is superior to LISA. The reason is that the number of sparse grid cells becomes larger with deletions of data points, which is applicable for LIDUSA. In stage of AI, the data points in the initial grid cells become denser with continuously inserting of data points, then LIDUSA does not need to merge grid cells. So, compared with LISA, the advantages of LIDUSA do not exhibit in experiment results.

KNN Query. In KNN query experiments, we also test the query performance on these three indexes for stages of Init, AI, and AD. Because the experimental results of these three stages are similar, only results of Init are described in Fig. 10. From these figures, we can find that R-tree works much better than LISA and LIDUSA in Normal_30, while LIDUSA works the best in other two datasets: Normal_60 and Normal_100. These results embody a similar phenomenon to that of range query, where the learned index works better for uniform datasets. Compared with LISA, LIDUSA could mitigate the harmful influence of data skewing by merging grid cells, then the results are better. In Normal_30, this advantage drops down. And LIDUSA and LISA are worse than Rtree, which indicates that when the degree of data skewing is too large, the improvement gained by merging and splitting grid cells also declines. So, in this case, it is necessary to introduce a new method of grid cell adjustment.

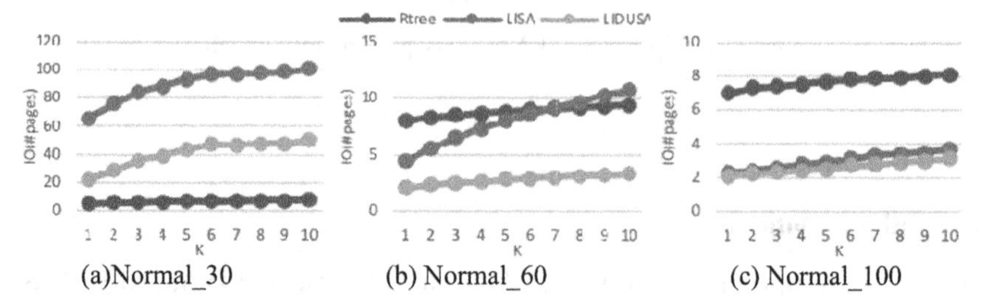

(a)Normal_30 (b) Normal_60 (c) Normal_100

Fig. 10. KNN query performance comparison of Init phase on synthetic datasets

6 Conclusion and Future Work

Based on the problem that existing learned indexes are difficult to adjust dynamically with data changes, a Learned Index structure for Dynamical Uneven Spatial dAta (LIDUSA) is presented in this paper. To handle with the problem of poor performance of KNN queries on sparse regions, LIDUSA could dynamically adjust data layout by merging and splitting corresponding grid cells, and relearn mapping function of this region to make data points stored in adjacent sparse grid cell also stored in neighboring disk pages. It combines the advantage of tree-shaped indexes, which could be adjusted dynamically, and that of learned indexes. In this paper, extensive experiments are conducted on a real-world dataset and synthetic datasets. From experiment results, it could be seen that LIDUSA is twice as fast as other existing indexes in the scenario of KNN query, which will greatly extend the applicable scope of learned indexes.

In the future, there are two directions to improve LIDUSA. The first is to make LIDUSA more intelligent, with introducing of reinforcement learning. Then, LIDUSA could merge or split grid cells automatically adapt to data distribution changes. The second is to introduce a new mapping function that could deal with data points in non-convex grid cells, which will make the merging or splitting of grid cells more flexible.

Acknowledgement. This work is partially supported by Science and Technology Planning Project of Fujian Province under Grant No. 2020H0023.

References

1. Li, P., Lu, H., Zheng, Q., Yang, L., Pan, G.: LISA: a learned index structure for spatial data. In: Proceedings of the 2020 ACM SIGMOD International Conference on Management of Data, pp. 2119–2133. ACM (2020). https://doi.org/10.1145/3318464.3389703
2. Bentley, J.L.: Multidimensional binary search trees used for associative searching. Commun. ACM **18**(9), 509–517 (1975)
3. Finkel, R.A., Bentley, J.L.: Quad trees: a data structure for retrieval on composite keys. Acta Informatica **4**(1), 1–9 (1974)
4. Meagher, D.: Geometric modeling using octree encoding. Comput. Graph. Image Process. **19**(2), 129–147 (1982)
5. Guttman, A.: R-trees: A dynamic index structure for spatial searching. In: Proceedings of the International Conference on Management of Data, pp. 47–57. ACM (1984)
6. Beckmann, N., Kriegel, H., Schneider, R., Seeger, B.: The R*-tree: an efficient and robust access method for points and rectangles. In: Proceedings of the International Conference on Management of Data, pp. 322–331. ACM (1990)
7. Kamel, I., Faloutsos, C.: Hilbert R-tree: an improved R-tree using fractals. In: Proceedings of the International Conference on Very Large Data Bases, pp. 500–509. Morgan Kaufmann (1994)
8. Sellis, T., Roussopoulos, N., Faloutsos, C.: The R+-tree: a dynamic index for multi-dimensional objects. In: Proceedings of the International Conference on Very Large Data Bases, pp. 507–518. Morgan Kaufmann (1987)
9. Sagan, H.: Space-Filling Curves. Springer, New York (1994)
10. Ramsak, F., Markl, V., Fenk, R., Zirkel, M., Elhardt, K., Bayer, R.: Integrating the UB-Tree into a database system kernel. In: Proceedings of the International Conference on Very Large Data Bases, pp. 263–272. Morgan Kaufmann (2000)
11. Kraska, T., Alizadeh, M., Beutel, A., Chi, E.H., Ding, J., Kristo, A., et al.: SageDB: a learned database system. In: Proceedings of the Biennial Conference on Innovative Data Systems Research (2019)
12. Wang, H., Fu, X., Xu, J., Lu, H.: Learned index for spatial queries. In: Proceedings of the IEEE International Conference on Mobile Data Management, pp. 569–574. IEEE (2019)
13. Royden, H.L., Fitzpatrick, P.M.: Real Analysis (2010)
14. Nathan, V., Ding, J., Alizadeh, M., Kraska, T.: Learning multi-dimensional indexes. In: Proceedings of the International Conference on Management of Data, pp. 985−1000. ACM (2020)
15. Breiman, L.: Random forests. Mach. Learn. **45**(1), 5–32 (2001)
16. Kraska, T., Beutel, A., Chi, E.H., Dean, J., Polyzotis, N.: The case for learned index structures. In: SIGMOD, pp. 489–504. ACM (2018)
17. Davitkova, A., Milchevski, E., Michel, S.: The ML-index: a multidimensional, learned index for point, range, and nearest-neighbor queries. In: Proceedings of the International Conference on Extending Database Technology, pp. 407–410 (2020)
18. Ho, D., Ding, J., Misra, S., Tatbul, N., Nathan, V., Vasimuddin, et al.: LISA: towards learned DNA sequence search. arXiv: Databases (2019)
19. Kirsche, M., Das, A., Schatz, M.C..: Sapling: accelerating suffix array queries with learned data models. bioRxiv (2020)
20. Kristo, A., Vaidya, K., Çetintemel, U., Misra, S., Kraska, T.: The case for a learned sorting algorithm. In: Proceedings of the International Conference on Management of Data, pp. 1001–1016. ACM (2020)

Why is Your Trojan NOT Responding? A Quantitative Analysis of Failures in Backdoor Attacks of Neural Networks

Xingbo Hu[1,2], Yibing Lan[1,2], Ruimin Gao[3], Guozhu Meng[1,2(✉)], and Kai Chen[1,2]

[1] SKLOIS, Institute of Information Engineering, Chinese Academy of Sciences, Beijing, China
mengguozhu@iie.ac.cn
[2] School of Cyber Security, University of Chinese Academy of Sciences, Beijing, China
[3] Mathematics and Statistics, University of Victoria, Victoria, Canada

Abstract. Backdoor has offered a new attack vector to degrade or even subvert deep learning systems and thus has been extensively studied in the past few years. In reality, however, it is not as robust as expected and oftentimes fails due to many factors, such as data transformations on backdoor triggers and defensive measures of the target model. Different backdoor algorithms vary from resilience to these factors. To evaluate the robustness of backdoor attacks, we conduct a quantitative analysis of backdoor failures and further provide an interpretable way to unveil why these transformations can counteract backdoors. First, we build a uniform evaluation framework in which five backdoor algorithms and three types of transformations are implemented. We randomly select a number of samples from each test dataset, and then these samples are poisoned by triggers. These distorted variants of samples are passed to the trojan models after various data transformations. We measure the differences of predicated results between input samples as influences of transformations for backdoor attacks. Moreover, we present a simple approach to interpret the caused degradation. The results as well as conclusions in this study shed light on the difficulties of backdoor attacks in the real world, and can facilitate the future research on robust backdoor attacks.

Keywords: Deep learning · Backdoor attack · Robustness · Transformation · Interpretability

1 Introduction

Deep learning has gained tremendous success in a variety of fields, such as image classification, speech recognition, natural language processing, and gaming. Moreover, its superior performance motivates the application in the security-critical areas including autonomous driving, face payment, and identity verification. However, deep learning has proved to be vulnerable and poses a great risk

Y. Lai et al. (Eds.): ICA3PP 2021, LNCS 13157, pp. 754–771, 2022.
https://doi.org/10.1007/978-3-030-95391-1_47

to its users. Since Szegedy *et al.* [33] first proposed the existence of adversarial examples in deep learning, researchers and practitioners have fleetly payed attention to issues of security and privacy in deep learning. It reveals that deep learning is suffering from adversarial attacks, model inversion, model extraction [11] and backdoor attacks [12]. Compared to other attacks, backdoor is more like an intentional attack initialized by a miscreant while the others are more like a special vulnerability of deep learning models. In a typical backdoor attack, training data is poisoned with well-crafted samples [8,22]. If an innocent developer trains a classification model with poisoned data, a backdoor is consequently implanted and the attacker can make the model output a chosen result as expected.

Backdoor attacks can incur severe damages and even threaten people's safety. If one face recognition model is implanted with a backdoor, Bob with a sticker on his face may deceive the model and buy a lunch on Alice's bill [3]. Even worse, an object detection system in a self-driving car may misclassify one STOP sign as a 30km/h speed limit due to an unconscious backdoor inside. This error can cause a serious traffic accident. To counteract this attack, prior studies have developed a number of techniques to detect trigger samples [5,7], reverse engineer backdoors in models [13,21,36], and harden models blindly [20]. However, it is unclear and still unexplored whether backdoor attacks are effective as claimed in prior studies and what difficulties will be confronted to trigger a backdoor in reality.

There are many uncertainties in triggering a backdoor of deep neural networks so that the implanted trojan may not respond to the attacker. First, these influencing factors can come from the physical world [38]. Taking face recognition as an example, a facial image is photographed by an on-site camera and it is heavily impacted by the shooting distance, angle, focus position and illumination conditions. Every time the face recognition system is used in the physical world, the images are likely varying. Second, there are uncertainties influencing the success rate of backdoor attacks in the digital world. The image may go through pre-processing and transformations like cropping, scaling, and rotation. These transformations are attributed to either the defensive measures employed by the target model, or an adaption of the size of the model input. Given that, it is intriguing and important to explore why backdoor attacks fail and evaluate the robustness of backdoor attacks.

In this study, we conduct a quantitative analysis of failures of backdoor attacks in deep neural networks. To be more specific, we aim to transform model inputs and determine the influence to the prediction results. To this end, we first build a uniform evaluation framework that integrates two vanilla deep neural networks – LeNet [14] and ResNet-34 [9]. Five backdoor algorithms are implemented and we obtain 8 trojaned models as the test subject. We then employ three transformations on both input samples and backdoor triggers, and create a number of test samples as input. The robustness of backdoor attacks are quantified by attack success rate, through which we shed light on the different resilience to transformations. Last, we leverage the interpretability algorithm SmoothGrad [31] to explain how models make a right or wrong prediction under transformations.

Contributions. To sum up, we make the following contributions.

- *A uniform evaluation framework.* It implements five backdoor algorithms and contains 8 trojaned models. We develop three types of transformations for both input samples and backdoor triggers that can simulate the uncertainties in the physical and digital worlds.
- *A quantitative analysis of backdoor failures.* We have created 165,000 samples in total as model input and measured the influence of data transformations on backdoor attack success rate.
- *Explanation on backdoor robustness.* Through the interpretability analysis on backdoor robustness, we unveil how the poisoned samples are recognized by models and their prediction limits in the context of data transformations.

2 Related Work

Backdoor Attacks in Deep Learning. Gu *et al.* [8] introduce a backdoor attack called BadNets for the first time, BadNets pollutes the training set, and achieves nearly 90% attack success rate in the traffic sign recognition. In order to enhance the concealment of injected backdoor, Chen *et al.* [3] mix backdoor triggers with a benign image. However, the attack success rate is proved to be related to the blending ratio. Besides, Li *et al.* [16] aim to regularize the disturbance trigger using p-norm so that the noise can be generated in a small range. Except for changing the trigger, Bagdasaryan *et al.* [2] claim that the loss computation was poisoned in the model-training code. On the one hand, Liu *et al.* [23] polluted the samples of reflected trigger image under common natural reflection phenomenon. Cheng *et al.* [4] define the trigger as style conversion, and train a generative adversarial network (GAN) model to generate polluted samples. Li *et al.* [18] plug the trigger into the image invisibly by image steganography. The above methods change the target labels of samples, so the attack can still be detected by checking the labels and samples. Therefore, a new attack strategy called clean label backdoor attack is proposed. Turner *et al.* [35] study the backdoor attack of clean labels at the beginning. They apply adversarial interference as a trigger to benign samples in the target category. Zhao *et al.* [42] extend this idea by using general perturbations in video classification. Saha *et al.* [29] minimize the distance of the target class in the feature space and inject the poisoned information into the image. Moreover, Quiring *et al.* [27] hide the trigger by zooming attack [39]. Apart from the deliberately designed triggers, some of the studies also use semantic shapes as backdoor triggers. For example, Bagdasaryan *et al.* [2] first explore this kind of backdoor attack named the semantic backdoor attack. Lin *et al.* [19] design hidden backdoor which can be activated by the combination of certain objects. In addition, some non-poisoning attacks have also been researched. For instance, Dumford *et al.* [6] explore non-poisoning backdoor attack and focus on modifying the parameters of models. Besides, Rakin *et al.* [28] consider to insert a target trojan during the training process. Tang *et al.* [34] introduce malicious backdoor module as trigger. As for the defense of backdoor attacks, some solutions have been proposed like unlearning [10].

Robustness Evaluation of Backdoor Attacks. The current research about the robustness of backdoor is basically sketchy. Weng *et al.* [37] analyze the relationship between the robustness of backdoor attacks and adversarial attacks. Xue *et al.* [40] demonstrate that the attack with a static trigger is vulnerable, and much less effective in the physical world. Furthermore, Li *et al.* [17] summarize that when triggers in testing images are not consistent with another trigger used for training, the attack may be unstable. Therefore, The transformation-based-pre-processing (e.g., flipping and scaling) on the testing image before prediction will sharply decrease attack success rate. Pasquini *et al.* [26] transform the triggers with typical image processing operators of varying strength, and discuss the results of the backdoored DNN. The response of geometric and color transformations suggests that the change of the trigger geometry and partial occlusion of trigger can lower the success rate. *Differently, our study considers a variety of backdoor algorithms and employs three types of transformations to measure their robustness. Moreover, we try to explore how backdoor models view trigger samples and the transformed.*

3 Preliminary and Overview

3.1 Backdoor Attacks

Deep learning can be interpreted as a process to learn an abstraction of massive amounts of data via multi-layer neural networks. For the supervised learning, it acquires a well-labeled data set $\{\mathcal{X}, \mathcal{Y}\}$ and computes an optimal parameter θ for the neural network F, that is, $F_\theta : \mathcal{X} \rightarrow \mathcal{Y}$. The model F_θ is correct under a certain probability where $F_\theta(\mathcal{X}) = \mathcal{Y}^*$ and the model accuracy can be computed with the portion of different elements between $\mathcal{Y}$ and $\mathcal{Y}^*$. Prior studies [3,8] show that the training data can be maliciously crafted to introduce a backdoor in a neural network, i.e., backdoor attack. As such, the trojaned model can output the attacker-chosen label via the trigger. Without loss of generality, we define backdoor attacks in neural networks as follows.

Definition 1 ***(Backdoor).*** *Given a trojaned neural network F_{θ^-}, there exists a set of samples X that will be classified as a fixed class (e.g., y_t) if they are decorated with a specific trigger t. That is, $F_{\theta^-}(x \oplus t) = y_t$.*

To evaluate the attack success rate (ASR) of backdoor attacks, an attacker can randomly choose a set of clean samples $\mathcal{X}$ that are decorated with the fixed trigger t, and determine how many percent of samples are classified as y_t. More rigorously, the clean samples $\mathcal{X}$ should not be of the factual class y_t, i.e., $\forall x \in \mathcal{X}, F_\theta(x) \neq y_t$. For ease of understanding, all the notations in this paper have been summarized in Table 1.

3.2 Approach Overview

In this study, we aim to evaluate the robustness of mainstream backdoor attack methods in the scenario of the physical world. The samples that serve as intentional triggers may be affected by the realistic environment that can undermine

Table 1. Notations in this paper

Notation	Description
$\mathcal{X} \times \mathcal{Y}$	The labeled data with input space $\mathcal{X}$ and label space $\mathcal{Y}$
$\mathcal{X} \times \mathcal{Y}^*$	The factual labels $\mathcal{Y}^*$ for a given data $\mathcal{X}$
θ	Model parameters
F_θ	A neural network with the parameter θ
F_{θ^-}	A backdoored neural network with the parameter θ^-
$\mathcal{X}^b$	The poisoned data for training
$\mathcal{Y}^b$	The model's predictions for the poisoned data
$x \oplus t$	A data point with trigger t attached
y_t	An attacker-chosen class for backdoor attacks
$\mathcal{S}$	The rendering space for a trigger

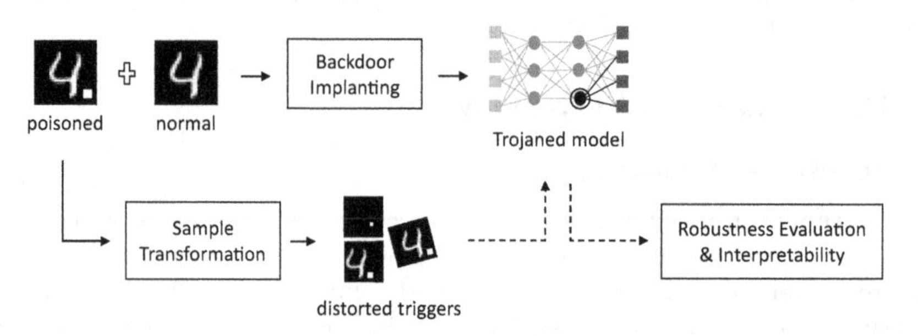

Fig. 1. System overview

the efficacy of trojaned models. Figure 1 shows the overview of this study. In particular, we first create a poisoned data set with diverse backdoor triggers and insert a backdoor into the trained model with five backdoor methods (i.e., *backdoor implanting*). All the poisoned samples undergo transformations such as scaling and rotation (i.e., *sample transformation*). The distorted samples are then passed to the trojaned model. Last, we evaluate the robustness of trojaned models in front of distorted samples and provide an interpretability analysis (i.e., *robustness evaluation & interpretability*).

4 Methodology

In this section, we present the details for the methodology.

4.1 Backdoor Implanting

In a conventional process of backdoor implanting, the attacker needs to first design a trigger, determine the optimization objective, and train a model with poisoned and non-poisoned samples.

Trigger Design. A trigger is the pattern used to poison training data and activate backdoors in a neural network [12]. In a backdoor attack, the original image I can be represented as $I : \{\langle x, y, z\rangle\}$, where x and y are the x-coordinate and y-coordinate, respectively, and z is the RGB value for the pixel at $\langle x,\ y\rangle$. The poisoned image can be created by performing a pixel-wise computation with the trigger t in the same coordinate, i.e., $I \oplus t$. There are several intriguing properties of a trigger, such as *size*, *color*, *texture*, *naturalness*, *detectability*, etc. A line of work has developed different types of triggers to inject a backdoor. In this paper, we select five representative trigger designs in terms of these properties. In particular, the selected triggers are from the following studies.

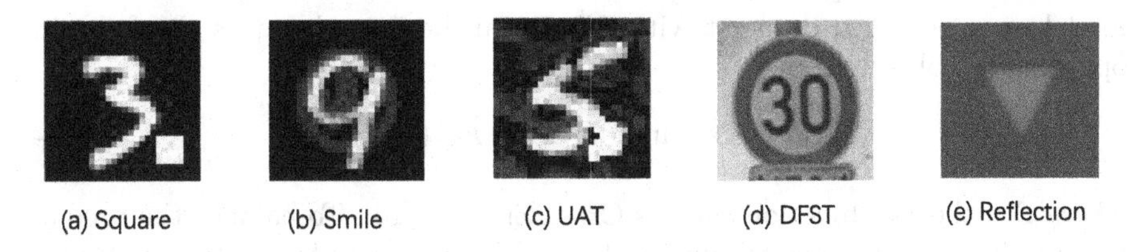

Fig. 2. Examples of backdoor trigger

1. BadNets. The trigger is a monochromatic square usually put at the corner of the original image as shown in Fig. 2 (a) [8]. As the background color of MNIST images is black, we render the trigger as a white square as $\{\langle x, y, z\rangle | \langle x, y\rangle \in \mathcal{S}\ \wedge z = 255\}$, where $\mathcal{S}$ is the rendering space of the trigger. Therefore, the poisoned image can be obtained by overlaying the trigger on image I.

2. Blended Injection in [3]. As shown in Fig. 2(b), it blends an image with the trigger in the following manner.

$$z_i = \lambda z_i + (1 - \lambda)z_t,\ \text{where}\ \langle x_i, y_i\rangle \in \mathcal{S} \wedge \lambda \in [0,\ 1] \tag{1}$$

Noted that λ is a parameter to control the blending ratio, and the trigger does not exist when $\lambda = 1$ and it is a BadNets trigger if $\lambda = 0$.

3. Deep Feature Space Trojaning (DFST). The trigger is not a fixed pattern in DFST, but generated dynamically depending on the original image. In particular, the attacker first trains a CycleGAN [4] that incorporates styles (e.g., sunset). The CycleGAN can then generate a stylized image as poisonous samples.

4. Refool. It proposes a new type of triggers using the natural reflection phenomenon [23] as indicated in Fig. 2 (d). Reflection is very common in reality when an object has a smooth surface. Assume the reflection image is x_R, the poisoned data x^b can be computed as

$$x^b = x + x_r \otimes \lambda \tag{2}$$

where λ is a convolution kernel that controls the reflection effect. Three effects are: reflection from the same-depth layer, out-of-focus layer and ghost effect.

5. Universal Adversarial Trigger (UAT) [42]. Similar to DFST, UAT is an optimized trigger based on the training data. It is initialized with random values for a fixed-size area, i.e., $t : \{\langle x, y, z\rangle | \langle x, y\rangle \in \mathcal{S} \wedge z \in \mathbb{R}^N\}$. The poisoned sample can be represented as $x^b = x \oplus t$ and trigger t can be optimized through:

$$t = \arg\min_t \Sigma_{x_i \in \mathcal{X}} - \frac{1}{l} \Sigma_{j=1}^{l} y_j \log(h_j(x_i \oplus t)) \tag{3}$$

where l is the number of output labels, y_j here is the probability of being class j, and $h(\cdot)$ is the softmax output of the model. In this manner, the attacker can generate a trigger that is universally workable for the training data.

Backdoor Training. Generally, a backdoor is installed by (re-)training the model on a poisoned data set with $\mathcal{X}^b$. It can be formulated as the following optimization objective.

$$\theta^- = \arg\min_{\theta^-} \Sigma_{x \in \mathcal{X}} \mathcal{L}(F_{\theta^-}(x), y) \tag{4}$$

where $\mathcal{L}$ is the loss function such as Cross Entropy [24]. To balance the predication accuracy for normal samples and attack success rate for backdoor samples, we choose 10% training samples randomly and put the trigger on them.

4.2 Sample Transformation

Backdoor attacks may not perform as effectively as claimed in prior studies due to data transformation in reality [26]. As aforementioned, there are many uncertainties from one sample with trigger to the target model [41,43]. These uncertainties can significantly affect attack success rate of backdoor attacks, i.e., one distorted trigger sample may fail to set the model on fire. Therefore, transforming input data can better illustrate how robust deep neural networks are when triggering a backdoor. In this study, we consider the following transformations.

Translation. In geometry, translation is to move a subject with a certain distance without rotating it. It often happens before an image is passed to the model. This transformation is eligible for both trigger and image. For instance, the focal point of one photographed image shifts by a small distance. The digital image has different sizes with the dimensions of model input and so it has to be cropped that can incur a translation of backdoor trigger in the range of image. To evaluate the stability of backdoor triggering, we translate the trigger in the image canvas by a certain distance (e.g., r) and create a number of patched images where the triggers' center is scattered on the ring of a circle with radius r. We also translate the entire image to simulate the process of image transformation in model defense [1]. There will be blanks in transformed images and we fill the blanks with the shift-out parts. To determine the largest translation distance, we employ a trial-and-error method to translate the unpatched image gradually unless it is wrongly classified by the model.

Scaling. One image and its trigger can be scaled outward or inward to imitate a varying focal length. We are then able to determine how backdoor attacks are affected by scaling. As for the image, we can cut out an area of same sizes as the original when the image is scaled out. There also exists the blank problem when shrinking images, and we fill the blanks with black color (i.e., *(0, 0, 0) in RGB images and 0 in grey-scale images*). We call the function "PIL.Image.resize()" and use nearest neighbor interpolation (i.e., PIL.Image.NEAREST) to scale the image. Similar to translation, we have to determine what the largest scaling ratio is in the trial-and-error method.

Rotation. Image rotation is an image processing routine, commonly used in data augmentation [30] during model training. Moreover, the images captured from the physical world oftentimes suffer such transformations considering the photographer is not at the right front of the photo subject (i.e., with a varying angle). As a consequence, the image has a certain angle horizontally. In this study, we intend to explore whether a rotated image or trigger can affect the performance of backdoor attacks. Given one image, we invoke the function "PIL.Image.rotate()" in Python library PIL to rotate the target image with a certain angle. Similar to scaling, we also fill the blanks due to rotation with black color. It is noted that some triggers are rotational symmetry. For example, the white square in Fig. 2(a) is 4-fold rotational symmetry so we only rotate the trigger by less than 90° with a stride of 15° in our experiment.

4.3 Robustness Evaluation and Interpretability Analysis

Given the poisoned data $\mathcal{X}^b$ of n dimensions, the model produces n-dimensional $\mathcal{Y}^b$. The attack success rate (ASR_1) can be computed as: $|\{y|y \in \mathcal{Y}^b \wedge y = y_t\}|/n$. We can perform transformations, i.e., $T(\cdot)$, on $\mathcal{X}^b$ and obtain data $\mathcal{X}^{\tilde{b}}$. Similarly, the output labels for $\mathcal{X}^{\tilde{b}}$ are represented as $\mathcal{Y}^{\tilde{b}}$. Then, the ASR_2 is: $|\{y|y \in \mathcal{Y}^{\tilde{b}} \wedge y = y_t\}|/n$. As a result, the attack success rate of backdoors is dropped by ASR_1 - ASR_2 because of transformation $T(\cdot)$. To gain a finer influence function of transformations on ASR, we parameterize these transformations such as rotating triggers by 15° for one checkpoint. The details for the parameterized transformations are described in each experiment at Sect. 5.

With the decorated samples and their distorted ones, we intend to explain why it sometimes fails to trigger a backdoor in an interpretable way. In particular, we employ SmoothGrad to visualize the important regions in an image that are responsible for the decision. In this way, we attempt to explain how a trigger image has a label flip, i.e., from y_t to $\neg y_t$, after transformations, and whether the important regions stay unaltered with the unchanged output label.

5 Evaluation

5.1 Experimental Setup

In the experiment, we take two datasets in account: MNIST [15] and GTSRB [25]. More specifically, MNIST is an image dataset of handwritten digits from 0 to 9, and contains 60,000 samples. Each data point in the set is a

Table 2. Performance of the original and five backdoor models on the two datasets. "ACC. (%)" denotes the accuracy of the model's main task, "ASR (%)" is the attack success rate of backdoor methods in our framework.

Model	Pattern	MNIST		GTSRB	
		ACC. (%)	ASR (%)	ACC. (%)	ASR (%)
Original	NA	99.13	–	97.05	–
BadNets	Square	99.17	99.95	97.32	95.79
Blend	Smile	99.02	99.78	96.58	96.63
UAT	UAT	97.01	90.06	94.30	82.60
DFST	Cezanne	–	–	95.27	98.30
Reflection	Apple	–	–	94.01	93.91

greyscale image of size 28×28. GTSRB is an image dataset of 43 kinds of German traffic signs, and contains about 40,000 samples. Each data point is a RGB image of difference size, and it is resized to size 96×96 before fed to the network. We train a LeNet [14] model on dataset MNIST and a ResNet-34 [9] model on dataset GTSRB since these two network structures achieve state-of-the-art performance on these datasets. As shown in Table 2, the accuracies of vanilla models are 99.13% and 97.05%, respectively.

For these two benign models, we apply five mainstream backdoor algorithms, i.e., BadNets, Blend, DFST, Refool and UAT, to introduce a backdoor. The BadNets method is the first method to propose the concept of deep learning backdoor attacks, which is also the main attack idea based on data pollution-based backdoor attacks. It mainly prints trigger patterns such as small white squares into the sample. Since the trigger by BadNets is obvious and easy-to-detected, the subsequent backdoor attacks are dedicated to increase the concealment of trigger. In particular, the Blend method mixes the trigger and the sample in a certain ratio. When the mixing ratio is small, the image will be close to the original sample. The UAT method is to generate the general disturbance trigger in the original image that is different from other methods. It does not need to maliciously change the label of the poisoned sample when the backdoor is implanted. The Refool method is inspired by the natural reflection phenomena and implant a backdoor with a reflected image. The DFST method basically trains a style transfer model with CycleGAN, with which it transfers the style of the current image to another, for example in a painting style of cezanne.

Because DFST and Reflection backdor attacks are limited in RGB images, we only implement BadNets, Blend and UAT attacks on dataset MNIST and set their poisoning rates to 10% for all. For GTSRB dataset, we implement all 5 backdoor attacks and set poisoning rates of DFST and Reflection to 20%.

Experiment Parameters. To preserve as many characteristics of triggers as possible during transformation, we make a proper design for the position and size of original triggers. Specifically, we use a 5×5 white bottom-right square as

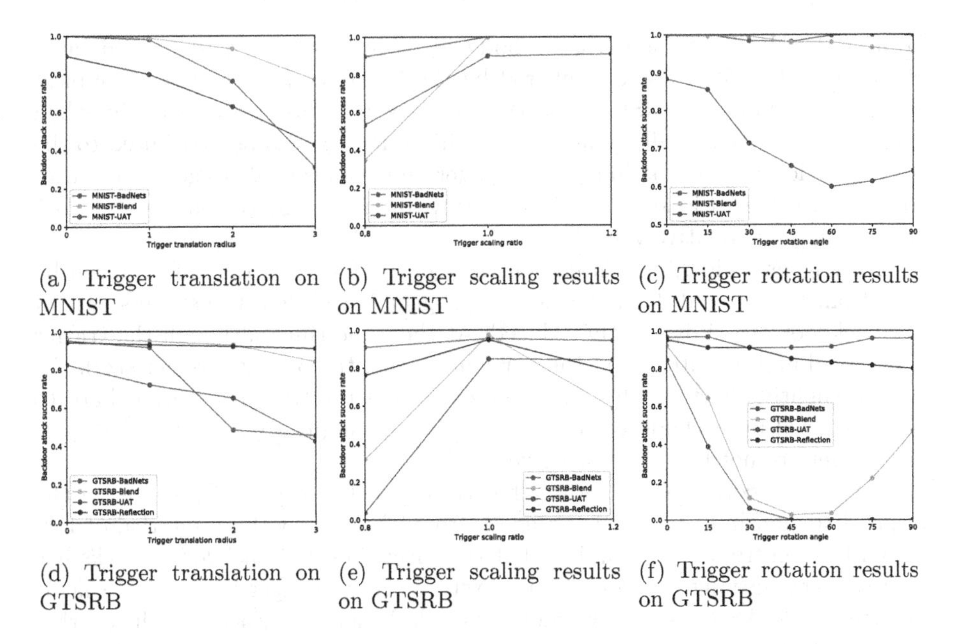

(a) Trigger translation on MNIST

(b) Trigger scaling results on MNIST

(c) Trigger rotation results on MNIST

(d) Trigger translation on GTSRB

(e) Trigger scaling results on GTSRB

(f) Trigger rotation results on GTSRB

Fig. 3. Success rate of backdoor attacks under trigger transformations

BadNets triggers. For UAT attack, we use a 8×8 perturbation square on MNIST, and a 28×28 perturbation block on GTSRB. To avoid exceeding the boundary of an image, we leave 3 bottom-right pixels for BadNets and UAT triggers on MNIST and BadNets triggers on GTSRB, but 6 pixels for UAT triggers on GTSRB. We use a smile meme as Blend trigger and stylize the images with the painting style of cezanne in attack DFST. Since a cezanne-styled trigger does not have a specific shape, we do not consider trigger transformations for this attack. For all experiments, we choose the first class as the target label y_t (e.g., 0 in MNIST and speed 20 limit sign in GTSRB).

5.2 Backdoor Robustness Under Trigger Transformation

In this experiment, we employ translation, scaling and rotation on triggers to evaluate the robustness of backdoor. For each transformation, we select a suitable transformation range to inspect backdoor attack changes from a small sample set. Figure 3 shows the results for applying these three transformations. The horizontal axis of each plot represents specific transformation parameters, and the vertical axis is backdoor attack success rate.

For trigger translation, we measure ASRs when the trigger is moved to a circle with a radius of 1, 2, 3 pixels, respectively. It is observed that as the translation distance increases, the attack success rate of all backdoor attacks decreases in Fig. 3a and 3d. However, the decline of the attack success rate varies from attack methods. BadNets have the fastest attenuation and this is partly because BadNets is the simplest method which overlays triggers to a certain area of samples. The features remembered by the backdoor model are the simplest and most obvious. The ASR of simple features will be dropped drastically when the trigger moves a small distance. In addition, it may also be attributed to the overlap ratio between the implanted trigger and the original trigger. It is easier to make overlap ratio smaller within a tiny amount of movement as the 5×5 square trigger is relatively small.

For trigger scaling, we amplify the trigger by 0.8, 1, or 1.2 times. It can be seen from Fig. 3b and 3e that when the trigger is reduced, the success rate of the backdoor attack is greatly reduced. At the same time, although the trigger is enlarged and deviated from the benchmark (i.e. the original trigger size), the ASR is slightly affected. This result indicates that the trigger feature is hard to be recognized by backdoored models after being compressed, and the details of the trigger are not lost after being zoomed in.

For trigger rotation, we rotate the trigger clockwise with the center of the original trigger and the rotation angle is from 0° to 90° with an interval of 15°. It can be seen from Fig. 3c and 4f that the success rate of backdoor attacks has decreased by varying degrees. For BadNets, since its trigger is a square that is symmetric, its ASR curve is symmetric as well. When it reaches 45°, the overlap between the transformed trigger and the original is the smallest, so that its ASR is the lowest. For UAT, the decline is most obvious, because the trigger of UAT is generated by the general perturbation value of an area in the lower right corner so that the model can maximize the output probability of the target label. When the UAT trigger is rotated, the value of the pixel is more likely to be different after the nearest neighbor algorithm, leading to the original neurons related to the target label cannot be activated as expected.

5.3 Backdoor Robustness Under Image Transformation

Similarly, we perform translation, scaling, and rotation transformations on the backdoor samples with triggers. Then we measure the impact of these transformations on backdoor attack success rate. For each transformation, we also select an appropriate transformation range, and generate 1000 samples under different transformation configurations to test the robustness of the five backdoor attack methods. In order to maintain uniformity with trigger transformation and ease the comparison of the results, we have adopted the same transformation settings for these three attacks. However, the transformation may lead a sample to be an invalid one, i.e., a wrong predication is not due to the backdoor, but the heavily distorted image. So we also perform these transformations on clean samples and test the prediction accuracy for them as a benchmark. Figure 3 reports the results of applying these three transformations. The horizontal axis of each plot represents the converted value of backdoor samples or clean samples, and the vertical axis represents the success rate of backdoor attacks or main attack accuracy, respectively. The black dotted line represents the accuracy of clean samples.

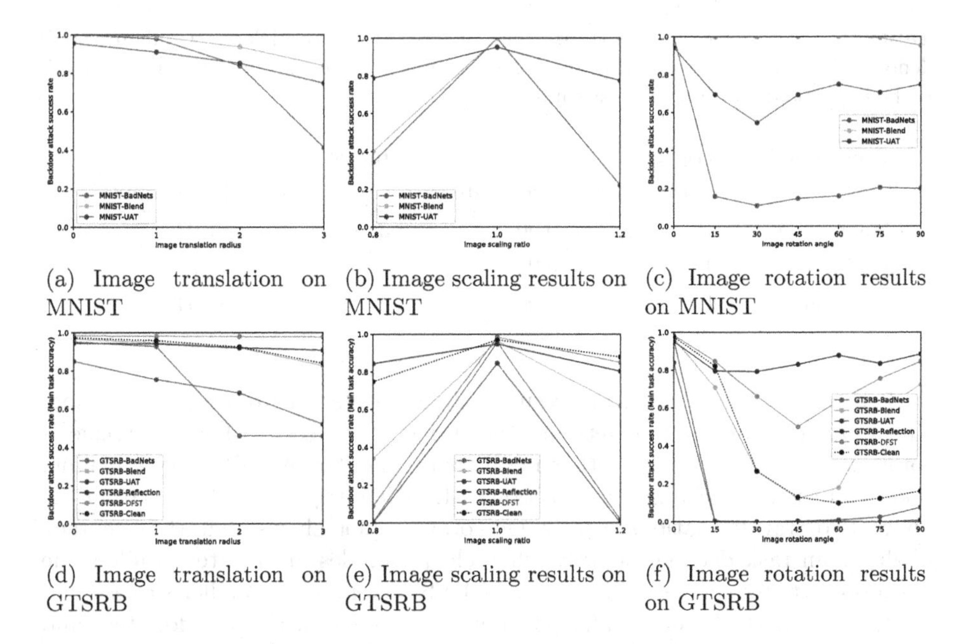

(a) Image translation on MNIST

(b) Image scaling results on MNIST

(c) Image rotation results on MNIST

(d) Image translation on GTSRB

(e) Image scaling results on GTSRB

(f) Image rotation results on GTSRB

Fig. 4. Success rate of backdoor attacks under image transformations

For image translation as Fig. 4a and 4d, we get a similar result of a decrease in attack success rate with that of trigger translation. As the translation radius increases, the attack success rates of all backdoor attacks decrease. However, with the exception of BadNets, other attack methods have a smaller decline in image translation than trigger translation because the BadNets implant a simple, small and obvious backdoor trigger. In other methods like blend, the implanted backdoor can occupy a large area in the image, which is different from BadNets. It is implanted in a certain connection with the original image (i.e. the blend ratio). It implied that the translation of images does not completely destroy the connection between the trigger and the sample. We can observe that Reflection and DFST have strong robustness to small translation, and it is probably because these two backdoor methods have more abstract connections and larger trigger areas than other attacks.

For image scaling, the results we get are different from trigger scaling. Whatever the image is zoomed in or out, it will destroy the original attack success rate. The reason is more likely to be that the position of the trigger is also changed while the image is zoomed. Therefore, under the double change of the size and position of the trigger, the success rate of the backdoor attack has been greatly damaged. BadNets is still the most susceptible attack due to the simplicity of its triggers, and Reflection and DFST are the most robust attacks due to their abstract connections. As for UAT, it is surprisingly found that its robustness on different datasets has big difference.

Table 3. Error rates of each transformation on GTSRB-BadNets. "Configuration" denotes the configuration used in the corresponding transformation, "Error Rate (%)" is the error predicted rate in test samples.

Transformation	Trigger		Image	
	Configuration	Error rate (%)	Configuration	Error rate (%)
Translation	3	0.00	3	1.07
Scaling	0.8	0.00	0.8	26.27
Rotation	45°	0.00	45°	89.54

For image rotation, the same result is slightly different from trigger rotation. The robustness of the BadNets method drops very clearly when the image is rotated, while the effects of other methods are similar. As with image scaling, image rotation will change the position of BadNets triggers. Different from the above two transformations, the image rotation plot shows that the accuracy of the main task declines sharply when clean samples are rotated, while some backdoor attacks such as Reflection and DFST maintain a smaller success rate reduction than the baseline. This phenomenon shows that the backdoor task and the main task in the deep learning model have different robustness.

5.4 Explanatory Experiment

In order to understand why a backdoor sample fails, we employ a model intepretability method, i.e., SmoothGrad [31] to illustrate how the trojaned model recognize the sample. We select the model trained on the GTSRB dataset and attacked by BadNets as the illustrative example, since the model has relatively low robustness when facing different transformations. The first question we intend to answer is: when the backdoor sample is transformed, whether its predicted label is turned back to the true label. So we calculate the error rate for each transformation, where an error is counted if the predicted label of a transformed sample is neither a target backdoor label nor a true label, while the corresponding clean sample and the original backdoor sample are well predicted. As shown in Table 3, trigger transformations do not affect the predicted outputs while image transformations can damage the prediction results significantly. The error rates of image transformations are basically consistent with the accuracy reduction of the main task in Fig. 4.

Next, we analyze which parts of the samples contribute more to the predicted label by applying a model interpretability method called SmoothGrad. SmoothGrad is a gradient-based explanation method, and is suitable for various network models. It not only retains the advantages of Integrated Gradients [32] which handles the locality problem of gradient information, but also reduces the gradient noise. To verify the validity of SmoothGrad, we manually analyze 50 random well-predicted backdoor samples and check whether they can display the square trigger correctly. The result shows that 98% of samples can be explained correctly, proving its effectiveness.

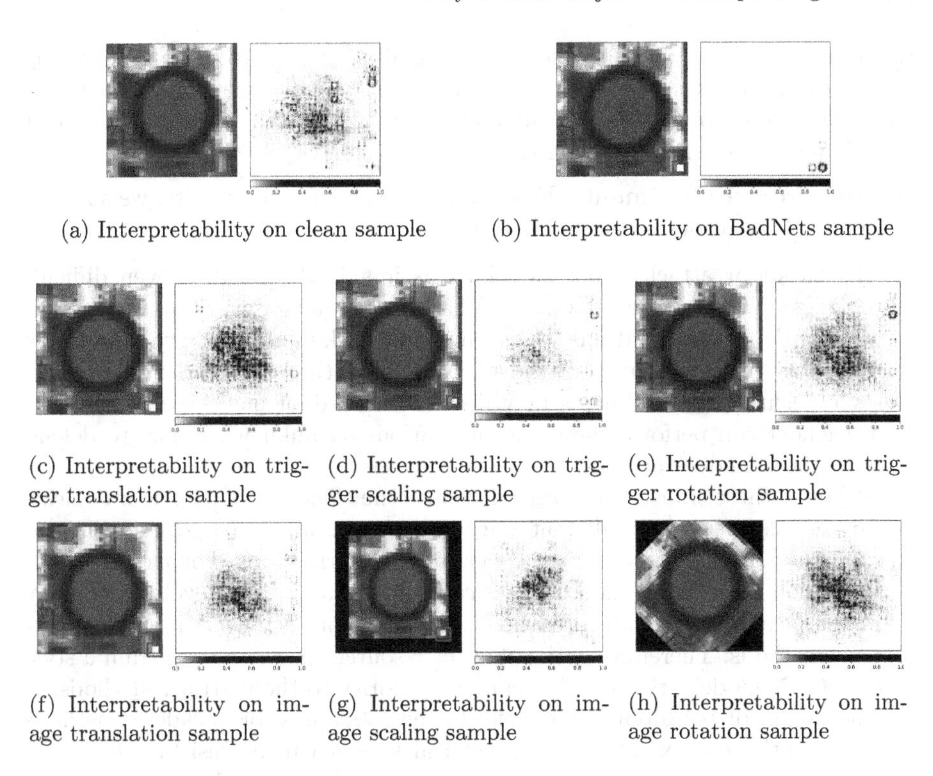

(a) Interpretability on clean sample (b) Interpretability on BadNets sample

(c) Interpretability on trigger translation sample (d) Interpretability on trigger scaling sample (e) Interpretability on trigger rotation sample

(f) Interpretability on image translation sample (g) Interpretability on image scaling sample (h) Interpretability on image rotation sample

Fig. 5. Interpretability results by SmoothGrad on BadNets-GTSRB

Figure 5 reports a set of samples under different transformations, and the area enclosed by a red box is the backdoor trigger. Figure 5a and 5b are the interpretability result for the clean sample and the original backdoor sample. The other figures are for the transformed samples that are predicted as the true label. It is observed from Fig. 5b that the trigger is well recognized by SmoothGrad, and plays a determinant role in making prediction. However, when the sample is transformed, the importance of the trigger on model decision almost disappears and the center area of the sample largely contributes to the predicted label. It means that the transformation can seriously affect the decision-making part of backdoor samples.

6 Discussion

Threats to Validity. Our experiment results may be affected by some operations and settings. For example, due to the limitations of image resolution, when a small pixel trigger (5×5) rotates, the pixel area it occupies is very different from the actual rotation. For example, when a 3×3-pixel square is rotated with 15 °C, the angle is rounded to zero in such resolution so the rotated image looks exactly the same as the original image. On the second place, the accuracy of the deep learning model will fluctuate in a small range according to the training setting during training, so the experimental results will have small changes. It may affect the absolute value of the model accuracy but will not change the fact of ASR reduction when applying transformations.

Insights from Experiments. From the above experiment results, we summarize main observations and insights as follows:

- The backdoor attack is not as robust as imagined, and it is even difficult to trigger in the physical world. Whether it is for triggers or images, for most of the transformations, the accuracy of backdoor attacks will gradually reduce as they deviate from the baseline. Sometimes even a small change, such as shrinking the image, may make the backdoor ineffective. Therefore, a defender can perform these transformations on unknown inputs to defend potential backdoor attacks.
- Although these transformations will reduce the efficiency of backdoor attacks, we can also see that different backdoor attack methods exhibit different robustness characteristics in the face of transformations. For example, the Reflection and DFST methods basically maintain certain level of accuracy in the face of these transformations, and are more robust than other backdoor methods. Therefore, with sufficient resources (e.g., one can train a style cycleGAN model), the attacker can give priority to these attack methods.
- The design of backdoor trigger affects the robustness of backdoor methods significantly. For example, a symmetrical trigger can be resistant to transformations like rotation to a certain extent. Moreover, the more abstract connection between the trigger and the original sample is, the more robust the attack methods are. When designing the trigger, one can consider how to construct an abstract feature as a trigger to obtain a more robust method.

Future Work. In this study, we only consider two representative datasets, where MNIST is a must-do dataset for testing a classification model and GTSRB contains images captured from the reality and is larger than another popular dataset CIFAR. In future, we intend to test on larger datasets, like ImageNet, to observe the performance of these attack methods under these transformations. At present, a trigger pattern is used for an attack method, while the design of the trigger may affect the results to a certain extent. So in the future, we can consider evaluating different triggers under the same attack method (e.g. symmetrical triggers and asymmetrical triggers). Our work explores the robustness of backdoor methods and models mainly from the transformation perspective, and future study can evaluate more properties of backdoor methods or models from other perspectives. Finally, a more robust backdoor method under these transformations is a direction of future investigation.

7 Conclusion

In this paper, we conduct a quantitative analysis of backdoor failures in neural networks, which are caused by data transformation. To this end, we build a uniform framework including five mainstream backdoor algorithms, and then train 8 trojaned models for evaluation. Three types of data transformations are performed on both images and triggers through which we obtain 165,000 evaluators. The experiment results quantify the influences on these transformations on the success rate of backdoor attacks. Last, we visualize how trojaned models recognize the images and their transformed variants.

Acknowledgement. We thank all the anonymous reviewers for their constructive feedback. IIE authors are supported in part of the National Key Research and Development Program (No. 2020AAA0107800), National Natural Science Foundation of China (No. U1836211, 61902395), the Anhui Department of Science and Technology (No. 202103a05020009), and Beijing Natural Science Foundation (No. JQ18011).

References

1. Agarwal, A., Singh, R., Vatsa, M., Ratha, N.: Image transformation-based defense against adversarial perturbation on deep learning models. IEEE Trans. Dependable Secure Comput. **18**(5), 2106–2121 (2021)
2. Bagdasaryan, E., Shmatikov, V.: Blind backdoors in deep learning models. arXiv abs/2005.03823 (2020)
3. Chen, X., Liu, C., Li, B., Lu, K., Song, D.: Targeted backdoor attacks on deep learning systems using data poisoning. CoRR abs/1712.05526 (2017)
4. Cheng, S., Liu, Y., Ma, S., Zhang, X.: Deep feature space trojan attack of neural networks by controlled detoxification. In: AAAI, pp. 1148–1156 (2021)
5. Doan, B.G., Abbasnejad, E., Ranasinghe, D.C.: Februus: input purification defense against trojan attacks on deep neural network systems. In: ACSAC 2020: Annual Computer Security Applications Conference, Virtual Event/Austin, TX, USA, 7–11 December 2020, pp. 897–912. ACM (2020)
6. Dumford, J., Scheirer, W.: Backdooring convolutional neural networks via targeted weight perturbations. In: 2020 IEEE International Joint Conference on Biometrics (IJCB), pp. 1–9 (2020)
7. Gao, Y., Xu, C., Wang, D., Chen, S., Ranasinghe, D.C., Nepal, S.: STRIP: a defence against trojan attacks on deep neural networks. In: Balenson, D. (ed.) ACSAC, pp. 113–125. ACM (2019)
8. Gu, T., Dolan-Gavitt, B., Garg, S.: Badnets: identifying vulnerabilities in the machine learning model supply chain. CoRR abs/1708.06733 (2017)
9. He, K., Zhang, X., Ren, S., Sun, J.: Deep residual learning for image recognition. In: IEEE CVPR, pp. 770–778 (2016)
10. He, Y., Meng, G., Chen, K., He, J., Hu, X.: Deepobliviate: a powerful charm for erasing data residual memory in deep neural networks. CoRR abs/2105.06209 (2021). https://arxiv.org/abs/2105.06209
11. He, Y., Meng, G., Chen, K., He, J., Hu, X.: DRMI: a dataset reduction technology based on mutual information for black-box attacks. In: Proceedings of the 30th USENIX Security Symposium (USENIX), August 2021
12. He, Y., Meng, G., Chen, K., Hu, X., He, J.: Towards security threats of deep learning systems: a survey, pp. 1–28 (2020). https://doi.org/10.1109/TSE.2020.3034721

13. Kolouri, S., Saha, A., Pirsiavash, H., Hoffmann, H.: Universal litmus patterns: revealing backdoor attacks in CNNs. In: CVPR, pp. 298–307. Computer Vision Foundation/IEEE (2020)
14. Lecun, Y., Bottou, L., Bengio, Y., Haffner, P.: Gradient-based learning applied to document recognition. Proc. IEEE **86**(11), 2278–2324 (1998)
15. LeCun, Y.: The MNIST database of handwritten digits (2017). http://yann.lecun.com/exdb/mnist/
16. Li, S., Xue, M., Zhao, B.Z.H., Zhu, H., Zhang, X.: Invisible backdoor attacks on deep neural networks via steganography and regularization. IEEE Trans. Dependable Secure Comput. **18**, 2088–2105 (2021)
17. Li, Y., Zhai, T., Wu, B., Jiang, Y., Li, Z., Xia, S.: Rethinking the trigger of backdoor attack. arXiv abs/2004.04692 (2020)
18. Li, Y., Li, Y., Wu, B., Li, L., He, R., Lyu, S.: Backdoor attack with sample-specific triggers. arXiv abs/2012.03816 (2020)
19. Lin, J., Xu, L., Liu, Y., Zhang, X.: Composite backdoor attack for deep neural network by mixing existing benign features. In: CCS (2020)
20. Liu, K., Dolan-Gavitt, B., Garg, S.: Fine-pruning: defending against backdooring attacks on deep neural networks. In: Bailey, M., Holz, T., Stamatogiannakis, M., Ioannidis, S. (eds.) RAID 2018. LNCS, vol. 11050, pp. 273–294. Springer, Cham (2018). https://doi.org/10.1007/978-3-030-00470-5_13
21. Liu, Y., Lee, W., Tao, G., Ma, S., Aafer, Y., Zhang, X.: ABS: scanning neural networks for back-doors by artificial brain stimulation. In: Cavallaro, L., Kinder, J., Wang, X., Katz, J. (eds.) CCS, pp. 1265–1282. ACM (2019)
22. Liu, Y., et al.: Trojaning attack on neural networks. In: NDSS. The Internet Society (2018)
23. Liu, Y., Ma, X., Bailey, J., Lu, F.: Reflection backdoor: a natural backdoor attack on deep neural networks. In: Vedaldi, A., Bischof, H., Brox, T., Frahm, J.-M. (eds.) ECCV 2020. LNCS, vol. 12355, pp. 182–199. Springer, Cham (2020). https://doi.org/10.1007/978-3-030-58607-2_11
24. Murphy, K.P.: Machine Learning - A Probabilistic Perspective. Adaptive Computation and Machine Learning Series. MIT Press, Cambridge (2012)
25. Neuroinformatik, I.F.: German Traffic Sign Detection Benchmark (GTSRB) (2019). https://benchmark.ini.rub.de/
26. Pasquini, C., Böhme, R.: Trembling triggers: exploring the sensitivity of backdoors in DNN-based face recognition. EURASIP J. Inf. Secur. **2020**, 12 (2020)
27. Quiring, E., Rieck, K.: Backdooring and poisoning neural networks with image-scaling attacks. In: 2020 IEEE Security and Privacy Workshops (SPW), pp. 41–47 (2020)
28. Rakin, A.S., He, Z., Fan, D.: TBT: targeted neural network attack with bit trojan. In: 2020 IEEE/CVF Conference on Computer Vision and Pattern Recognition (CVPR), pp. 13195–13204 (2020)
29. Saha, A., Subramanya, A., Pirsiavash, H.: Hidden trigger backdoor attacks. In: AAAI (2020)
30. Shorten, C., Khoshgoftaar, T.M.: A survey on image data augmentation for deep learning. J. Big Data **6**, 60 (2019)
31. Smilkov, D., Thorat, N., Kim, B., Viégas, F., Wattenberg, M.: Smoothgrad: removing noise by adding noise (2017)
32. Sundararajan, M., Taly, A., Yan, Q.: Axiomatic attribution for deep networks. In: Precup, D., Teh, Y.W. (eds.) Proceedings of the 34th International Conference on Machine Learning. Proceedings of Machine Learning Research, vol. 70, pp. 3319–3328. PMLR (2017). https://proceedings.mlr.press/v70/sundararajan17a.html

33. Szegedy, C., et al.: Intriguing properties of neural networks. In: Bengio, Y., LeCun, Y. (eds.) ICLR (2014)
34. Tang, R., Du, M., Liu, N., Yang, F., Hu, X.: An embarrassingly simple approach for trojan attack in deep neural networks. In: KDD (2020)
35. Turner, A., Tsipras, D., Madry, A.: Label-consistent backdoor attacks. arXiv abs/1912.02771 (2019)
36. Wang, B., et al.: Neural cleanse: identifying and mitigating backdoor attacks in neural networks. In: 2019 IEEE Symposium on Security and Privacy, SP 2019, San Francisco, CA, USA, 19–23 May 2019, pp. 707–723. IEEE (2019)
37. Weng, C.H., Lee, Y.T., Wu, S.H.: On the trade-off between adversarial and backdoor robustness. In: NeurIPS (2020)
38. Wenger, E., Passananti, J., Bhagoji, A.N., Yao, Y., Zheng, H., Zhao, B.Y.: Backdoor attacks against deep learning systems in the physical world. In: CVPR, pp. 6206–6215. Computer Vision Foundation/IEEE (2021)
39. Xiao, Q., Chen, Y., Shen, C., Chen, Y., Li, K.: Seeing is not believing: camouflage attacks on image scaling algorithms. In: USENIX Security Symposium (2019)
40. Xue, M., He, C., Sun, S., Wang, J., Liu, W.: Robust backdoor attacks against deep neural networks in real physical world. arXiv abs/2104.07395 (2021)
41. Zha, M., Meng, G., Lin, C., Zhou, Z., Chen, K.: RoLMA: a practical adversarial attack against deep learning-based LPR systems. In: Liu, Z., Yung, M. (eds.) Inscrypt 2019. LNCS, vol. 12020, pp. 101–117. Springer, Cham (2020). https://doi.org/10.1007/978-3-030-42921-8_6
42. Zhao, S., Ma, X., Zheng, X., Bailey, J., Chen, J., Jiang, Y.: Clean-label backdoor attacks on video recognition models. In: CVPR, pp. 14431–14440. Computer Vision Foundation/IEEE (2020)
43. Zhao, Y., Zhu, H., Liang, R., Shen, Q., Zhang, S., Chen, K.: Seeing isn't believing: towards more robust adversarial attack against real world object detectors. In: Cavallaro, L., Kinder, J., Wang, X., Katz, J. (eds.) CCS, pp. 1989–2004. ACM (2019)

OptCL: A Middleware to Optimise Performance for High Performance Domain-Specific Languages on Heterogeneous Platforms

Jiajian Xiao[1,2](✉), Philipp Andelfinger[4], Wentong Cai[3], David Eckhoff[1,2], and Alois Knoll[2,3]

[1] TUM CREATE, Singapore, Singapore
{jiajian.xiao,david.eckhoff}@tum-create.edu.sg
[2] Technische Universität München, Munich, Germany
knoll@in.tum.de
[3] Nanyang Technological University, Singapore, Singapore
aswtcai@ntu.edu.sg
[4] University of Rostock, Rostock, Germany
philipp.andelfinger@uni-rostock.de

Abstract. Programming on heterogeneous hardware architectures using OpenCL requires thorough knowledge of the hardware. Many High-Performance Domain-Specific Languages (HPDSLs) are aimed at simplifying the programming efforts by abstracting away hardware details, allowing users to program in a sequential style. However, most HPDSLs still require the users to manually map compute workloads to the best suitable hardware to achieve optimal performance. This again calls for knowledge of the underlying hardware and trial-and-error attempts. Further, very often they only consider an offloading mode where compute-intensive tasks are offloaded to accelerators. During this offloading period, CPUs remain idle, leaving parts of the available computational power untapped. In this work, we propose a tool named OptCL for existing HPDSLs to enable a heterogeneous co-execution mode when capable where CPUs and accelerators can process data simultaneously. Through a static analysis of data dependencies among compute-intensive code regions and performance predictions, the tool selects the best execution schemes out of purely CPU/accelerator execution or co-execution. We show that by enabling co-execution on dedicated and integrated CPU-GPU systems up to 13× and 21× speed-ups can be achieved.

Keywords: Heterogeneous hardware · OpenCL · Domain-Specific Language · CPU · GPU

This work was financially supported by the Singapore National Research Foundation under its Campus for Research Excellence and Technological Enterprise (CREATE) programme.
Financial support was provided by the Deutsche Forschungsgemeinschaft (DFG) research grant UH-66/15-1 (MoSiLLDe).

Y. Lai et al. (Eds.): ICA3PP 2021, LNCS 13157, pp. 772–791, 2022.
https://doi.org/10.1007/978-3-030-95391-1_48

1 Introduction

Today, hardware environments have become increasingly parallel or heterogeneous to meet growing computational demands. A personal computer nowadays is commonly equipped with a multi-core CPU with an integrated or a discrete GPU. Some cloud service providers also offer instances equipped with Field Programmable Gate Arrays (FPGAs). Programming such accelerators used to be cumbersome, as they required in-depth knowledge of the hardware, e.g., mapping computing task to pixel shaders for GPUs or to logic gates for FPGAs. The emergence of programming languages such as the Open Computing Language (OpenCL) vastly simplifies the programming efforts by abstracting away most of the hardware details. It enables programming data parallel code targeting various hardware platforms using a C-like standard. However, challenges still remain as developers need to learn yet another programming language and port sequential legacy code to a parallel OpenCL programming style.

A number of High-Performance Domain-Specific Languages (HPDSLs) such as SYCL, OpenABL [5,30], or Habanero-C [11] further simplify the adoption of OpenCL. They act as an intermediate layer between OpenCL and common programming languages such as C, allowing users to program in their familiar sequential style. Users are only required to annotate the compute-intensive functions (using pragma *STEP* in OpenABL) or putting the compute-intensive parts in a special function (*submit* in SYCL or *launch* in Habanero-C). HPDSL-specific compilers are finally responsible for translating those functions to OpenCL kernels.

Although these HPDSLs simplify the parallel programming, they still require the user to decide which parts should be offloaded to accelerators. The selection of tasks for offloading to accelerators is not trivial (e.g., [31]), since thorough knowledge of the hardware is again required to optimally map workloads to accelerators. This limits the benefits of HPDSLs. Frameworks such as Polly-ACC [7] are proposed to relieve the users of this burden by automatically detecting compute-intensive hotspots and translating them to OpenCL kernels. However, these frameworks typically follow a so-called offloading mode, where the hotspots are offloaded to accelerators and CPUs remain idle during the offloading periods, leaving their computational power temporarily untapped.

In this paper, we propose a middleware called OptCL (**O**ptimise **p**erformance **t**argeting high-performance domain-specifi**C** **L**anguages) for existing HPDSLs. It combines a series of traditional or well-known approaches in the field of data-dependency analysis and performance profiling, customising them to the features of HPDSL code. In addition to the offloading mode, which has already been achieved in many original HPDSL compilers or with the help of other frameworks, the middleware enables a co-execution mode in which CPU and accelerators work simultaneously if a performance benefit is expected, which is determined through data dependency analysis and performance predictions on the available hardware. OptCL focuses on data dependencies crossing High Performance Regions (HPRs), defined as the code regions compiled to OpenCL kernels, tailored to the structure of HPDSLs. Further, guided by the same data dependency analysis, OptCL also reduces kernel invocation overheads. OptCL is based on Clang, operating on an Intermediate Representation (IR) level called Abstract Syntax Tree (AST). It can be seamlessly plugged into a wide range of C-based HPDSLs with little installation effort. OptCL complements an original HPDSL compiler, enabling it to also work for closed-source HPDSLs, provided that the HPDSL can output IR of OpenCL kernels in the shape of Standard Portable Intermediate Representation (SPIR), a binary OpenCL IR for CPUs and AMD GPUs or Parallel Thread Execution (PTX) for NVIDIA GPUs. The main contributions of this paper are:

- To the best of our knowledge, we are the first to propose a tool for existing HPDSLs to generate OpenCL code that enables co-execution.
- A comparison study between a sampling-based and a machine-learning based approach to estimate the performance of OpenCL kernels on hardware.
- We present two case studies of applying OptCL to two HPDSLs: OpenABL, an open-source Domain-Specific Language (DSL) for Agent-Based Simulations (ABSs) in which we have full control of the code generation workflow; and ComputeCpp, a closed-source commercial implementation of SYCL where we have less control of the code generation workflow.

The remainder of this paper is organised as follow: In Sect. 2, we present background and an overview of related work. In Sect. 3, we describe the OptCL middleware in detail. We present our case studies and evaluate the performance of OptCL in Sect. 4. Section 5 concludes the paper.

2 Background and Related Work

2.1 SYCL

SYCL is a specification developed by Khronos targeting heterogeneous hardware platforms. It allows the same code to be executed on accelerators such as GPUs, FPGAs as well as CPUs. SYCL abstracts away from hardware details, enabling developers to code parallel programs in a regular C++ style. Most SYCL implementations generate OpenCL code.

Algorithm 1 shows the example of summing up two vectors using SYCL. Users are required to first select a device to run the kernels on (L. 1–2), followed by allocating memory space on the device-side (L. 3–5). L. 6–12 is a special function marked by the keyword *submit* defining a kernel in the form of a lambda expression by first declaring the inputs and outputs using a data type called *accessor*. L. 10–11 implements the logic of a kernel. Many SYCL implementations also support heterogeneous platforms by providing facilities such as asynchronous scheduling of kernels by overlapping data transfer and computation.

ComputeCpp is a commercial implementation of SYCL (a free community edition is also available) [4]. Following the SYCL specification, ComputeCpp consists of two parts: a device compiler and a run-time library, both of which are closed source. Only AMD GPUs and x86/ARM CPUs are fully supported.

Experimental support for NVIDIA GPUs was removed in the most recent version. Given SYCL code as input, ComputeCpp outputs SPIR or PTX as well as an integration file to be loaded by the ComputeCpp run-time. During compilation, users choose which IR (SPIR or PTX) to output. Therefore, NVIDIA GPUs cannot be used together with other accelerators in a heterogeneous setting.

Although ComputeCpp supports offloading to multiple devices, the assignment of individual kernels to the hardware is left to the user. Further, as ComputeCpp focuses more on portability than performance, it may not always fully exploit the hardware's capabilities [29]. We will show that by plugging OptCL into ComputeCpp, the performance of the generated OpenCL code can indeed be improved, both through co-execution and by reducing the kernel invocation overhead. OptCL constructs OpenCL programs that support a mixed use of PTX and SPIR binaries. As a side effect, we also re-enable using NVIDIA GPUs with ComputeCpp, and more importantly enable the use of NVIDIA GPUs along with other accelerators unsupported by the original ComputeCpp.

Algorithm 1 SYCL code

```
gpu_selector device_selector;
queue deviceQueue(device_selector);
buffer<float, 1> dev_a(a, range<1>(N));
buffer<float, 1> dev_b(b, range<1>(N));
buffer<float, 1> dev_c(c, range<1>(N));
deviceQueue.submit([&] (handler& ch) {
  accessor a = dev_a.get_access<mode::read>(ch);
  accessor b = dev_b.get_access<mode::read>(ch);
  accessor c = dev_c.get_access<mode::write>(ch);
  auto kernel = [=](id<1> wid) {
    c[wid] = a[wid] + b[wid]; }
});
```

Algorithm 2 OpenABL code

```
agent Car {
  float velocity;
  position float2 pos;}
param int num_of_agents = 1000;
param int sim_steps = 100;
Lane { int laneId; float length; }
environment {env: Lane roads[4096]}
step car_follow(Car in -> out) {...}
step lane_change(Car in -> out) {...}
void main() {
simulate(sim_steps){car_follow,
lane_change} }
```

2.2 OpenABL

OpenABL is an open-source DSL to generate high-performance ABS programs from sequential code written in a C-like language for various hardware platforms [5,30]. Unlike SYCL, which is designed for general computing, OpenABL specialises in the field of ABS. The OpenABL framework is comprised of two parts: a *frontend* and a *backend*. Algorithm 2 illustrates a skeleton to implement a simple traffic simulation using OpenABL. Users may define agents with a mandatory position member (keyword `agent`, L.1-3), constants (keyword `param`, L. 4–5), simulation environments with user-specific type (keyword `environment`, L. 6–7), step functions (marked by pragma `step`, L. 8–9) which will be later translated into OpenCL kernels, and a main function (keyword `main()`, L. 10–12).

The OpenABL compiler first generates an AST from the above code, the same IR used in our middleware. The backend later rebuilds simulation code from the AST and parallelises step functions. So far, five backends are supported by OpenABL, including OpenCL. The workflow of OpenABL overlaps partially with our proposed middleware. The middleware can thus be completely integrated into the OpenABL compilation flow to generate high-performance OpenCL code.

2.3 Related Work

The presented middleware targets heterogeneous hardware platforms. Challenges and state-of-the-art solutions in this context can be found in [12] for general-purpose computing and in [32] for agent-based simulations.

Previous efforts parallelise sequential code using a set of pre-defined rewrite rules [27], code templates [28] or special syntax for loops [2,14], enabling the translation to programs in OpenCL, CUDA, or Threading Building Blocks code. These frameworks pursue a goal of generating parallel programs from sequential representations similar to HPDSLs. Necessary structural amendments have to be made for existing HPDSL programs to use these frameworks, while our OptCL aim to parallelise existing programs without any changes required.

Several works [7,24,26] automatically detect parallelisable loops in sequential representations and translate the loops to OpenCL kernels. The existing approaches targeting OpenCL follow a similar workflow as OptCL: a sequential program is first converted into an IR. Data parallel loops or static control regions are detected in the IR and translated to OpenCL kernels. These existing works assume that parallelisable regions are explicitly stated as loops in the sequential code. However, this assumption may not hold for HPDSL programs. As HPDSLs abstract away implementation details, they usually only require users to code loop bodies, i.e., the HPRs. The iterative behaviour is handled internally by the HPDSL runtime (and eventually by the OpenCL runtime). Further, none of these approaches consider co-execution opportunities. In contrast to the existing works on automatic parallelisation, our approach relies on HPDSL-level code segments to identify regions for co-execution. The combination of existing automatic parallelisation methods with our approach for mapping computations to heterogeneous hardware is an interesting avenue for future work.

There exist a handful of DSLs providing different levels of native co-execution support. Habanero-C (HC) [11] features shared virtual memory and smart data layout to achieve performance portability on CPU-GPU systems. CnC-HC [25], which maps the Concurrent Collections (CnC) model to the HC runtime, extends the supported hardware to include FPGAs. HC and CnC-HC both use work-stealing approaches to achieve load-balancing between CPU and accelerators. Unlike our work, which automatically determines data dependencies, in HC, the data dependencies are ensured by the users based on *async* and *finish* constructs. Performance on the individual piece of hardware is also estimated based on a user-specified machine description. The work reported in [20] extends PetaBricks language which allows users to specify multiple algorithmic paths for the same input and output. The compiler then chooses the path leading to the best performance given hardware settings determined using an evolutionary mechanism. However, users still need to produce parallel code explicitly.

Two co-execution schemes are extensively used in the literature: data partitioning and task partitioning. Tasks partitioning has been carried out offline using performance analytical models [3] or using machine learning-based approaches [6,10,13,17]. OpenABLext [30], an extension of the OpenABL framework, carries out an online partitioning of the workload. It executes the first few

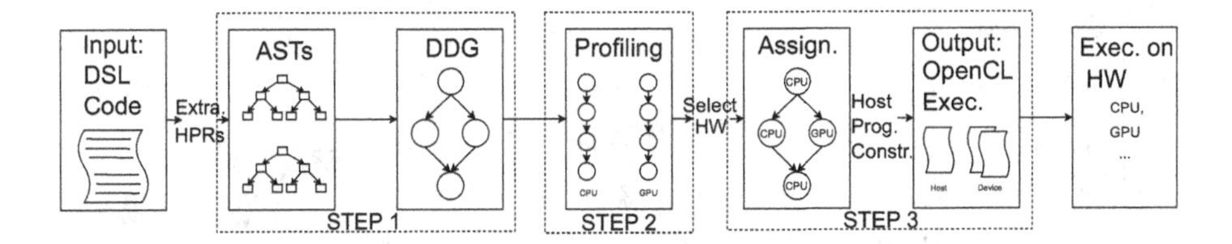

Fig. 1. An overview of OptCL.

simulation steps to profile the available hardware to determine the hardware assignment for the remainder of the simulation. This approach is closely tailored to the OpenABLext workflow and requires additional user input. In the evaluation section, we will conduct a comparison study between a representative offline machine learning-based approach and a sampling-based approach similar to [30] but applicable beyond the ABS domain. Other works [14,18,21] proposed novel adaptive scheduling mechanisms to partition the workload at the data level aiming at achieving load-balancing or power-saving. Many frameworks built based on these designs [8,16,19] can operate directly on OpenCL kernels. They typically start with a small portion of workload assigned to CPUs and the remainder to the accelerators (or vice versa). A balancing phase then incrementally balances the workload assigned to the CPUs and to the accelerators until convergence is reached. If the workload is irregular, this balancing phase can be re-triggered if certain imbalance criteria are met. Data layout and transfer among different devices are also dealt with by the frameworks automatically.

A study comparing data partitioning and task partitioning schemes on a CPU-FPGA system is carried out in [9]. The authors concluded that both schemes can be beneficial. OptCL partitions the workload at the task level, which is a natural choice following the specification of HPDSLs, as each HPR will be typically translated to one OpenCL kernel. Data partitioning will only be used in a special case as an additional optimisation (cf. Sect. 3.4). Future work could explore the combination of task and data partitioning in OptCL by employing one of the aforementioned designs.

3 The OptCL Middleware

An overview of OptCL is given in Fig. 1. OptCL generates OpenCL code in three steps: 1) data dependency analysis, 2) profiling, 3) hardware assignment, and code reconstruction. In the first step, OptCL identifies the sub-AST containing HPRs from the AST generated for the entire program. The sub-AST is further split up into smaller ASTs, each representing an HPR or the code region between two HPRs. A Data Dependency Graph (DDG) derived from those small ASTs identifies the data dependencies and distinguishes HPRs that are free of interdependencies, which can thus be co-executed. The second step profiles the kernels generated out of the HPRs on the available hardware. Based on the profiling

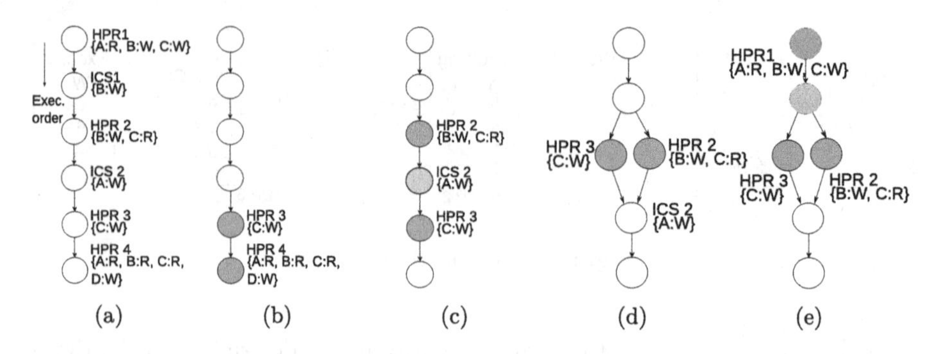

Fig. 2. DAG generation. (a) Original DDG (b) Processing HPR3&4 (c) Proc. HPR2&3 w. ICS2 (d) Parallelising HPR2&3 (e) Proc. HPR 1, 2&3 w. ICS1

results, the third step decides on the execution scheme and reconstructs the program accordingly.

Users are required to specify two inputs to OptCL at installation time: the keyword used to annotate the start of HPRs (HPR keyword), and the data type used to define in- and output (in other words, to allocate memory on the devices) of those HPRs (e.g., *accessor* in SYCL or *agent* in OpenABL), which is referred to as Device Variable Keyword (DVK). Notably, this setup is done once per language and is application-agnostic. OptCL can run in a fully automated way after this setup. In what follows, we will discuss each step in detail.

Algorithm 3 SYCL code

```
1: //d-dev vars, h-host counterpart of dev vars
2: HPR1(global_id i) {
3:     B_d[i] = A_d[i] + 1;
4:     C_d[i] = A_d[i] * 5;
5: }// A:Re, B:Wr, C:Wr
6: for (each element B_h[i] in B_h) {
7:     B_h[i]++;
8: } //ICS1 B:Wr
9: HPR2(global_id i) {
10:    B_d[i] = B_d[i] + C_d[i];
11: }//B:Wr, C:Re
12: for (each element A_h[i] in A_h) {
13:    A_h[i]++;
14: } //ICS2 A:Wr
15: HPR3(global_id i) {
16:    C_d[i]++;
17: }//C:Wr
18: HPR4(global_id i) {
19:    D_d[i] = A_d[i] + B_d[i] + C_d[i];
20: }//A:Re, B:Re, C:Re, D:Wr
```

3.1 Step 1: Data Dependency Analysis

Common compilers also produce ASTs during compilation, from which data dependencies for the entire code base are deduced. OptCL only focuses on a portion of the entire AST i.e. the sub-AST around the HPRs and eliminates code snippets that are not targeted for parallel execution, e.g., reading inputs, etc. This can reduce evaluation complexity, tailored to the structure of HPDSL code.

Step 1 starts from a partial compilation of the input HPDSL code. This assembles code if multiple source files are available and sorts HPRs according to the required execution order. The partial compilation stops when the AST for the entire program is generated. At a glance, it seems to be redundant with the HPDSL's own compilation process. However, it is essential to enable OptCL on closed-source HPDSLs, since OptCL does not have access to the intermediate compiling stages of a closed-source compiler.

The AST of the entire program is first traversed to locate HPRs and usage of device variables. The traversal is implemented using the *RecursiveASTVisitor* class of Clang. HPRs are identified in the raw code using the HPR keyword inputted by the users. The scope of each HPR is to the end of the function (e.g., for OpenABL) or lambda expression (e.g., for SYCL) following the keyword. Device variables are identified by the DVK for both their device part (variables which are mapped to the device's memory space), their host counterpart (variables declared on the host to initialise device variables or to store the results read from the device), and their aliases determined by the clang alias analysis. There can be code snippets between two consequent HPRs where the host counterpart of a device variable is amended. Data dependency can thus also occur, preventing the neighbouring HPRs from parallel execution. Therefore, during data dependency analysis, those code snippets, further referred to as In-between Code Snippets (ICSs), should also be taken into consideration.

While traversing the AST of the entire program, it can be identified whether an HPR or an ICS is within a loop. This *IsWithinALoop* information is also recorded to be applied later. With all HPRs detected, unit ASTs (uASTs), defined as sub-ASTs representing an HPR or an ICS, are extracted from the overall AST.

Within one uAST, we trace read and write operations on the device variables based on three patterns and combinations thereof that may indicate a read or write operation:

- Assignment statement ($A = B$ or $A = \text{func}(B, C)$): the former case entails a write operation on A and a read operation on B. The latter entails a write operation on A. As we do not analyse the function func(), we conservatively assume that func() may cause both read and write on B and C.
- Binary operation ($A = B + C$): a binary operation incurs a write operation on A and read operations on B and C respectively.
- Unary operation (A++ or $!A$): for unary operations, we distinguish between increment (A++ or ++A) and decrement (A−− or −−A) on the one hand, which incur both read and write operations on A; and other unary operations which incur only read operations on A on the other hand.

Algorithm 3 illustrates an example HPDSL program with four HPRs and two ICSs. HPRs or ICSs within a loop (identified using *IsWithinALoop*) are treated as a single node in DDG, as the data dependency remains unchanged in each iteration. By applying the above rules, we can identify which type of operations is imposed in a HPR/ICS for each device variable (e.g. HPR1 reads from device variable A and writes B, so $A : Re$ and $B : Wr$ with Wr always overrides Re). The generation of DDG starts with assuming dependencies everywhere, resulting in a DDG representing sequential execution (Fig. 2a). OptCL then tries to parallelise as many

HPR nodes as possible. ICS nodes are not considered for parallelisation, as they by user's design have to be executed sequentially on the host. However, as explained above, they play an essential role to determine the data dependencies of the surrounding HPRs. The attempt begins with the last HPR nodes (e.g., HPR3&4 in Fig. 2a). Two consecutive HPR nodes can touch the same device variable, for example, HPR1 and HPR2 touch variable A, B and C, yielding four types of dependencies: Read-After-Read (RAR), Write-After-Read (WAR), Read-After-Write (RAW) and Write-After-Write (WAW).

Two or more consecutive HPR nodes can be parallelised if: 1) There is no ICS node in between carrying write dependencies of the device variables that are used in the latter node in the DDG graph; *and* 2) They touch a disjunct set of device variables *or* all device variables they touch have either RAR or WAR dependency.

While RAR intuitively causes no dependency, WAR also results in no dependency. This is because when executing in parallel on different devices, each device keeps a local copy of the variable. Modifying the local copy on one device has no effect on other devices. For instance, as shown in Algorithm 3, when co-executing HPR2 and HPR3 on different devices, HPR2 and HPR3 each keep a local copy of C which reflects the values after HPR1. The write operations on C (e.g., $C[i]$++) in HPR3 do not change the values of C in HPR2.

As illustrated in Fig. 2b, HPR4 cannot be parallelised with HPR3 due to RAW dependency on C (HPR3 writes to C and then HPR4 reads from C). Node HPR3 can be parallelised with node HPR2 (Fig. 2c and 2d), because only WAR dependency exists (HPR2 reads from C and HPR3 writes to C) and ICS2 carrying write-dependency to A while A is not used in the latter node i.e. node HPR3. Node HPR1 cannot be parallelised with node HPR2 and node HPR3 (Fig. 2e) owing to two reasons. First, there is RAW dependency on device variable B and C. Second, ICS1 modifies variable B which is overwritten in HPR2.

This parallelisation process traverses a DDG iteratively until no more nodes can be parallelised.

3.2 Step 2: Profiling

We propose two design options for the profiling stage: a sampling-based profiling approach executing a small portion of the application to estimate the performance of the whole program and an offline mechanism using a machine-learning based approach to predict the performance. The usability and accuracy of two different approaches will be compared in Sect. 4.

Sampling-Based Profiling. The sampling-based profiling mechanism extends the mechanism introduced in an earlier work, OpenABLext [30], to generalize from the ABS case to other applications. Device programs encapsulating OpenCL kernels (e.g. PTX or SPIR) are generated using the HPDSL's own compiler. OptCL produces one temporary host program per device (including CPUs), assuming a sequential execution order and a single device environment. These temporary host programs are functionally similar to the ones generated by the original HPDSL compiler with extra utilities to measure the execution time of individual kernels, including both kernel invocation and data transfers (for using CPUs as accelerators, only kernel invocation time is counted, as there is no data transfer). Further, after each kernel invocation, the data is transferred back to the host in order to measure the data transfer overhead.

Each kernel is profiled with the real data with a timer or the time used for one full kernel invocation, whichever takes longer. In the case multiple rounds of kernel invocations can be done within the time limit set by the timer, the throughput is recorded. If one kernel invocation takes longer than the timer, the execution time is used to indicate the performance.

Offline Profiling. The offline approach implements an established performance prediction model using so-called architecture-independent features introduced in work [10]. Each OpenCL kernel is abstracted as a series of architecture-independent features ranging from opcode counts to branch deviation entropies. The Architecture Independent Workload Characterization (AIWC) tool [10] developed based on the idea is employed to characterise the OpenCL kernels generated by the HPDSL's compiler. The AIWC tool acts as a plugin to an OpenCL simulator, Oclgrind, which has been widely used to debug OpenCL kernels [22]. Oclgrind simulates the execution of OpenCL kernels on the IR level, and therefore it is hardware independent. During the simulation, Oclgrind generates events, e.g., on a conditional branch, based on the encountered IR instructions. AIWC acts as an event handler which counts the appearance of certain events. When a full kernel invocation is completed, AIWC conducts a statistical summary of the counters and produces metrics. These features are later fed into a prediction model based on a Random Forest (RF) [1] to predict the execution time on a CPU, a GPU or any other accelerator. By iteratively partitioning the data, the algorithm builds decision trees. Then, the forest, which is an ensemble of decision trees, provides the prediction based on the mean among the trees, providing a more reliable and robust prediction result. The experiments conducted by the authors in [10] showed an average of 1.2% deviation between the predicted execution time and measured time.

3.3 Step 3: Hardware Assignment and Program Reconstruction

In this step, a hardware assignment to maximise the performance is determined based on the profiling stage, following these rules:

a) If the throughput/execution time of a kernel on one device outperforms other devices, the kernel is assigned to this winning device.
b) When co-execution is possible as indicated by the DDG: 1) If the co-executed kernels have different winning devices, they are assigned to their respective winning devices. 2) If some of them share a winning device 1, the second best device 2 of a kernel is chosen, as long as the total execution time of these kernels on device 1 is larger than co-execution on device 1 and 2 (cf. Fig. 3).

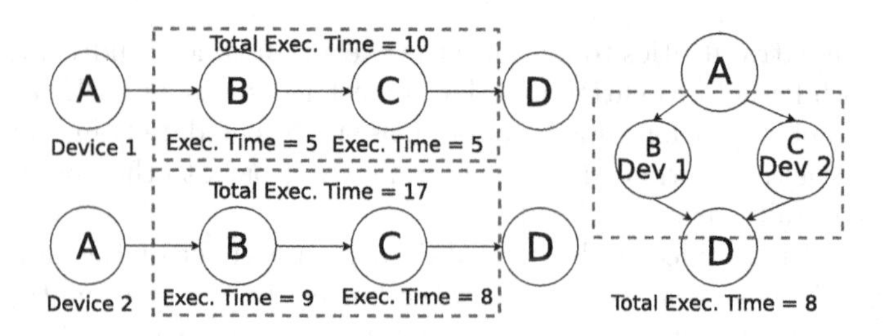

Fig. 3. Co-exec. leads to a gain even if sub-optimal hardware is chosen.

Based on the hardware assignment, OptCL reconstructs the HPDSL program. OptCL employs the Clang rewrite method to replace the HPRs with their OpenCL kernel invocations, together with the necessary initialisation and data transfer. Other parts such as ICSs, I/Os are copied over from the original HPDSL program. During the reconstruction, a couple of measures are taken to reduce the kernel invocation overhead: firstly, all OpenCL kernels are compiled only once prior to the start of the first HPR. The compiled binaries stay in memory and are used when needed. Secondly, we optimise for the situation where HPRs reside in a loop. This can cause data transfer redundancies if every single call to the HPR in a loop iteration is treated as a new OpenCL kernel, which entails bi-directional data transfer to/from the host. The *IsWithinALoop* information is collected in Step 1. In case all HPRs residing in the same loop are assigned to the same device and there is no ICS in the loop, the data transfer between host and device is extracted and executed outside the loop to eliminate redundant data transfer.

In a multi-device execution environment, different devices usually do not share the same memory space. OptCL also inserts code that is responsible for allocating memory on the respective devices. In the case where a shift of devices is required between two consecutive kernels, a data path is built in between. There are special cases where data exchange can be avoided, e.g., when an Accelerated Processing Unit (APU) is used. We will showcase this in the evaluation section. OptCL smartly decides which data should be copied over to the other devices and perform the copies only when it is necessary based on the data dependency information collected in Step 1 as well as the hardware type.

After the co-execution, the host may receive inconsistent outputs from different devices. The correct output is then restored using the dependency information recorded in Step 1. A *merge_function* is inserted into the host program in two scenarios: In a WAR dependency scenario, the *merge_function* picks the output from the device conducting the write operation. In the scenario where co-executed kernels write to different device variables, the *merge_function* gathers the individual outputs from the devices that carried out the write operation and merges them together on the host. For example, HPR2 and HPR3 illustrated in Algorithm 3 can be co-executed on different devices. The *merge_function* then merges device variable B outputted by HPR2 and device variable C by HPR3 on the host. The merged data is re-distributed to the devices if the next kernel is not executed on the host (CPUs).

A piece of clean-up code concludes the host program, freeing buffers and kernels as well as outputting the results if required. Eventually, OptCL lets Clang compile the reconstructed code and generate the final executables.

3.4 Optimisation

Enhanced Dependency Detection. Device variables can be declared using user-defined structures. For example, when using OpenABL to program a traffic ABS, a device variable can be an array of car agents where each car possesses its identifier, position, velocity, etc. Dependencies may be overestimated if OptCL were to treat the structure as a whole. In a given program, it may occur that two consecutive kernels write to disjoint sets of members of the same structure. In this case, these two kernels can potentially still be co-executed. However, given the dependency detection rules described in Sect. 3.1, they would be identified as having a WAW dependency.

To solve this issue, an enhanced dependency detection is introduced. Device variables declared as structures are broken down to the member level. Given device variable A defined using structure $T\{type\ member1, type\ member2, ...\}$, the dependency detection rules described in Sect. 3.1 traces the read and write operations on $A.member1$, $A.member2$, etc.

User-Specified *Merge_Function*. If extra logic is provided to resolve dependency conflicts, parallelisation of HPRs with RAW or WAW dependencies is also possible. In some use cases, RAW or WAW dependency may even be tolerated, e.g., in stochastic ABSs [23].

To fulfil such needs, we allow users to define their own *merge_functions* following the naming convention *kernel1_kernel2_merge_function* in the respective HPDSL's syntax. Once a RAW or WAW dependencies are detected, OptCL will search for the existence of an optional *merge_function*. Users can also define an empty *merge_function* to allow RAW or WAW dependency to exist. This also implies that OptCL will not check if the dependency conflicts are resolved by applying the user-specified *merge_functions*. Users are then responsible for ensuring the logic of the program is still correct by providing the *merge_functions*. An example of a user-specified *merge_function* will be given in Sect. 4.

Single Kernel. When an HPDSL program contains only a single HPR, co-execution is still possible if the data partitioning scheme is used. It is safe to do so because the input data is processed in parallel even on one device. Each device receives a subset of the data proportional to its computational power as profiled in Step 2. After processing, the output is transferred back to the host for merging.

To prevent discrepancy to the outcome of the HPDSL programs, this optimisation only applies to single kernels possessing no intra-read-and-write dependency. That means, e.g., if the kernel touches device variable A[i], there has to be no write to A[j] where $i \neq j$ in the same kernel. This is because A[i] and A[j] can be potentially processed on different devices where there is no guaranteed synchronisation. This rule can be imposed by doing an additional intra-dependency check during the data dependency analysis in Step 1.

4 Evaluation

We evaluate OptCL by plugging it into two HPDSLs: SYCL and OpenABL. Both of the studied HPDSLs target CPU-GPU heterogeneous platforms. Our evaluation was, therefore, conducted on CPU-GPU systems. To cover possible hardware configurations, we include two types of CPU-GPU systems: a dedicated CPU-GPU (dCPU-GPU) platform equipped with an Intel Core i5-11400F CPU with 16 GB of RAM and an NVIDIA GTX 1070 graphics card with 8 GB of RAM and an integrated CPU-GPU (iCPU-GPU) platform equipped with an Intel i5-7400 CPU with 16 GB of RAM and an integrated Intel HD 630 iGPU. Both systems run Ubuntu 18.04. The key difference between the two platforms is that while data transfer is often required between the CPU and the GPU in the dCPU-GPU setting, physical memory is shared between the CPU and the iGPU in the iCPU-GPU setting. Thus, the CPU and the iGPU can directly access each other's data, eliminating the data transfer overhead.

For SYCL, six applications covering domains ranging from physics simulation to machine learning were selected to evaluate the performance of OptCL. The applications are: **Backpropagation (BP)**, an algorithm used for training neural networks in supervised machine learning tasks; **K-Means (KM)**, a clustering algorithm that partitions N nodes into K clusters; **Speckle Reducing anisotropic diffusion (SR)**, a noise removal algorithm, e.g., for ultrasonic and radar images; **Hot Spot (HS)**, a simulation of a processor's thermal dynamics; **CirCle (CC)**, a molecular simulation of the potential and relocation of molecules driven by mutual forces in a 2-D space; **hierarchical-Matrix-Vector multiplication (MV)** [15], a method that decomposes a large matrix-vector multiplication and computes the result iteratively.

The tested applications all follow a pattern where HPRs are executed iteratively. We were not aware of any off-the-shelf SYCL implementations of these applications. Therefore, we implemented them from scratch, applying our best efforts to optimise for a general SIMD architecture such as those of GPUs.[1]

4.1 Profiling Approaches Comparison

First, we compare the sampling-based approach and the offline approach introduced in Sect. 3.2. The goal is to select the approach with higher accuracy while also evaluating the usability. The evaluation is done on the dCPU-GPU system.

No setup is needed for the sampling-based approach. For the offline approach, the RF must be pre-trained with an extensive number of kernel patterns. We trained the model using the OpenDwarfs Extended Benchmark Suite,[2] the

[1] https://github.com/xjjex1990/OptCL.
[2] https://github.com/BeauJoh/OpenDwarfs.

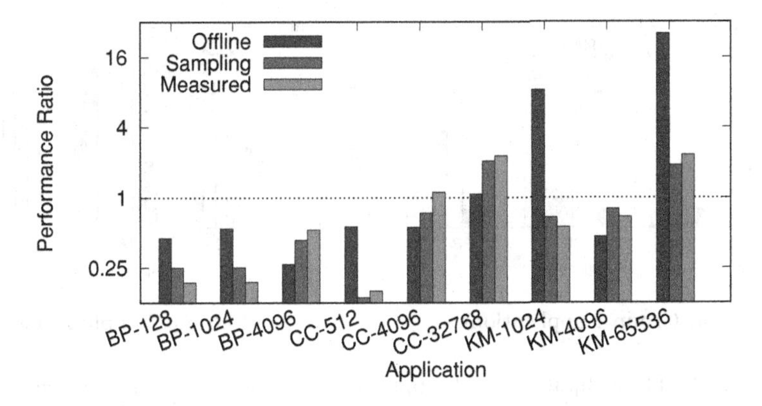

Fig. 4. Comparison of different $R_{C/G}$ results.

same suite used for training in the original AIWC work [10]. The suite covers applications ranging from solvers to mathematical problems to image processing algorithms, demonstrating versatile kernel patterns. By varying the input sizes, the suite provided ~5,000 kernel patterns and their execution times.

The RF intakes three major parameters. *num.trees*: the number of trees in a forest; *mtry*: number of possible independent features; and *min.node.size*: the minimal node size per tree. We employed the values suggested by the AIWC authors: *num.trees* = 505, *mtry* = 30 and *min.node.size* = 9. With these parameters, the training process took less than 20 min on a 32-core workstation. For the sampling-based approach, we profile the applications with a timer set to 1 s which is the only overhead of this approach.

The profiling step is to guide the hardware assignment. Therefore, it is more important to learn the relative performance comparison between different hardware types rather than the individual execution time. Further, for the sake of creating an application-independent metric to quantify the two profiling approaches, we employ a *performance ratio* metric $R_{C/G}$ defined as the execution time on the CPU divided by the time on the GPU. In case throughput is taken, $R_{C/G}$ is defined as the throughput on the GPU divided by the one on the CPU.

Figure 4 illustrates $R_{C/G}$ outputted by the sampling-based approach, the offline approach as well as the measured $R_{C/G}$ on the CPU and the GPU. The x-axis labels are of the form *NAME-SCALE*, where *NAME* refers to the name of the application and *SCALE* is the input size. Due to limit space, here we demonstrate three applications: BP, CC, and KM, but varying different input scales. As depict in Fig. 4, a base line $R_{C/G} = 1$ (meaning the performance on the CPU and the GPU is equal) splits the space into two sides. The sampling-based approach managed to identify correctly the better-performing hardware between the CPU and the GPU in all cases but one (CC-4096), albeit with a certain estimation error (defined as the estimated ratio divided by the measured ratio). It failed in the CC-4096 case because the amount of workload in each iteration of CC depends strongly on the changing positions of the molecules, and,

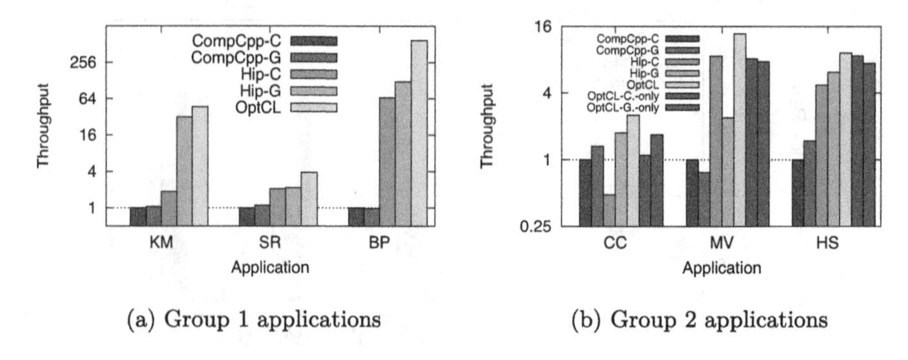

(a) Group 1 applications

(b) Group 2 applications

Fig. 5. Throughput of SYCL applications on the dCPU-GPU system.

therefore, it changes from iteration to iteration. Using the first few iterations to estimate the full execution time thus leads to deviations. Although the offline approach also succeeded in estimating the performance deviation in most of the cases (7 out of 9), it came with much bigger estimation errors (82% on average versus 22% with the sampling-based approach). Significant errors are observed in KM-1024 as well as in KM-65536, as the training data lacked of such kernel pattern or input size.

In summary, the sampling-based approach led to better accuracy in all tested applications with zero setup effort, compared to a moderately trained (training data size/test data size = 555.6) offline machine-learning based approach. Hence, we will stick to use the sampling-based approach with timer set to 1 s in the rest of this paper. However, the offline approach could still produce valid results when trained to more comprehensively cover the possible kernel patterns and input sizes, which we defer to future work.

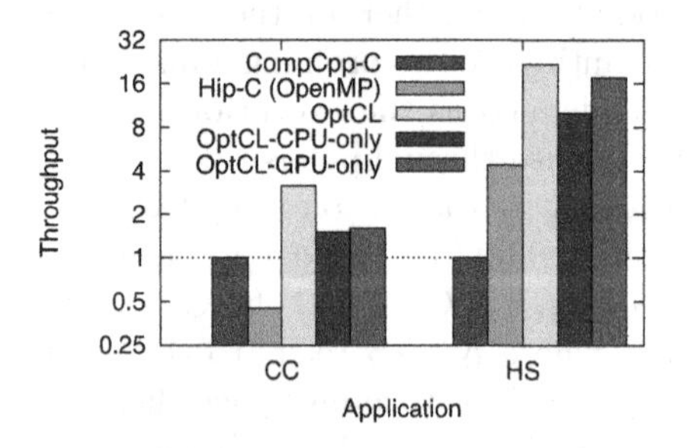

Fig. 6. Throughput of SYCL applications on the iCPU-GPU system.

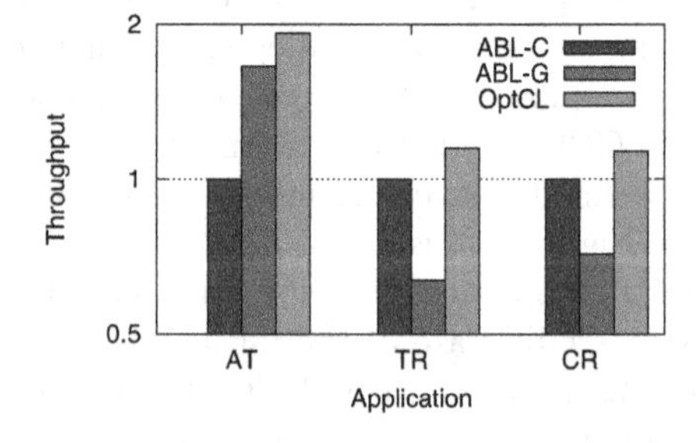

Fig. 7. Throughput of OpenABL applications.

4.2 Case Study 1: SYCL

We relied on ComputeCpp version 2.21, which still supports NVIDIA GPUs. All the tested applications were compiled with `-O3` optimisation.

Six applications are categorised into two groups based on the observed speed up types: **Group 1: BP, KM and SR.** These applications consist of several kernels. Owing to the dependencies between kernels, co-execution is not feasible. However, performance improvements are still possible through kernel invocation overhead reduction introduced in Sect. 3.3 as well as by executing them on the best suitable hardware. **Group 2: HS, CC and MV.** Co-execution is feasible. In addition to the performance benefits achieved by less invocation overhead, further performance improvements are observed due to co-execution. We report the performance of these two groups separately.

To put the performance of OptCL into perspective, we provide another baseline. The performance of the same SYCL code compiled by hipSYCL, an open-source SYCL implementation using OpenMP for CPU and CUDA for GPU as the backends. hipSYCL supports NVIDIA graphics cards (via clang-CUDA), and for CPUs, it uses OpenMP, which incurs only little invocation overhead.

Figure 5a and 5b show the throughput of CompCpp-C/G (the throughput of the SYCL code compiled by ComputeCpp and executes on CPU/GPU), Hip-C/G (the same code compiled by hipSYCL and runs on CPU/GPU) with each application normalised to the throughput of its respective CompCpp-C. As depicted in the two figures, OptCL achieved the best performance in all settings. The performance was not optimal for the ComputeCpp variants as shown in Fig. 5a, due to the incomplete support of NVIDIA GPUs and the kernel invocation overheads as the HPR in each iteration was treated as a new kernel. The runtime freshly compiles the kernel and redundantly transfers data to accelerators each time an HPR is invoked, which can be verified by running the executables in the Oclgrind simulation. By compiling the same SYCL code using hipSYCL, substantial speed-ups up to 120× were achieved. The performance was improved further by employing the OptCL middleware, thanks to the kernel invocation overhead reduction and using the most suitable hardware.

To better illustrate the power of co-execution, in Fig. 5b we also show the 'otherwise' scenarios, where we suppose OptCL would assign the kernels to only CPU (OptCL-C.-only) or GPU (OptCL-G.-only). Notably, Hip-C (OpenMP) performed worse than the ComputeCpp executables in the CC application. This is because in CC, each molecule traverses the global memory for other nearby molecules, causing large numbers of cache misses. As can be seen in the 'otherwise' scenarios, while running the OptCL variants on a single accelerator (OptCL-CPU-only and OptCL-GPU-only) already produced similar or even better performance than other variants, co-execution further boosted the throughputs. A maximum speed-up of 1.67× over the 'otherwise' scenarios and 13× over the baseline CompCpp-C was observed for the MV application.

The performance on the iCPU-GPU system is displayed in Fig. 6. As both ComputeCpp and hipSYCL do not well support iGPU (the support is experimental in hipSYCL), we exclude them from the evaluation. The iGPUs are usually not as powerful as the dedicated ones due to fewer cores and thermal concerns. As a consequence, co-execution for the MV application was not feasible in the iCPU-GPU setting, because of the large throughput deviation between the CPU and the iGPU, causing long execution time on the iGPU even with a small amount of data. However, for the applications (CC and HS) that are eligible for co-execution, more significant performance benefits were obtained owing to the zero-copy technology reducing the data transfer overhead. We achieved a speed-up of 7× and 5× over the Hip-C executables and 3× and 21× over the CompCpp-C executables in CC and HS, respectively.

4.3 Case Study 2: OpenABL

Independent kernels are common in the domain of ABS where multiple models operate on different attributes of the same agent. Different types of agents in the same simulation can perform individual models that do not rely on the states of other types which presents ample opportunity for co-execution. However, not all of them can benefit from co-execution, as the performance on one type of hardware can dominate the others, resulting in the same situation as the MV case in the iCPU-GPU setting. Due to limited space, we only demonstrate applications that benefit from co-execution.

Three ABS models from three domains were selected: social science, transportation, and biology: **CRowd (CR)**, modelling the flocking behaviour of people following fire wardens to escape from a single-entrance room in case of a fire accident; **TRaffic (TR)**, a traffic simulation comprised of a car following model and a lane changing model. A user-specified *merge_function* is added to enable co-execution of these two models. For experimentation purposes, we set the *merge_function* such that the car following model always overwrites the output of the lane changing model; **AnTs (AT)**, a simulation of the foraging behaviour of ants. We based the evaluation on an extended version of OpenABL which supports generating OpenCL code.[3]

Despite the fact that data transfer overhead was avoided, co-execution was not feasible for the tested applications on the iCPU-GPU system. Similarly to the MV application, this is caused by the large performance deviation between the CPU and the iGPU. Therefore, in what follows, we only report the performance using the dCPU-GPU system. As shown in Fig. 7, compared to running on a single device, co-execution led to the best performance in all three applications. In the CR application, the path finding is slow on the GPU owing to the heavy memory operations, causing an overall performance reduction on the GPU as indicated by the ABL-G bar in Fig. 7. Similarly, for the TR application, the car following model requires memory-intensive search for nearby vehicle which is again slow on the GPU. By co-execution, i.e., assigning the memory-intensive kernel to the CPU and the rest to the GPU, a speed-up of 1.15× and 1.13× over running on the CPU was achieved by OptCL for the TR and CR application.

Although all kernels run faster on the GPU than the CPU in the AT applications due to GPU's massive parallelism, overlapping execution of kernels on the CPU and the GPU can still lead to performance benefits. This is because of the reason explained in Fig. 3. For the AT applications, co-execution was also 1.15× faster than running on the GPU.

[3] https://github.com/xjjex1990/OpenABL_Extension.

5 Conclusion and Future Work

In this paper, we presented OptCL, a middleware to generate efficient co-execution-enabled OpenCL code from existing high-performance Domain-Specific Languages (HPDSLs) code. Through a data dependencies analysis among High Performance Regions (HPRs) and performance predictions, OptCL assigns each kernel to the most suitable hardware device and selects the best execution strategy out of purely CPU-based execution, offloading to an accelerator, or co-execution. Kernel invocation and data transfer overheads are also minimised in the generated code.

The workflow of OptCL consists of three steps. Starting from HPDSLs code, OptCL triggers a partial compilation, where an Abstract Syntax Tree (AST) is generated. After identifying all High Performance Regions (HPRs), we can then only focus on the sub-AST that is related to HPRs. A data dependency graph is built using the information gathered from analysing the sub-AST, revealing the dependencies between HPRs and thus possibilities for co-execution. In Step 2, kernels converted from HPRs are profiled on the available hardware devices. The profiling results are used to assign each kernel to its best suitable hardware and to enable co-execution where possible. Two profiling approaches are studied to estimate the power of each device: a sampling-based approach by executing the applications with a small amount of data and an offline approach using a prediction model. In our comparison study, the sampling-based approach enabled higher performance than the offline approach. Finally, in Step 3, OpenCL executables are generated reflecting the hardware assignment.

We demonstrated the versatility of OptCL by using it with two existing HPDSLs, SYCL and OpenABL in two different hardware settings: a system using a CPU and a discrete GPU as well as an integrated CPU-GPU system. In an extensive study using various applications at different scales, OptCL outperformed existing solutions and exhibited significant speed ups. We showed that OptCL can be used to enable high-performance execution on heterogeneous hardware environments without in-depth knowledge of programming paradigms for the underlying hardware. Maximum speed-ups of 13× and 21× over the original compiler were achieved on the dCPU-GPU system and iCPU-GPU system respectively.

OptCL assumes users are aware of all the parallelisable opportunities and code them in the HPRs accordingly. A possible direction to improve our middleware is to automatically detect parallelisable code snippets. We will further consider extending the hardware support to more accelerators such as FPGAs not yet targeted by most existing HPDSLs.

References

1. Breiman, L.: Random forests. Mach. Learn. **45**(1), 5–32 (2001)
2. Brown, K.J., et al.: Have abstraction and eat performance, too: optimized heterogeneous computing with parallel patterns. In: 2016 IEEE/ACM International Symposium on Code Generation and Optimization (CGO), Barcelona, Spain, pp. 194–205. IEEE (2016)

3. Chikin, A., Amaral, J.N., Ali, K., Tiotto, E.: Toward an analytical performance model to select between GPU and CPU execution. In: 2019 IEEE International Parallel and Distributed Processing Symposium Workshops (IPDPSW), Rio de Janeiro, Brazil, pp. 353–362. IEEE (2019)
4. Codeplay: Codeplay: ComputeCpp. https://www.codeplay.com/products/computecpp/. Accessed 30 July 2020
5. Cosenza, B., et al.: Easy and efficient agent-based simulations with the OpenABL language and compiler. Future Gener. Comput. Syst. **116**, 61–75 (2021)
6. Grewe, D., O'Boyle, M.F.P.: A static task partitioning approach for heterogeneous systems using OpenCL. In: Knoop, J. (ed.) CC 2011. LNCS, vol. 6601, pp. 286–305. Springer, Heidelberg (2011). https://doi.org/10.1007/978-3-642-19861-8_16
7. Grosser, T., Hoefler, T.: Polly-ACC transparent compilation to heterogeneous hardware. In: Proceedings of the 2016 International Conference on Supercomputing, Istanbul, Turkey, pp. 1–13. ACM (2016)
8. Guzman, M.A.D., Nozal, R., Tejero, R.G., Villarroya-Gaudo, M., Gracia, D.S., Bosque, J.L.: Cooperative CPU, GPU, and FPGA heterogeneous execution with EngineCL. J. Supercomput. **75**(3), 1732–1746 (2019)
9. Huang, S., et al.: Analysis and modeling of collaborative execution strategies for heterogeneous CPU-FPGA architectures. In: Proceedings of the 2019 ACM/SPEC International Conference on Performance Engineering, Mumbai, India, pp. 79–90. ACM (2019)
10. Johnston, B., Falzon, G., Milthorpe, J.: OpenCL performance prediction using architecture-independent features. In: 2018 International Conference on High Performance Computing & Simulation (HPCS), Orleans, France, pp. 561–569. IEEE (2018)
11. Majeti, D., Sarkar, V.: Heterogeneous Habanero-C (H2C): a portable programming model for heterogeneous processors. In: 2015 IEEE International Parallel and Distributed Processing Symposium Workshop, Hyderabad, India, pp. 708–717. IEEE (2015)
12. Mittal, S., Vetter, J.S.: A survey of CPU-GPU heterogeneous computing techniques. ACM Comput. Surv. (CSUR) **47**(4), 1–35 (2015)
13. Moren, K., Göhringer, D.: Automatic mapping for OpenCL-programs on CPU/GPU heterogeneous platforms. In: Shi, Y., et al. (eds.) ICCS 2018. LNCS, vol. 10861, pp. 301–314. Springer, Cham (2018). https://doi.org/10.1007/978-3-319-93701-4_23
14. Navarro, A., Corbera, F., Rodriguez, A., Vilches, A., Asenjo, R.: Heterogeneous parallel_for template for CPU-GPU chips. Int. J. Parallel Program. **47**(2), 213–233 (2019)
15. Ohshima, S., Yamazaki, I., Ida, A., Yokota, R.: Optimization of hierarchical matrix computation on GPU. In: Yokota, R., Wu, W. (eds.) SCFA 2018. LNCS, vol. 10776, pp. 274–292. Springer, Cham (2018). https://doi.org/10.1007/978-3-319-69953-0_16
16. Pandit, P., Govindarajan, R.: Fluidic Kernels: cooperative execution of OpenCL programs on multiple heterogeneous devices. In: Proceedings of Annual IEEE/ACM International Symposium on Code Generation and Optimization, Orlando, FL, USA, pp. 273–283. ACM (2014)
17. Pereira, A.D., Rocha, R.C., Ramos, L., Castro, M., Góes, L.F.: Automatic partitioning of stencil computations on heterogeneous systems. In: 2017 International Symposium on Computer Architecture and High Performance Computing Workshops (SBAC-PADW), Campinas, Brazil, pp. 43–48. IEEE (2017)

18. Pérez, B., Bosque, J.L., Beivide, R.: Simplifying programming and load balancing of data parallel applications on heterogeneous systems. In: Proceedings of the 9th Annual Workshop on General Purpose Processing Using Graphics Processing Unit, Barcelona, Spain, pp. 42–51. ACM (2016)
19. Pérez, B., et al.: Auto-tuned OpenCL kernel co-execution in OmpSs for heterogeneous systems. J. Parallel Distrib. Comput. **125**, 45–57 (2019)
20. Phothilimthana, P.M., Ansel, J., Ragan-Kelley, J., Amarasinghe, S.: Portable performance on heterogeneous architectures. In: Proceedings of the 18th International Conference on Architectural Support for Programming Languages and Operating Systems, Houston, Texas, USA, pp. 431–444. ACM (2013)
21. Pérez, B., Stafford, E., Bosque, J., Beivide, R.: Sigmoid: an auto-tuned load balancing algorithm for heterogeneous systems. J. Parallel Distrib. Comput. **157**, 30–42 (2021)
22. Price, J., McIntosh-Smith, S.: Oclgrind: an extensible OpenCL device simulator. In: Proceedings of the 3rd International Workshop on OpenCL, Palo Alto, CA, USA. ACM (2015)
23. Rao, D.M., Thondugulam, N.V., Radhakrishnan, R., Wilsey, P.A.: Unsynchronized parallel discrete event simulation. In: 1998 Winter Simulation Conference. Proceedings (Cat. No. 98CH36274), Washington, USA, vol. 2, pp. 1563–1570. IEEE (1998)
24. Riebler, H., Vaz, G., Kenter, T., Plessl, C.: Transparent acceleration for heterogeneous platforms with compilation to OpenCL. ACM Trans. Archit. Code Optim. (TACO) **16**(2), 1–26 (2019)
25. Sbîrlea, A., Zou, Y., Budimlíc, Z., Cong, J., Sarkar, V.: Mapping a data-flow programming model onto heterogeneous platforms. In: Proceedings of the 13th ACM SIGPLAN/SIGBED International Conference on Languages, Compilers, Tools and Theory for Embedded Systems, Beijing, China, pp. 61–70. ACM (2012). https://doi.org/10.1145/2248418.2248428
26. Sotomayor, R., Sanchez, L.M., Blas, J.G., Fernandez, J., Garcia, J.D.: Automatic CPU/GPU generation of multi-versioned OpenCL kernels for C++ scientific applications. Int. J. Parallel Program. **45**(2), 262–282 (2017)
27. Steuwer, M., Fensch, C., Lindley, S., Dubach, C.: Generating performance portable code using rewrite rules: from high-level functional expressions to high-performance OpenCL code. ACM SIGPLAN Not. **50**(9), 205–217 (2015)
28. Tillet, P., Rupp, K., Selberherr, S.: An automatic OpenCL compute kernel generator for basic linear algebra operations. In: Proceedings of the 2012 Symposium on High Performance Computing, Orlando, FL, USA, pp. 1–2. ACM (2012)
29. Trigkas, A.: Investigation of the OpenCL SYCL programming model. Master's thesis, The University of Edinburgh, UK (2014)
30. Xiao, J., Andelfinger, P., Cai, W., Richmond, P., Knoll, A., Eckhoff, D.: OpenABLext: an automatic code generation framework for agent-based simulations on CPU-GPU-FPGA heterogeneous platforms. Concurr. Comput. Pract. Exp. **32**, e5807 (2020). https://doi.org/10.1002/CPE.5807
31. Xiao, J., Andelfinger, P., Eckhoff, D., Cai, W., Knoll, A.: Exploring execution schemes for agent-based traffic simulation on heterogeneous hardware. In: Proceedings of the International Symposium on Distributed Simulation and Real Time Applications, Madrid, Spain, pp. 1–10. IEEE (2018)
32. Xiao, J., Andelfinger, P., Eckhoff, D., Cai, W., Knoll, A.: A survey on agent-based simulation using hardware accelerators. ACM Comput. Surv. (CSUR) **51**(6), 1–35 (2019)

Author Index

Printed in the United States
by Baker & Taylor Publisher Services

Printed in the United States
by Baker & Taylor Publisher Services